Infection:

Microbiology and Management

ONE WEEK LOAN

Infection
Microbiology and Management

Barbara Bannister

MSc, FRCP
Consultant in Infectious and Tropical Diseases
Royal Free Hospital
London

Stephen Gillespie

MD, FRCP (Edin), MRCPath
Professor of Microbiology and Regional Microbiologist
Royal Free and University College Medical School
Royal Free Hospital
London

Jane Jones

MFPH
Consultant Epidemiologist
Health Protection Agency Centre for Infections
London

Third Edition

© 2006 Barbara Bannister, Stephen Gillespie, Jane Jones
Published by Blackwell Publishing Ltd
Blackwell Publishing, Inc., 350 Main Street, Malden, Massachusetts 02148-5020, USA
Blackwell Publishing Ltd, 9600 Garsington Road, Oxford OX4 2DQ, UK
Blackwell Publishing Asia Pty Ltd, 550 Swanston Street, Carlton, Victoria 3053, Australia

First edition 1996
Second edition 2000
Third edition 2006
1 2006

Library of Congress Cataloging-in-Publication Data
Bannister, Barbara A.
  Infection : microbiology and management / Barbara Bannister, Stephen
Gillespie, Jane Jones. -- 3rd ed.
    p. ; cm.
  Rev. ed. of: Infectious disease. 2nd ed. 2000.
  Includes bibliographical references and index.
  ISBN-13: 978-1-4051-2665-6 (pbk.)
  ISBN-10: 1-4051-2665-5 (pbk.)
  1. Communicable diseases.   I. Gillespie, S. H. II. Jones, Jane.
III. Bannister, Barbara A. Infectious disease. IV. Title.
  [DNLM: 1. Communicable Diseases.   WC 100 B219i 2006]
RC111.B364 2006
616.9--dc22

2006005836

ISBN-13: 978-1-4051-2665-6
ISBN-10: 1-4051-2665-5

A catalogue record for this title is available from the British Library

Set in 9.5/11.5 Minion by Sparks, Oxford – www.sparks.co.uk
Printed and bound in India by Replika Press Pvt Ltd

Commissioning Editor: Martin Sugden
Editorial Assistant: Eleanor Bonnet
Development Editors: Mirjana Misina, Hayley Salter
Production Controller: Kate Charman

For further information on Blackwell Publishing, visit our website:
http://www.blackwellpublishing.com

Contents

Preface

Infection: Microbiology and Management encompasses the management of infections in both the hospital and the community. It has been designed as a problem-solving text, which will be equally useful both to those preparing for examinations and to those working in the clinical setting.

The introductory section sets out the important definitions required for the understanding of infectious diseases, their origins and routes of transmission to humans, their diagnosis, management and control. The presentation, diagnosis and management of individual diseases are described in the systematic chapters. Each chapter introduces the range of diseases that can affect the relevant system, and lists the pathogens responsible for each presentation in approximate order of importance. For each individual pathogen, the epidemiology and microbiology, clinical presentations and diagnosis, and strategies for prevention and control are described.

This textbook is designed to be used either as a learning text or as a practical textbook in the clinical setting. It should be most useful for senior medical and pharmacy students and for doctors preparing for examinations and assessments at the completion of general medical and surgical training programmes. It would be a useful text for those undertaking specialist training in clinical microbiology or public health, and would provide background information relevant to their everyday work as specialists. It contains much information that will be useful for infection control and for public health nurses, community nurses and environmental health officers.

Since the publication of the second edition, there has been an enormous expansion in knowledge of pathogens, in diagnostic technology and in the management of many infections. Molecular and genetic techniques have enhanced our understanding of pathogens and pathogenesis. The ribosomal RNA of bacteria or viruses can be detected in a range of clinical specimens; individual pathogens can be recognized by their unique nucleic acid signatures. Many new pathogenic species have been recognized, even during the preparation of this book. New antiviral agents have been developed, especially for the treatment of chronic infections. Novel vaccines have been added to routine childhood and adult vaccination programmes, while some older vaccines have been abandoned. Emerging threats, such as SARS, the risk of an influenza pandemic or the deliberate release of dangerous pathogens, have led to enhanced worldwide collaboration in public health. Systems for rapid diagnosis, surveillance and communication are increasingly shared between countries. To take account of these developments, many chapters in his edition have been extensively revised and reorganized. New data and diagrams have been included, and a new chapter on emerging infections has been added.

Infection is an exciting and ever-evolving specialty. We hope that this new edition will provide you with the information and insight required for addressing infection-related problems in the varied and challenging setting of modern medicine.

Part 1: Infection, Pathogens and Antimicrobial Agents

The Nature and Pathogenesis of Infection

1

Introduction

The nature of pathogens and infections

The terms used in discussion of infectious diseases are constantly changing to keep pace with changes in our knowledge and understanding. We will start by defining some of the common terms used in this book.

A **pathogen** is an organism capable of invading the body and causing disease. Such an organism is termed **pathogenic**.

Robert Koch developed criteria for the definition of a pathogen when he isolated and identified pathogenic bacteria such as *Mycobacterium tuberculosis* and *Bacillus anthracis*.

Koch's **'postulates'** indicate when an identified bacterium is a pathogen:
• the bacterium can always be identified in cases of the disease;
• the bacterium is only found in the presence of the disease;
• when the bacterium is cultured in pure growth outside the body and then re-introduced into a healthy host, it will produce the same disease.

Koch's definition works well for many bacteria, but does not fully define the host–pathogen interaction with agents, including viruses or prions, which have now been described. For instance, *Escherichia coli* is found in huge numbers in the healthy human bowel, and could therefore be defined as non-pathogenic. However, some strains of *E. coli* can produce potent enterotoxins and other pathogenicity determinants and cause significant diarrhoeal diseases. *E. coli* can therefore behave as a pathogen or as a colonizer, depending on various circumstances.

A broader definition of a 'biological agent' is used in European Union legislation: any microorganism, cell culture or toxin capable of entering the human body and causing harm.

An **infectious disease** is an illness caused by a pathogen, which invades body tissues and causes damage.

Numerous microorganisms colonize the skin and mucosal surfaces, to form the normal flora of the human body. The mere presence of multiplying microorganisms does not constitute an infection (Fig. 1.1). Indeed, colonizing organisms cause no damage, but often provide benefit to the host, by competing with potential pathogens for attachment sites and nutrients, and by producing antimicrobial substances toxic to pathogens. It is only when there is **associated tissue damage** that an infectious disease exists. Even potential pathogens can act as colonizers. *Staphylococcus aureus*, which is capable of causing severe disease, commonly exists on the surface of healthy skin. It is only when it invades the skin tissues or the blood that it causes an infectious disease. Tetanus occurs when *Clostridium tetani* multiplies in a wound, elaborating the neurotoxin tetanospasmin. Because the organism is multiplying in the host's tissues, tetanus can be called an in-

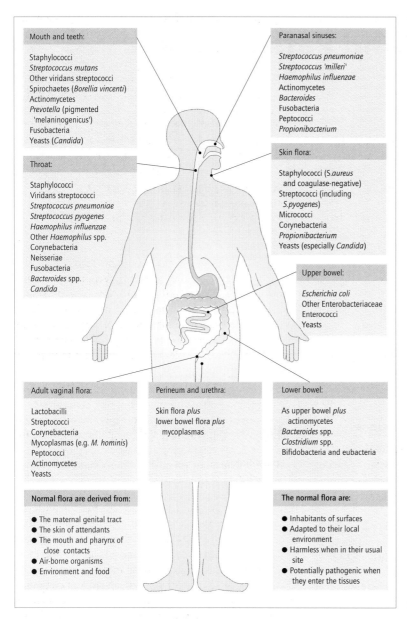

Figure 1.1 Normal human flora.

fectious disease. In contrast, adult botulism is caused by the ingestion of food in which *C. botulinum* has grown and produced a neurotoxin. The organism itself does not replicate in the human host, so botulism is defined as an intoxication (poisoning), rather than an infectious disease. *C. difficile* is present in the faeces of 5–20% of normal individuals. The organism only produces its toxin, causing pseudomembranous colitis, when conditions within the large bowel are altered by antibiotic therapy. In this case, colonization, the harmless presence of microorganisms, has developed into an infectious disease. In this book the term 'infection' will often be used as shorthand for 'infectious disease'.

Increasing numbers of patients are immunocompromised by natural disease or medical treatments, and therefore have increased susceptibility to infections. For example, around one-third of the world's population is infected with the bacteria that cause tuberculosis. Between 10 and 20% of immunocompetent people will develop tuberculosis disease at sometime in their life, as a result of this. In patients with co-existing HIV infection the risk increases to 10% per year. Furthermore, immunocompromised patients cannot resist infection by organisms such as environmental bacteria or saprophytic fungi, which are usually considered non-pathogenic. Even routine medical treatments can make otherwise immunocompetent people more susceptible to infectious disease. For instance, intravascular cannulae can provide a pathway for staphylococci, normally part of the skin flora, to enter the blood and behave as a pathogen.

> A **communicable disease** is an infectious disease that is capable of spreading from person to person.

Not all infectious diseases are communicable. A patient with pneumonia caused by *Legionella pneumophila* is suffering from an infectious disease. This is not, however, a communicable disease as it is unable to spread from this patient to another. Communicable diseases may be transmitted by many routes: direct person-to-person transfer; respiratory transmission; sexual or mucosal contact; parenteral inoculation; by insect vectors; or by means of fomites (inanimate objects).

> A **parasite** is an organism that lives on or in another organism, deriving benefit from it but providing nothing in return.

Not all 'parasites' are harmless: *Entamoeba dispar*, a protozoan, lives in the human gut without causing disease and is thus a colonizer. The closely related species *E. histolytica* is capable of invading the tissues, causing colitis and abscesses in the liver, brain and other tissues. It is thus a pathogen, and this term will be applied to it, and to other pathogenic parasites, in this book. Multicellular parasites such as schistosomes may also be pathogens. In the past, diseases caused by metazoan parasites, such as schistosomiasis, were sometimes called *infestations*. Nowadays all parasitic diseases are called infectious diseases.

> **Pathogenicity** is the ability to cause disease.

Neisseria gonorrhoeae is the causative organism of gonorrhoea. It is a small Gram-negative diplococcus, some strains of which bear surface projections called pili. Those organisms with pili can attach to the urethral epithelium and cause disease. Those that lack this feature cannot, and are non-pathogenic. In this example, pili confer pathogenicity. Mechanisms of pathogenicity are numerous, and will be discussed more fully later (see pp. 13–16).

> **Virulence** is a pathogen's power to cause severe disease.

When a pathogen causes an infectious disease, the resulting illness may be asymptomatic or mild, but is sometimes very severe. This variation may be due to host factors or to virulence factors possessed by the organism. Influenza virus is constantly able to modify its antigenic structure, on which its virulence depends. The difference in the attack rate and the severity of disease in succeed-ing epidemics is related to the antigenic structure of the causative virus.

Pathogenicity and virulence are not necessarily related. For instance, *Streptococcus pneumoniae* cannot cause disease if it does not possesses a polysaccharide capsule. However, the biochemical nature of the capsular polysaccharide determines the virulence of the organism. Pneumococci of capsular type 3 and type 30 both produce much capsular material, and are therefore pathogenic. Infection with type 3 usually causes severe disease, whereas infection with type 30 is rarely severe. For a further discussion of pneumococcal pathogenicity, see p. 153.

> **Infectiousness** is the ease with which a pathogen can spread in a population.

Some organisms always spread more readily than others. For example, measles is highly infectious and mumps is much less so. For communicable diseases, a measure of infectiousness is the intrinsic reproduction rate (IRR), which is the average number of secondary cases arising from a single index case in a totally susceptible population. The IRR for measles is 10–18, while for mumps it is 4–7.

Epidemiology of infections

> **Epidemiology** is the study of the distribution and determinants of diseases in populations.

The distribution of diseases may be described in terms of time (day, month or year of onset of symptoms), person (age, sex, socio-economic circumstances) or place (region or country). Determinants of diseases are those factors that are associated with an increased or decreased risk of disease. Their effects are usually identified by analytical studies such as case-control or cohort studies. For example, the epidemiology of meningococcal meningitis is characterized by its distribution (commonest in winter, peak incidence in young children, worldwide occurrence but especially in sub-Saharan Africa) and its determinants (close contact with a case, passive smoking).

The variation over time in the number of new cases of infectious disease occurring in a given time period (or incidence) is often represented graphically. When viewed in this way cyclical phenomena can often be observed. Seasonal cycles are common (e.g. peaks of respiratory illness in the winter) or cycles occurring over several years (e.g. measles has a typical two-year cycle, while mycoplasma pneumonia peaks typically every three or four years). Gradual changes in incidence over many years are called 'long term secular trends' and may be due to demographic, social, behavioural or nutritional changes in the host population, to climatic or environmental changes or to public health intervention.

Outbreaks and epidemics

The terms **outbreak** and **epidemic** have the same definition.

> An **epidemic** or **outbreak** is an incidence of a disease clearly in excess of normal experience or expectancy.

The distinction between these two terms is somewhat arbitrary. 'Outbreak' may be applied to smaller or more geographically localized increases in disease incidence than 'epidemic'. Since the latter often has ominous connotations in public perception, the term 'outbreak' tends to be used most commonly by public health officials so as not to cause unnecessary public alarm.

There are three main types of outbreaks: point-source, extended-source and person-to-person (Fig. 1.2).

> A **point-source outbreak** occurs when a group of individuals is exposed to a common source of infection at a defined point in time.

An example of this is a group of wedding guests who consume a contaminated food item at the reception. All those affected develop symptoms within a few days of each other.

> An **extended-source outbreak** occurs when a group of individuals is exposed to a common source over a period of time.

An example is an outbreak of hepatitis B associated with a tattoo parlour using contaminated equipment that is inadequately sterilized between customers. Extended-source outbreaks may occur over long periods of time.

> A **person-to-person outbreak** is a propagating outbreak with no common source: the outbreak is maintained by chains of transmission between infected individuals.

Examples of person-to-person outbreaks are a continuing *Shigella sonnei* outbreak in a school, or a persisting norovirus outbreak on a cruise ship.

> Outbreaks occurring in animal populations are called **epizootics**.

Two other terms that sometimes cause confusion are 'endemic' and 'pandemic'.

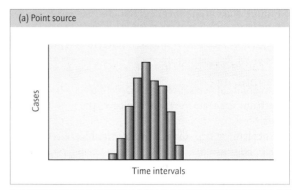

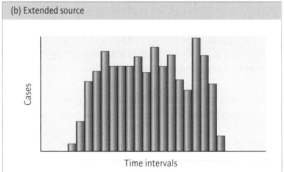

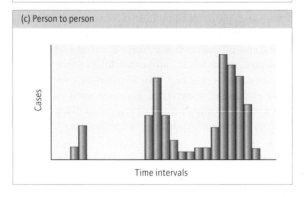

Figure 1.2 Three types of outbreak. (a) Point-source cases occur in a cluster after a single exposure, for example to a contaminated meal. (b) Extended-source cases occur over a period of time after continuing exposure, for example to a commercial distributed food. (c) Person-to-person cases occur in clusters, separated by an incubation period.

> • **Endemic** refers to a disease that occurs commonly all the year round, for example malaria in West Africa.
> • **A pandemic** is an epidemic that affects all or most countries in the world at the same time.

Four communicable diseases have caused pandemics: influenza, plague, cholera and acquired immunodeficiency syndrome (AIDS). A pandemic of severe acute respiratory syndrome was threatened but did not occur when the

SARS coronavirus spread rapidly to several countries via air travel from China in 2003.

Interaction between host, agent and environment

The behaviour of a pathogen in a population depends upon the interaction between the pathogen, host and environment. Changes in any one of these three factors will affect the likelihood of transmission occurring, and of disease resulting. For communicable diseases, population factors are also important.

Host factors

Host factors affect both the chance of exposure to a pathogen and the individual's response to the infection. Important host factors include socioeconomic circumstances, nutritional status, previous immunity and underlying disease, and behavioural factors (e.g. hygiene, sexual behaviour and travel).

Agent factors

Agent factors include infectiousness, pathogenicity, virulence and ability to survive both in human and animal hosts and under different environmental conditions. Other important factors, such as the ability to resist vaccine-induced immune responses, or drugs, also play a large part in the effect of some diseases.

Environmental factors

Environmental factors such as temperature, dust and humidity, and the use of antibiotics and pesticides affect the survival of pathogens outside the host.

Malaria as an example

The spread of malaria is a good example of the interaction between host, pathogenic agent and environment. *Plasmodium* sp. is a protozoan parasite transmitted by the bite of an infected female anopheline mosquito. Subsequent asexual development of the organism takes place within the human hepatocytes and erythrocytes. In some forms of malaria (e.g. *P. vivax*) organisms may remain dormant in hepatocytes, to mature months later and produce relapses. This does not occur in infections due to *P. falciparum*.

Many host factors affect the transmission of malaria: individuals who live in endemic areas develop partial immunity from repeated exposure and rarely suffer severe disease. This immunity is lost after 1 or 2 years away from endemic exposure. Newcomers to endemic areas will usually suffer severe disease if infected. Certain genetic factors also affect the outcome of infection. For example, individuals with sickle-cell trait have a relatively low parasitaemia when infected with *P. falciparum*, because the organism cannot derive effective nutrition from haemoglobin S.

The agent of *P. falciparum* malaria has developed resistance to an increasing range of prophylactic drugs, making it harder for travellers to protect themselves from infection. The differing antigenic structures of the several stages of the parasite life cycle have, so far, prevented the development of an effective vaccine.

Environmental factors are particularly important in the spread of malaria. Transmission occurs predominantly (although not exclusively) in tropical zones, especially during the rainy season. The anopheline mosquito breeds in stagnant freshwater environments, and malaria is particularly common in these areas. Drainage of ponds and tanks is an effective means of reducing malaria transmission. Residual insecticides have been used to control adult mosquito vectors; however, this measure has had limited success due to the emergence of insecticide-resistant mosquitoes.

Population factors

For communicable diseases, the spread of a pathogen through a population depends on certain population characteristics. It is, for example, possible to define the minimum size of host population required for the continued survival of a pathogen in the population. This critical population size varies according to the pathogen involved, the demographic structure and environmental conditions of the host population. For an outbreak or epidemic to occur there must be a threshold number of susceptible people in the population. The proportion of the population that is immune to an infection is termed 'herd immunity'. If herd immunity is high, e.g. as a result of an effective immunization programme, the population of susceptibles may be too small to maintain the pathogen, or at least insufficient to support large-scale epidemics. The presence of immune individuals protects those who are not themselves immune because the cycle of transmission is broken when the pathogen encounters an immune individual.

Sources and reservoirs of infection

Pathogens are either endogenous, arising from the host's own flora, or exogenous, arising from an external source.

The **reservoir of infection** is the human or animal population or environment in which the pathogen exists, and from which it can be transmitted.

Infections can be transmitted from carriers of an organism as well as from those suffering active disease.

Person-to-person transmission is the most common method of spread. Horizontal spread is between individuals in the same population, as in the case of influenza. Vertical spread is also possible, from mother to fetus during gestation or birth, as in the case of hepatitis B virus infection. Many pathogens can cross the placenta, but only a few cause fetal damage. The consequences of vertical transmission are usually, but not always, most serious when infection occurs during early pregnancy (see Chapter 17).

Sometimes a normal infectious cycle occurring between animals is accidentally entered by humans, usually where there is close contact between humans and animals, for instance in occupations such as farming or veterinary work. Recreational activities involving contact with animals or their excretions can also be important. For instance, leptospirosis is primarily an occupational disease in those who work with animals since the causative agent is commonly excreted by rodents or cattle. However, it is increasingly associated, both in the UK and overseas, with recreational activities such as canoeing, windsurfing and swimming, which involve exposure to freshwater that can become contaminated with rodent/cattle urine.

> A **zoonosis** is an animal disease which can spread to humans.

Many pathogens are environmental organisms, for example *Listeria monocytogenes*, *Legionella pneumophila* and *Clostridium tetani*. Spread from environment to humans can occur by ingestion (*Listeria monocytogenes*), inhalation (*Legionella pneumophila*) or inoculation (*C. tetani*).

Routes of transmission of infection

> **Fomites** are inanimate environmental objects that passively transfer pathogens from source to host.

Fomites such as towels or bedding may transmit *Staphylococcus aureus* between hospital patients. Make-up applicators, towels and ophthalmic equipment have all been shown to carry bacterial or viral pathogens from eye to eye when shared without adequate cleaning between uses.

> A **vector** is a living creature that can transmit infection from one host to another.

Many arthropod species are able to transmit pathogens (see Chapter 25 and Table 25.3). Some vectors are damaged by the organisms they carry: fleas are killed by the *Yersinia pestis* bacteria, which cause plague, and ticks are killed by *Rickettsia prowazeckii*, the agent of typhus.

Direct contact

Where pathogens are present on the skin or mucosal surfaces, transmission may occur by direct contact. Infectious diseases of the skin such as impetigo spread by this means. More fragile organisms cannot survive in a dry, cool environment, but can spread via sexual contact. In children, among whom direct contact is greater than in adults, pathogens in respiratory secretions may also be transmitted by direct contact.

A few environmental pathogens can penetrate the skin and mucosae directly. An example of this type of spread is leptospirosis, in which organisms contaminating freshwater can penetrate the mucosae or broken skin of a human. Another example is schistosomiasis, a tropical parasitic disease in which larvae (cercariae) can directly penetrate human skin that is exposed in freshwater inhabited by the intermediate snail host.

Inhalation

Droplets containing pathogens from the respiratory tract are expelled during sneezing, coughing and talking. Droplet nuclei (1–10 μm in diameter) are formed by partial evaporation of these droplets, and they remain suspended in air for long periods of time. Inhalation of droplet nuclei is a major route of transmission for many human respiratory pathogens, e.g. influenza viruses or *Mycoplasma* spp. Transmission of pathogens by inhalation can also occur from animals to humans, as with *Chlamydia psittaci*, which is present in the droppings and secretions of infected birds. Inhalation of the organism usually occurs when infected birds are kept in a confined space.

Environmental pathogens can also be transmitted by inhalation. The most important example is *Legionella pneumophila*, the causative agent of legionnaires' disease, which is present in aerosols generated from air-conditioning cooling towers, cold-water taps, showers and other water systems. Depending upon wind speed, these aerosols can travel up to 500 m and infect large numbers of individuals.

Ingestion

Enteric pathogens are usually transmitted via contaminated food, milk or water. Many foods are produced from animals, thus ingestion commonly results in animal-to-person transmission. The two most important pathogens causing bacterial food poisoning, *Salmonella* and *Campylobacter*, are both zoonotic pathogens readily transmitted to humans. Food-borne transmission is most likely if food is eaten raw or undercooked, as the pathogens are killed by heat. Many milk-borne infectious diseases are also zoonoses acquired by ingestion of unpasteurized milk products from infected cows, sheep or goats.

Spread by ingestion can occur when pathogens discharged in faeces, vomit, urine or respiratory secretions contaminate the hands of an infected individual or fomites such as handkerchiefs, clothes and cooking and eating utensils. Subsequent spread to food or water is favoured by conditions of poor sanitation.

Faecal–oral transmission occurs through direct contact between faecally contaminated hands and oral mucosa. This is a common form of transmission by ingestion.

Salmonella typhi can spread by all of these means – a few organisms deposited in food will multiply to achieve an infective dose. Transmission from environment to humans by ingestion is less common. However, the soil- and sewage-borne bacterium *Listeria monocytogenes* is an example, as it can contaminate food and cause invasive disease following ingestion.

Inoculation

Transmission can occur when a pathogen is inoculated directly into the body via a defect in the skin. Contaminated transfusions, blood products or materials from non-sterile needles and syringes can transmit viruses such as hepatitis B and human immunodeficiency virus (HIV). Malaria may also be transmitted by contaminated blood transfusions.

Animal-to-person transmission occurs when an infected animal bites, scratches or licks an individual. Rabies is usually spread by this route. Alternatively, the skin may be broken by sharp animal bristles or rough bone-meal containing pathogens such as *Bacillus anthracis*.

Environment-to-person spread by inoculation also occurs: *Clostridium tetani* is usually introduced through a puncture wound contaminated with soil, dust or animal faeces.

Dynamics of colonization and infection

When a microbe encounters a potential host, a sequence of events takes place. On making contact with the host's mucosa or skin, an organism may be able to adhere to and colonize this surface. If successfully established, colonization often continues without ill effect for a variable length of time. During this period the host may develop immunity to the organism. This is a common means of development of immunity to a number of pathogens such as *Haemophilus influenzae* and *Neisseria meningitidis*. This process is also important since organisms that have little capacity to cause disease may share some antigenic markers with human pathogens. Antibodies developed to these agents of low pathogenicity may provide immunity to powerful pathogens. This effect of cross-immunity is exploited when bacillus Calmette–Guérin (BCG) vaccine is given to confer protection against tuberculosis or leprosy.

Alteration of the host–microbial interaction may permit a change from colonization to invasion of local tissues or the whole body. For many infectious agents the majority of interactions are restricted to colonization.

For other agents, such as poliomyelitis viruses, invasion of the host may take place as part of the life cycle of the organism. In an unimmunized population, newly exposed to the virus, many individuals will be infected. The great majority suffer only a mild bowel or throat infection and become immune to further attacks. In a few cases this infection is complicated by self-limiting viral meningitis, and a minority of meningitis cases develop anterior horn cell infection and paralysis. Paralysis is most likely to affect older children and adults. Almost all adults are immune in populations where the virus is common, so most infections occur in young children, who rarely develop paralysis. The disease therefore exists in equilibrium with the population, where the combination of host factors and microbial pathogenicity does not favour the occurrence of symptomatic or severe disease. This situation is often described as 'the iceberg of infection' where the majority of host–microbial interactions are colonization–clearance episodes and only a small proportion result in morbidity or mortality (Fig. 1.3).

The manifestation of an infectious disease is a complex balance between the direct effect of the pathogen or its toxins, and the response of the affected patient. The patient's response depends on several factors, including immune competence, previous experience of the same or similar pathogens and his or her own genetic structure.

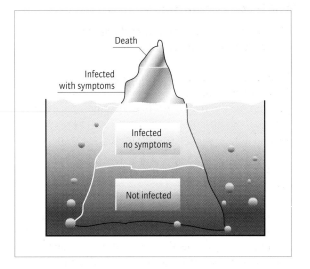

Figure 1.3 The iceberg of an infection.

Mechanisms of resistance to infection

The defences of the human host against infection can be classified into three parts:
- local defences against pathogens,
- non-specific (innate) resistance to infection, and
- the specific (adaptive) immune system (Fig. 1.4).

Each part is vital for survival against the continuous pressure of microorganisms. Innate resistance to infection does not depend on the development of specific reactions to individual pathogens. It therefore provides an early, non-specific defence that will often 'buy time' to permit the development of an efficient and specific immune response.

Local defences against infection

Many components of this function are normal mechani-

cal and physiological properties of the host. They include the skin, the mechanical flushing activity of urine and intestinal contents, ciliary removal of mucus and debris, the enzymatic action of lysozyme in tears, the phagocytes, the alternative complement pathway and the normal flora.

The skin

Natural defences of the skin
1 Keratinous surface.
2 Antibacterial effects of sebum.
3 Effect of normal flora.

The skin forms an impermeable mechanical barrier to pathogens. Sebum secreted by sebaceous glands inhibits the multiplication of many microorganisms. The skin's resident bacterial flora competes with potential invaders, and may produce metabolic products inhibitory to other species. This combination of effects is called **colonization**

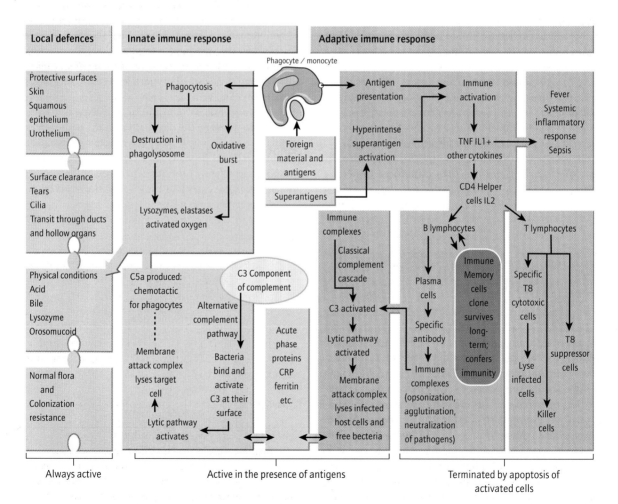

Figure 1.4 Overview of body defences against infection.

resistance and is also important in the pharynx and the bowel.

Scratches, ulcers and other defects in the skin surface can bypass its protective mechanisms and permit the entry of skin pathogens such as staphylococci or herpes simplex viruses, as well as environmental organisms such as *Leptospira* spp. Parasites such as hookworm larvae and schistosome cercariae are capable of penetrating intact skin. Cercariae, which hatch from infected water snails, swim towards potential hosts. The head part of the cercaria secretes proteolytic enzymes, which break down the skin (and, incidentally, cause a local dermatitis or 'swimmer's itch'). The physical barrier of the skin can also be breached by biting arthropods. Many types of pathogen are transmitted by this route.

Intravenous access devices enable coagulase-negative staphylococci and corynebacteria from the skin to enter the bloodstream and cause septicaemia and endocarditis. Hepatitis B and C virus infections are also transmitted by this route, through the use of contaminated needles.

Mucosal defences against infection

Natural defences of mucosae
1 Mechanical washing by tears or urine.
2 Lysozyme in surface fluid.
3 Surface phagocytes.
4 Ciliary action moving mucus and debris.

Many bacteria and viruses can invade through intact mucosal surfaces, which are not keratinized and are often only one cell thick. Nevertheless, mucosal surfaces also have natural defences. They are usually moistened by tissue fluids, which contain lysozymes capable of destroying microbial peptidoglycan. Phagocytic neutrophils are often expelled at mucosal surfaces and can ingest foreign material, including pathogens. The mucosae of the gut and the urinary tract are 'washed' by the constant transit of liquid contents. In the respiratory tract material is moved upwards towards the pharynx by the action of mucosal cilia. These defence mechanisms are so effective that the urinary tract and respiratory tract are bacteriologically sterile, except for areas near the exterior, such as the mouth and the lower urethra. Sites such as the bowel that possess a colonizing flora are also protected by the colonization resistance that this resident flora confers.

Obstruction of a bronchus, ureter or bile duct will interfere with many of these defences, permitting pathogens to accumulate at sites of stagnation, thus predisposing to infection. Similarly, insertion of a urinary catheter or an endotracheal tube will alter clearance mechanisms, encourage stagnation of mucus secretions around the tube and provide an inanimate substrate, encouraging the entry and establishment of a bacterial flora.

Defences against infection via the gut

Natural defences of the gut
1 Gastric acid.
2 Chemical environment produced by normal flora.
3 Mucosal phagocytes and lymphocytes.
4 Bacteriocins.

The gut provides an important route for the acquisition of microbes. The first barrier to infection is gastric acid, which inhibits the survival of many intestinal pathogens. Patients with achlorhydria are more susceptible to infections transmitted by ingestion. Phagocytes will migrate through the mucosa, in response to chemotaxins released at sites of tissue damage. The normal flora of the gut is important in competing for nutrients and attachment sites, and in inhibiting the action of pathogens. Facultative and obligate anaerobes produce potent inhibitors of bacterial growth called bacteriocins, which inhibit the growth of competing organisms. Many obligate anaerobes secrete free fatty acids, which alter the local redox potential, making the environment less supportive to other microorganisms. This delicate competitive balance can be upset by disease or by antimicrobial therapy. The most dramatic example of this is the overgrowth of *Clostridium difficile* in the intestine of patients treated with antibiotics, when *C. difficile* produces a toxin that causes severe ulcerative disease (pseudomembranous colitis) in the large bowel.

Innate resistance to infection

Phagocytosis
Phagocytosis of invading organisms is an important innate defence mechanism (Fig. 1.4). Neutrophils and macrophages are attracted to the site of inflammation by mediators (chemotaxins) such as complement components. The efficiency of phagocytosis is enhanced when organisms are 'opsonized' by attached complement or specific antibody, which provide receptors for the attachment of phagocytes. Organisms are taken up into phagosomes, which fuse with the lysosomes containing free radicals and lytic enzymes, resulting in killing.

Patients with deficiencies in phagocyte function suffer repeated pyrogenic infections, and develop chronic suppurative granulomata (see p. 432).

Classical and alternative complement systems

The **complement system** is a complex of plasma enzymes.

Complement component C3 is cleaved to produce C3b. This active product initiates a reaction involving complement components C6–C9, producing a complex of proteins called the 'membrane attack complex', which damages cell membranes and leads to cell lysis.

The **alternative complement pathway** is an extremely important component of the innate immune response. It depends on spontaneous cleavage of C3 at bacterial cell surfaces and, although it is relatively slow, it can act in the absence of specific antibody and it provides early defence against such severe infections as meningococcal septicaemia. Complement can act to enhance resistance to bacterial and parasitic infection by the action of breakdown products such as C3a and C5a, which promote capillary permeability and are chemotactic to neutrophils and macrophages. C3b deposited on the surface of bacteria will opsonize them for phagocytosis.

Patients with congenital deficiencies of the early complement components are more susceptible to pneumococcal infections, in which activation of the alternative complement pathway is important in resistance to infection. Deficiencies in components of the alternative pathway, such as properdin, render the individual highly susceptible to invasive meningococcal infection.

Sialic acid inhibits the natural breakdown of C3. Successful pathogens, such as meningococci, have sialic acid on their surface, which probably reduces the effectiveness of the alternative complement pathway.

Activation of the **classical pathway** depends on the activation of a preliminary cascade of active proteins, initiated in the presence of immune complexes. It therefore depends on the presence of antibodies specific to the pathogen. These bind to C3 and initiate the breakdown process that activates it. The preliminary cascade of proteins serves to amplify the rate of complement activation. The classical complement cascade is therefore much more rapidly acting than the alternative pathway, but it depends on the existence of an adaptive immune response, which provides the necessary antibodies. This pathway can also be initiated by C-reactive protein, an acute-phase plasma protein that is elevated during acute inflammation.

The adaptive (specific) immune response

The adaptive immune response is a series of changes whereby the host develops specific defensive responses to individual microorganisms. This is based on the selection from the body's repertoire of antigen-recognizing cells – cells with receptors best fitted to recognize and bind to unique antigens that the pathogen possesses. These cells are then amplified to produce clones of cells, which perform functions such as antibody production, cytotoxicity and immunological memory.

Antigens are peptide or sugar molecules unique to the pathogen, and exposed at the surface of the pathogen or infected host cells, to which the immune responsive cells of the host can bind and initiate an adaptive immune response.

Antigens are displayed in an array on the pathogen's surface, based on the tertiary structure of the surface proteins or polysaccharides. Short sequences of the antigens, which are binding sites for immune responsive cells, are called epitopes. Immunogenic epitopes may also be parts of toxin molecules, or of abnormal surface proteins displayed on virus-infected host cells. Different epitopes stimulate T and B cell immunity. Organisms each contain many antigens, within which are many different epitopes.

The adaptive immune response comprises two main components. In the humoral response, B-lymphocytes develop into clones of antigen-producing cells, elaborating antibodies against antigens of the invading organism. In the cellular response, clones of cytotoxic lymphocytes develop, which can destroy cells bearing the foreign antigens. Both the humoral and the cellular immune response depend for their amplification on CD4 helper T-lymphocytes. In the absence of effective CD4 cells, new virus infections cannot be overcome by cellular immune responses, and new antibody responses are inefficient.

T cell-independent antigens

These antigens induce a humoral immune response without the involvement of T helper cells. They are usually large polymeric molecules. It is thought that they bind to many adjacent antigen receptors on B cell surfaces, accidentally influencing intervening receptors concerned with recognizing helper function. The resulting immunoglobulin response is a rather small and short-lived IgM response. The absence of a mature IgG response results in poor and short-lasting immunological memory. This is important as many polysaccharide antigens, such as pneumococcal capsular antigens, act as T cell-independent antigens, especially in young children.

Superantigens

These antigens activate large numbers of lymphocytes by activating V-beta receptors, which are possessed by up to 30% of all lymphocytes. Superantigen binding does not activate the process of apoptosis, so that activated cells are not programmed to die, as normally activated cells are. Superantigens therefore cause an intense and destructive immune reaction that is not limited or terminated in the usual way. Staphylococcal enterotoxins, toxic shock syndrome toxins (TSSTs), pyrogenic exotoxins of *Streptococcus pyogenes* and some viral antigens act as superantigens.

Pathogenesis of infection

While the spread of a disease is influenced by interaction between the host, the agent and the environment, and by population factors, the effects of disease in the host depend on a number of factors particular to the agent or pathogen.

> **Pathogenicity factors** are attributes of a pathogen that are important determinants of pathogenicity.

The pathogen exploits its host to best advantage if it achieves optimum levels of survival and multiplication. The death of the host is not an advantage, unless this contributes to the transmission of microbial genes. Pathogenicity factors are genetically maintained, by natural selection, because they facilitate survival or transmission via the host.

Characteristics of a successful pathogen

> **Characteristics of a successful pathogen**
> **1** Survival and transmission in the environment.
> **2** Attachment to the surface of the host.
> **3** Overcoming the body defences against infection.
> **4** Ability to damage the host, e.g. by toxin production.
> **5** Ability to replicate in the host, producing progeny able to infect others.

Survival in the environment

Many microorganisms are killed by drying, ultraviolet light and variation from their optimum temperature for growth. To overcome these difficulties, organisms have developed many strategies. Organisms that are predominantly environmental have developed survival mechanisms such as the bacterial endospore. A spore is a structure that contains a single copy of the bacterial DNA in a keratinous protective 'shell', which has little retained water and a very low metabolic rate. Adverse environmental conditions are the stimulus for sporulation, for example the rise in pH in the duodenum for *Clostridium perfringens*. Bacterial spores are capable of survival for many years, 'germinating' to form vegetative cells when conditions are favourable. The Scottish island of Gruinard was contaminated with anthrax spores early in the Second World War, and was only declared free of infection 50 years later.

Other organisms have found protected ecological niches in the environment: *Legionella pneumophila* inhabits freshwater and can survive in the protected environment within the cytoplasm of free-living amoebae.

Some organisms have such a close relationship with an animal host that they can exist reversibly, as either a pathogen or a commensal. To increase the chances of survival, some organisms will infect a wide range of species, for example the rabies virus is able to infect many mammalian species. This diversity provides a large and adaptable reservoir of infection, which will ensure survival of the pathogen if one host group is eliminated.

Transmission

The problem of transmission between hosts is related to survival. Spore-forming organisms can survive and be transmitted by many routes. Organisms with moderate survival potential can be transmitted by spreading in the air on droplet nuclei (see p. 8).

Bacteria such as *Neisseria gonorrhoeae* are extremely delicate and are unable to survive outside the host. Sexual transmission overcomes this difficulty by depositing the pathogen directly on to the genital mucosa of the new host, and in addition ties the organism's life cycle into an essential part of the host's life cycle, ensuring the survival of the pathogen.

Attachment of organisms to body surfaces

For organisms to gain access to the body via the mucosal surfaces they must first attach themselves. They have to overcome the natural defence mechanisms present in each area.

Organisms may gain attachment by specialized organelles of attachment, or more simply with attachment molecules. Uropathic *Escherichia coli*, which must overcome the flushing action of urine, uses fimbriae to attach to the urinary epithelium. These fimbriae are pathogenicity determinants. Influenza virus adheres to the host's respiratory mucosal cells via its haemagglutinin molecule.

Microbial defence against immunological attack

From the moment the pathogen enters a new host, it must avoid the host's defence mechanisms. Host secretory IgA is an important defence mechanism against organisms that invade via the mucosal surfaces. Many respiratory tract pathogens, including *Streptococcus pneumoniae*, *Haemophilus influenzae* and *Neisseria meningitidis*, elaborate a protease that selectively destroys IgA.

Bacterial capsules are an important defence against phagocytosis by neutrophils or macrophages. This probably depends on the negative charge on the capsular polysaccharide molecules. Capsulate organisms resist phagocytosis unless opsonized by the attachment of specific antibody. For organisms such as the pneumococcus, which activate the alternative complement pathway at their cell wall, the capsule acts as a physical barrier, preventing attached C3b on the cell wall being recognized by phagocytes.

Some organisms are able to exploit phagocytes to enhance their life cycle. Once they have entered the phagocyte they are protected from antibodies, and can survive for very long periods. *Mycobacterium tuberculosis* is an intracellular pathogen that can survive inside macrophages, and re-emerge to cause disease if the host's immune defences become compromised. Intraphagocytic survival depends on safe entry into the phagocyte and subsequent avoidance of enzymic degradation by lysozymes. Phagocytic ingestion of a particle is usually accompanied by a 'respiratory burst', which produces intensely toxic oxygen radicals. *Leishmania* spp. overcome this by utilizing alternative mannose/fucosyl and C3b receptors to attach to and enter phagocytes. These, unlike the Fc receptors, do not trigger a respiratory burst.

Within the phagocyte, there are three main mechanisms by which pathogens can survive. Organisms may prevent phagolysosomal fusion, avoiding contact with lysozyme, and continuing to multiply within the phagosome. *Toxoplasma gondii* and *Chlamydia* act in this way. *Leishmania* survives inside the phagolysosome by metabolic adaptation to the hostile environment, and by excreting a factor that scavenges the normally lethal oxygen radicals. Mycobacteria escape from the phagolysosome into the cytoplasm where they are partly protected from digestion by their high lipid content. This effect has been demonstrated by coating staphylococci with the phenolic glycolipid of *M. leprae*. Although successfully ingested, these coated staphylococci are not killed, while uncoated control staphylococci are destroyed.

Many viruses contain genes that encode cytokine-like molecules or cell receptors, thus interfering with the natural functions of cells. Epstein–Barr virus produces proteins that switch off programmed cell death in infected lymphocytes, helping to maintain the population of productively infected cells.

Antigenic variation

For organisms that are obliged to live extracellularly, antibody attack poses a major problem. Some organisms are able to evade the humoral immune system by varying the antigenic make-up of their surface. The major surface antigen of *Trypanosoma brucei* var. *rhodesiense* is the variable surface glycoprotein (VSG). As a result of a complex series of molecular events, the trypanosomes are able to express a different VSG every few days. Thus, as a humoral immune response is produced and parasite numbers are falling, a new clone of trypanosomes emerges with a different VSG and is able to multiply unhindered by the immune system. This process is continued through a pre-programmed set of variations, which are reflected by the episodic nature of symptoms in the early phases of the infection. *Borrelia recurrentis* and *B. duttoni* also undergo antigenic variation, producing a characteristic, relapsing fever. A variation of this approach is adopted by adult schistosomes, which absorb host proteins to their surfaces, thus evading detection by the immune system.

Influenza virus survives as a pathogen by antigenic variation because its genome can undergo antigenic 'drift' and 'shift'. Drift is the process of gradual changes in the genes coding for viral surface haemagglutinin, enabling the virus partly to escape the effects of population immunization by previous epidemics. Shift is a major change in the antigenic structure of the virus, producing a novel strain to which nobody has any immunity. Antigenic shift may initiate a worldwide epidemic (pandemic).

Immune suppression

Pathogens employ various strategies to evade or impair the host's immune response.

Many parasitic infections cause overstimulation of the humoral immune system. High concentrations of ineffective antibodies are produced at the expense of normal antibody responses. African trypanosomiasis and leishmaniasis are examples of this; not only is the parasitic infection uncontrolled, but many sufferers die of intercurrent bacterial infections such as acute pneumonia.

The cellular immune response can also be depressed. This occurs in severe tuberculosis and in lepromatous leprosy, where the infection induces a specific cell-mediated immune defect, limiting T cell responses to the mycobacteria. Acute viral infections, such as infectious mononucleosis, cause temporary suppression of cell-mediated immune responses.

Immune suppression can be broad spectrum, when a whole arm of the system is impaired by the action of a pathogen. HIV causes a selective depletion of CD4 cells, resulting in susceptibility to tuberculosis, *Pneumocystis* infections and toxoplasmosis. Additionally, reduced T helper-cell function causes an increased susceptibility to many other pathogens, including bacteria such as pneumococci and salmonellae.

Ability to damage the host
Toxin production

Toxins are responsible for many of the damaging effects of infection. They are also excellent vaccine targets, as chemically modified toxin vaccines (toxoids) stimulate strong immune responses. Such vaccines have helped in the virtual elimination of diseases such as tetanus and diphtheria. Toxins are often essential for the life cycle of the pathogen, and their pathogenic effects may be coincidental. Diphtheria toxin, for example, mediates the pharyngeal, cardiac and neurological damage in diphtheria. The gene coding for diphtheria toxin exists in a beta-phage, and only organisms carrying this lysogenic phage are toxigenic. In this symbiosis, the phage requires *Corynebacterium diphtheriae* as a host and *C. diphtheriae* is given a biological advantage in colonization of the human host by possession of the toxin gene.

Bacterial toxins are conventionally classified as exotoxins and endotoxins. Exotoxins are toxic substances excreted by organisms. The word endotoxin is usually used to describe the lipopolysaccharide antigen of Gram-negative bacterial cell walls. However, it is now known that many bacterial structural antigens can have toxic effects.

Endotoxin

The lipopolysaccharide of Gram-negative bacteria is an important pathogenicity factor. Lipopolysaccharide is made up of three main parts:
• the core region, lipid A, which is responsible for the main toxic effects;
• an oligosaccharide region, which contains heptoses (Hep) and hexoses linked to lipid A via the unusual sugar ketodeoxyoctanoic acid (KDO); and
• attached to this a long polysaccharide chain, which is the somatic antigen ('O' antigen) of the individual organism. This polysaccharide partly protects Gram-negative bacteria such as salmonellae against the bactericidal activity of serum (see Fig. 1.5).

Lipid A acts by stimulating cells of the macrophage series to produce cytokines, such as interleukin 1 (IL-1), which eventually leads to the production of a cytokine called tumour necrosis factor (TNF). TNF is a powerful activator of the complement and clotting cascades, and causes endothelial damage, leading to fever, hypotension, intravascular coagulation, organ failure and other metabolic and physiological changes. These changes produce the syndrome of sepsis.

> **Sepsis** is the manifestation of the systemic inflammatory response to infection.

Toxic activity also resides in the cell wall of some Gram-positive bacteria. The C-polysaccharide and F (Forssmann) antigen of *Streptococcus pneumoniae* are released into the host's tissues, and activate the alternative complement pathway. The products of the complement cascade are responsible for increased capillary permeability and leucocyte migration to the site of infection.

Exotoxins

Bacterial exotoxins are diverse in function and clinical effect. They may be classified either according to the body system that they affect (for instance enterotoxin, neurotoxin or cytotoxin), or according to their mode of action, when this is known (Table 1.1).

Effects of microbial toxins

Some toxins cause important features of a disease, and it is convenient to consider a few examples here.

Streptococci and staphylococci can produce pyrogenic exotoxins (SPEs), toxins that damage the skin and sometimes the internal organs. The erythrogenic toxin of *Streptococcus pyogenes* can cause the rash of scarlet fever. *Staphylococcus aureus* may produce toxic shock syndrome toxins (TSSTs), the cause of toxic shock syndrome with a scarlet fever-like rash. TSSTs also behave like the entero-

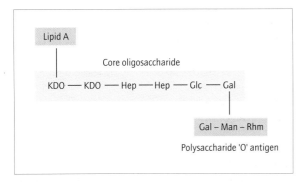

Figure 1.5 An example of the structure of endotoxin.

Table 1.1 Actions of bacterial exotoxins

Type	Examples
Extracellular cytotoxins (directly poison cells)	Streptococcal hyaluronidase *Pseudomonas aeruginosa* exotoxin A
Transmembrane cytotoxins (enter cells via implanted receptor/transporting molecule)	*Escherichia coli* verotoxin Shiga toxin Diphtheria toxin
Membrane-damaging toxins (cause haemolysis or cytolysis)	Streptolysin O *Clostridium perfringens* alpha toxin *Staphylococcus aureus* P-V leukocidin
Deregulating toxins (cause overactivity of secretory mechanisms)	*E. coli* heat-labile toxin Cholera toxin
Competitive inhibitors (competitive blockers of natural transmitters)	Botulinum toxin Tetanus toxin

toxins of *S. aureus* and cause diarrhoea. Some organisms produce haemolytic toxins; *Clostridium perfringens* can cause severe haemolysis by this mechanism. Diphtheria toxin directly damages myocardial cells and causes demyelination of nerve axons, while botulinum toxin inhibits neuromuscular conduction of nerve impulses, and causes paralysis.

Microbial synergy

Microbes can act together to establish infection, facilitate tissue invasion, reduce the host's immune response and enhance the virulence of pathogens. *Streptococcus pneumoniae* cannot bind to intact respiratory epithelium, but can bind to basal membrane. Influenza virus causes damage to, and shedding of, respiratory epithelium, exposing the underlying basement membrane to attack.

Many infections are polymicrobial, with obligate and facultative pathogens multiplying together and creating the conditions for each other to survive. In synergistic gangrene the metabolic products of facultative organisms reduce the redox potential sufficiently to enable obligate anaerobes to multiply and cause extensive tissue necrosis.

Infection with HIV is an example where one infectious agent reduces the immune response of the host, allowing other organisms to invade. Many parasitic infections (leishmaniasis, trypanosomiasis and malaria) can trigger polyclonal activation of B cells with overproduction of antibody, impairing the host's response to intercurrent bacterial infections.

Chronic schistosomiasis is associated with recurrent *Salmonella* infection. Salmonellae can bind to schistosome eggs, which provide a niche for salmonellae to cause persisting colonization and recurrent infection.

Manifestations of infectious disease

Fever is the most common manifestation of the systemic inflammatory response, the clinical expression of sepsis, which accompanies infection. Fever occurs in all but the most trivial or unusual cases.

The body temperature of a healthy person is set and maintained by the hypothalamus. It follows a circadian cycle in which the temperature is lowest in the early morning and highest at about 10 p.m. local time, varying by 0.5 °C or more. Also, in women who ovulate, a monthly variation in temperature can be detected, with an abrupt step at the time of ovulation.

Infection, in common with a number of other events, can cause a resetting of the hypothalamus to a higher body temperature. This change is initiated by the release of cytokines, particularly IL-1, TNF and alpha-interferon, by activated mononuclear phagocytes. The cytokines act on specialized endothelial cells in the hypothalamic blood vessels, causing the release of prostaglandins, which act on the hypothalamic cells (Fig. 1.6).

A raised body temperature is probably useful in combating infection. Many pathogens replicate best at temperatures at or below 37 °C. These include respiratory viruses, pneumococci and other bacteria, and many agents of tropical skin infections. Such pathogens are adversely affected by higher temperatures. Even if the pathogen is unaffected by temperature change or, like some campylobacters, replicates well at higher temperatures, fever can still help by accelerating immune reactions such as phagocytosis, antibody and cytokine production, and the complement cascades.

Adverse effects of fever

Delirium
Delirium is an organic confusional state. It is most common in children and the elderly, occurring when the temperature is at its highest, especially at night. It may cause agitation, drowsiness, distressing dreams or visual hallucinations. Although sometimes caused directly by the disease process (toxaemia or cerebral infection), it can often be improved or cured simply by reducing the temperature.

Febrile convulsions
Febrile convulsions affect children between the ages of 6 months and 6 years. They are rarely a sign of true epilepsy, and usually cease spontaneously as the child reaches the age of 4–6. The convulsions most often happen as the temperature is rising.

Treating fever and its complications

Treatment of fever
1 Tepid bathing.
2 Paracetamol.
3 Ibuprofen.
4 Aspirin (should not be given to children under the age of 16 years).

Fever can be reduced by sponging or washing the patient with tepid (not cold) water, or by giving antipyretic drugs. Paracetamol is the drug of choice. Ibuprofen is an alternative for both adults and children. Aspirin is an effective antipyretic, but should not be given to children under 16, because of the association of aspirin treatment with Reye's syndrome.

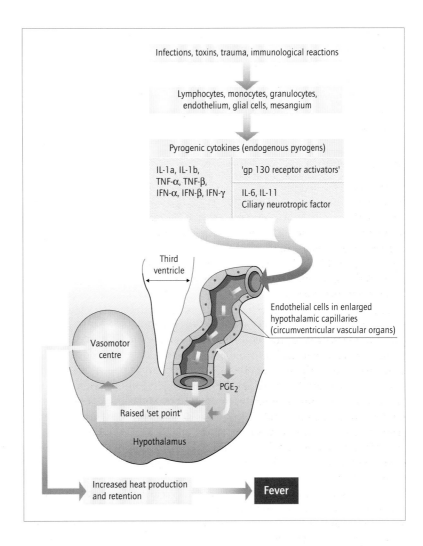

Figure 1.6 The mechanism of fever.

Febrile convulsions can usually be terminated by lowering the body temperature. If this is unsuccessful anticonvulsant treatment is given. Often a once-only dose of lorazepam or diazepam is enough to avert further attacks during a brief illness. On rare occasions a short course of regular anticonvulsant dosage is required; in this case sodium valproate or carbamazepine is the drug of choice, as in true childhood epilepsy.

Pre-existing disorders

Pre-existing disorders may be adversely affected by fever. Epilepsy may become poorly controlled, and is then best managed, if possible, by treating fever, rather than by altering established drug routines. The neurological deficits of multiple sclerosis are reversibly exacerbated by fever. Cerebral ischaemic symptoms may recur or worsen during fever.

Inflammation

Inflammation is a complex combination of events, whose pathogenesis is still poorly understood.

> **Key pathological features of inflammation**
> 1 Vasodilatation at the affected site.
> 2 Exudation of tissue fluid from dilated capillaries.
> 3 Accumulation of neutrophils and macrophages at the site.
> 4 Release of active chemicals from neutrophils (Fig. 1.7).

These events combine to cause local heat and redness, sometimes with the advantage of adversely affecting the responsible pathogen. The tissue exudate contains complement components, an important part of the innate defence against pathogens. Some complement activation products are chemotaxins and attract phagocytes to the site.

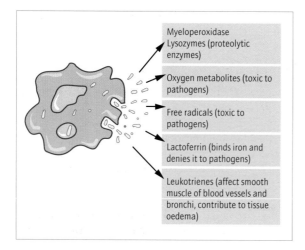

Figure 1.7 Active substances released by neutrophils.

The phagocytes ingest bacteria and debris, becoming activated and releasing chemicals that both attack the pathogens and contribute to inflammation. These include enzymes that promote rapid synthesis of prostaglandins, a variety of chemicals that are vasoactive and also affect platelet activation. Prostaglandins are important initiators of inflammation. The effects of non-steroidal anti-inflammatory drugs are due to their strong inhibition of prostaglandin synthesis.

Useful anti-inflammatory drugs
1 Ibuprofen (adult and child preparations).
2 Aspirin (avoid in under-16s).
3 Intramuscular or rectal diclofenac.

Detecting inflammation
The signs and symptoms of inflammation are pain, heat, redness and swelling. They are helpful in indicating the site of a localized infection, for instance an abscess or an infected joint, and should always be sought during clinical examination when assessing a patient with suspected infection.

Neutrophils, which collect at sites of inflammation, may be shed, e.g. in the urine, or discharged from the tissues as pus. Microscopical examination of the appropriate specimen can therefore reveal evidence of infection, even when the patient cannot indicate the affected site.

The swelling of inflammation may be deep in the body and undetectable by surface examination. An X-ray may reveal a soft-tissue shadow (Fig. 1.8), isotope scans may demonstrate the site of hyperaemia, and other imaging procedures such as computed tomography or nuclear magnetic resonance scans are excellent for demonstrating oedema.

Acute phase proteins
Several plasma proteins show a large rise in concentration in the presence of inflammation. Notable among these are caeruloplasmin, ferritin, haptoglobin, $alpha_1$-antitrypsin, $alpha_1$-glycoprotein (orosomucoid) and C-reactive protein (CRP). Levels of transferrin, fibronectin and albumin tend to fall.

The function of these changes, which can be induced by prostaglandins, interferon-alpha or IL-1, is unknown. Caeruloplasmin and haptoglobin bind to oxygen radicals, perhaps inhibiting their damaging potential in blood. Alpha$_1$-acid glycoprotein can inhibit platelet aggregation, possibly protecting against platelet activation and thrombus formation.

C-reactive protein
C-reactive protein is produced in the liver, and synthesis is greatly increased in acute inflammation. It is a disc-shaped pentameric molecule that readily binds a number of substances, including the C fraction of pneumococcal lysates (from which it gets its name). In its bound form it strongly activates the classical complement pathway, acting as an innate defence against infection. It is elevated in many acute bacterial and viral infections and other inflammatory

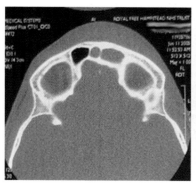
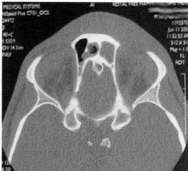

Figure 1.8 CT head scans of a young teenager with acute sinusitis. The opacified frontal and ethmoid sinuses are seen, with a large soft-tissue swelling, due to inflammatory oedema, over the frontal and orbital areas on the affected side.

conditions. Because of its rapid response to inflammation C-reactive protein is useful for monitoring responses to treatment in conditions such as endocarditis.

Plasma viscosity and erythrocyte sedimentation rate

The protein changes in inflammation alter the viscosity of the plasma. This can be measured directly, but is more often inferred from changes in the erythrocyte sedimentation rate (ESR: the rate at which red blood cells settle in anticoagulated blood on standing). The normal ESR is not more than about 20 mm/h. This rises to 30–50 mm/h in acute infections, but may reach 70–100 mm/h in some atypical pneumonias and chronic conditions such as abscess formation or immunological disease. The ESR has non-specific diagnostic value, like CRP levels, but re-sponds less rapidly than CRP to changes in the degree of inflammation.

Rashes

Rashes are a particular form of inflammation or tissue damage, affecting the skin. The causative pathogen may be present in the lesions. The rash of an infectious disease may be generalized or local, and often evolves in a predictable way, starting at a particular site, spreading in a particular direction and containing typical types of skin lesions. Some of the skin lesions that occur in rashes are illustrated in Fig. 1.9.

The rashes of infectious diseases, unlike those of hypersensitivity reactions, are rarely painful or irritating. The

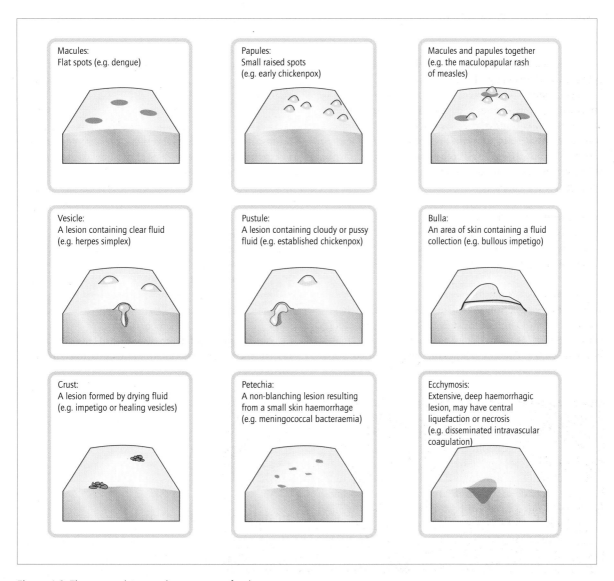

Figure 1.9 The nomenclature and appearance of rashes.

lesions of chickenpox may itch quite severely, but this is not so in every patient. However, in rashes caused by severe tissue damage, e.g. the meningococcal rash caused by intravascular coagulation, the more necrotic lesions can be painful.

Harmful effects of immune responses

Immune reactions can be clinically detectable as part of the acute disease, or as a late effect of the disease. This is described as the immunopathology of the disease.

The rashes of some viral infections such as scarlet fever are the manifestation of an immune vasculitis of the skin. The lung damage of respiratory syncytial virus infection is immunopathological, and can be made worse in experimental conditions by immunization against the virus.

Antibodies that accidentally damage human tissues may be manufactured in the course of an infection. Examples include immune thrombocytopenia after rubella and other viral infections, and red-cell agglutination in *Mycoplasma pneumoniae* infections.

Interferon is a lymphokine with many effects, including the inhibition of viruses and reduction of the metabolic activity of virus-infected cells. It also causes the symptoms of fatigue, malaise and myalgia that are typically seen in acute viral infections. High concentrations may contribute to the neutropenia of some viral diseases by a toxic effect on the bone marrow.

The primary function of cell-mediated immunity is to destroy infected cells. Occasionally a very vigorous response can cause severe tissue damage, such as hepatic necrosis in viral hepatitis. It is thought that a similar but slower-onset mechanism is responsible for post-viral encephalitis.

Antibody–antigen complexes (immune complexes) often form during immune reactions to infection. Most are harmlessly destroyed or cleared, but some may lodge in tissues such as glomerular capillaries or synovial membranes. If they combine with complement there is a risk that the complement will be activated, causing local inflammation and tissue damage. This takes time to develop, but the late effects can produce autoimmune-like post-infectious disorders. Rheumatic fever after *Streptococcus pyogenes* infections is a classic example of this, but post-infectious arthritis, nephritis and neuritis are nowadays more common (see Chapter 21).

These relatively rare complications of infection probably depend also on genetic factors in the patient. A good example of this is the predisposition of HLA B27-positive individuals to develop Reiter's syndrome. Other recognized factors include secretor status. Rare individuals who do not secrete blood-group antigens in their body fluids appear to be at greater risk of acquiring disease from organisms that they carry and also of developing some post-infectious disorders.

> A **post-infectious disorder** is a condition arising as a result of an infectious disease, but whose features tend to appear some time after the acute illness, and are markedly different from those of the acute infectious syndrome. Common post-infectious conditions include erythema multiforme (an immunologically-mediated vasculitis), reactive arthritides (often immune-complex-mediated), glomerulonephritis and encephalitis.

An immunocompromised patient may lack the immunopathological features of a disease. Thus, a child with leukaemia may have severe respiratory features of chickenpox with little or no rash, or a patient with AIDS may fail to develop granulomata in organs infected by mycobacteria (see Chapter 18).

Clinical assessment of a feverish patient

Most acute, short-lasting fevers are caused by infections. However, only a little over half of prolonged fevers are infectious in origin (see Chapter 20).

Commonest causes of community acquired infections
- Acute upper respiratory infections.
- Lower respiratory infections.
- Urinary tract infections.
- Gastrointestinal infections.
- Skin and soft-tissue infections.
- Bacteraemic infections (less than 5%).

Causes of fevers of unknown origin (fevers lasting over 2 weeks, with no diagnosis after initial investigation)
- Infections.
- Malignancies (especially lymphomas).
- Connective tissue and auto-immune diseases.
- Hypersensitivity to drugs or allergens.
- Recurrent pulmonary emboli.
- Metabolic disorders.
- Endocrine disorders.

Principles of investigation

Clinicians should remember that screening procedures work more efficiently if the population to be screened is preselected to include a high proportion of likely positives.

Clinicians select patients by obtaining a history of epidemiological exposure, or susceptibility to diseases, a clinical history of the evolution of the condition and the results of clinical examination. The results of these indicate which investigations and tests are likely to be useful. Otherwise an almost infinite range of tests could be performed, each with a different sensitivity and specificity, presenting the investigator with a hugely complex task when interpreting results.

History of exposure or susceptibility to diseases

A patient may have been exposed to infection by known contact with other cases, by travel, food, water, occupation or recreation, or by association with animals, including farm animals or pets.

Protection or resistance may be the result of natural immunity following previous infection, or it can be induced by immunization. Temporary resistance can also be obtained by the use of chemoprophylaxis, as for malaria, or passive immunization, as for hepatitis A. While none of these confers absolute protection from a condition, they reduce the likelihood of a particular disease and allow the investigator to choose priorities in the differential diagnosis of the fever.

Exposure to drugs and allergens may be important; this could be iatrogenic exposure to antibiotics or other drugs, or to environmental agents at work, home or play. Allergens might include bird proteins (as in pigeon-fancier's lung), organic dusts such as cotton or contaminated hay (byssinosis and farmer's lung) or industrial dusts and vapours, including vinyl chloride monomer or beryllium, which can both cause inflammatory or granulomatous lung disease.

Predisposition to non-infectious fevers may be indicated by a family history either of rare disorders, such as relapsing serositis, or Reiter's syndrome, and connective tissue diseases, including systemic lupus erythematosus and rheumatoid arthritis.

The patient may have a history of exposure to carcinogenic agents, such as radiation, including intensive radiotherapy. A history of sustained immunosuppressive therapy, for instance with cyclosporin, also indicates an increased likelihood of lymphomas or other malignant diseases because of impaired 'immune surveillance'.

Evolution of the feverish condition

Although the current complaint may be fever, this might have been preceded by symptoms either of an earlier stage of the disease, or by a recent precipitating condition. Such a history can point to appropriate diagnostic tests at an early stage in investigation.

The severity or pattern of the fever itself are not often helpful. The tertian fever of benign malaria is an exception, occurring when disease is well established or in re-

lapse. Similarly, the undulant fever of chronic brucellosis, the escalating fever of early typhoid and the relapsing fever of *Borrelia recurrentis* infection can be diagnostically helpful (Fig. 1.10) but they do not always occur in their classic form.

Acute viral infections are often marked by prostration, myalgia, arthralgia and shivering attacks. Transient diarrhoea, constipation, sore throat or cough could hint at the systemic site of the problem. Bacterial infections may similarly produce transient localizing symptoms. Abscesses and loculated sepsis are often accompanied by intermittent bacteraemias, indicated by rigors – severe shaking chills that make speech and other movement difficult.

In post-infectious conditions and connective tissue diseases transient rashes can occur and recur. They may be visible only when the temperature is highest (Fig. 1.11) or when the skin has been warmed by bathing.

If fever is due to a post-infectious disorder, the precipitating illness probably occurred 10–14 days earlier. Typically it would be a sore throat or a viral-type infection with respiratory symptoms or a rash. Next most likely would be a gastrointestinal complaint.

Physical examination

Sometimes subtle physical signs can be helpful in making a diagnosis; they should always be sought, and acted upon when found.

Essential investigations in a feverish patient

In general practice
- Epidemiological exposure history.
- Clinical history.
- Clinical examination (lymph nodes, skin, chest, ears, throat – and check the spleen if tonsillitis found).
- Rapid urine (dipstick) test.

In hospital
- Epidemiological exposure history.
- Clinical history.
- Full clinical examination.
- Chest X-ray.
- Rapid urine test.
- Urine microscopy and culture.
- Full blood count, with differential white cell count.
- Blood film examination.
- ESR and/or CRP.
- Liver function tests.

Localized bone or joint pain

Bone or joint pain can be extremely mild at the onset of bone and joint infections, appearing as discomfort, stiffness or, in children, reluctance to move the affected part. When it affects the leg or lumbar spine it is often more

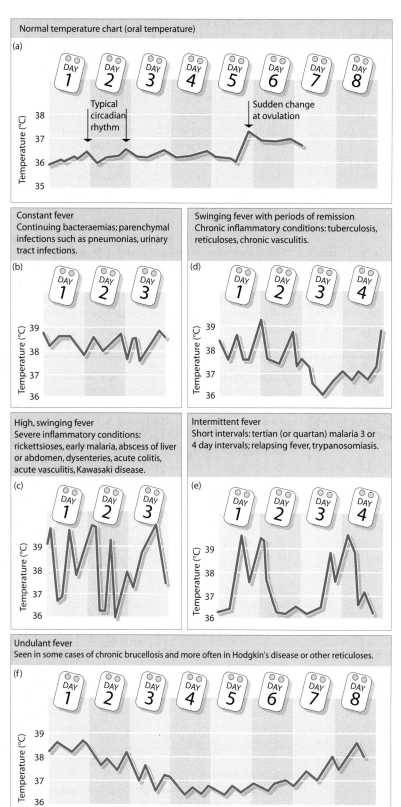

Normal temperature chart (oral temperature)

(a) Typical circadian rhythm / Sudden change at ovulation

Constant fever
Continuing bacteraemias; parenchymal infections such as pneumonias, urinary tract infections.

(b)

Swinging fever with periods of remission
Chronic inflammatory conditions: tuberculosis, reticuloses, chronic vasculitis.

(d)

High, swinging fever
Severe inflammatory conditions: rickettsioses, early malaria, abscess of liver or abdomen, dysenteries, acute colitis, acute vasculitis, Kawasaki disease.

(c)

Intermittent fever
Short intervals: tertian (or quartan) malaria 3 or 4 day intervals; relapsing fever, trypanosomiasis.

(e)

Undulant fever
Seen in some cases of chronic brucellosis and more often in Hodgkin's disease or other reticuloses.

(f)

Figure 1.10 A normal temperature chart, and some examples of patterns of fever.

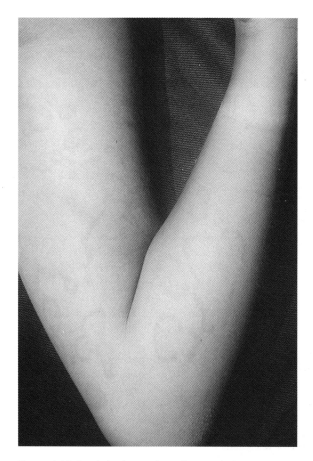

Figure 1.11 Post-infectious rash: erythema marginatum appearing during an episode of fever.

obvious on standing or walking. X-ray changes may not be visible for some weeks. Other imaging techniques are more likely to show early signs of infection (see Chapter 14).

Soft systolic murmurs
Soft systolic murmurs may be flow murmurs in feverish patients, but may also be signs of endocarditis, pericarditis or myocarditis. They should be reassessed regularly to detect changes, and investigated further at an early stage.

Subtle skin rashes
Subtle skin rashes may be the only sign of embolic or vasculitic phenomena. Showers of petechiae, Osler's nodes or simply small, vasculitic lesions of the digits can be signs of endocarditis or immune vasculitis. Splinter haemorrhages, small retinal haemorrhages and cytoid bodies have a similar significance.

Meningism
Mild meningism is an important warning sign of central nervous system disease. It is important to suspect such disease before more definite (and often more irreversible) signs develop. If there is doubt, early investigation by imaging and cerebrospinal fluid examination is essential.

Mild localized abdominal tenderness
Mild localized abdominal tenderness is easily dismissed. Some patients may be excessively sensitive, or trying to help by pointing up every possible sign, but it is difficult to describe vague visceral discomfort. Trivial abdominal signs may indicate peritoneal or subphrenic pathology. If vague signs persist, they should be followed up.

Chest X-ray
The chest X-ray is so important in assessing fever that initial examination is not complete without it. Extensive pulmonary consolidation or cavitation can exist with few or absent physical signs. Granulomas, infiltrations, small pleural effusions and mediastinal swellings are other important abnormalities immediately detectable by X-ray.

Initial laboratory investigations
Granulocyte counts
High neutrophil counts often occur in infections caused by capsule-bearing bacteria, in which the white cell count may reach $15–25 \times 10^9$/l. Intermediate elevations of around $12–16 \times 10^9$/l are common in many bacterial infections and in the presence of abscesses. Most bacterial infections are accompanied by neutrophilia, but important exceptions include early *Staphylococcus aureus* infections, even bacteraemias. Typically, enteric fevers and brucellosis induce neutropenia but in early infection there may be a neutrophilia as focal inflammation develops.

Eosinophilia may indicate an invasive helminth infection. It is a helpful sign in early and late schistosomiasis, filariasis, strongyloidiasis and liver fluke infection.

Lymphocyte count
The lymphocyte count is usually normal or low in viral infections, with the exception of the mononucleoses. The proportion of lymphocytes may appear raised because of mild neutropenia, possibly caused by the toxic effects of interferon. Scanty atypical or activated mononuclear cells are often seen in acute viral infections such as hepatitis A or rubella, but these do not approach the numbers seen in the mononucleoses. Pertussis toxin causes marked lymphocytosis ($15–25 \times 10^9$/l), which is an important diagnostic feature of pertussis.

Platelet count
The platelet count is rarely altered. Exceptions are severe Gram-negative sepsis in which intravascular coagulation may deplete platelet numbers, and malaria in which a similar mechanism operates. A reduced platelet count is

seen in established dengue fever, and is usual in dengue haemorrhagic fever. A high platelet count is usual in established Kawasaki disease and other severe vasculitides, and is a rare finding in disseminated tuberculosis.

Red cells

Obvious parasitization, distortion and granule formation affect the red cells in malaria. Typical inclusion bodies are seen in rare conditions such as babesiosis (red blood cells) or erlichiosis (monocytes or granulocytes). Abnormalities of red cell production or survival can alter the blood film in many infections. In parvovirus B19 infection, erythropoiesis is arrested by viral invasion of red cell progenitors. This causes a fall in the reticulocyte count. Conversely, the reticulocyte count will rise, causing an increase in the mean corpuscular volume, if haemolysis occurs, as in *Mycoplasma pneumoniae* infections.

The stained blood film can be diagnostic if it contains tropical parasites such as microfilariae or trypanosomes. It may be worth examining films taken in special circumstances, e.g. at night or after a dose of filaricide, to increase the likelihood of detecting parasitaemia.

Non-specific tests for inflammation (see above)

The erythrocyte sedimentation rate (ESR) is moderately raised (e.g. 35–50 mm/h) in acute infections and other inflammatory conditions. It is often markedly raised (e.g. over 70 mm/h) in the presence of persisting abscesses, in some pneumonias, such as legionnaires' disease, and in severe connective tissue diseases and hypersensitivity reactions. The C-reactive protein rises rapidly in bacterial infections, viral infections, parasitic diseases and non-infectious conditions. It can be high when the ESR is near normal, often in non-infectious conditions, or in the presence of fibrinogen consumption. Its best use is in demonstrating the presence of an inflammatory response and in monitoring the response to treatment (as in the management of endocarditis).

Blood biochemical tests

Aspartate transaminase and alkaline phosphatase levels

Slight elevations of transaminases are common in acute viral infections, but rarely to levels above 60–100 IU/l.

Higher levels point to specific liver inflammation or occasionally to severe and widespread tissue damage. Elevated alkaline phosphatase levels are usually of liver origin. They can indicate space-occupying lesions in the liver, but do not distinguish many small lesions, such as granulomata, from larger lesions, such as abscesses or tumours.

Other enzymes

Blood amylase levels may be high in pancreatitis or inflammation of the salivary glands. Creatine kinase elevations are seen in myositis, including myocarditis, and in toxic shock syndromes.

Urine biochemical tests

Urine biochemical tests for protein, blood, nitrite and excess neutrophil lactate dehydrogenase (LDH) may give early indication of infection (see Chapter 2). Products of haemolysis may be present in severe malaria, and bilirubin may be present when jaundice is too mild for clinical detection. Screening tests for porphyria are carried out on urine samples (preferably taken during episodes of fever).

Initial microbiological investigations (see Chapter 3)

Microscopy of easily obtained specimens

This is a rapid diagnostic procedure that is easy to overlook. As well as a search for neutrophils and pathogens in urine and stools (see Chapter 2), or the examination of the cerebrospinal fluid obtained by lumbar puncture, investigation of aspirates from abscesses, enlarged lymph nodes or the bone marrow can aid early diagnosis. Smears and fine-needle aspirates of lymph nodes may be stained to demonstrate granulomas, malignant cells or acid–alcohol-fast bacilli. Bronchial aspirate can be examined by various stains or polymerase chain reaction (PCR) techniques to reveal fungal, herpesvirus or respiratory virus infections. Actinomycosis can be diagnosed by the Gram-stained appearance of 'sulphur granules' from pus.

In many cases, the early findings from history-taking and simple investigation will indicate the patient's diagnosis. More details of the natural history and diagnosis of specific conditions will be found in the following chapters of this book.

Case study 1.1 Tuberculosis and the 'iceberg of infection'

History
Two 6-year-old classmates developed tuberculous meningitis within 3 weeks of one another. Both were English children, born and brought up in a prosperous town in East Anglia, where tuberculosis is extremely uncommon.

Question
How should this unusual event be followed up?

Epidemiological investigation
Both cases were notified to the Proper Officer of the Local Authority, as required by statute. The consultant in communicable disease control visited the school to gather more information about the teacher and classmates. The children belonged to a class of 25 6- and 7-year-olds. None had received BCG vaccination, as it would not normally be offered in this community. On enquiry, it was found that their teacher had a 10-week history of persistent cough, and was found on subsequent investigation to have infectious pulmonary tuberculosis.

The 23 classmates all had tuberculin skin testing, and 17 of them tested tuberculin positive (usual prevalence in this community < 2%). Four of these had X-ray evidence of pulmonary or pleural tuberculosis.

Question
What action should be taken for the different groups of children?

Management and progress
The six children with tuberculosis were all treated and recovered. The tuberculin-positive children received prophylactic chemotherapy with daily isoniazid for 6 months, as recommended for tuberculin-converters under the age of 16 years (who are at high risk of developing disease); non-infected children were offered immunization with BCG.

This clearly demonstrates the iceberg of infection – 8% of exposed children had severe infection; 16% had milder clinical disease; 52% had immunological evidence of infection without demonstrable clinical disease; 28% escaped infection and remained susceptible.

Structure and Classification of Pathogens

2

Introduction

Taxonomy is the process of classifying living organisms into groups (taxa) of related individuals or species. Taxonomists weigh individual characteristics equally but clinical microbiologists consider ability to cause disease more important than, say, the ability to ferment a particular sugar or possession of a particular 16S rRNA gene sequence. The medical microbiologist aims to classify microorganisms into groups that are relevant to, and can predict, their behaviour in patients. DNA sequencing is increasingly available in routine microbiology laboratories, so microbiologists must be familiar with molecular taxonomy. More accurate identification of an organism will aid our study of its pathogenic potential.

Pathogens can be classified into five main groups: viruses, bacteria, fungi, protozoa and metazoa (usually helminths). Viruses are smaller than bacteria and consist of a piece of either DNA or RNA supported by nucleoprotein, some enzymes for replication and a 'casing' of structural protein. Some viruses possess an envelope, derived from host cells (Fig. 2.1). Viruses cannot replicate independently but grow inside host cells, taking control of cellular biochemical processes and subverting them for virus production. Some bacteria, e.g. *Chlamydia*, and protozoa, such as the microsporidia, are also obligate intracellular pathogens.

Classification of microorganisms

- Viruses.
- Bacteria.
- Protozoa.
- Fungi.
- Metazoa.

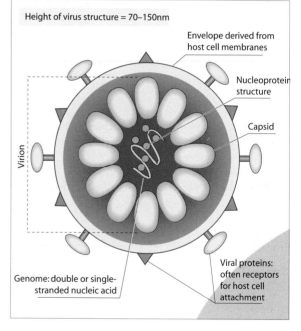

Height of virus structure = 70–150nm

Envelope derived from host cell membranes

Nucleoprotein structure

Capsid

Virion

Viral proteins: often receptors for host cell attachment

Genome: double or single-stranded nucleic acid

Figure 2.1 General structure of a virus.

Bacteria are single-celled organisms with a single circular DNA chromosome that is not enclosed in a nucleus. Such organisms are called prokaryotes. Bacteria have a plasma membrane and a cell wall of characteristic composition. Bacteria replicate by binary fission. They have specialized surface structures that serve special functions: pili and fimbriae are for attachment and flagella for motility (Fig. 2.2), but they have no internal organelles. Some bacteria can form extremely durable spores. The detailed structure of bacteria is discussed below.

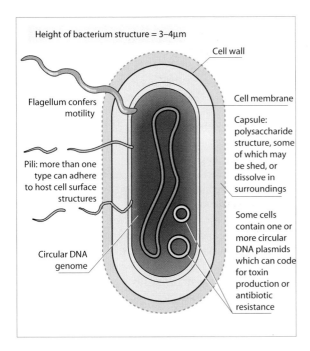

Height of bacterium structure = 3–4µm

Cell wall

Cell membrane

Flagellum confers motility

Capsule: polysaccharide structure, some of which may be shed, or dissolve in surroundings

Pili: more than one type can adhere to host cell surface structures

Some cells contain one or more circular DNA plasmids which can code for toxin production or antibiotic resistance

Circular DNA genome

Figure 2.2 General structure of a bacterium.

Mycoplasmas are small bacteria lacking peptidoglycan-containing cell walls. They are the smallest organisms able to live and replicate independently.

Rickettsiae and chlamydiae are also small, and structurally resemble Gram-negative bacteria. They lack some enzymes needed for independent existence, and must replicate intracellularly, borrowing host enzyme systems.

Protozoa are single cells; their name is derived from the Greek words for 'first animals'. Protozoa are diploid, possessing paired chromosomes. As they also possess a nucleus they are eukaryotes. They have several specialized organelles (see below). Protozoan lifecycles may be simple, involving only one species, or require intermediate hosts or vectors. Fungi are also eukaryotes, with cell walls containing chitin, cellulose or both. They reproduce by sexual or asexual processes, forming germinative spores.

The final group of organisms that must be considered are the metazoa or multicellular organisms. Virtually all metazoa pathogenic to humans are helminths.

Structure and classification of viruses

The classification of pathogenic viruses developed later than the classification of other organisms. Original classifications of viruses simply described the diseases caused (e.g. rubella), the site of viral shedding (e.g. enteroviruses) or the means of transmission (arthropod-borne; arboviruses). Modern classification is based on genetically determined features (Table 2.1), particularly:

1 the type of nucleic acid and its means of transcription;
2 the structure and symmetry of the structural proteins (capsids); and
3 the presence or absence of an envelope.

Table 2.1 Classification of viruses

Name of family	Type of nucleic acid	Genome size (kb)	Envelope	Examples: genera of medical importance
Poxviridae	ds-DNA	130–280	No	Molluscum contagiosum
Herpesviridae	ds-DNA	120–220	Yes	Herpes simplex, Epstein–Barr virus, etc.
Adenovirus	ds-DNA	36–38	No	Adenovirus
Papovaviridae	Circular ds-DNA	8	No	Human wart virus
Parvoviridae	ss-DNA	5	No	Parvovirus
Hepadnaviridae	ds-DNA with ss portions	3	Yes	Hepatitis B virus
Picornaviridae	ss(+)RNA	7.2–8.4	No	Poliovirus
Togaviridae	ss(+)RNA	12	Yes	Rubella virus
Flaviviridae	ss(+)RNA	10	Yes	Yellow fever virus
Caliciviridae	ss(+)RNA	8	No	Norwalk agent
Rhabdoviridae	ss(–)RNA	13–16	Yes	Rabies virus
Paramyxoviridae	ss(–)RNA	16–20	Yes	Measles virus
Orthomyxoviridae	ss(–)RNA	14	Yes	Influenza virus
Reoviridae	ds segmented RNA	16–22	No	Rotavirus
Arenaviridae	ss(–)RNA	10–14	Yes	Lassa
Retroviridae	ss(+)RNA	3–9	Yes	HIV-1

and an RNA-dependent RNA polymerase that is used to replicate the viral RNA.

If the RNA is negative, or antisense, a different strategy must be adopted. These viruses encode an RNA polymerase (or transcriptase) that transcribes the viral genome into positive RNA, which then acts either as a template for further viral genomic (negative-strand) RNA or as mRNA for translation into proteins (Fig. 2.3).

The third RNA-based strategy is that of the retroviruses, which have single-stranded positive (sense) RNA that cannot act as mRNA. The RNA is transcribed into DNA by an RNA-dependent DNA polymerase (reverse transcriptase, or RT). The DNA enters the host nucleus and is inserted into host DNA. In the host DNA molecule, it is under the control of host transcriptase enzymes, which make mRNA and viral genomic RNA.

The reoviruses, which include rotaviruses, possess a segmented, double-stranded RNA genome. Their complex reproductive strategy includes the use of a double-

---

Genomic structure in viruses

DNA viruses
- Double-stranded DNA.
- Single-stranded DNA.

RNA viruses
- Positive single-stranded RNA (sense).
- Negative single-stranded RNA (antisense).
- Double-stranded RNA.
- Positive single-stranded RNA that cannot act as messenger (retroviruses).

Nucleic acid

Both the type of nucleic acid and its method of transcription are considered important, as there is considerable variation in mechanisms of viral reproduction.

DNA viruses

Viral DNA can be either double-stranded or single-stranded (ds or ss). Double-stranded DNA viruses include a number of important human pathogens. The poxvirus family, possessing the largest viral genomes, includes the agents of smallpox, molluscum contagiosum and also vaccinia virus (an essential factor in the smallpox eradication programme, and now a potential vaccine vector). The herpesvirus family includes herpes simplex, varicella zoster, cytomegalovirus, Epstein–Barr virus and the more recently described human herpes virus 6 and 7. Adenoviruses also possess double-stranded DNA. Hepatitis B virus is unusual in that it possesses double-stranded DNA with a single-stranded portion. The papilloma and polyomaviruses are small viruses containing double-stranded DNA. These viruses are associated with benign tumours (warts) and malignant tumours (cervical, genital and laryngeal cancer).

There are two single-stranded DNA viruses: parvovirus, responsible for 'fifth disease' or 'slapped cheek syndrome', and TT virus, a blood-borne virus transmissible by transfusion.

Viral DNA is usually replicated in the nucleus of host cells, using viral DNA polymerase. Newly formed viral DNA is not inserted into host chromosomal DNA, but is assembled with viral, nucleoprotein and capsids to form new virus particles.

RNA viruses

Single-stranded RNA viruses adopt one of three reproduction strategies, depending on whether the RNA is positive or negative (sense or antisense). If it is positive, or sense, it can serve directly as messenger RNA (mRNA) and be translated into protein, which includes structural proteins

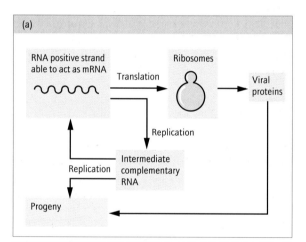

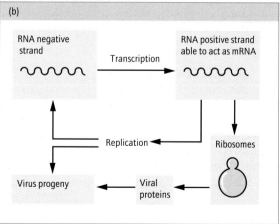

Figure 2.3 (a) Replication of RNA virus with positive polarity (able to be read as messenger (m)RNA). (b) Replication of RNA virus with negative polarity.

(a)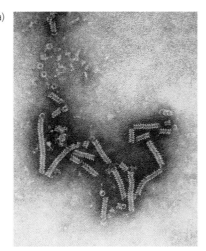

(b)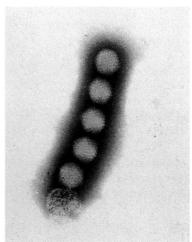

Figure 2.4 (a) Electron micrograph of a virus with helical symmetry. Parainfluenza type 3 virus × 100 000. (b) Electron micrograph of a virus with icosahedral symmetry. Adenovirus × 100 000.

stranded RNA–single-stranded RNA polymerase, which produces positive (sense) RNA from the double-stranded portion of viral RNA, using the antisense strand as a template. The positive (sense) RNA is extruded from the virus, serving both as mRNA and also as a template to make further antisense RNA, which is then annealed with the complementary strand to form double-stranded RNA.

Capsid symmetry

Viral nucleic acid is contained within a protein coat made up of repeating units known as capsids, arranged in either icosahedral or helical structures. In RNA viruses with helical symmetry the capsids are bound around the helical nucleic acid (Fig. 2.4a). Icosahedral symmetry occurs in both DNA and RNA viruses, the capsids forming an approximately spherical polyhedral structure (Fig. 2.4b). Using simple, repeating structures minimizes the amount of nucleic acid devoted to viral coat production and simplifies the process of viral assembly.

Envelopes

In some viruses the nucleic acid and capsid (the nucleocapsid) are surrounded by a lipid envelope derived from the membrane of the host cell or nucleus (Fig. 2.5). The host membrane is altered by virus-encoded proteins or glycoproteins that may fuse several host cells together, which facilitates the passage of viruses from cell to cell. Viruses possessing an envelope are sensitive to ether and other substances that dissolve lipid membranes.

Virus attachment

Virus attachment to host cell membrane is a critical step in the process of infection, and in determining tissue tropism. Viruses have evolved specific antigens that target

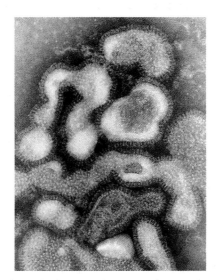

Figure 2.5 Electron micrograph of an enveloped virus. Influenza virus × 100 000.

receptors on the host cells. These are sometimes known as virus attachment proteins or VAPs. The VAP of influenza virus is a haemagglutinin, whereas that of human immunodeficiency virus (HIV) is a glycoprotein that binds to the CD4 antigen of T cells.

Structure and classification of bacteria

Classification

For descriptive purposes bacteria are often grouped by

four main characteristics: the Gram reaction, shape, atmospheric requirements for respiration and the presence of spores.

Classification of bacteria
- Gram reaction.
- Shape.
- Atmospheric requirement.
- Presence of spores.

Gram reaction
The Gram reaction or Gram stain uses the ability of stains such as crystal violet to bind strongly to the cell wall teichoic acids of Gram-positive cells, and thereby resist decoloration by alcohol or acetone. It is a simple way of identifying bacteria with different cell wall structures (Fig. 2.6). This is important, as Gram-positive and Gram-negative organisms have different pathogenic potential and different antibiotic susceptibilities.

Shape
Bacterial cell walls determine the organisms' shape, which is consistent for individual genera. Cocci are spherical

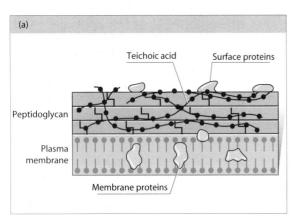

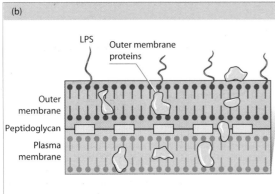

Figure 2.6 Structures of (a) Gram-positive and (b) Gram-negative bacterial cell walls. LPS, lipopolysaccharide.

and include important human pathogens such as streptococci and staphylococci. Bacilli are rod-shaped and may be short (coccobacilli) or long. They may also vary their shape throughout their length, as do *Fusobacterium* spp. Spiral bacteria are a diverse group, including *Treponema pallidum*, the causative organism of syphilis, which has a very short 'wavelength', and *Borrelia* spp., which have a longer wavelength, and include the organisms of Lyme disease and relapsing fever. Other spiral organisms, such as the intestinal pathogens *Campylobacter* and *Helicobacter*, have only a few turns.

Atmospheric requirements
Bacteria may be classified by their atmospheric requirements into five groups.
- **Obligate aerobes**, e.g. *Bordetella pertussis*, are obliged to use oxygen as a terminal electron acceptor.
- **Microaerophilic** organisms, e.g. *Campylobacter* spp., use oxygen as the terminal electron acceptor but only grow under conditions of reduced oxygen tension.
- **Capnophiles**, e.g. *Brucella* spp., are organisms that grow optimally in an atmosphere with an increased carbon dioxide concentration.
- Many human pathogens are **facultative anaerobes**, capable of growing aerobically or anaerobically, e.g. staphylococci, streptococci and enteric organisms.
- **Obligate anaerobes** only grow in the absence of oxygen. They are divided into aerotolerant and strict anaerobes, depending on their sensitivity to oxygen. This difference has practical importance as strict anaerobes do not survive long in specimens and are much more difficult to isolate in the laboratory.

Endospores
Two genera of bacteria found in humans, *Clostridium* and *Bacillus*, possess a bacterial endospore. The shape, size and position of this spore may be helpful in identifying species, especially in the genus *Clostridium* (Fig. 2.7). The ability to form spores helps organisms to survive in the environment and has an important role in the epidemiology of several diseases, including tetanus and gas gangrene.

Bacterial structures

The cell wall
The bacterial cell wall is essential for survival, as it must withstand the immense osmotic pressure difference between the interior and exterior of the cell. Its strength and rigidity depends on peptidoglycan, a polymer of muramic acid and *N*-acetylglucosamine, cross-linked by peptide bridges.

The Gram-positive cell wall has two layers, consisting of the plasma membrane and a thick peptidoglycan layer. In contrast, the Gram-negative wall has three layers: the

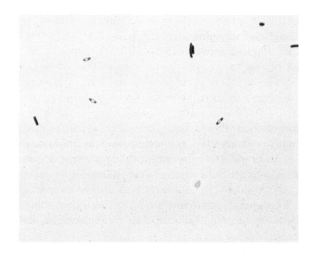

Figure 2.7 Structure of *Clostridium* sp. showing endospore.

plasma membrane, a thinner peptidoglycan layer and an outer, lipid membrane (see Fig. 2.6). Cell walls are important targets for beta-lactam and glycopeptide antibiotics (see Chapter 4).

Many bacterial antigens present in the cell wall have an important role in pathogenesis (see Chapter 1). Important examples are lipopolysaccharide (endotoxin), which is an integral part of the Gram-negative outer membrane, and teichoic acid from *Streptococcus pneumoniae*, which causes both complement activation and attraction of neutrophils to the site of infection. The cell wall of mycobacteria contains approximately 40% lipid, including lipoarabinomannan, mycolic acid and phenolic glycolipids. It will not stain satisfactorily with conventional Gram stain. Vigorous staining and decolorizing methods are required to demonstrate it (see Chapter 18). Several of the surface lipids of mycobacteria are implicated in inhibiting macrophage function (e.g. lipoarabinomannan) or phagocytosis (phenolic glycolipid from *Mycobacterium leprae*).

The plasma membrane

The bacterial plasma membrane, as in other cells, consists of a trilaminar membrane with two outer hydrophilic layers and an inner lipid core. In the absence of internal membrane-bound structures, the plasma membrane is an important site of bacterial metabolism.

Bacterial capsules

These are polysaccharide surface structures possessed by some of the most important pathogens, including *Streptococcus pneumoniae*, *Haemophilus influenzae*, *Neisseria meningitidis*, *Streptococcus agalactiae*, *Salmonella typhi*, *Bacillus anthracis* and *Klebsiella pneumoniae*. *Pseudomonas aeruginosa* sometimes elaborates such abundant alginate capsular material that pus appears gelatinous at infected sites or on culture plates (Table 2.2). Capsules protect pathogens from phagocytosis and humoral immune attack, and inhibit activation of the alternative complement pathway (see Chapter 1).

A capsule is not, in itself, a pathogenicity determinant as many organisms not usually pathogenic for humans are capsulate. The bacterial capsule may have evolved as a mechanism to aid survival during conditions of desiccation.

Immune response to polysaccharide capsular antigens

The human host does not respond efficiently to polysaccharide antigens. Such antigens are called T cell-independent, as T cells are bypassed in the development of antibody to them, and effective immunological memory is not therefore established. T cells are, however, involved in the modulation of humoral responses to these antigens. Antibodies are mainly of the immunoglobulin G_2 (IgG_2) isotype. The multiplicity of capsular types expressed by some organisms makes repeated infections possible, and complicates the design of vaccines.

The spleen has an important role in the control of infection with capsulate organisms, as intrasplenic phagocytes can ingest particles that have not been opsonized; thus patients with splenectomy are especially susceptible to pneumococcal infection (see Chapter 22).

Protein 'capsules'

The M antigen of *Streptococcus pyogenes* is a fibrillar pro-

Table 2.2 Bacterial pathogens whose capsules are important in pathogenesis, and used to identify pathogenic strains

Species	Associated diseases
Streptococcus pneumoniae	Meningitis, pneumonia, septicaemia
Neisseria meningitidis	Meningitis, septicaemia
Haemophilus influenzae	Meningitis, septicaemia
Streptococcus agalactiae	Neonatal septicaemia and meningitis
Klebsiella pneumoniae	Pneumonia, septicaemia, wound infection
Pseudomonas aeruginosa	Pneumonia in cystic fibrosis patients
Escherichia coli	Meningitis, septicaemia
Bacillus anthracis	Anthrax

tein forming an extracellular capsule-like structure, which protects against phagocytosis by polymorphonuclear leucocytes. Strains of *S. pyogenes* cannot be ingested unless opsonized with specific anti-M antibody. There are more than 80 antigenically distinct M types, and immunity to one does not protect against strains carrying different M antigens. This accounts for the recurrent attacks of streptococcal tonsillitis that some individuals may experience throughout their life. The M antigen may cross-react with a number of human tissue antigens, with important pathological consequences. Cross-reaction with myocardial antigens is found in many different M types and is thought to be responsible for the pancarditis of rheumatic fever. Infection with a second type of *S. pyogenes* may cause a recrudescence of rheumatic fever. Post-streptococcal glomerulonephritis is confined to a few M types, including 4, 12 and 49, resulting from cross-reaction with the glomerular basement membrane. Infection with another strain of *S. pyogenes* will not result in a recrudescence unless it too is a 'nephritogenic' strain (Table 2.3).

Glycolipid capsules

Many species export extracellular glycolipid material, which appears to be important to the organism as a pathogenicity determinant. *Mycobacterium leprae* can survive inside macrophages. It has been shown that the phenolic glycolipid capsule inhibits the activity of the macrophage myeloperoxide–halide microbicidal system, possibly by scavenging free radicals.

Bacterial extracellular material

1 Polysaccharide capsules.
2 Extracellular slime.
3 M protein.
4 Phenolic glycolipid.

Bacterial adhesins

1 Capsular polysaccharide.
2 Extracellular slime.
3 Fimbriae.
4 Lectins.

Extracellular slime

Many organisms of low virulence, such as *Staphylococcus epidermidis*, colonize intravascular prosthetic devices. This causes fever or septicaemia in patients with intravenous cannulae or long-term intravascular devices such as Hickman catheters. Slime-producing organisms adhere better to the cannulae and more readily colonize and cause infection.

Pseudomonas aeruginosa is an important pathogen of children with cystic fibrosis. It can express a mucoid phenotype in which copious amounts of exopolysaccharide alginate is produced. Alginate is a non-repeating copolymer of beta-D-mannuronate and its C5 epimer, alpha-D-glucuronate, similar to an alginate produced by brown seaweed. Alginate production assists the organism because it enables the formation of microcolonies protected from humoral and cellular immune responses. Paradoxically, alginate-producing strains are often exquisitely sensitive to antimicrobial agents.

Fimbriae or pili

Fimbriae (pili) have an important role in adhesion and assist bacteria in allowing access to a new host or locating the organism close to a nutrient supply. Pili are filamentous proteins capable of binding to host antigens by acting as lectins (proteins that bind to carbohydrate residues). One bacterial strain may express several different pili and may up- and down-regulate different types as required. For some organisms, such as *Neisseria gonorrhoeae*, pili

Table 2.3 Number of *Streptococcus pyogenes* isolates from different sites from 1980–1990 by M type

M type	Number of isolates			
	Scarlet fever	Acute glomerular nephritis	Rheumatic fever	Puerperal infections
1	21	24	5	10
2	10	3	0	2
3	85	6	3	11
4	75	10	2	9
9	0	2	1	20
12	9	20	0	10
R28	8	3	0	48
49	0	8	0	2
75	3	0	0	15

Data courtesy of the Streptococcal Reference Laboratory, Health Protection Agency, formerly Public Health Laboratory Service.

are necessary for pathogenesis. Possession of pili may assist the adhesion of organisms to epithelia. Uropathic *Escherichia coli* express not only type 1 fimbriae, in common with many Enterobacteriaceae, but also P fimbriae, which bind to a digalactoside found in the urinary tract. The role of fimbriae is discussed in more detail on p. 176.

Flagella

The flagellum is the main organ of bacterial motility. It is made up of a globular protein, flagellin, arranged in a multistrand helix. The flagellum is bound to the plasma membrane via the motor unit, where energy provided by an ion gradient across the membrane is converted into rotary movement. Motility *per se* is not often a pathogenicity determinant, but *Vibrio cholerae* use their motility to burrow through intestinal mucus to the epithelium and strains lacking flagella are non-pathogenic.

Flagella are useful in the laboratory as an identification feature. Flagellar antigens are frequently used for serological classification of enteric Gram-negative bacteria such as *Salmonella* and *Shigella*.

Pathogenicity islands

These are collections of genes that code for pathogenicity and virulence determinants. These genes may include toxins, adhesins, invasins or other virulence factors. Pathogenicity islands are found in both Gram-positive and Gram-negative pathogens and are large genomic regions often containing up to 200 kb of DNA. The G + C content (proportion of guanine plus cytosine in the DNA bases) often differs from the rest of the genome, indicating that the organism has acquired this section of DNA from another organism. The genetic segments are often flanked by specific DNA sequences such as direct repeats or insertion sequence (IS) elements. They may also contain bacteriophage attachment sites and cryptic genes coding for bacteriophage integrases or origins for plasmid replication. These features indicate that pathogenicity islands are transmissible among bacterial populations by horizontal gene transfer, and probably contribute to rapid evolution in bacterial populations.

Structure of protozoa

Protozoa are unicellular microorganisms. They exist in almost every type of environment, over a wide range of pH (3–9.5), temperature, salinity and redox potential.

Classification

The classification of protozoa can be very complicated but may be simplified by subdivision into four main groups: spore forming, flagellate, amoeboid and ciliate.

> ### Classification of protozoa
> 1 Sporozoa (e.g. *Microsporidium*, *Plasmodium*, *Toxoplasma*).
> 2 Flagellates (e.g. *Leishmania*, *Trypanosoma*, *Trichomonas*).
> 3 Amoeboid (*Entamoeba histolytica*, *Acanthamoeba* spp.).
> 4 Ciliates (e.g. *Balantidium coli*).

Spore-forming protozoa

Spore-forming protozoa can be divided into two groups.
1 Microspora have recently been recognized as human pathogens, particularly in acquired immunodeficiency syndrome (AIDS) patients (see Chapter 15). *Encephalitozoon cuniculi*, *Enterocytozoon beneusi*, *Encephalitozoon intestinalis*, *Pleistophora*, *Nosema connori* and *Vittaforma corneum* are the most important species reported.
2 Sporozoa. This group includes many important human pathogens. The sporozoa possess a complex of organelles, known as the apical complex, that is important in cell invasion. In the Eimeridia (which include *Toxoplasma gondii*), sexual reproduction occurs in the intestine of the definitive host (e.g. cat species for *Toxoplasma*). In contrast, the sexual stage of haemosporidians, such as *Plasmodium* spp., occurs in mosquitoes, or in ticks for piroplasms such as *Babesia*.
   - Eimeridia: *Isospora belli*, *T. gondii*, *Cryptosporidium parvum*, *Sarcocystis* sp.
   - Haemosporidians: *Plasmodium* spp.
   - Piroplasms: *Babesia* spp.

Flagellate protozoa

These comprise the intestinal/genital flagellate protozoa and the blood flagellates. Intestinal flagellates include *Giardia intestinalis* and *Trichomonas* spp.

The blood flagellates are the Trypanosomatidae and are characterized by the possession of a kinetoplast (a dense body at the base of the flagellum). They are transmitted to humans by the bite of arthropod vectors: sandflies in the case of *Leishmania*, *Glossina* (tsetse) flies for African trypanosomiasis, and triatomid bugs for South American trypanosomiasis. *Leishmania* spp. exist in a flagellate form in the arthropod vector (Fig. 2.8) but after injection into the human host they live intracellularly as non-flagellate amastigotes.

Amoeboid protozoa

The most important genus of amoebae in human pathology is *Entamoeba*. The main recognized pathogenic species, *E. histolytica*, could be subdivided by electrophoresis of certain enzymes into pathogenic and non-pathogenic subgroups (or zymodemes). These groups can now be distinguished by monoclonal antibodies, lectin binding and

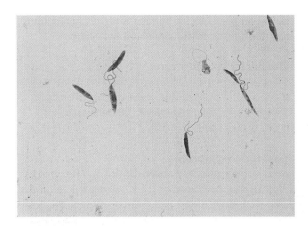

Figure 2.8 Flagellate appearance of *Leishmania* sp.

serum sensitivity, and are assigned to two different species: *E. histolytica*, which includes the pathogenic organisms, and *E. dispar*, which contains the non-pathogenic strains.

Naegleria fowleri causes a rare but usually fatal form of meningitis. It has an amoeboid form in the tissues, and a cyst and a flagellate form in the environment. It could therefore be classified with either group. *Acanthamoeba* can cause meningitis, and also a severe form of conjunctivitis associated with contaminated contact lenses.

Blastocystis hominis is an organism of uncertain pathogenic potential but it has been associated with diarrhoeal disease. Its taxonomic position is also uncertain but it is often classified with the amoebae. *Dientamoeba fragilis*, traditionally classified as an amoeba, has now been reclassified as a trichomonad.

Ciliate protozoa

Balantidium coli is the only ciliate protozoan regularly found in human specimens. It is sometimes identified in faeces, but is rarely implicated in disease.

Nuclear structure of protozoa

Protozoa are eukaryotes, whose DNA exists as chromosomes within a nucleus. The nucleus is surrounded by a tough nuclear membrane, which has multiple channels connecting the nucleoplasm with the endoplasm. Some organisms have a single nucleus, and others may be binucleate (e.g. *Giardia*) or form multinucleate cysts (e.g. *Entamoeba histolytica*). The nucleus may contain single or multiple linear chromosomes. Reproduction is by both sexual and asexual mechanisms.

Cytoplasmic structure of protozoa

The endoplasm is the inner portion of the cytoplasm

immediately surrounding the nucleus. It contains chromatidoid bodies, endoplasmic reticulum, Golgi bodies, mitochondria, food vacuoles and microsomes. The ectoplasm is the metabolically active portion of the cell, and is involved in locomotion, respiration, osmoregulation and phagocytosis. The cell is limited by the plasma membrane, which controls the intake and output of food, secretions and metabolic waste products. The plasma membrane may vary in shape and can often form pseudopodia for locomotion.

Protozoa have a number of structures external to the plasma membrane. These include an external glycocalyx (as in *Leishmania donovani*), or a tough cyst wall to enable the organism to survive in the external environment (as in *Giardia intestinalis*). Encystation is triggered by alteration in food or oxygen supply, excess of catabolic components, pH changes or desiccation. Some protozoa have specialized organelles: *Leishmania* spp. possess flagella, which arise from a specialized kinetoplast containing its own DNA and metabolic processes. Others, such as *B. coli*, express cilia, which beat in formation to propel the organism.

Life cycle

Protozoa have reproductive cycles that may involve a sequence of different hosts and environments. These may include vectors, acting to transmit the infection between hosts and reservoirs of infection. An example of this is the life cycle of *Plasmodium* sp., the cause of malaria (Fig. 2.9). Vectors can also act as semipermanent reservoirs of infection by sustaining vertical transmission, for example certain viral encephalitides in generations of ixodid ticks.

Examples of protozoa that may be transmitted by vectors

1 Mosquitoes (*Plasmodium*).
2 *Glossina* spp. flies (*Trypanosoma brucei*).
3 Sandflies (*Leishmania*).
4 Ticks (*Babesia*).
5 Bugs – triatomid (*Trypanosoma cruzi*).

Protozoa may also be transmitted by water, air, sexual intercourse, food and the faecal–oral route (Table 2.4).

Sexual stages of protozoal life cycles

Unlike bacteria, which have only limited sexual function based on various means of transferring genome fragments, protozoa have well-developed sexual reproduction. In amoebae, simple exchange of DNA is thought to take place, and in ciliates microgametocytes are exchanged during conjugation. Sporozoa have asexual and sexual generations in their life cycle; in *Toxoplasma*, schizogony, gametocyte formation and zygote formation all occur in

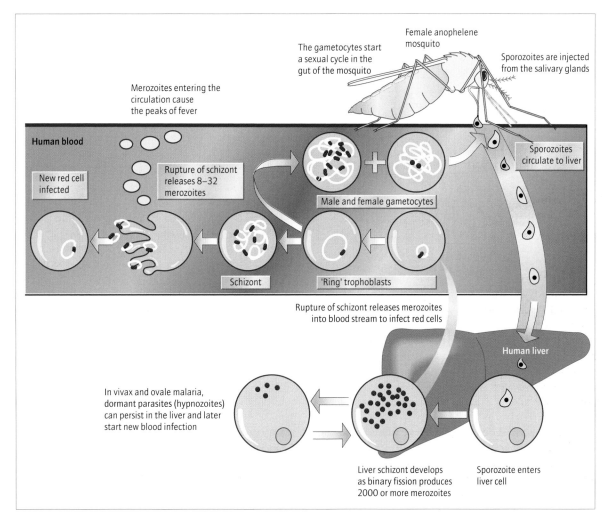

Figure 2.9 Life cycle of *Plasmodium* sp.

Table 2.4 Transmission of protozoa

Route	Example
Faecal–oral	*Entamoeba histolytica*
Water	*Cryptosporidium parvum*
Food	*Toxoplasma gondii*
Air	*Pneumocystis carinii*
Sexual	*Trichomonas vaginalis*
Vector	*Plasmodium vivax*

the same host. In sporozoa, including malarial parasites, schizogony and gametocyte production occur in the vertebrate host and gametogeny is completed in the invertebrate host.

Classification of fungi

Clinical classification of fungi of medical importance
- Cutaneous, e.g. dermatophytes.
- Systemic, e.g. *Coccidioides*, *Histoplasma*.
- Fungal infections of immunocompromised hosts, e.g. *Aspergillus*.

The formal classification of fungi is based upon means of reproduction and morphology of sexual and asexual stages (Fig. 2.10). Unfortunately, this mode of classification has little clinical relevance, so a simplified clinical classi-

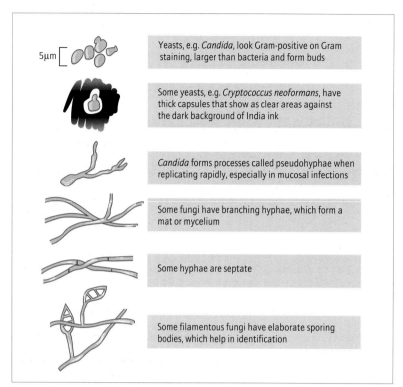

5μm

Yeasts, e.g. *Candida*, look Gram-positive on Gram staining, larger than bacteria and form buds

Some yeasts, e.g. *Cryptococcus neoformans*, have thick capsules that show as clear areas against the dark background of India ink

Candida forms processes called pseudohyphae when replicating rapidly, especially in mucosal infections

Some fungi have branching hyphae, which form a mat or mycelium

Some hyphae are septate

Some filamentous fungi have elaborate sporing bodies, which help in identification

Figure 2.10 The morphologies of pathogenic fungi.

fication is described here. This divides pathogenic fungal species into three groups:

1 the superficial and subcutaneous fungi;
2 the systemic fungi; and
3 the fungi associated with immunocompromised patients.

Superficial fungal infections are common. They include infection with dermatophytes, such as *Microsporum*, *Trichophyton*, and the yeast-like fungus *Malassezia furfur*, the causative organism of pityriasis versicolor. Rarer cutaneous fungal pathogens include *Sporothrix schenckii*, the cause of sporotrichosis, and the agents of piedra.

Systemic fungi include *Histoplasma capsulatum*, *Coccidioides immitis* and *Paracoccidioides braziliensis*. These are dimorphic fungi, which have both yeast-like and filamentous forms. They are environmental organisms that usually enter the human body via the respiratory tract. Infection is geographically localized and often clinically mild. Severe disease can occur, however, particularly in immunocompromised subjects.

The main fungi that infect immunocompromised patients are the yeasts (*Candida albicans*) and related species, such as *C. krusei* and *Torulopsis glabrata*. *Aspergillus* species are important filamentous fungi, usually causing pulmonary or disseminated infection. The commonest species implicated in human infection are *A. niger*, *A. fumigatus* and *A. flavus*. The yeast *Cryptococcus neoformans* was a rare cause of chronic lymphocytic meningitis in patients with deficient cell-mediated immunity, and is now an important problem in HIV-infected patients.

Introduction to helminths

The term helminth is derived from a Greek word meaning 'worm'. It was initially applied to roundworms, but now encompasses all metazoan internal parasites (Tables 2.5 & 2.6). Helminths are complex multicellular organisms with developed organs. Adult worms may possess an alimentary canal whose morphology can be helpful in identification (e.g. hookworms). Female helminths have uteri and produce large numbers of eggs daily. Many pathogenic helminths, especially the hookworms and the tapeworms, possess specialized organs of attachment.

In some helminth diseases, humans are the only host. The pathogen produces eggs, which are excreted and survive in the environment; eggs, or free-living larvae from excreted eggs, are later ingested by a new human host and a new infection develops. Such infections include those caused by threadworms and roundworms.

Some helminths have complex life cycles, involving an intermediate host (e.g. the pig or cow in tapeworm diseases) or a vector (e.g. the mosquito in filariasis). The

Table 2.5 Nematodes that cause human disease

Family	Species	Disease
Trichuroidae	*Trichinella spiralis*	Trichinosis
	Trichuris trichuria	Trichuriasis
	Capillaria hepatica	
	C. phillipinensis	
Rhabditoidae	*Strongyloides stercoralis*	Strongyloidiasis
	S. fuelleborni	
Ancylostomatoidae	*Ancylostoma duodenale*	Hookworm disease
	A. ceylonicum	Ancylostomiasis
	Necator americanus	
Metastronguloidae	*Angiostrongylus cantonensis*	Eosinophilic meningitis
Oxyuridae	*Enterobius vermicularis*	Threadworms
Ascarididae	*Ascaris lumbricoides*	Ascariasis
	Toxocara canis	Visceral larva migrans and ocular disease
	T. cati	
Dracunculoidae	*Dracunculus medinensis*	Guinea worm
Filaroidae	*Wuchereria bancrofti*	Lymphatic filariasis
	Brugia malayi	
	B. timori	
	Onchocerca volvulus	River blindness
	Loa loa	Loiasis

Table 2.6 Most important platyhelminths causing human infection

Family	Species	Disease
Schistosomatoidae	*Schistosoma mansoni*	Bilharzia
	S. haematobium	
	S. japonicum	
Echinostomatoidae	*Fasciola hepatica*	Liver flukes
	Taenia solium	Tapeworms and cysticercosis
Taeniidae	*T. saginata*	
	Echinococcus granulosus	Hydatid disease
	E. multilocularis	

eggs of schistosomes are deposited by humans into water, where they hatch, invade snails, are released as infectious cercariae, then reinfect humans by penetrating intact skin. An infected human is an essential part of the life cycle of filariasis or ascariasis, but hydatid disease is a zoonosis, in which humans replace the sheep in a dog–sheep–dog life cycle.

Classification

Helminths are classified into the nematodes (roundworms) and the platyhelminths (flatworms), which include cestodes (tapeworms) and flukes. Nematodes are classified according to the main site of infection: intestinal nematodes such as *Ascaris* and hookworms; blood nematodes such as *Filaria*; tissue nematodes, e.g. *Trichinella*; and skin nematodes such as *Onchocerca volvulus*. The flukes are subdivided by the place where the adult fluke is found, i.e. blood, liver or lung.

Pathogenic helminths

Nematodes (roundworms)
- *Ascaris lumbricoides, Ancylostoma duodenale, Strongyloides stercoralis* (intestinal)
- *Wuchereria bancrofti, Brugia malayi* (blood filariae)
- *Onchocerca volvulus, Trichinella spiralis* (tissue forms)

Platyhelminths (flatworms)

Cestodes (tapeworms)
- *Taenia solium, Diphyllobothrium latum* (intestinal form)
- *Echinococcus granulosus* (tissue cyst form)

Trematodes (flukes)
- *Fasciola hepatica, Opisthorchis sinensis* (liver flukes)
- *Schistosoma mansoni, S. japonicum* (blood flukes)
- *Paragonimus westermani* (lung fluke)

Laboratory Techniques in the Diagnosis of Infection

3

Introduction

The microbiology laboratory plays a crucial role in the diagnosis of infectious diseases. Many different types of tests are available for the detection of infection. There are therefore several ways of making a microbiological diagnosis.

1 Microscopical methods.
2 Cultural methods.
3 Serological methods.
4 Molecular methods.

Collection of specimens

Body fluids, secretions and biopsy material can all be examined to detect pathogens, antigens or products or the immune response to them. Samples from the environment, e.g. water, food or soil, may also be examined. Some samples must be collected at a particular time; for example malaria parasites are best sought at the peak of fever and a short time afterwards, whereas blood for bacterial culture should be taken as the fever begins to rise. Special precautions must often be taken to ensure survival of the pathogen and exclude contaminants, e.g. cleaning the perineum before collecting a midstream specimen of urine. Anaerobic species may die if exposed to atmospheric oxygen and survive better in samples of pus, rather than in swab specimens. Many pathogens, such as *Neisseria* and most anaerobic species, die quickly outside the body and must be transported to the laboratory without delay. *Neisseria gonorrhoeae* is susceptible to drying, so specimens likely to contain this organism should be inoculated onto microbiological medium near to the patient: many genitourinary medicine clinics are equipped to do this.

Universal precautions and laboratory safety

Specimens may contain hazardous pathogens and must be handled with care. A system of 'universal precautions' is employed to reduce the risk of transmitting blood-borne viruses. These are personal protective measures, taken in collecting and examining specimens irrespective of their source. The idea is to handle specimens on the assumption that they may contain a transmissible pathogen, rather than relying on clinical suspicion or written clinical details, which may be faulty or absent. For example, any sputum specimen is assumed to carry the risk of tuberculosis even if this is not the test requested.

Direct microscopic examination

Antony van Leeuwenhoek first saw microscopic 'animalcules', and Alexander Ogston, a surgeon, described the characteristic microscopical morphology of staphylococci in pus and discovered their role in pyogenic sepsis. The light microscope has since been indispensable in the study of microorganisms. The equipment required is cheap, reagent costs are low and early results can be obtained. The diagnosis of malaria or vaginal trichomoniasis, for instance, can be made while the patient waits at the clinic. The organism sought need neither multiply nor even be alive. Microscopy is especially useful for detecting organisms that are difficult or dangerous to grow.

Types of microscopy for the diagnosis of infections

1 Unstained preparations.
2 Simple stains: Gram, Giemsa.
3 Special stains: Ziehl–Nielsen, Gomori–Grocott, India ink.
4 Immunofluorescence: direct and indirect.
5 Electron microscopy.

Microscopy of unstained preparations

Direct microscopy of unstained preparations

- Faecal protozoa and helminths.
- Vaginal discharge.
- Urine for bacteria and pus cells.

Direct examination of unstained 'wet' preparations is suitable for rapid diagnosis in the laboratory and the outpatient setting. Many pathogens have a characteristic appearance, e.g. parasites in the faeces (Fig. 3.1), or may appear in diagnostic circumstances, for instance bacteria,

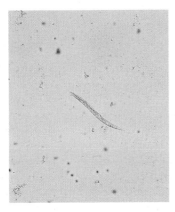

Figure 3.1 Unstained wet preparation of faeces showing larva of *Strongyloides stercoralis*. × 94 approximately.

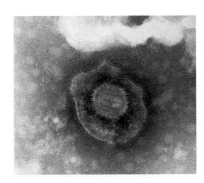

Figure 3.2 Negatively stained electron micrograph of herpesvirus from a mouth swab. Courtesy of Dr David Brown, Health Protection Agency, Centre for Infections.

together with white cells, in the urine from a case of acute urinary tract infection.

Viruses can also be detected directly by electron microscopy, using a negative-staining technique. The viruses are concentrated by vigorous centrifugation and are then suspended in a heavy metal solution. As it dries, the heavy metal salt fills in the spaces between the viruses, providing an electron-dense background that outlines the virus. This technique is useful for detecting viruses with distinctive morphology, such as poxviruses in scrapings from the lesions of orf and molluscum contagiosum, herpesviruses in vesicle fluid, or one of the many gastrointestinal viruses such as rotavirus or calicivirus in diarrhoea stools (Fig. 3.2).

Microscopical examination of preparations stained with simple stains

Gram stain

Dried preparations of specimens that are fixed to kill the organisms can be examined using simple stains, such as Gram stain, that dye the bacteria. This technique can demonstrate the shape of the bacteria and the capability of 'Gram-positive' bacteria to retain the methylene blue dye (Fig. 3.3). The Gram stain provides a rapid answer to the clinical question: 'are there any organisms present?' It is most useful when sterile fluids such as cerebrospinal fluid (CSF) or pleural fluid are examined. However, the sensitivity of Gram stain is relatively low and at least 10 000 organisms per millilitre must be present for a diagnosis to be made.

Gram stain

- Sterile fluids (cerebrospinal fluid, ascites, pleural fluid).
- Sputum (to exclude poor-quality specimens).
- Pus from any site.
- Urethral discharge.

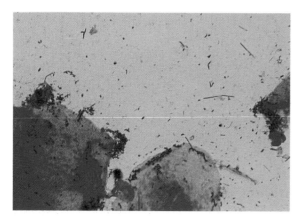

Figure 3.3 Gram stain of a smear from the mouth, showing a mixture of Gram-positive (blue) and Gram-negative (pink) cocci and bacilli.

Gram staining alone can rarely answer the question 'what organism is present?', because the morphology of bacteria is rarely diagnostic. Important exceptions to this include the finding of Gram-positive or Gram-negative diplococci in CSF from a patient with meningitis, or the characteristic appearances of *Borrelia* and *Fusobacterium* in Vincent's angina. Similarly, a Gram-stained preparation of urethral pus showing Gram-negative intracellular diplococci is sufficiently characteristic to allow a presumptive diagnosis of gonorrhoea.

Other simple stains

Other simple stains include acridine orange (which is more sensitive in demonstrating organisms than Gram stain, but more prone to confusing artefactual effects), lactophenol blue to demonstrate the morphology of fungi, or India ink, which is used to detect the presence of *Cryptococcus neoformans* in the CSF by negative staining (Fig. 3.4).

Special stains

Stains such as Ziehl–Nielsen (ZN) are used to demonstrate specific features of organisms that simple stains will not demonstrate. Specimens are stained with carbolfuchsin, destained with an acid–alcohol solution and then counter-stained with methylene blue. The lipid-rich mycobacterial cell wall retains the pink dye and organisms are seen as pink bacilli against the blue background (Fig. 3.5). The number of acid-fast species is limited and this technique is therefore useful in the diagnosis of mycobacterial infection, including tuberculosis and leprosy, and parasitic infections such as cryptosporidiosis.

A variation of this technique uses the naturally fluorescent substance auramine to stain the organisms. The specimen is processed in a similar way to the ZN method, and acid-fast organisms fluoresce bright yellow under an ultraviolet light. Auramine microscopy is used for screening large numbers of specimens but, because it lacks the specificity of the ZN stain, all positive specimens must be overstained by the ZN method, and re-examined to confirm the findings.

Romanowsky stains

Romanowsky stains, which colour cytoplasm and chromatin, are widely used to demonstrate blood cells. Stains of this type are very useful for revealing blood parasites. In malaria or filariasis, Giemsa-stained smears not only demonstrate the presence of the organisms, but permit speciation by displaying morphological details (Fig. 3.6).

Special stains used in microbiology

1 Ziehl–Nielsen: *Mycobacterium* spp.
2 Gomori–Grocott: Fungi, *Pneumocystis carinii*.
3 Giemsa: Malaria, *Filaria*.

Figure 3.4 India ink-stained preparation of cerebrospinal fluid, showing *Cryptococcus neoformans* with a thick, clear capsule.

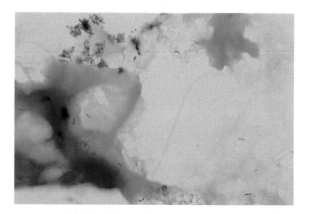

Figure 3.5 Ziehl–Nielsen-stained smear, showing acid-fast bacilli.

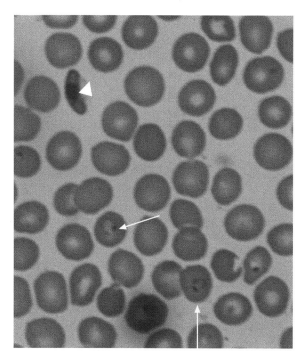

Figure 3.7 Immunofluorescence-stained preparation of legionellae. Courtesy of Dr Tim Harrison, Central Public Health Laboratory.

Figure 3.6 Romanowsky-stained thin blood film, showing morphologically typical trophozoites of *Plasmodium falciparum* (arrows) and a gametocyte (arrowhead).

Immunofluorescence

Direct immunofluorescence techniques detect organisms by their binding with fluorescence-labelled antibodies. Specimens are dried on a multiwell slide, together with positive and negative 'control' specimens. A specific fluorescein-labelled antibody is then added. The slides are washed and examined microscopically under ultraviolet illumination. Where the antibody has bound to the pathogen there is an apple-green fluorescence (Fig. 3.7). This technique is both sensitive and specific and provides a rapid, presumptive diagnosis. It can be applied to a wide range of specimens and is used in the diagnosis of upper respiratory tract virus infections including influenza, parainfluenza and respiratory syncytial virus, and also measles and rabies.

> **Some organisms detectable by direct immunofluorescence**
> **1** Viruses: Parainfluenza viruses, respiratory syncytial virus.
> **2** Bacteria: *Legionella*, *Treponema pallidum*.
> **3** Protozoa/fungi: *Giardia intestinalis*, *Pneumocystis jiroveci* (*carinii*).

Immunofluorescence techniques are also used for the detection of specific antibody. This will be described in more detail below.

Cultural methods

Culture can aid diagnosis in bacterial, parasitic and viral diseases. In bacteriology, culture permits amplification of the number of bacteria. Isolating them on solid media makes identification and susceptibility testing possible. Bacterial culture is made possible by the use of agar, a gelatinous substance derived from seaweed, which melts at 90 °C but solidifies at 50 °C. It is highly stable, is rarely affected by organisms in cultures and can be mixed with nutrients such as blood, serum and protein digests to make solid media. Koch introduced agar to microbiology to replace gelatin, which melts at a lower temperature, near to that used for incubation.

Bacteriological culture on solid agar is usually performed in Petri dishes: these are plastic dishes, 90 mm in diameter, with a vented lid. When prolonged culture is necessary, as in the diagnosis of mycobacterial or fungal infection, it is usually performed in sealed containers to prevent desiccation and the entry of contaminating organisms. The choice of medium and the conditions of incubation depend on knowledge of the organism's requirements for optimum growth (see Chapter 2).

Media

Three types of bacteriological media are used: enrichment, selective and indicator.

> **Bacteriological media**
> **1** Enrichment.
> **2** Selective.
> **3** Indicator.

Enrichment media

Enrichment media are used to ensure that small numbers of fragile pathogens can multiply sufficiently to be detected. They are useful when seeking to identify fastidious organisms such as *Streptococcus* spp., *Haemophilus influenzae* or *Bacteroides fragilis*.

Enrichment media are made by adding blood, yeast extracts, tissue infusions, meat, etc. to simple base media. Some enrichment media include materials to neutralize toxic bacterial products that would otherwise inhibit growth. An example of this is charcoal in the enrichment medium for *Bordetella*. The media may be solid, for example blood agar, or liquid, as with Robertson's cooked meat broth. Liquid media are especially valuable for investigating body fluids that are normally sterile. Very small numbers of organisms in such fluids will multiply in the highly nutritious medium, and they can then be subcultured on to solid media for identification and susceptibility testing.

> **Enrichment medium** aims to amplify the organism by growth, e.g. for *Neisseria* spp., *Streptococcus* spp. or *Vibrio* spp.

Selective media

Selective media are used to identify a pathogen existing in a mixture of organisms. Many body sites, such as the upper respiratory tract or the gut, have a normal resident flora, and pathogens must be isolated from this bacterial competition. Selective media contain compounds such as chemicals (e.g. selenite F), dyes (crystal violet) or mixtures of antibiotics (such as lincomycin, amphotericin B, colistin and trimethoprim in New York City medium) that selectively inhibit the normal flora, enabling the pathogen to grow through (Table 3.1). These complex media often require careful preparation.

Although selective agents have their maximum effect on the unwanted organisms, some inhibition of the target organism inevitably occurs. Therefore, an enrichment medium should also be inoculated so that small numbers of pathogens can be detected.

> **Selective cultures**, e.g. sputum, stool or throat-swab specimens, aim to identify, for instance, *Haemophilus* spp., *Salmonella* spp. or *Streptococcus* spp. from among normal flora. Antibiotics, dyes, antiseptics, chemicals and bile salts are examples of substances used in selective bacteriological media.
>
> Culture of pathogens from sites with normal flora requires selection, and from sterile sites requires enrichment.

Indicator media

Indicator media are used to identify colonies of pathogens among the mixture of organisms able to grow on the selective medium. Many indicator media are also selective. An example of a selective indicator medium is MacConkey's agar, which uses bile salts to select for bile-tolerant enteric organisms. It also contains lactose and the indicator neutral red. Colonies of lactose-fermenting organisms produce lactic acid, and are coloured red by the neutral red indicator. Some enteric pathogens such as *Salmonella* and *Shigella* are usually non-lactose fermenters, which produce easily identified, colourless colonies in this medium (Fig. 3.8).

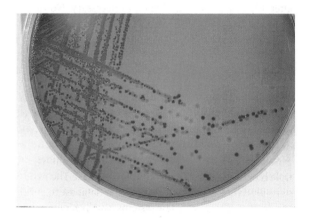

Figure 3.8 Mixed growth of Enterobacteriaceae on MacConkey's agar plate, showing pink, lactose-fermenting colonies of *Escherichia coli* among the colourless non-lactose fermenters.

Table 3.1 Examples of selective media for bacteriological culture from sites containing a normal flora

Medium	Selective agent	Specimen	Organism/s sought
Crystal violet–blood agar	Crystal violet	Throat swab	*Streptococcus pyogenes*
Desoxycholate citrate	Desoxycholate	Faeces	*Salmonella, Shigella*
New York City medium	Lincomycin, colistin, amphotericin, trimethoprim	Urethral or cervical smear	*Neisseria gonorrhoeae*
Selenite broth	Selenite F	Faeces	*Salmonella*
Sabouraud's agar	High dextrose content	Many	Fungi

Indicator medium (often combined with selective function): e.g. MacConkey
• Selects with bile salts.
• Indicates lactose fermentation with pH indicator.

There are many liquid indicator media, containing different sugars or other substrates such as urea and citrate. They also contain indicator dyes, which change colour when bacterial metabolic products of the substrates alter the pH of the medium. Thus an organism that ferments a particular sugar lowers the pH, or one that metabolizes urea produces ammonia and raises the pH. These indicator media are usually used in banks of tests that allow identification of an organism by its biochemical profile. Preprepared sets of miniculture vials are commercially available (Fig. 3.9).

Automation

Bacterial culture is time-consuming and labour-intensive. Before automation of blood cultures became widely available it was necessary to subculture each bottle manually, after 12, 24 and 48 h, and later, as necessary. Nowadays, bacterial growth is detected by the microbial production of carbon dioxide, or changes in the electrical impedance of the medium, indicating the need for subculture. These detection systems provide continuous monitoring of blood culture bottles, allowing earlier subculture, identification and testing of pathogens. Such methods have been adapted to mycobacteria, reducing the time taken to detect a positive culture from a mean of 5 weeks to 2 weeks or less.

Some automated systems can analyse the results of multiple biochemical tests on duplicated isolates. The pattern of results is compared with a computer database to permit species identification. These systems usually include a battery of antibiotic susceptibility tests. All of these 'intelligent' systems allow the operator to choose results profiles that, for example, permit identification of expanded spectrum beta-lactamase producers.

Limitations of culture

The main limitation of bacterial culture is the time required for incubation. For ill, infected patients requiring immediate treatment, decisions on initial therapy must be made based on the clinical features of the case, plus rapid microscopical examination and haematological or biochemical results. Culture is valuable for confirmation or rebuttal of a diagnosis, and particularly for revealing unusual or unexpected organisms, or unusual antibiotic susceptibility patterns. The initial therapy in such cases may need modification when the culture result is available.

Even in dangerously ill patients, the bacterial load in easily obtained specimens may be very small. Treating the patient with antibiotics before taking cultures may lead to falsely negative culture results.

The main reason for performing culture is to identify the infecting pathogen. The likely course of disease, and treatment response, can then be predicted from our knowledge of the organism's behaviour. As antibiotic resistance is becoming more common, culture allows the susceptibility pattern of the infecting organism to be determined. However, molecular methods are now increasingly used to predict bacterial antibiotic susceptibility, and are already the standard way of predicting viral susceptibility.

Culture has an important epidemiological purpose, not related to the treatment of individual patients. Most cases of acute diarrhoea could be managed properly without microbial culture. The isolation of the same *Salmonella* sp. from several individuals, however, might prompt a search for a common source, such as a contaminated food. Similarly, the isolation of a toxigenic strain of *Corynebacterium diphtheriae* from a throat swab should prompt immediate surveillance and control measures among the patient's contacts.

Screening

Microbiological culture can be used to screen patients and healthcare workers for colonization with pathogens such as *Streptococcus pyogenes* or epidemic methicillin-resistant *Staphylococcus aureus* (EMRSA). It can allow a prompt response to the presence, in a hospital setting, of dangerous or difficult-to-treat organisms such as multiple drug resistant *Acinetobacter baumanii*.

Antimicrobial susceptibility testing

Testing antimicrobial susceptibility is an important func-

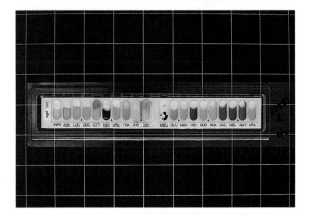

Figure 3.9 A typical bank of biochemical tests for the identification of species of Enterobacteriaceae.

tion of the microbiology laboratory. Apart from confirming expected results, it is valuable when infecting pathogens have unpredictable antimicrobial susceptibilities. This is especially true of Enterobacteriaceae, among which multidrug resistance easily develops and spreads in the hospital environment. In some geographical areas *Salmonella* may be resistant to most of the commonly used therapeutic agents. Strains of *Streptococcus pneumoniae*, previously sensitive to many antimicrobials, have recently developed resistance to penicillin and other drugs, making empirical therapy difficult.

Almost all currently available methods of antimicrobial susceptibility testing depend on isolation of the organism, followed by reculture of the pure growth in the presence of antimicrobial agents. This is described in more detail on pp. 57–60.

Typing microorganisms

Typing is the use of further identification methods to distinguish between strains of organisms within the same species.

For most clinical purposes it is sufficient to identify the genus or species of an infecting organism. Further characterization is desirable when, for example, increased numbers of *Salmonella typhimurium* infections could be caused by different strains of this common pathogen. Showing that the strains are identical, by typing, proves that the cases represent an outbreak.

An effective typing system divides a species of microorganism into sufficient different groups to be epidemiologically discriminating. For example, a system separating a species into only six groups would not be useful, nor would a system that placed all the common strains in the same group. Typing methods should be reproducible so that results from one laboratory can be compared with those of another. They should also be simple to perform, as delay in receiving results hinders early control of an outbreak. Finally, typing should be inexpensive. The results of typing should also be 'portable' so that they can be used by anyone with access to computers (see below).

Methods of typing microorganisms

Methods of typing are numerous, ranging from simple biochemical to complex genetic characterization.

Simple laboratory typing

Simple phenotypic markers can indicate that isolates are probably identical but are not sufficiently discriminating to ensure certainty. There are a wide range of phenotypic typing methods, including biochemical activity, serology

and phage typing, which are often used in combination. Such methods are being increasingly superseded by genetic methods (see below).

Serological typing

Many laboratories still type enteric pathogens such as *Shigella flexneri* or *Salmonella* using sets of antisera raised in mice or guinea-pigs. Serological typing is performed by suspending a pure growth of the organism in a drop of antiserum on a glass slide and mixing them with a rocking movement of the slide. The development of agglutination indicates that the organism carries the antigen specific to the antiserum used.

Phage typing

Many bacteriophages lyse the bacteria they infect as part of their life cycle (this is known as a lytic cycle). Different strains of an organism have different sensitivities to the phages that commonly infect the species. This property is still used to type *Salmonella* spp.

Molecular methods

Many molecular typing methods are now used in clinical and research practice, much improving our ability to follow outbreaks and to study the epidemiology of infection.

Nucleic acid typing

Nucleic acid typing methods use the activity of endonuclease enzymes to split the genome into a characteristic range of different-sized fragments. Genomic or plasmid DNA, or ribosomal RNA, is harvested from the test organisms and digested with restriction endonucleases. The resulting fragments are then separated by electrophoresis or pulsed-field gel electrophoresis (PFGE), to produce a pattern of bands. The variation in band patterns is called restriction fragment length polymorphism (RFLP; Fig. 3.10). This method can be varied by using a frequently cutting enzyme, and using electrophoresis to demonstrate the presence of repetitive sequences of DNA. Many organisms have characteristic sequences, or numbers of repeats. This approach has proved to be especially valuable in typing *M. tuberculosis*.

Larger DNA fragments can be investigated by Southern blotting. In this process the DNA is bound to a membrane; after washing, a nucleic acid probe is added and if homologous sequences are present on the membrane it will bind to them. Detecting the presence of bound probe will confirm the presence of the DNA fragments from the pathogen.

The problem of gel-to-gel variation is reduced by computer programs that 'normalize' the gels, then store and

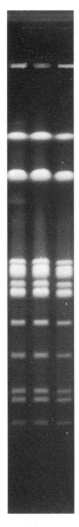

using PCR-based methods. Primers are designed to permit amplification of the repeated element, and characterize the organism by the number of copies that it possesses. The number of each of the identified repeating elements possessed by the strain is collated into a score (strain number) for that strain. The strain number can be compared with numbers for other known strains in a relational database.

Multi-locus sequence typing

In multi-locus sequence typing a number of genes related to physiological bacterial functions are chosen. These genes should usually be 'stable', and not under selective pressure, so their sequence changes slowly. Approximately 450 bp segments of the genes are amplified and sequenced. The sequences of the selected genes are compared with an international central database using the internet. Any single base pair change in any of the genes is defined as a new allele: only those strains with exactly the same sequence are considered identical.

> **Methods of typing bacteria**
> 1 Serotyping.
> 2 Phage typing.
> 3 Restriction fragment length polymorphism typing.
> 4 PCR-based methods.
>   • Outer membrane protein genes.
>   • Repetitive elements.
>   • Multi-locus sequence typing.

Figure 3.10 Electrophoresis of DNA restriction fragments of three cytomegalovirus isolates, showing that the infected patients had acquired the same strain.

compare the data, making it possible to build up enormous international databases that can be used to track the transmission of organisms across the globe. (see Chapter 18).

Polymerase chain reaction-based methods

Polymerase chain reaction methods (PCR, see p. 50) can be used to generate DNA, from a variable gene or genes, for RFLP typing. Genes encoding the variable outer membrane proteins of *Neisseria* may be analysed in this way. Bacterial genomes contain many repetitive nucleotide sequences. PCR protocols can be designed to amplify parts of these, allowing organisms that possess them to be typed

Culture of protozoa and helminths

Culture of protozoa and helminths tends to be difficult and often produces no better diagnostic yield than microscopy. For many organisms, such as *Cryptosporidium parvum*, only one stage in the life cycle may be culturable. In some, such as *Strongyloides stercoralis*, the conditions of the natural environment can be reproduced in the laboratory. Using activated charcoal as a culture medium, larvae in faecal specimens will mature and multiply.

Trophozoites of *Plasmodium falciparum* can be cultured in banked red blood cells, enabling direct testing of sensitivity to antimalarial agents. Although relatively simple, this technique is inappropriate for routine diagnosis, as direct microscopy is sufficiently sensitive to detect clinically significant parasitaemias. Similarly, amoebae can be cultured in solid media if certain bacteria are included, but simple microscopy provides accurate and rapid diagnosis, and culture remains a research procedure.

The promastigote stage of *Leishmania* can be cultured in artificial media. To achieve culture of the amastigote, or tissue stage, material must be inoculated into isolated peripheral blood macrophages or macrophage cell lines.

Tissue culture for viruses and other intracellular organisms

Viruses and chlamydiae are obligatory intracellular pathogens, so culture for these organisms must be performed in living cells or tissues. Cells are usually prepared as monolayers on to which the prepared specimen is inoculated. Some cells used for culture are 'primary cultures', extracted directly from an organ, as with fetal lung cells. Primary cell lines can be propagated for up to 50 generations, and are essential for isolation of some viruses. Alternatively, continuous cell lines can be sub-cultured indefinitely. Cell cultures are maintained at 37 °C in an atmosphere of increased humidity and carbon dioxide. Cultured viruses infect the cells and take over their metabolic processes as they do in the human host. The cell monolayer is inspected at intervals for the presence of viral damage to the cells (the cytopathic effect: CPE). Some viruses may be presumptively identified by their distinctive CPE; for example, measles virus produces multinucleated giant cells (Fig. 3.11). Some viruses produce no visible CPE but growth of the virus is detected by, for example, adherence of red cells to viral haemagglutinin molecules expressed on the cells. Viral antigens can be directly demonstrated by fixing the cell monolayer and using immunofluorescence staining techniques (Fig. 3.12).

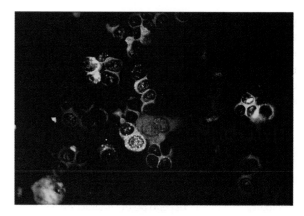

Figure 3.12 Fluorescence micrograph of Lassa virus-infected vero cells, stained with immunofluorescent-labelled anti-Lassa serum. Courtesy of Dr David Brown, HPA Centre for Infections.

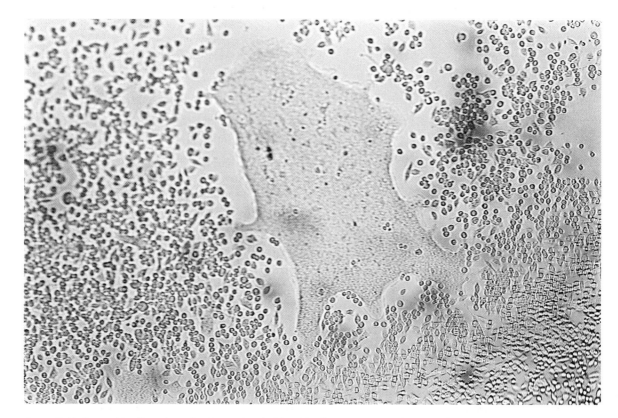

Figure 3.11 Cytopathic effect of measles virus on lymphoblastoid cells; progressive coalescence of infected cells into a syncytium has formed a giant cell. Such giant cells may contain up to 50 nuclei.

Evidence of virus growth in cell cultures

1 A cytopathic effect.
2 Haemagglutination.
3 Antigen detection.
4 Electron microscopy.

Virus neutralization

Identification of virus isolates can be achieved by lysing the cells of the monolayer and visualizing the virus under the electron microscope. Alternatively, inhibition of subculture by specific antibodies can be used to identify the species. This is known as virus neutralization.

Serology in the detection of infection

Serological techniques depend on the interaction between antigen and specific antibody. They are of particular value when the pathogen is difficult or impossible to culture, or is dangerous to handle in hospital laboratories.

The process can be divided into two parts:
1 the antigen–antibody interaction; and
2 the demonstration of this interaction by a testing process.

The antigen–antibody reaction depends on the specific binding between epitopes on the pathogen and the antigen-binding sites on the immunoglobulin molecules. The sensitivity of a serological test depends partly on the specificity and strength of the antigen–antibody reaction, but mostly on the ability of the test system to detect the reaction.

In older tests antibody–antigen binding was detected by observing a natural consequence of this interaction: precipitation, agglutination or the ability of the antigen–antibody complex to bind and activate (fix) complement. Some laboratory tests based on these reactions are still in daily use. Newer tests use 'labelled' immunoglobulin molecules to facilitate detection. The main methods employed are labelling with fluorescein or enzymes (see below).

Agglutination tests

Slide agglutination

Agglutination tests are used to identify the species or serotype of an infecting organism, by observing the aggregation of a suspension of bacteria in the presence of specific antibody. They can be performed on glass slides and are in daily use in speciating faecal pathogens such as *Salmonella*, *Shigella* or *Streptococcus pneumoniae*.

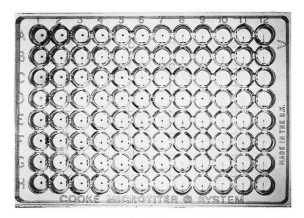

Figure 3.13 Rapid microagglutination test for the detection of antibodies to *Legionella pneumophila*. Each row is composed of doubling dilutions of an individual patient's serum, mixed with dead organisms. The microtitre plate has conical wells, so that centrifugation 'jams' the agglutinated organism at the bottom of the well. Unagglutinated organisms form a streak when the plate is tipped on its edge: a dot is therefore positive and a streak is negative.

Tube agglutination

In tube agglutination, particulate antigens form a lattice with specific antibody in the serum being tested, and fall to the bottom of the test vessel as a fine mat (a positive result). In a negative test, the particulate antigen falls quickly to the bottom to form a condensed button. This technique has been widely used in *Brucella* standard agglutination tests. A rapid microagglutination test (RMAT) is still the standard test for antibodies to *Legionella pneumophila* (Fig. 3.13).

Coagglutination

Specific antibodies can be attached to uniform latex particles, or killed protein A-possessing staphylococci. These particles will be agglutinated when they attach in large numbers to antigen molecules. In this way, otherwise soluble immune complexes may be detected in an agglutination reaction. Such latex and coagglutination techniques are used to detect the presence of the polysaccharide antigens of *Streptococcus pneumoniae*, type b *Haemophilus influenzae*, *Neisseria meningitidis* and *Cryptococcus neoformans* in CSF, or *S. pyogenes* in throat swabs.

Complement fixation tests

In a complement fixation test, defined antigen is added to a patient's serum (which has been heated to destroy naturally occurring complement) and supplemented with a measured amount of guinea-pig complement. When specific antibody is present in the serum, it interacts with the antigen, which activates (or fixes) complement. Sheep red cells sensitized with rabbit anti-sheep red cell antibody are added and if complement has already been fixed none remains to lyse the sensitized sheep cells, so no lysis will occur (a positive result). In contrast, if no antigen–antibody reaction has taken place in the first reaction, the complement is still available to lyse the red cells (a negative result). The test is technically difficult and requires careful standardization. It is therefore being superseded by those described below.

Indirect fluorescent antibody tests

Specific antigen is fixed on to a multiwell microscope slide and patient's serum added. The slide is incubated, then washed. Fluorescein-labelled anti-human immunoglobulin is then added, followed by further incubation. After a final wash the slides are examined under ultraviolet illumination. Where antibodies from the patient's serum have bound to the antigen, the anti-human globulin will bind, and is indicated by apple-green fluorescence. Individual positive sera may be titrated. Indirect immunofluorescence is both sensitive and specific, but rather time-consuming. It is used in the diagnosis of a number of infections, especially where the throughput of specimens is small, for instance in the diagnosis of syphilis (FTA, fluorescent treponemal antibody test) or parasitic disease such as leishmaniasis or amoebiasis.

Enzyme-linked immunosorbent assay

In this technique either the antigen or antibody in the reaction is allowed to bind to a solid phase, such as the walls of microtitre wells. There are many variations of enzyme-linked immunosorbent assay (ELISA) but four will be described here and illustrated in the figures.

Antibody-detecting ELISA tests

In an antibody-detecting ELISA, specific antigen is coated on to the wells of a microtitre ELISA plate. The patient's serum is added and any specific antibody binds to the antigen ('antibody-capture'). The plates are then washed and enzyme-labelled anti-human immunoglobulin is added. Plates are washed again and substrate for the enzyme is added. The enzyme–substrate reaction generates a colour, which indicates the specific antibody interaction (Fig. 3.14).

The optical density of the wells is measured by an ELISA reader. A positive result can be determined by reference to control values: for example, a positive result may be defined as a well with an optical density three standard deviations above the mean of a series of negative controls.

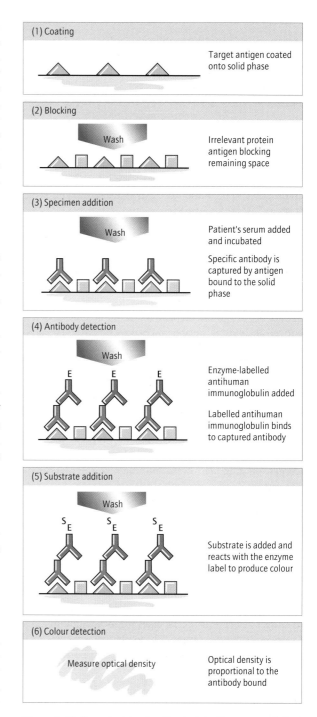

Figure 3.14 The principle of antibody-detection enzyme-linked immunosorbent assay.

Alternatively, control positive and negative sera can be examined in parallel and a positive result is reported if the optical density is significantly higher than that of the control negative.

As in other serological techniques, sera can be titrated but this is time-consuming and defeats the main advantages of ELISA, which are simplicity of performance and automation. It is difficult to use an antibody capture assay to answer any question other than: 'is there antibody present?' If a diagnosis is to be made on a single specimen, specific immunoglobulin M (IgM) must be detected. This can be achieved by purifying an IgM fraction from the serum and retesting this in an antibody detection test. A simpler alternative is the IgM antibody-capture ELISA described below.

IgM antibody-capture ELISA

By placing anti-human IgM antibody on the solid phase, it is possible to capture all IgM antibody. After washing, labelled antigen is placed in the well so that if the serum contains antigen-specific IgM the labelled antigen will bind. A positive result is detected by adding substrate, when a colour will be produced by the enzyme-labelled antigen.

Antigen-capture ELISA

Antibody to a specific antigen is bound on the solid phase. If antigen is present in the specimen, it will bind to the antibody. After washing, the bound antigen is then detected by enzyme-labelled specific antibody (Fig. 3.15). The amount of antigen present can be quantified by reference to an antigen standard curve.

Competitive ELISA

Competitive ELISA is a technique that is particularly useful for the measurement of antigen concentrations and is also used to detect antibodies or antibiotic concentrations. Microtitre plates are coated with antibody and the patient's serum is added together with a known quantity of labelled antigen. Labelled antigen and any unlabelled antigen existing in the specimen compete for binding sites on the solid phase. The quantity of labelled antigen bound is then determined by the addition of substrate and measurement of the optical density as before. If antigen is present in the specimen, little labelled antigen will bind and there will be no colour change. Results are computed in comparison with a control antigen curve.

ELISA techniques utilize relatively inexpensive reagents and simple, inexpensive detection systems. They readily lend themselves to automation and the reagents have a long shelf-life. ELISAs are therefore widely applied in the diagnosis of bacterial, parasitic and viral infections.

Western blotting (immunoblotting)

Microbial proteins can be separated by SDS-PAGE and transferred electrophoretically to a nitrocellulose membrane. Strips of the membrane are exposed to the patient's

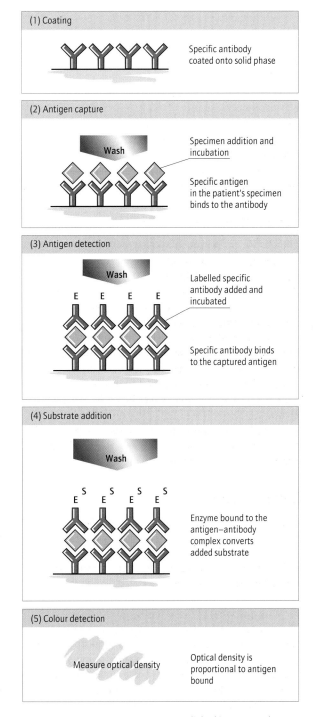

(1) Coating

Specific antibody coated onto solid phase

(2) Antigen capture

Wash

Specimen addition and incubation

Specific antigen in the patient's specimen binds to the antibody

(3) Antigen detection

Wash

Labelled specific antibody added and incubated

Specific antibody binds to the captured antigen

(4) Substrate addition

Wash

Enzyme bound to the antigen–antibody complex converts added substrate

(5) Colour detection

Measure optical density

Optical density is proportional to antigen bound

Figure 3.15 Antigen-capture enzyme-linked immunosorbent assay, e.g. used for demonstrating the presence of hepatitis B surface antigen in the test serum.

diluted serum, so that any antibodies specific to microbial proteins are bound, and can be detected using enzyme-labelled anti-human antibodies. The pattern of antibody recognition can be used to confirm a diagnosis, and to demonstrate the stage of disease by the repertoire of antibody specificities that the patient has developed. This is useful in the diagnosis of Lyme disease, in which simpler serological techniques are unreliable. It is also used in the study of human immunodeficiency virus (HIV) seroconversion illnesses.

Molecular amplification methods

Polymerase chain reaction

In a polymerase chain reaction (PCR), a reaction mixture consists of the specimen together with a pair of primers (short sequences of nucleotides specific for a nucleic acid sequence of the pathogen sought). Nucleotides and Taq polymerase (an enzyme that catalyses the construction of DNA but is stable at high temperatures) are added. The reaction mixture is heated, to separate all DNA strands. The primers bind to their target sequences, if they exist in the specimen, and are then annealed by cooling the mixture. The Taq polymerase adds nucleotides to make double-stranded DNA fragments. At the end of one cycle there are therefore two copies of the nucleic acid sequences bound by the primers. The cycle of temperature manipulations is repeated, allowing exponential multiplication of these sequences. In positive reactions, the amplified sequences are detected by a variety of methods including the characteristic size of the product, or by hybridization.

A range of different nucleic acid amplification tests (NAAT) have followed the development of PCR. These use different reaction procedures, which all result in amplifying the target nucleic acid DNA. NAAT methods can be modified to demonstrate viral RNA by using reverse transcriptase to make DNA, which is then amplified in the usual way. It has been applied to the diagnosis of many viral diseases, including HIV, cytomegalovirus and some, such as hepatitis C, whose agent cannot be cultivated. NAAT methods have transformed viral diagnosis, allowing rapid detection of many pathogens. It is also transforming other disciplines in microbiology.

A most important development has been the introduction of 'real time' NAAT. This uses an amplification process that incorporates labelled primers that, although varying between different test platforms, permit measurement of the product as it forms. Once the signal level passes a pre-set threshold the test is positive. The specificity can be checked by testing the 'melting point' of the product. As well as significantly increasing the speed of testing, the time at which the specimen signals positive is directly proportional to the bacterial or viral number in the original specimen. This allows rapid, accurate monitoring of antiretroviral therapy and is helpful in deciding the sig-

nificance of positive viral NAAT. For many pathogens NAAT are significantly more sensitive than conventional culture methods, improving the diagnostic rate. An example of this is *Chlamydia* diagnosis, where the increased sensitivity of NAAT allows effective screening, using non-invasive specimens such as urine. Rapid positive detection could be used to select specimens for culture, as set out in Figure 3.16, also allowing the microbiologist to determine susceptibility of the organism. Other advantages are the availability of a rapid negative for many specimens, and species identification for many positive samples.

Applications of molecular amplification methods

Detection of pathogens that grow slowly or that cannot be cultured

A major advantage of NAAT is that they avoid the need to await growth of the organism in culture. This is especially important for virology where tissue culture is complex, expensive and slow and many common viruses cannot readily be cultured. *Tropheryma whipellii*, the causative bacterium of Whipple's disease, was first identified by the use of 16S rRNA gene NAAT, which remains the diagnostic method of choice. Tuberculosis can be reliably diagnosed by NAAT in one day, compared to the many weeks required for culture. NAAT methods obviate the need for culture of dangerous organisms in the diagnosis of Lassa fever or anthrax. They have also been adapted for detection of bio-terrorist attacks by allowing workers to detect suspected organisms in minutes using hand-held devices.

Use of NAAT in routine bacteriology

One of the disadvantages of NAAT is that it only detects the organisms that match the chosen test. In clinical practice, one of a range of organisms may be causing the patient's illness, and it would be impractical to apply numerous specific NAAT to search for them all. Ribosomal RNA genes (16S in bacteria, 18S in fungi) are highly conserved with stretches of variability. Primers designed for the constant regions can allow amplification of segments including the interspersed variable regions. Sequencing the resulting products and comparing them with databases allows the infecting species to be identified. However, if more than one organism is present the mixed sequence is jumbled and unreadable. When this happens the initial reaction products can be cloned and the 'nonsense' rRNA gene inserted into an *E. coli* vector. The *E. coli* will 'sort out' any functional rRNA sequences. On culturing the *E. coli*, random colonies can be picked for re-examination of the rRNA, and the new PCR products can be sequenced for comparison with the rRNA database (Fig. 3.17).

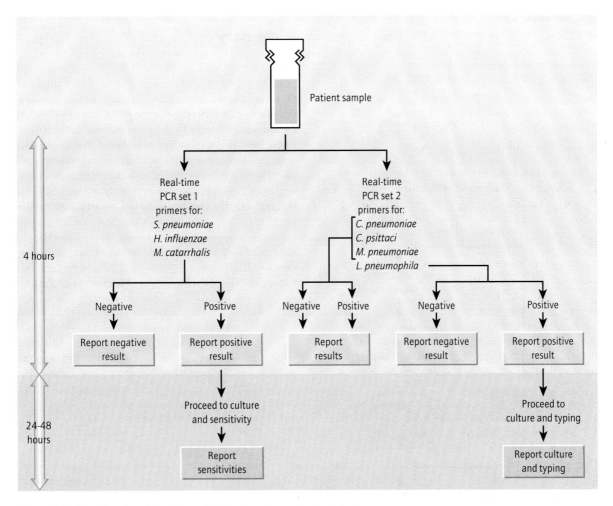

Figure 3.16 Algorithm describing diagnostic techniques for respiratory infections.

Susceptibility testing

The molecular basis of drug resistance is now becoming understood. For some organisms the presence of a particular gene is associated with resistance to antimicrobial agents, for example the *Pfmdr* gene in multidrug-resistant *Plasmodium falciparum*. HIV readily mutates to resistance, and the nucleotide sequence of the affected genes (protease and reverse transcriptase genes) can confirm the resistance and indicate which other drugs may also be affected (see Chapter 4).

High density DNA 'arrays' can identify patterns of relevant genes in a microorganism, in a single procedure. Several research groups are studying their use for detecting an organism's range of drug resistance genes.

Response to treatment

Amplification techniques can be useful in showing the early response to treatment. Examples include their use in measuring the response to antituberculosis therapy; the number of organisms in a respiratory sample can be estimated, using limiting-dilution PCR (as the number of organisms falls, negative results are obtained after fewer dilutions). A quicker alternative is to detect bacterial mRNA, which is short lived and only present in viable bacteria. The presence of specific mRNA implies the presence of viable organisms allowing the success or failure of therapy to be judged.

In managing HIV and hepatitis C infections, PCR is used to quantify the number of viral genome copies in the serum (viral load). The progress of treatment can be followed by serial measurements.

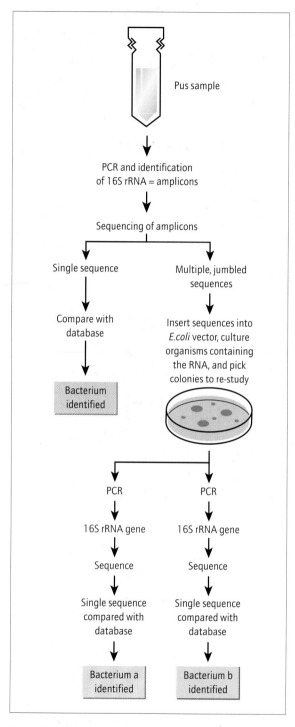

Pus sample

↓

PCR and identification
of 16S rRNA = amplicons

↓

Sequencing of amplicons

Single sequence

↓

Compare with
database

↓

Bacterium
identified

Multiple, jumbled
sequences

↓

Insert sequences into
E.coli vector, culture
organisms containing
the RNA, and pick
colonies to re-study

PCR   PCR

↓   ↓

16S rRNA gene   16S rRNA gene

↓   ↓

Sequence   Sequence

↓   ↓

Single sequence   Single sequence
compared with   compared with
database   database

↓   ↓

Bacterium a   Bacterium b
identified   identified

Figure 3.17 Use of 16S rRNA sequencing to identify bacterial species from contaminated specimens.

Summary

An important skill in clinical infectious diseases and microbiology is the choosing of appropriate diagnostic investigations. Molecular diagnostic techniques now provide microbiologists with the possibility of providing a diagnosis more rapidly than was possible with culture-based methods. It must be remembered that timely and accurate diagnosis depends on obtaining the correct specimen, at the optimum time of collection. Specimens must be transported to the laboratory quickly and in conditions that maintain the viability of the organisms present or the integrity of the antigen or DNA sought. In the laboratory the correct diagnostic method must be used and the results effectively communicated to the doctor managing the case. This process depends on close co-operation between clinician, microbiologist and laboratory scientists.

Antimicrobial Chemotherapy

Introduction

Many medicines owe their origins to the use of herbs. The use of cinchona bark originated in the theory that where hazards existed, in this case marshy land (the bringer of malaria), the natural remedy could be found. This led to the development of quinine, which is still obtained from its natural source. More recently, artemesinin antimalarial drugs have been developed from species of wormwood, which have been used to treat fever in China for two millennia.

Ehrlich thought that the selective staining of microorganisms by dyes might be used to target and kill pathogens. From this he developed salvarsan, the first specific antimicrobial. All modern antimicrobial agents depend for their effect on the ability to severely damage microorganisms while having much less effect on human metabolism.

Antimicrobials have numerous modes of action. Some interrupt unique microbial metabolic pathways, not possessed by humans. These are often false substrates or competitive inhibitors of microbial enzymes. Sulphanilamide, developed by Domagk in 1935, was the first of the sulphonamides. These compounds competitively inhibit the conversion of para-aminobenzoic acid (PABA) into dihydropteroic acid, an essential precursor of folate, which is a substrate in the synthesis of DNA. Bacteria rely on this metabolic pathway, as they cannot absorb preformed folate from the host. Further along the folate synthesis pathway, trimethoprim and pyrimethamine inhibit the activity of dihydrofolate reductase, and prevent the conversion of dihydrofolate to tetrahydrofolate (Fig. 4.1). The two effects can be combined in the treatment of bacterial infections (trimethoprim–sulfamethoxazole) or protozo-

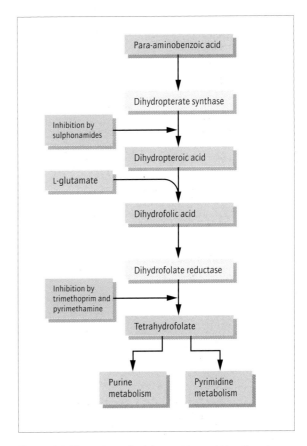

Figure 4.1 The actions of sulphonamides and trimethoprim.

an infections such as toxoplasmosis (pyrimethamine and sulfadiazine).

The bacterial cell wall protects the organism from the osmotic gradient between the interior and exterior of the cell. Its major structural component is peptidoglycan, composed of long polysaccharide chains containing N-acetylglucosamine and muramic acid molecules. The peptidoglycan chains are cross-linked between short peptide side-chains by an amide linkage (Fig. 4.2). Beta-lactam antibiotics (penicillins, cephalosporins, monobactams and penems) work by inhibiting transpeptidation and preventing cross-linking (Fig. 4.3). These antibiotics each possess a beta-lactam ring, which mimics the shape of the amide bond. Inhibition of cross-linking weakens the cell wall, and the bacteria are killed by osmotic lysis. Vancomycin and teicoplanin, glycopeptide antibiotics, inhibit cross-linking by binding to the D-alanine of the peptide chain and the lipid II precursors. This activity is limited to Gram-positive cell walls.

To package DNA in the bacterial cell, its long molecule must be supercoiled. It must then be uncoiled and recoiled to permit replication and transcription. Quinolone antimicrobials, which include nalidixic acid, ciprofloxacin and moxifloxacin, work by interfering with enzymes of the topoisomerase or DNA gyrase series, which excise aggregated bacterial DNA or remove supercoiling (Fig. 4.4).

Rifamycin antibiotics such as rifampicin and rifabutin act by inhibiting DNA-dependent RNA polymerase, preventing transcription of the genetic code to mRNA.

Bacterial protein synthesis (Fig. 4.5) can be inhibited by several different mechanisms. Tetracycline inhibits the binding of transfer RNA (tRNA) to the 30S ribosome, whereas macrolides inhibit RNA-dependent protein syn-

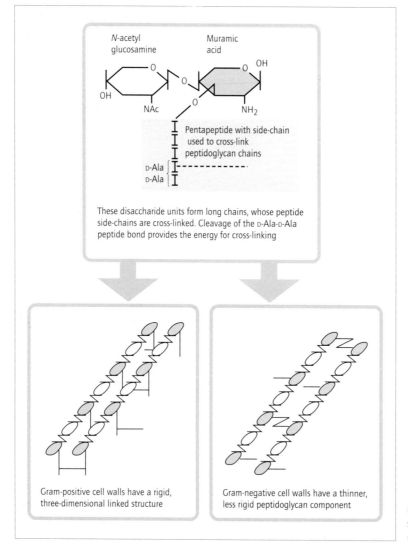

N-acetyl glucosamine

Muramic acid

Pentapeptide with side-chain used to cross-link peptidoglycan chains

D-Ala

D-Ala

These disaccharide units form long chains, whose peptide side-chains are cross-linked. Cleavage of the D-Ala-D-Ala peptide bond provides the energy for cross-linking

Gram-positive cell walls have a rigid, three-dimensional linked structure

Gram-negative cell walls have a thinner, less rigid peptidoglycan component

Figure 4.2 The peptidoglycan cell wall structures in Gram-positive and Gram-negative organisms.

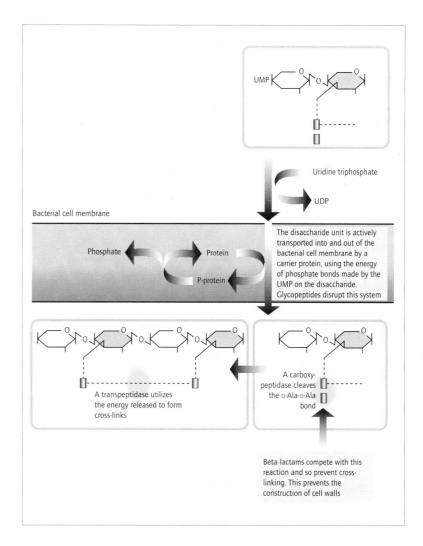

Figure 4.3 The actions of beta-lactams and glycopeptides. UDP, uridine diphosphate; UMP, uridine monophosphate.

thesis at the 50S ribosome. Oxazolidinones also act at this site by inhibiting the initiation phase of protein translation. Chloramphenicol prevents binding of tRNA to the 50S subunit of the ribosome. Aminoglycosides bind to both subunits of the ribosome, causing the genetic code to be misread and interfering with protein synthesis.

Antimicrobial action

1 Antimetabolites interrupt microbial chemical pathways.
2 Cell wall agents prevent the construction of bacterial cell walls.
3 Protein synthesis inhibitors interrupt the transcription and/or translation of microbial genes.
4 DNA gyrase inhibitors damage the tertiary structure of bacterial DNA.

Antimicrobial pharmacology

Effective treatment of infection depends on:
1 knowledge of the likely infecting organisms
2 probable antimicrobial susceptibilities
3 the site of infection
4 the spectrum of action of antimicrobial agents considered for use
5 absorption and distribution of the antimicrobials within the body.

Understanding the pharmacology of antimicrobial agents is important in ensuring that adequate antibiotic concentrations are achieved at the site of infection (bioavail-

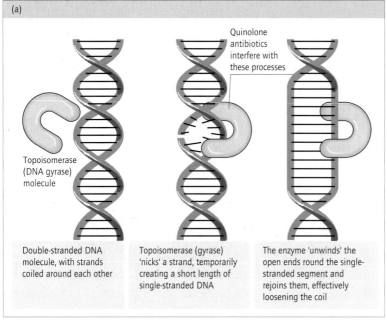

Double-stranded DNA molecule, with strands coiled around each other

Topoisomerase (gyrase) 'nicks' a strand, temporarily creating a short length of single-stranded DNA

The enzyme 'unwinds' the open ends round the single-stranded segment and rejoins them, effectively loosening the coil

Quinolone antibiotics interfere with these processes

Topoisomerase (DNA gyrase) molecule

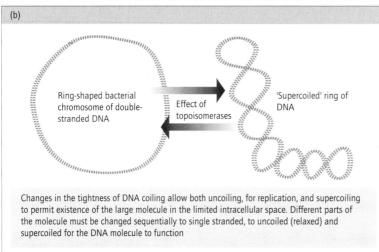

Ring-shaped bacterial chromosome of double-stranded DNA

Effect of topoisomerases

'Supercoiled' ring of DNA

Changes in the tightness of DNA coiling allow both uncoiling, for replication, and supercoiling to permit existence of the large molecule in the limited intracellular space. Different parts of the molecule must be changed sequentially to single stranded, to uncoiled (relaxed) and supercoiled for the DNA molecule to function

Figure 4.4 The actions of quinolone antimicrobials. (a) Topoisomerase tightens (supercoils) or loosens (relaxes) the coiling of the DNA molecule by opening and reclosing small gaps in a DNA strand. DNA must be relaxed for its strands to part during replication or the production of RNA. (b) A bacterial DNA molecule is approximately 1 m long. It must be coiled many times to enable it to fit inside a bacterial cell. It must then be uncoiled, part by part, to allow its code to be read. Quinolone antibiotics prevent these changes in configuration, disrupting the function of the DNA.

ability). Infection in special sites such as the meninges can only be treated if effective amounts of antibiotics cross the blood–brain barrier. Similarly, penetration of antibiotics into abscesses is poor and likely to result in treatment failure.

The well-established principles of pharmacology apply to the behaviour of antibiotics within the body. Non-polar (lipid-soluble) agents such as chloramphenicol are well absorbed and cross the blood–brain barrier easily, whereas the more polar agents, penicillins and cephalosporins, are less well absorbed and are confined to extracellular compartments. Protein binding also affects the duration of action and the bioavailability of antimicrobial agents.

The pharmacokinetic properties of antibiotics, such as absorption, distribution and elimination of the drug, can be calculated by measuring blood levels at different times. The maximum concentration (C_{max}) or the area under the pharmacokinetic curve (AUC) are important in measuring the effective exposure of organisms to the drug. Taken together with the minimal inhibitory concentration (MIC) of the drug against the pathogen, e.g. the ratio of AUC/MIC, a direct indicator of patient outcome can be calculated.

Conditions at the site of infection may greatly modify the effect of an antimicrobial agent. A concentration gradient of antimicrobial may be set up in an abscess, with

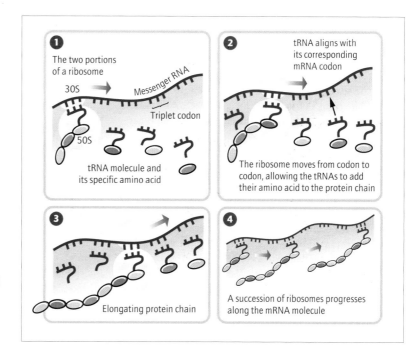

Figure 4.5 The mechanisms of bacterial protein synthesis. Various aspects of this process are disrupted by tetracyclines, aminoglycosides, macrolides and rifampicin.

lower concentrations at its centre than in surrounding serum. The presence of cellular or bacterial debris, and metabolic products, lower the local pH and redox potential, interfering with the action of antimicrobials such as gentamicin and erythromycin. An infected site may be devascularized, or an abscess with a fibrous cavity may develop, making drug penetration difficult. Often, organisms are not 'free' in the tissues but exist inside macrophages or other cells. Organisms at the site of infection do not multiply exponentially, as they do in the test-tube where nutrients are not limited. Infecting bacteria may be in a stationary phase because of limited nutrients or immunological suppression. In some cases, such as tuberculosis, they may be dormant. In active infection the immune system is a vital contributor to recovery and may enhance the apparent activity of antimicrobials.

The route of excretion of a drug also influences the effect of antimicrobial therapy. Agents excreted by the kidney are likely to be effective in pyelonephritis and cystitis as high concentrations will be found in renal tissue, the bladder and urine. However, they will accumulate in cases of renal failure, and the dose must be reduced to avoid toxicity. Similarly, antibiotics excreted in the bile are likely to be effective in acute cholecystitis or cholangitis.

Many antimicrobials are metabolized by the liver, and hepatic disease may interfere with their excretion, causing an increased risk of side-effects and often necessitating reduction in dosage.

Factors affecting the effect of antimicrobials at the sites of infections

1 The concentration of the antimicrobial (influenced by its ease of access and mode of excretion).

2 Local pH and redox potential.

3 Ability of the pathogen to destroy the antibiotic.

4 Destruction of antimicrobial by host lysozymes and proteases.

5 Renal and/or hepatic damage, which can impair antibiotic excretion.

Antimicrobial sensitivity testing

Concepts of susceptibility testing

Definitions

An organism is defined as sensitive or resistant to an antimicrobial, depending on whether treatment with the usual dose is likely to be successful. A moderately resistant organism is likely to respond to an increased dose. An organism that is resistant is unlikely to be successfully treated with a given antibiotic, irrespective of dosage.

These definitions depend on the assumption that the results of laboratory tests reflect what is happening in the patient. Laboratory tests measure directly or indirectly, under controlled conditions, the inhibitory or killing effect of the antimicrobial on the pathogen isolated from the patient. Usually, the pure drug is tested against bacteria growing exponentially in artificial culture. This is unlike the situation at the site of an active infection (see above). The results of the tests, together with the known pharmacokinetics of the antibiotic can assist the clinician in choosing the most appropriate therapy.

Minimal inhibitory concentration and minimal bactericidal concentration

The minimal inhibitory concentration (MIC) is the lowest antibiotic concentration at which growth of the organism is completely inhibited. The minimal bactericidal concentration (MBC) is the lowest concentration at which the organism is killed (defined as a 99.9% kill). In planning chemotherapy, the aim is to achieve a concentration of antimicrobial exceeding the MIC at the site of the infection. In some cases, where the multiplying organisms are inaccessible to the additive effect of phagocytosis and antibody, it is highly desirable that the MBC should be exceeded; for example in endocarditis where bacteria are buried in thrombus. Artificial materials, such as intravenous cannulae or prosthetic implants, also 'protect' bacteria and pose an especially difficult problem for therapy.

The MIC can be determined by cultivating organisms in broth cultures or on agar plates that incorporate a range of concentrations of the desired antimicrobial. By subculturing the broth cultures that show no bacterial growth, the death of the organisms can be confirmed, and the MBC can be determined (Fig. 4.6).

Methods of testing antimicrobial susceptibility

MIC testing can be too cumbersome for a clinical laboratory although commercial automated systems have made it more widely available (see Chapter 3). However, using rapid, indirect methods, a large number of isolates can be tested against many antimicrobials. The most popular methods employ filter-paper discs impregnated with antimicrobials. These are placed on agar plates that have been seeded with bacteria. The antimicrobials diffuse from the disc into the medium, inhibiting bacterial growth. At a certain distance from the edge of the disc, the concentration falls below that which will inhibit the growth of the organism, producing a visible zone of demarcation. The diameter of this zone is related to the MIC of the organism. The size of the zone may be altered by a number of factors unrelated to the susceptibility of the organisms

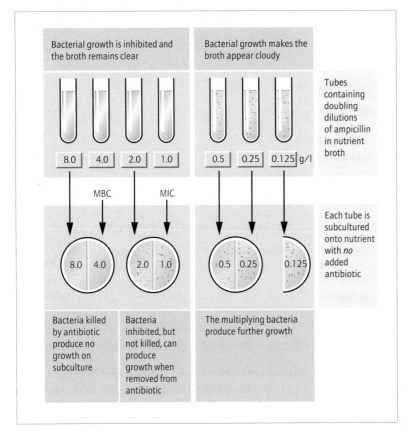

Figure 4.6 The principle of minimum inhibitory concentration (MIC) and minimum bactericidal concentration (MBC) determination.

tested. These include the composition of the medium, particularly the concentration of divalent cations; the pH; the molecular size of the antimicrobial; and the incubation conditions. These variables must be controlled if reproducible results are to be obtained. The different methods are outlined below.

Disk susceptibility methods
NCCLS method
The National Committee for Clinical Laboratory Standards (NCCLS) method is the standard technique used in the USA. The medium, inoculum (which must be confluent) and incubation conditions are strictly defined in this method. The zone sizes are directly related to MIC using a regression line, a graph that plots the known MIC of a series of organisms against zone size. The zones surrounding the test organisms are measured and compared with the previously defined ranges for susceptible organisms. Control strains are regularly tested and must be within defined limits of variation if the results are to be accepted as valid.

BSAC standardized disc sensitivity testing method
The British Society of Antimicrobial Chemotherapy (BSAC) uses a standard medium, Iso-Sensitest Agar, that can be supplemented with 5% whole horse blood with or without 20 mg/l beta-nicotinamide adenine dinucleotide (NAD) if required. The only exception to this is the use of Mueller–Hinton agar for detection of methicillin resistance in staphylococci. A semiconfluent inoculum is used, defined discs are applied and the plates incubated within 15 min. Plates are incubated for 18–20 h (except for enterococci, which require 24 h, and coagulase negative staphylococci, which may require incubation for 48 h when testing for methicillin resistance). Zone sizes are measured and compared with preprepared tables that relate zone size to MIC.

MIC determinations
Agar incorporation method
In the agar incorporation method, concentrations of antibiotic are incorporated into a solid medium in Petri dishes. The technique has the advantage that many tests can be performed on the same plate, using a multipoint inoculator. It is difficult, however, to obtain an estimate of the MBC in this test. The agar incorporation MIC method can be simplified for routine use by using only two carefully chosen antibiotic concentrations or 'breakpoints'. Organisms inhibited by the lower concentration of antibiotic are considered 'sensitive' to conventional doses of the antimicrobial. If inhibition occurs only at the higher concentration, successful therapy would require higher doses and this result is reported as 'moderately sensitive'. Organ-

isms not inhibited by either concentration are considered resistant, and therapy would be predicted to fail.

High- and low-concentration antibiotic-impregnated discs are sometimes used to test organisms in this way. *Neisseria gonorrhoeae*, for example, is often sensitive to benzylpenicillin, and shows a large zone of inhibition around a low-concentration penicillin disc. Some gonococci have altered penicillin-binding characteristics, and are inhibited only by high-concentration penicillin discs. Penicillinase-producing *Neisseria gonorrhoeae* (PPNG) are not inhibited even by high-concentration penicillin discs. Placing the two types of disc on the original selective medium plate therefore provides an early indication of antibiotic sensitivity (Fig. 4.7), but this screening test must be confirmed by formal testing of the purified organism.

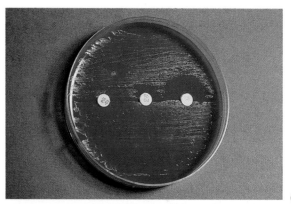

(a)

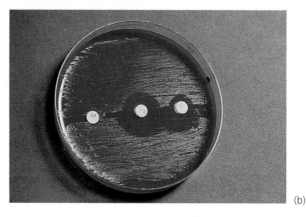

(b)

Figure 4.7 (a) The organism on the upper half of the plate is a penicillin-sensitive gonococcus with zones to the penicillin 0.25 µg and 1.0 µg disc. Below is a strain resistant to penicillin. (b) The organism on the upper half of the plate is a non-beta-lactamase-producing penicillin-resistant gonococcus as there is only a small zone to co-amoxiclav (right-hand disc). The strain in the lower half is also penicillin-resistant: no zone to the ampicillin disc on the left but a wide zone to co-amoxiclav, which negates this bacterium's beta-lactamase.

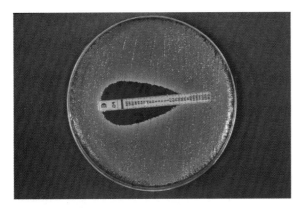

Figure 4.8 Use of the E-test, showing loss of inhibition of the organism at lower antibiotic concentrations; MIC is indicated at the point where growth meets the strip.

E-test

The E-test uses a semiconfluent inoculum and a strip impregnated with varying concentration of antibiotic throughout its length. The point at which growth meets the calibration on the strip is equivalent to the MIC (Fig. 4.8). A wide range of organisms can be tested against most antimicrobials (even including difficult-to-test mycobacteria).

Automated methods

Many laboratories use automated systems that detect growth for identification and susceptibility testing. These provide a report that includes a judgement on susceptibility versus resistance and can often provide an MIC value also. These methods are described in Chapter 3.

Methods of routine sensitivity testing

1 NCCLS method.
2 BSAC method.
3 Agar incorporation method.
4 Breakpoint method.
5 The E-test.
6 Automated methods.

Molecular and other susceptibility test methods

Beta-lactamase testing

Neisseria gonorrhoeae commonly becomes resistant to penicillin by elaborating a beta-lactamase enzyme. Penicillin resistance can therefore be detected by demonstrating the action of the beta-lactamase enzyme. This is done by using a chromogenic cephalosporin that changes colour when the beta-lactam bond is hydrolysed by beta-lactamase. This test takes only a few seconds to perform.

Molecular methods of susceptibility testing

It is possible, for some organisms, to detect the presence of gene mutations associated with resistance phenotypes. This is done by amplifying the target gene and determining the DNA (or RNA) sequence. This method is very useful in investigating the antiviral susceptibility of human immunodeficiency virus (HIV) as the gene sequence can also indicate which drugs might be successful. A similar approach can indicate susceptibility of *Mycobacterium tuberculosis*, especially for rifampicin. *M. tuberculosis* becomes resistant to rifampicin through mutations in the RNA polymerase gene, which can be amplified by polymerase chain reaction (PCR) and resistance-associated mutations detected. Among the various techniques for this are direct nucleotide sequencing, or hybridization, of *rpoB* NAAT products to a series of oligonucleotides.

Plasmodium falciparum becomes resistant to several antimalarial agents by acquiring the *Pfmdr* gene. A PCR that amplifies this gene enables multidrug-resistant organisms to be identified much more simply than by *in vitro* cultivation. The *mecA* gene of *S. aureus* coding methicillin resistance or the *VanA* gene of enterococci encoding vancomycin resistance can also be detected in this way. Adapting these techniques to real-time amplification platforms enables susceptibility data to be available in a few hours.

Checking the effectiveness of antimicrobial chemotherapy

Therapeutic monitoring

Antibiotic concentrations are nowadays rarely monitored except for the following reasons.

1 To ensure that adequate concentrations of antibiotic are found in the serum (or in the cerebrospinal fluid (CSF) in difficult cases of meningitis).

2 To ensure that antibiotic concentrations have not reached toxic levels. This is important for antibiotics with a narrow therapeutic index such as aminoglycosides or vancomycin, or when the patient is in renal failure.

3 To monitor patient compliance. This is important in managing infections that require long-term outpatient therapy, e.g. tuberculosis.

4 To confirm adequate absorption of oral antibiotics after a shift from parenteral administration, for example in the case of infective endocarditis.

5 To study the pharmacokinetics of new antibiotics.

Several different methods are used in the laboratory. The simplest are the commercial automated immunoassays available for many antibiotics, notably aminoglycosides and vancomycin. For other antibiotics, high-performance liquid chromatography (HPLC) must be used. Specimens must be taken at appropriate times: specimens for trough concentration should be taken just before the next dose is given, and peak concentrations should be taken 1 hour after the parenteral dose.

Adverse effects of antimicrobials

Antimicrobial chemotherapy depends on selective toxicity, which damages the pathogen's metabolism with a minimal effect on the host. Most antimicrobial agents have very favourable therapeutic indices, with a high ratio of toxic level to therapeutic level. Adverse effects fall into two major groups: those that are dose-dependent, as in the bone marrow toxicity of sulphonamides or renal toxicity of aminoglycosides; and idiosyncratic reactions, such as penicillin-induced anaphylaxis or chloramphenicol-induced aplastic anaemia.

> **Adverse antimicrobial effects**
> **1** Dose-dependent toxicity.
> **2** Idiosyncratic reactions.

Renal toxicity

Renal damage can occur for a number of reasons during antimicrobial chemotherapy. Not all are direct effects of the antimicrobial drug; for instance, renal failure can follow an anaphylactic reaction, because the accompanying hypotension causes acute tubular necrosis. Obstructive nephropathy can occur if high concentrations of drugs, such as the antiretroviral drug indinavir, or first-generation cephalosporins, form crystalline deposits in the renal tubules or collecting ducts.

Direct renal cell damage

Renal cell damage can be caused by nephrotoxic antimicrobials. Aminoglycosides are toxic to proximal renal tubules, causing accumulation of membrane structures within the cell. This occurs at serum levels close to those necessary for effective antimicrobial action; the therapeutic index is narrow. There is evidence of mild, self-limiting tubular effects in all patients, and small reductions in the glomerular filtration rate occur in approximately 80% of patients. Toxicity is more likely in elderly patients and those with pre-existing renal compromise, including renal failure caused by sepsis. It is increased by co-administration of diuretics, other renal toxic drugs or recent aminoglycoside therapy. Renal function can decline rapidly in patients treated with aminoglycosides for infective endocarditis, because of the additive effects of aminoglycoside toxicity, uncontrolled sepsis and the nephritis accompanying the infective endocarditis syndrome.

In the ear aminoglycosides may damage the hair cells of the organ of Corti, or the type I hair cells at the summit of the ampullar cristae, leading to impaired hearing. To prevent these consequences the serum aminoglycoside

concentration must be closely monitored, and the dose adjusted if toxic levels are approached.

Renal toxicity is common in patients treated with amphotericin. The mechanism is not known but may be associated with distal tubular lesions and a decrease in glomerular filtration rate. This problem has been largely overcome by using the less toxic liposomal preparations.

Tetracycline can cause renal damage by several different mechanisms. Its catabolic action produces an increase in nitrogen degradation products, including urea. Some tetracycline degradation products may be directly toxic to the kidneys.

Sulphonamides are a diverse group of agents, some of which can crystallize in the renal tubules, causing obstructive uropathy. Occasionally they cause direct renal tubular cell damage. Co-trimoxazole, which contains sulfamethoxazole, may accumulate in renal failure, producing a fall in the glomerular filtration rate. Patients with a small urinary volume may develop crystalluria, even with this modern sulphonamide preparation.

Liver toxicity

Damage to the liver may take the form of acute or chronic hepatitis, cholestasis, fatty degeneration or a granulomatous hepatitis.

Hepatocellular damage

This is the commonest form of liver damage. It is often associated with the antituberculous agents isoniazid and rifampicin. Among patients given isoniazid prophylaxis, acute hepatitis occurs in approximately 1%. The risk is age-related, being much greater in patients aged over 35, and is also more likely in patients with previous liver disease, or a history of alcohol abuse. Toxicity is probably caused by an intermediate metabolite, acetylhydrazine.

With rifampicin, transient elevation of transaminases is common when dosing is started, but levels usually return to normal despite continued dosing. Severe hepatitis with jaundice is occasionally seen. Rifampicin may also produce transient hyperbilirubinaemia due to competition for hepatic excretion. The combination of rifampicin and isoniazid in the treatment of tuberculosis causes hepatitis in about 5% of patients. Hepatitis can complicate the use of erythromycin estolate, parenteral tetracycline, pyrazinamide and ethionamide.

Rare cases of granulomatous hepatitis have been reported following high-dose ampicillin or flucloxacillin treatment, or prolonged quinine therapy.

Cholestatic jaundice

Cholestatic jaundice is a dose-related effect of fusidic acid, induced by doses above 2 g/day. Tetracycline may induce fatty change in the liver if high doses are given intravenously or if patients are predisposed by pregnancy or renal failure.

Effects on the haemopoietic system

Many different agents can cause bone marrow toxicity, due to either the normal action of the drug, an idiosyncratic reaction, or by immune mechanisms. Toxicity may affect granulocytes alone or the whole haemopoietic system.

Aplastic anaemia

Aplastic anaemia can follow chloramphenicol therapy from 10 days after therapy until up to 6 months. This idiosyncratic reaction must be distinguished from a mild, reversible dose-dependent bone marrow depression, which occurs in many patients on higher doses. True aplasia has an incidence of 1 in 40 000 to 1 in 100 000, and is often irreversible.

Metabolic bone marrow depression

Metabolic bone marrow depression can complicate treatment with antimicrobials affecting nucleic acid synthesis, either directly, as with ganciclovir and zidovudine, or by reducing folate availability, as with sulphonamides. These effects can be made worse by concurrent bone marrow suppressive therapy for the treatment of leukaemia or other malignancy.

Granulocytopenia

Granulocytopenia is usually idiosyncratic. It is commonest after sulphonamide or sulphone therapy, but rarely follows high-dose benzylpenicillin therapy (>12 MU/day). Penicillin-associated agranulocytosis is always associated with the same reaction to cephalosporins, and will recur following even small oral doses of these drugs.

Immunologically mediated cytopenias

Immunologically mediated cytopenias are rare. Antibiotics, like other drugs, may act as haptens and induce antibodies to red blood cells or platelets, resulting in a Coombs-positive haemolytic anaemia or immune thrombocytopenia.

Cutaneous adverse reactions

These vary from fixed drug eruptions, urticarial and maculopapular eruptions to erythema multiforme or even life-threatening Stevens–Johnson syndrome. The commonest causes of severe skin reactions are sulphonamides, and the risk increases dramatically in the elderly. Mild, itching rashes are common with sulphonamides, penicillins (Fig. 4.9) and cephalosporins. Cross-reactivity between different types of beta-lactam is rare (less than 5%). Although not strictly an allergic response, 95% of patients with infectious mononucleosis develop a rash if treated with ampicillin, but this does not recur on re-exposure to the drug after convalescence.

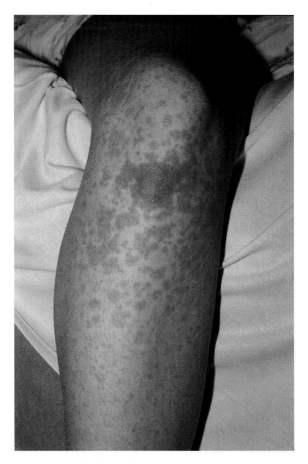

Figure 4.9 Rash induced by ampicillin; such rashes are usually painful or irritating.

Anaphylaxis occurs when an antibiotic generates a type I hypersensitivity reaction, usually as a result of hapten–drug sensitization. This complication is uncommon, perhaps affecting 1 in 100 000 treatments. Penicillins are most often responsible. Cross-reaction with cephalosporins is more likely than for skin reactions. Monobactams have much lower cross-sensitivity rates.

Mechanisms of resistance to antibacterial drugs

Each antimicrobial agent is effective against a limited range of organisms, usually because the metabolic process with which they interfere does not occur in all species. Vancomycin prevents peptidoglycan cross-linking only in Gram-positive organisms, but Gram-negative organisms are naturally resistant. Metronidazole is active at such low redox potentials that only anaerobic bacteria are susceptible. Amphotericin is active against fungi because it inhibits ergosterol synthesis, necessary for cell wall production in fungi, but not in bacteria.

Reasons why organisms may be naturally resistant to antimicrobials:

1 They are naturally impermeable to the antimicrobial agent.
2 They lack the target binding site.
3 They lack the target metabolic pathway.
4 They naturally produce antibiotic-destroying enzymes.
5 They produce more of the target to overcome inhibition.

Factors affecting antibiotic penetration

Reduced permeability of bacterial outer membranes may result in high-level resistance, often affecting all antibiotics of a class. This is common in *Pseudomonas* spp. Some bacteria become resistant to tetracyclines by actively transporting the antibiotic out of the cell (the efflux mechanism). This results in resistance to all compounds in this class.

The major resistance mechanism for tetracycline among Gram-negative organisms is an alteration of an inner membrane protein that inhibits accumulation of tetracycline within the bacterial cell.

Glycopeptide-resistant *S. aureus* has a much thicker cell wall, with subtle alteration to its peptide structure. Increased absorption of glycopeptide to this surface prevents the antibiotic reaching the target site.

Altered antibiotic-binding sites

An antibiotic's effect may be inhibited by preventing its binding to the target site. Organisms resistant to macrolides and aminoglycosides have small alterations in their ribosomal binding sites. Resistance to streptomycin, for example, may develop in a one-step mutation that causes methylation of the ribosomal binding site, preventing the binding of the antibiotic. Resistance to quinolones develops through mutations in the topoisomerase and DNA gyrase genes. Penicillin resistance in *Streptococcus pneumoniae* depends on altered affinity of penicillin-binding protein (PBP). This has developed in a multiple stepwise fashion under the selective pressure of frequent exposure to antibiotics. Low frequency (10^{-13}) recombination between PBP genes in *S. pneumoniae* and commensal organisms has eventually resulted in penicillin-binding proteins with reduced affinity for penicillin, but which still support cell wall synthesis (Fig 4.10). Resistance to sulphonamides and trimethoprim occurs due to point mutations in the relevant gene, which alter the target enzymes, dihydropterate synthetase or dihydro-

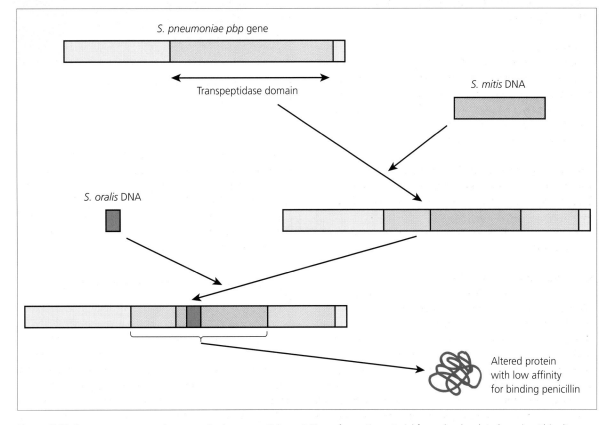

Figure 4.10 Pneumococcus acquires new pbp by sequential acquisition of genetic material from closely related species. This alters the pbp gene, which now codes for an altered protein with low affinity for binding penicillin.

folate reductase. Rifampicin resistance arises by alteration of the peptide sequence of RNA polymerase. Many viruses develop resistance to antiviral agents by point mutations in the genes encoding antiviral target proteins; for example, HIV becomes resistant to lamivudine by a single point mutation in the reverse transcriptase gene.

Antibiotic-inactivating enzymes

Some organisms naturally produce enzymes that inactivate certain antimicrobials. When penicillin was first introduced, 95% of *Staphylococcus aureus* were sensitive, but 5% produced a beta-lactamase that destroyed the beta-lactam bond needed for activity. Over a few years an increasing proportion of strains acquired beta-lactamase. Now almost all *S. aureus* isolated from patients are penicillin-resistant. Selective pressure applied by exposure to antibiotics favours the survival of beta-lactamase-producing strains. Other organisms expressing a beta-lactamase may also be favoured by the same selection pressure. The genes coding for these enzymes may be transferred to other organisms, passing on the selective advantage. By these means, antibiotic-destroying enzymes can become widespread. For example, a beta-lactamase that first emerged in enteric organisms has spread widely to include *H. influenzae* and *N. gonorrhoeae*.

Other antibiotics are inhibited by different bacterial enzymes. Bacteria become resistant to aminoglycosides by developing inactivating enzymes which adenylate, hydroxylate or acetylate the aminoglycoside molecule at various sites. Resistance to chloramphenicol depends on possession of an acetyltransferase enzyme.

Methods by which bacteria can acquire resistance to antimicrobial agents:

1 An alteration in permeability to the antibiotic.
2 An alteration in the target binding state.
3 Use of an alternative metabolic pathway.
4 Switching on a gene for an antibiotic-destroying enzyme.
5 Acquiring a new gene for an antibiotic-destroying enzyme.

Genes that encode antibiotic resistance features can be transmitted not only between organisms of the same species but even between different genera. Thus, resistance to commonly used antibiotics can become widespread in previously sensitive organisms. Bacterial resistance genes, like all other DNA fragments, are transmitted by four main mechanisms: transformation, transduction, conjugation and transposons (Fig. 4.11).

• Transformation, the process whereby bacteria take up segments of naked DNA and incorporate them into their own genome, is the mechanism for emergence of penicillin resistance in *S. pneumoniae* and *N. gonorrhoeae*.

• Transduction occurs when DNA fragments are taken up by a bacteriophage and transmitted when the phage infects a new bacterial cell. Aminoglycoside and macrolide resistance can be transmitted in this way.

• Conjugation involves transfer of chromosomal and/or plasmid DNA between conjoined bacterial cells.

• Transposons and integrons are genetic elements capable of transferring from one bacterial strain to another, between plasmids, and between plasmids and the bacterial chromosome.

Some Gram-negative pathogens may possess multiple resistance determinants, gathered together on one of these transmissible elements. This makes the organisms both difficult to treat in the patient and to eradicate from the hospital environment. Multiple drug resistant strains of *Klebsiella pneumoniae* and *Actinobacter baumanii* circulating in the hospital environment can be resistant to almost all currently available antibiotics. Control of antibiotic prescribing may be effective in limiting emergence of these strains, but to eradicate existing strains all positive selective pressure must be avoided by excluding key antimicrobials to which the organism is resistant from the hospital environment.

Mechanisms for gene transfer between bacteria:

1 Transformation (transfer of naked DNA).
2 Transduction (where the gene is acquired with a bacteriophage infection).
3 Conjugation (DNA transfer between conjoined bacteria).
4 Transposons and integrons (transposition).

Agents used in treating infection

The main types of anti-infective agents will be reviewed here briefly. Antiviral, antifungal and antiparasitic agents will also be discussed with their main therapeutic targets in the systematic chapters.

Antibacterial agents

Beta-lactam antibiotics

Beta-lactam antibiotics include penicillins and cephalosporins, which have similar structures.

Penicillins can be conveniently divided into classes on the basis of their antibacterial activity:

1 Natural penicillins (penicillin G).
2 Penicillinase-resistant penicillins (cloxacillin).
3 The aminopenicillins (ampicillin-like agents).
4 Expanded-spectrum penicillins.
5 Carbapenems.
6 Monobactams.
7 Beta-lactamase inhibitors.

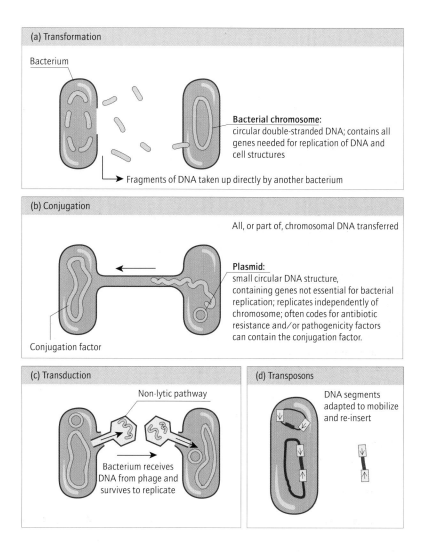

(a) Transformation

Bacterium

Bacterial chromosome:
circular double-stranded DNA; contains all
genes needed for replication of DNA and
cell structures

Fragments of DNA taken up directly by another bacterium

(b) Conjugation

All, or part of, chromosomal DNA transferred

Plasmid:
small circular DNA structure,
containing genes not essential for bacterial
replication; replicates independently of
chromosome; often codes for antibiotic
resistance and/or pathogenicity factors
can contain the conjugation factor.

Conjugation factor

(c) Transduction

Non-lytic pathway

Bacterium receives
DNA from phage and
survives to replicate

(d) Transposons

DNA segments
adapted to mobilize
and re-insert

Figure 4.11 The mechanisms of DNA
transfer between bacteria.

Spectrum of activity

Penicillin G is derived from cultures of *Penicillium chrysogenum*. It is active against Gram-positive bacteria, including streptococci, staphylococci (excluding most *Staphylococcus aureus*), clostridia and corynebacteria, and some Gram-negative cocci, including *Neisseria meningitidis*. *Bacteroides fragilis* is resistant, but *Prevotella* sp., *Porphyromonas* sp. and anaerobic cocci are susceptible.

Isoxazolyl penicillins such as flucloxacillin, oxacillin and cloxacillin were developed to combat penicillin-resistant *S. aureus*. Their bulky side-chain sterically hinders the binding of penicillinase to these drugs (Fig. 4.12). These agents retain activity against streptococci and *Neisseria*, and are well absorbed orally.

The aminopenicillins include ampicillin, its esters and amoxicillin. They are well absorbed orally. Their expanded spectrum includes *Haemophilus influenzae* and many species of the Enterobacteriaceae, including *Escherichia*

coli, *Salmonella* sp. and *Shigella* sp. Following the introduction of ampicillin, hospital cross-infection occurred due to Gram-negative pathogens naturally resistant to this agent. These included *Klebsiella* spp. and *Pseudomonas aeruginosa*. Newer penicillins, developed to address this problem, included the acylureidopenicillins (azlocillin and piperacillin), and the carboxypenicillins (carbenicillin and ticarcillin) (Fig. 4.13). All of these agents must be administered parenterally. They are active against some ampicillin-resistant Enterobacteriaceae and *P. aeruginosa* but are susceptible to plasmid-mediated beta-lactamases, which limits their effectiveness.

The carbapenems, imipenem (always given with cilastatin) and meropenem, have a broad spectrum of activity against Gram-positive and Gram-negative bacteria, including beta-lactamase-producing Gram-negative organisms and *Bacteroides fragilis*. Orally available carbapenems are in the late stages of clinical development. The

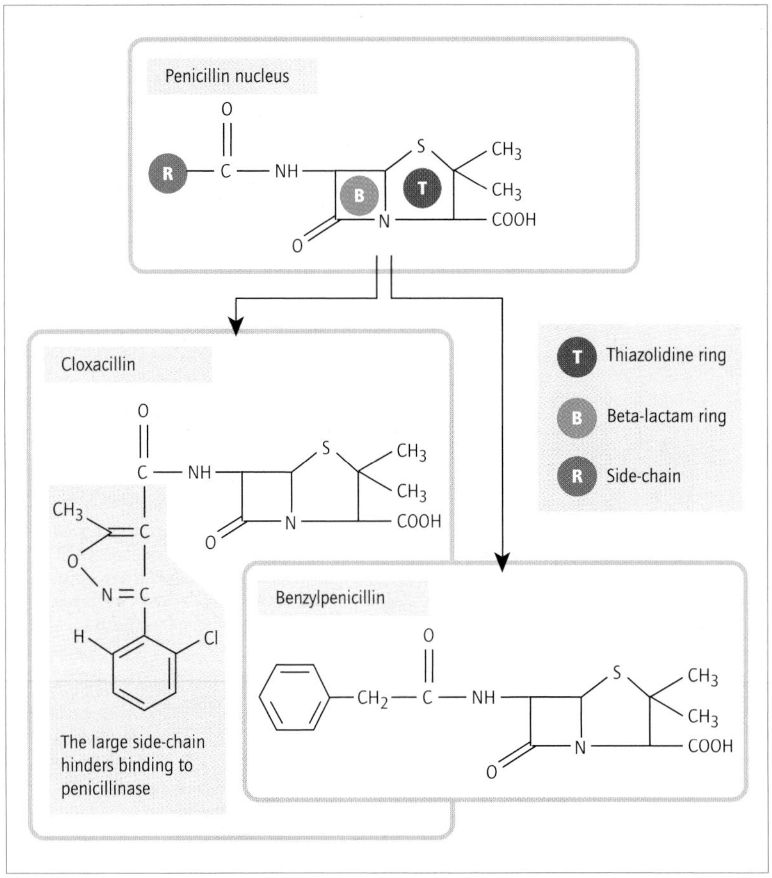

Figure 4.12 The structure of penicillin G and cloxacillin.

monobactams, of which aztreonam is the best example, are synthetic compounds that have little anti-Gram-positive activity but are very active against Gram-negative species, including beta-lactamase-producing strains and *Pseudomonas* sp.

As the elaboration of beta-lactamases is a major mechanism of resistance to penicillins, co-administration of a beta-lactamase inhibitor is a logical step. These are beta-lactamase-stable beta-lactam compounds with minimal antimicrobial activity. The compounds used must have similar pharmacokinetic properties to their companion active agent. Several combinations are available for clinical use, including clavulanic acid, used with amoxicillin or ticarcillin, and sulbactam, used with ampicillin or piperacillin.

Pharmacology

Penicillins vary markedly in their oral absorption. Penicillin G is not stable to gastric acid and must be given intravenously. Long-acting derivatives of benzylpenicillin such as procaine and benzathine penicillin are available for intramuscular administration in special outpatient situations. Penicillin V (phenoxymethylpenicillin), in contrast, is sta-ble to gastric acid and can be given orally. The aminopenicillins and isoxazolylpenicillins are well-absorbed orally. All other penicillins must be administered intravenously.

All orally active penicillins reach peak serum levels at 1–2 h after ingestion but this peak is delayed by taking the dose with food. Penicillins are excreted and secreted by the kidney. The consequence of rapid excretion is a very short half-life, ranging from 30 to 72 min. Protein binding is variable, from 17% for aminopenicillins to 97% for dicloxacillin. Penicillins are distributed in extracellular fluid to most of the body tissues, including lung, liver, kidney, muscle, bone and placenta. They do not cross the blood–brain barrier unless the meninges are inflamed. High concentrations of penicillin are found in the urine. Penicillins are actively secreted into the bile. Imipenem is broken down by renal dihydropeptidase I so, to be effective it must be co-administered with cilastatin, a specific enzyme-inhibitor. Meropenem is not broken down, and can be given alone.

Unwanted actions

Penicillins are widely used because of the infrequency of their adverse effects. The most serious side-effect is acute

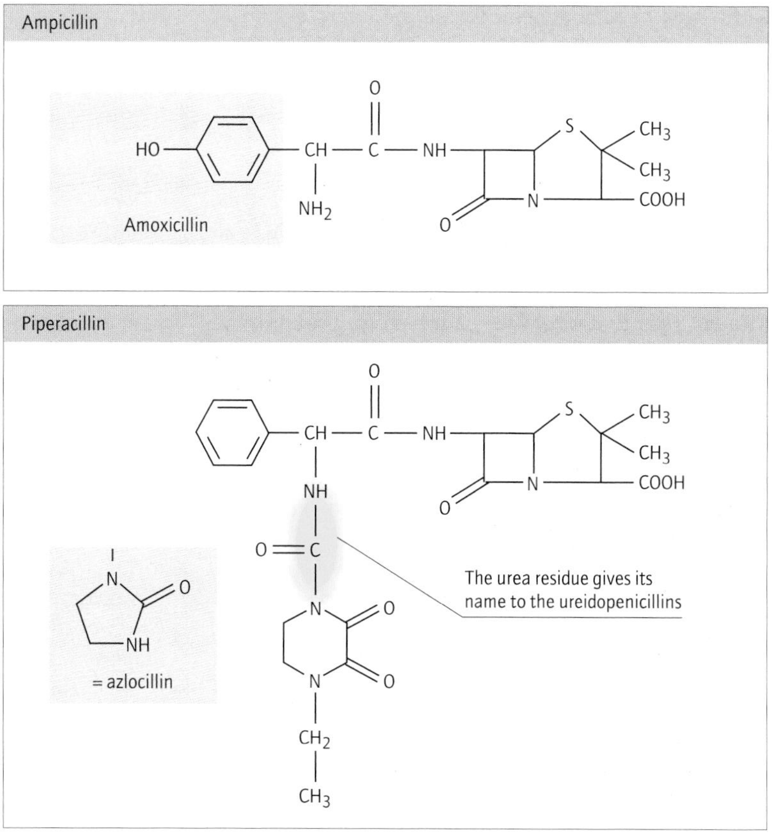

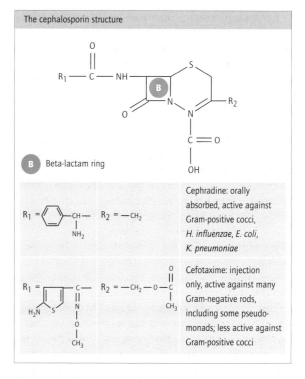

Figure 4.13 The structure of ampicillin and piperacillin.

anaphylaxis, which is uncommon but is most likely to complicate treatment with benzylpenicillin. Less severe allergic manifestations include angioneurotic oedema, pruritus and urticaria. Occasionally, patients suffer a delayed 'serum sickness' reaction with urticaria and arthralgia. Other side-effects include Stevens–Johnson syndrome, dermatitis, morbilliform eruptions and allergic vasculitis. Neutropenia occasionally occurs with prolonged high dosage of benzylpenicillin. A high salt load is associated with penicillin sodium salts, especially the disodium salts of expanded-spectrum penicillins. In patients with renal failure, this may result in hypernatraemia, hypokalaemia and convulsions.

Ampicillin can precipitate antibiotic-associated diarrhoea and, occasionally, pseudomembranous colitis due to its activity against obligate anaerobes of the intestinal flora, permitting the emergence of *C. difficile* infection.

Cephalosporins

Cephalosporins are naturally occurring compounds closely related to penicillins (Fig. 4.14). They are usually classified according to the route of administration and spectrum of activity.

The first group is the orally active cephalosporins, which all have a strong Gram-positive spectrum, initially

Figure 4.14 The structure of cephalosporins.

introduced to combat penicillinase-producing *S. aureus*. They are, however, less effective than isoxazolyl penicillins against this organism. Some agents such as cefaclor also have useful activity against *H. influenzae*, making it useful in the treatment of community-acquired respiratory infections.

The second group contains older, injectable agents, such as cefazolin, cefamandole and cefuroxime. These all retain good activity against Gram-positive organisms, and are also highly effective against *E. coli* and some species of *Proteus*. They have established a role in surgical prophylaxis (e.g. cefazolin plus metronidazole before large bowel surgery). Esters of cefuroxime, such as cefuroxime axetil, are absorbed orally.

Third-generation cephalosporins are newer injectable cephalosporin agents. Cefotaxime is active against most Gram-negative organisms, retains good activity against streptococci but is less active against *S. aureus*. Ceftriaxone has an extended half-life, making single-dose treatment of penicillinase-producing *N. gonorrhoeae* and once-daily therapy of Gram-negative infections, including typhoid, possible. These agents are widely used in the management of septicaemia, serious community-acquired pneumonia and neonatal meningitis. Some cephalosporins have only weak activity against *P. aeruginosa*. However, ceftazidime and cefepime are active against this species. More recently, orally active cephalosporins such as cefixime have become available, with a spectrum of activity similar to injectable third-generation cephalosporins.

Cephalosporins are eliminated by the kidney with a half-life of 1–2 h (note the exception of ceftriaxone). They are distributed widely in the extracellular fluid, and penetrate well into tissues. Earlier agents did not cross the blood–brain barrier but the newer injectable cephalosporins such as cefotaxime and ceftriaxone do. Hepatic metabolism is important for some compounds, including cefotaxime.

Cephalosporins have a low incidence of anaphylaxis, but cross-reactivity occurs in approximately 10% of penicillin-allergic patients. Rashes may also occur, as with the penicillins. Cephaloridine and cephalothin are toxic to the kidney but more recent agents have not had this side-effect.

Glycopeptides

Vancomycin is a glycopeptide antibiotic first isolated from *Streptomyces orientalis*. Glycopeptides inhibit Gram-positive cell wall construction by blocking cross-linking of peptide bridges and inhibiting chain-lengthening. They are large polar molecules and cannot penetrate the outer membrane of Gram-negative bacteria. Resistance among Gram-positive organisms, initially rare, is increasingly reported in enterococci and, recently, staphylococci. Vanco-

mycin and teicoplanin are not absorbed orally and must be administered parenterally, or occasionally locally into the peritoneal cavity. Vancomycin is distributed in the extracellular fluid. It does not cross the blood–brain barrier unless there is meningeal inflammation. More than 80% is excreted unchanged by the kidney.

Vancomycin causes thrombophlebitis, so administration via a central venous line is preferred. An idiosyncratic vasodilatation–hypotension ('red man') syndrome, probably related to histamine release, develops in up to 15% of patients given bolus doses. To avoid this, the drug is always given by slow infusion. Teicoplanin may be given by a peripheral line or by intramuscular injection.

Renal toxicity is important and is more likely in patients receiving concomitant aminoglycosides. Deafness may develop, particularly in elderly patients or those with renal impairment. These toxic effects can be minimized by careful attention to dosage schedules and regular monitoring of serum levels.

Vancomycin is indicated for the treatment of severe Gram-positive infections, including methicillin-resistant *Staphylococcus aureus* (MRSA) infections, unresponsive to beta-lactam antibiotics, or for patients who are sensitive to beta-lactams. It has an important role in the therapy of invasive infections caused by methicillin-resistant staphylococci and ampicillin-resistant enterococci. Oral therapy is indicated for the treatment of pseudomembranous colitis. Gram-positive infections in chronic ambulatory peritoneal dialysis may be treated by administration of the drug in the dialysis fluid. The dosage of vancomycin should be modified in patients with renal failure, to avoid accumulation and toxicity.

Vancomycin-resistant enterococci (VRE) have emerged in many countries, causing significant problems in the hospital environment (see above for mechanism), particularly as enterococci, notably *E. faecium*, which are naturally resistant to many commonly used antibiotics. Vancomycin resistance has emerged in staphylococci in two forms: the glycopeptide intermediate *S. aureus* (GISA) in which the bacterium expresses excess peptidoglycan, which 'mops up' the glycopeptide, and fully glycopeptide-resistant *S. aureus* strains, which possess the glycopeptide resistance mechanism of enterococci.

Teicoplanin

Teicoplanin is a new glycopeptide antibiotic isolated from *Actinoplanus teichomyceticus*. It is similar to vancomycin in its spectrum of activity. However, some strains of *Staphylococcus haemolyticus* are naturally resistant. Teicoplanin does not cause the severe adverse reactions reported with vancomycin and serum monitoring is unnecessary. The clinical indications for the use of teicoplanin are similar to those of vancomycin.

New agents
Several new glycopeptide antibiotics are in clinical development, including oritavancin, dalbavancin and the glycolipodepsipeptide, ramoplanin.

Quinolones
All members of the quinolone group of antimicrobial agents (Fig. 4.15) have a similar action. When DNA is transcribed, the supercoiled molecule must be unwound for transcription to occur. Quinolones interfere with the action of the responsible bacterial enzyme, DNA topoisomerase A. The first clinically used quinolone was nalidixic acid, active mainly against Enterobacteriaceae. It reaches high concentrations in the urine, and was used to treat Gram-negative urinary tract infections. *Pseudomonas* spp. and *Serratia* spp. are resistant to nalidixic acid, as are staphylococci and streptococci.

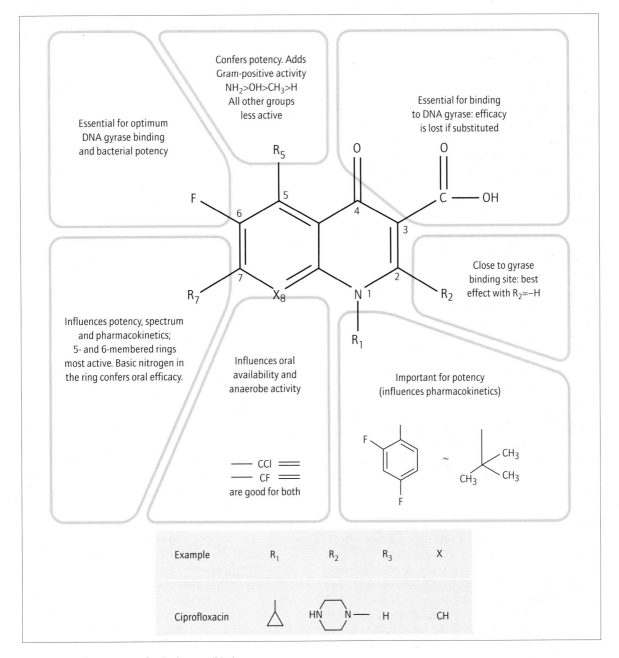

Figure 4.15 The structure of quinolone antibiotics.

Quinolones interfere with DNA function by preventing the uncoiling and recoiling of the molecule necessary for polymerization and transcription and untwisting bacterial DNA: DNA gyrase and topoisomerase.

Modification of the quinolone nucleus by the introduction of a fluorine atom produced the fluoroquinolones, with much greater antibacterial activity than nalidixic acid. Fluoroquinolones such as ciprofloxacin and ofloxacin are highly active against the Enterobacteriaceae, *Pseudomonas* spp., *Haemophilus* spp., *Neisseria* spp., *Chlamydia* spp. and also mycobacteria. Some newer quinolones have excellent activity against anaerobic bacteria and streptococcal species.

Fluoroquinolones are well absorbed orally, reaching peak serum concentrations 90 min after oral administration. Protein binding is low and they are distributed widely in the tissues, including 'difficult' areas such as the prostate. They do not cross the blood–brain barrier. Quinolones penetrate cells, including macrophages and polymorphs. They vary in their degree of metabolism and renal excretion. Ciprofloxacin is partly metabolized and excreted in the urine, so dosage should be reduced in patients with renal insufficiency.

Quinolone antibiotics are generally well tolerated but nalidixic acid often causes nausea, vomiting, diarrhoea, abdominal pain and skin reactions, including photosensitivity. Fluoroquinolones are associated with mild gastrointestinal side-effects, dizziness, tiredness, restlessness and depression. They decrease the seizure threshold in epilepsy, and should be avoided or used with caution in known epileptics. Important interactions with some drugs, including aminophylline, increase their toxic effects.

Fluoroquinolones are valuable in managing many systemic bacterial diseases, including hospital-associated Gram-negative septicaemia, especially in neutropenic patients when *P. aeruginosa* is a likely pathogen. They readily enter bronchial secretions, and are useful in treating hospital-acquired Gram-negative pneumonia, and chest infections in children with cystic fibrosis (who are subject to recurrent *Pseudomonas* infection). They are widely used in the management of invasive *Salmonella* infections and sexually transmitted diseases, and are valuable in managing 'difficult' mycobacterial infections. The progressive emergence of resistance, particularly in *Salmonella typhi* and *Neisseria gonorrhoeae*, has limited their usefulness, however. Newer fluoroquinolones such as gemifloxacin and moxifloxacin possess enhanced activity against Gram-positive pathogens and are increasingly used for treating respiratory tract infections. Ciprofloxacin, ofloxacin and moxifloxacin are active against *M. tuberculosis* and have a role in the managing of patients with tuberculosis who have resistant organisms or intolerance of standard drugs.

Aminoglycosides

The first aminoglycoside, streptomycin, was isolated from *Streptomyces griseus* in 1943. A number of different agents were subsequently developed, including kanamycin, gentamicin, tobramicin, netilmicin and amikacin (Fig. 4.16). Aminoglycosides possess an aminocyclitol structure to which amino sugars are linked by glycoside bonds. They inhibit bacterial protein synthesis by preventing ribosomes from translating messenger RNA codes. They are active against aerobic and facultatively anaerobic Gram-negative bacilli, *S. aureus* and also *Mycobacterium* and *Brucella* species.

Aminoglycosides are not absorbed from the gastrointestinal tract and must be administered parenterally. They are polar and are distributed in the extracellular fluid. They are excreted unchanged in the urine. Although they

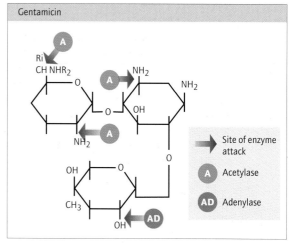

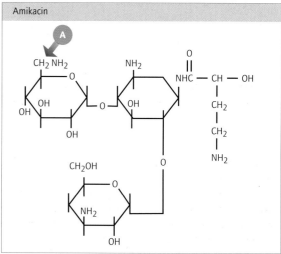

Figure 4.16 Structure of gentamicin, indicating sites of antimicrobial enzyme attack. Amikacin, by contrast, lacks all but one of these vulnerable sites.

have a low therapeutic index, they are valuable for their effectiveness in treating Gram-negative septicaemia.

The three main adverse events associated with aminoglycoside use are ototoxicity, renal toxicity and neuromuscular paralysis.

Bacteria become resistant to aminoglycosides by three main mechanisms. Methylation of the ribosome prevents streptomycin from binding to its substrate, but does not affect other aminoglycosides. The bacterial cell membrane may become impermeable to aminoglycoside antibiotics, and this affects all aminoglycosides. Bacteria can also become resistant by developing aminoglycoside-inactivating enzymes. There are three main enzyme groups: *N*-acetyltransferases, *O*-phosphotransferases and *O*-nucleotidyltransferases. Despite this repertoire of enzyme-mediated resistances, aminoglycoside-resistant bacterial strains are uncommon in hospitals with well-organized infection control and antibiotic policies.

Tetracyclines

Tetracyclines were first isolated from *Streptomyces* species and are based around four fused rings (a hydronaphthacene nucleus). They act by binding to the 30S ribosomal subunit, locking band A of transfer RNA to the septal site on the messenger RNA–ribosomal complex, and arresting the sequence of transfer RNA molecules. Many tetracycline compounds have been described, all with similar spectra of activity. Tetracyclines are active against *S. aureus*, streptococci, including *Streptococcus pneumoniae*, *N. gonorrhoeae*, *N. meningitidis*, *H. influenzae*, some species of Enterobacteriaceae, and also *Bacillus* species and many species of anaerobes. They are also effective against mycoplasmas, rickettsiae, coxiellae, chlamydiae, and spirochaetes, including *Treponema pallidum*. Some species of mycobacteria are also susceptible. The activity of tetracyclines is not limited to bacteria but also includes protozoa, such as *Plasmodium* spp. and *Entamoeba histolytica*.

Tigecycline, a glycylcycline tetracycline that must be administered parenterally, is now entering clinical practice. It may find a role in treatment of multiple drug resistant Gram-positive and Gram-negative bacteria including *Acinetobacter baumanii*.

Tetracyclines are well absorbed orally but vary widely in their half-life. Some agents, such as doxycycline, have a long half-life, and adequate therapeutic levels may be obtained by once-daily dosage. All are widely distributed in body tissues including the lung, liver, kidney, brain and respiratory tract. They are concentrated in the bile but do not cross the blood–brain barrier into the CSF. With the exception of doxycycline and minocycline, they should be avoided in renal insufficiency. They cause increased protein breakdown, with a rise in blood urea. High-dose tetracyclines can cause hepatocellular disorder, as excretion is via the liver; they should be avoided in severe liver disease. Fatty degeneration of the liver, with fatal liver failure, has been reported in pregnant patients.

Gastrointestinal intolerance is the commonest adverse event. Skin reactions include photosensitivity and exfoliation. Tetracyclines are concentrated in developing bones and teeth, and are therefore contraindicated in children and pregnant women.

Because of their wide range of side-effects, tetracyclines are first-line treatment for relatively few infections. They are the treatment of choice for severe rickettsial infections, and for Q fever.

Metronidazole

Metronidazole is a nitroimidazole drug active against anaerobic bacteria, and the parasites *Giardia intestinalis*, *Trichomonas vaginalis* and *Entamoeba histolytica*. The drug acts as an electron acceptor at low redox potentials, allowing the formation of toxic metabolites that damage bacterial DNA.

Metronidazole is indicated for the treatment of serious infections in which anaerobic bacteria may play a part: abdominal sepsis, brain abscess, lung abscess, anaerobic lung infections and dental infections. It may also be effective in oral or parenteral therapy of pseudomembranous colitis. It is the treatment of choice for giardiasis, intestinal amoebiasis and amoebic liver abscess and in the treatment of vaginal trichomoniasis. It is also valuable for surgical prophylaxis for abdominal and gynaecological surgery (see Chapter 23).

Metronidazole is well absorbed orally and rectally, and can also be administered intravenously. It is not strongly protein bound. It is widely distributed throughout all tissues and body compartments. It crosses the blood–brain barrier and achieves high concentrations in brain abscesses. Approximately 30% of the dose is metabolized in the liver and 70% is excreted unchanged in the urine. Initial therapy of serious sepsis and abscesses is via the intravenous route but oral or rectal administration can be commenced as soon as practicable.

Metronidazole is usually well tolerated, but many patients complain of minor side-effects, including fever, dizziness, dry mouth and a metallic taste. More serious side-effects include encephalopathy and peripheral neuropathy, more likely with extended therapy. A disulfiram (Antabuse) reaction may occur in patients who drink alcohol while taking metronidazole.

Macrolides

The macrolides are a group of related antimicrobials, erythromycin, clarithromycin, azithromycin and spiramycin, that possess a macrocyclic lactam ring with 14, 15 or 16 atoms. All the agents have a similar antimicrobial spectrum, including Gram-positive organisms, *Neisseria*, *Haemophilus*, *Bordetella*, and some Gram-negative anaerobes. They are also active against *Mycoplasma* sp., *Rickettsia* sp. and *Toxoplasma gondii*. They bind to the 50S ribosome, but their mechanism of activity is uncertain, occurring in the early stages of RNA-dependent protein synthesis. Bacteria become resistant to macrolides by mechanisms that decrease cell permeability and by the alteration of the ribosomal binding site. Resistance may be inducible or noninducible (the latter confers resistance to both macrolides and the lincosamide, clindamycin).

Erythromycin

Erythromycin is derived from *Streptomyces erthythreus*. Erythromycin base is poorly soluble and inactivated by gastric acid. To improve absorption it may be protected by enteric coating but this may lead to delayed or incomplete absorption. Erythromycin stearate and ethyl succinate esters are better absorbed. A water-soluble lactobionate preparation is available for intravenous use. Erythromycin is absorbed orally and distributed throughout the total body water. It is concentrated in polymorphs and alveolar macrophages, also in the liver, and it is excreted in the bile. It does not cross the blood–brain barrier but does cross the placenta.

Erythromycin is well tolerated but some patients complain of nausea and gastric irritation. Thrombophlebitis may follow intravenous use. Cholestatic jaundice may develop, but was most common with older oral preparations, such as the estolate salt. Allergic reactions are rare. It is often indicated as an alternative to penicillin in allergic subjects. It is a first-choice treatment of primary atypical pneumonia, including legionellosis, and is the treatment of choice in neonatal chlamydial infections. It may shorten the clinical course of *Campylobacter enteritis*. It is also used in treating diphtheria and whooping cough, and may be valuable in some mycobacterial infections.

Other macrolide antibiotics, including clarithromycin and azithromycin, have a similar antimicrobial spectrum to erythromycin, but improved pharmacokinetic properties. Azithromycin may be given in a single daily dose for the treatment of respiratory tract infections. It reaches very high intracellular concentrations and may be particularly useful in treating persisting chlamydial infections, such as trachoma. Clarithromycin is often used for the treatment of *Mycobacterium avium-intracellulare* infection.

Telithromycin

Telithromycin is a ketolide, a new class of antibiotics closely related to the macrolides. It is active against pathogens of reduced susceptibility to beta-lactams, macrolides and fluoroquinolones, having efficacy equal or superior to comparator agents in numerous studies. It is active against many common respiratory pathogens, including macrolide-resistant *Streptococcus pneumoniae*. It penetrates rapidly into bronchopulmonary tissue, with peak levels obtained in 1 to 2 h and has the advantage of oral, once-daily dosing.

Streptogramins

The streptogramin class of antibiotic contains the synergistic combination agent quinupristin/dalfopristin produced by *Streptomyces pristinaespirilis*. The two drugs bind to the 50S ribosome at different sites and cooperate to inhibit protein synthesis. They are active against streptococci, *Mycoplasma*, *Chlamydia* and *Legionella*. They are effective against *Staphylococcus aureus* (including methicillin-resistant *S. aureus*, MRSA), and resistant enterococci. Its major use is for the treatment of resistant Gram-positive respiratory and soft tissue infections. Various resistance mechanisms have already been described including target modification, efflux and modifying enzymes. Resistance encoded by the *erm* gene provides cross-resistance to quinupristin but does not affect the bacteriostatic activity of the streptogramins.

Oxazolidinones

Oxazolidinones act by binding to the 50S ribosome and inhibiting the initiation phase of protein translation. They are a novel class of antibiotic, active against Gram-positive organisms and some Gram-negative organisms. They show no cross-resistance with other classes of antibiotic. They are useful in the management of multiple drug resistant Gram-positive infections. Linezolid has bacteriostatic activity against both vancomycin resistant *E. faecium* and *E. faecalis*, and methicillin-resistant and -susceptible *S. aureus*. It can be administered intravenously and is also rapidly and completely absorbed after oral dosing. It is cleared by both renal and non-renal routes. Studies in patients with renal failure show that metabolites do accumulate but their clinical significance is unknown: a decrease in dose may therefore not be necessary. Both linezolid and its metabolites are removed during dialysis, so post-dialysis dosing is recommended. Adverse effects have been predominantly gastrointestinal symptoms, headache and taste alteration. In addition there are reports of thrombocytopenia, but this appears to affect mainly patients receiving treatment for longer than 14 days. Linezolid resistance has been reported in a small number of *E. faecium* and *S. aureus* strains and is secondary to a base-pair mutation in the genome encoding for the bacterial 23S ribosome binding site.

Chloramphenicol

Chloramphenicol was first isolated from *Streptomyces venezuelae*. It inhibits protein synthesis by reversibly binding to the 50S ribosomal subunit, preventing the attachment of aminoacyl transfer RNA, and subsequent peptide bond formation. It is active against most Gram-positive and Gram-negative aerobic bacteria, and many obligate anaerobes, including *Bacteroides*. It is also active against spirochaetes, rickettsiae, chlamydiae and mycoplasmas. Against most microorganisms it is bacteriostatic, but it is bactericidal against *H. influenzae* and *Neisseria* spp.

Chloramphenicol may be given parenterally but is well absorbed orally. It is widely distributed in the tissues and crosses the blood–brain barrier. It is metabolized in the liver and excreted by the kidney. In neonates insufficient hepatic enzymes to conjugate chloramphenicol efficiently allow toxic levels to build up, eventually causing circulatory collapse; dosage should be adjusted to achieve blood levels of 15–25 mg/l (peak) and not more than 15 mg/l (trough). In adults, irreversible aplastic anaemia is the most feared toxic effect. It occurs in approximately 1 in 40 000 patients even after the completion of therapy and is not dose-related. For this reason, chloramphenicol is rarely used in clinical practice.

Fusidic acid

Fusidin or fusidic acid is a fermentation product of *Fusidium coccineum*. It is active against Gram-positive bacteria, including most *S. aureus* strains, and against Gram-negative cocci. Streptococci and pneumococci are relatively resistant and Gram-negative bacilli are highly resistant. It has some activity against a variety of protozoa, including *Giardia intestinalis*. Resistance arises naturally and readily develops during therapy, unless a second agent is co-administered to prevent this. It is well absorbed orally and widely distributed. It does not reach the CSF but does penetrate into cerebral abscesses and bone. When given orally, sodium fusidate is well tolerated, although mild gastrointestinal upset and rashes have been reported. With intravenous use, cholestasis and jaundice can develop but usually resolve following withdrawal of therapy.

Fusidic acid is indicated for the treatment of severe staphylococcal infection, especially infections of bones and joints. It is also used for the treatment of recurrent furunculosis and for topical therapy of some superficial staphylococcal infections.

Antiviral therapy

The scope of antiviral therapy is rapidly broadening, in parallel with increasing understanding of the mechanisms of viral replication.

Amantadine

Amantadine is a symmetrical 10-carbon tricyclic amine. It interferes with the replication of influenza type A virus, probably by an effect on cell penetration and viral uncoating. It is used for the prophylaxis and treatment of influenza A during outbreaks. Prophylaxis is usually given to patients with underlying chronic conditions who are at increased risk from severe disease. Resistance to amantadine often develops when the drug is given as post-exposure prophylaxis but occurs less commonly after treatment courses.

Amantadine is well absorbed orally and concentrations in respiratory secretions equal those in plasma. Side-effects include insomnia, poor concentration, nervousness, dizziness and headaches. In overdose convulsions, psychoses and cardiac dysrhythmias can develop. Amantadine is contraindicated in pregnancy, as teratogenicity has been observed in animals.

Nucleoside analogues

These drugs are chemical analogues of natural purine or pyrimidine nucleosides. They require phosphorylation to an active form in host cells, and then interact with viral DNA or RNA polymerase, either becoming incorporated into the nucleotide chain and distorting it, or terminating further elongation.

Aciclovir

Aciclovir is a synthetic acyclic purine nucleoside (Fig. 4.17) active against herpes simplex virus (HSV) types 1 and 2 and varicella zoster virus (VZV). It is activated by monophosphorylation by virally encoded thymidine kinases (TKs) followed by further phosphorylation by cellular thymidine kinases to form aciclovir triphosphate (Fig. 4.18). The need for viral thymidine kinase to perform the first phosphorylation means that the drug only becomes active in virally infected cells. The activated drug inhibits HSV DNA polymerase and is a DNA chain terminator. Resistance arises by alteration of the viral thymidine kinase.

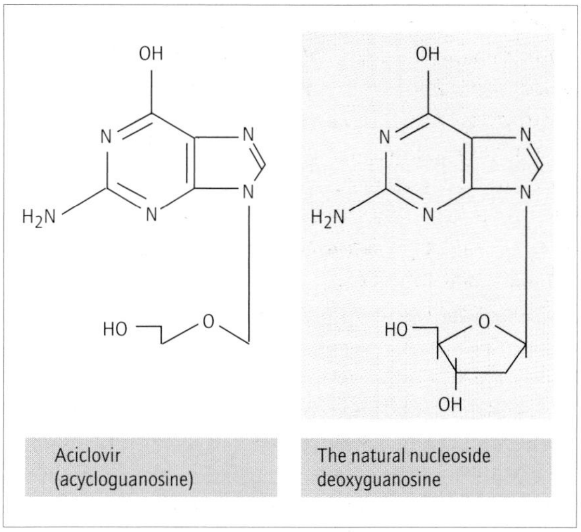

Figure 4.17 The structure of aciclovir.

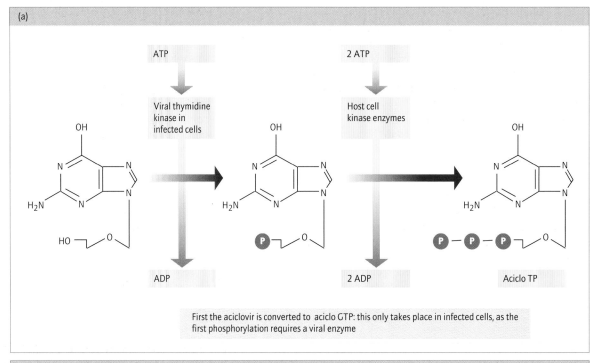

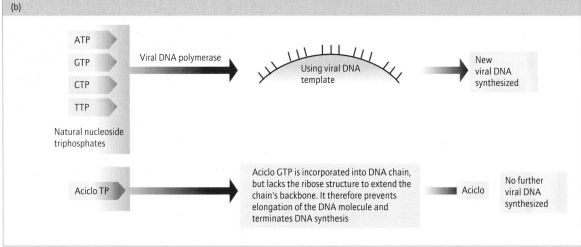

Figure 4.18 The mechanism of action of aciclovir. (a) Stages in the phosphorylation of aciclovir to a 'false' nucleoside triphosphate. (b) Incorporation of aciclovir into the DNA molecule prevents further chain extension.

Resistance occurs naturally and is readily induced in the laboratory but resistant viruses are not highly pathogenic and rarely cause clinical problems.

Aciclovir is absorbed orally, with 15–20% bioavailability, and can be administered parenterally. It is widely distributed and adequate concentrations are found in the CSF. It is well tolerated, although renal function declines if the drug is given in a rapid bolus or in overdosage. It is excreted by the kidneys, so the dose must be reduced in renal impairment.

Aciclovir is used parenterally for the treatment of severe HSV and VZV infections in normal and immunocompromised subjects. Topical preparations are available for the treatment of herpes simplex infections of the skin and mucous membranes but these have only a weak, local action.

Valaciclovir, the valyl ester of aciclovir, is well absorbed orally, then rapidly and completely converted to aciclovir by hepatic enzymes, increasing bioavailability approximately threefold compared with aciclovir. Another ester compound is famciclovir, a prodrug of penciclovir. It has excellent bioavailability after oral dosing. Like aciclovir it depends on virally encoded TK for activation. These oral drugs are used for treating cutaneous and mucocutaneous herpesvirus infections.

Ganciclovir

Ganciclovir (Fig. 4.19) is more active than aciclovir against cytomegalovirus (CMV) (which does not contain the same viral thymidine kinase as herpes simplex and varicella zoster). It is not well absorbed orally and must usually be given intravenously. Oral valganciclovir, the valyl ester of ganciclovir, provides serum drug levels comparable with intravenous ganciclovir. Both drugs are used in the treatment of life- or sight-threatening CMV infections in immunocompromised patients, but are too toxic to the bone marrow to be used for less severe indications.

Cidofovir

Cidofovir is an acyclic nucleoside phosphonate that is phosphorylated only by cellular enzymes. Consequently it is active against strains of herpesvirus, such as cytomegalovirus, with altered or absent TK. It also has some action against poxviruses. Mild or moderate cutaneous adverse reactions are commonly reported and renal tubular toxicity is an important adverse effect. Adefovir is a similar drug, active against hepatitis B virus.

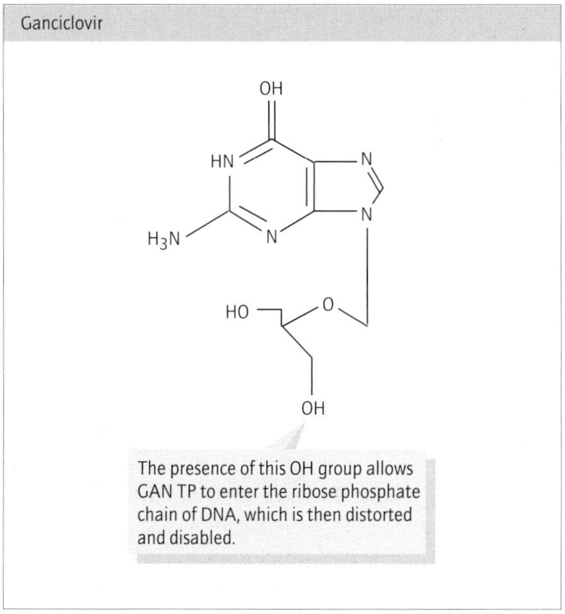

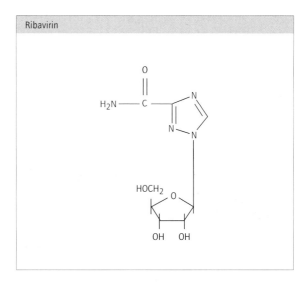

Figure 4.20 The structure of ribavirin.

Ribavirin

Ribavirin is a synthetic nucleoside derivative (Fig. 4.20). It is active against DNA viruses such as the herpesviruses and some human RNA viruses such as the orthomyxoviruses, paramyxoviruses, arenaviruses including Lassa virus, Bunyaviridae including Rift Valley virus, and hantaviruses. It is absorbed orally and can be administered by aerosol for the treatment of respiratory syncytial virus (RSV) infections in compromised infants or immunosuppressed patients. It is well tolerated, although a dose-related haemolytic anaemia is common. It is used, combined with interferon-alpha, for the treatment of chronic hepatitis C virus infection and is effective in the treatment of Lassa fever, Crimean-Congo haemorrhagic fever and hantavirus infections, if given early in the course of disease.

Other types of antiviral drug

Foscarnet

This is a non-nucleoside pyrophosphate analogue. It inhibits the DNA polymerases of herpesviruses, including CMV. It must be given by intravenous infusion. It is indicated in CMV retinitis in acquired immunodeficiency syndrome (AIDS) patients, and mucocutaneous herpes simplex unresponsive to aciclovir in immunocompromised patients. It is relatively toxic, causing impaired renal function, including acute renal failure, hypokalaemia, granulocytopenia and thrombocytopenia. It should be used with care in renal impairment and electrolyte disturbance.

Neuramidase inhibitors

Zanamivir and oseltamivir are anti-influenza agents that inhibit the neuraminidase of influenza A and B. This in-

The presence of this OH group allows GAN TP to enter the ribose phosphate chain of DNA, which is then distorted and disabled.

Figure 4.19 The structure of ganciclovir.

terferes with virus attachment, and with the release of new virus particles from the membranes of host cells. It is effective in clinical trials in reducing the duration and severity of infection and in prophylaxis of disease. Zanamivir must be given by the respiratory route using an inhaler device and this has tended to limit its application in clinical practice. Oseltamivir has excellent oral bioavailability, but is more likely to cause nausea.

Protease inhibitors for other viruses

Protease inhibitors are being investigated as therapeutic agents in a number of viral infections including hepatitis C, CMV, enterovirus and rhinovirus.

Drugs in development

Drugs are being developed that inhibit the binding of enteroviruses to their host-cell receptor. They are given orally and are widely distributed in the body, including the CSF. They are intended for the treatment of viral meningitis.

Antiretroviral drugs

The introduction of highly active antiretroviral therapy (HAART) has transformed the benefits of HIV care by producing dramatic reductions in the HIV viral load and improvement in the CD4 count. This has led to a fall in the incidence of opportunistic infections, with much improved survival and quality of life. Three main classes of antiretroviral drugs are available and these are described below (Table 4.1).

Nucleoside reverse transcriptase inhibitors (NRTIs) or nucleoside analogues

These are nucleoside analogues that inhibit reverse transcriptase, the enzyme responsible for the conversion of viral RNA into a DNA copy. These include the longest established antiretroviral drugs zidovudine (AZT), lamivudine (3TC, also effective against hepatitis B virus), stavudine (d4T), zalcitabine (ddC), and abacavir. They are the mainstay of antiretroviral therapy and are used in combination in initial therapy (see below). Lamivudine is highly active but selects for resistance to other NRTIs, and should only be used in completely suppressive regimens.

Zidovudine

Zidovudine (azidodeoxythymidine or AZT) is a dideoxynucleoside analogue originally developed as an anticancer agent (Fig. 4.21). It is phosphorylated to AZT monophosphate by viral thymidine kinase and then converted to diphosphate and triphosphate by cellular thymidylate kinase. It inhibits retroviral reverse transcriptase to which it binds preferentially.

It is well absorbed orally and crosses the placenta and the blood–brain barrier. It is metabolized in the liver. The main side-effects are reversible anaemia and neutropenia, which usually occur after about 6 weeks of therapy, particularly in patients with pre-existing cytopenias. Other adverse events reported include nausea, headache, rash, abdominal pain, fever, myalgia and, rarely, metabolic acidosis.

Protease inhibitors

Protease inhibitors are the antiretroviral compounds that produce the greatest fall in viral load when used as

Table 4.1 Currently approved antiretroviral drugs

Method of action	Name
HIV reverse transcriptase inhibitors	
Nucleoside analogues	Zidovudine (ZDV, AZT)
	Didanosine (ddl)
	Zalcitabine (ddC)
	Stavudine (d4T)
	Lamivudine (3TC)
	Abacavir (ABC)
Non-nucleoside analogues (NNRTI)	Nevirapine (NVP)
	Delavirdine (DLV)
	Efavirenz (EFV)
Nucleotide analogues	Tenofovir (TDF)
HIV protease inhibitors	Saquinavir (SQV)
	Ritonavir (RTV)
	Indinavir (IDV)
	Nelfinavir (NFV)
	Amprenavir (APV)
	Lopinavir/ritonavir (LPV/r)
Fusion inhibitor	Enfuvirtide (T-20)

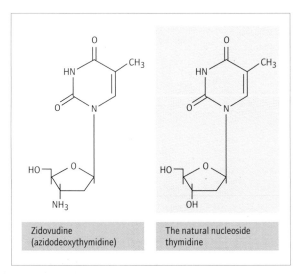

Figure 4.21 The structure of zidovudine.

single agents. They are central to HAART. They include indinavir, ritonavir, saquinavir and nelfinavir, which all have similar activity. Lopinavir is used only in a synergistic combination with ritonavir (Kaletra).

Protease inhibitors can cause significant toxicity. This includes unusual complications such as renal stones, crystalluria, a 'buffalo hump' and gynaecomastia. Lipodystrophy is a recently recognized complication of protease inhibitor therapy and is characterized by lowered body fat, insulin resistance and raised cholesterol and triglyceride concentrations. Fat loss occurs in all regions other than the abdomen.

Protease inhibitors act variably as substrates and inhibitors of cytochrome P 3A4 enzymes. They have many important interactions with other such drugs, and with one another, requiring caution and expertise in their use.

Non-nucleoside reverse transcriptase inhibitors (NNRTIs)

These drugs inhibit reverse transcriptase by an alternative mechanism to NRTIs. They have been shown to be effective in combination regimens. Because of their susceptibility to viral resistance after a single mutation event, they should only be used in regimens designed to be maximally suppressive. For new drugs see Table 4.2.

Fusion inhibitors

Fusion inhibitors are a new class of antiretroviral drugs (ARVs) of which enfuvirtide is the first to enter clinical practice. These drugs block fusion, the last step in the process of viral capsid entry into the host cell. Clinical trials demonstrated high efficacy and low toxicity. Enfuvirtide is a peptide mimetic of a segment of viral envelope glycoprotein gp41. It functions by blocking gp41 structural rearrangements necessary to complete fusion. Resistance occurs by mutation in a 10 amino acid motif in gp41, which forms part of the enfuvirtide binding site. As this motif is critical for viral fusion, resistant mutants grow more slowly compared with wild type, and reversion to a wild type has been reported following enfuvirtide withdrawal.

Integrase inhibitors

Investigations are ongoing into new compounds that interfere in various ways with the binding of proviral DNA to integrase.

Resistance to antiviral drugs

RNA viruses have relatively unstable genomes because of the lack of 'proof-reading' mechanisms to preserve

Table 4.2 Some new antiretroviral agents in various stages of development

Method of action	Name
HIV reverse transcriptase inhibitors	
Nucleoside analogues	Emtricitabine (FTC)
	Amdoxovir (DAPD)
Non-nucleoside analogues	DPC-083
	Capravirine (Ag 1549)
	TMC 125
	Calanolide A
Nucleotide analogues	GS-7340 (tenofovir pro-drug)
HIV protease inhibitors	Atazanavir (ATV, BMS-232 632)
	Tipranavir (TPV, PNU-140690)
	Mozenavir (DMP-450)
	TMC 114
HIV entry inhibitors	CD4 attachment inhibitors
	PRO 542
	BMS-806
Chemokine receptor inhibitors – CXCR4	AMD-3100*
Chemokine receptor inhibitors – CCR5	Schering C (SCH-C, SCH 351125)
	PRO 140
	Schering D (SCH-D)
	UK 427,857
Fusion inhibitors	T-1249
HIV integrase inhibitors	S-1360

*Development discontinued

the fidelity of genome replication. Many genetic variants (quasi-species) exist in a virus population. Mutations may be neutral, having no effect on fitness, or deleterious, where the variants replicate less well than the 'wild' type. Selective mutations are those where the mutation provides advantage in the face of immunological or therapeutic pressure. In HIV infection around 10 billion virus particles are produced daily, containing one mutation in 9200 nucleotides. Thus every possible single drug-resistant mutant is likely to arise each day. Spontaneous double mutations are extremely unlikely. Resistant virus clades can be detected in patients not previously treated. Selective pressure applied by therapy allows pre-existing minor quasi-species to emerge as the dominant type. Resistance to some drugs can arise by a single mutation event (e.g. lamivudine and certain NNRTIs). When these are given in incompletely suppressive regimens, resistance will emerge within weeks. For other drugs (e.g. zidovudine and certain protease inhibitors), three or more mutations must occur before a resistant phenotype is expressed.

Emerging resistance is increasingly encountered in the antiviral therapy of chronic hepatitis B, a DNA virus infection. Treatment with multiple drugs, or drugs that require multiple mutations to resistance, should be preferred. To prevent the accumulation of resistant mutants, viral replication should be kept as low as possible. Previous treatment may have selected mutants that are no longer detectable but will reappear rapidly if treatment is restarted with that drug.

Antifungal therapy

Many antifungal drugs exist, but most are highly toxic, and only suitable for topical application. The polyenes are the oldest drugs, but among them, amphotericin B is still the drug of first choice for many systemic mycoses. Modern formulations are designed to maximize its efficacy and minimize its significant toxicity.

Polyenes

Polyenes possess a macrolide ring with a variable number of hydroxyl groups along the hydrophilic side and numerous repeating double bonds on the hydrophobic side. They bind to sterols and eukaryotic cell membranes, causing leakage of cellular contents leading to cell death. Some polyenes are toxic only to fungal membranes but others are equally toxic to mammalian cells.

Nystatin

This drug is used as topical treatment, mostly for mucosal *Candida* infections, and for intestinal candidiasis. It is not absorbed by the oral route and is too toxic for parenteral use. It has largely been replaced by the more convenient and highly active imidazole drugs in the treatment of vaginal candidosis.

Amphotericin B

This important polyene drug is a heptaene, with seven double bonds. It is active against most fungi pathogenic for humans and is used as parenteral treatment for systemic mycoses. It also has activity against *Leishmania* spp. and can be used in the treatment of kala-azar and cutaneous leishmaniasis. Acquired resistance is rare, but has been described in some *Candida* species, notably *C. tropicalis*, *C. parapsilosis*, *C. lusitaniae* and *C. krusei*. Amphotericin B is administered parenterally and is highly protein-bound, and penetrates poorly into the CSF. Treatment causes fever, rigors, headache, vomiting and thrombophlebitis, which may be ameliorated by concurrent administration of corticosteroids. Hyperkalaemia, hypomagnesaemia, anaemia and renal impairment are also common. Amphotericin's toxicity has been reduced by incorporating the drug into liposomes or other lipid carriers, producing liposomal AMB (ambisome) or amphotericin B lipid complex. This allows much higher doses to be administered with fewer toxic reactions.

Azoles

The azoles are a large group of synthetic agents possessing an imidazole or triazole ring with an N-carbon substitution. These agents act by blocking 14 alpha-D-methylation in the biosynthesis pathway of ergosterol, leading to ergosterol depletion and fungal membrane dysfunction. The azoles have a wide spectrum of activity against *Candida* and dermatophytes, and some are active against *Aspergillus* sp.

Imidazoles

Topical imidazoles

Topical imidazoles are powerful topical agents used for treating oral, vaginal and skin candidosis, and dermatophyte infections. They include clotrimazole, econazole, fenticonazole, isoconazole, miconazole, sulconazole and tioconazole. Miconazole can be given by mouth for treatment of intestinal candidosis. Adverse effects are rare; irritation caused by the pharmaceutical vehicles or preservatives in creams is relatively common.

Ketoconazole

Ketoconazole can be given orally, as in the treatment of histoplasmosis, but is more toxic than the triazoles (see below). In oral use, side-effects include severe hepatitis (1 case per 15 000 patients). Topical formulations are used to treat seborrhoeic dermatitis and severe, resistant fungal infections of the skin or vagina.

Triazoles

Fluconazole

Fluconazole is active against yeast-like fungi, including *Candida*, *Histoplasma* and *Cryptococcus* spp. It is effective in dermatophytosis but not against aspergillosis. It is well absorbed orally and is well tolerated. Elevation of transaminases may occur, and warfarin activity may be increased. It is useful in the oral treatment of vaginal candidosis, and treatment of mucosal and systemic candidosis in immunocompromised patients.

Itraconazole

Itraconazole, closely related to fluconazole, additionally has some activity against *Aspergillus* spp. It is orally absorbed, achieving high concentrations in the stratum corneum and hair. It is used to treat superficial mycoses including dermatophytosis, pityriasis versicolor, and oral and vaginal candidosis. It may also be useful as an alternative to amphotericin in the treatment of aspergillosis, histoplasmosis and cryptococcosis.

Voriconazole

Voriconazole, a second-generation triazole, is a synthetic derivative of fluconazole that additionally inhibits 24-methylene dihydrolanosterol demethylation of certain yeasts and filamentous fungi. This results in an extended spectrum of activity, compared with older triazoles. It has increased potency against fluconazole-resistant *Candida* species, including *C. krusei* and *C. glabrata*.

Echinocandins

Echinocandins are recently developed antifungal agents that inhibit the synthesis of the fungal cell wall by noncompetitive blockade of the enzyme (1,3)-beta-D-glucan synthase, preventing the formation of (1,3)-beta-D-glucan, an essential cell wall component. This leads to cell wall weakness and subsequent cell lysis. Caspofungin was the first of this class to be licensed. It can only be administered intravenously and is useful in the management of resistant and unresponsive systemic fungal infections, including those in neutropenic patients.

Allylamines

Terbinafine is an allylamine drug. It is used topically for treating *Candida* and dermatophyte infections of the skin. By oral dosage it is used for treatment of ringworm, and *Candida* or dermatophyte nail infections. It is useful for prolonged treatment of toenail infections that are unlikely to respond to other drugs. It acts by inhibition of squalene epoxidase, resulting in accumulation of aberrant and toxic sterols in the fungal cell wall. Stevens–Johnson syndrome, toxic epidermal necrolysis, loss of the sense of taste and hepatic toxicity are reported adverse effects.

Other antifungal drugs

Flucytosine (5-fluorocytosine)

Flucytosine is a synthetic fluorinated pyrimidine with activity limited to *Candida* spp., *Cryptococcus neoformans* and some fungi causing chromomycosis. It is incorporated into fungal mRNA in the place of uracil, thus inhibiting protein synthesis. It also blocks thymidylate synthetase causing interruption of DNA synthesis. It is given either orally or parenterally and is excreted by the kidney. Dosage must be reduced when the creatinine clearance falls. Resistance is quite common and may arise during treatment. The main side-effects are bone marrow aplasia, enteritis and hepatotoxicity. Flucytosine is used in combination with amphotericin for the treatment of systemic candidosis and cryptococcal infections, when it allows reduction of amphotericin dosage, thereby reducing side-effects.

Griseofulvin

Griseofulvin is an antibiotic derived from *Penicillium* species, such as *P. griseofulvum*. Its mechanism of action is unknown but it interferes with nucleic acid synthesis. Its activity is restricted to the dermatophytes. It is absorbed orally and appears in keratinized tissues at concentrations sufficient to prevent further fungal invasion. It is useful as an alternative to terbinafine for treating nail infections, but must be given for extended periods until all infected keratinized structures are shed and replaced. Serious adverse events are uncommon but headache and rashes occasionally occur. Griseofulvin reduces the anticoagulant effect of warfarin.

Antiprotozoal drugs

Antimalarials

Quinine

Quinine is a 4-aminoquinoline drug, a potent schizonticide thought to act by inhibiting haem polymerase. It can be given orally or parenterally, and crosses the blood–brain barrier with CSF concentrations reaching 7% of serum levels. It is indicated for the treatment of *P. falciparum* malaria. Quinine has a narrow therapeutic index: tinnitus, headache and nausea are common with normal doses. The drug also aggravates hypoglycaemia in severe malaria. Patients with cardiac diseases should be closely monitored during therapy, as quinine depresses cardiac function. Overdosage, or intravenous bolus dosing, cause blindness and encephalopathy.

Chloroquine

Chloroquine, a 4-aminoquinoline drug, is schizonticidal against all four human species of *Plasmodium* (though chloroquine resistance is now widespread in *P. falciparum*) and against amoebae. It is probably acts by inhibiting the enzyme that polymerizes and detoxifies ferriprotoporphyrin IX. It is approximately 90% bioavailable after oral administration, and has a long half-life. It is highly bound to thrombocytes, granulocytes and plasma proteins, and also to melanin-containing cells in the retina. Retinal damage can follow prolonged high dosage. Chloroquine is the treatment of choice for non-falciparum malaria, but is not now used for treating *P. falciparum* infections, though it retains a role in prophylaxis. Minor adverse effects are common, including nausea and vomiting, diarrhoea, dizziness, and pruritus in dark-skinned individuals. It exacerbates psoriasis and is contraindicated in this condition.

Mefloquine

Mefloquine was developed after over 300 000 compounds were synthesized and tested by the American Army, following their experience of drug-resistant malaria in Vietnam. It is a quinolone carbinol possessing a quinine-like ring structure. Mefloquine is administered orally and absorbed from the gut, with peak concentrations 2–12 h later. It is 98% protein-bound. The average half-life is 21 days. Weekly dosage with 250 mg gives a steady-state serum concentration of approximately 1 mg/l (2 mg/l inside red cells). The drug is mainly excreted in the bile and faeces.

Mefloquine is effective in treatment and prophylaxis of all species of human malaria, and is recommended for prophylaxis of chloroquine-resistant *P. falciparum*, but mefloquine resistance already exists in some areas of the Far East. Adverse effects may limit its tolerability. Agitation, vivid dreams, dizziness, altered balance, vomiting and diarrhoea are relatively common with prophylactic doses. Depression or psychosis occur rarely, particularly with therapeutic doses. Mefloquine should be avoided in patients with renal compromise or a past history of psychiatric disease or epilepsy. It is teratogenic to rats and mice in early pregnancy and relatively contraindicated for pregnant and lactating women.

Primaquine

Primaquine, an 8-aminoquinoline drug, is the only effective agent for eradicating long-term liver infection with the hypnozoite forms of *P. vivax* and *P. ovale*. It is given after treatment of the feverish acute infection with other antimalarials. It causes haemolysis in individuals with glucose-6-phosphate dehydrogenase (G6PD) deficiency, and must not be given to patients whose G6PD status is unknown. This problem prevents its routine use for malarial prophylaxis.

Artemisinins

Artemisinin is derived from the plant *Artemis annua* or qinghaosu, used in traditional Chinese medicine. It is a sesqueterpinoid, whose mode of action is believed to depend on cleavage of its endoperoxidase bridge, generating free radicals that bond covalently with parasite proteins. All artemisinin derivatives are converted *in vivo* into the active antimalarial compound dihydroartemisinin.

In the Far East and many African countries, artemesinin has replaced quinine in treatment of *P. falciparum* malaria. It shortens fever clearance time by 8 h compared with quinine, and is therefore useful for treating serious and complicated malaria. It causes mild cerebral irritation and does not terminate coma as rapidly as quinine, but this is not clinically important.

This drug can be administered intravenously, rectally, intramuscularly and orally. Artusenate is a water-soluble form for intravenous and oral use. Artemether is a methyl derivative for intramuscular injection or oral use. In severe cases of malaria an effective dose is 2 mg/kg followed by 1 mg/kg 12 h later and each 12 h until the patient is able to take oral medication. A high relapse rate follows monotherapy so artemesinin derivatives are now usually given in combination with other drugs, often a tetracycline or mefloquine.

Other antimalarial drugs

Proguanil is used in combination with chloroquine for malarial prophylaxis, although increasing resistance has rendered this combination less useful. It may be used alone for prophylaxis of non-falciparum malaria. Other drug combinations used against malaria are: Fansidar®, a combination of pyrimethamine plus sulfadoxine; and Malarone®, atovaquone plus proguanil.

Hexadecylphosphocholine (Miltefosine)

This compound is active against *Leishmania*, through interference with protein kinase C, which is found on the *Leishmania* membrane. It has better bioavailability after oral rather than intravenous administration, and is lethal against cutaneous and visceral leishmaniasis species. Clinical trials indicate its utility in practice but the relapse rate is higher than for liposomal amphotericin, more adverse events were reported and some resistant strains emerged on treatment.

Other antiparasitic drugs
Pyrimethamine

Pyrimethamine is an antifolate drug that is usually used as part of a drug combination (as in Fansidar®). It is used in combination with sulfadimidine for treating toxoplasmosis.

Atovaquone

Hydroxynaphthoquinones are inhibitors of mitochondrial electron transport, competing with the biological electron carrier ubiquinone. They inhibit the ubiquinone-linked dihydro-oroate dehydrogenase, essential for pyrimidine biosynthesis. Atovaquone binds to the ubiquinoyl–cytochrome c reductase region (complex III).

It is active against chloroquine-sensitive and chloroquine-resistant *P. falciparum*, *Toxoplasma gondii* and *Pneumocystis carinii*. It is a second-line drug for the treatment of parasitic infections in HIV patients and is combined with proguanil in Malarone®, which is licensed for the treatment and prophylaxis of *P. falciparum* infections.

Antihelminthic drugs

Benzimidazoles

Three members of this group, mebendazole, thiabendazole and albendazole, are commonly used in the treatment of nematode infections. They interfere with beta-tubulin function, causing failure to form cytoplasmic microtubules and leading to paralysis of the worm.

Mebendazole

Mebendazole is the drug of choice for *Enterobius*, *Ascaris*, *Trichuris* and hookworm infections. Although a single dose of 100 mg is sufficient to eradicate threadworm infections, *Ascaris*, *Trichuris* and hookworm infections require a dose of 100 mg twice daily for 3 days. The drug is usually well tolerated but abdominal pain, diarrhoea and headache may occur, particularly with a heavy worm load. Mebendazole is not sufficiently well absorbed for use in systemic parasite infections. Serum concentration is lowered by phenytoin or carbamazapine, but increased by cimetidine. An oral suspension is available for the treatment of children.

Thiabendazole

Thiabendazole is absorbed after oral dosage and is used worldwide for the treatment of strongyloidiasis, visceral larva migrans, and sometimes for intestinal helminth infections. It can be made into a topical paste for treating cutaneous larva migrans. It can cause significant gastrointestinal irritation and skin hypersensitivity rashes.

Albendazole

Albendazole achieves better tissue concentrations and is less toxic than thiabendazole. It crosses the blood–brain barrier to achieve a CSF concentration 20% that of plasma, and also achieves useful concentrations inside parasitic cysts. It is metabolized quickly and completely, with an approximate plasma half-life of 9 h.

Albendazole is active against *Enterobius*, *Ascaris*, *Trichuris* and hookworms, and also against *Strongyloides*, *Capillaria*, *Trichinella* and the tissue stages of hydatid disease and cysticercosis (*Taenia solium*). It is well tolerated, with only transient side-effects such as epigastric pain and diarrhoea following single-dose treatment. Surprisingly, albendazole has activity against some species of microsporidia and other protozoa.

Other antihelminthic drugs

Piperazine

Piperazine causes paralysis of *Ascaris* and *Enterobius*. Bioavailability data are scanty, though it is thought to be well absorbed and eliminated rapidly. It has a success rate of over 95% but side-effects, including nausea, diarrhoea and urticaria, are described. Severe neuropsychiatric reactions occur rarely.

Levamisole

Levamisole (not currently available in the UK) is well absorbed and widely distributed through the body tissues. It is metabolized in the liver and mainly excreted by the kidney, giving a plasma half-life of 4–5 h. It is active against *Ascaris* but less effective against hookworm, and is therefore used only for *Ascaris* mono-infections. Nausea, vomiting, abdominal pain and headache have been reported. Influenza-like reactions, blood disorders, photosensitivity and renal failure have occurred at higher doses used for immunomodulation in some non-infectious diseases.

Ivermectin

Ivermectin is an antiparasitic agent used for single-dose treatment of systemic filariasis. It is the treatment of choice for suppression and cure of onchocerciasis. It is also useful in strongyloidiasis, particularly in the immunosuppressed.

Niclosamide

Niclosamide is a chlorinated salicylanilide that acts against *Taenia saginata* and *T. solium* by interfering with the energy metabolism of the worm. The pharmacokinetics are unknown but the drug achieves a high cure rate (90–100%). Rare side-effects include nausea, vomiting and diarrhoea. An antiemetic should be given before treatment, as retrograde ingestion of eggs can cause cysticercosis if vomiting occurs. For this reason, patients with intestinal tapeworms should preferably be referred to a specialist for treatment with praziquantel (see below).

Praziquantel

Originally developed for the treatment of schistosomiasis, praziquantel is effective against blood, lung and liver flukes and tapeworms. It acts by altering the calcium permeability of membranes, resulting in tetanic contraction and disruption in the parasite surface allowing the host's immune system to attack the parasite. This is most important in killing adult schistosomes, which evade host defences by 'cloaking' their surface antigens with adsorbed host antigens.

Praziquantel is well absorbed and distributed to all the body tissues. It crosses the blood–brain barrier, reaching CSF levels around 25% of serum concentrations. It is metabolized in the liver and excreted rapidly in the urine, having a half-life of 1–1.5 h. Serum praziquantel levels are reduced by co-administration of phenytoin, carbamazepine and dexamethasone. There is no information on human teratogenicity; clinically insignificant amounts are excreted into breast milk. Side-effects are uncommon, mainly dose related, and include dizziness, headache, lassitude and limb pain. A single dose of 40 mg/kg is adequate for treating beef and pork tapeworms and simple schistosomiasis. Higher or multiple doses are required for liver flukes and *Schistosoma japonicum*.

Part 2: Systematic Infectious Diseases

Infections with Skin, Mucosal and Soft-tissue Disorders

5

Introduction

Structural considerations

The skin consists of the superficial epidermis and the deeper dermis (Fig. 5.1).

The epidermis is composed of layers of squamous cells, arising from a basal cell layer, with a thick acanthocyte layer overlaid by the increasingly keratinized and flattened granular, lucid and horny layers. The keratinous surface is further protected by a film of sebaceous secretion, which resists penetration by fluids and microorganisms.

Beneath the basal cell layer lies the dermis. This is a sensitive and vascular structure supported by loose connective tissue, which is well hydrated and supple, allowing movement of the skin surface over deeper structures.

The epidermis is penetrated by hair follicles and the ducts of sweat glands arising in the dermis. These can provide a niche for colonizing organisms and, if occluded by keratinous debris or dried secretions, they can become sites of loculated or spreading infection. The hair and nails are modified epidermal structures, consisting mainly of keratin.

The epidermis, hair and nails can be locally invaded by organisms, usually fungi, which digest keratin or sebum.

They cause flaking and discoloration, sometimes with a minimal inflammatory reaction. Bacterial infection of the epidermis is more invasive and inflammatory. Some infections involve the epidermal surface, producing a weeping macerated lesion, which advances to adjacent areas. Such open lesions are infectious, often spreading by contact to other sites or individuals. Intra-epidermal infections are less common. They advance within the epidermal cell layers, sometimes separating them to form fluid-filled bullae. Erysipelas is an example of this.

Infections of the dermis spread widely in the loose connective tissue, and the advancing inflammation has an indistinct edge. This is the typical appearance of cellulitis. The dermis can be greatly expanded by the oedema and cellular infiltrate that accompany infection. Scarring or fibrosis can tether the skin to underlying structures such as fascia or muscle. In severe cases the underlying tissue can become infected (for instance, when a deep skin ulcer erodes underlying bone; see Chapter 14).

Skin changes in systemic disease

The skin often displays abnormalities in systemic diseases. Rashes, jaundice, and vasculitic, embolic and haemorrhagic lesions are valuable physical signs, and may be typi-

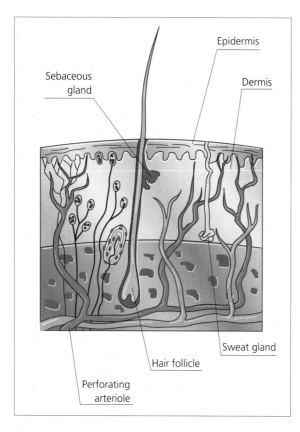

Figure 5.1 Structure of the skin.

cal and diagnostic. They may be due to infectious lesions, as in the chickenpox rash, to a vasculitic reaction, as in erythema nodosum, or to damage by toxins, as in scarlet fever or toxic shock syndrome. Vasculitis or intravascular coagulation may produce petechial or haemorrhagic lesions, seen in rickettsiosis or meningococcal disease. Chemical infiltration with bilirubin or methaemalbumin may cause typical discolorations.

Natural defences of the skin

The surface structure and acidic sebaceous secretions are hostile to many pathogens. A dense population of normal flora inhabits the surface, ducts and follicles. These organisms are adapted to the skin environment, which they may further modify by adjusting the pH, redox potential or local concentration of bacteriostatic substances. The rich blood and lymphatic supply of the dermis ensure that both innate and adaptive immune responses can quickly be recruited when pathogens enter the skin.

Important normal skin flora
1 Coagulase-negative staphylococci.
2 *Staphylococcus aureus.*
3 *Streptococcus pyogenes.*
4 Many species of *Corynebacterium* (diphtheroids).
5 *Propionibacterium* spp.
6 *Candida* spp.

Skin defences are compromised if the surface is penetrated by injury or thinned and excoriated by inflammatory processes. Sustained wetness of the skin causes maceration of the keratinous epidermis, which can then be invaded by pathogens. Patients who suffer from eczema are highly susceptible to skin infections caused by herpes simplex or *Staphylococcus aureus.*

Natural defences of the skin
1 Keratinous surface.
2 Antibacterial effects of sebum.
3 Effect of normal flora.

Mucosal structure and defences
Mucosae do not have a keratinized surface; they are always moist and some are only one cell thick. They are more susceptible than skin to invasion by surface pathogens. For defence they rely on the washing action of secretions, which may contain lysozymes or specific secretory immunoglobulin A (sIgA). Mucosae adjacent to skin, as in the nose and mouth, are colonized and protected by their own normal flora, which merges into that of the skin at the mucocutaneous margin.

Natural defences of mucosae
1 Mechanical washing by tears or urine.
2 Lysozyme or antibody in surface fluid.
3 Surface phagocytes.
4 Ciliary action moving mucus and debris.
5 Secretory IgA at the mucosal surface.

Viral infections of the skin and mucosae

Organism list
- Papillomaviruses
- Herpes simplex virus type 1 (HSV 1)
- Herpes simplex virus type 2 (HSV 2)
- Human herpesvirus type 6 (HHV6) (see Chapter 11)
- Human herpes virus type 7 (HHV 7) (see Chapter 11)

- Varicella zoster virus (VZV) (see also Chapter 11)
- Coxsackie A viruses
- Molluscum contagiosum
- Cowpox
- Orf/paravaccinia
- Vaccinia

Papillomavirus infections (warts)

Introduction

There are many types of human papillomaviruses (HPVs), of which the commonest are those associated with common warts, plantar warts (verrucae), genital warts and, occasionally, laryngeal warts. HPV infections of the skin are rarely associated with malignancy. Exceptions include rare squamous cell transformation of condylomata acuminata (genital warts), and the uncommon Bowenoid papulosis and epidermodysplasia verruciformis. Some papillomaviruses associated with genital warts, particularly HPV-16 and HPV-18 but including other types, are associated with cervical interstitial neoplasia (see Chapter 15).

Common (plane) warts

These are usually seen in children, as immunity to the viruses is gained by adulthood. They spread by contact, the viruses entering tiny fissures, and infecting cells of the acanthocyte layer. The infected cells hypertrophy and multiply, producing a keratinized, nodular papilloma. Curetting the lesion confirms its identity by revealing the small bleeding points of dermal capillaries supplying the lesion. In involuting lesions, thrombosed capillaries appear as black dots. While solitary on the palms, plane warts may be multiple in the looser skin on the wrists or dorsa of the hands.

Plantar warts may become large, a centimetre or more in diameter. They are pressed into the foot by the body weight and often develop an eroded central area. Local pressure can cause severe pain on weight-bearing. They can be distinguished from corns by the bleeding points revealed by paring the lesion; corns are composed entirely of keratin, with no capillaries.

Treatment of warts is rarely required in children, as immunity gradually develops and the lesions involute. Warts in adults can be curetted, or ablated by freezing with liquid nitrogen. Plantar warts may be reduced in size by paring. Salicylate and lactic acid ointment, painted on to the lesion, allowed to dry and covered with an adhesive dressing, softens the keratin and inhibits the viruses.

Plantar warts are infectious, especially during swimming, when the uncovered skin is softened by immersion. If the lesions are temporarily covered with a waterproof plaster or latex sock, however, swimming need not be forbidden.

Herpes simplex infections

Introduction and epidemiology

Herpes simplex viruses are colonists and invaders of mucosae. They can also infect skin if entry is available via small punctures, fissures or macerated areas. Typically, infections of the mouth and upper body have been caused by herpes simplex type 1, while genital infections have been due to the genetically distinct type 2 virus. Nowadays this distinction is less clear, and either virus may be found in either site.

Like other herpesviruses, herpes simplex has the property of latency; after causing a primary infection it persists in the immune host's sensory nerve ganglia in a dormant form. Subsequent reactivations may cause either asymptomatic virus excretion or a clinical lesion, usually a localized papulovesicular eruption on the mucocutaneous border or skin in the same body area that was affected by the primary infection.

Herpes simplex infection is spread by direct contact with mucosa, skin or secretions from where virus is being shed. In infancy primary infection is often buccal, acquired by oral contact or shared eating utensils. Nailfold or finger pulp infections may follow contact with saliva, and often occur in infants who suck their thumb or nurses who provide mouth care for debilitated patients. Inoculation into the eye may cause conjunctival or corneal infection.

Reactivation lesions may be precipitated by acute feverish illness and are therefore called cold sores. Severe cold sores often accompany pneumococcal or meningococcal infections. Exposure to strong sunlight is also a common precipitating event.

Sites affected by herpes simplex infection

1 Buccal mucosa (primary gingivostomatitis).
2 Lip margin (cold sore).
3 Nailbed (felon, or whitlow).
4 Finger pulp.
5 Facial skin.
6 Eczema-affected skin (eczema herpeticum).
7 Cornea (dendritic ulcer).
8 Conjunctiva.
9 Genitalia (see Chapter 15).
10 Acute facial nerve paralysis ('Bell's palsy'); up to 70% of cases have some evidence of HSV infection (PCR-positive in saliva, tears or blood mononuclear cells).
11 Central nervous system infections (see Chapter 13).

Effective cell-mediated immunity is essential for the control of herpes simplex infection. Patients treated with cytotoxic drugs or high-dose corticosteroids, those undergoing bone marrow grafting, or those with human immunodeficiency virus (HIV) infection are at risk of frequent reactivation events and severe infections.

Pathology

Herpes simplex is a cytolytic virus that causes degeneration and 'ballooning' of epidermal prickle cells, leading to vesicle formation. Keratinized cells form the vesicle roof, but in the mouth and in other mucous membranes the roof is unstable and sloughed away. The vesicle contains cellular debris with fused keratinocytes containing typical inclusion bodies (Tzanck cells) (Fig. 5.2).

Virology

The genome of HSV consists of a linear double-stranded DNA genome that codes for 84 proteins. There are two serotypes that differ in the repeats of sequences from both terminal ends of the genome. The homology between HSV-1 and HSV-2 is approximately 50%. Restriction endonuclease digestion can be used to distinguish between strains and subtypes.

Virus spreads from cell to cell and also up sensory axons, establishing latent infection in neurones of the sensory nerve ganglia. HSV DNA can be demonstrated in the trigeminal ganglia of approximately 40% of patients tested. It is not integrated into the host DNA. The lesions of mucocutaneous herpes simplex are caused by tissue destruction by replicating virus, and visceral disease such as encephalitis by virus-induced cell damage. Immune-mediated complications include erythema multiforme, haemolytic anaemia and thrombocytopenia.

Latency is a dynamic equilibrium with low-grade viral infection that does not cause cell lysis. Only a few of the HSV genes are expressed. Intermittently, replication of all virus genes resumes and viral particles travel down the sensory neurone to cause a reactivation lesion. The mechanisms of latency and reactivation are not clearly understood, but specific antisense DNA sequences from HSV are detectable in latently infected cells. Some HSV latency genes interfere with host immune recognition and others prevent apoptosis of the infected cell.

Clinical features

Primary herpes simplex: skin and mucosal lesions

Primary herpetic gingivostomatitis is common in infants, though some cases present in their teens or later. After a day or two of fever, with tender enlarged cervical lymph nodes, painful, oval, whitish vesicles appear in the mouth. They may cover the tongue and are common on the soft palate, the gums and inside the lips. Soreness often prevents suckling or taking solid food. Dribbling of infectious saliva often causes secondary lesions on the fingers, chin or chest (Fig. 5.3a,b). The fingers may be infected by contact with infectious saliva or lesions, virus can then be transferred to the conjunctiva or genitalia (Fig. 5.3c,d). Few new lesions appear after 3 to 5 days of eruption, and most have deroofed or scabbed and healed within 7–10 days.

Herpes simplex is the name given to primary and reactivation lesions on the skin or mucocutaneous border. A localized patch of itching or tingling papules progresses to form round vesicles, then pustules, which crust from the centre outwards. The affected area may overlap the midline, and is not dermatomal in distribution. The natural history of the lesions averages one to two weeks (Fig. 5.3e).

⚠️ **Caution:** Eczema herpeticum is herpes simplex infection of eczema-affected skin. Even if the eczema is currently inactive, affected skin is highly susceptible to HSV. Extensive, severe, spreading skin lesions can cause severe, occasionally life-threatening illness, especially in infants with severe eczema, or those on corticosteroid therapy (Fig. 5.3f). Antiviral treatment is always indicated. Secondary *Staphylococcus aureus* infection of affected areas may be difficult to distinguish from crusted herpes simplex lesions, and the two conditions may also coexist. Additional antibiotic therapy with an antistaphylococcal agent is often necessary.

Diagnosis

This is usually clinically obvious. Herpes zoster is unilateral, and dermatomal in distribution. Oral blisters and a vesicular cutaneous rash occur in hand, foot and mouth disease, but these have a typical distribution (see below). Oral candidiasis sometimes appears as oval plaques on mucosal surfaces, but without associated fever and lymphadenopathy.

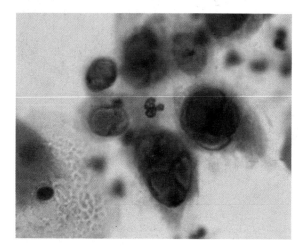

Figure 5.2 Microscopical appearance of cells from a herpetic vesicle.

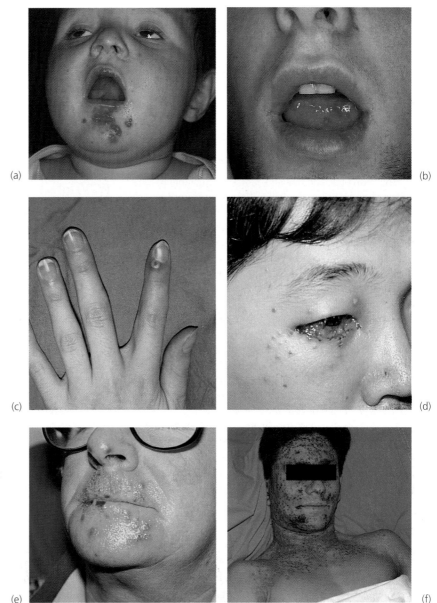

Figure 5.3 (a) Primary herpetic gingivostomatitis in an infant, herpetic lesions on skin macerated by infected saliva. (b) Primary herpes simplex lesions on the tongue (unusual in an adult). (c) Herpetic whitlow affecting an intensive care nurse. (d) Herpes simplex conjunctivitis. (e) Herpes simplex reactivation lesion (cold sore). (f) Eczema herpeticum originating from a cold sore.

Laboratory diagnosis

Light microscopy of scrapings from the base of lesions may reveal cells with typical intranuclear inclusions surrounded by a clear halo (Tzanck cells) and, in skin specimens, fused cells (polykaryocytes). Electron microscopy demonstrates herpesvirus particles in scrapings from the base of the lesions and is a useful investigation, if available, in early cases. None of these investigations can distinguish between HSV and VZV infections. Herpes simplex virus is readily cultivated. Fresh vesicle fluid, saliva or scrapings should be inoculated as soon as possible. A cytopathic effect can often be seen after 24 h but may be delayed for up to 7 days. It is characterized by grape-like clusters of refractile cells, usually more pronounced with herpes simplex virus type 2. Herpes simplex antigen can be detected by immunofluorescence staining of cells from herpetic lesions. Increasingly, polymerase chain reaction (PCR) is used for the diagnosis of herpes simplex infection, especially encephalitis.

Techniques for diagnosing herpes simplex infections of skin and mucosae

1 Antigen detection.
2 PCR for herpes simplex DNA.
3 Electron microscopy.
4 Light microscopy (cytology).
5 Cell culture.
6 Serology.

Serology

Herpes virus-specific IgG and IgM are detectable by immunofluorescence or enzyme-linked immunosorbent assay (ELISA). Specific IgM antibodies in serum indicate active infection but do not reliably differentiate between primary infection and reactivation. Seroconversion occurs in primary infection. A fourfold rise in IgG antibody titres demonstrated by these methods often accompanies reactivation.

Management

In primary gingivostomatitis, maintenance of hydration is important. Mild analgesia with paracetamol may help. In severe cases an analgesic gel can be applied sparingly to the oral mucosae. Older children and adults usually cope by taking frequent cool or tepid drinks. Antiviral drugs have little effect on the duration of the lesions.

Skin and mucocutaneous lesions will also heal spontaneously. Treatment with oral antiviral drugs shortens the duration of the condition by a day or two. Topical antiviral creams have little effect on the course of the lesions, and cannot affect viral replication in nerve cells.

Treatment of herpes simplex

For skin and mucosal infections

1 Aciclovir orally 200 mg five times daily for 5 days.
2 Valaciclovir orally 500 mg twice daily for 5 days.
Doses may be doubled in severe cases or immunosuppressed patients.

For severe infections or extensive eczema herpeticum

1 Aciclovir i.v. 5 mg/kg 8-hourly.
Dose may be doubled in immunosuppressed patients.

⚠ **Caution**: Aciclovir is associated with dose-dependent renal impairment, and is also excreted mainly by the kidney. Intravenous aciclovir should be given by infusion in 0.9% saline solution, over a period of at least 1 hour, to avoid the achievement of toxic blood levels. Doses should be reduced and/or the dose-interval increased in patients with creatinine levels above 150 mg/l or a creatinine clearance below 50 ml/min (see British National Formulary).

Prevention and control

Transmission may be reduced by avoiding direct contact with herpetic lesions, although asymptomatic infections and reactivations, with transient viral shedding, are probably common. To avoid herpetic whitlow those attending dental and tracheostomy patients should always wear gloves. Patients with herpetic lesions should avoid direct contact with immunosuppressed patients, newborns and patients with severe eczema or burns.

Herpes zoster and other varicella-zoster virus (VZV) infections

Introduction and epidemiology

Primary VZV infection causes varicella (chickenpox), a systemic disease with a generalized vesicular rash (see Chapter 11). When the acute infection is terminated, VZV establishes latency, with viral genome existing in an undetectable form in the dorsal root ganglia of the central nervous system. In contrast to latent HSV, specific latency sequences of VZV have not been demonstrated.

Herpes zoster is the condition caused by reactivation of latent varicella zoster virus (VZV). Reactivation is related to declining cell-mediated immunity to VZV, and may be spontaneous or may follow a debilitating physical or emotional insult, such as a fever, injury or bereavement. It is common in adulthood and old age, but children and teenagers can develop zoster if they had chickenpox as babies, and neonatal zoster occasionally follows intrauterine varicella infection.

Herpes zoster can also be an early feature of pathological cell-mediated immunodeficiency, including that of clinical AIDS.

The virus has been implicated in the aetiology of some granulomatous vasculitides (such as temporal or cranial arteritis, in which viral DNA can be demonstrated in affected tissue). VZV can also cause a range of uncommon neurological conditions, some of which may be associated with underlying vasculitis.

The importance of herpes zoster is the severe morbidity that it can cause. It may immobilize and incapacitate the elderly by causing severe pain and extensive central nervous system inflammation. Patients over age 50 years suffer persistent pain as a result of herpes zoster, though this can be reduced by antiviral therapy. The burden of ill health caused by herpes zoster is therefore many times greater than that caused by chickenpox. As with chickenpox, herpes zoster can be severe or life threatening in patients with cell-mediated immunodeficiency.

Spectrum of clinical presentations of VZV infection

1 Chickenpox (varicella; primary VZV infection: see Chapter 11).

2 Herpes zoster (VZV reactivation).

3 Lymphocytic meningitis.

4 Acute facial nerve paralysis ('Bell's palsy'): 20% of cases have possible evidence of active VZV infection.

5 Post-infectious cerebellitis or encephalitis.

6 Meningoencephalitis (rare: more common in immunosuppression).

7 Necrotizing retinitis (very rare: more common in immunosuppression).

Pathogenesis

VZV shares structural similarities with other Herpesviridae. It is a 150–200 nm enveloped virus with glycoprotein spikes. The genome encodes approximately 75 proteins with a similar genomic organization to other herpes viruses. Five families of glycoprotein are the primary markers of humoral and cell-mediated immune responses and antibodies against some of these antigens can neutralize viral infectivity. Only enveloped viral particles are infectious.

As viral replication progresses, infected cells undergo a characteristic ballooning degeneration, eventually forming giant cells and eosinophilic intranuclear inclusions.

The disease begins with inflammation in the affected dorsal root ganglion and the appearance of virus particles in the neurones. Inflammation affects the surrounding meninges and the courses and connections of the neurones in the central and peripheral nervous system. On reaching the skin the virus causes blistering lesions in the dermatome served by the affected nerve root.

Clinical features

Illness begins with pain and hyperaesthesia in the dermatome served by the affected dorsal root ganglion. Viruses spread centrally, causing intramedullary and meningeal inflammation. A few patients have clinically apparent meningism, and children in particular may present with viral meningitis and cerebrospinal fluid pleiocytosis before they develop a rash (Fig. 5.4).

The rash begins as small, pink patches at the site of perforating cutaneous nerves. Vesicles quickly develop in these areas, and in severe cases they spread until the dermatome is filled. A few lesions may appear in other areas of the body. The rash is painful, with a disturbing burning, stabbing quality. Unlike chickenpox vesicles, those of zoster may coalesce to form large, flaccid bullae that easily de-roof, forming ulcerated areas (Fig. 5.5). The vesicles heal by drying and scabbing, rarely leaving a scar. Some

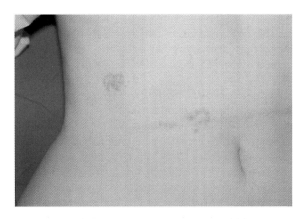

Figure 5.4 Herpes zoster lesions appearing 2 days after this 8-year-old presented with lymphocytic meningitis; the rash did not extend further.

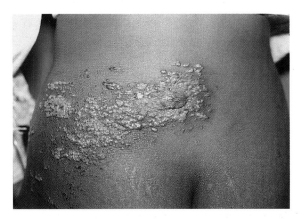

Figure 5.5 A close-up view of herpes zoster lesions, showing their tendency to coalesce.

dermatomes are more commonly affected than others; the supraclavicular, thoracic and lumbar dermatomes and the ophthalmic branch of the trigeminal nerve are often affected. Zoster of the arm, or of the lower face, is rare. Zoster sine herpete (herpes zoster with dermatomal pain and paraesthesiae, but without a rash) can also occur.

Herpes zoster of the ophthalmic branch of the trigeminal nerve may involve the ciliary nerves, leading to lesions on the tip of the nose and inflammation of the uveal tract. This causes a painful, red eye, with an inflamed iris and the risk of synechiae. Inflammation in the extensive central connections of the trigeminal nerve is particularly likely to cause debility and confusion in the elderly.

Herpes zoster of the geniculate ganglion of the seventh nerve causes so-called Ramsay-Hunt syndrome, with vertigo (due to disturbance of the adjacent vestibular nerve), facial paralysis and loss of taste sensation in the anterior part of the tongue. Less extensive involvement of the ganglion is probably one cause of VZV-related facial paralysis.

Multidermatomal and bilateral zoster occur rarely, when more than one sensory ganglion is affected. This is more common in the immunosuppressed. Dissemination of virus, with the development of numerous lesions in a chickenpox-like rash, also indicates immunodeficiency (Fig. 5.6).

Deficit of motor function is occasionally seen in herpes zoster, usually occurring as weakness or loss of tendon reflexes at the affected dermatomal level, or diaphragmatic paralysis in cervical zoster. This is probably due to extensive segmental inflammation, affecting the anterior spinal cord.

> ⚠ **Warning signs of immunodeficiency in herpes zoster**
> 1 Bilateral or multidermatomal rashes.
> 2 Unusually severe, deep or scarring lesions.
> 3 Recurring herpes zoster.
> 4 Association with extensive chickenpox-like rash.
> 5 Association with progressive central nervous system features.

The course of the illness depends on the patient's age. In children the rash may be minimal; in young adults the average duration of rash is 5–8 days. In elderly people there can be weeks of illness, with bullous rash and severe

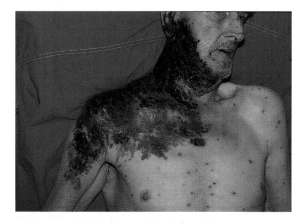

Figure 5.6 A severe and haemorrhagic herpes zoster rash, with disseminated chickenpox lesions in a patient with chronic lymphocytic leukaemia (the patient recovered fully on treatment).

central nervous system disruption causing confusion and debility.

Diagnosis
Diagnosis is usually made clinically. Virus particles can be demonstrated by electron microscopy, if facilities are available, in material scraped from the base of vesicles. The virus grows readily with a typical cytopathic effect in cell culture. PCR is now routinely used for diagnosis and is especially valuable in investigation of CSF. Testing of paired sera usually shows a large rise in IgG antibodies (secondary response; see Chapter 1).

> **Laboratory diagnosis of herpes zoster**
> 1 Demonstration of VZV DNA by PCR.
> 2 Electron microscopy.
> 3 Cell culture.
> 4 Serology: IgG ELISA.

Management
Rest and analgesia are sufficient in younger patients and mild cases, in whom antiviral therapy only marginally shortens the course of the disease. Ophthalmic zoster and Ramsay-Hunt syndrome should always be actively treated, as should patients over the age of 50. The illness can be significantly shortened in older people by early treatment with valaciclovir, famciclovir or aciclovir. Oral aciclovir has poor bioavailability and may be less effective. The prodrugs, valaciclovir (an ester of aciclovir) and famciclovir (an ester of penicyclovir), are well absorbed before being hydrolysed to the parent drug. They can be given in lower and less frequent dosage than aciclovir. The blood levels reached by all three of these drugs may not be high enough to affect all strains of VZV. Severe cases benefit from intravenous aciclovir treatment for 4 or 5 days (10–14 days in the immunosuppressed).

> **Treatment of herpes zoster**
>
> **For uncomplicated infections**
> 1 Oral valaciclovir 1 g three times daily for 7 days.
> 2 Oral famciclovir 250 mg three times daily for 7 days or 750 mg once daily for 7 days.
> 3 Oral aciclovir 800 mg five times daily for 7 days.
>
> **For severe infection**
> 4 Aciclovir infusion 5 mg/kg 8-hourly for 4–7 days.
>
> **For immunocompromised patients**
> 5 Aciclovir infusion 10 mg/kg for 7–14 days.
>
> **Caution**: intravenous aciclovir (see box on page 90).

Complications

ZAP; zoster-associated pain (post-herpetic neuralgia)

Post-herpetic neuralgia is causalgic (pain unrelated in quality or severity to its cause), and severely affects the elderly, lasting an average of 60 days in the over-50s. It may be related to inflammatory damage to small nerve fibres and/or persisting low-level viral activity. ZAP is approximately halved in duration in those commencing antiviral treatment within 48 h of rash onset. It is often noticed when the rash is healing, and responds poorly to opiate analgesics, phenytoin, carbamazepine and vitamin B$_{12}$ injections. Gabapentin often gives relief. Tricyclic antidepressants may be an effective alternative, but can cause unwanted sedation or confusion. Either can be combined with simple analgesics.

Secondary bacterial infection

Secondary bacterial infection, most often with *Staphylococcus aureus*, can affect severe or denuded rashes, especially in the elderly or immunosuppressed. Oral treatment with flucloxacillin, an oral cephalosporin or a macrolide is usually effective.

Ascending myeloencephalitis

Ascending myeloencephalitis is an exceptionally rare complication that should be treated with high-dose intravenous aciclovir. Immunosuppressed patients are at higher risk, and respond less well to treatment.

Hand, foot and mouth disease

Hand, foot and mouth disease is a highly infectious disease of toddlers caused by coxsackie A viruses, most often A16, but also A5 and enterovirus type 71. Family outbreaks are common, and can affect adults and children, though most adults are immune. A brief feverish prodrome is followed by the simultaneous eruption of several blisters on the hands and feet and in the pharynx. The palms and soles are most affected (Fig. 5.7). There is also a papular rash on the buttocks. A mild sore throat is common. Illness is short and self-limiting in most cases. In a recent epidemic in the Far East caused by a strain of enterovirus type 71, some cases were complicated by systemic disease with encephalopathy and/or shock and acute respiratory distress syndrome.

If microbial diagnosis is needed, the viruses can be identified by cell cultures of vesicle fluid, throat swab or faeces but increasingly PCR is the main method of diagnosis. The laboratory diagnosis of enterovirus infections is discussed in Chapter 6.

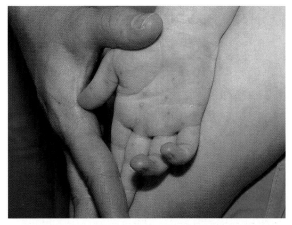
(a)

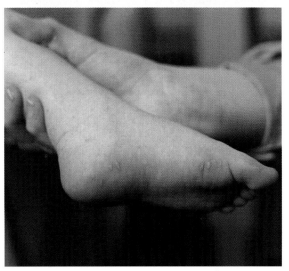
(b)

Figure 5.7 Hand, foot and mouth disease: the typical blisters.

Cutaneous poxvirus infections

The poxviruses that infect the skin are molluscum contagiosum, cowpox, orf and vaccinia. Monkeypox and tanapox cause systemic infections with cutaneous pocks.

Poxviruses are divided into three groups:
1 orthopoxviruses, which include monkeypox, cowpox, vaccinia, and variola (smallpox: declared extinct in 1977);
2 parapoxviruses, which include orf; and
3 unclassified viruses, which include molluscum contagiosum and tanapox.

The viruses are large, with a double-stranded DNA genome coding for more than 100 viral proteins. They have complex symmetry with a brick-like shape 200–

250 × 250–300 nm. Virus uncoating is initiated by host enzymes but completed by viral enzymes. Viral replication takes place in the cytoplasm of host cells, producing large inclusion bodies.

Specimens of vesicle fluid are suitable for investigation. The viruses are stable and survive well while being transported. In recent monkeypox cases in Zaire the diagnosis was made by a combination of electron microscopy, virus isolation and serology. Distinctive pocks are produced on chorioallantoic membrane culture in hens' eggs. Monkeypox, vaccinia and variola virus can be distinguished by PCR techniques. Antibody to specific monkeypox antigen can be detected in serum absorbed with smallpox and vaccinia antigens. Surveillance of systemic monkeypox is maintained because of the remote possibility of an increase in virulence.

Molluscum contagiosum

This common wart-like condition is caused by a poxvirus infection of the acanthocyte layer of the epidermis. The infected cells proliferate, vacuolate and enlarge, protruding above the surface of the skin as typical, pearly lesions up to 3 mm in diameter. They are umbilicated, with a small central cavity containing whitish, pulpy material. The vacuolated cells are shed into the lesion, which is highly infectious. Infection is transmitted by skin contact, and by fomites such as clothing and towels. Autoinoculation to other skin sites is common.

Lesions typically occur in groups, often on the face or arms. Children are frequently affected. The spread of lesions is limited by cell-mediated immunity. HIV-positive patients may have numerous lesions, particularly on the face. Their appearance is diagnostic but, if necessary, poxviruses can be demonstrated by the presence of characteristic 'molluscum bodies' in scrapings stained with, for example, Giemsa or iodine, or by electron microscopy of expressed material (Fig. 5.8).

Spontaneous resolution is uncommon, but the lesions can be removed by curetting, ablated by freezing or 'killed' by inserting the point of an orange-stick dipped in 80% phenol solution into the umbilicated centre. Extensive skin lesions in HIV patients often respond well to cidofovir 10% topical ointment.

Cowpox and orf

These are zoonoses caused by poxviruses. Cowpox produces lesions on the udders of cattle, and causes systemic disease with pneumonitis in cats. Human cases have occurred after contact with cattle, domestic cats and big cats. Transmission is by direct contact, and causes large, volcano-shaped pocks with vesicular or necrotic centres, usually on the hand (Fig. 5.9). The central crater may appear black, and a history of animal contact may lead to suspicion of anthrax. Cowpox lesions have no halo of vesicles, however, and the surrounding oedema is rarely as severe or extensive as in anthrax.

Orf is a parapoxvirus whose natural host is sheep. Human infections follow the handling of sheep or their carcasses. The hand or wrist is usually affected by one or

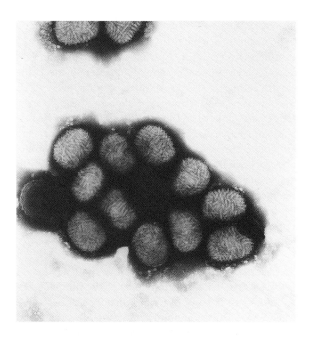

Figure 5.8 Virion of molluscum contagiosum virus. Courtesy of Professor P. D. Griffiths and Ms G. Clewley, Department of Virology, Royal Free Hospital School of Medicine.

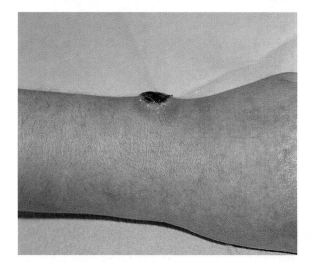

Figure 5.9 Lesion of cowpox on the wrist of a stable-girl (possibly contracted from the farm cat).

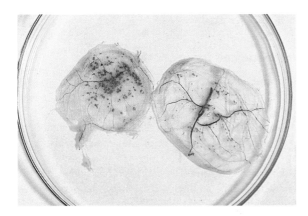

Figure 5.10 Pocks produced by cowpoxvirus on the chorioallantoic membrane of a hen's egg.

more indurated papules 1–2 cm in diameter. There may be a central vesicle or crater, or simply a depression.

Viruses can be identified by electron microscopy, or culture of biopsy material or curettings from the edge or base of the lesions. Culture of cowpox on the chorioallantoic membrane of hens' eggs produces characteristic lesions (Fig. 5.10).

No specific treatment is required, as the conditions are self-limiting, with a natural history of up to 3 or 4 weeks. Aciclovir and penciclovir are not effective against poxviruses.

Vaccinia

Vaccinia exists in many strains whose origins are uncertain. It may be descended from subcultures of cowpox, horsepox or variola (the agent of smallpox). It was used for three centuries as an effective vaccine against smallpox (which was declared extinct from nature in 1977). Vaccinia has been used for developing hybrid vaccines, though canarypox and other animal poxviruses are now more favoured. It is rarely given as a vaccine as the indication no longer exists, though some countries retain stocks of vaccine, in case it should be needed to protect from accidental or deliberate release of smallpox. However, vaccinia is a potentially pathogenic virus (Fig. 5.11). The vaccine lesion is accompanied by fever and headache. Accidental inoculation to another site in the patient or a close contact is common. Vaccinia has a predilection for eczematous skin. It is occasionally transmitted by the trans-placental route. A post-infectious myocarditis or encephalitis can follow vaccination. It can cause severe necrotizing or progressive systemic infection in patients with altered immunity.

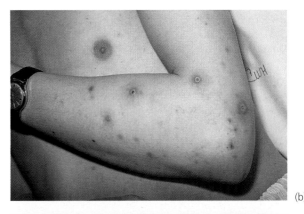

(b)

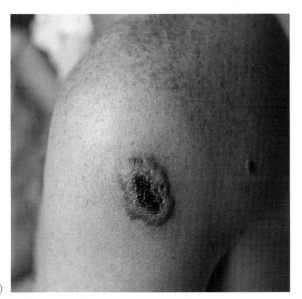

(a)

(c)

Figure 5.11 (a) Large vaccinia lesion following smallpox vaccination. (b) Autoinoculation lesions of vaccinia on the skin. (c) Accidental inoculation of vaccinia into the conjunctiva.

Aciclovir is not effective, but cidofovir shows significant antivaccinia activity in laboratory settings.

Monkeypox

Monkeypox is a zoonotic orthopoxvirus infection that causes smallpox-like disease in humans in central and West Africa, where the case-fatality rate is about 12%. The diagnosis may be suspected clinically because of the typical, smallpox-like rash, lymphadenopathy and recent monkey contact. Some rodents can be infected by monkeypox. In 2003, Gambian giant rats, imported as pets to the USA, infected pet prairie dogs. These became ill, and infected humans through inoculation via biting or scratching, causing local lesions followed by a systemic infection. Human-to-human spread is uncommon.

Tana

Tana is a poxvirus infection probably spread by insect bites, and usually acquired in Africa. It is unusual to see cases with more than a single skin lesion. The diagnosis is suggested by the travel history. The virus can be demonstrated by electron microscopy or cultivated in cell culture. Neutralizing antibodies are produced in convalescence.

Diagnosis

In many instances the clinical circumstances will permit a presumptive diagnosis to be made, e.g. recent travel to an endemic area or occupational exposure. Electron microscopy of vesicle fluid reveals the characteristic ball of wool appearance of pox viruses. The viruses can be grown readily in cell culture. Concerns about bioterrorism have led to the development of rapid DNA amplification-based techniques for diagnosis.

Bacterial infections of the skin and mucosae

Organism list

- *Staphylococcus aureus*
- *Streptococcus pyogenes*
- *Corynebacterium* spp., including *C. minutissimum* and *C. diphtheriae*
- *Borrelia burgdorferi*, *B. garinii*, *B. afzelii*
- *Pasteurella* spp.
- *Bartonella henselae*
- *Francisella tularensis*
- *Erysipelothrix rhusiopathiae*
- Mycobacteria, including *Mycobacterium tuberculosis*, *M. marinum* and environmental mycobacteria

- *Actinomyces* spp.

Impetigo and furunculosis

Introduction

Impetigo and furunculosis are caused by pyogenic bacterial infection affecting surface structures of the skin.
- Impetigo is an infection of the surface of the epidermis.
- Furunculosis is infection of the sebaceous glands or sweat glands.

Both are usually caused by *Staphylococcus aureus*, and are characterized by an intense local inflammatory response and the production of pus.

Pathology and epidemiology

Impetigo is a common infection of children, who suffer frequent mild skin trauma, and have little immunity to strains of *S. aureus*, but adults can also be affected, particularly by secondary 'impetiginization' of skin lesions. Pus from skin lesions is highly infectious. Staphylococcal skin infections therefore spread easily, contiguously to adjacent skin sites, by autoinoculation to distant sites or, by contact, to the skin of other individuals. Sweating or dampness of the skin macerates the epidermal surface, increasing susceptibility to infection. Swimmers and divers, who are exposed to untreated water, may develop furunculosis caused by waterborne *Pseudomonas* spp.

Microbiology

Staphylococci are Gram-positive, catalase-producing organisms in the family Micrococcaceae. *Staphylococcus* spp. are facultative, but *Micrococcus* spp. (which rarely cause human disease) are obligate aerobes. The organisms are non-motile and rarely produce capsules.

There are at least 30 *Staphylococcus* species of varying pathogenicity to humans. They can be divided into the pathogenic coagulase-positive *S. aureus*, and the less invasive coagulase-negative staphylococci, on the basis of the coagulase test. *S. aureus* colonies on modern media are rarely gold-coloured, and are therefore not morphologically distinguishable from some coagulase-negative staphylococci. *S. aureus* is distinguishable by phage-typing into many groups, some of which produce powerful toxins. Some toxins act locally, contributing to the pathogenesis of lesions. Others are systemically active, causing severe systemic disease as a consequence of relatively limited local infection. The most important example of this is toxic shock syndrome (see below).

Pathogenesis of *Staphylococcus aureus* infection
(Fig. 5.12)

S. aureus expresses many potential pathogenicity determi-

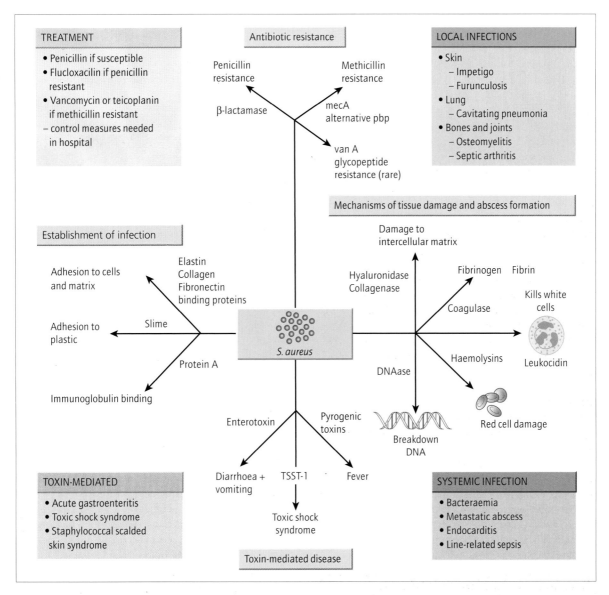

Figure 5.12 Pathogenicity factors of *S. aureus*.

nants. Different stages of staphylococcal infection correspond with the expression of different bacterial antigens. Initially, matrix-binding proteins favour colonization and, later, exoproteases facilitate spread through the tissues. Global regulatory genes, which control expression of groups of staphylococcal genes, have been studied, including *agr*, which suppresses expression of surface protein through a bacterial density-sensing peptide. Methicillin-resistant *S. aureus* are now a particular problem where methicillin resistance has become coupled with the ability to be transmitted readily.

Capsules

Most clinical isolates of *S. aureus* express one of 11 capsular types. Serotypes 1 and 2 elaborate profuse capsular material, producing mucoid colonies, but are rare in clinical practice. Most strains possess much smaller capsules (micro-capsules). Over 75% of all clinical isolates are serotypes 5 or 11, and both express a micro-capsule that appears to be antiphagocytic. Non-capsulate type 5 organisms have reduced virulence in animal models. Experiments with mucoid capsular types suggest that immunization with the polysaccharide protects from lethal infection.

Surface proteins

Many *S. aureus* surface proteins possess a secretory signal sequence, a hydrophobic membrane-spanning domain, and a cell wall anchoring region. A ligand-binding domain is exposed on the bacterial cell surface, enabling some of these proteins to function as adhesins. An example is protein A, which has antiphagocytic properties, related to its ability to bind the Fc portion of immunoglobulin. Several surface-exposed proteins (microbial surface components recognizing adhesive matrix molecules; MSCRAMM) bind extracellular matrix molecules. These include fibronectin-binding proteins and collagen-binding proteins.

Toxins

The main toxins of *S. aureus* are the pyrotoxic superantigens (pyrogenic exotoxins, or SPEs: a group of exotoxins that share several biological characteristics, secreted by either *S. aureus* or *Streptococcus pyogenes*; see pp. 128–9). They include TSST-1 and most of the staphylococcal enterotoxins (SEA, SEB, SECn, SED, SEE and SEH), which each exhibit at least three biological properties: pyrogenicity, superantigenicity and the capacity to enhance the lethality of endotoxin in rabbits up to 100 000-fold. SPE A binds tightly to endotoxin and is then toxic to immune system cells. TSST-1 increases the lethal effects of endotoxin on renal tubular cells. The staphylococcal enterotoxins (SEs) are potent emetic agents whereas the other pyrotoxic superantigens are not. Antibody to pyrotoxic superantigens has an important role in active and passive immunity.

Other staphylococcal enterotoxins (SEG, SEH and SEI) or genes (*sej* and *sek*) exhibit superantigenic activity or sequence homology to known superantigens.

S. aureus possesses other exotoxins such as Panton Valentine leukocidin and α-haemolysin, which are thought to act in the pathogenesis of cutaneous infection.

Matrix lytic enzymes

These are extracellular enzymes, such as hyaluronidase, collagenase and lipase, that break down host tissues and may facilitate invasion. Cytolytic toxins, conventionally named haemolysins, are also produced. Their contribution to the pathology of the staphylococcal disease remains uncertain.

Coagulase

The enzyme coagulase, which converts fibrinogen to fibrin, is the conventional marker of pathogenicity for *S. aureus*. It was thought to help in protecting the organism from non-specific immune attack, but there is no clear association between coagulase and virulence.

Cell wall components

Staphylococcal peptidoglycan and lipoteichoic acid stim-

ulate the release of tumour necrosis factor-alpha (TNF-α), interleukin (IL)-1, IL-6 and IL-8, leading to shock and coagulopathy.

Factors affecting pathogenicity of staphylococci

1 Coagulase.
2 Capsules.
3 Fibronectin receptor.
4 Extracellular enzymes.
5 Haemolysins.
6 Pyrogenic exotoxins.
7 Enterotoxins A–E.
8 Toxic shock syndrome toxins 1 and 2.

Clinical features

Impetigo

Impetigo often occurs around the mouth or nose, where skin is easily damaged, or in superficial skin lesions such as scratches, insect bites or broken chickenpox vesicles. Beginning as a small spot (Fig. 5.13), it extends to form a plaque-like inflamed lesion on which a yellowish exudate forms, and dries into thick scabs. The lesions are irritating and sore; scratching or rubbing contributes to infection of other sites (Fig. 5.14).

Furunculosis

Furunculosis tends to occur in stagnant sebaceous material, and therefore affects children and men more than women. Poor personal hygiene predisposes to furunculosis by allowing blockage of sebaceous ducts. Lesions may be single (often called boils), or multiple. They are common on the back, or the thighs, and begin as small papules, which increase in size and tenderness to a variable degree. Some will 'point' and discharge yellowish pus before gradually resolving, while others may remain 'blind', gradually healing by becoming less indurated and inflamed.

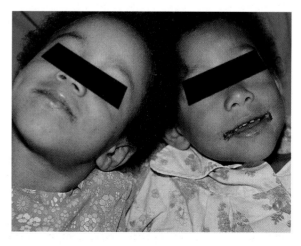

Figure 5.13 Impetigo in a typical site: both sisters are affected.

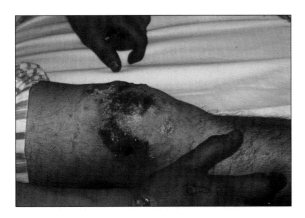

Figure 5.14 Staphylococcal impetigo developed on the scratched knee of a builder; a satellite lesion then appeared on his hand.

Concurrent infection of several neighbouring glands causes a composite lesion with multiple discharging sinuses at the skin surface. These lesions are called carbuncles. They tend to occur in men, often affecting the axilla or the hairline at the back of the neck, and can reach several centimetres in diameter.

Diagnosis

The diagnosis is usually clinically evident. *S. aureus* can readily be recovered from cultures of exudate, allowing confirmation of the antibiotic sensitivities of the organism. Microbiological examination is important in detecting the minority of impetiginous lesions caused by *Streptococcus pyogenes* or *Pseudomonas* spp. Streptococcal lesions may develop bullae, but this is not a reliable distinguishing feature.

Microbiological diagnosis

Staphylococci grow readily on simple media. Colonies of *S. aureus* on blood agar are opaque, butyrous golden-yellow to creamy-white, and may elaborate a haemolysin, producing a zone of complete haemolysis. Coagulase-negative colonies are smaller and white. To identify *S. aureus* from sites contaminated by other organisms, selective media are used. Staphylococci can tolerate high salt concentrations, so 5% salt agar is selective for them. Mannitol and an indicator are incorporated, as mannitol is fermented by almost all *S. aureus*, allowing selection of positive colonies for further study. This screening technique is used to identify carriers of *S. aureus* in outbreaks of infection. Another selective medium is Baird–Parker medium. Adding antibiotics, such as ciprofloxacin, allows selection of ciprofloxacin-resistant epidemic MRSA strains. The identity of *S. aureus* is confirmed by characteristic morphology on Gram staining, catalase production and the presence of coagulase. Strains of MRSA can now be identified from screening specimens within a few hours, by real time PCR tests. The use of primers within both the *mecA* gene and another gene, *orf3*, from the same gene cassette, prevents false positive reactions from mixtures of coagulase-negative staphylococci (which might contain *mecA*) and *S. aureus*. A slide agglutination test, using a monoclonal antibody to PBP2, together with a coagulase test, will then confirm a diagnosis of MRSA.

S. aureus strains with intermediate glycopeptide sensitivity cannot be distinguished from susceptible strains by disc diffusion methods. Both minimal inhibitory concentration (MIC) and E-test are satisfactory methods.

Coagulase is either bound to the bacterial cell wall (when it is known as clumping factor) or expressed extracellularly. Isolates are usually screened by slide agglutination, to detect clumping factor. This identifies more than 89% of *S. aureus* strains. Negative isolates are then tested using a tube technique that detects free coagulase. This identifies more than 99% of *S. aureus*. More than 95% of *S. aureus* produce deoxyribonuclease (but so do up to 10% of coagulase-negative staphylococci). Once an isolate is identified as a *S. aureus* no further identification is usually necessary. Speciation of coagulase-negative staphylococci may be indicated, to confirm that multiple isolates from the same patient are the same, and therefore likely to be clinically significant. Useful commercial testing kits are available for use in routine laboratories. Most automated detection systems will identify the species of coagulase-negative staphylococci.

Typing

Several molecular typing methods are used to demonstrate the transmission of resistant staphylococci in a hospital environment: these include pulsed field gel electrophoresis and multi-locus sequence typing (see page 45).

Antimicrobial susceptibility

S. aureus is susceptible to erythromycin, clindamycin, fusidic acid, ciprofloxacin, aminoglycosides, chloramphenicol, the first-generation cephalosporins and the glycopeptides vancomycin and teicoplanin. Individual strains may acquire resistance to any of these agents. Intermediate glycopeptide resistance has been reported and may be an increasing clinical problem.

S. aureus strains were once almost universally susceptible to penicillin but when penicillin was used widely, penicillinase-producing strains soon predominated. The introduction of methicillin and, later, flucloxacillin initially solved this problem. However, strains of methicillin-resistant *S. aureus* (MRSA) have become a major problem in many hospitals throughout the world. Methicillin resistance develops by acquisition of the gene *mecA*, which codes for a low-affinity penicillin-binding protein PBP2. This enables the organism to synthesize its cell wall despite the presence of the antibiotic. MRSA is commonly resistant to several other antimicrobials, and some MRSA, known as epidemic ('E') strains, spread readily in hospitals. Most hospitals have policies to control the spread of MRSA (see Chapter 23).

Laboratory tests used to identify *S. aureus*

1 Slide coagulase test.
2 Tube coagulase test.
3 Tests for deoxyribonuclease.
4 Slide agglutination tests to detect PBP2s (detects MRSA).

Management

Impetigo is a superficial infection that responds readily to modest doses of antibiotics. Oral antistaphylococcal agents flucloxacillin or cloxacillin are usually used, and are also effective in cases caused by *Streptococcus pyogenes*. In penicillin-allergic individuals oral cephalosporins may be given (unless anaphylaxis was the problem, in which case there is approximately a 10% chance of anaphylactic reaction to cephalosporins also). Macrolides are also often effective. For methicillin-resistant staphylococci, treatment must be determined by sensitivity testing of the organism.

Topical antimicrobial agents such as fusidic acid or mupirocin cream are also effective, but carry a risk of skin sensitization and encourage the emergence of resistant skin flora. They should not be used for large lesions or for periods greater than 10 days. Aminoglycosides may be absorbed from the lesions, with risk of toxicity if used on large areas. Silver sulfadiazine 1% cream also has a wide spectrum and is used for prophylaxis and treatment of infection in burns and some leg ulcers.

Treatment of staphylococcal skin infections

First choice
• Oral flucloxacillin 250 mg for 5–7 days.

Alternatives
Oral cephalosporins 250–500 mg 6–8-hourly (see data sheet) for 5–7 days *or* erythromycin 250 mg 6-hourly *or* clarithromycin 250 mg 12-hourly *or* trimethoprim 200 mg 12-hourly for 5–7 days.

Note that mild furunculosis, with small or moderate pustular lesions, will often heal spontaneously if good skin hygiene is maintained. Large lesions and carbuncles are painful and may be accompanied by spreading inflammation and fever. They should be treated with antistaphylococcal antibiotics, which may need to be given parenterally in severe cases.

⚠ **Surgical management**
Once an abscess has formed, antibiotic treatment may limit local inflammation and abolish systemic features but infected pus remains, with risk of recurrent infection. Large pustules and abscesses should therefore be aspirated or incised. This is usually followed by rapid healing and relief of pain. Surgical drainage is the most effective treatment for large abscesses and carbuncles.

Complications

Streptococcal impetigo can be complicated by surgical scarlet fever (see Chapter 6) and, in young children, by post-streptococcal nephritis (see Chapter 21).

Failure to respond to treatment

Skin infections may be caused by organisms resistant to commonly used antibiotics. Non-healing lesions should be investigated microbiologically. Ciprofloxacin or a tetracycline may be affective if the patient is infected by a marine organism or a pseudomonad.

Scalded skin syndrome or toxic epidermal necrolysis

Scalded skin syndrome, common in children, is caused by the pyrogenic exotoxins of *S. aureus*; it is also called Ritter's or Lyell's syndrome when it affects neonates. Each infected staphylococcal lesion is surrounded by an area of epidermal damage. Affected epidermis is loose and can easily be rubbed off the underlying layers (Nikolsky's sign). Early lesions look pale, and often form flaccid, shallow bullae, which may be very extensive. In severe cases the lesions become confluent and the surface of the skin separates, leaving a typical scalded appearance (Fig. 5.15).

Prompt treatment with antistaphylococcal agents will abort the lesions. The surface epidermal layers are shed,

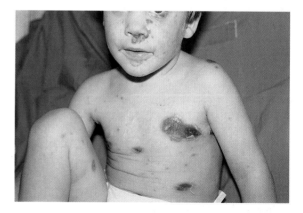

Figure 5.15 Scalded skin appearance: the child has healing chickenpox, the lesions of which have become infected with *Staphylococcus aureus*.

leaving regenerating skin areas. Severe cases must be nursed in warm, humid conditions to avoid excessive loss of heat and moisture from denuded areas. This is especially important for infants, because of their relatively large surface area.

⚠️ Toxic epidermal necrolysis is an immunologically mediated skin disorder, sometimes caused by drug reactions. The lesions resemble those of scalded skin syndrome but there is no evidence of staphylococcal infection. A history of drug ingestion may be obtainable. Commonly implicated drugs include sulphonamides, sulphonylureas, phenytoin, indometacin, allopurinol and nevirapine. Antistaphylococcal treatment is not indicated unless secondary bacterial infection occurs.

Staphylococcal toxic shock syndrome

This is a systemic disease caused by the toxic shock syndrome toxins (TSSTs) of *S. aureus*. The patient often has a trivial staphylococcal infection, commonly a skin abscess or, in women, a vaginal infection associated with the use of highly absorbent tampons during menstruation. Rarely, toxic shock syndrome complicates staphylococcal bacteraemia or endocarditis.

The main features of the illness are fever, diarrhoea, myalgia, rash, hypotension and confusion. The rash is similar to that of scarlet fever, but without the rough, punctuate effect. It particularly affects the peripheries, where it may contain petechiae, and is exaggerated in the flexures. Conjunctival injection is often prominent (Fig. 5.16). The white cell count may be normal or high, usually with a predominance of neutrophils.

Definition of toxic shock syndrome

1 Temperature 39 °C or greater.
2 Rash: diffuse macular erythema.
3 Desquamation: 1–2 weeks after onset.
4 Systolic blood pressure 90 mmHg or less.
plus involvement of three or more of the following organ systems:
- Gut: vomiting or diarrhoea.
- Muscles: creatine kinase twice upper limit or more.
- Mucosae: vagina, mouth, conjunctiva inflamed.
- Renal: creatinine twice upper limit or more.
- Hepatic: bilirubin or transaminases twice upper limit or more.
- Central nervous system: altered consciousness without focal signs.
- Platelets: 100 000/mm^3 or less.

The diagnosis can be suspected clinically. It is important to seek the focus of infection and take pus or swab specimens to identify the organism and check its sensitivity spectrum.

Treatment of toxic shock syndrome

1 Fluid resuscitation and haemodynamic support (see Chapter 19).
2 Intravenous antistaphylococcal antibiotics; first choice: flucloxacillin: 1 to 2 g 6-hourly (child under 2: 0.25 times adult dose, child 2 to 10: 0.5 times adult dose), rifampin or fusidic acid may be added, as they readily penetrate into inflamed tissues; for penicillin-allergic patient or for MRSA: teicoplanin 400 mg 12-hourly for 3 doses, then 200 to 400 mg daily, child over 2 months: 10 mg/kg 12-hourly for 3 doses, then 6 mg/kg daily.
3 Remove any precipitating cause: drain abscesses; search for and remove infected tampon or pessary.
4 Consider intravenous immunoglobulin; as lack of antibodies to TSSTs may contribute to the development of the disease.

It may take some days for diarrhoea and hypotension to resolve. The erythema often heals by desquamation, with characteristic shedding of the whole nailfold and skin overlying the finger pulp.

Cellulitis and erysipelas

Introduction and pathology

Cellulitis and erysipelas are infections that cause spreading red lesions of the body surface. The infection is usually endogenous, caused by bacteria from the skin surface entering through breaks in the epidermis.

Cellulitis is a spreading infection of the lower epidermis and dermis. It is often caused by *S. aureus*, but *Streptococcus pyogenes* is also common. Coliforms or enterococci may be a cause in immobile or incontinent patients; *Pasteurella multocida* may complicate dog or cat bites. Marine vibrios can enter via scratches or cuts from rocks and coral. In hospitals and care homes, healthcare-acquired and opportunistic organisms such as *Acinetobacter*, *Bacillus* or *Pseudomonas* spp. may be a cause.

Erysipelas, often called 'streptococcal cellulitis', is a distinct intradermal infection caused by *Streptococcus pyogenes*. It can often be clinically distinguished from typical cellulitis, and this is important both in choosing immediate treatment and in determining which cases may have severe underlying disease, such as bacteraemia. Properties of *S. pyogenes* predisposing to virulence are discussed in Chapter 6.

Clinical features

Cellulitis appears as a spreading, tender, red lesion, usually with an indistinct margin. It often arises from an obvious point of entry for infection, such as a wound, or the insertion site of a needle or cannula (Fig. 5.17). It can occur in any body site: commonly the arm, the lower leg, the foot, or

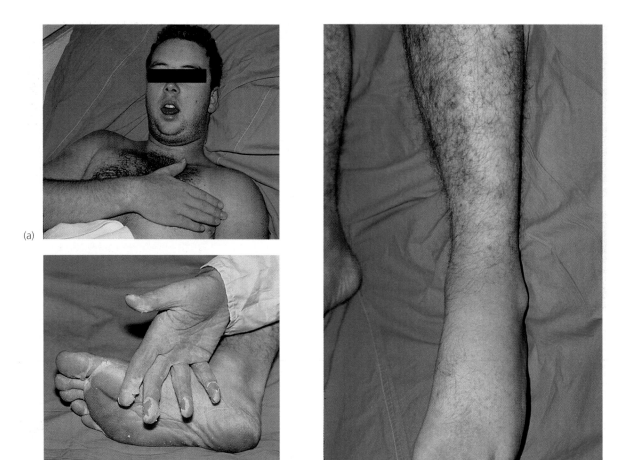

Figure 5.16 (a) Toxic shock syndrome: erythema and conjunctival injection in a confused and hypotensive patient (the causative infection was a small abscess on the occiput). (b)Toxic shock syndrome: petechial component in rash on legs. (c) Toxic shock syndrome: typical desquamation of the digits.

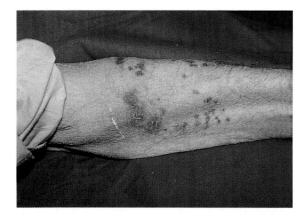

Figure 5.17 Small area of cellulitis surrounding a venepuncture site (this was caused by a methicillin-resistant *Staphylococcus aureus*).

surgical sites on the trunk. Elderly or diabetic patients are at risk of leg and foot cellulitis; those who self-inject drugs may develop cellulitis at multiple sites, due to attempted intravenous injections or to 'skin popping'. Pus may form in the infected tissues, and may produce abscesses, or ooze from the entry point of the infection. Lymphangitis and lymphadenitis of draining nodes is common.

Erysipelas almost always affects the face, the shin or the dorsum of the foot, where the skin is easily traumatized; a small fissure between the toes or at the angle of the nose or mouth may afford entry for virulent streptococci (Fig. 5.18). Aching, throbbing or tenderness of the skin is followed in a few hours by an indurated, hot, tender, erythematous lesion with distinct demarcation from the normal skin, both visually and by palpation; the patient can easily indicate this boundary. In severe infection, fluid-filled bullae may form in the epidermal layers, but

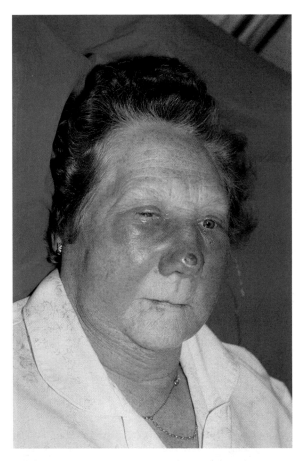

Figure 5.18 Erysipelas: this rash spread from a tiny fissure in the nose.

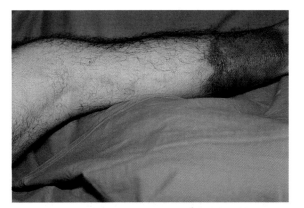

Figure 5.19 Erysipelas: lymphangitis ascending from the skin lesion (the patient had tender, enlarged inguinal lymph nodes).

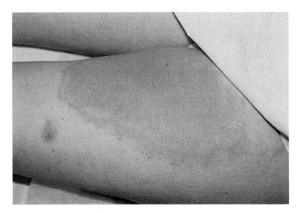

Figure 5.20 Erysipelas: spreading infection surrounding the draining inguinal lymph node.

pus is rarely seen. There is a variable fever, moderate malaise and often a neutrophilia with a total white cell count of $11–13 \times 10^6$/l. If untreated, the lesion spreads rapidly by separating the epidermal cells in an expanding 'bleb'. Tender enlargement of draining lymph nodes is common, and can progress to suppurative lymphadenitis (Fig. 5.19) with surrounding erythema. Breakdown of lymph nodes indicates a danger of secondary bacteraemia (Fig. 5.20).

> ⚠ **Caution: Necrotizing fasciitis and systemic sepsis**
> • Severe pain, out of proportion to the appearance of the skin lesion, may be an early sign of necrotizing fasciitis (see below); it should prompt early vigorous treatment and urgent investigation.
> • Erysipelas, other than on the face or shin, is extremely uncommon and is an important marker of streptococcal bacteraemia; this useful warning feature often precedes the appearance of the signs of sepsis.

Diagnosis

Clinical suspicion is aroused by the site and appearance of the lesions. In cellulitis, examination of swabs or aspirated pus can indicate the identity of the causative organism. In true erysipelas, the infection is 'enclosed' in the epidermis; *S. pyogenes* is rarely recovered from swabs, and blood cultures are usually negative. A serological diagnosis can be made by detecting rising anti-streptolysin O titres (ASOT), but the rise is often insignificant or delayed. Anti-DNase or anti-hyaluronidase titres (reference laboratory tests) may show greater rises.

Management

Early treatment is important to limit extension of the infection. Many cases are successfully treated in family practice with oral antibiotics. Severe infection, especially in debilitated patients, or failure to respond within 36–

48 hours, should prompt admission for intravenous therapy. Treatment of hospital-acquired cellulitis may need to be guided by culture and sensitivity data.

Short-course treatment is often adequate for cellulitis. However, it is important to treat erysipelas until healing is established and erythema resolves, as recrudescence and recurrence are common.

Treatment of cellulitis and erysipelas

Oral treatment
In mild cases, treat for 5–10 days with *either* amoxicillin 500 mg 8-hourly (or ampicillin 500 mg 6-hourly) plus flucloxacillin 500 mg 6-hourly *or* co-amoxiclav (expressed as amoxicillin) 250–500 mg 8-hourly *or* cefradine 250–500 mg 6-hourly *or* 0.5 to 1 g 12- hourly *or* erythromycin 500 mg–1 g 6-hourly.

Intravenous treatment
In severe cases, treat for 5–10 days (possibly longer for erysipelas) with *either* benzylpenicillin 2.4 g 4–6-hourly plus (for cellulitis) flucloxacillin 1.0–1.5 g 6-hourly *or* cefuroxime 1.5 g 6–8-hourly *or* erythromycin 1 g 6–8-hourly (infuse over 30–60 minutes in 250 ml, to avoid phlebitis) *or* (oral alternative for severe infections) clindamycin 300–450 mg 6-hourly (discontinue if significant diarrhoea occurs, and test for *C. difficile* toxin in the stools).

⚠ **Caution**: the skin lesion of erysipelas usually spreads for 12–24 h after treatment is commenced, before the lesion flattens and inflammation subsides – this does not indicate the need to change the antibiotic; desquamation often accompanies healing.

⚠ **Caution**: cellulitis caused by marine organisms (vibrios or Chromobacteriaceae), or tropical organisms (*Acinetobacter* spp.) may not respond to the above antibiotics; ciprofloxacin orally, 500 mg 12-hourly, *or* intravenously, 400 mg 12-hourly, may be effective.

Complications
Complications of these localized infections are rare. Tissue necrosis occasionally occurs even after apparently prompt and effective treatment. Full-thickness sloughing of skin may require referral for grafting (Fig. 5.21).

Recurrence of erysipelas is common, and may occur weeks or months after the first attack. A few patients suffer repeated attacks, usually in the same site. Prophylactic oral penicillin or erythromycin may prevent further attacks, but is not effective in all cases. If no attack occurs after a year of prophylaxis, it may be possible to discontinue the antibiotic.

Figure 5.21 Erysipelas: this patient made a rapid recovery on penicillin treatment, but the affected site sloughed, requiring a full-thickness skin graft.

Rare toxic complications include post-streptococcal glomerulonephritis in small children, and toxic shock syndrome in children and adults.

Necrotizing infections of skin and soft tissue

Introduction
Necrotizing infections of skin and soft tissue may be endogenously acquired, or follow penetrating skin lesions, surgical incisions and wounds. They occasionally affect decubitus or diabetic ulcers.

Localized gas-forming infections
These are often polymicrobial infections, affecting devitalized tissue. This type of infection is relatively common in diabetic ulcers and decubitus ulcers, and can complicate diabetic or ischaemic gangrene of the toes or feet.

The infection is frequently caused by facultative Gram-negative organisms (which often produce gas during carbohydrate metabolism), together with Gram-negative anaerobes, and also enterococci or anaerobic cocci. Hypoxic tissue enables the growth and multiplication of facultative organisms, which further lower the redox potential, allowing obligate anaerobes to multiply. There is inflammation and moderate gas formation at the affected site, but extension beyond the devitalized area, or systemic toxaemia, are rare.

Clinical examination may reveal slight crepitus in the inflamed area, and X-rays may show streaks of gas in tissue planes. In spite of these appearances, treatment with broad-spectrum antibiotics is often successful. Suitable treatment should include cover for Enterobacteriaceae, enterococci and anaerobes.

Treatment of localized gas-forming infections

Oral
Co-amoxiclav or amoxicillin (the latter may be combined with metronidazole).

Intravenous
High-dose amoxicillin *or* piperacillin/tazobactam (Tazocin). These penetrate tissues and are effective in anaerobic environments and against anaerobes. For beta-lactam-allergic patients: ciprofloxacin *plus* clindamycin.

Surgery
Surgery is occasionally necessary, to remove non-viable tissue, digits or limbs: it should be limited to that demanded by the pre-existing condition.

The diabetic foot
The feet of diabetics are often compromised by vascular disease and/or peripheral neuropathy. Small lesions easily develop ulcerating infection, which readily extends in the skin and underlying tissues. Infection is often polymicrobial, but the commonest predominating pathogens are *S. aureus*, *Pseudomonas* spp., anaerobes and, less frequently, enterococci or Gram-negative rods.

Management of diabetic foot ulcers
1 Send swabs or pus for bacteriological examination.
2 Empirical antibiotic treatment: initially; clindamycin *plus* ciprofloxacin.
3 Treat localized gas-forming infections as above.
4 Assess and optimize diabetes control.
5 Assess vascular function and perfusion of the affected site; seek vascular surgery opinion if indicated.

⚠ **Caution**: non-healing or recurrent breakdown of ulcers may indicate underlying invasion of bone; MR scan is the best investigation. Antibiotic treatment alone may succeed for localized infection of distal bones; digit or ray amputation are often successful if this fails. Infections of the tarsus or calcaneum may only be treatable by below-knee amputation.

Gas gangrene
This is a clostridial infection of subcutaneous tissue, particularly of muscle (clostridial myonecrosis). Most cases follow the inoculation of *Clostridium perfringens* organisms or spores into a wound or incision but several other clostridial species may also cause gas gangrene. They all produce copious amounts of gas from the metabolism of either saccharides or proteins. Mainly saccharolytic organisms include *Clostridium perfringens, C. septicum* and *C. tertium*. Mainly proteolytic organisms include *C. oedematiens* and *C. histolyticum*.

Clostridia, which are spore-bearing, anaerobic Grampositive rods, are common in faeces and soil, so that wounds acquired out of doors, in wars or in field sports are at risk. Operations at or near the perineum or, rarely, gallbladder surgery can also lead to gas gangrene. Rare, apparently spontaneous cases, especially in unusual sites such as the arm or trunk, are often the result of bloodborne seeding from a malignancy of the colon or genital system. Initiation of infection is also facilitated by the inoculation of foreign material, such as soil or surgical implants. Clostridial infections have been associated with the inoculation of contaminated heroin and other 'street' drugs.

C. perfringens, in particular, is a highly toxic organism. It produces an alpha toxin, which is a strongly haemolytic lecithinase; a necrotizing beta toxin; a similarly necrotizing epsilon toxin; and a theta toxin, which is strongly haemolytic, producing large clear zones around colonies on blood agar plates. Many other toxins are also produced (including a delta toxin, which causes rare cases of necrotizing jejunitis).

Clinical features
There is rapidly advancing swelling and mottled discoloration of a limb or other affected tissue, with obvious crepitus. The skin often contains fluid- and gas-filled blisters. The infection may produce a characteristic sickly sweet smell. The patient is feverish and hypotensive, with obvious sepsis, and may also be severely anaemic, due to toxin-mediated haemolysis.

Diagnosis
The diagnosis is mainly performed clinically and the management of the patient should not be delayed to make a laboratory diagnosis. The presence of gas in the tissues is easily demonstrated by clinical and X-ray examination. Microscopy of vesicle fluid or wound swabs shows plentiful Gram-positive rods, with surprisingly few neutrophils. The white cell count is often raised, there is often haemolysis with methaemalbuminaemia, and haptoglobins are often reduced. Fluid and wound cultures readily produce a growth of clostridia on anaerobic culture. Blood cultures should be performed, as bacteraemia can coexist.

Treatment of gas gangrene

Intravenous antibiotics

Benzylpenicillin 2.4 g 4- to 6-hourly *plus* metronidazole 500 mg 6-hourly (a broad-spectrum cephalosporin can be substituted for penicillin if Gram-negative rods may be involved); *or*, for penicillin-allergic patients: teicoplanin 400 mg 12-hourly for 3 doses, then 400 mg daily (gentamicin may be added for Gram-negative cover).

Early, aggressive surgery is important

Wide debridement, or amputation, with several revisions in the following days, may be necessary to halt the spreading infection, and to remove all infected, necrotic tissue.

Hyperbaric oxygen therapy

This has been strongly advocated but has never gained a place in routine therapy. It may save critically devitalized tissue and/or make demarcation between salvageable and necrotic tissue more evident to the surgeon.

Note that anti gas gangrene serum (AGGS) is no longer available.

Prevention

Prevention of clostridial myonecrosis is important. Benzylpenicillin or metronidazole is given prophylactically in at-risk operations such as high amputations of the leg, and after contaminated traumatic or military wounds. Adequate cleaning of wounds, removal of debris and devitalized tissue, and avoiding primary closure of severely contaminated wounds all decrease the risk of anaerobic infection.

Necrotizing fasciitis and synergistic gangrenes

Necrotizing fasciitis is a rapidly spreading infection of subcutaneous and perimuscular fat. Necrotic liquefaction of fatty tissue, due to infective thrombosis of the perforating skin arterioles, is the characteristic pathology. Many cases are caused by a mixed bacterial infection, which may include pathogens derived from the skin and bowel. A significant proportion of reported cases are due to infection with *Streptococcus pyogenes* (group A streptococcus or GAS). Injuries in marine environments can lead to necrotizing *Vibrio vulnificus* infections. Immunosuppressed patients may have lesions caused by *Pseudomonas* spp.

Severe pain in the infected site and developing features of sepsis or septic shock are accompanied by spreading inflammation of the overlying skin, which may become livid, mottled and necrotic. The diagnosis can be suspected clinically, but surgical incision and inspection of the tissues is important in confirming an early diagnosis, and should be carried out urgently. Microscopy and culture of

tissue samples can provide early indication of a streptococcal or other aetiology.

Fournier's gangrene (a full-thickness necrosis of the perineal skin that can leave the testicles denuded), and Meleney's synergistic gangrene (which affects wounds of the abdominal wall) have a similar aetiology in which infection with facultative organisms causes lowered oxygen tension and reduced redox potential, allowing obligate anaerobes to invade the tissues. These types of gangrene affect the full thickness of the skin.

Treatment of necrotizing fasciitis and synergistic gangrenes

Vigorous antibiotic therapy

• For necrotizing fasciitis due to GAS: intravenous clindamycin 600 to 1200 mg 6-hourly (first choice) *or* benzylpenicillin 2.4 g 4-hourly *plus* metronidazole 500 mg 8-hourly.
• For polymicrobial fasciitis and synergistic gangrenes or when causative organism is not known *add* ceftazidine 2 g 8- to 12-hourly *or* meropenem 1 to 2 g 8-hourly (for Gram-negative cover); adjust therapy when the aetiology is confirmed.

For all of these infections

Resuscitation and haemodynamic support as indicated (see Chapter 19).

Early, wide debridement

All affected tissue must be completely removed, to halt the advance of the gangrenous process; daily revisions are performed until no further necrosis is detected; split-skin grafting can be carried out when tissues are viable, to protect denuded areas; reconstructive surgery may be required when the infection is controlled.

Other therapies

Hyperbaric oxygen therapy and intravenous immunoglobulin have both been used in small numbers of cases; firm evidence for their benefit is currently lacking.

Clostridial infections in injecting drug abusers

Clostridia may be present in adulterated or unhygienically manufactured 'street' drugs, or in dirty spoons, syringes and needles. They replicate well in the anaerobic environment of infected lesions produced by subcutaneous and intramuscular injections ('popping'). Affected drug users may present with severe clostridial necrotizing infections or toxin-mediated disease.

An outbreak of *C. novyi* infection with necrotizing lesions and severe sepsis occurred in Britain in 2000. In 2003, between 20 and 30 cases of tetanus were treated, and there has been a steady incidence of wound botulism.

Necrotizing infections present, and are treated, in the same way as any type of necrotizing fasciitis or gangrene.

Tetanus presents with neck, back and jaw stiffness, and increasing localized and generalized spasms (see page 301). Botulism presents with blurred vision, dry mouth, increasing weakness of speech and swallowing and progressive descending paralysis (see Chapter 8). These toxin-mediated diseases can progress extremely rapidly. The precipitating tissue lesion may be trivial, but should be sought, incised and debrided. Treatment should be given with metronidazole, plus any other indicated antibiotic. Tetanus or botulism antitoxin should be given without delay.

Acute pyomyositis

Acute pyomyositis is a rare infection usually seen in the tropics. It is a pyogenic *Staphylococcus aureus* infection of muscle, frequently affecting the thigh or sometimes the spinal muscles. CT or MRI imaging, with contrast, will often show muscle oedema or later abscess formation, and permit needle aspiration of material for microscopy and culture. Pyomyositis is difficult to treat with antistaphylococcal drugs alone. Surgical drainage and debridement are usually required.

Otitis externa

This is a superficial inflammation of the skin of the external auditory meatus, often a complication of eczema. It presents with erythema, weeping and local pain, or may be similar to impetigo or furunculosis. The poorly ventilated auditory canal can be colonized by common skin-infecting bacteria, by fungi, including *Aspergillus* spp., and in swimmers and divers, whose ears are often wet, pseudomonal otitis externa is caused by water-borne organisms.

Simple cleaning and drying of the ear canal under direct vision may be sufficient to allow resolution. Short courses of corticosteroid drops may eradicate the eczematous reaction. Mild infection is often treated topically with neomycin, framycetin or clioquinol drops, but courses should be no longer than 7 days, to avoid skin sensitization or the establishment of fungi. Topical clotrimazole is useful if fungal infection is present. Staphylococcal infection is best treated with oral flucloxacillin, as for other skin infections.

Erythema chronicum migrans (early cutaneous Lyme disease)

Erythema chronicum migrans (ECM) is the cutaneous manifestation of early localized Lyme disease, caused by tick-borne borrelias (*Borrelia burgdorferi* commonest in the USA, *B. afzelii, B. garinii,* and *B. valaisiana,* more common in Europe, and probably others: see Chapter 24). A circular or discoid lesion begins and expands from the site of an infecting tick bite (Fig. 5.22). Approximately 75% of individuals with early *Borrelia* infection display ECM, 1

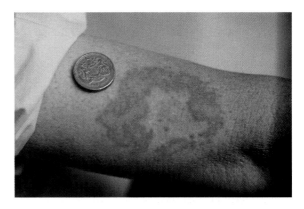

Figure 5.22 Erythema chronicum migrans expanding from the site of a tick bite on the arm. Courtesy of Dr M. G. Brook.

or 2 weeks after exposure. Some lesions are large, disappearing when they have traversed a whole limb, while occasional patients have atypical or multiple lesions.

Typical ECM is pathognomonic of Lyme disease: just over half of patients have serum IgM ELISA antibodies to *Borrelia* spp. but antibodies take up to 8 weeks to become detectable. Early treatment may prevent the appearance of antibodies. False positives occur in VDRL-positive patients, and those with autoimmune disorders. Western blotting will confirm true positive reactions. Diagnosis therefore rests on combined clinical and serological evidence. Borrelias can be cultured from blood or biopsies of lesions, but this technique is confined to reference laboratories. *Borrelia* DNA can be demonstrated in lesions by PCR, and is useful in early cases, but the combination of ELISA and western blot is the usual means of diagnosis.

Treatment is important. Although untreated ECM is self-limiting, this leaves the risk of later manifestations of Lyme disease (see Chapter 24). The treatment of choice (especially in countries where the same ticks may transmit rickettsial diseases or babesiosis) is oral doxycycline 200 mg, followed by 100 mg daily for 3 weeks. Alternatives, suitable for children are: amoxicillin 250–500 mg (child up to 10: 125–250 mg 8-hourly); erythromycin 125–250 mg 6-hourly. Treatment should be continued for 2 to 4 weeks. Azithromycin is not as effective as other macrolides.

Erythrasma

This is a superficial inflammation of skin flexures, usually caused by *Corynebacterium minutissimum* infection. The advancing flexural erythema can be mistaken for a fungal infection or for erysipelas, but it is not painful, and antifungal treatment is ineffective. Under ultraviolet illumination the lesion shows characteristic, salmon-pink fluorescence.

In the laboratory, *C. minutissimum* may be distinguished from *C. jeikeium* and *C. bovis*, which also do not utilize nitrate, hydrolyse urea or digest gelatin, by its lack of dependence on lipid in the culture medium. It produces small colonies that fluoresce red-orange under Wood's light when cultivated on serum-containing medium.

The organism is sensitive to erythromycin and tetracycline. Infections may fail to respond to oral antibiotics, but can often be eradicated by topical treatment with compound benzoic acid ointment.

Erysipeloid

Erysipelothrix rhusiopathiae is a zoonotic organism that causes erysipeloid in pigs. Human infections result from inoculation injuries, such as puncture wounds from bone splinters. A dull red erythema advances, often spreading from one finger to another via the web. Underlying joints may become sore. Systemic manifestations rarely occur.

Clinical diagnosis is usually possible from the history of exposure (often occupational), and clinical features. The infection responds rapidly to 5–7 days' treatment with oral penicillins or tetracyclines.

E. rhusiopathiae is a facultative, catalase-negative, non-sporing Gram-positive rod, which produces alpha haemolysis on blood agar. Unlike *Listeria*, it is non-motile and it produces hydrogen sulphide in Kligler's triple sugar–iron medium. The mechanisms of pathogenesis are uncertain but are thought to be related to neuraminidase production.

Cat-scratch disease

This is a lymphocutaneous disease caused by *Bartonella henselae*, a commensal of the mouth of cats. A history of cat scratch or cat exposure is common. After 5–10 days' incubation, a nodular or indurated swelling appears at the site of the scratch, and may discharge a little pus. The local-draining lymph nodes enlarge and become tender, occasionally suppurating. The disease is self-limiting, but may last for 3 weeks or more. Treatment with an oral tetracycline or macrolide may shorten the course.

B. henselae can cause systemic or bacteraemic infection, pelious hepatitis or indolent dermatoses in the immunosuppressed (see Chapter 22).

Histology of affected lymph nodes shows non-specific inflammatory reactions including granulomata and stellate necrosis. Bacilli may be demonstrated by Warthin–Starry silver staining (Fig. CS.1, p. 116) and less effectively by Gram staining. *Bartonella henselae* has been isolated in culture on freshly prepared brain–heart infusion agar containing 5 or 10% rabbit or horse blood, incubated in a humid atmosphere for up to 3–4 weeks. *Bartonella* spp. grow best on semisolid media containing rabbit heart infusion. Unfortunately they do not produce turbidity or convert enough carbon dioxide for ready detection in automated systems. Colonies on blood agar are pleomorphic. Plates must be sealed and incubated for up to 30 days in a humid atmosphere with enhanced carbon dioxide. Real-time PCR-based techniques are now sufficiently sensitive to be very useful in diagnosis.

The diagnosis is often made serologically: immunofluorescence and enzyme immunoassay techniques for the detection of IgM and IgG antibodies to *Bartonella henselae* are available through reference laboratories.

Actinomycosis

This is an infection of skin and subcutaneous tissue caused by *Actinomyces israelii* or, rarely, by other *Actinomyces* spp. The organism may invade from underlying mucosa of the mouth, pleura, genital tract or peritoneum. The commonest lesion is an abscess of the cheek.

The lesion begins as a hard, enlarging, nodule, which suppurates and discharges greyish pus containing tiny pale or yellow dots. These 'sulphur granules' are spherical colonies of the branching bacteria; they can be crushed between a slide and a cover slip and stained to demonstrate their morphology and variable degree of Gram positivity (Fig. 5.23). The organisms can be isolated by anaerobic culture but the laboratory must be informed of the suspected diagnosis as incubation should be continued for at least 2 weeks. Isolation allows accurate speciation by sequencing of the 16S gene, and sensitivity testing. The *Actinomyces* sp. is usually accompanied by coexisting bacteria, such as *Actinobacter actinomycetecomitans* or Gram-negative bacteria of intestinal origin, which may

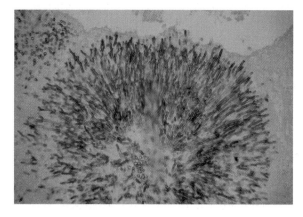

Figure 5.23 Gram-stained crush preparation of 'sulphur granule', showing a Gram-variable, branching appearance.

require concurrent treatment with additional antibiotics to ensure eradication of the infection.

Penicillin or erythromycin are effective, but high doses should be given for 6 or 8 weeks to eradicate infection. Co-amoxiclav is useful monotherapy for both *Actinomyces* spp. and the combination of accompanying organisms often found in lesions.

Cutaneous mycobacterial infections

Some 'environmental' mycobacteria and, rarely, *Mycobacterium tuberculosis* can cause skin infections that fail to respond to simple antibiotic treatment and that seem sterile on standard bacterial culture.

Mycobacterium marinum is found in pools, rivers and aquaria. Infection, probably of small skin defects, produces indolent, nodular lesions, almost always on the hand or wrist. These can spread to affect subcutaneous tissue, fascia and tendons. Extensive lesions may ulcerate and/or discharge pus. Mycobacteria can be seen and cultured in curettings or biopsy material (see Chapter 18). Treatment with rifampicin plus ethambutol or clarithromycin is often successful; debridement is sometimes required.

M. chelonei and *M. fortuitum* occasionally infect inoculation sites, for instance in diabetics taking insulin. The infection usually causes a 'sterile' subcutaneous abscess. Mycobacteria are recovered by culture of pus. Treatment with co-trimoxazole or ciprofloxacin is often curative without the need for drainage.

M. tuberculosis causes lupus vulgaris, a granulomatous lesion that appears slightly nodular, lichenified and sometimes scaling, often with an atrophic centre. It may have a natural history of years, leading to misdiagnosis as chronic dermatitis. Pressing a glass slide on the lesion to blanch it may demonstrate the granulomata as groups of translucent granules (called the apple-jelly appearance). Mycobacteria may be demonstrated, and cultured from, biopsy material.

Many patients also have pulmonary or other foci of tuberculosis. Standard antituberculosis treatment will cure the skin lesion as well as any other focus.

Propionibacterium acnes and acne

Acne is a multifactorial skin disorder, in which excessive sebaceous secretions are produced in response to strong androgenic stimulation. Sebaceous glands become engorged and blocked, causing pustules and comedones. Secondary infection may cause severe inflammation or furunculosis and contribute to later scarring. *P. acnes* can be recovered from the lesions. It is not known how the organism contributes to the pathology of acne, or whether other skin flora are also involved, but broad-spectrum antibiotic treatment can produce considerable improvement. Doxycycline is given once daily in courses of several months. Erythromycin is also often effective.

Fungal infections of the skin and its appendages

Organism list

- *Candida albicans*
- Dermatophytes
  Microsporum spp.
  Trichophyton spp.
  Epidermophyton spp.
  Malassezia furfur
- Rarities
  Sporothrix schenckii
  Blastomyces
  Histoplasma capsulatum
  Cryptococcus neoformans

Introduction

Common fungal infections (mycoses) seen in Europe are superficial, affecting only the skin, with inflammation confined to the site of infection. They are usually recognizable by their typical skin lesions, and the organisms are easily identified in swabs, scrapings or culture.

Deep mycoses are rare in temperate countries. Although some are respiratory infections (see Chapter 7), several are infections of subcutaneous tissue, acquired by inoculation through the skin. They tend to produce granulomatous lesions, which sometimes invade, either by spread to adjacent tissue or by metastasis to lymph nodes and other body sites. The lesions of deep mycoses must be distinguished from infections, autoimmune granulomata and tumours.

Candidiasis

Introduction and epidemiology
Candida albicans is a yeast found among the normal flora of the skin, mucosae and bowel. Other local flora and the health of the tissues are important in preventing *Candida* infection. Normal skin is rarely affected, but wetness and maceration of the epidermis readily favour the establishment of candidiasis. Antibiotic treatment increases the likelihood of infection by altering local flora. Mild degrees of immunosuppression, including the effects of corticosteroid or cytotoxic therapies, pregnancy, diabetes, other endocrine diseases and HIV infection can all predispose to candidiasis.

Predisposing conditions to candidiasis

1 Antibiotic treatment.
2 Corticosteroid treatment.
3 Cytotoxic therapy.
4 Diabetes mellitus.
5 Pregnancy.
6 Cell-mediated immune deficiency (including HIV-induced CMI).

Clinical features

The flexural skin under the breasts, in the natal cleft or under the abdominal 'apron' of the obese are most often affected. Candidiasis can affect the perineum of infants or incontinent adults, but is largely prevented by the use of highly absorbent pads, which dry the skin. Individuals whose hands are always damp or sweaty may develop candidal paronychia.

Flexural infection (intertrigo) produces reddening and slight thickening of the skin, with clearly demarcated, dull plaque-like lesions, which may produce a slight, sticky exudate. Drier lesions often appear shiny or flaky, may have circular satellite lesions nearby, and must be distinguished from psoriasis. Both types of lesion are often irritating or itchy.

Nailfold lesions (paronychia) cause swollen, thickened rolls of skin, which may bulge over the nail. A 'cheesy' exudate is sometimes seen in the cleft under the swelling. Inflammation of the nailbed causes ridging of the nail and rare infection of the nail itself produces an opaque greenish or brownish discoloration.

Diagnosis

The clinical features are often typical. The differential diagnosis includes erythrasma, dermatophyte infections, contact dermatitis and flexural psoriasis. Swabs from lesions or exudate can be Gram stained to demonstrate the diagnostic presence of budding yeasts. Inoculation on to Sabouraud's agar allows cultural identification of yeasts in cases of doubt. Newly introduced chrome agars can provide a presumptive identification with a characteristic coloration of the colonies – *C. albicans* develops a characteristic green colour on this agar. *Candida* will also grow on blood agar or heated blood agar, but produces tiny colonies that are easily overlooked in mixed cultures of skin organisms. Other useful identification tests include germ tube formation, specific for *C. albicans*, and auxanograph testing, which allows speciation of most isolates. Alternatively isolates can be speciated by sequencing of the r18S gene.

Management

Removing the predisposition, when possible, will aid healing and reduce recurrences. Most infections respond readily to topical treatment. Water-soluble creams are recommended, as they do not have the occlusive, macerating effect of ointments. The imidazoles – miconazole, clotrimazole and econazole – are effective; the last two are available as sprays, solution or lotion for application to large or hairy areas. The polyenes, nystatin and amphotericin, are effective and cheap.

In severe or extensive disease, or persisting predisposition, topical treatment may fail. Orally administered triazoles – itraconazole or fluconazole – may then be effective. They are contraindicated in liver disease and should be given with caution during pregnancy and lactation; they have important interactions with protease-inhibiting anti-HIV agents, terfenadine, warfarin and some other drugs. Itraconazole can precipitate cardiac failure in the elderly and those with cardiac disease or on negative inotropic therapies.

Treatment of severe *Candida* infections of skin or mucosae

1 Itraconazole 100 mg daily for 15 days (200 mg daily in immunosuppressed or neutropenic cases); *or*
2 Fluconazole 50 mg daily for 14 days (100 mg daily in refractory or immunosuppressed cases).
(Up to 4 or 6 weeks for foot infection or severe intertrigo.)

Dermatophytoses (tinea infections)

Introduction

Dermatophytes are filamentous fungi that digest keratin. Different species have varying affinities for skin, hair and nails. Although they cannot invade living tissues, their presence in the epidermis can induce an inflammatory reaction in the affected site. Some species are exclusive to humans; others are acquired by contact with infected animals, and these often cause the most severe inflammation.

Organism list

See Table 5.1.

Laboratory identification of dermatophytes

Specimens from infected sites, skin scrapings, hair and nail clippings should be sent dry to the laboratory. These are clarified by gently heating in a solution of potassium hydroxide, and examined under the microscope for the presence of typical branching hyphal elements.

Dermatophytes grow readily on many microbiological media but a useful medium is Sabouraud's dextrose agar. As all are resistant to the action of cyclohexamide, which is incorporated as a selective agent. Chloramphenicol and gentamicin can be added when bacterial contamination is likely. A specialized dermatophyte test medium incorpo-

Table 5.1 Summary of clinical and epidemiological characteristics of dermatophyte infections

	Organism	Host	Fluorescence
Scalp infections	*Microsporum canis*	Cat, kitten, dog	Positive
	M. audouinii	Human	Positive
	Trichophyton sulphureum	Human	Negative
	T. violaceum	Human	Negative
	T. schoenleinii	Human (favus)	Positive
Body infections	*T. mentagrophytes*	Animal	Negative
	T. verrucosum	Animal	Negative
Groin and foot infections	*T. rubrum*	Human	Negative
	T. interdigitale	Human	Negative
	Epidermophyton floccosum	Human	Negative
Nail infections	*T. rubrum*	Human	Negative

rates all three of these agents, with an indicator to detect the rise in pH that accompanies dermatophyte growth. Cultures are incubated at 30 °C for up to 4 weeks.

Dermatophytes are identified based on colonial morphology, microscopic appearance of the fungal hyphae and conidia, and on physiological and biochemical testing. Slide preparations of mycelia can be stained with lactophenol cotton blue and examined microscopically for the morphology of the conidia and chlamydospores. These are often characteristic for different species; an example is seen in Fig. 5.24. Other tests include the ability of the fungal isolate to penetrate an uninfected hair, to hydrolyse urea and to produce characteristic growth on rice grains.

Clinical features

The usual appearance of dermatophyte infection is an expanding lesion with a scaly or inflamed advancing edge.

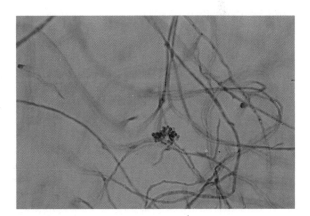

Figure 5.24 Lactophenol blue-stained preparations of dermatophytes, showing the characteristic morphology of conidia.

This is a typical tinea or ringworm infection. The eruption is described by its position on the body, e.g. tinea capitis, tinea corporis or tinea cruris.

Scalp ringworm (tinea capitis)

Scalp ringworm is common in children, usually caused by a *Microsporum* species. One or more oval patches of hair loss, and skin scaling, expand steadily and can affect the whole scalp. The damaged hairs are broken off near to the skin (unlike alopecia, in which there is no scaling, and hairs are absent in the acute stage).

Trichophyton species can infect both adults and children. Scalp swelling is often marked and hairs broken off at the skin surface appear as black dots. A purulent exudate from the follicles may cause hairs to be completely shed. This terminates the infection, but can leave scarring of the scalp and follicles, with permanent hair loss.

The clinical features strongly suggest the diagnosis. *Microsporum* infections cause a greenish-blue fluorescence of affected hairs and skin under ultraviolet light (Wood's light). Scales and plucked hair stumps can be examined by microscopy and culture.

Topical treatment is difficult to apply and rarely succeeds. Oral terbinafine in a 2-week course is effective; adult dose: 250 mg once daily, child over 1 year (10–20 kg): 62.5 mg daily; body weight 20–40 kg: 125 mg daily. Griseofulvin is licensed for children; dose 10 mg/kg (up to 500 mg) daily; a course of 6 weeks or more is usually needed. The need for further treatment may be reduced by applying miconazole or clotrimazole cream. Oral itraconazole and fluconazole are licensed for treating adults.

Re-infection can be prevented by seeking and treating any reservoir of infection, which may be another child or adult, a family pet or a farm animal.

Ringworm of the body (tinea corporis)

Ringworm of the body is usually an obvious round or oval

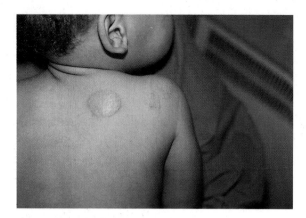

Figure 5.25 A typical lesion of tinea corporis (the family cat was also affected).

expanding lesion with a scaly, slightly inflamed periphery (Fig. 5.25). It must be distinguished from discoid eczema and isolated lesions of psoriasis. Examination and culture of scrapings will confirm the diagnosis.

Treatment with topical antifungals such as clotrimazole or miconazole cream is adequate for most mild lesions. Oral itraconazole is effective in a dose of 100 mg daily for 15 days or 200 mg daily for 7 days. Extensive or severely inflamed lesions may be best treated with a 3- or 4-week course of oral griseofulvin.

Ringworm of the groin (tinea cruris)

Ringworm of the groin is often intertriginous, affecting the inguinal folds and adjacent thighs. *T. rubrum* infection, however, can cause extensive lesions spreading down the thighs and posteriorly to the natal cleft and buttocks. Inadvertent treatment with topical corticosteroids partly inhibits the inflammation, producing an ill-defined papulopustular lesion (tinea incognita). This can also occur on the face in similar circumstances. Microscopy and culture of scrapings will make the diagnosis. The important differential diagnoses are intertriginous candidiasis or psoriasis, and erythrasma.

Oral terbinafine 250 mg daily for 2 to 4 weeks is effective; itraconazole is an alternative – the dose is as for tinea corporis. Topical antifungal treatment often causes skin irritation, especially if the skin is intensely inflamed; if topical treatment is necessary, dilute potassium permanganate soaks can be applied twice a day for 3 or 4 days before commencing a weak corticosteroid cream and topical antifungal. The steroid can be discontinued when the inflammation subsides.

Ringworm of the hands, feet and nails

Dermatophyte infections of thick keratin produce extremely scaly lesions that resist treatment with topical an-

tifungals. The interdigital webs may become fissured and macerated, with scaly infection extending on to the digits or the dorsum of the foot (or hand). *Trichophyton* spp. can cause extensive, slipper-shaped infection on the dorsum of the foot. Nails become opaque, discoloured and brittle, starting at the tip, gradually affecting the lateral margins and then the whole nail plate, which may flake away. The nailfolds do not swell, as they do in candidiasis. The main differential diagnoses are psoriasis and contact dermatitis.

Treatment must be systemic and prolonged. Terbinafine 250 mg daily for 2–4-weeks is effective in palmar and plantar disease, as well as early nail infections. It must be given for 6–12 weeks for severe nail infections (longer in some toenail infections). It is well tolerated, apart from occasional gastrointestinal side-effects. Rare allergic rashes and liver toxicity have been reported. Itraconazole 100 mg daily for 30 days may be effective in treating the hands or feet; for toenail infections the dose is 200 mg daily for 3 months (pulses of 200 mg twice daily for 7 days can be used, repeated after 3 weeks; two pulses for fingers, three for toes). Oral griseofulvin for 6 months is the minimum for nail infections; 12 months or more is usual for toenails. As griseofulvin enhances the effects of alcohol, and is teratogenic, many people prefer not to use it for toenail infections.

Treatment for fungal infection of the nail

1 Terbinafine 250 mg daily for 6 weeks–3 months (not recommended for children).
2 Itraconazole 200 mg daily for 3 months *or* 200 mg twice daily for 7 days, repeated after 21 days; two repeats for fingers, three for toes.
3 Griseofulvin 500 mg (child 10 mg/kg) daily for several weeks or months until cure is complete. Beware of its Antabuse-like effect.

⚠ **Caution**: some adverse effects of antifungal drugs
• Itraconazole: liver disorder – check liver function tests if there is history of liver disease or for courses longer than 1 month; rare heart failure in the elderly or those on negative inotropic therapy; enhances effect of warfarin, acenocoumarol, cyclosporin, sirolimus, tacrolimus and indinavir; increased risk of severe dysrhythmias with QTc-prolonging drugs; increases risk of myopathy with simvastatin and atorvastatin; some reports of oral contraceptive failure.
• Fluconazole: less liver toxicity than itraconazole: enhances effects of warfarin, acenocoumarol and rifabutin (with risk of uveitis); enhances action of sulphonylureas; rare reports of oral contraceptive failure.
• Griseofulvin: enhances the effect of alcohol and increases the degradation of warfarin and oral contraceptives. Women should avoid conception for 1 month and men for 6 months after therapy.

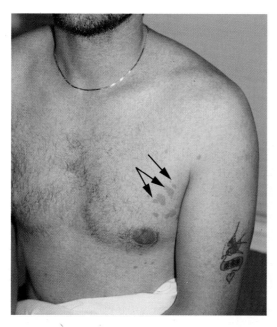

Figure 5.26 Pityriasis versicolor: this appears red because the patient has a slight fever.

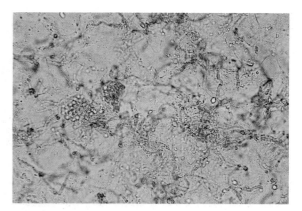

Figure 5.27 Potassium hydroxide preparation of scraping from a case of pityriasis versicolor (stained with blue ink (Quink)).

Pityriasis versicolor

Pityriasis versicolor is a superficial skin infection caused by the filamentous fungus *Malassezia furfur*. Pale brown, fine scaly macules develop on the upper chest or back, forming an irregular pattern, which appears pale brown in a white-skinned person, or slightly pale in a dark skin. There is little or no inflammation, and sensation is normal in the affected areas. When the skin is warm after a bath or during a feverish illness, the plaques may become red or pink (Fig. 5.26).

The clinical appearance is characteristic, but the diagnosis can be confirmed by microscopy. Skin scrapings are placed on a microscope slide, and mixed with a drop or two of 5% potassium hydroxide. This reveals a mycelium in which are scattered groups of rough, round sporing bodies. This typical 'meat balls and spaghetti' appearance is diagnostic (Fig. 5.27).

Almost any topical antifungal cream will clear the lesions, but recurrence is common. Washing all clothes and shampooing the hair with selenium sulphide (Selsun) shampoo may help to remove a reservoir of infection. Oral itraconazole 200 mg daily for a week is also effective.

Sporotrichosis

Sporotrichosis is a localized, nodular skin infection caused by *Sporothrix schenckii*, a ray fungus that produces characteristic stellate microcolonies *in vitro*. The fungus exists in wood, soil and vegetation. Infection is usually by inoculation, and leads to painless nodules, abscesses or ulcers, which often itch and may expand locally. Satellite lesions often appear along lymphatic pathways, and draining lymph nodes may be affected. Haematogenous spread occasionally occurs in debilitated or immunosuppressed individuals.

Differential diagnoses include mycobacterial infections (fishtank granulomata) or rare syphilitic lesions. Histology and culture of biopsy material are diagnostic.

The treatment of choice for sporotrichosis is oral itraconazole 100–200 mg daily for 3–6 months. Terbinafine 125 mg daily has also be shown effective. Where these are not available, saturated potassium iodide solution is effective; starting dose 1 ml 8-hourly, increasing to 4 to 6 ml 8-hourly and continuing for 1 month after resolution (total courses of 2 to 5 months are usual). For systemic disease, itraconazole is the treatment of choice.

Pityriasis rosea

Pityriasis rosea is associated with seroconversion to human herpesvirus type 7 (HHV-7). A 'herald patch' appears as an inflamed, scaly, rather scabby lesion up to 4 cm in diameter on the trunk or upper leg (Fig. 5.28). After 3 to 14 days a symmetrical eruption of oval plaques appears, with smaller macules and papules, which are sometimes itchy. The plaques affect the trunk and proximal limbs ('vest and pants' distribution), and have their long axes aligned with the skin creases, giving the rash a typical 'Christmas tree' appearance (Fig. 5.29).

Each plaque, or medallion, has an indistinct, slightly raised margin with central arrays of pointed scales, with the points orientated towards the edge of the lesion. The lesions expand up to 5 cm long in about 2 weeks, and then fade over about 2 months.

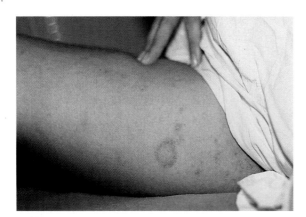

Figure 5.28 Pityriasis rosea: herald patch on the thigh.

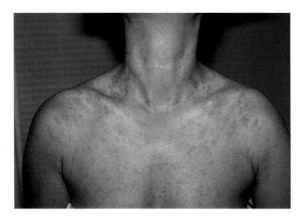

Figure 5.29 Pityriasis rosea: typical 'vest and pants' rash, with 'Christmas tree' orientation of lesions.

The differential diagnosis is of tinea or discoid eczema or, rarely, secondary syphilis or HIV seroconversion. There is no specific treatment. The condition is not contagious, and second attacks are extremely rare.

Rarities

Fungal infections of the skin are more common in the tropics than in temperate countries. Several exotic fungi can produce nodular or granulomatous lesions on the skin, typically of the lower legs. Diagnosis is by histology and culture of biopsy material. A positive complement-fixation test is usual in histoplasmosis, which is often a systemic disease with risk of lymph-node, buccal mucosa and lung involvement. Expert advice should be sought about management, which may be difficult and is always prolonged.

Mycetoma and 'Madura foot'

A chronic granulomatous infection caused by fungi (eumyc-etoma) or bacterial actinomycetes (actinomycetoma). Rarely seen in Western countries, but endemic in the tropics and subtropics.

- Begins with traumatic subcutaneous inoculation of organisms; slowly extends from an indurated nodule into a mass invading subcutaneous tissues and even bones, with many draining sinuses discharging pus, blood and 'fungal' grains. Can affect the hand, arm, buttock, perineum, leg or foot, rarely the chest or back. Satellite lesions can follow haematogenous or lymphatic spread.
- True fungal grains may be black, white or yellow (*Aspergillus flavus* can produce green grains), bacterial grains may be white, yellowish, pinkish or red.
- Diagnosis is by histopathology of deep biopsies, examination of grains from pus (see Fig. 5.23), or PCR of tissue preparations for *Madurella mycetomatis* (culture and speciation are difficult for fungi: anaerobic actinomycetes may not grow in cultures).
- Management: eumycetoma needs combined surgery (to remove isolated lesions or aggressively excise and debride larger, complex lesions) plus antifungal drugs such as itraconazole 400 mg daily; treatment may need to continue for years, with liver function monitoring: actinomycetoma can often be treated with antibiotics alone (dapsone plus streptomycin, rifampin or pyrimethamine *or* amikacin plus co-trimoxazole).

Parasites of the skin

Scabies

Scabies is caused by the mite *Sarcoptes scabiei*, which burrows in the epidermis, the female mites laying eggs along their tracks. It is infectious by skin-to-skin contact; mites being attracted by warmth. It is easily transmitted between sexual partners, those who share beds, or to the carers of infected individuals.

The infection itself is asymptomatic, but hypersensitivity to mites, their eggs or surface proteins eventually causes intense irritation of affected skin. The soft skin of the flexures, digital webs, perineum and axillae is most affected; the face is spared, except in small children and severe infections. So-called Norwegian scabies is caused by an aggressive strain of mite that causes severe lesions, even affecting the face in adults. Scabies can cause widespread chronic lesions in immunosuppressed HIV patients.

The diagnosis is suggested by the typically distributed, itchy rash. Burrows may be visible, with a tiny, pearly nodule at the advancing end; this can be teased out and shown to be a mite. Burrows are often destroyed by scratching, but skin scrapings may still contain round, black dots of mite faeces.

Treatment is usually with topical acaricides. All members of a household should be treated, and the medication should be applied to the whole skin excluding the face. Lindane 1% lotion or cream is inexpensive; malathion 0.5% lotion or liquid may succeed if lindane fails. Permethrin 5% cream is also available, and can be applied to the face if necessary. Sulfiram solution may be used for treating children. Most cases will respond to one application of acaricide, washed off after 12–24 h. If lindane is used, a second application should be given after 1–3 days.

Itching may persist for many days after successful treatment. It is often ameliorated by topical crotamiton cream. Ivermectin, orally, 200 μg/kg, combined with topical treatment is often curative in immunosuppressed patients and in 'Norwegian' scabies.

Case study 5.1 Complicated cat bite

History

A 31-year-old woman was bitten on the left first metacarpophalangeal joint while removing a stray cat from her garden shed. Rabies is not endemic in Britain, and she had recently had a tetanus toxoid booster immunization, so she cleaned the puncture wounds with proprietary disinfectant solution and covered the site with a waterproof adhesive dressing. Thirty-six hours later she attended the emergency department because of increasing redness, aching and irritation at the site. She was previously well, and took no regular medications.

Physical findings

The finger was swollen from the dorsum of the hand to the first interphalangeal joint, and the overlying skin was shiny and dusky red. Four puncture wounds were present on the dorsal and medial aspect of the metacarpophalangeal joint, and yellowish pus was visible in three of them. There was no associated lymphangitis or lymphadenopathy.

Questions

- Are cat bites common?
- Do they commonly become infected?
- Which bacterial pathogens are likely to be identified?
- Are any of these pathogens potentially invasive?
- What antimicrobial chemotherapy can be recommended?

Management and progress

Animal bites are extremely common; over 250 000 accident and emergency consultations yearly are caused by them, and many more general practitioner consultations take place. Over 90% are for cat and dog bites. About 80% of cat bites are puncture wounds and two-thirds affect the hand. About half of dog bites are puncture wounds, the remainder are lacerations or mixed injuries, and about half affect the hand. Children are more likely to suffer severe bites, and bites on the face and head. Various studies show that 30–80% of cat bites become infected, compared with 5–20% of dog bites. A mixed aerobic and anaerobic flora can be identified in half of infected bites, in 40% only aerobic or facultative organisms are found, and the remaining infections are purely anaerobic (see Table CS.1).

Many of these organisms are sensitive to benzylpenicillin or phenoxymethylpenicillin, but *S. aureus*, *Moraxella* and some *Pasteurella* species produce beta-lactamase enzymes. Bites from humans are often infected with a wide range of organisms, including anaerobes. The treatment of choice for infected animal and human bites is therefore co-amoxiclav. Clarithromycin or cefuroxime axetil are suitable second choices.

Pasteurella multocida can cause invasive infection in debilitated individuals. Bacteraemias and endocarditis have been described in patients with alcoholic liver cirrhosis, those receiving chemotherapy or transplant immunosuppression and, rarely, as an opportunistic infection in AIDS. *Capnocytophaga canimorsus* is also recognized as potentially invasive.

Table CS.1 Bacteria commonly isolated from infected dog and cat bites.

Type	Details
Aerobic and facultative organisms	
Pasteurella species	*P. septica* and *P. multocida*, most often from cat bites; *P. canis*, mostly from dogs
Streptococci	*S. pyogenes*, *S. milleri* types, and viridans streptococci
Staphylococci	Half of these are *S. aureus*, probably intrinsic to the host, the remainder are coagulase-positive and -negative staphylococci of animal and human origin
Neisseria species	Several species that are pharyngeal commensals of animals
Moraxella species	Pharyngeal commensals of animals
Capnocytophaga species	Pharyngeal commensals of animals
Anaerobic organisms	A range of organisms including *Fusobacterium*, *Bacteroides*, *Prevotella*, *Porphyromonas* and *Propionibacterium* species of animal or human origin

The patient was given a 5-day course of oral co-amoxiclav 250 mg three times daily. She re-presented 2 weeks later, saying that after initial slight improvement, the finger had remained inflamed, and she had developed a tender swelling in the left axilla. Examination revealed improvement in finger swelling, but persistent discharge of pus from the puncture sites and a 3 × 2.5 cm tender, mobile lymph node in the left axilla.

Questions

- What less common organism could cause ulceroglandular infection?
- Are debilitated and immunocompromised individuals at high risk from this infection?
- What treatment is available?

Further management and progress

Lymph-node biopsy (Fig. CS.1) showed typical inflammatory changes with cords of active mononuclear cells, and intracellular rod-shaped bacteria demonstrated by silver staining (immunological staining techniques are also available). Although rare, cat-scratch disease classically follows a cat scratch or bite. It is caused by *Bartonella henselae*, an obligately intracellular Gram-negative rod, which is a member of the family of Rickettsiaceae. A serum enzyme-linked immunosorbent assay (ELISA) test, available from reference laboratories, is positive in over 85% of infected patients. Ulceroglandular tularaemia, caused by *Francisella tularensis*, can cause a similar presentation. It results from rodent exposure and occasionally from cat bites in wooded or forested terrain, but is not endemic in

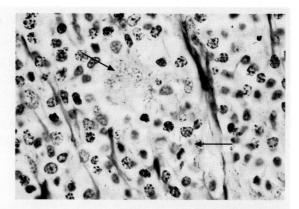

Figure CS.1 Lymph-node biopsy examined with Warthin–Starry silver staining, showing many active cell nuclei, and groups of intracellular rod-shaped bacteria (arrows).

Britain. It tends to affect hunters and trappers, particularly in North America.

B. henselae infection can cause indolent skin and reticuloendothelial disease in the immunosuppressed, and has particularly been described in AIDS, in which an angiomatous dermatitis or hepatic peliosis (bacillary angiomatosis) may occur.

This patient was treated with doxycycline 100 mg daily for 2 weeks, resulting in gradual healing of the skin lesions. The lymphadenitis resolved over the following 3 weeks. Serological tests showed antibodies to *B. henselae* in a titre of 1 : 256, considered to indicate recent infection.

Upper Respiratory Tract Infections

6

Introduction

The upper respiratory tract comprises the conjunctiva, nose, paranasal sinuses, middle ear, nasopharynx, oropharynx and laryngopharynx. It is largely covered with ciliated columnar epithelium. Exceptions are the oropharynx, vocal cords, upper posterior epiglottis and mastoid antrum of the middle ear, which are lined with stratified squamous epithelium. The conjunctiva is also composed of stratified squamous epithelium, continuous with and similar to the epithelium of the cornea.

The adenoids and tonsils are important structures of the upper respiratory tract. They are lymphoid organs whose surfaces are marked by many deep clefts, both macroscopic and microscopic.

The whole upper respiratory tract is colonized by a variety of normal flora.

Normal upper respiratory tract flora
1 *Streptococcus pneumoniae.*
2 Anaerobic and microaerophilic streptococci.
3 *S. 'milleri'* group of streptococci (found in the sinuses).
4 *Haemophilus influenzae.*
5 Other *Haemophilus* species.
6 Diphtheroids.
7 Coagulase-negative staphylococci.
8 *Staphylococcus aureus.*
9 *Moraxella catarrhalis* and *Neisseria* spp.
10 *Prevotella melaninogenicus* and related species.

Temporary colonization of the pharynx, nose or eye by potential pathogens is also common, and may provide an important reservoir of infection, for instance with *Neisseria meningitidis* or *Corynebacterium diphtheriae*. Similarly, highly transmissible viruses, such as rhinoviruses, paramyxoviruses, enteroviruses, adenoviruses and myxoviruses, can infect the nasopharynx, mildly or asymptomatically.

Latent viruses may be intermittently shed from the pharynx. Herpes simplex virus, other human herpesviruses, such as HHV-6, Epstein–Barr and cytomegalovirus virus are the most important of these.

The environment of the upper respiratory tract is varied, and different areas are susceptible to infection with different pathogens. While most infections are of surfaces, the middle ear and the paranasal sinuses are hollow structures with narrow outlets (the ostia of the sinuses and the Eustachian tubes of the middle ears) whose obstruction leads to loculated infection and abscess formation. The soft tissues of the fauces, surrounding the tonsils, are susceptible to abscess formation if severely inflamed.

Conjunctivitis and keratoconjunctivitis

Introduction

The conjunctiva is often inflamed during infections of the

respiratory tract, such as colds, influenza and measles. It is exposed to many air-borne infections, but the washing action of the tears discourages the establishment of infection. Tears contain a number of substances, including lysozymes and immunoglobulins, that inhibit pathogens. Nevertheless, a number of organisms commonly cause primary conjunctivitis. Conjunctival infections are easily transmitted directly from eye to eye by fingers, by fomites such as ophthalmological instruments, or shared face towels and, in conditions of poor hygiene, by flies. When the cornea is involved, the condition is called keratitis or keratoconjunctivitis.

Occlusion of the conjunctiva by contact lenses increases the likelihood of infection, and poor lens hygiene can lead to severe pseudomonal or amoebic infections with the risk of severe corneal damage.

Organism list

- Adenoviruses
- Enteroviruses (especially type 30)
- Herpes simplex virus
- *Staphylococcus aureus*
- *Moraxella lacunata*
- *Streptococcus pneumoniae*
- *Haemophilus influenzae*
- *Neisseria gonorrhoeae* and *N. meningitidis*
- *Chlamydia trachomatis*
- *Pseudomonas aeruginosa**
- *Acanthamoeba* spp.*
- *Naegleria* spp.*

*These organisms are particularly associated with contact lens use.

Clinical features

The eye feels sore and itchy, and there is a discharge of watery, mucoid or purulent material, which may dry, especially during sleep, and glue the eyelids together. Severe infection may cause swelling of the eyelids, which further inhibits eye-opening and drainage of secretions.

The conjunctiva appears red, often with a thin, clear outline surrounding the iris. Occasionally marked swelling causes it to bulge through the palpebral fissure.

Contact lenses exacerbate the inflammation, and may increase susceptibility to conjunctival and corneal infections, especially if not hygienically maintained. Soft or hydrophilic contact lenses may interfere with the access of drops to the eye, and can be damaged by the chemicals in eye drops. Their use should be discontinued during treatment for eye infections.

Childhood conjunctivitis

Childhood conjunctivitis (pink eye) mainly affects small children, and easily spreads in families and school com-

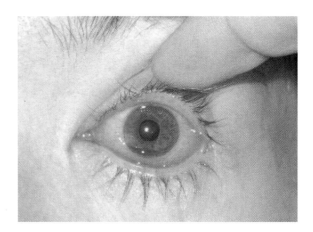

Figure 6.1 Conjunctivitis, showing red eye with sparing of an area around the cornea. (Courtesy of Dr K. Dansingani, Royal Free Hospital.)

munities. It begins unilaterally and often spreads to the other eye, but is usually mild, with a natural history of a few days. Respiratory strains of adenovirus (often type 3 or 7) are the common causes (Fig. 6.1).

'Shipyard eye'

Shipyard eye is acute keratoconjunctivitis spread by ophthalmological equipment. Often caused by adenovirus type 8, 19 or 37, it was once common in occupational settings such as shipyards, where minor eye trauma and frequent clinic visits occurred. Hand-washing by staff, and use of sterile or disposable equipment, control its spread in modern clinics.

Haemorrhagic conjunctivitis

Haemorrhagic conjunctivitis caused many epidemics worldwide in the early 1980s. The agent was enterovirus type 30. The disease was abrupt in onset, moderate to severe, and associated with intense, haemorrhagic inflammation of the conjunctiva.

Diagnosis of conjunctivitis

The clinical diagnosis is usually evident. Important differential diagnoses for a red, painful eye include herpes simplex keratitis (dendritic ulcer: see below) and acute glaucoma. Both of these can be sight-threatening, and should be considered when a red eye is severely painful.

In infants the lacrimal sac may drain poorly, causing swelling and a mucus discharge at the inner canthus. The condition is non-infectious, harmless and self-limiting. Lacrimal sac drainage by digital compression abolishes the 'sticky eye' and can be discontinued after 1 or 2 weeks.

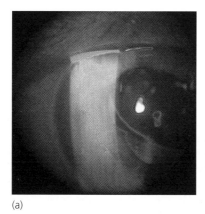
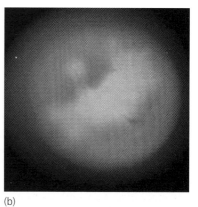

Figure 6.2 (a) Dendritic ulcer viewed by slit lamp with fluorescent contrast: the linear, meandering ulcer is seen at the edge of the cornea. (b) Large dendritic ulcer with many linear branches, occurring in an immunosuppressed patient. (Courtesy of Dr K. Dansingani, Royal Free Hospital.)

(a)   (b)

Management

Viral conjunctivitis usually resolves spontaneously in a few days. Failure to respond should prompt investigation with swabs for bacterial and viral culture, a search for chlamydial infection, and arrangements for slit-lamp examination with and without fluorescein staining, to exclude herpetic keratitis.

> **⚠ Cautions**
> **1** A red eye unresponsive to antibiotic treatment should not be treated with corticosteroids until the possibility of herpes simplex keratitis has been ruled out.
> **2** A red eye accompanied by severe pain or headache could indicate acute glaucoma: seek an ophthalmological opinion without delay.

Adenoviruses

Adenoviruses are double-stranded linear DNA viruses (70 nm) that are classified into five subgenera and at least 51 serotypes. Infection can produce either a lytic cycle, with cell destruction and release of infectious virus, or latent or chronic infection, usually in tonsillar or lymphoid tissue. Alternatively, in 'oncogenic transformation' only the early replicative stages occur and the viral DNA is integrated into host cell DNA without production of infectious virions. Replication-incompetent adenoviruses are used in research as vectors for gene insertion.

Herpes simplex keratitis (dendritic ulcer)

This is a progressive infection of the corneal epithelium that presents as a red, painful eye. It produces an ulcer, which is often complex or branching.

The diagnosis is suggested by a persistently red eye, unresponsive to topical antibiotic treatment *and that progresses rapidly if treated with topical corticosteroids.* The linear, meandering ulcer can be seen on slit-lamp ex-

amination, or by inspection after the instillation of fluorescein drops (Fig. 6.2). Untreated herpes simplex ulcers erode the corneal epithelium, leading to scarring (Fig. 6.3) and visual impairment. Diagnosis is confirmed by demonstration of herpes simplex virus antigen (by direct fluorescence staining) or DNA (by polymerase chain reaction, PCR) in corneal scrapings, or by viral culture.

The treatment of choice is topical aciclovir ointment. Treatment is continued until healing is complete. Follow-up by an ophthalmologist is important, both to monitor healing and to detect residual corneal scarring.

> **Treatment of herpes simplex keratitis**
> **1** Topical aciclovir 3% eye ointment five times daily (continue for at least 3 days after complete healing) *or*
> **2** Topical ganciclovir 0.15% in gel basis, five times daily until complete corneal re-epithelialization, then three times daily for 7 days.

Bacterial conjunctivitis

Bacterial conjunctivitis may occur alone, or complicate

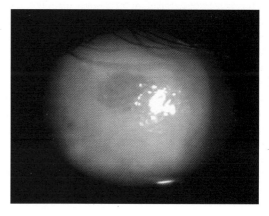

Figure 6.3 Loss of corneal epithelium, and scarring following herpes simplex keratitis. (Courtesy of Dr K. Dansingani, Royal Free Hospital.)

upper respiratory infections; common causes are *Staphylococcus aureus*, *Haemophilus influenzae* and *Streptococcus pneumoniae*. Purulent exudate is produced, often forming crusts at the inner canthus. The condition is usually mild and often self-limiting.

Unusual bacteria affecting the conjunctiva include *Moraxella lacunata*, which causes indolent or subacute infections (often in outbreaks where spread is by fomites or unwashed hands), and *H. aegyptius*, more common in tropical climates: this causes an aggressive infection and may also spread by the droplet route. *Pseudomonas aeruginosa* can cause keratoconjunctivitis, with blurred vision, in contact lens wearers. Infection is derived from unsterile cleaning fluids or from inappropriate use of stored tap water.

Chloramphenicol eye drops are the treatment of choice for mild and moderate bacterial conjunctivitis. Chloramphenicol ointment can be used at night. The course should rarely be longer than a week. The risk of agranulocytosis from topical use is now thought to be negligible. Specific bacterial infections may be treated with other appropriate antibiotics, for instance, *Pseudomonas* infections with ciprofloxacin drops, or staphylococci with fusidic acid. Gentamicin drops are also available for treating *Pseudomonas* conjunctivitis.

Severe, unresponsive or ulcerating eye infections require specialist management, which may include subconjunctivally injected and/or parenteral antibiotics.

Treatment of bacterial conjunctivitis

1 Chloramphenicol 0.5% eye drops at least 2-hourly, then four times daily when infection is controlled. Continue for 48 h after healing. Chloramphenicol 1% eye ointment may be used instead of drops at night, or alone in a dose of three or four times daily.

2 For staphylococcal infections: fusidic acid 1% drops in a gel basis twice daily.

3 For *Pseudomonas* infections: ciprofloxacin 0.3% *or* ofloxacin 0.3%, eye drops, in the same regimen as for chloramphenicol drops (*for corneal ulcer*, use ciprofloxacin every 15 min for 6 h, then every 30 min for that day, every hour the second day, then every four hours for days 3–14, *maximum course: 21 days*).

4 *Alternatives for superficial conjunctivitis:* gentamicin 0.3%, neomycin 0.5%, or framycetin 0.5%, may be used in the same regimen as for chloramphenicol *or* lomefloxacin 0.3%, every 5 minutes for five doses, then twice daily.

⚠️ **Caution**: Ciprofloxacin, ofloxacin and lomefloxacin are well-absorbed: in pregnancy their safety is not established; they should be used only if an expert risk assessment has been made.

Neonatal conjunctivitis

Neonatal conjunctivitis is an intrapartum infection, derived from the colonized or infected maternal birth canal.

Causes of neonatal conjunctivitis

- Onset at 2 to 4 days: *Neisseria gonorrhoeae*.
- Onset at 3 to 10 days (often unilateral): *Chlamydia trachomatis*.
- Onset at 2 to 16 days: herpes simplex.

Although these infections will respond to topical treatment, there may be coexisting infection of the respiratory, alimentary or genital tracts of the infant (see Chapter 17). Chlamydial pneumonia commonly occurs in neonates presenting with chlamydial conjunctivitis. Systemic treatment is therefore given.

Prophylaxis of neonatal bacterial conjunctivitis is possible, using chlortetracycline or chloramphenicol eye ointment, in a single application.

Treatment of neonatal conjunctivitis (see also Chapter 17)

1 Gonococcal: ceftriaxone 25–50 mg/kg; single intravenous dose.

2 Chlamydial: erythromycin 50 mg/kg daily, orally, in four divided doses for 14 days.

3 Herpes simplex: aciclovir 50–100 mg orally four times daily for 5 to 10 days (may be supplemented by topical treatment; then use lower oral dose).

 In all cases, both parents should be offered investigation and treatment for urogenital infections.

Chlamydial conjunctivitis and trachoma

Oculogenital strains of *Chlamydia trachomatis* commonly cause conjunctivitis, as well as colonizing the genital tract. Subacute conjunctivitis, unresponsive to chloramphenicol treatment, spreads between sexual contacts and from eye to eye. A rapid diagnosis is made from eye swabs, either by detecting chlamydial lipopolysaccharide (LPS) group antigen, using enzyme immunoassay (EIA), or detecting chlamydial DNA by PCR or ligase chain reaction. *Chlamydia trachomatis* inclusion bodies can be seen by microscopy of epithelial cells in conjunctival scrapings, but this is now rarely done. Oculogenital chlamydiae can be distinguished from one another and from the serotypes that cause trachoma and lymphogranuloma venereum, by serotyping. This is based on outer membrane proteins, nowadays identified by molecular methods.

Chlortetracycline eye ointment is the treatment of choice. Oral treatment, e.g. erythromycin or tetracycline, may be added. Investigation and treatment of the patient and sexual partner for genital infection are also necessary (see Chapter 15).

Trachoma

Trachoma is a disease of crowding and poor hygiene that predominantly affects poor, marginalized and displaced communities. It is precipitated by persisting or repeated infection with *C. trachomatis*, spread from eye to eye by unwashed hands, and possibly by flies. Untreated infection and repeated super-infections may lead to the formation of a plaque of vascular inflammatory tissue (pannus), which deforms the eyelid. Scarring, leading to entropion and trichiasis, is an important cause of corneal damage, scarring and blindness.

For the treatment of trachoma, community wide application of single dose oral azithromycin has been found as effective as 6 weeks of once- or twice-daily tetracycline eye ointment administered under supervision. Azithromycin is therefore more likely to be effective, if the use of tetracycline eye ointment cannot be supervised.

> ### World Health Organization (WHO) SAFE strategy for combating trachoma
> **S**urgery for in-turned lids.
> **A**ntibiotics for active disease.
> **F**ace washing.
> **E**nvironmental improvement, to reduce transmission.

Amoebic keratoconjunctivitis

Amoebic keratoconjunctivitis is a rare condition associated with using contact lenses or with environmental exposure (e.g. to water in hot springs). Contact lens cleaning fluid can become colonized by free-living amoebae, usually *Acanthamoeba*, but occasionally *Naegleria* spp., which are then repeatedly inoculated into the eye. The resulting severe, ulcerating keratitis is difficult to treat and often damages vision. The treatment of choice is propamidine isetionate (Brolene) 0.1% eye drops four times daily or 0.15% dibromopropamidine isetionate ointment once or twice daily. These may be used in combination with chlorhexidine and neomycin eye drops (specialist supervision required). Control is by the use of only sterile cleaning fluids, which are discarded after use.

Infections of the middle ear

Acute otitis media

Organism list
- Many respiratory viruses
- *Streptococcus pneumoniae*
- *Haemophilus influenzae*
- *S. pyogenes*
- *Staphylococcus aureus*
- *Chlamydia pneumoniae*
- *Mycoplasma pneumoniae*
- *Moraxella catarrhalis*

Introduction
Childhood otitis media
Infection of the cavity of the middle ear (otitis media, or OM), is a common condition in small children. It causes pain, reddening and opacification of the tympanic membrane, sometimes with mild fever. Many cases of OM are probably viral. *Chlamydia pneumoniae* is also recognized as a cause. Symptomatic treatment with simple analgesics, such as paracetamol, is usually adequate, as the disorder resolves spontaneously in 48 to 72 hours. Several studies have shown that the time to recovery is rarely shortened by antibiotic treatment. Myringotomy and culture of middle ear contents do not lead to more effective treatment, and are not recommended.

Acute suppurative otitis media
Acute suppurative otitis media (ASOM) can affect individuals of any age. It is an aggressive infection, with fever, livid inflammation of the eardrum, a visible, whitish fluid level, or early tympanic rupture and frankly purulent discharge. It may be spontaneous, but often complicates a respiratory infection, in which respiratory tract organisms ascend the Eustachian tube to infect the obstructed or virus-inflamed cavity. ASOM is a common complication of influenza, and a major complication of measles. Causative organisms include *Streptococcus pneumoniae*, *H. influenzae*, *S. pyogenes* and *Staphylococcus aureus*. Swabs should be obtained of any discharge from the ear. Antibiotic therapy should be commenced without delay, to minimize damage and scarring of the eardrum.

> ### Management of otitis media
> **1** For mild pain and reddening of drum, with no fluid level or other features, or with general upper respiratory symptoms: analgesics, and decongestant if indicated. Review if persistent or worsening.
> **2** Evidence of *C. pneumoniae* or *M. pneumoniae* infection (see Chapter 7): systemic erythromycin (or tetracycline in an adult) may be beneficial.
> **3** ASOM, with severe pain, discharge or important precursor such as measles or influenza: ideally, obtain pus for bacterial culture; oral antibiotic treatment should include an antistaphylococcal spectrum. Co-amoxiclav, cephradine, clarithromycin or azithromycin are appropriate. Analgesia is essential.

Complications
Mastoiditis
Mastoiditis is the extension of pyogenic middle ear infec-

tion into the mastoid antrum. If treated early with antibiotics this may resolve, but loculated pus in the air cells of the antrum makes the infection difficult to eradicate. There is severe pain behind and within the ear, and often a high fever. Surgical treatment, with opening and debridement of the air cells (mastoidectomy) is curative and removes the risk of advancing intracranial infection.

Attic infection

Attic infection involves the high roof of the middle ear cavity, which includes the course of the facial nerve. It is more common in chronic or neglected ear infections. A cholesteatoma (tumour of waxy inflammatory tissue) may form, and can erode the temporal bone, predisposing to intracranial infection. Treatment of cholesteatoma is surgical. Rare cases of chronic middle ear infection can be complicated by the presence of anaerobic pathogens.

Paranasal sinusitis

In this condition the paranasal sinuses fill with exudate, and the draining ostia become blocked. Like otitis media, it is often secondary to a catarrhal infection and may be caused by *S. pneumoniae*, *H. influenzae*, *S. 'milleri'*, anaerobes or *S. aureus*.

Clinical features are pain and tenderness over the affected sinus, usually a maxillary or frontal sinus. There is a loss of transillumination, and X-rays show thickening of the soft-tissue wall of the cavity, often with a fluid level (see Fig. 1.8).

Treatment includes elevation of the head, and decongestants to aid drainage. An oral broad-spectrum antibiotic such as co-amoxiclav, cefuroxime axetil or tetracycline will reduce the purulent exudate. In relapsing or chronic cases, the ostia of the frontal sinuses can be surgically enlarged, or false ostia can be made to drain the maxillary sinuses to the buccal cavity.

Complications of sinusitis

The ethmoid sinus has thin lateral and superior walls, which can rupture if infection causes raised pressure in the sinus cavity. Lateral spread of infection causes orbital cellulitis, while superior spread may lead to meningitis.

Cavernous sinus thrombosis is a grave complication of severe, untreated sphenoid sinusitis, or posterior extension from ethmoid sinusitis. Warning signs are high fever, severe headache, periorbital oedema and sometimes altered consciousness. It is best diagnosed by computed tomography or magnetic resonance scan. High-dose parenteral antibiotic treatment is mandatory. A third-generation cephalosporin, such as ceftriaxone, 2–4 g daily by intravenous infusion, should be given. Flucloxacillin 1.5–2.0 g 6-hourly may be added to improve antistaphylococcal cover. Clindamycin 600–900 mg 6-hourly by intravenous infusion provides antistaphylococcal and anti-anaerobe cover with excellent tissue penetration.

Viral infections of the throat and mouth

Organism list

- Rhinoviruses
- Coronaviruses
- Enteroviruses
- Adenoviruses
- Epstein–Barr virus
- Herpes simplex virus

The common cold (coryza)

Introduction

The common cold is caused by numerous strains of rhinoviruses and sometimes by coronaviruses or other respiratory virus infections. Colds are extremely infectious by the droplet route, and are also transmitted by contaminated hands or fingers. Mild fever, swelling of the mucosae of the nose, throat and conjunctiva, and often sore throat are followed by a copious mucoid nasal exudate. Otitis media in children and sinusitis in adults are common complications.

Virology and pathogenesis of common colds

Rhinoviruses are members of the Picornaviridae and exist in over 100 different serotypes. They are small RNA viruses (28–34 nm), possessing a single, approximately 7.2 kb strand of positive RNA and expressing icosahedral symmetry. A single polypeptide is produced and cleaved. There are four capsid proteins, VP1–4. The viral shell consists of 60 capsids organized as 12 pentamers. These each contain a distinct valley, within which is a 'pocket', which is thought to play an important role after attachment, leading to release of viral RNA. In most strains this contains a strongly hydrophobic molecule known as pocket factor. This pocket is a potential site for specific inhibitors of the virus. The virus binds via its VP1 protein to host cell intercellular adhesion molecule 1 (ICAM-1) to gain attachment and entry. A single point mutation confers protection from virus neutralization, but new serotypes of rhinovirus are not emerging rapidly.

Coronavirus is the only genus in the family Coronaviridae. It is a pleomorphic, non-enveloped, positive sense, single-strand RNA virus of variable size (60 to 220 nm). The virus has characteristic club-shaped surface projections 20 nm in length (Fig. 6.4). These are composed of a high-molecular-weight glycoprotein (180 kDa) and dis-

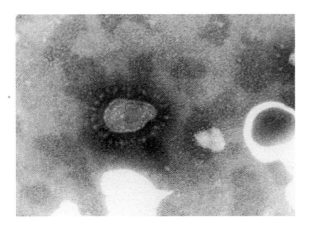

Figure 6.4 Negatively stained electron micrograph of a coronavirus.

play strain-specific antigens. They are readily removed by protease activity. The three main structural proteins are the nucleocapsid proteins, which are species-specific, the surface projection protein and the transmembrane (matrix) protein. These coronaviruses differ from the SARS coronavirus, which is discussed in more detail in Chapter 26.

The main mechanism of pathogenesis is probably a direct cytotoxic effect on respiratory epithelial cells.

Laboratory diagnosis

The principal approach to respiratory virus diagnosis is the identification of viral RNA by RT-PCR. Cell culture is rarely attempted due to the frequency and trivial nature of the infection. Nasal washings may be inoculated into human embryonic lung fibroblasts. An enterovirus-like cytopathic effect (CPE) often develops within 8 days but may require a second passage to become evident. Serotype-specific antibodies can be detected, but serological tests are too cumbersome for routine use.

Coronaviruses are difficult to isolate, and serological investigations are not readily available. Virus can be isolated from nasal and throat swabs and nasopharyngeal aspirates by inoculation in human embryonic lung fibroblasts. The CPE consists of small granular round cells in the monolayer. RT-PCR provides the simplest approach to diagnosis (see Chapter 3).

Management

Treatment is symptomatic and includes antipyretic analgesics, mild decongestants such as pseudoephedrine tablets, and bed rest in severe cases.

Enteroviral pharyngitis

Epidemiology

Enteroviral pharyngitis is common, with an annual epi-

demic peak in summer and autumn. Echovirus strains and Coxsackie type A10 are most often implicated. Humans are the only known reservoir of infection and transmission occurs by direct contact or droplet spread. Epidemics frequently affect young children in nurseries, playgroups and schools. Crowding and poor hygiene increase the risk of transmission.

Virology and pathogenesis of enteroviral infections

Enteroviruses all belong to the family Picornaviridae, which comprises four genera – of which rhinoviruses and enteroviruses cause disease in humans. More than 70 enterovirus serotypes have been isolated from human sources, and these belong to five main groups:

1 polioviruses;
2 group A Coxsackieviruses;
3 group B Coxsackieviruses;
4 ECHO (enteric cytopathogenic human orphan) viruses; and
5 five recently characterized human enteroviruses types 68–72 (type 72 is hepatitis A virus).

Each of these has different tissue tropisms (see also Chapter 13).

The viruses are 30 nm in diameter, with icosahedral symmetry. The virion contains four proteins: VP1–4. The genome is a single strand of positive-sense RNA. Variations in VP1–VP3 are responsible for serological diversity; antibodies generated by infection neutralize only the homologous virus strain. VP4 mediates binding to host cell receptor sites: different tissue tropisms among enteroviruses probably depend on the specificity of these receptors. The VP4 protein structure of two poliovirus serotypes has been fully identified. Both possess a deep cleft through which the virus binds to the host cell membrane.

> **Groups of enteroviruses**
> 1 Polioviruses.
> 2 Coxsackieviruses group A.
> 3 Coxsackieviruses group B.
> 4 Echoviruses.
> 5 Other recently characterized human enteroviruses.

Pathology

Enteroviruses enter the body through the pharynx and alimentary tract. The virus multiples in the tonsils, Peyer's patches and other bowel-associated lymphoid tissue. Viraemia often occurs, and may be followed by disease in different organs, for example meninges, myocytes, brain or skin. In poliomyelitis the virus multiplies in the anterior horn cells and the resulting cell damage leads to flaccid paralysis (see Chapter 13).

Clinical features

The incubation period of about 1 week is followed by sore

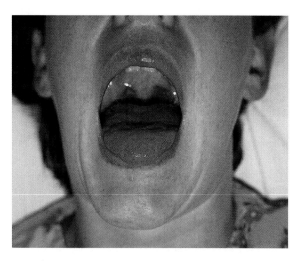

Figure 6.5 Faucial blisters in a case of herpangina.

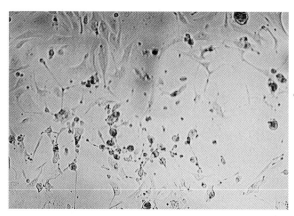

Figure 6.6 Cytopathic effect of enteroviruses, showing rounding, shrinkage and loss of contiguousness of the cell monolayer. Courtesy of Dr M. Zambon, HPA Centre for Infections.

throat and fever, whose severity and duration vary greatly. In severe cases, headache, stiff neck or meningism may occur. Symptoms rarely last more than 5–7 days.

Reddening of the fauces is usual, but does not parallel the intensity of symptoms. Groups of moderately enlarged lymph nodes are often palpable in the anterior and posterior triangles of the neck. Rarely, both echovirus and Coxsackievirus infections produce a rash of sparse macules or small papules, concentrated on the cheeks and trunk.

Coxsackie A infections may cause a pharyngeal rash of blisters with inflamed haloes (herpangina; Fig. 6.5). Hand, foot and mouth disease of toddlers causes similar lesions in the mouth, accompanied by blisters on the palms and soles, and a papular rash on the buttocks (see Chapter 5).

Coxsackie B infections can cause pleurodynia (Bornholm disease), with high fever, sore throat and tender, painful orchitis and, rarely, myocarditis.

Management

There is no specific antiviral treatment for enteroviral pharyngitis. Analgesics, especially non-steroidal anti-inflammatory agents, are helpful. In patients over the age of 16 years, soluble aspirin gargles may help, and can be swallowed for their systemic effects.

Laboratory diagnosis

Detection of enteroviral RNA by RT-PCR is a useful rapid test for the diagnosis of enteroviral meningitis or encephalitis. By culture, enteroviruses are most easily recovered from faeces, but throat swabs and cerebrospinal fluid should also be examined in cases of meningitis. Culture of an enterovirus from a sterile site is diagnostic, whereas isolation from faeces is less certainly so. Infected human embryonic lung (HeLu) cells become rounded and refrac-

tile before separating from the monolayer (Fig. 6.6). Isolates are typed by neutralization using pooled antisera.

Coxsackie group A viruses grow poorly in cell cultures, but can be replicated by intracerebral inoculation of mice. Serodiagnosis is available using immunoglobulin M (IgM) antibody capture methods.

Complications

The spectrum of enteroviral infections includes lymphocytic meningitis, myositis, pericarditis and acute myocarditis. Patients with significant meningism, precordial pain, dysrhythmias or heart failure require further investigation.

Adenoviral sore throats and pharyngoconjunctival fever

Introduction and pathology

Adenoviruses of types 1–10 are capable of infecting the respiratory tract. Typically, they cause severe, inflamed sore throat, high fever and painful enlargement of cervical lymph nodes. Some also infect the conjunctiva (leading to pharyngoconjunctival fever), or the lower respiratory tract. The illness may last 7–10 days and is often followed by debility in the convalescent period. Neonates and immunocompromised patients may suffer severe pneumonia (types 1–7), urethritis (type 37) and hepatitis in liver allografts.

Pathology of adenovirus infections

Human adenoviruses are unenveloped, icosahedral viruses with a double-stranded DNA genome of approximately 35kDa. They belong to the genus Mastadenovirus, from the family Adenoviridae. Forty-two adenovirus serotypes are recognized, divided into subgenera A–F. Group A strains cause asymptomatic enteric infection, groups B and C respiratory disease, group D keratoconjunctivitis, group E conjunctivitis and respiratory disease, and group F infantile diarrhoea. These infections occur throughout the year. They are transmitted by the faecal–oral or droplet routes. Eye infections are also transmitted by hand–eye contact, particularly in crowded swimming pools. Epidemic keratoconjunctivitis is highly infectious.

Adenoviruses multiply inside the nuclei of epithelial cells. They are cytopathic for human cells, and this is probably the mechanism of tissue damage associated with infection. Different target specificity of serotype-specific surface 'fibres' contributes to the different tissue tropisms. A toxin-like activity has been associated with the vertex capsomeres.

Clinical syndromes associated with adenovirus serotypes

- acute febrile pharyngitis: (serotypes 1–7).
- pharyngoconjunctival fever (serotypes 3, 7).
- acute respiratory infection and pneumonia (serotypes 3, 4*, 7*).
- pertussis-like syndrome (serotype 5).
- conjunctivitis (serotypes 3, 7).
- epidemic keratoconjunctivitis (serotypes 8, 19, 37).
- gastroenteritis (serotypes 40, 41).
- acute haemorrhagic cystitis (serotypes 11, 21).

⚠ A bivalent adenovirus type 4 and 7 vaccine is available for prevention of acute respiratory syndromes (see Chapter 7).

Laboratory diagnosis

Adenovirus is most easily isolated from stool but can be recovered from conjunctival swabs, nasopharyngeal aspirates and cerebrospinal fluid. Human cell lines, e.g. Hep-2, are suitable for virus isolation. A cytopathic effect, seen in 48 h, is characterized by rounded-up cells with refractile intranuclear inclusion bodies. The identification can be confirmed by electron microscopy. In patient's serum, antibodies to a group-specific antigen are detected by complement fixation test or EIA. PCR diagnosis is available.

Management

There are currently no specific anti-adenoviral drugs. Treatment of adenoviral infections is therefore supportive and includes adequate fluid intake, analgesia and rest.

Infectious mononucleosis

Introduction

Infectious mononucleosis is caused by primary Epstein–Barr virus (EBV) infection. It is a systemic disease, but in most clinically detected cases it presents as a sore throat, so that the differential diagnosis and main complications fall into this chapter. However, 15% present as hepatitis and 10% as fever alone (see Chapters 9 and 20).

Clinical presentations of Epstein–Barr virus infection

1 Sore throat (anginose infectious mononucleosis) 75%.
2 Hepatitis 15%.
3 Fever alone 10%.
4 Rare: viral-type meningitis, mononeuritis or polyneuritis, perisplenic pain.

Although primary disease can cause considerable morbidity, late effects of infection may also be important. Both Burkitt's lymphoma and nasopharyngeal tumours are consequences of EBV infection in early infancy. Cofactors such as early infection with malaria, or ingestion of toxins in, for example, preserved vegetables, may also be important. In the immunosuppressed, EBV infection can cause interstitial pneumonitis, precipitate a haemophagocytic syndrome or trigger EBV-driven B-cell lymphomas.

Epidemiology

EBV is shed in pharyngeal secretions, and transmission occurs via close oral contact, shared eating utensils or, in some cultures, by a mother chewing food for her infant.

Seroconversion is commonest in young children, when it is usually asymptomatic. Clinical disease affects mainly teenagers and young adults. The estimated annual incidence of clinical disease in the UK, based on the last 20 years' statistics from the Royal College of General Practitioners', is 1.6 per 100 000 population.

Virology and pathogenesis of EBV infections

EBV was first identified in Burkitt's lymphoma tissue. It is morphologically identical to other herpesviruses, with icosahedral symmetry and an envelope derived from the host-cell plasma membrane. The double-stranded DNA is 172 kb: large enough to code for 100–200 proteins. These include the latent membrane protein (LMP), the terminal protein, the membrane antigen complex, the early antigen (EA) complex, the viral capsid antigen (VCA) and the Epstein–Barr nuclear antigen (EBNA) complex. The EBNA complex contains at least six proteins, probably important in maintaining the virus in the infected cell. EBNA-1 antigen is expressed on all infected cells but may be lost as infected cells die. The early antigen (EA) complex depends on genes encoding thymidine kinase, DNA polymerase and late structural genes including viral capsid antigens, and marks cells that have entered a lytic phase. Host antibodies to EBNA and EA appear early in infection and are transient. Antibodies to capsid antigens appear in IgM by the onset of illness, and may persist for weeks or months.

IgG anti-capsid antibodies indicate immunity, and persist throughout life.

Virus binds to the CD21 receptor of host B-cells (the receptor for the C3d component of complement) via the major viral glycoprotein gp350/220. Adherence is followed by endocytosis, fusion of virus with the endocytotic vesicle membrane via viral gp85, and viral release into the cytoplasm. Another viral antigen, gp42, binds the major histocompatibility complex class II molecules, which serve as co-receptors for B-cells.

During viral latency, EBV DNA persists in host cells as double-stranded episomes, organized into nucleosomes similar to chromosomal DNA: 10% of EBV genes are expressed including six EBNAs (1, 2, 3A, 3B, 3C and LP). The function of these genes in latency and in the immortalization of infected cells is now known. For example, LMP-1 mimics the growth and survival signals given to B cells by CD4 T cells. It mediates EBV-induced transformation, and is found in many EBV-related lymphomas.

Virus first invades pharyngeal cells, from where B lymphocytes become infected and carry the virus throughout the body. EBV-mediated activation and immortalization of infected B cells is the probable pathogenic mechanism for the many 'immunological' effects and complications of infection.

Clinical features

After 6–8 weeks' incubation, fever, sore throat and widespread lymphadenopathy develop more or less simultaneously. A white creamy exudate appears on the tonsils, and becomes confluent within 24–36 h. The exudate may become bulky but is rarely discoloured and it does not involve the pharyngeal mucosa (Fig. 6.7). The pharyngeal, conjunctival and nasal mucosae are congested and swollen. Gross pharyngeal swelling may make it impossible to swallow saliva and can threaten the airway.

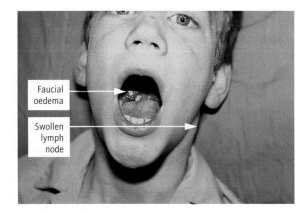

Figure 6.7 Exudative tonsillitis with marked cervical lymphadenopathy in a case of infectious mononucleosis.

Labels: Faucial oedema; Swollen lymph node

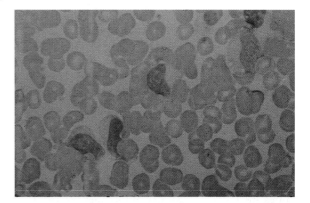

Figure 6.8 Atypical mononuclear cells in the peripheral blood of a case of infectious mononucleosis.

The spleen is palpable in 25–40% of cases; the liver edge is often palpable. Chest X-ray occasionally shows enlarged mediastinal lymph nodes and, surprisingly, as many as 20–25% of patients have a lung opacity suggesting segmental pneumonitis. These abnormalities resolve spontaneously as the fever abates.

Infected B cells remain within reticuloendothelial tissues, while numerous activated T cells appear in the bloodstream. These are seen as 'atypical mononuclear cells' and may constitute 40% or more of the total lymphocyte count during the acute infection (Fig. 6.8).

The liver function tests show raised transaminases, usually in the range of 60–500 IU/ml. The alkaline phosphatase may rise during convalescence, sometimes approaching 1000 IU/ml, but this usually resolves uneventfully.

The activated B cells produce various antibodies, including the diagnostic 'heterophile antibodies', which agglutinate horse and sheep red blood cells. Others include haemolysins, platelet antibodies, anti-nuclear antibodies, rheumatoid factors and anti-cardiolipin antibodies. These antibodies are occasionally associated with immune cytopenias or autoimmune-like diseases, but this affects fewer than 1 in 20 000–30 000 clinical cases.

The duration of fever and exudative pharyngitis varies from a few days to 3 weeks or more. Most patients subsequently convalesce steadily, but a minority suffer prolonged post-infectious fatigue. Persisting chronic fatigue syndrome (CFS) is a rare complication of EBV infection.

Diagnosis

The clinical picture is often sufficient for diagnosis. The exudate seen in streptococcal pharyngitis is follicular and rarely confluent. The pseudomembrane of diphtheria is sometimes confluent, but it is discoloured, and tends to spread beyond the tonsillar margin. In both bacterial diseases there is a low-grade neutrophilia, unlike the atypical lymphocytosis of EBV infection.

The Monospot slide agglutination test demonstrates the presence of heterophile antibody, utilizing horse red cell agglutination, and is positive in about 90% of patients at presentation. Children below age 5 years, however, do not produce high levels of heterophile antibodies, and about 50% of these patients have a negative test. PCR is emerging as a sensitive method of detecting EBV infection.

IgM anti-EBV capsid antibodies are detectable by ELISA at presentation, and are diagnostic of acute infection. Rising IgG titres to VCA cannot be detected, as concentrations are already high when the patient presents. Although EBV can be cultivated, this approach is not used as cultures are likely to be positive in specimens from latently infected asymptomatic individuals.

Management

There is no specific treatment; most cases recover uneventfully. In older children and adults soluble aspirin in standard doses may be gargled and swallowed. In the under-16s paediatric formulations of ibuprofen are useful if paracetamol is not effective. Stronger analgesics may be indicated if pharyngeal pain is severe.

Complications

Threatened respiratory obstruction

Threatened respiratory obstruction is the commonest reason for hospital admission. It often improves with elevation of the head (to encourage drainage of oedema from the pharyngeal tissues) and anti-inflammatory analgesics. Danger signs are inability to swallow saliva and a rapidly increasing pulse rate; increasing respiratory rate and cyanosis occur later. Obstructing pharyngeal oedema can often be reduced by intravenous treatment with a bolus of 100–200 mg hydrocortisone. This is effective within 20–30 minutes. As in corticosteroid treatment of croup and bronchiolitis, adverse effects are rare and dosing can be repeated if non-steroidal anti-inflammatory agents do not maintain the improvement. Emergency intubation or tracheostomy are rarely necessary.

Corticosteroid treatment does not affect the duration of symptoms.

Effects of abnormal antibodies

Effects of abnormal antibodies are rarely clinically important. Thrombocytopenia is the least uncommon, followed by haemolytic anaemia. A handful of cases of systemic lupus erythematosus-like disease are reported, with joint pains, and with anti-DNA positive antibodies, which may persist for some months. Symptoms resolve during convalescence when B-lymphocyte activation ceases and the antibodies disappear.

Suppurative complications

Peritonsillar abscess, pharyngeal abscess, ethmoiditis,

infection of other intracranial sinuses and periorbital cellulitis, sometimes with severe sepsis, can result from secondary infection of inflamed mucosae and obstructed sinuses. Severe pain and swelling affecting these sites should be actively investigated and appropriately treated.

Rupture of the spleen

Rupture of the spleen is extremely rare. It may present with left upper quadrant and shoulder pain during acute disease or be precipitated by apparently trivial trauma, such as a blow during play, a sudden movement or a cough. A peritoneal tap may yield blood-stained fluid. Imaging studies may demonstrate free fluid in the peritoneal cavity, or disrupted splenic anatomy. Prompt surgical intervention is needed to terminate bleeding. Contact sports or combat sports should be avoided until lymphadenopathy (and therefore splenomegaly) have subsided.

Neurological complications

Neurological complications include lymphocytic meningitis, mononeuritis or brachial plexitis (Fig. 6.9). These are benign and self-limiting. Occasional cases of encephalopathy are reported, of which some are progressive or even fatal.

Haemophagocytic syndrome

Haemophagocytic syndrome is a rare, life-threatening complication, with a high case-fatality rate. It results from inability to terminate EBV viral replication in activated lymphocytes. It is commonest in individuals with recognized disorders of cell-mediated immune (CMI) responses. However, some apparently normal young people develop the disorder, probably due to a subtle cell-mediated immunodeficiency that only becomes apparent when they contract EBV infection. Management should be undertaken by an expert clinical immunologist, and may

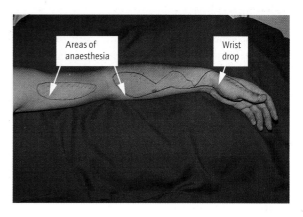

Figure 6.9 Brachial plexitis complicating infectious mononucleosis: complete recovery occurred within 3 weeks.

include the use of antiviral drugs, etoposide or other immunosuppressive regimens.

Bacterial throat infections

Organism list

- *Streptococcus pyogenes*
- *Haemophilus influenzae*
- *Corynebacterium diphtheriae*
- Other bacteria, including *Neisseria meningitidis*, *H. haemolyticum*, *Chlamydia pneumoniae* and *Staphylococcus aureus*.

Streptococcal tonsillitis

Introduction
Streptococcal tonsillitis is caused by *S. pyogenes* (group A streptococcus, or GAS). It is common worldwide, and can affect both adults and children. It causes considerable short- and medium-term morbidity, and can be recurrent, as *S. pyogenes* is a tenacious colonist of the throat, and exists in many serotypes. It is important because streptococcal throat infections can be complicated by scarlet fever, post-streptococcal nephritis, post-streptococcal reactive arthritis or rheumatic fever (see Chapter 21).

Epidemiology
Streptococcal pharyngitis is common in temperate climates, occurring mainly in the winter. Transmission is mainly by direct contact with respiratory secretions, and less commonly by air-borne droplets or indirect contact by hands. Rarely, outbreaks have resulted from consumption of contaminated food and milk.

The infection presents mainly in children, up to 20% of whom may be asymptomatic carriers. Disease is commoner in crowded settings such as children's homes and military camps.

The incidence and severity of scarlet fever (and other manifestations of group A streptococcal infection) have declined steadily over the past 50 years. In 1936, when yearly notifications began, there were 104 862 notifications and 440 deaths from scarlet fever in England and Wales; a case fatality ratio of 0.42 per 100. By 1986, this had dropped to 6888 notifications and 3 deaths (0.05 per 100). The incidence (but not the case fatality ratio) increased during 1988 and 1989, but has subsequently declined to around 2000 cases per annum. A similar fall was seen in the USA; however, a resurgence of severe infections complicated by rheumatic fever occurred in the late 1980s, associated with mucoid strains of *S. pyogenes*.

Pathogenesis of *S. pyogenes* infections (Fig. 6.10)
Inhibition of host defences
The M protein of *S. pyogenes* is a fibrillar protein possessing conserved and variable domains. It binds complement control proteins. It also binds fibrinogen, which inhibits alternative complement binding and reduces the ability of neutrophils to recognize and phagocytose GAS. The organism has multiple M types and several M protein homologues that bind IgG, IgA and other host proteins, all contributing to avoidance of phagocytosis. The polysaccharide capsule, a hyaluronic acid polymer, is a poor immunogen and also contributes to inhibiting phagocytes. Surface expression of C5a- and immunoglobulin-binding proteins may interfere with chemotaxis and humoral immune responses.

Host attachment
S. pyogenes expresses at least 17 adhesion molecules including fibronectin, vitronectin, collagen binding proteins and lipoteichoic acid.

Tissue damaging toxins
Streptolysin S is critically important in the pathogenesis of necrotizing fasciitis. Streptolysin O activates complement and destroys neutrophils, lymphocytes and tissue cells by inserting pores into the plasma membrane. C5a peptidase, encoded by *scpA*, destroys C5a, the complement component that recruits neutrophils to the site of infection. Hyaluronidase and collagenase may aid tissue invasion by breaking down collagen and hyaluronic acid in connective tissue. Streptokinase acts as a plasminogen activator, producing clot lysis, possibly enhancing spread of the organism. Surface-expressed enolase is a metabolic enzyme that acts as a plasminogen activator and enhances invasiveness across tissue boundaries. Other lytic enzymes include four serologically different DNAses (A–D).

The ability to disrupt connective tissues, kill cells and reduce tissue redox potential is probably important in the pathogenesis of synergistic gangrene, mixed infection with Gram-positive cocci, including *S. pyogenes* and/or *Staphylococcus aureus*, and obligate anaerobes, which can invade and destroy tissue planes.

Three streptococcal pyrogenic exotoxins (SPEs) have been identified: A, B and C. A and C are structurally similar: they cause the rash of scarlet fever (erythrogenic toxins). They are superantigens, closely homologous with some *Staphylococcus aureus* exotoxins. They stimulate the production of tumour necrosis factor, a major mediator of sepsis, and are thought to mediate the shock syndrome in severe GAS infections.

SPE B has a different structure. It is secreted as a zymogen and converted to a proteinase that is mitogenic and cardiotoxic.

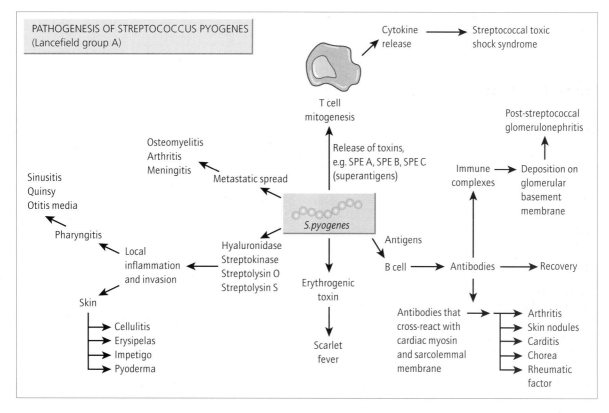

Figure 6.10 Pathogenicity of *Streptococcus pyogenes*.

Intracellular survival

Some strains of *S. pyogenes* that possess the *PRTF1* gene encoding fibronectin binding protein, are able to invade epithelial cells. They can survive within cells of the respiratory tract, possibly explaining why beta-lactam antibiotics cannot eradicate carriage.

Inappropriate immune responses

Through as yet uncharacterized cross-reacting antigens, *Streptococcus pyogenes* may trigger rheumatic fever and acute glomerulonephritis.

Pathogenicity factors for *Streptococcus pyogenes*

1 M proteins.
2 Streptolysin S.
3 Streptolysin O.
4 Hyaluronidase.
5 Collagenase.
6 DNAase.
7 Streptokinase.
8 Pyogenic exotoxins A, (B) and C.
9 Fibronectin binding protein.
10 Cross-reacting antigens.
11 Enolase.

Clinical features

The usual features are fever, pain in the throat, enlargement of the tonsils and tender swelling of the tonsillar lymph nodes at the angles of the jaw. The severity of the symptoms is variable; some patients with severe pain have only mild pharyngeal inflammation, while some with moderate pain have alarming tonsillar enlargement.

The classical appearance is of follicular tonsillitis, in which enlarged, red tonsils are dotted with patches of soft white exudate (Fig. 6.11). The throat is painful, rather than sore, and a large, tender lymph node swells downwards from beneath the angle of the jaw. One tonsil may be predominantly affected, with proportionately greater swelling of the lymph node on that side.

Scarlet fever

Scarlet fever is a hypersensitivity response to SPE produced by the infecting *Streptococcus pyogenes*. It presents with severe malaise and often vomiting, followed by the appearance of erythema, first on the chest but quickly becoming generalized. Erythema is increased in the folds and valleys of the skin (Pastia's sign). The papillae of the skin are swollen, forming tiny conical papules, and roughening the skin ('punctate' erythema). There may be a less affected area around the mouth or 'snout' area, but this is neither a constant nor a diagnostic feature (Fig. 6.12).

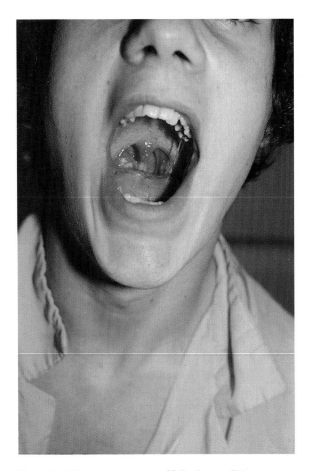

Figure 6.11 Typical appearance of follicular tonsillitis.

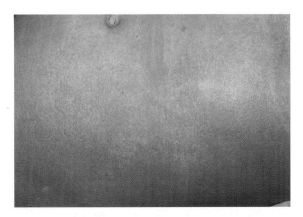

Figure 6.12 Typical rash of scarlet fever.

Initially the tongue is furred and white (white strawberry tongue). Over about 3 days this clears from the tip backwards, leaving a reddened (red strawberry tongue) appearance (Fig. 6.13).

Figure 6.13 Red strawberry tongue of scarlet fever.

Streptococcal toxic shock syndrome

This is a severe version of scarlet fever, with shock, thrombocytopenia and multisystem involvement. Pleural and peritoneal effusions are common, and renal failure is often seen but haemorrhage is rare.

Diagnosis

Mild streptococcal tonsillitis is clinically anonymous and often self-limiting. Follicular tonsillitis with tender enlargement of tonsillar lymph nodes is typical of severe *S. pyogenes* infection, and is clinically diagnostic. Neutrophilia, in cases of sore throat, is a predictor of bacterial aetiology (with a sensitivity of 75%), and supports a decision to give antibiotic treatment. Recovery of *S. pyogenes* from throat swabs supports the diagnosis, but streptococcal colonization can coexist with viral and other sore throats. Also, *S. pyogenes* does not transfer well from swab to culture, so that there is a 20–30% false negativity rate in the results of swab cultures.

S. pyogenes is fastidious, growing only on rich nutrient media, which usually contain blood. Haemolytic toxins produce the characteristic zone of complete (beta) haemolysis, revealing suspect colonies for identification. Both growth and haemolysis are enhanced by incubation in an anaerobic atmosphere. For screening multiple specimens, selective media can be used. They may contain antibiotics such as ofloxacin or nalidixic acid, or dyes such as crystal violet, which inhibit Gram-negative and Gram-positive commensal organisms, respectively.

For routine laboratories, identification of *S. pyogenes* is based on colonial morphology (large clear colonies surrounded by a zone of beta haemolysis) and identification of the group A Lancefield antigen. Antigen is extracted from a bacteria suspension with an enzyme and detected by latex agglutination, using particles coated with an-

Table 6.1 Species of beta-haemolytic streptococci and their pathogenicity

Group	Species	Diseases caused
A	*Streptococcus pyogenes*	Acute pharyngitis, quinsy, otitis media, erysipelas, synergistic gangrene, rheumatic fever, post-streptococcal glomerulonephritis, puerperal fever
	S. 'milleri' (minute colony)	Metastatic suppurative infection
B	*S. agalactiae*	Neonatal septicaemia, meningitis and pneumonia
C	*S. dysgalactiae*	Rare cause of skin sepsis and endocarditis. Post-infectious glomerulonephritis has been reported
	S. equi	
	S. equisimilis	
	S. zooepidemicus	
D	*S. bovis*	Endocarditis and bacteraemia associated with colonic neoplasm
F	*S. 'milleri'*	Metastatic suppurative disease, dental sepsis

tibodies to the different Lancefield antigens. Lancefield grouping is useful, as it relates to streptococcal species and pathogenicity (Table 6.1). Rapid diagnostic tests exist, based on latex agglutination of throat-swab material, and can be used in a general practitioner surgery. They are highly specific but of variable sensitivity (as low as 55% in some studies).

Laboratory identification of *Streptococcus pyogenes*

1 Chains of Gram-positive cocci, growing on blood agar, plus beta haemolysis, plus Lancefield group A antigen.
2 Rapid antigen detection in throat swabs.
3 Antibody detection: high or rising titres of anti-streptolysin O, anti-hyaluronidase and/or anti-DNAse.

Management

The treatment of choice is penicillin. Early and mild cases often respond to oral ampicillin, but severe infections require inpatient treatment with intravenous benzylpenicillin. Suitable alternative drugs are erythromycin, azithromycin or clindamycin. However, a small proportion of highly virulent strains are resistant to macrolides, and fail to respond to these drugs.

S. pyogenes is difficult to eradicate from the throat. Research shows that 2 weeks of vigorous (often parenteral) penicillin therapy is needed and erythromycin treatment is unreliable. This is important in dormitory- or barracks-associated outbreaks of streptococcal sore throat, especially those associated with rheumatic fever. In a domiciliary setting a course of a least 7–10 days' treatment is probably advisable.

Treatment of streptococcal sore throat

1 Early and mild cases: oral ampicillin 250–500 mg 6-hourly for 10 days, or erythromycin in the same dosage and schedule.
2 Severe cases: benzylpenicillin 1.2–2.4 g 4–6-hourly (child over 1 month, 150–300 mg/kg daily in 4 to 6 divided doses); *alternative:* clindamycin 600–900 mg 6-hourly by IV infusion (child over 1 month 20–40 mg/kg daily in 3 or 4 divided doses).

Complications

Peritonsillar abscess

Peritonsillar abscess (quinsy) is a common complication of tonsillar sepsis. The fauces and soft palate on the affected side become oedematous and pendulous, and may completely envelope the tonsil. The throat, and the draining lymph node, are intensely painful and tender. A small proportion of cases have bilateral quinsy, which carries a high risk of airway obstruction (Fig. 6.14).

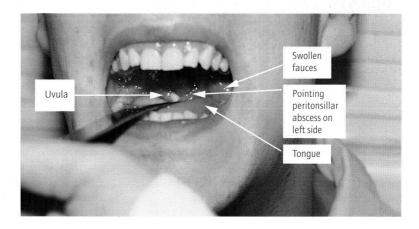

Figure 6.14 Bilateral quinsies (peritonsillar abscesses).

Prompt, vigorous treatment can avert the need for surgical drainage. The addition of metronidazole to high-dose penicillin therapy, or the substitution of clindamycin, which readily penetrate oedematous tissue, may speed improvement by inhibiting streptococcal growth in this rather anaerobic environment. In an emergency, oedema can be rapidly reduced by giving an intravenous bolus of 100–200 mg hydrocortisone. Many early abscesses will resolve; others drain into the throat, with rapid relief of symptoms. A few progress and enlarge, requiring urgent surgical drainage.

Streptococcal bacteraemia

Streptococcal bacteraemia is a rare complication, with a mortality rate reaching 25–30%. Warning signs are high fever, extreme pain (rarely, even suppuration or sloughing) in the draining lymph nodes (Fig. 6.15) or the appearance of erysipelas-like lesions on the skin. Patients with these features should be treated intravenously while blood culture results are awaited (see Chapter 19).

Post-streptococcal disorders

Post-streptococcal disorders include rheumatic fever, reactive arthritis, nephritis, erythema multiforme and erythema nodosum. These are discussed in Chapter 21.

Acute epiglottitis

Introduction and epidemiology

This is a severe throat infection that causes massive oedema of the epiglottis and threatens the airway. It is a rare but important disease. Before vaccination against *H. influenzae* type b (Hib) was included in national programmes (see Chapter 13), Hib caused many cases of epiglottitis in pre-school children. Most cases nowadays are caused by *S. pyogenes*, and affect mainly adults. The disease must be considered in cases of severe sore throat with stridor or

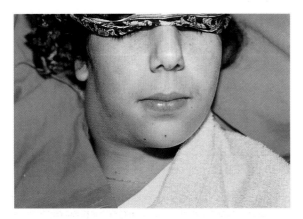

Figure 6.15 Suppuration of cervical lymph node following tonsillitis (300 ml of pus was drained at surgery).

inability to swallow saliva. The airway is vulnerable in this condition; respiratory obstruction can develop suddenly, or be precipitated by throat examination.

Clinical features

Illness develops rapidly, with high fever, severe throat pain, swelling and tenderness of the neck and hyoid region, and great difficulty in swallowing. As epiglottic swelling increases the patient drools and then develops stridor. The tonsillar lymph nodes may be tender and enlarged.

On examining the throat with the tongue depressed, the red, swollen epiglottis can be seen protruding upwards like a cherry. However, manipulation of the throat can precipitate complete respiratory obstruction and should be avoided if urgent X-ray diagnosis is available. The swollen epiglottis is visible on a lateral X-ray of the soft tissues of the neck, in which it looks like the rounded tip of the thumb, filling the lower oropharynx (Fig. 6.16). The blood count often shows a neutrophilia, with a total white count of $12–16 \times 10^9/l$.

Diagnosis

The diagnosis should be considered in any patient with severe throat pain, features of sepsis, and drooling or stridor. Diagnosis is often delayed because epiglottitis is not considered.

Confirmation is provided by a lateral X-ray of the soft tissues of the neck, or by direct inspection of the throat if there is no alternative. Blood culture should be obtained for microbiological diagnosis. Throat swabs should be deferred until the airway is secure.

Differential diagnoses

Differential diagnoses include infectious mononucleosis, bilateral peritonsillar abscesses, diphtheria, retropharyngeal abscess and Ludwig's angina. A differential white count and heterophile antibody test will exclude infectious mononucleosis (PCR or IgM antibody tests for EBV are more reliable in small children). Cautious examination of the mouth and fauces, avoiding the use of a tongue depressor, helps to exclude the gross sublingual swelling of Ludwig's angina, faucial swelling of quinsy and the membrane of diphtheria. It may be delayed until emergency measures have secured the airway. The X-ray will show the site of swelling within, rather than behind, the pharynx.

Management

Acute epiglottitis is a medical emergency. The patient should be allowed to sit up, which helps to keep the airway open. High-dose antibiotic treatment should be begun immediately. The treatment of choice is a broad-spectrum cephalosporin such as cefotaxime or ceftriaxone. This can be changed to penicillin if *S. pyogenes* infection is confirmed. A course of 10 days is adequate for most cases.

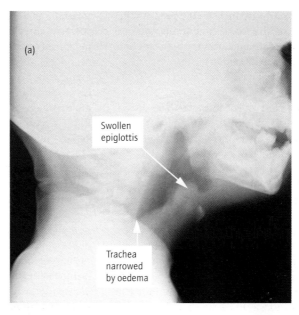

Figure 6.16 Acute epiglottitis: (a) enlarged epiglottis demonstrated by lateral X-ray of the soft tissues of the neck; (b) oedema tracking downwards narrows the trachea.

If the airway is critically obstructed, oxygen should be given by mask while urgent tracheostomy is considered. An intravenous bolus dose of hydrocortisone, 100 to 200 mg, may reduce oedema and avoid the need for tracheostomy. It will not compromise the response to treatment. Many ear, nose and throat departments keep a supply of heliox (80% helium and 20% oxygen), which has an extremely low viscosity and will pass much better than pure oxygen through a tiny airway or a small cannula in the trachea.

Treatment of acute epiglottitis

First choice: cefotaxime – child i.v. 150–250 mg/kg daily in two to four divided doses; adult 2–4 g 8-hourly, *or* ceftriaxone – child 50 mg/kg as a single daily dose; adult 2–4 g as a single daily dose. Treatment continued for 10 days.

 Consider a bolus dose of corticosteroid to reduce epiglottic oedema.

Complications

Once the patient reaches medical care the mortality is low

and the complications are mainly those of intubation and ventilation: hypostatic lung infections, pneumothorax and infected tracheostomy sites.

Prevention and control

Many cases are prevented by childhood immunization against *H. influenzae* type b.

Diphtheria

Introduction

Diphtheria is a rare infection of the respiratory mucosa, and sometimes of broken skin, caused by toxigenic *Corynebacterium diphtheriae*. Although controlled in most communities by immunization, elderly unvaccinated travellers, or refugees and immigrants from rural areas may suffer from the disease or carry the organism in the nose and throat. Vaccine-induced immunity declines in adult life. In tropical countries where hygiene is poor, diphtheria is still common. *C. ulcerans* infection from unpasteurized milk occasionally presents as classical diphtheria. Areas where diphtheria is transmitted include former USSR, the Indian subcontinent, south east Asia and south America. Western tourists may unwittingly pass through endemic areas, and be unexpectedly infected.

Epidemiology

Humans are the only reservoir of infection. The disease is spread by direct contact with cases or carriers. Patients with cutaneous diphtheria are more infectious than those with other forms, though the less extensive lesions of cutaneous disease may confer immunity without causing severe toxic illness. Nasal diphtheria also tends to be mild, but the infected discharge is important in the spread of disease among children. Between the introduction of routine immunization in Britain in 1942, and the 1960s, diphtheria was almost eliminated. Since 1985, one death has been reported. Sporadic cases, sometimes with limited indigenous transmission, occur occasionally.

Non-toxigenic *C. diphtheriae* isolates are reported in about 10 cases per year, mostly in adults who were infected abroad. There has been a recent resurgence of these strains in European regions, isolated from cases of pharyngitis, septic arthritis, endocarditis and other conditions.

Pathogenesis of diphtheria

The consequences of infection with *C. diphtheriae* are twofold: the effects of the potent exotoxin; and obstruction of the airway by necrotic debris, which forms a tough pseudomembrane on infected respiratory mucosa (Fig. 6.17).

Diphtheria toxin is the major pathogenicity determinant of *C. diphtheriae*. Its genetic code resides on a beta-phage. Bacteria not infected by the phage are non-toxigenic and do not cause diphtheria. The protein toxin possesses three domains. A receptor portion binds to the target cell, and the central portion, which is highly hydrophobic, dissolves in the cell membrane, carrying the toxin portion into the cell. The toxin itself is an adenosine diphosphate ribosylase, which ribosylates an amino acid diphthamide present in elongation factor 2. Elongation factor 2 is essential for protein synthesis in the host cell, and its inhibition leads to cell death.

Clinical features

After 3–5 days incubation, the infected site becomes inflamed and gradually acquires a spreading, tough, adherent slough (often called the membrane). The toxic effects of diphtheria are proportional to the extent of the membrane.

Pharyngeal diphtheria is the commonest respiratory form. This presents with fever, sore throat and marked oedema of the cervical lymph nodes, which may produce a bull-neck appearance (Fig. 6.18). The membrane may affect the tonsils but, unlike the exudate of infectious mono-

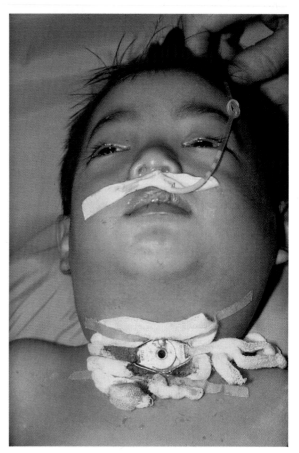

Figure 6.18 Diphtheria: bull-neck appearance and tracheostomy. Courtesy of the World Health Organization.

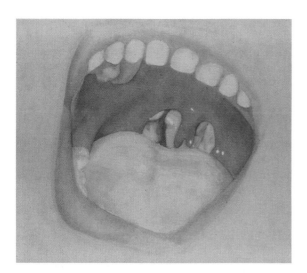

Figure 6.17 Diphtheria: the off-white, smooth pseudomembrane affects the tonsils and pharynx.

nucleosis, also spreads across the pharyngeal mucosa. It is off-white or greyish and semitransparent, opacified where it contains areas of altered blood, and sometimes blackened and necrotic. Attempts to scrape it away cause pain and bleeding. Diphtheria can also affect the larynx and trachea, and then presents as croup. The true diagnosis is often suspected late in such cases, as the membrane is not visible in the pharynx.

Threatened airway obstruction causes stridor. Children often assume a characteristic posture, leaning forward with the neck extended, to hold the airway open.

When diphtheria affects the nares, the conjunctiva or skin abrasions or ulcers, infected sites are inflamed, often produce a serosanguineous exudate and may have small adherent patches of membrane. C. diphtheriae can colonize sites of other infections and has been found in impetigo, cellulitis and broken chickenpox lesions.

During acute diphtheria there is a modest fever, but disproportionate prostration. There is a neutrophilia in the peripheral blood, but minimal disturbance of renal and liver function.

Effects of diphtheria toxin

The diphtheria toxin causes cardiac damage in the first week. Heart failure and conduction defects are common, and profound heart block is a risk, but the myocardium recovers completely when convalescence is established.

The neurological damage appears from the second week. It is caused by demyelination, and occurs earliest and most severely near to the site of the membrane. Palatal and ocular palsies are common after throat infections. After 3 or 4 weeks some patients develop a generalized weakness or paralysis similar to Guillain–Barré syndrome, but this also is fully reversible.

In rare cases a late nephritis causes impaired renal function.

Fatalities are usually due to irreversible heart failure. Intrabronchial or tracheal membrane can cause respiratory obstruction, which is difficult to treat if it cannot be by-passed by tracheostomy. In the recovery phase, these membranes may be sloughed, and can lodge in the trachea

and bronchi. Very severe cases occasionally die from a Waterhouse–Friderichsen syndrome of adrenal failure and haemorrhagic features.

Diagnosis

Clinical suspicion should be aroused by severe throat or pharyngeal swelling, typical membrane, and modest fever with severe prostration and neutrophilia.

Laboratory diagnosis

Specimens from the throat, larynx, nose or skin may be examined. Swabs are adequate. It is very important to inform the laboratory that diphtheria is suspected, otherwise special media will not be inoculated and the pathogen may be discarded as a diphtheroid. On tellurite media, corynebacteria reduce tellurite to metallic tellurium. The colonies have a black shiny appearance, aiding selection for identification. The Loeffler's slope contains a rich serum medium on which organisms grow rapidly. Sufficient growth is usually available after 6 h to allow staining by Albert's method, which demonstrates the volutin granules found in this species. The organism should then be subcultured on to blood agar for biochemical confirmation using sugar tests adapted for corynebacteria (Table 6.2).

Confirmation of toxigenicity is made by polymerase chain amplification, which allows rapid detection of the toxin gene in C. diphtheriae isolates.

The laboratory's role is to confirm the presence of toxigenic C. diphtheriae and to alert the public health services. The identification of non-toxigenic C. diphtheriae requires no public health control measures.

Management
Antibiotic treatment

Intravenous benzylpenicillin is the treatment of choice. Erythromycin or a cephalosporin are alternatives. As inflammation subsides in 24–36 h the membrane loosens. Casts of the upper airway or bronchi may be shed, sometimes needing assisted removal by suction or bronchoscopy.

Table 6.2 *Corynebacterium diphtheriae*: tests to identify and distinguish *gravis*, *intermedius* and *mitis* biotypes

Organism	Sucrose hydrolysis	Mannitol hydrolysis	Starch hydrolysis	Glycogen hydrolysis	Urease	Nitrate production	D toxin	Cp toxin	Both toxins
C. diphtheriae var. gravis	+	+	+	+		+	+		
C. diphtheriae var. intermedius	−	+	−	−		+	+		
C. diphtheriae var. mitis	+	+	−	−		+	+		
C. ulcerans	−	+			+	−	+	+	+

D, diphtheria toxin; Cp, toxin produced only by the animal pathogen C. *pseudotuberculosis* and by C. *ulcerans*.

Table 6.3 Antitoxin treatment of diphtheria

Type of diphtheria	Dose (units)	Route
Nasal	10 000–20 000	Intramuscular
Tonsillar	15 000–25 000	Intramuscular or intravenous
Pharyngeal or laryngeal	20 000–40 000	Intramuscular or intravenous
Combined sites or late diagnosis	40 000–60 000	Intravenous
Cutaneous disease	Not widely recommended: wound toilet and antibiotics preferred	

Antitoxin

Antitoxin is given to neutralize circulating toxin and prevent further damage to myocardium and myelin. Dosage ranges from 10 000 IU for nasal disease to 120 000 IU for aggressive nasolaryngeal diphtheria.

> **Antitoxin treatment in diphtheria**
>
> See Table 6.3.
>
> Human immunoglobulin is not effective, so diphtheria antitoxin is prepared from horse serum. Precede full dose by subcutaneous test dose of anti-toxin, e.g. 50–100 U, 30 min before main dose.
>
> (From WHO (1994) Manual for the Management and Control of Diphtheria in the European Region. World Health Organization.)

Elective tracheostomy

Elective tracheostomy avoids possible emergency tracheostomy, which is difficult when the tissues are very oedematous. It also provides airway protection in case palatal palsy develops later.

Follow up

Diphtheria cases require prolonged observation to detect cardiographic changes, rhythm disturbances and late neurological complications, which may require airway protection or other support. When patients have recovered, they are likely to be carriers of *C. diphtheriae*. Carriage is difficult to eradicate, often requiring two 14-day cycles of oral erythromycin.

Clinical diphtheria does not always induce effective levels of antitoxin. Patients should therefore be immunized or receive reinforcement 'booster' immunization after recovery with a diphtheria toxoid vaccine appropriate to their age.

Prevention and control

All forms of diphtheria, including cutaneous disease, are notifiable in Britain, and many other countries.

Cases and carriers of toxigenic *C. diphtheriae* should remain isolated until two consecutive nose and throat cultures, taken 24–48 hours apart, have proved negative. Cultures may be taken from 24 hours after completion of antibiotic courses (or at least 2 weeks after, for cutaneous diphtheria).

Household, healthcare and other close contacts of cases should have surveillance swabs taken, have a complete or booster course of vaccination, as appropriate, and should be offered chemoprophylaxis.

> **Chemoprophylaxis of diphtheria**
>
> 1 Oral erythromycin 500 mg 6-hourly for 7 days; child up to 2 years: 125 mg 6-hourly; 2–8 years: 250 mg 6-hourly.
> 2 Bacteriological clearance is confirmed by two consecutive negative nose and throat swabs, the first at least 24 hours after completion of chemoprophylaxis.
>
> If surveillance swabs are positive, continue treatment for a further 10 days.

Diphtheria vaccine is a formalin-inactivated toxoid preparation. A standard paediatric dose contains at least 30 IU of antigen. In adults and children over 10 years a low-dose (1.5 IU) vaccine is used because of the risk of hypersensitivity reactions. Three doses of vaccine, given at monthly intervals, starting at 2 months of age, are recommended for primary immunization in the UK, with a booster dose 3 years later and again before leaving school (see Chapter 26). Nowadays over 95% of children in the UK receive a full course of vaccine. Adults born before 1942, when routine immunization was introduced, have only naturally acquired immunity. Up to 25% of people in this age group have no measurable antibody.

> **Measuring immunity to diphtheria by antitoxin concentration**
>
> 1 < 0.01 IU/ml: no protection.
> 2 0.01–0.10 IU/ml: partial protection.
> 3 > 0.10 IU/ml: reliable protection.

Ludwig's angina

This is a suppurative infection of the hypoglossal tissue planes. It can become an emergency because the oedema and exudate push the tongue upwards and backwards, potentially threatening the airway.

The origins of the infection are probably the mouth and teeth. Most infections are polymicrobial. Typical implicated organisms include *Streptococcus pyogenes*, *Staphylococcus aureus*, oral streptococci and 'mouth' anaerobes such as *Prevotella melaninogenicus* and *Fusobacterium* spp.

Clinical features develop rapidly. They include pain, fever, difficulty in swallowing and increasing stridor. Examination shows a bull-neck appearance with tenderness of the neck and throat. It is difficult to open the mouth and, when it is open, the tongue is elevated so that its underside is visible above the lower teeth. The fauces may also be swollen.

Treatment should be prompt. Antibiotics should include one active against *S. aureus*. Satisfactory therapy includes penicillin plus flucloxacillin, or a cephalosporin such as cefuroxime or cefotaxime. Although these agents are effective against mouth anaerobes, metronidazole penetrates oedematous tissues well and may be useful additional treatment. Treatment should usually last for about 10 days.

Oedema may be reduced medically with a bolus dose of corticosteroid, e.g. 200 mg hydrocortisone or 10 mg dexamethasone, intravenously. If this does not provide relief, drainage of pus from the sublingual space may be effective. This is performed by passing perforated drains through the floor of the mouth and out through the skin anterior to the hyoid bone.

Retropharyngeal abscess

This is a suppurative infection in the tissue spaces behind the pharynx. Normally there is only a narrow space between the posterior pharyngeal wall and the anterior ligaments of the spinal column. If oedema and pus expand this space, the posterior pharyngeal wall is pushed forwards, obstructing the airway.

The abnormal position of the pharyngeal wall is difficult to see on inspection, especially if the abscess is low in the throat. Many patients therefore present as emergencies, with neck or throat pain and difficulty in breathing. The diagnosis can be revealed by showing a wide soft-tissue space between the vertebrae and the air-filled pharynx on a lateral X-ray of the neck.

Emergency tracheostomy may be life-saving in urgent cases. Medical treatment is identical to that for Ludwig's angina, as the infection is almost always of mouth origin (and only rarely from a spinal infection). Pus can be released by incising the posterior pharyngeal wall. Spinal infection should be excluded by appropriate imaging of the spinal tissues. Rare cases of retropharyngeal abscess result from cervical infection with tuberculosis or (in endemic areas) brucellosis, so pus should be obtained for culture if the spine is involved.

Vincent's angina

This is a synergistic infection of the mouth that particularly affects the gums. It is associated with poor dental health and poor oral hygiene. The patient presents with extreme soreness of the mouth and gums, accompanied by an offensive halitosis.

The microbial cause is the synergistic action of the mouth spirochaete *Borrelia vincenti* and the anaerobe *Fusiformis* spp. If laboratory confirmation of the diagnosis is needed, the organisms can be demonstrated in large numbers by making a Gram stain of the material on a mouth swab.

Treatment with metronidazole will quickly eradicate the anaerobic organisms. Penicillin or ampicillin is also effective. Improved mouth care may be needed to avoid recurrence.

Lower Respiratory Tract Infections

7

Introduction

Respiratory infections are the commonest community-acquired and also healthcare-acquired infections of humans.

The lower respiratory tract includes all structures below the vocal cords. The trachea, bronchi and bronchioles are lined with ciliated columnar respiratory epithelium, within which are distributed mucus-producing goblet cells. The airway walls contain smooth-muscle cells, and are elastic, dilating on inspiration and narrowing during expiration. Any pathological narrowing is therefore increased during expiration.

The smallest bronchioles have no muscular coating, and their epithelium is flat and non-ciliated. The alveoli are lined with two types of cells; smooth, flat cells, and slightly thicker cells with a granular cytoplasm. The thicker cells are related to the alveolar macrophages, which are mobile within the alveoli. The alveolar cells are separated from the underlying capillary endothelium by a film of interstitial fluid, in hydrodynamic equilibrium with the alveoli (which usually contain no fluid) and the capillary blood.

The normal lower respiratory tract is bacteriologically sterile. Inhaled particles, including bacteria, are trapped in the mucus that lines the airways and are moved towards the pharynx by the beating of the epithelial cilia. The mucus is then swallowed and the bacteria are destroyed by gastric acid. Particles that reach the alveoli are phagocytosed by alveolar macrophages, which are also expelled by the 'ciliary escalator'. Secretory immunoglobulin A (IgA) in the respiratory epithelium helps to inhibit or destroy invading pathogens.

Common mechanical problems in the respiratory tract include:
1 Paralysis of the cilia by cigarette smoke.
2 Excessive volumes of mucus that cannot be effectively cleared.
3 Mucus too thick to be effectively cleared (in cystic fibrosis).

4 Paroxysmal narrowing of the airways by asthma attacks.

5 Immobilization or damage of alveolar macrophages by particles that they cannot destroy (e.g. silica or asbestos particles).

6 Loss of ability to cough, due to coma or paralysis.

7 Endotracheal intubation, which introduces microorganisms and inhibits effective coughing.

These factors impair the clearance of bacteria and other particles. Larger airways can be cleared by coughing, but this cannot maintain the patency of smaller airways, even when supplemented by vigorous physiotherapy. Thus, chronic infection and exudation lead to disruption of these structures, eventually resulting in bronchiectasis.

The lower respiratory tract is exposed to a variety of inhaled pathogens, which it can often expel before infection becomes established. Particles above 5 μm diameter (such as liquid droplets or dust) usually lodge in the nose and pharynx; smaller particles may reach the bronchi, and particles of 1 μm and smaller (such as droplet nuclei or spores) can reach the alveoli.

Streptococcus pneumoniae and *Haemophilus influenzae*, from the adjacent pharynx, are common colonists, and are also potential pathogens.

Their rich blood supply makes the lungs vulnerable to infection by blood-borne organisms, of which *Staphylococcus aureus* is the most important.

Non-infectious conditions can mimic lower respiratory infections, by presenting with fever and cough or shortness of breath. These include:

1 Autoimmune disorders such as systemic lupus erythematosus.

2 Granulomatous conditions such as sarcoidosis.

3 Vasculitides, particularly Wegener's granulomatosis or polyarteritis nodosa (both of which can also produce nodular opacities on chest X-ray).

4 Hypersensitivity reactions such as farmer's lung or the more severe Goodpasture's syndrome.

5 Malignancies, such as infiltrating lymphomata.

Laboratory diagnosis

Introduction

Identifying pathogens in respiratory specimens is difficult because of the problems of obtaining uncontaminated specimens. Diagnostic efforts are therefore directed at the common important pathogens. Adequate diagnosis can usually be achieved by a combination of sputum micro-scopy and culture. In severe pneumonia or suspected tuberculosis samples from the lower respiratory tract can be obtained by bronchoalveolar lavage.

Sputum examination

Many sputum samples contain mostly saliva and buccal epithelial cells, and are not representative of lower respiratory tract secretions. Effective coughing, sometimes assisted by physiotherapy, is necessary to obtain lung secretions. This can be further aided by prior inhalation of nebulized hypertonic saline, which induces a strong cough reflex (providing an 'induced sputum' specimen).

The specimen is first examined by microscopy of a Gram-stained smear. Specimens containing few polymorphs and more than 25 epithelial cells per low-power field are largely salivary, and likely to yield only upper respiratory tract flora on culture. Alternatively, specimens containing many neutrophils and few epithelial cells are largely of bronchial origin.

The common respiratory pathogens are derived from the normal upper respiratory tract flora so that care must be taken in interpreting culture results. The numbers of any potential pathogen in the specimen are estimated as a guide to significance. Organisms present in numbers of 10^6 CFU/ml or more, are considered more likely to be acting as pathogens.

Bronchoalveolar lavage

Bronchoscopy is valuable in the diagnosis of lower respiratory infections with unusual pathogens or in immunocompromised patients (who often produce little sputum because of their reduced inflammatory responses). The bronchoscope is passed into a lower airway and 100–200 ml of sterile saline is injected and re-aspirated. Although the material has come directly from the alveoli, semiquantitative methods are advisable, as contamination can still occur from the advancing end of the bronchoscope. Organisms present at more than 10^3 CFU/ml are likely to be pathogens. Bronchoalveolar lavage is of particular value for the diagnosis of infections such as tuberculosis, legionellosis, fungal infections and *Pneumocystis jiroveci* infection.

Obtaining specimens from the lower respiratory tract

1 Sputum obtained by coughing.
2 'Induced sputum' specimens.
3 Bronchiolar lavage.
4 Direct bronchial aspiration.

Viral infections of the lower respiratory tract

Introduction

Viral respiratory infections are among the most common of all infections. Many are part of a systemic disease such as influenza or varicella. Nevertheless, some viral infections specifically cause severe pulmonary disease.

Laboratory diagnosis

Nasopharyngeal secretions, which always contain respiratory cells, should be obtained from children under the age of 5 years, and from adults with suspected myxovirus or paramyxovirus infections. A fine-bore catheter is passed through the nostril into the nasopharynx and secretions are aspirated using gentle suction via either a 20- or 50-ml syringe or a low-pressure suction apparatus.

Direct immunofluorescence is a useful rapid laboratory test for some viral antigens, particularly RSV, influenza, parainfluenza or measles. Nasopharyngeal secretions are dried on a Teflon-coated slide, acetone-fixed and incubated with rabbit antibody to the expected virus. If viral antigen is present in the respiratory cells, the rabbit antibody will bind to it, and is demonstrated by staining with fluorescein-labelled anti-rabbit antibodies.

Organism list

- Influenza virus
- Parainfluenza viruses
- Human metapneumovirus (HMPV)
- Respiratory syncytial virus (RSV)
- Adenoviruses
- Measles, varicella and Epstein–Barr viruses (EBV)
- SARS coronavirus

Croup

Introduction

Croup is a syndrome of laryngotracheobronchitis, usually affecting children but occasionally seen in adults. Its importance is that almost every young child has at least one episode, often the child's first significant illness. The physical signs are distressing and respiratory obstruction can occur in severe cases, due to gross swelling of the aryepiglottic folds. The common aetiologies of croup must be distinguished from rarer causes of partial respiratory obstruction such as epiglottitis or diphtheria.

Epidemiology

Croup is common in infants. The principal causal agents are parainfluenza viruses types 1 and 2, which circulate during the late autumn and winter months. They are highly infectious by the airborne route. Prodromal measles causes a similar cough; influenza can cause croup in both adults and children.

Organisms that may cause croup

1 Parainfluenza virus.
2 Influenza virus.
3 Respiratory syncytial virus.
4 Measles virus.
5 *Haemophilus influenzae*.
6 *Corynebacterium diphtheriae*.

Pathology of parainfluenza virus infections

Parainfluenza viruses belong to the genus *Paramyxovirus* within the family Paramyxoviridae. Four parainfluenza virus types (types 1–4) are pathogenic to humans. The virus is pleomorphic with a roughly spherical shape, 120–300 nm in diameter, and possesses an envelope. The envelope contains glycoprotein projections of haemagglutinin, and a neuraminidase. The virion contains a helical nucleocapsid, which consists of viral negative-sense RNA attached to the large (L) protein and the P protein (which are both polymerase complexes), and the matrix (M) protein. Viral RNA is transcribed by virally encoded transcriptase into positive (sense) RNA.

The virus invades epithelial cells throughout the tracheobronchial tree, via the binding of haemagglutinin and neuraminidase proteins to sialic acid-coated cells. The haemagglutinin–neuraminidase complex antigen activates the fusion F protein by a serine protease that may be unique to the host cells that are targeted by paramyxoviruses. The columnar ciliated cells are damaged, and lose their ciliary function. There is mucosal oedema, with an acute neutrophilic infiltrate. A minority of cases suffer severe pneumonia, with alveolar exudate containing polymorphs and macrophages, and destruction of the ciliated epithelium. Fatal haemorrhagic bronchiolitis is occasionally seen.

Clinical features

The syndrome develops rapidly, with bursts of harsh, barking coughs interspersed with noisy breathing. Many cases are mild, with no other features. In more severe cases there is fever and increasing stridor, sometimes with intercostal recession on inspiration. Restlessness and tachycardia increase as respiratory obstruction develops. Auscultation of the chest rarely reveals abnormal sounds, unless the bronchi are sufficiently affected to be loaded with mucus. Cyanosis is a late and grave sign. Most cases resolve within a week, but a croupy cough may return if there are further respiratory infections in following weeks.

Diagnosis

Diagnosis is usually clinical, based on the typical cough. A low or normal white cell count supports a probable viral aetiology. In patients with neutrophilia, other causes of croup or pharyngeal obstruction should be considered, including epiglottitis and diphtheria.

A parainfluenza or influenza virus may be demonstrable by direct immunofluorescence staining or ELISA testing of nasopharyngeal aspirate. Increasingly, laboratories use multiplex ELISA, or reverse transcriptase polymerase chain reaction (RT-PCR) tests to detect the major viral pathogens. Virus may be recovered by cell culture of respiratory secretions although this is too slow for clinical use. Serodiagnosis is rarely undertaken.

Management

Symptoms are improved by surrounding the patient with a warm, steamy atmosphere, which helps to relieve pain and reduce cough. The heat may inhibit viral replication.

In more severe cases, measures may be required to reduce airway swelling.

Treatment of airway swelling in croup

1 Dexamethasone, orally 150–600 µg/kg, or intravenously 300–600 µg/kg.
2 Budesonide, 2 mg by nebulizer.
3 If 1 and 2 are insufficient: nebulized adrenaline solution, 1 mg/ml; up to maximum 5 mg can be given, repeated if necessary after 30 minutes (the effect will last for 2–3 hours: *expert supervision of this treatment is needed*).

Bronchiolitis

Introduction and epidemiology

This condition shares similar pathology with croup, but mainly affects the small bronchioles instead of the upper airways.

The infection principally affects infants up to the age of 2 years. Peaks in numbers of cases occur during late winter and early spring though the size of the peak varies from winter to winter (Fig. 7.1). These are usually due to respiratory syncytial virus (RSV), although other viruses, particularly human metapneumovirus, may cause the same syndrome. Bronchiolitis causes considerable morbidity in infants, but with modern treatment mortality is very low in normal children. In England, 2.8% of hospital admissions in children under 12 months old are attributed to RSV each year. Outbreaks often coincide with increases in the incidence of sudden infant death syndrome, suggesting a possible role of RSV in its aetiology. Preterm infants and those with congenital heart disease are at particular risk of severe infection. Serious outbreaks have been reported among newborn babies on intensive care units.

Humans are the only source of infection. The incubation period is 5–8 days and the patient remains infectious for about 7 days after the onset of symptoms.

Pathology of RSV infection

RSV belongs to the genus *Pneumovirus* in the family Paramyxoviridae. It is an enveloped virus of 120–300 nm diameter. The helical, single-stranded, positive-sense RNA genome codes for at least 10 polypeptides including F, G

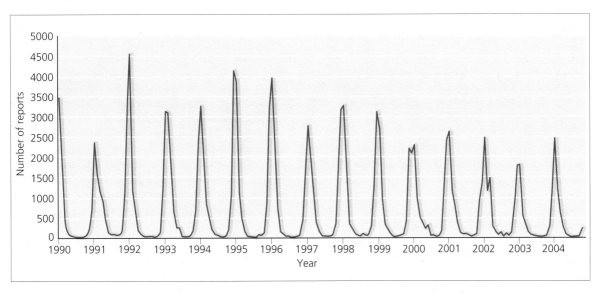

Figure 7.1 Epidemic curve of respiratory syncytial virus infections, showing annual winter/spring epidemics.

and Sh envelope-associated glycoproteins. The F (fusion) protein is associated with penetration of the virus into cells, and its spread from cell to cell, but these processes depend upon the co-expression of all three glycoproteins. The larger G protein is responsible for initial attachment of the virus to host cells. There is antigenic variation among strains of RSV but this is not easy to demonstrate, as human sera contain antibody that neutralizes all viruses of this genus. Monoclonal antibodies allow the distinction of two major groups that are further divided into subgroups. The antigenic variation between the two groups resides in the G, F, Sh and NS1 antigens. Strain heterogeneity is found in the G and Sh glycoproteins.

RSV is acquired through the upper respiratory tract and spreads throughout the respiratory epithelium. Infection is characterized by peribronchial inflammation, with oedema of the submucosa and adventitium. There is necrosis of the bronchiolar epithelium and plugging of the airways. When the bronchiolar lumen is incompletely obstructed a ball-valve effect occurs, which results in air-trapping, causing the characteristic patchy hyperinflation seen in this condition.

Clinical features

Illness begins with fever, mucoid secretions and a loose cough. After 3 or 4 days the fever rapidly declines, the respiratory rate increases and intercostal recession appears. Breathing is wheezy, and suckling is often impossible, as crying or feeding causes cyanosis. Respiratory difficulties can develop suddenly in a child with an apparently mild illness.

Chest examination reveals widespread wheeze, particularly in expiration. Patchy areas of dullness, reduced breath sounds or adventitious sounds are common and may change with coughing, which is repetitive and often constant.

Bronchiolar obstruction causes atelectasis in some lung areas, with air-trapping and hyperinflation in others. The chest X-ray therefore shows varying areas of opacity and lucency.

Slow improvement begins after 5 or 6 days. It may take a week or two for the child to return to normal, but the capillary oxygen saturation is often subnormal for twice as long.

Diagnosis

Clinical diagnosis is usually obvious.

Laboratory diagnosis

Laboratory diagnosis of RSV infection can be made by direct immunofluorescence, or enzyme immunoassay (EIA) performed on nasopharyngeal secretions. RT-PCR performed on the same sample has proved more sensitive and specific. RSV can also be cultured in Hep-2 or HeLa cells. The characteristic cytopathic effect is the formation of syncytia, generally seen between 2 and 7 days after inoculation.

A retrospective diagnosis can also be achieved by detecting rising titres of RSV antibodies, by complement fixation, virus neutralization, EIA and indirect immunofluorescence tests.

Laboratory diagnosis of RSV infection

1 Direct immunofluorescence or EIA of respiratory secretions.
2 Reverse transcriptase PCR to demonstrate viral RNA.
3 Cell culture.
4 Serological: complement fixation, enzyme immunoassay or indirect immunofluorescence.

Management

Oxygenation and adequate hydration are the immediate requirements. Humidified oxygen can be given in a cot-sized oxygen tent or an infant head-box. The inspired oxygen concentration is adjusted to maintain as near normal arterial oxygen saturation as possible. Pulse oximetry is a useful means of monitoring saturation.

Bronchospasm sometimes plays a part in the bronchiolar narrowing; a trial of nebulized salbutamol is worthwhile, particularly if there is a past or family history of atopy.

Nebulized ribavirin shortens the duration of hypoxia and fever in bronchiolitis. Its effect is small, but morbidity from bronchiolitis is great in those with pre-existing cardiac or lung disorders, and it is licensed for use in these children. A 20 mg/ml solution of ribavirin is nebulized, and administered for 12–18 h daily, for at least 3 days.

Complications

Secondary bacterial infection is uncommon, but can threaten life. High fever, the development of purulent bronchial secretions, sustained deterioration in oxygen saturation and rising neutrophil count are warning signs. The chest X-ray may show a segmental opacity or an air bronchogram, suggesting consolidation.

Blood and aspirated secretions should be cultured and intravenous antibiotic treatment started without delay. A cephalosporin such as cefuroxime or cefotaxime will be effective against *Streptococcus pneumoniae* or *Haemophilus influenzae*, which are the likely pathogens, and will also act against occasional *Staphylococcus aureus* infections.

Prevention

No vaccine is currently available. Passive immunization with the monoclonal antibody preparation Palivizumab, 15 mg/kg monthly by intramuscular injection, is used for prevention of RSV infections in susceptible infants. These are infants born at up to 35 weeks gestation and who are less than 6 months old at the onset of the RSV season, or children aged less than 2 years who have been treated for bronchopulmonary dysplasia within the last 6 months. Cardiopulmonary resuscitation facilities must be available. Side-effects include fever and agitation, sometimes cough, wheeze, diarrhoea, vomiting, rash, leucopenia and liver function abnormalities.

Influenza

- Influenza A
- Influenza B
- Influenza C
- Avian influenza H5N1 and others, including H7N3

Introduction

Influenza is a moderate to severe illness most often caused by influenza A or B viruses. It is highly infectious, and with its short incubation period it can rapidly cause large epidemics and pandemics. Sudden absence of staff adversely affects commerce, industry and public services. Influenza cases can be seriously ill, and severe secondary bacterial infections cause as much morbidity and mortality as the influenza itself.

Virology and pathogenesis of influenza

Influenza virus belongs to the genus *Orthomyxovirus* in the family Orthomyxoviridae. Virus particles are 80–120 nm in diameter. The RNA genome is segmented, with four fragments. The RNA is closely associated with nucleoprotein (NP) in a helical structure. The NP is specific to the three types of influenza virus: A, B and C. The matrix (membrane) protein, with a mass of 2 kDa, surrounds the nucleocapsid, has M1 and M2 components, and is the major protein of the virus particle. This is enclosed by the envelope, which contains two virus-encoded glycoproteins: haemagglutinin (H), responsible for attachment of the virus to host cell receptors, and neuraminidase (N), of molecular weight 200–250 kDa, important for release of virus from infected cells (Fig. 7.2). Both the H and the N glycoproteins are antigenically variable, and this is useful in the classification of the viruses.

The virus adsorbs to host cell sialic acid-containing surface structures (a process that can be blocked by specific secretory IgA). After internalization the haemagglutinin molecule is transformed into a fusogenic form, and facilitates viral fusion with the endosome. The M2 matrix protein forms an ion channel that permits influx of protons, releasing the core from the M1 matrix protein. Viral progeny are released by budding from the host-cell membrane. Host-cell death eventually results from virally mediated degradation of newly synthesized host mRNA, blocking protein production. Also, an influenza virus protein (PA) degrades host-cell proteins and stimulates apoptosis. The virus can also inhibit the functioning of polymorphs,

This is a typical influenza A or B virus (influenza C does not possess neuraminidase structures)

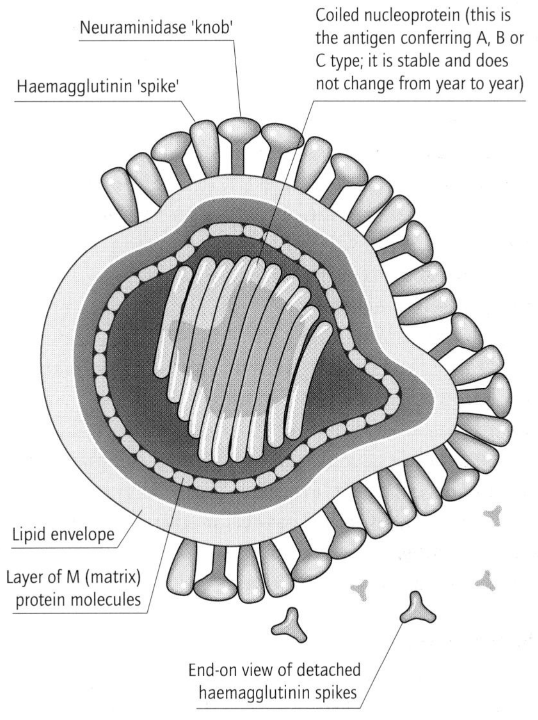

Neuraminidase 'knob'

Haemagglutinin 'spike'

Coiled nucleoprotein (this is the antigen conferring A, B or C type; it is stable and does not change from year to year)

Lipid envelope

Layer of M (matrix) protein molecules

End-on view of detached haemagglutinin spikes

Width of virus structure = 100 nm

Figure 7.2 Structure of influenza viruses.

lymphocytes and monocytes. Virus shedding commences just before the onset of illness and continues at a high level for around 48 hours, before declining rapidly and ceasing within 5 to 10 days. Despite our knowledge of influenza pathogenicity, reasons for the severity of the 1918 influenza pandemic remain unknown.

Influenza virus infects ciliated respiratory cells, killing many, and causing shedding of the majority. This leaves denuded airways, susceptible to colonization and invasion by bacterial pathogens. There is a vigorous immune response, with much production of interferon and temporary impairment of cell-mediated immunity (the tuberculin reaction cannot be elicited until convalescence is complete). Although influenza viraemia is difficult to demonstrate, viral genome is present in large quantities within the mononuclear cells of the blood during acute infection, and many body tissues can be invaded.

Epidemiology

Increased transmission of influenza occurs in most years during the winter months and sporadic cases occur throughout the year. There are two main virus types that cause disease, influenza A and influenza B (influenza C is an uncommon type that infrequently causes infection). Influenza A is responsible for most cases and is usually a more severe infection than influenza B. Influenza viruses are easily transmitted by droplet spread. Attack rates during epidemics seldom exceed 10% in the general community, but may approach 50% in closed communities such as boarding schools, nursing homes and military barracks. An epidemic is defined in the UK when the weekly family practice consultation rate for influenza-like illness exceeds 200 cases per 100 000 population. Pandemics are much less common, and in the last century have been recorded in 1918, 1957, 1968 and 1977.

Peaks of transmission, epidemics and pandemics are the results of continuing changes in the antigenic structure of the virus.

Antigenic drift is responsible for the yearly winter peak of influenza activity as new subtypes of haemagglutinin (H) slowly evolve. Infection with the circulating strain provides antibody that cross-reacts with closely related H types, conferring partial immunity to infection by slowly evolving types from year to year. Antigenic drift is driven by selection pressure imposed by the partial immunity of the population. Influenza A drifts more than influenza B. Periodically a major change (shift) occurs in the haemagglutinin or neuraminidase antigen, often as a result of reassortment of genome segments from two different influenza A strains, one of which may be an animal- or bird-adapted virus (see Chapter 26). The population has little or no cross-reacting antibody to the 'new' virus, and an epidemic or pandemic results.

Age-specific attack rates during an epidemic reflect existing immunity from exposure to previously prevalent strains. Attack rates are often highest among school-age children. In contrast, complication rates and mortality are greatest in the elderly and those with underlying chronic conditions. Numerous and severe infections occur during pandemics, as few members of the population possess immunity to the novel virus strain.

Excess mortality during influenza epidemics

1 Deaths due to respiratory disease may increase by as much as 50%.
2 Deaths due to cerebrovascular and cardiovascular disease also increase.
3 Excess mortality is not usually followed by a deficit during the following year.
4 There were an estimated 26 080 excess deaths attributable to influenza in England and Wales during the 1989/90 epidemic.
5 Over 80% of excess deaths are in people aged 65+ years.

Clinical features

After 3 or 4 days incubation there is a sudden onset of se-

vere malaise, arthralgia, myalgia and prostration. Fever may be high, with shivering and sweating. Associated symptoms include vomiting, headache, sore throat, loose stools and shortness of breath.

Clinical definition of influenza-like illness (ILI)

An acute feverish illness, accompanied by at least two of:
• cough
• sore throat
• myalgia
• headache

⚠ In an influenza epidemic, these features are strongly correlated with a virological diagnosis of influenza.

Uncomplicated influenza usually persists for 4–7 days before the fever decreases and gradual convalescence begins. Children are often less severely affected than adults, but are still an important source of infection (Fig. 7.3).

The respiratory tract is a major target of the infection. Respiratory features include tracheitis, croupy cough, laryngitis and sometimes sinusitis and conjunctivitis. Influenza C infections are uncommonly recognized, and often mild, with conjunctivitis as a prominent feature. Pneumonitis may be widespread, but is rarely clinically detectable, except by a slight reduction in the capillary oxygen

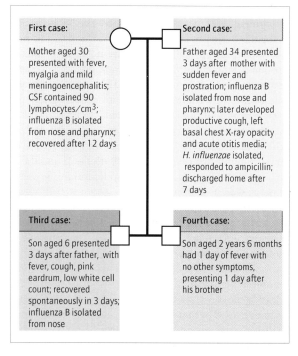

First case:
Mother aged 30 presented with fever, myalgia and mild meningoencephalitis; CSF contained 90 lymphocytes/cm³; influenza B isolated from nose and pharynx; recovered after 12 days

Second case:
Father aged 34 presented 3 days after mother with sudden fever and prostration; influenza B isolated from nose and pharynx; later developed productive cough, left basal chest X-ray opacity and acute otitis media; *H. influenzae* isolated, responded to ampicillin; discharged home after 7 days

Third case:
Son aged 6 presented 3 days after father, with fever, cough, pink eardrum, low white cell count; recovered spontaneously in 3 days; influenza B isolated from nose

Fourth case:
Son aged 2 years 6 months had 1 day of fever with no other symptoms, presenting 1 day after his brother

Figure 7.3 A family outbreak of influenza B, illustrating the range of presentations and their relationship to the patients' ages.

saturation on pulse oximetry. The chest X-ray is usually normal in uncomplicated influenza. Severe interstitial or segmental pneumonia is rare, but severe pneumonitis can be life-threatening, especially in high-risk groups.

Other body systems are often affected. Myocarditis is common, as shown by a slightly altered electrocardiogram. Clinical features, such as variable extrasystoles, and postural or exercise-related hypotension, are rare. Skeletal myositis is sometimes seen in early convalescence.

There is a variable rate of encephalitis or meningoencephalitis with different influenza strains. This coincides with viral excretion from the respiratory tract. Typical features are headache, meningism, irritability, altered personality and drowsiness, with slow recovery. The cerebrospinal fluid (CSF) contains excess lymphocytes and slightly raised protein levels.

Avian influenza affects individuals, including children, who have very close contact with poultry. The disease is severe, with multisystem involvement, particularly gastrointestinal symptoms, and a high case-fatality rate. Human to human transmission is usually rare.

Patients considered at increased risk from influenza infection

1 Patients aged over 65 years.
2 Patients with chronic respiratory, cardiovascular (excluding hypertension) or renal disease.
3 Patients with diabetes mellitus.
4 Immunosuppressed patients.

Diagnosis of influenza

Cases may be identified clinically in epidemics. Mild and sporadic cases may be indistinguishable from other acute viral respiratory infections. Diagnosis is then confirmed by virus identification in nose- or throat-swab specimens, nasopharyngeal aspirate, bronchial washings or CSF.

Laboratory findings reflect tissue damage and the effects of interferon. The white cell count is low, often 3 or $4 \times 10^9/l$. Transaminase levels of both liver and tissue origin are raised. Amylase levels are also occasionally raised. Oxygen saturation, measured by pulse oximetry, is often reduced, even when there is no clinical evidence of significant pneumonitis. Some patients with severe myalgia and weakness have elevated creatine kinase levels. Troponins may be raised in myocarditis.

Laboratory diagnosis

A rapid diagnosis can be made by demonstrating type-specific antigen in nasopharyngeal cells by EIA or direct immunofluorescence; RT-PCR techniques can demonstrate viral RNA. Virus isolation permits definition of the antigenic structure of the virus for future vaccine prepa-

ration. Nasopharyngeal secretions are inoculated on to continuous human diploid fibroblasts incubated at both 33 °C and 37 °C, as some viruses replicate optimally at a lower temperature. The growth of virus in the cell culture is detected by haemadsorption using guinea-pig and chicken erythrocytes (but guinea-pig erythrocytes do not adhere to cells infected with type C). Virus can also be recovered from nose or throat swabs, bronchial secretions and occasionally from CSF. After primary isolation, the virus is typed in dedicated reference laboratories.

Rising titres of neutralizing, complement-fixing or haemagglutinin antibodies can be demonstrated in paired sera.

Laboratory diagnosis of influenza

1 EIA, direct immunofluorescence, or RT-PCR of respiratory secretions.
2 Cell culture of respiratory secretions or cerebrospinal fluid.
3 Serologically by complement fixing or haemagglutinating antibodies.

Management

Rest, warmth, adequate hydration and analgesia all give considerable relief, and are adequate for managing mild cases. The myocardium is protected from excess work, and oxygen desaturation is limited, while the patient rests. Reduction of fever is not usually necessary; indeed virus replication in mucosae may be reduced when the local temperature rises above 35 °C.

Antiviral drugs

Antiviral drugs, given within 48 hours of first symptoms, can shorten the duration of illness by 24–36 hours. The UK National Institute for Clinical Excellence recommends treatment for influenza A or B in at-risk patients when indicators of epidemic influenza activity are recorded. Treatment of healthy individuals is not recommended.

The World Health Organization indicator of epidemic influenza activity is the occurrence of more than 200 cases of acute influenza-like syndrome per 100 000 population, in 1 week.

Zanamavir may cause gastrointestinal side-effects, or bronchospasm: short-acting bronchodilators should be available. Oseltamivir may cause nausea and other gastrointestinal symptoms, ear problems or dizziness and rare hypersensitivity rashes. Dosage must be reduced or avoided in renal impairment. Amantadine is active against clinical influenza A. Side-effects include drowsiness, confusion and reticulate rash, which can be difficult for elderly patients to tolerate.

Antiviral treatment of influenza

1 Zanamavir, by powder inhalation: 10 mg 12-hourly for 5 days (not recommended for children under 12).

2 Oseltamivir, orally: 75 mg 12-hourly (child doses all orally, 12-hourly: over 1 year, body weight up to 15 kg: 30 mg; 16–23 kg: 45 mg; 24–40 kg: 60 mg) treat for 5 days.

⚠ Amantadine (for influenza A only; adults and children over age 10 years: not recommended in the UK, because of easy emergence of resistance) 100 mg daily for 4–5 days.

Intravenous ribavirin has been used to treat severe influenza pneumonia, but is not validated or licensed for this.

Complications

Secondary bacterial infection

Secondary bacterial infection is common and should be suspected in any patient with deteriorating respiratory function or features of sepsis. *Streptococcus pneumoniae* and *Haemophilus influenzae* secondary infections are common, but severe *Staphylococcus aureus* infection also occurs, with deteriorating lung function, purulent sputum and slow or late increase in the white cell count. Nodular, segmental or lobar opacities are seen on chest X-ray; abscess cavities may develop; and staphylococcal bacteraemia may occur.

Blood and sputum cultures should be obtained, and antibiotic treatment promptly commenced, including drugs that are effective against *S. aureus*.

Bacterial sinusitis or otitis media can complicate influenza in both children and adults. The expected pathogens are the same as for pneumonia.

Post-viral complications

Post-viral complications occur with variable frequency, depending on the characteristics of both the causative virus and the affected population. Prolonged convalescence with fatigue is common, often lasting from 6 to 12 weeks. Neurological effects include Guillain–Barré syndrome, mononeuropathies and occasionally encephalopathy. Persisting encephalopathy is rare, but the encephalitis lethargica cases in 1918 may have been related to the influenza pandemic of that year.

Prevention and control

Influenza vaccines are non-replicating virus vaccines, whose efficacy is around 70%. They are less efficacious in the elderly, debilitated or immunosuppressed, but still reduce influenza-related morbidity and mortality in these groups. Different strains of antigen are used in different years, to match the antigenic structure of the prevalent influenza virus strains (see Chapter 26). Usually the vaccine contains antigens of two type A and one type B virus. Some vaccines are composed of purified viral surface antigens; others, called split vaccines, are made by separating viral core material from surface structures and using the partly purified surface structures as the vaccine antigens.

Vaccination is generally recommended in the UK for all people over 65 years of age, and for all those over 6 months of age with chronic respiratory and cardiac disease, chronic renal failure, diabetes, other endocrine disorders and immunosuppression. It should also be given to residents of nursing homes and other institutions for the elderly. In some countries healthcare workers are routinely vaccinated. The vaccine is contraindicated in patients with known anaphylactic hypersensitivity to egg products.

Oseltamivir is effective for post-exposure prophylaxis of influenza A and B.

UK guidelines on influenza prophylaxis

Where it is known that either influenza A or B is circulating in the community, oseltamivir may be used for the post-exposure prophylaxis of influenza in at-risk people aged 13 years or over who are not effectively protected by vaccination, have been exposed to someone with influenza-like illness and are able to begin prophylaxis within 48 hours of exposure. People at risk are defined in the same way as the group recommended for vaccination. For those at risk over 13 years of age and living in a residential care establishment, oseltamivir prophylaxis is recommended irrespective of prior vaccination, provided it can be started within 48 hours of exposure. (Guidelines available at www.nice.org.uk.)

Oseltamivir dose for prophylaxis: 75 mg daily.

⚠ Amantadine hydrochloride has been used for prophylaxis of influenza type A. Resistant influenza A strains quickly emerge during its use, and it is no longer recommended in the UK.

Other viral pneumonias

Many viruses are capable of causing localized or generalized pneumonitis. Segmental pneumonia, with a distinct chest X-ray opacity, can be seen in infections with human metapneumovirus, adenovirus, RSV, parainfluenza virus, and EBV. Pneumotropic hantaviruses, excreted by rodents, cause severe and fatal pneumonias in various parts of the world (see Chapter 24), but are not endemic in the UK.

SARS coronavirus pneumonia

In February to May 2003, an increasing number of cases of severe viral pneumonia were seen, first in mainland China, then in Hong Kong and rapidly spread to other countries on direct air routes from Hong Kong. Clinical disease mainly affected adults, with a case-fatality rate of over 30% in the elderly. A worldwide programme of travel restrictions, and quarantine of cases and contacts, helped to terminate the pandemic. Apart from isolated cases of laboratory-acquired SARS, no transmission of the virus has been detected by intense surveillance in subsequent years (see Chapter 26).

Human metapneumovirus

Virology

This virus was first described in Holland in 2001. It has paramyxovirus-like morphology and is most closely related to the pneumovirinae. With a single strand, non-segmented, negative sense RNA genome, it is closely related to avian pneumovirus. Its structure includes a matrix protein, fusion protein, nucleoprotein and phosphoprotein. Gene sequencing demonstrates that human metapneumoviruses currently cluster into at least two groups.

Diagnosis

Human metapneumovirus grows poorly in cell culture but can be identified by its cytopathic effect (small round granular refringent cells) after 21 days growth in monkey kidney cells. Confirmation of the isolate is by immunofluorescence. In the clinical context the most reliable means of diagnosis is RT-PCR, using primers directed against the matrix protein. Serology is unhelpful, as antibodies are universal by the age of five.

Adenovirus

Adenoviruses, particularly types 4, 5 and 7, tend to affect children and young adults. Chest infection may be accompanied by pharyngitis, conjunctivitis, tender lymphadenopathy or, occasionally, a morbilliform rash. Epidemics can affect closed communities in barracks or institutions. Vaccines are used to prevent epidemics in military recruits.

Respiratory syncytial virus

RSV can cause outbreaks of segmental pneumonia, particularly among the elderly. The illness is mild to moderate, with a low mortality.

Pneumonias in other viral infections

Parainfluenza and EBV tend to cause subclinical lung disease, though there may be a cough. A segmental lung opacity is often discovered by chance on chest X-ray. EBV can also cause lymphocytic lung nodules or pneumonitis in immunosuppressed patients.

Diffuse pneumonitis can be caused by parainfluenza virus, measles and chickenpox. The lung disease is rarely clinically significant in immunocompetent individuals. Varicella pneumonia is the exception, being potentially severe or life-threatening, but it is accompanied by the characteristic chickenpox rash (see Chapter 11).

Some hantavirus strains can cause a severe inhalational pneumonia, with a high fatality rate. This may or may not be accompanied by a haemorrhagic renal syndrome, typical of other hantavirus diseases (see Chapter 24).

'Other' viruses causing pneumonitis

1 Parainfluenza virus.
2 Varicella zoster virus.
3 Epstein–Barr virus.
4 Pneumotropic hantaviruses.
5 SARS coronavirus.

Pyogenic bacterial respiratory infections

Chronic obstructive pulmonary disease (COPD, chronic bronchitis)

Introduction

Chronic obstructive pulmonary disease is characterized by excessive mucus production and poor clearance of the bronchi, with varying degrees of accompanying bronchospasm. The abnormal bronchi easily acquire a colonizing flora, and are also susceptible to intercurrent infections with respiratory viruses, 'atypical' infections and recrudescences of *Bordetella pertussis* infection, which are now recognized to be common in adults. Repeated infections cause progressive bronchial damage and bronchospasm, and an inexorable decline in lung function.

Epidemiology

COPD is among the commonest causes of hospital admission. The incidence is greatest in adults, although children are also affected. Patients who smoke and those with dust diseases are particularly affected, probably due to impaired ciliary clearance of secretions. Alpha-1 anti-trypsin deficiency and inherited defects of ciliary function (Kartagener's syndrome) also predispose to severe infections. Patients with cardiac disease have poor alveolar clearance due to oedema. Most infective episodes occur during the winter. There is evidence that the incidence is related to levels of atmospheric pollution, particularly sulphur dioxide levels.

Organism list

Common
- *Streptococcus pneumoniae*
- *Haemophilus influenzae*
- *Chlamydia pneumoniae*
- *Mycoplasma pneumoniae*
- Respiratory viruses
Uncommon
- *Moraxella catarrhalis*

- *Klebsiella pneumoniae*
- *Escherichia coli* and other Gram-negative rods
- *Bordetella pertussis*

Note that infections caused by mixed flora are probably common.

Pathology

The infections are superficial, affecting the mucosa of the bronchi and the mucus-producing goblet cells. There is oedema, excess mucus and acute inflammatory exudate, all of which acutely exacerbate chronic obstruction and bronchospasm.

Clinical features

The illness develops with cough, raised respiratory rate, and increased expectoration and sputum purulence. The tendency to bronchospasm increases. Systemic effects are few, but the body temperature and the white cell count may be slightly raised. The chest X-ray rarely shows any definite change unless bronchial obstruction has precipitated a complicating segmental infection.

Diagnosis

This is clinical. The sputum is always colonized with mixed upper respiratory tract flora, and may contain neutrophils. This does not change, except in degree, during acute exacerbations.

Management

General measures

Measures to optimize bronchial patency, drainage of secretions and oxygenation are as important, or more so, than therapy aimed at reduction or eradication of microbial flora. Measures include: optimum bronchodilator treatment with inhaled or nebulized selective beta-agonists, or antimuscarinic bronchodilators; adding or increasing the dose of oral corticosteroids, chest physiotherapy, oxygen therapy sufficient to maintain adequate arterial oxygen saturation (with care not to inhibit respiratory function) and, in suitable patients, non-invasive ventilatory support.

A short course of antibiotic treatment

Oral treatment is often adequate. Agents suitable for the commonest associated organisms include tetracyclines, amoxicillin, co-amoxiclav or clarithromycin. Almost all *Moraxella catarrhalis* isolates produce a beta-lactamase, but are sensitive to co-amoxiclav and most non-beta-lactam antibiotics. Courses of antibiotic should be just long enough to terminate infection, but insufficient to encourage the emergence of new colonizing flora; 5 days is ideal. Patients who have received repeated courses of antibiotics may become colonized with Gram-negative rods such as *Escherichia coli* or *Klebsiella pneumoniae*. These

may need treatment with antibiotics such as co-amoxiclav or cefuroxime.

In severe episodes, parenteral antibiotics may be indicated. Co-amoxiclav or cefuroxime is often effective. The use of very broad-spectrum antibiotics, such as third-generation cephalosporins, predisposes to antibiotic-associated diarrhoea.

Non-invasive ventilation (NIV)

Non-invasive ventilation is the provision of positive-pressure triggered respiratory support, using a sealed face mask (or, occasionally, a nasal mask, which is less efficient). It reduces the work of breathing and assists in reducing carbon dioxide retention and acidosis. It is increasingly used to provide temporary respiratory support to patients with type 2 respiratory failure (low oxygen saturation with increased carbon dioxide levels in the blood) during exacerbations of COPD. It prevents or shortens admission to intensive care, as it assists weaning from invasive ventilation, and is valuable in supporting patients who can return to a life of significant quality after an episode of exacerbation.

Indications for considering NIV (non-invasive ventilation)

If, using optimum inspired oxygen therapy by Venturi mask, PaO_2 cannot be maintained above 8.0 kPa without the $PaCO_2$ rising above 6.5 kPa and the arterial pH falling below 7.35, when cardiovascular status is stable.

⚠ NIV is useful if a patient has coexisting chest deformity or neuromuscular impairment that exacerbates temporary impairment of respiratory function.

⚠ NIV is not indicated for type 1 respiratory failure (when the $PaCO_2$ is normal or low), or if acute severe asthma coexists; it may be unsuitable for a patient with impaired consciousness, with difficulties in protecting the airway, or with excessive secretions or uncontrolled vomiting.

Bronchiectasis

In bronchiectasis the bronchial walls are damaged by severe or repeated infections, sometimes resulting from congenital immunoglobulin deficiency or ciliary abnormalities in conditions such as Kartagener's syndrome. Saccular or fusiform spaces derived from dilated bronchi affect one or more areas of the lungs (Fig. 7.4). Ciliary clearance is ineffective in these wide spaces, which contain pools of stagnant secretions. The respiratory epithelium may become squamous after many infectious insults. Patients have a productive cough, and produce copious purulent sputum, especially on rising in the morning. Severe, recurrent infectious episodes are common.

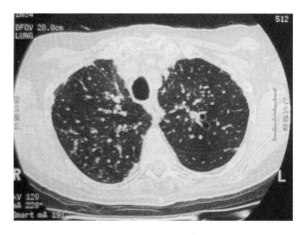
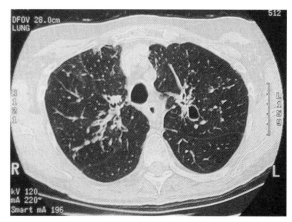

Figure 7.4 Bronchiectasis: these CT scans show cystic dilatation of bronchi entering the left upper lobe; they are not currently infected – in chronic infection the walls of the ectatic bronchi lose their ciliated epithelium, thicken and become rigid, hindering the clearance of secretions.

Long-term complications include amyloidosis, hypertrophic pulmonary osteoarthropathy (HPOA) and late malignant change.

Cystic fibrosis

Cystic fibrosis (CF) is a congenital condition in which abnormal, tenacious mucus accumulates in the bronchial tree. The ciliary clearance process is ineffective, and segments of the bronchial tree are chronically obstructed and inflamed, producing large volumes of mucus and mucopus. There is colonization of the bronchi with a resident flora that causes repeated infections. CF sufferers also have poor phagocyte function, which additionally predisposes to chronic infection with *S. aureus*, *Pseudomonas aeruginosa* and *Burkholderia cepacia*. *P. aeruginosa* sometimes produces large quantities of alginate capsule-like material, enclosing microcolonies that are protected from the host's immune defence. Furthermore, it produces a variety of enzymes and toxins, including elastases, proteases, DNase, lecithinase and exotoxin A. These are damaging to both bronchial structures and defence systems, and contribute to deteriorating lung function. Some strains of *Burkholderia cepacia* can colonize the lungs of CF sufferers and cause severe infective episodes. The rate of decline in lung function is directly related to the acquisition of these organisms. Debility and failure to thrive in infants is exacerbated by coexisting abnormalities of bowel and digestive function.

Management depends on vigorous physiotherapy and postural drainage, usually a lifelong twice-daily burden. This is combined with early and vigorous antibiotic treatment of infections. In CF patients it is usual to treat exacerbations immediately with intravenous drugs that have an antipseudomonal action, e.g. ceftazidime. Many CF patients have implanted cannulae with subcutaneous access so that domiciliary treatment can be commenced without delay.

Localized bronchiectasis can be treated by removal of the damaged and infected lung segment or lobe. This often permits a marked improvement in general health. Heart–lung transplant may be offered to CF patients. This cures respiratory failure and normalizes lung mucus, but the denervated lungs lack a cough reflex, so physiotherapy must be continued to avoid aspiration, and minimize infection of the transplanted lung.

Community-acquired pneumonia (CAP)

Organism list
- *Streptococcus pneumoniae* (35 to 40% of all identified pathogens)
- *Haemophilus influenzae* (10% of identified causes)
- *Chlamydia pneumoniae*
- *Mycoplasma pneumoniae*
- *Staphylococcus aureus*
- *Moraxella catarrhalis*
- *Legionella pneumophila*
- *Klebsiella pneumoniae*
- *Escherichia coli* and other Gram-negative rods
- *Fusobacterium necrophorum*
- Rare bacterial lung infections: *Streptococcus pyogenes*, enteric fevers (see Chapter 24), anthrax, plague (see Chapter 25), leptospirosis (see Chapter 9), respiratory viruses (about 10% of identified causes).

Note that a causative organism cannot be identified in nearly half of investigated CAP cases.

Introduction

Community-acquired pneumonia (CAP) is a common acute infection characterized by fever and respiratory symptoms such as cough, shortness of breath, chest pain, or sputum expectoration, combined with clinical features of lung consolidation and radiographic evidence of lung infiltrate.

CAP affects all age groups, with an annual incidence of 5–6 per 1000 in adolescents and working adults, but reaching 35 per 1000 in the over-75s. The predominant causative organisms are similar in different groups. Although some clinical features are particularly associated with individual organisms, it is not often possible to determine aetiology on clinical grounds. Even with modern diagnostic methods, no causative organism is identified in nearly half of all investigated cases.

Patients at high risk of CAP include those with lung damage due to heart failure, dust diseases or autoimmune disorders, patients with chronic cardiac, respiratory, liver and renal disease, diabetes mellitus, alcohol abuse and immunosuppression, including young patients with poor CD4 T-cell function (such as immunosuppression due to HIV disease), who are not old enough to have acquired antibodies protective against respiratory pathogens.

Very young or elderly patients, and those with coexisting diseases have a worse outcome from CAP than otherwise healthy individuals.

Management of CAP

Initial management is aimed at confirming the diagnosis, assessing the grade of severity and, in the hospital setting, obtaining diagnostic specimens. The patient should be supported with immediate measures such as oxygen, intravenous hydration or analgesia.

In the community

• Obtain a full history and examination, to demonstrate signs of pulmonary infection, to exclude additional pathology (e.g. heart failure), and to seek epidemiological information (e.g. exposure to animals, birds or potentially legionella-infected aerosols, or recent hospital admission).
• Review the severity of the illness and arrange hospital admission for severe or complicated cases.

In hospital

• Obtain a chest X-ray to confirm the diagnosis of CAP.
• Take two sets of blood cultures.
• Perform baseline laboratory tests to assess severity and coexisting conditions (full blood count, CRP, renal and liver profile, blood glucose; blood gas analysis if indicated by clinical signs).
• Obtain urine for pneumococcal and legionella antigen tests.
• Collect sputum (if obtainable) for laboratory examination.
• Obtain a baseline serum specimen for later serodiagnosis.
• Consider other investigations, including examination of induced sputum specimen and/or nasopharyngeal aspirate.

Table 7.1 summarizes the assessment of the severity of CAP.

Choice of antibiotic treatment for CAP

In most cases, initial antibiotic treatment must be empirical, as initial assessment, and sputum microscopy rarely reveal a causative pathogen. The antibiotic regimen can be adjusted in the light of later laboratory results (Table 7.2).

Table 7.1 Assessment of the severity of CAP

Grade of severity	Criteria	Other comment
Uncomplicated	Acute illness with fever, respiratory symptoms and a chest X-ray opacity consistent with infective lung infiltrate	
Severe (CURB-65):	2 or more of: confusion, urea >7 mmol/l, respirations >30/min, blood pressure <60 mmHg diastolic, age >65	Additional findings indicating severity: hypoxia (PaO_2 < 8.0 kPa); white cell count < 4×10^9/l or > 20×10^9/l; multilobar opacities on X-ray
Indications for intensive therapy (invasive ventilatory support)	Blood pH <7.3; systolic blood pressure <90 mmHg; multiorgan failure; severe disseminated intravascular coagulation	Non-invasive ventilation (NIV) is relatively contraindicated in the presence of focal consolidation, as it preferentially ventilates the normal lung areas in this situation
Significant coexisting conditions	Chronic heart (excluding managed hypertension), lung or renal disease, diabetes mellitus, alcohol abuse, immunosuppression	

Table 7.2 Recommendations for antibiotic treatment of CAP

Grade of severity	Recommended treatment
Uncomplicated	Adult orally: doxycycline 100 mg daily, *or* azithromycin 500 mg daily for 3 days orally **or** i.v.: amoxicillin 500 mg to 1.0 g 8-hourly *or* erythromycin 500 mg 6-hourly, *or* clarithromycin 500 mg 12-hourly (**there is no evidence that adding a macrolide to a beta-lactam regimen provides significant benefit, even if an 'atypical' organism is identified**) Child orally: azithromycin, 3-days course; over age 6 months, 10 mg/kg daily; or 15–25 kg: 200 mg daily; 26–35 kg: 300 mg daily; 36–45 kg: 400 mg once daily, *or* orally or i.v.: erythromycin 50 mg/kg daily in 6-hourly divided doses (may be given by continuous infusion) *or* amoxicillin 50–100 mg/kg daily in 3 or 4 divided doses
With coexisting conditions	Co-amoxiclav 625 mg orally or 1.2 g i.v., both 8-hourly *or* oral clarithromycin 500 mg 12-hourly *or* i.v. cefuroxime 750 mg 8-hourly
Severe	Adult: co-amoxiclav 1.2 g 8-hourly i.v., *or* cefuroxime 1.5 g 8-hourly i.v., *or* cefotaxime 1 g 8-hourly i.v.; *plus* erythromycin 500 mg 6-hourly *or* clarithromycin 500 mg 12-hourly (rifampin 600 mg 12-hourly, orally or i.v., *or* ciprofloxacin 400 mg 12-hourly i.v., can be added if there is evidence of *Legionella* infection) Child: co-amoxiclav i.v.; up to 3 months: 25 mg/kg 8-hourly; 25 mg/kg 8- or 6-hourly, *or* cefuroxime i.v. 20–30 mg/kg 8-hourly *plus* erythromycin i.v. 50 mg/kg daily as continuous infusion or in 4 divided doses. (**NB rifampin i.v.; under 1 year: 5 mg/kg 12-hourly; over 1 year: 10 mg/kg 12-hourly, may be added**)
Severe CAP with coexisting conditions or requiring intensive care	Cefotaxime i.v. 1–2 g 12-hourly (child: 100–150 mg/kg daily in 2-4 divided doses) *or* ceftriaxone i.v., 2–4 g daily (child 50–80 mg/kg daily by i.v. infusion) *or* meropenem i.v. 500 mg–1 g 8-hourly (child 3 months to 12 years 20 mg/kg 8-hourly, adult dose from weight 50 kg upwards) (**NB: i.v. rifampin or ciprofloxacin, as above, may be added if legionellosis is suspected**: Ciprofloxacin is not licensed for use in children; if no alternative is available, an appropriate dose is 8–16 mg/kg daily in 2 divided doses)

⚠ Co-amoxiclav and broad-spectrum quinolones are more likely than amoxicillin or macrolides to produce adverse effects such as intestinal disorders or candidiasis. There is a small risk that broad-spectrum fluoroquinolones may predispose to infection with methicillin-resistant *S. aureus*.

⚠ Repeated, unexplained community-acquired pneumonias may be a sign of underlying disease: neonates and infants may have undiagnosed cystic fibrosis; patients of any age may have a congenital or acquired disorder of immunity, and should be investigated (see Chapter 22).

Complications of CAP
Sterile pleural effusion
A common problem, this is a sterile exudate, containing few white cells. It usually occurs during treatment, and often responds to aspiration. Loculated, or repeatedly accumulating effusions may require drainage by one or more chest drains. Rarely, persistent re-accumulation of fluid may require operative drainage and pleurodesis.

Extension of infection
Extension of infection to adjacent pleura or pericardium is a particular risk with *S. pneumoniae* and *S. aureus* infec-

tions. It is more likely in late or incompletely treated cases. It is marked by swinging fever and typical localizing pain, sometimes with increasing shortness of breath or, rarely, signs of pericardial tamponade.

Empyema should be drained as completely as possible, to avoid loculation. Withdrawing large amounts of fluid may lead to hypotension or reactive pulmonary oedema, requiring appropriate supportive treatment; this is rare when volumes below 1.5 litres are removed. An indwelling chest drain can be used if fluid repeatedly reaccumulates (Fig. 7.5). High-dose intravenous antibiotic treatment should be given. Intrapleural treatment offers no advantage, as the antibiotic is trapped at the site of injection, or destroyed by proteases or lysozymes.

Pericardial drainage is hazardous, but must be carried out if there are signs of tamponade. The procedure is best performed by an experienced operator.

Bacteraemia
Bacteraemia is a serious complication, which should always be considered. In *S. pneumoniae* infections, it increases the case fatality rate to 20%. It is more likely in infections with type 3, 6 or 5 pneumococci. Metastatic infections, such as meningitis, peritonitis or brain abscess, may further complicate bacteraemia.

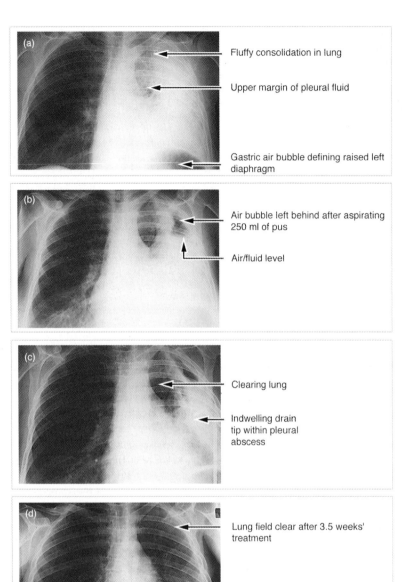

(a)

Fluffy consolidation in lung

Upper margin of pleural fluid

Gastric air bubble defining raised left diaphragm

(b)

Air bubble left behind after aspirating 250 ml of pus

Air/fluid level

(c)

Clearing lung

Indwelling drain tip within pleural abscess

(d)

Lung field clear after 3.5 weeks' treatment

Left diaphragm still raised, indicating some loss of lung volume

Figure 7.5 Chest X-ray series showing management of pneumococcal empyema complicating lobar pneumonia: the empyema was first aspirated, but later required drainage with an indwelling catheter before resolution was achieved.

Notes on the organisms that can cause pneumonia

Streptococcus pneumoniae

Streptococcus pneumoniae (or pneumococcus) is the commonest bacterial cause of pneumonia. It attacks previously healthy individuals as well as those with predispositions, and causes severe morbidity. Certain occupational groups and geographical populations are at increased risk, for example South African gold miners and highlanders of Papua New Guinea. Predispositions include sickle-cell

disease, anatomical or functional asplenia, chronic cardiac, respiratory, liver and renal disease, diabetes mellitus, alcohol abuse and immunosuppression. The case fatality rate is 5–10% (20% in bacteraemic cases), and is highest in elderly or compromised patients.

In Britain, less than 5% of pneumococci are moderately resistant to benzylpenicillin, but can be treated with adequate doses of this drug (1.8–2.4 g 6-hourly); in other Western countries, around 15% are moderately resistant and 10% are fully resistant. The same situation applies to amoxicillin and co-amoxiclav. In the USA, cefuroxime resistance varies from 3 to 12%, but cefotaxime and ceftri-axone are effective in 95%. Resistance to macrolides and tetracyclines is seen in 3–10%. A number of isolates with vancomycin tolerance have been described.

Pathogenesis of *S. pneumoniae* infections (Fig. 7.6)

S. pneumoniae is a Gram-positive coccus belonging to the *S. oralis* group together with *S. oralis* and *S. mitis*. There are more than 90 different serotypes based on polysaccharide capsular antigens. Type-specific antibodies confer resistance to infection by the homologous pneumococcus.

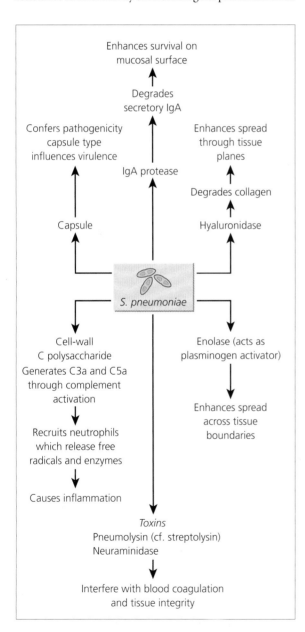

Figure 7.6 The pathogenicity of *Streptococcus pneumoniae*.

Ongoing infection is terminated by production of these antibodies.

The capsule is a major pathogenicity determinant. Virulence is related to the capsular polysaccharide type rather than the quantity produced. Possession of a capsule reduces the lethal infective dose for a mouse by a factor of 10^8. Type 3 pneumococci, which produce copious amounts of a linear polysaccharide, cause severe and fatal infections, but type 30 strains, which produce similar amounts of capsule, rarely cause clinical disease.

The alternative complement pathway is important in limiting early infection. Cell wall C polysaccharide activates the production of vasoactive chemotaxins C3a and C5a, which recruit polymorphs to the infected site.

Toxins produced by *S. pneumoniae* include pneumolysin, a cytotoxin analogous to streptolysin O (in *S. pyogenes*) and to listeriolysin (in *Listeria monocytogenes*). It inserts a ring-shaped molecule into plasma membranes, producing a hole, and appears to be important in establishing bacteraemia. *S. pneumoniae* possesses two forms of neuraminidase, an enzyme that removes sialic acid from host glycoproteins. Immunization against these toxins partially protects mice from lethal pneumococcal infection.

S. pneumoniae, like *S. pyogenes*, produces a hyaluronidase and a surface-exposed enolase. Surface protein A is a surface-exposed protein with several serological variants, whose role is not yet established. Like other mucosal pathogens, *S. pneumoniae* possesses an IgA protease, important in maintaining mucosal colonization.

Pathogenicity factors for *Streptococcus pneumoniae*

1 Capsular polysaccharide.
2 Cell wall (C) polysaccharide.
3 Toxins: pneumolysin, neuraminidase.
4 Surface protein A.
5 Immunoglobulin A protease.
6 Hyaluronidase and enolase.

Clinical features

Pneumococcal (lobar) pneumonia is an infection of the alveoli. There is sudden onset of high fever and non-specific symptoms including pleuritic chest pain, headache, vomiting and loose stools. After a variable interval a dry cough begins, and signs of consolidation develop. Sweating is not usually prominent. There is often a neutrophilia, of 15–$25 \times 10^9/l$. The chest X-ray typically shows a dense opacity filling all or most of a lobe (Fig. 7.7).

Bloodstained alveolar exudates may be expectorated as rust-coloured sputum, though early treatment may achieve resolution of the pneumonia before sputum is produced.

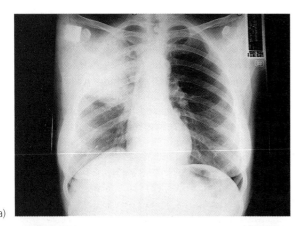

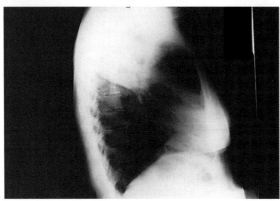

(a)

(b)

Figure 7.7 Lobar pneumonia: chest X-ray showing typical consolidation in the right upper lobe: (a) posteroanterior and (b) right lateral views.

With recovery, the alveoli resume their normal appearance and the chest X-ray changes slowly resolve.

Diagnosis

Sputum examination, if available, may demonstrate sheets of Gram-positive cocci but this is not sufficiently specific for diagnosis. Sputum culture is useful to allow confirmation of antibiotic susceptibilities, as reduced sensitivity to penicillin, and resistance to macrolides and quinolones, are increasing problems. Blood cultures are essential as bacteraemia is an important prognostic sign. Antigen detection in urine is highly sensitive (and occasionally signals positive in patients who are only carrying pneumococci). PCR-based methods are available for diagnosis and are more sensitive that culture-based techniques. With extensive microbiological study the aetiology of CAP can be determined in over 60% of cases. Amplification of antibiotic resistance genes can indicate the organism's antibiotic susceptibilities.

Prevention and control

Pneumococcal polysaccharide vaccine is a non-replicating vaccine. It contains purified capsular polysaccharide from each of 23 types of pneumococcus, which together cause over 85% of invasive pneumococcal disease in most Western countries. The efficacy is 60–70%, but is lower in those with immunosuppression. It is indicated for individuals above age 2 with predisposing conditions (see above), and is offered to people above 65 years of age in the UK (and those over age 60 in many other countries).

For patients undergoing elective splenectomy, the vaccine should be given 2 weeks before the operation.

Revaccination is not normally advised, because of the risk of adverse reactions, except for patients with asplenia or nephrotic syndrome, whose antibody levels tend to decline more rapidly.

Polysaccharide vaccines are not effective for children below age 2 years. However, pneumococcal conjugate vaccine, containing antigens of seven common *S. pneumoniae* serotypes, is effective in preventing pneumonia, bacteraemia and otitis media. It is offered to children under the age of 2 who are at increased risk from pneumococcal infections. Like Hib conjugate vaccine, it reduces the carriage of vaccine serotypes. It will be included in the UK infant vaccination schedule from December 2006.

Long-term antibiotic prophylaxis with penicillin or amoxicillin is also indicated for patients with splenic dysfunction, including sickle-cell disease, as they can become infected with uncommon types of pneumococci.

Klebsiella pneumoniae

Klebsiella pneumoniae can cause lobar pneumonia in patients with long-standing asthma, debilitating disease or alcoholism. Unlike pneumococcal infection, there is copious sputum containing many neutrophils and large, capsulated Gram-negative rods, permitting presumptive diagnosis. This organism is resistant to amoxicillin, and is usually treated with an injectable cephalosporin (see also Table 7.1).

Haemophilus influenzae

There are six capsular serotypes (a–f) of *H. influenzae*, differing in their polysaccharide composition. Type b (Hib) readily colonizes unimmunized infants below the age of 5 years, but is rare since Hib immunization became routinely available. Manifestations of Hib disease include pneumonia, meningitis, epiglottitis, septic arthritis and facial cellulitis, all commonly accompanied by bacteraemia. The differential diagnoses are pneumococcal and *Mycoplasma pneumoniae* infection, both of which are common in toddlers.

Non-encapsulated (untypable) *H. influenzae* strains are commensals and secondary pathogens of the upper and lower respiratory tract. They are commonly involved in exacerbations in COPD and other chronic chest conditions. Non-encapsulated *H. influenzae* strains are susceptible to most antibiotics used for community-acquired pneumonia. However, Hib isolates are increasingly resistant to amoxicillin and erythromycin. Cefuroxime or cefotaxime are treatments of choice. Clarithromycin is effective against 99% of isolates. Quinolone antibiotics are effective against almost all *H. influenzae* strains.

Clinical features of Hib pneumonia

Illness develops over a few days, with fever and increasing respiratory rate. Cough may not be prominent, but chest examination usually reveals features of consolidation. Chest X-ray shows a segmental, lobar or more widespread opacity, typically with an air bronchogram. There is usually a neutrophilia of $15–20 \times 10^9$/l. Blood cultures are often positive. Sputum for culture may be obtainable by pharyngeal aspiration (which stimulates coughing).

Laboratory identification

Haemophilus is a fastidious genus whose growth requires the presence of nicotinamide adenine dinucleotide (NAD) and haematin normally found in blood. These are not available in blood agar, but are released by adding the blood to the medium at 80 °C, producing a chocolate appearance. If a clear medium is required, as for the study of capsulation, blood is replaced with a filtered red-cell extract or a peptic digest of meat. Dependence on NAD and haematin can be tested by inoculating the organism on to nutrient agar and placing paper discs containing the factors; growth will only occur around a disc containing both factors. The presence of a capsule, confirming type b infection, is demonstrated by slide agglutination with serotype-specific antiserum.

Antibiotic resistance is common in Hib, especially to ampicillin (about 15% of UK isolates). The beta-lactamase, TEM-1, which originated in *E. coli*, is usually responsible. It can be detected by rapid tests using a cephalosporin, nitrocephin, which changes colour when its beta-lactam bond is broken.

Prevention and control

Routine infant vaccination programmes in many countries include polysaccharide conjugate vaccines against Hib (see also Chapter 26). They have greatly reduced the incidence of all invasive Hib diseases, including pneumonia. Hib vaccines do not protect against pneumonia caused by non-type b, or non-encapsulated strains of *H. influenzae*.

Legionnaires' disease and legionelloses

Introduction

Legionnaires' disease is important because its recognition is often delayed, it responds poorly to conventional treatment for community-acquired pneumonia, and it can be life-threatening. It commonly occurs in clusters or outbreaks of cases.

Epidemiology

Infection is transmitted by inhalation of contaminated aerosols (particle diameters around 1 μm) generated from hot-water tap and shower outlets, water-cooling towers of office and industrial sites, and domestic or recreational whirlpool spas. The organism can also be isolated from water systems in buildings, from ponds, streams and from soil. Legionellae are found in water at 5–50 °C, and replicate at temperatures between 20–50 °C. They survive and are supported in the cytoplasm of free-living amoebae, which, with other organisms, form a biofilm, the major reservoir of infection. Within biofilms, legionellae are protected from biocides, which are often used to reduce contamination in water towers and air-conditioning systems.

Legionellosis occurs as sporadic cases or as localized outbreaks associated with a contaminated water source, usually in a large building. Illness preferentially affects the middle-aged and elderly, particularly males and those who smoke or have underlying chronic respiratory disease. In the UK, about 200 cases are reported annually, with a case fatality rate between 10% and 15%. Nearly 50% of UK cases are travel-associated, and mainly contracted within Europe (Fig 7.8). The implicated source is often the water system of a tourist hotel (Fig. 7.9).

The incubation period is 2–10 days for legionnaires' disease and 1–2 days for Pontiac fever, the milder non-pneumonic form of legionellosis. Person-to-person spread does not occur.

Microbiology and pathogenesis

Legionellae are Gram-negative aerobic non-sporing, fastidious bacteria that need L-cysteine for growth. They are closely related to *Coxiella*. At least 48 serotypes are described (and have previously been proposed as different species). Legionellae vary in their invasive potential but the reasons for this are not certain. They are taken up by macrophages via binding of their major outer membrane protein with the C3 receptor, which does not trigger a respiratory burst. They are able to survive within macrophages, remaining in a phagosome that does not fuse with the lysosome thus avoiding the effects of acidification and enzymic degradation. The bacterial type IV secretion mechanism is essential for promoting intracellular infection and inhibiting phagolysosome formation.

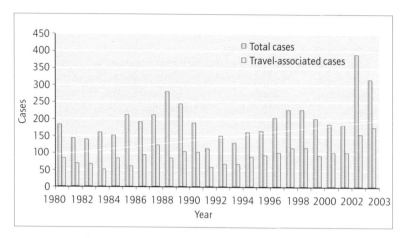

Figure 7.8 Reports of legionnaires' disease in England and Wales, 1980–2003. Source: Health Protection Agency.

Figure 7.9 Poorly maintained Turkish bath in a hotel that caused an international outbreak of legionnaires' disease: 11 people from four countries contracted the infection.

Legionellae can also suppress the effect of complement- and polymorph-mediated killing.

Legionella lipopolysaccharide has some toxic properties and also protects the organism against the effect of complement. Variants of this antigen are associated with variations in virulence. Macrophage infectivity promoter (Mip) is surface exposed on legionellae, and is required for intracellular infection and virulence. *L. pneumophila* secretes many potential toxins and degradative enzymes, including lipases, phospholipases and a zinc metalloprotease, via a type II secretion mechanism.

Clinical features

Many Legionellaceae are capable of producing an influenza-like illness, sometimes with a cough. When recognized as legionellosis, this is called Pontiac fever.

Legionnaires' disease is a severe systemic infection with pneumonia, usually caused by a serotype of *L. pneumophila*. After 2–10 days incubation, there is fever, prostration and cough, unresponsive to beta-lactam antibiotics and often misinterpreted as influenza or bronchitis. Loose stools are common, and the patient often becomes confused. Cough, breathlessness and purulent sputum production progress rapidly. Severe respiratory and/or renal failure can develop quickly.

Physical examination shows tachypnoea and purulent sputum in a morose or confused patient. There is usually a distinct area of consolidation or coarse crepitations in the chest, and one or more large opacities on chest X-ray (Fig. 7.10).

Laboratory findings include neutrophilia, high C-reactive protein and erythrocyte sedimentation rate, variably elevated urea and creatinine levels and raised transaminases. There may be marked hyponatraemia and/or normocytic anaemia.

Diagnosis

This may be suspected on epidemiological grounds. The presence of pneumonia with severe confusion and metabolic disturbance in a previously fit patient is highly suggestive.

Laboratory diagnosis

Gram-stained smears of sputum are not useful as legionellae take up the stain poorly.

Culture has a relatively low yield and is often negative in patients with mild infection. Sputum, or bronchoalveolar lavage specimens can be inoculated onto cystine-containing charcoal yeast extract agar, made selective by adding antibiotics that inhibit respiratory tract commensals. *Legionella* spp. grow slowly and are identified by their cystine dependence. They are serotyped by direct immunofluorescence antibody testing.

An enzyme-linked immunosorbent assay (ELISA) method, which detects serogroup 1 antigen in urine, is now the diagnostic method of choice. It provides rapid re-

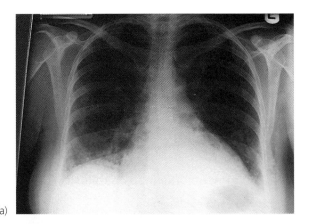

(a)

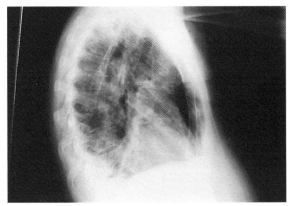

(b)

Figure 7.10 Legionnaires' disease: extensive lung opacity in a patient with severe symptoms (one of several people infected during a hotel-based group holiday): (a) posteroanterior and (b) right lateral views.

Treatment of legionnaires' disease

- Erythromycin i.v. 3–4 g daily in four divided doses *or* clarithromycin 1.0 g 12-hourly *plus* either ciprofloxacin i.v. 200 mg twice daily (ofloxacin may be substituted, in the same dosage) *or* rifampicin 600 mg twice daily.
- Oxygen, and/or ventilatory support may be needed.
- Renal failure in late or severe cases demands intensive management.

Complications

Patients who recover usually do so completely, though convalescence may be prolonged. A few develop lasting, patchy cerebral or bulbar deficit, often attributed to severe hypoxia, or inflammatory encephalopathy during the illness. Guillain–Barré syndrome is a rare but recognized complication.

Prevention and control

Effective maintenance of building water systems and other potential sources of infection is the key to prevention of legionellosis. This includes adequate chlorination of water supplies and stored cold water, and heating of hot water (to above 60 °C). The design of water systems should avoid stagnation of water in peripheral sections, and permit the achievement of a temperature of 55°C at all hot taps within 15 seconds of turning on. Whirlpool spas should be frequently cleaned and continuously disinfected. A European surveillance system for travel-associated legionnaires' disease works closely with member state health authorities and the travel industry, to manage outbreaks.

Aspiration pneumonia

Aspiration pneumonia results from overwhelming aspiration of contaminated liquid, or from failure to cough and protect the airways from accumulating secretions. It can occur in near-drowning, in unconscious or anaesthetized patients, or in those with neuromuscular disease affecting swallowing or breathing. When the chest is erect, the right main and lower lobe bronchi are nearly vertical, so the right lower lobe tends to be most affected. In a recumbent patient the apical bronchus of the right lower lobe is affected most.

Aspirated secretions contain mixtures of mouth and pharyngeal organisms, mainly penicillin-sensitive streptococci and spirochaetes mixed with anaerobes. These produce a patchy pneumonia, most dense in the vulnerable segment, sometimes leading to abscess formation. Aspirated vomitus is also highly acidic.

The diagnosis must be suspected in vulnerable patients, and treated with physiotherapy, a second- or third-gen-

sults with up to 90% sensitivity. However, it cannot detect other serotypes of *Legionella*, a serious drawback especially in immunocompromised patients, in whom non-serotype 1 disease is more likely. *L. pneumophila* serotype 1 can be identified in sputum by a direct fluorescent antibody method. It has relatively poor sensitivity but a high specificity. Detection of *Legionella* DNA by PCR has been described and is probably about as sensitive as culture.

Legionnaires' disease can be diagnosed serologically by detecting a fourfold rise in antibody titres or a single high value. Methods such as indirect immunofluorescence and commercial rapid microagglutination tests are available.

Management

Appropriate treatment should be given on suspicion of the diagnosis, to avoid prolonged hypoxia and respiratory or renal failure. The treatment of choice is intravenous erythromycin or clarithromycin. The addition of ciprofloxacin, ofloxacin or rifampicin may speed the response.

eration cephalosporin or broad-spectrum penicillin plus metronidazole. If gastric contents have been aspirated, the inflammatory effect of gastric acid can be inhibited by giving a corticosteroid such as methylprednisolone or prednisolone for the first 24–36 hours.

Patients who have aspirated contaminated water may suffer infection with Enterobacteriaceae or *Pseudomonas* spp, or with marine *Chromobacterium* or *Vibrio* spp. They may also have aspirated or ingested chemical toxins, on which advice should be sought through local public health sources.

Lung abscess

The commonest cause of a lung abscess is aspiration. Most lung abscesses are polymicrobial, with facultative and anaerobic organisms occurring equally commonly in most case-series. These abscesses can be treated as for aspiration pneumonia.

Primary lung infections or bacteraemic seeding are rarer causes. Staphylococcal bacteraemia is often accompanied by a nodular pneumonitis, which can progress to abscesses. Intravenous drug abusers are at risk from infected pulmonary emboli (usually seeded with skin-derived *S. aureus* – Fig. 7.11). Patients at risk of *S. aureus* infections such as these should be treated with flucloxacillin (or alternatively a glycopeptide).

Hospital-acquired lung abscesses may behave in the same way as community-acquired ones, but *Klebsiella pneumoniae, Pseudomonas aeruginosa*, MRSA and *Candida* spp. are additional causes, which carry a particularly poor prognosis. Rare infections such as nocardiosis and aspergillosis (more common in neutropenic or immunosuppressed patients) can cause multiple lung abscesses. Sputum, bronchial specimens and blood cultures may aid aetiological diagnosis.

The bronchi provide a natural route for drainage, and healing usually occurs with antibiotic treatment alone. If the abscess persists, percutaneous aspiration or surgical drainage, with culture and sensitivity testing of the pus, may be helpful.

Anaerobic pneumonia (necrobacillosis)

Introduction and epidemiology

Anaerobic pneumonia is a rare condition that tends to affect adolescents and young adults, particularly men. It is a bacteraemic disease, usually caused by *Fusobacterium necrophorum*, which is a colonist of the pharynx and mouth.

Clinical features and diagnosis

Illness begins abruptly with a high fever and severe sore throat. Suppuration of cervical lymph nodes with thrombophlebitis of the underlying jugular vein (Lemmière's syndrome) is a recognized presentation. When the lungs are involved, cough and often chest pain develop 2 or 3 days later. One or more areas of consolidation are detectable on examination and chest X-ray. Standard antibiotic treatment is unsuccessful, and progressive features of sepsis may develop. The disease may be fatal unless suspected and treated early.

The age and sex of the patient, and the severe sore throat should arouse suspicion. Blood cultures are often positive,

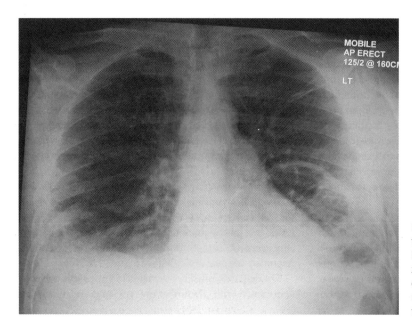

Figure 7.11 Staphylococcal disease in the lungs of an intravenous drug abuser: large abscesses are seen in both lower zones; these originate from pulmonary embolization by *S. aureus*-infected material. An empyema is also present on the right.

and must always be obtained. Sputum culture may also be positive, but the bacteria are strictly anaerobic, and survive poorly unless maintained in an anaerobic environment.

Management

Although the pathogen is penicillin-sensitive in laboratory conditions, metronidazole is more rapidly effective. It can be given together with penicillin or broader-spectrum drugs while the diagnosis is confirmed.

Atypical pneumonias

Introduction

The atypical pneumonias were so-called because they differed from classical lobar pneumonia, having a gradual onset, prolonged fever, profuse sweating and slow recovery with no point of crisis. Causative pathogens were not demonstrable by Gram-staining or conventional culture. They are important because some of them cause morbidity in children, but they also affect adults of working age and often occur as outbreaks.

Organism list

- *Mycoplasma pneumoniae*
- *Chlamydophila pneumoniae*
- *Chlamydia psittaci*
- *Coxiella burnetii*

Epidemiology

M. pneumoniae and *C. pneumoniae* are human pathogens whose reservoir of infection resides in infected members of the population.

Mycoplasma pneumonia affects all ages but is commonest in school-age children and young adults. Outbreaks occasionally occur, especially in closed institutions. In the UK, epidemics occur at 3- or 4-year intervals. During epidemic years, *M. pneumoniae* competes with *Streptococcus pneumoniae* as the commonest cause of community-acquired pneumonia.

Chlamydophila pneumoniae was distinguished from *C. psittaci* in the 1980s. The prevalence of seropositivity rises steadily with age until about 60% of individuals have evidence of past infection by the fifth decade. Large epidemics have been recognized in Denmark and Canada. Both primary infection and reinfection can be symptomatic. The suggestion that *C. pneumoniae* might be a factor in the development of atherosclerosis has never been proved.

C. psittaci, the agent of ornithosis, is widespread among birds, which excrete many organisms in their faeces. In-fection in humans is acquired by inhaling the organism from desiccated bird droppings and secretions of infected birds. Sporadic cases and limited outbreaks have resulted from contact with parrots and parakeets. Large outbreaks have been associated with duck or poultry farming, and with poultry packing plants. Person-to-person spread is rare. The incubation period is 4–15 days. A strain affecting sheep (that causes enzootic ovine abortion) can cause severe illness and abortion in pregnant women.

Coxiella burnetii is a pathogen mainly of sheep. Transmission to humans occurs by airborne spread of organisms in dust contaminated by infected placental tissues and faeces. Rarer sources of infection include parturient cats and unpasteurized milk. The infection mainly affects meat workers, farmers and veterinarians. Chronic infection can occur, often associated with endocarditis. Infection is severe in pregnant women, particularly affecting the placenta, and often precipitating abortion. No case of transmission by a parturient woman has been described.

Microbiology and pathogenicity of chlamydiae

The family Chlamydiaceae consists of two genera: *Chlamydia* and *Chlamydophila*. The *Chlamydia* are divided into two species: *Chlamydia trachomatis*, which affects the eye and genital tract, causing trachoma, non-specific urethritis and lymphogranuloma venereum; and *C. psittaci*, the cause of a zoonotic infection with pneumonia. *Chlamydophila pneumoniae* causes upper and lower respiratory tract infection.

The Chlamydiaceae are obligate intracellular bacteria that exist in two forms. The reticulate body is the intracellular vegetative form found within a cytoplasmic inclusion body inside host cells. The reticulate body divides by a series of fissions via intermediate forms to generate elementary bodies, which are released from the cell (Fig. 7.12). The elementary bodies are the transmissible form of the organism.

Chlamydiae express a genus-specific lipopolysaccharide (LPS) similar to the rough LPS of certain salmonellae. The serological classification of *C. trachomatis* is based on the major outer membrane protein (MOMP). The organism has a specific surface-exposed trisaccharide epitope that induces pro-inflammatory cytokine production by host cells, and stimulates scarring and fibrosis. Chlamydiae possess genes for a type III secretory system, which may contribute to pathogenicity. The 12, 60 and 75 kDa heat shock proteins are closely related to their counterparts in *Escherichia coli* (GroEL, GroES and DnaK), and also to human mitochondrial proteins Hsp10, Hsp60 and Hsp70. They are thought to play a major pathogenetic role by interacting with interferon gamma (IFN-γ) to upregulate Hsp60, which precipitates a host autoimmune response.

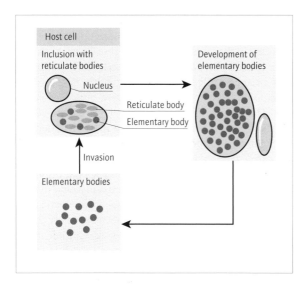

Figure 7.12 Appearance of intra- and extracellular forms of *Chlamydia* spp.

Clinical features

The acute illness is similar in all types of atypical pneumonia. After an incubation period of 4 to 14 days, there is a gradual onset of irregular, swinging fever, malaise, fatigue and sweating, especially at night. There is often a dry cough. Physical examination is rarely remarkable, though there may be localized crepitation or dullness in the lung fields. The chest X-ray may show anything from a faint, ill-defined segmental opacity to a large, dense consolidation. Some patients have a small or moderate pleural effusion, often with pleuritic pain on the affected side.

Laboratory findings are variable. There may be a modest neutrophilia, but this is rare in *C. pneumoniae* infections. Mild elevation of liver transaminases is common; the C-reactive protein and erythrocyte sedimentation rate are markedly raised. Most patients with *Mycoplasma* pneumonia develop cold agglutinins, but these tend to appear in convalescence, too late for diagnostic use. A few patients develop severe haemolysis in convalescence. Associated clinical features differ in the different infections.

Chlamydophila pneumoniae

C. pneumoniae usually causes acute self-limiting respiratory infections, often mild, frequently starting with sore throat and rarely presenting to hospital. Other syndromes include moderately severe atypical pneumonia, otitis media, acute-on-chronic bronchitis, persistent or relapsing pharyngitis and, less commonly, myocarditis. *C. pneumoniae* antigen and DNA have been demonstrated, and the organism has been recovered from the foamy cells of atheromatous plaques. There is an epidemiological association between *C. pneumoniae* seropositivity and acute cardiovascular events, but a causative relationship is not proved.

Mycoplasma pneumoniae

M. pneumoniae usually causes pneumonia, but this may be accompanied by tracheitis, bullous myringitis, mild hepatitis or sometimes by a lymphocytic meningitis. Pericarditis and myocarditis are rare but recognized features. Probably at least 25% of infections are feverish illnesses without pneumonia.

Microbiology and pathology of mycoplasmas

Mycoplasmas are the smallest form of life capable of extracellular existence and *M. genitalium* was one of the first bacterial genomes to be fully sequenced. They lack the rigid cell wall of other bacteria as they do not possess peptidoglycan. They are consequently completely resistant to beta-lactam antibiotics. Mycoplasmas are widely distributed in nature and cause many animal and plant diseases. Several species are important in human disease including *M. pneumoniae* (respiratory infections), *Mycoplasma hominis* and *Ureaplasma urealyticum* (genital infections), *M. amphoriforme* (respiratory infections in immunocompromised patients) and *M. fermentans* (joint infections in immunocompromised patients).

Mycoplasmas attach to both ciliated and non-ciliated host cells via a foot like projection. The P1 antigen appears to be essential in achieving adherence.

Chlamydia psittaci

C. psittaci causes pneumonia in about 60% of clinically infected individuals. The remainder have fever alone. Infection is debilitating and can be severe and accompanied by respiratory failure. It is a rare cause of culture-negative endocarditis. Occasional cases of myelopathy have been reported, with or without associated pneumonia.

Coxiella burnetii

C. burnetii causes Q fever. This may be an atypical pneumonia, but over half of cases are feverish illnesses with hepatocellular disorder. A small proportion of cases, probably less than 1%, become chronic with swinging fever, granulomatous hepatitis, often clinical jaundice and, usually, endocarditis. In endemic areas this is an important cause of morbidity. Mild to moderate myocarditis may occur in acute or chronic disease.

Diagnosis of atypical pneumonias

The diagnosis can sometimes be inferred from epidemiological circumstances and clinical presentation. RSV and influenza can cause similar illnesses, often in outbreaks.

Mycoplasmas

Although *M. pneumoniae* can be cultivated on serum-based artificial media this technique is not suitable for routine diagnosis as the organism is slow-growing and the diagnostic yield is low. PCR-based diagnostic methods are now the standard technique for diagnosis as they combine sensitivity with rapidity. ELISA antigen detection tests are also available. Serology is a reliable method of diagnosis and IgM-detecting ELISA tests are commercially available. Susceptibility testing can be performed, but it is slow and difficult. It is occasionally used for testing isolates from immunocompromised patients whose chronic infections (such as septic arthritis) fail to respond to antibiotic therapy. It is valuable in evaluating new drugs.

Chlamydiae

Chlamydial infections are often still detected by finding rising titres of complement-fixing antibodies to whole-cell antigen preparations. There is much cross-reaction between antibodies to different species. Species-specific antibodies are detected by microimmunofluorescence techniques, which employ elementary body antigens. All of these techniques lack sensitivity and are slow. Rapid diagnosis is possible using DNA amplification techniques. Commercial assays are available. DNA amplification techniques are now a standard of care for diagnosis of genital *C. trachomatis* genital infections. They are sufficiently sensitive to be used in screening programmes, as reliable diagnosis can be obtained from a urine sample. This is an important approach to combating the prevalence of genital *Chlamydia* infection.

Culture of chlamydiae is not useful in the diagnosis of pneumonias, but *C. trachomatis* grows readily in culture with cyclohexamide-treated McCoy cells. Culture is used to confirm the presence and serotype of *Chlamydia trachomatis* genital infection in cases of sexual abuse or rape.

Q fever

Q fever is diagnosed by finding rising titres of complement-fixing antibodies to phase 2 and (in chronic Q fever) phase 1 antigens. Coxiellae isolated from infected animals express phase 1 antigens, but after passage through embryonated eggs, these are replaced by phase 2 antigens.

In cases of hepatitis, there is a characteristic histological change, with many small granulomata, often with clear centres and always with a halo of eosinophils in their periphery (see p. 217). *Coxiella burnetii* can be demonstrated in silver- or Giemsa-stained smears, or by direct immunofluorescent staining of biopsy specimens.

Management of atypical pneumonias

The antibiotic of choice is a tetracycline. Macrolides are useful alternatives, particularly for children, in whom

they are drugs of choice but they may be less effective in Q fever and some cases of psittacosis. Treatment may be given orally in mild and moderate cases, but should be given parenterally in severe or complicated cases (this is easier with macrolides). The response is not always fast, so courses of 10–14 days are advisable to avoid relapse or recrudescence.

Chloramphenicol may still be a treatment of choice when Q fever occurs in pregnancy. Ciprofloxacin and newer fluoroquinolones are highly active in laboratory testing and may be valuable alternatives but treatment failures have been reported and there may be a risk of cartilage damage in the fetus. Ciprofloxacin is not reliably effective in chlamydial and mycoplasmal pneumonias; newer fluoroquinolones such as levofloxacin and moxifloxacin are more effective.

Treatment of *Mycoplasma pneumoniae* and *Chlamydia pneumoniae*

1 Erythromycin orally or i.v. 500 mg 6-hourly *or* clarithromycin 250–500 mg 12-hourly.
2 Alternative: oxytetracycline orally 250–500 mg 6-hourly *or* doxycycline 200 mg single dose, then 100 mg daily.
Both above regimens for 10–14 days.
3 Azithromycin 500 mg daily for 10 days.

Treatment of acute Q fever

1 First choice: a tetracycline orally – same dose as in note 2 above.
2 In pregnancy: chloramphenicol orally or i.v. 500 mg 6-hourly *or* ciprofloxacin orally 500 mg twice daily (i.v. 200 mg twice daily).
All above regimens for 14 days.

Treatment of psittacosis

1 First choice: a tetracycline orally – same dose as for *Mycoplasma pneumoniae* and *Chlamydia pneumoniae*.
2 Alternative: erythromycin or clarithromycin orally or i.v – same dose as for *M. pneumoniae* and *C. pneumoniae*.
Both above regimens for 10–14 days.

> ⚠ Care must be taken not to miss endocarditis in Q fever. The presence of phase 1 antibodies is a warning. Echocardiography, physical examination, repeated C-reactive protein estimations and follow-up assessments for 2 or 3 months are advisable in cases of doubt.

Pertussis

Introduction

Pertussis, or whooping cough, is a prolonged, severe and

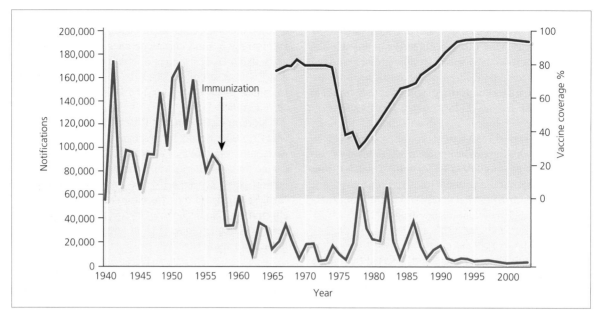

Figure 7.13 Decline and resurgence of pertussis in England and Wales with changing vaccine coverage.

distressing contagious disease of children caused by *Bordetella pertussis*. It is now rare, as immunizaton is widespread. Large epidemics quickly recurred in the 1980s after concerns about vaccine safety resulted in low rates of acceptance (Fig. 7.13).

Epidemiology and pathology

B. pertussis is readily transmitted by the airborne route, and is most often seen in infants too young to have been vaccinated or in families who have declined immunization. Morbidity is high in infants and in older children with respiratory or cardiac disease. The reservoir of infection is in adult carriers, also mild or subclinical cases, as immunity wanes after childhood. *B. pertussis* is a small Gram-negative rod that is typed by its cell wall proteins, or agglutinogens, designated 1, 2 and 3, antibodies to which are protective. Most clinical isolates are of type 1,3 or 2,3.

Clinical features

After 14–20 days incubation, illness begins with a simple cough and slight fever. Over 4 or 5 days coughs begin to be grouped into paroxysms, of up to 20 or 30 coughs with no inspirations between them. Thick mucus is expectorated, with difficulty, in ropey strands. Cyanosis and even apnoeic attacks can occur. Finally the paroxysm ends with a stridulous inspiratory cry (a 'whoop'), and often with

a vomit. Paroxysmal cough, whooping and vomiting may present separately. Adults and a few children may suffer sneezing paroxysms. The paroxysms continue for 2 or 3 weeks before gradually improving over 10–14 days.

During paroxysms, the venous pressure is raised, there is facial lividity and sometimes conjunctival haemorrhage or facial petechiae (Fig. 7.14). In infants there may be intracranial haemorrhage. Secondary or aspiration pneumonia often occurs, but the rare fatalities are usually caused by cerebral hypoxia and vascular damage.

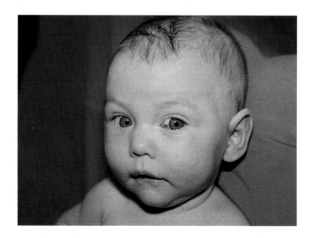

Figure 7.14 Pertussis in a 6-month-old: subconjunctival haemorrhage produced by severe cough.

Diagnosis

Diagnosis is often clinically obvious, but mild cases may be similar to croup, bronchiolitis or persisting asthma. A lymphocytosis of $15–25 \times 10^9/l$, affecting all lymphocyte subgroups, is strongly suggestive of pertussis.

Microbiology

B. pertussis is a small Gram-negative coccobacillus, related to *B. parapertussis*, which is rarely associated with the whooping cough syndrome, and *B. bronchoseptica*, which is primarily a pathogen of animals.

The pathogenicity of the organism is related to its toxins, including fimbrial haemagglutinin, pertussis toxin and a 67 kDa protein. Pertussis toxin mediates the paroxysmal cough and the lymphocytosis, which affects all lymphocyte subsets. The bacterial agglutinogens and fimbrial haemagglutin are the main adhesion antigens. The most effective vaccines contain all of these antigens. Other toxins include tracheal cytotoxin, which causes ciliostasis and destruction of ciliated epithelium, and adenyl cyclase, which destroys polymorphs and stimulates the cough reflex.

Laboratory diagnosis

A pernasal swab is the optimal specimen for culture. Cultivation of the organism is difficult because *B. pertussis* is fastidious and relatively slow-growing, but it may be detected by culture of swab specimens on Bordet–Gengou or charcoal yeast extract agar with antibiotic supplements. Few laboratories now perform *B. pertussis* cultures, and DNA amplification techniques are now used by reference laboratories. Serodiagnosis is based on detection of specific IgM by EIA, but this only provides a retrospective diagnosis and is not sensitive.

Management

Erythromycin appears to ameliorate cough, though it has not been shown to shorten the illness in controlled trials. Supportive treatment is important. The airway should be protected by holding infants face- and head-downwards during paroxysms. Cyanotic or apnoeic attacks can often be helped by suctioning the airways to remove tenacious mucus. A humidified atmosphere may help. Severely affected infants may be nursed in an oxygen atmosphere. Assisted ventilation is rarely needed.

Paroxysms are easily precipitated by feeding or by breathing cold air. Most children lose weight during the paroxysmal stage, and frequent feeds are important; food should be offered after coughing or vomiting. Small children may be fearful of paroxysms, and need sympathetic reassurance.

Prevention and control

Erythromycin treatment may reduce the infectiousness of cases, and is also given for post-exposure prophylaxis. Cases remain infectious for up to 3 weeks after paroxysms begin. An effective, five-component, acellular vaccine containing fimbrial haemagglutinin, pertussis toxin, 67 kDa antigen and agglutinogens, forms part of the infant vaccine programme in the UK.

Fungal infections of the lower respiratory tract

Introduction

Fungal respiratory infections are rare in immunocompetent patients in the UK, mainly because exposure to highly pathogenic fungi is rare. There is a limited list of pathogens, mostly from endemic areas overseas, that cause diseases requiring differentiation from unusual, nodular or cavitating chest infections.

Aspergillosis

Aspergillus spp. have a worldwide distribution. They inhabit dark areas, such as ventilation ducts, cavities in buildings and spaces behind wall panels. They produce millions of spores, which are inhaled and primarily infect the lungs.

Allergic bronchopulmonary aspergillosis

Indolent colonization of ectatic bronchioles or old lung cavities can produce a hypersensitivity reaction with bronchoconstriction, leading to deterioration in patients with pre-existing asthma or chronic bronchitis. Affected patients have bronchospasm, anti-aspergillus precipitin antibodies in the blood, changeable pulmonary infiltrates, coinciding with eosinophilia, and raised serum IgE (>150 IU/l). Prolonged treatment with itraconazole, 200 mg daily, often reduces symptoms, but systemic corticosteroid therapy (0.5 mg/kg for 2 weeks, followed by gradual reduction) is the mainstay of management.

Aspergilloma

In pre-existing lung cavities, fungal hyphae may produce a ball-shaped growth that irritates the cavity wall, causing cough and recurrent haemoptysis (a severe problem in up to 25% of sufferers). This does not induce hypersensitivity, and precipitins are not usually present. The X-ray or imaging appearance of this aspergilloma is typically of a round, space-occupying lesion surrounded by a narrow,

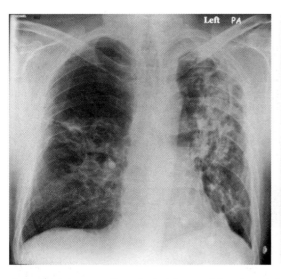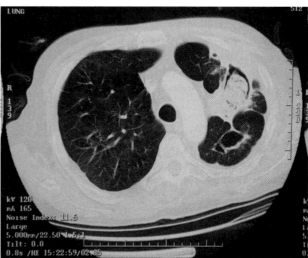

Figure 7.15 An aspergilloma in the left upper lobe of a patient who had episodes of haemoptysis after successfully-treated TB: the X-ray shows the elongated cavity containing opaque material; the CT scan clearly shows the 'fungus ball' surrounded by a clear space.

clear air space (Fig. 7.15). Haemoptysis can be treated by imaging-controlled embolization of the arterial blood supply, which often also kills the fungus. The fungus may be killed by injection of a ketoconazole paste into the cavity. Oral or parenteral antifungal drugs are rarely effective, as the fungus ball is not vascularized. Resection of the affected lung lobe or segment may be possible if other treatments are unsuccessful.

Invasive interstitial (necrotizing) aspergillosis

This is a rare condition with widespread infiltrating disease, often associated with immunodeficiency. Voriconazole has good activity against the fungus, and penetrates tissues well. It has proved effective in treatment, but very prolonged courses may be required. Itraconazole may also be effective. Amphotericin B preparations are not the first choice, because of their less certain effect and potential for toxicity.

Histoplasmosis

Histoplasma capsulatum is a yeast, widely distributed in dry, warm areas, and some caves. Acute clinical disease appears after 5–18 days, with fever, chills and myalgia. Physical examination is often normal, but mild hepatosplenomegaly, lymphadenopathy and maculopapular rashes or erythema nodosum sometimes occur. Chest X-ray may show midzone opacities and hilar adenopathy. Mild, self-limiting disease is common.

Antigen detection and ELISA antibody tests are useful for diagnosis. Bronchoscopic biopsies show interstitial pneumonia with eosinophilic infiltrates. Itraconazole, 200 mg once or twice daily for 10 days, is effective treatment. The chest X-ray becomes normal in 2–4 weeks; serology remains positive for many months.

Disseminated disease occasionally occurs, with metastatic foci or cavitary pneumonia, particularly affecting immunosuppressed patients. Itraconazole is the treatment of choice plus or minus amphotericin (see Chapter 22).

Blastomycosis

This is caused by the yeast *Blastomyces*, whose endemic area is confined to the USA. It causes a disease similar to histoplasmosis. Diagnosis and treatment are as for histoplasmosis.

Coccidioidomycosis

The cause of this disease, *Coccidioides immitis*, is virtually confined to the San Joachim valley of the western USA. It tends to cause a rapidly progressive bronchopneumonia or pneumonia unresponsive to antibiotics, but mild cases and rare, metastatic foci also occur. The fungal hyphae grow rapidly in standard cultures, and produce segmented chains containing millions of spores. The spores are highly infectious, and patients with suspected coccidioidomycosis should always be isolated.

Treatment is with amphotericin (see Chapter 4).

Case study 7.1: Chef's exotic cough

History

A 62-year-old chef of Indian origin had not worked for 5 years because of chronic obstructive pulmonary disease. He had a 1-week history of fever, increasing cough, shortness of breath, and increasing thick, greenish sputum. He also had a long history of itching of the lower legs. He had entered the UK 12 years previously and never travelled overseas since then. He used regular budesonide and ipatropium inhalers, and took nifedipine for control of moderate hypertension. For 6 weeks he had been taking oral prednisolone 5 mg daily for control of worsening asthma.

Physical findings

He was distressed, and short of breath at rest. Respiratory rate 32 per minute, pulse 110 bpm, blood pressure 130/85 mmHg, heart sounds normal with a precordial third sound. Chest expanded, with limited movement, especially on the left; widespread expiratory wheeze, coarse crepitations at both bases and the left mid-zone, with reduced breath sounds and bronchial breathing in the left axilla. Abdominal and neurological examination normal.

Investigations

Haemoglobin 15.6 g/dl, white cell count 16.7×10^9/l, with 12.2×10^9/l neutrophils and 2.2×10^9/l eosinophils. Blood urea 8.7 mmol/l, sodium 129 mmol/l, potassium 4.0 mmol/l, ESR 65 mm/h, alkaline phosphatase 255 IU/l, other liver function tests normal. Arterial PO_2 7.7 kPa, PCO_2 6.4 kPa. Chest X-ray showed widespread ground-glass appearance and a large, ill-defined opacity in the left mid-zone (Fig. CS.2).

Questions

- What diagnosis is fulfilled by the findings?
- What treatment is indicated?
- What other important pulmonary infection should be excluded?
- What unexpected finding should be followed up?

Management and progress

A diagnosis of severe pneumonia was made in this patient with significant co-morbidity (see Table 7.1). Treatment was commenced with intravenous cefotaxime, and nebulized salbutamol by inhalation with 30% oxygen. His prednisolone dosage was temporarily increased to 10 mg daily. Sputum and blood cultures were obtained, and tuberculosis was excluded initially by negative sputum examination. Moderate eosinophilia is unusual in chronic pulmonary disease; this was further investigated. The aspergillus precipitin test was negative.

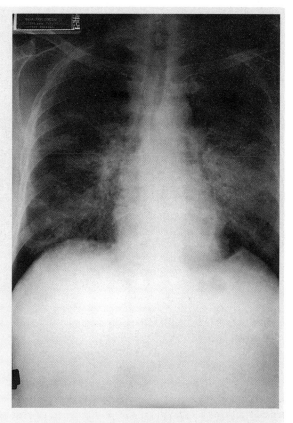

Figure CS.2 Chest X-ray of the chef on presentation, showing widespread ground-glass appearance and a large, ill-defined opacity in the left mid-zone.

Question

- What is a common cause of isolated eosinophilia in individuals from third-world countries?

Further management and progress

Strongyloidiasis is a persisting parasite infection in which adult worms in the bowel generate rhabditiform larvae, which may become invasive and reinfect by penetrating the colonic mucosa or perianal skin. The larvae then become filariform, and migrate through the tissues to the lungs, where they cause an inflammatory reaction and are coughed into the pharynx from where they reinvade the gut. Migrating parasites in the blood or tissues stimulate a protective eosinophilia. Serological tests were requested for strongyloidiasis and filariasis. Stool samples were obtained and charcoal culture performed to identify *Strongyloides* larvae. Sputum examination revealed many neutrophils, a few eosinophils and a heavy growth of *Escherichia coli*, sensitive to cefotaxime and co-amoxiclav. Scanty *Strongyloides* larvae were present in the sputum. He was treated with oral ivermectin 200 µg/kg daily for 2 days.

The fever and shortness of breath resolved after 3 days' therapy, and oral co-amoxiclav was substituted for intravenous therapy on the fourth day. Oral prednisolone dosage was stepped down and discontinued over 2 weeks.

Questions

- Why did *Strongyloides* larvae appear in the sputum?
- Is the finding of coliform pneumonia of interest in this case?
- Is a standard two-dose course of ivermectin adequate treatment for the strongyloidiasis?

Completion of management

In debilitated patients, *Strongyloides* larvae may migrate, carrying coliform organisms from the patient's bowel in their gut. The resulting pulmonary irritation may be mistaken for asthma and treated with increasing doses of prednisolone, resulting in further immunosuppression and escalating larval migration. Coliform pneumonia or even meningitis may result. In severe immunosuppression, eosinophilia does not occur, and larvae are found in the stool, sputum, liver biopsies and effusions. Fatal invasive disease and/or coliform infections may result. In this mildly immunosuppressed patient a two-day course of ivermectin was given. The course was repeated two weeks later, after ceasing prednisolone dosage.

Gastrointestinal Infections and Food Poisoning

8

Introduction

Infections of the gastrointestinal tract are among the commonest infections in all communities of the world. They cause major morbidity and mortality, especially in infants, children and the elderly. Even in the UK, where water supplies, sanitation and education reach high standards, surveys suggest that intestinal infections cause significant illness and loss of working days.

Structure and environment of the gastrointestinal tract

The gastrointestinal tract is a tube, open to the environment at both ends, into which much foreign material is introduced. Microorganisms, which may be pathogens, enter every time fingers, food, drink or utensils are put into the mouth. Local environments vary widely throughout the gastrointestinal tract, and this influences the nature of the local flora and of pathogens that may invade.

The stomach is highly acidic and contains variable amounts of air. Few bacteria can survive this environment for long, but many pass quickly through. Boluses of food, especially fats, can protect bacteria during this passage. Chocolate and cheese are thought to enhance survival of *Salmonella* in this way. *Helicobacter pylori* and mycobacteria resist the effect of gastric acid, and can persist in the stomach.

The upper small intestine is alkaline. Its vascular mucosa is attractive to adherent parasites. It contains bile, which inhibits the growth of many bacteria, and selects a bile-resistant local flora; almost all small bowel flora grow well on bile-containing media, e.g. MacConkey's agar. The absorptive mucosa of the jejunum and ileum affords entry for toxins, both those that target mucosal cells and systemic toxins that cause non-intestinal diseases such as botulism. Toxins may be elaborated by bacteria in food or may be components of the food itself.

The terminal ileum and colon contain large numbers of anaerobic bacteria, including Gram-positive and Gram-negative cocci, rods, and sporing organisms. Facultative organisms such as *Escherichia coli* and enterococci are also plentiful.

Some organisms, such as the yeast *Candida*, are able to reside throughout the bowel. Others may pass through, being excreted in the faeces for a while, but not causing long-term colonization.

Natural defences of the gastrointestinal tract

Gastric acid
Gastric acid is the first major defence of the gastrointestinal tract. Although *Helicobacter* and acid-fast organisms can survive in it, it greatly reduces the numbers of most bacterial species that reach the intestines. Patients with achlorhydria or who take strong antacids such as proton pump inhibitors or H_2 antagonists are easily infected by very small numbers of salmonellae or vibrios, and suffer severe illness as a result.

Bile salts
Bile salts in the duodenum inhibit many organisms, killing some by disrupting their cell surfaces. Exceptions are the Enterobacteriaceae, enterococci and other enteric organisms, which survive well in a medium of bile salts. Lower in the bowel, bile salts are reabsorbed and their effect is diminished.

Normal bowel flora
The normal bowel flora confers colonization resistance, a complex phenomenon depending on competition between microorganisms. Bacterial metabolic products can alter the local pH or redox potential. Some bacteria produce toxic hydrogen sulphide or volatile fatty acids, which inhibit other organisms. Many Enterobacteriaceae produce natural antibiotics, bacterocines, which harm other species. It has been shown that enterococci inhibit potential pathogens, such as *Clostridium difficile*. All of these factors combine to favour particular types of organisms in the environments of the bowel.

The protective effect of the normal flora can be reduced by antibiotic action, increasing the risk of infection if a pathogen is encountered. Thus, *Salmonella* infection in exposed people is more likely if they have recently taken streptomycin or tetracycline.

Immune responses
The bowel mucosa is rich in lymphocytes and contains much lymphoid tissue. It is known that some lymphocytes respond actively to antigens of recently active pathogens. T cells are critical to the defence against intestinal infection and HIV-infected individuals are especially susceptible to persistent *Cryptosporidium* and *Microsporum* infections. Humoral immunity is provided by secreted immunoglobulin A (IgA). Immunoglobulin-producing plasma cells

are often plentiful in established mucosal inflammation. Breastfed infants are protected from intestinal pathogens by IgA, which constitutes around 10% of the protein in breast milk. Breast milk also contains lysozyme, which can destroy bacteria.

The immune responses of the gut are exploited when live oral vaccines are given; after immunization the mucosa resists invasion by the 'wild' viruses, which are thus deprived of their usual means of entry into the body.

Motility
The motility of the gastrointestinal tract assists in clearing pathogens. This is demonstrated when obstruction or stasis alters the local environment. The stagnant stomach becomes colonized by organisms that ferment its contents, and reduce its acidity. Blind loops or diverticula in the bowel easily become the foci of abscesses containing bowel flora. Approximately 50% of the weight of faeces is composed of bacteria. The diarrhoea of bowel infections is probably an important mechanism for clearing pathogens from the gut.

Protection against toxins
Enzymes produced by hepatocytes detoxify many drugs and other substances. Toxins absorbed from the gut enter the portal circulation and are intercepted by the liver before reaching the general circulation. Gut flora naturally produce endotoxin, and portal venous blood contains significant endotoxin concentrations. However, after passage through the liver, hepatic venous blood contains only negligible amounts. Isolated Kupffer cells can inactivate endotoxin *in vitro*.

Normal defences of the bowel

1 Gastric acid.
2 Bile salts.
3 Lymphoid tissue.
4 Enterocines.
5 Normal bowel flora.
6 Secretory immunoglobulin A.
7 Motility.
8 Hepatic deactivation of toxins.

General principles of managing gastrointestinal infections

Diagnostic tests

Stool specimens
Stool specimens are easily collected when diarrhoea occurs, but it is important to include the most liquid part of the stool, where most excreted pathogens are found.

Samples of any mucus should also be obtained (a syringe is a convenient means of collecting liquid or viscous material).

Mucosal specimens

Some parasites and ova are only sparsely excreted and more direct specimens increase the likelihood of positive diagnostic results. Examples include *Entamoeba histolytica* trophozoites or schistosome ova, which may both be more readily detected in scrapings or small biopsies of the rectal mucosa obtained at proctoscopy (a small spatula or gloved finger may be used to obtain scrapings). Specimens should be taken from the edge of any ulcer or area of inflammation.

Intestinal fluids

In giardiasis or strongyloidiasis, parasites may not be detectable in stool samples, but are readily demonstrated in duodenal fluid or mucosal biopsy. Duodenal fluid may be obtained by a 'string test', for which the patient swallows a gelatin capsule containing a weighted length of absorbent string. The free end of the string is fixed at the mouth; the weighted end unwinds from the capsule and passes through the stomach into the duodenum. After 30–60 minutes, the string is withdrawn and the adherent material is examined for parasites.

Vomitus

Vomitus is rarely obtained but it may contain viruses in acute viral gastroenteritis, or bacteria in some types of toxic food poisoning. It is worth attempting to collect specimens, especially in patients who do not have coexisting diarrhoea.

Other specimens

Blood cultures are mandatory in patients with fever. Positive blood cultures are seen in a minority of patients with salmonellosis and occasionally in other gastrointestinal infections. Serum may contain enterotoxins, botulinum neurotoxin or antibodies to toxins.

Management of gastrointestinal diseases

Principles of oral rehydration

Few gastrointestinal infections disable the entire absorptive function of the bowel. In toxin-mediated diarrhoeas there is no damage to absorption; hypersecretion is the problem. In other diarrhoeas, a proportion of mucosal cells often remain sufficiently functional to absorb fluid and electrolytes. Absorption of sodium and water can be maximized by giving sufficient of each, with the optimum amount of glucose to 'drive' the active transport systems. A steady intake of this mixture can allow absorption to overtake diarrhoeal fluid loss, even in severely dehydrated patients. Potassium passes passively along concentration gradients. Oral rehydration solution can be given until diarrhoea ceases and normal hydration is restored, as determined by clinical condition or body weight.

> **Recommended composition of oral rehydration fluid** (Fig. 8.1)
> - Sodium: 50–60 mmol/l
> - Glucose: 200–220 mmol/l
> - Potassium: 4–5 mmol/l
>
> This solution can be given until diarrhoea ceases and normal hydration is restored.

(a)   (b)

Figure 8.1 (a) Oral rehydration salts sufficient for addition to one litre of safe, or boiled, water. (b) Simple scoop for measuring sugar and salt, to make one glass (200 ml) of rehydration solution (for an adult, to be consumed after each diarrhoea stool).

Table 8.1 A guide to fluid and electrolyte requirements

Age	Daily baseline fluid requirement (ml/kg)	Daily sodium (mmol/kg)	Daily potassium (mmol/kg)	Total daily calcium (mmol)
1–2 days	75–100	2.5	2.5	12.5–17.5
Up to 1 year	150	2.5	2.5	12.5–17.5
1–3 years	100	2.5	2.5	20–25
4–6 years	90	2.0	2.0	20–25
7–10 years	70	2.0	2.0	20–25
10–14 years	60	1.5	1.5	30–38

Intravenous rehydration

This is indicated for shock, exhaustion precluding oral feeding and failure of oral hydration therapy. The electrolyte solution of choice is half-normal sodium chloride solution (0.45%). This provides adequate sodium, but with less risk of sodium overload than normal saline. Potassium may be added if required. Other electrolyte supplementation is seldom necessary, as rehydration alone usually restores normal homeostasis. A guide to children's fluid and electrolyte requirements is given in Table 8.1.

Infant feeding in diarrhoeal illnesses

Infants who are fed during diarrhoea episodes recover sooner and lose less weight than those who are offered only fluids. Some work in Third World countries suggests that rehydration using rice-water leads to earlier improvement of diarrhoea than with plain water.

Secondary acquired lactose intolerance

This particularly affects infants and young children. It is commonest after rotavirus infection and enteropathogenic *Escherichia coli* infections, which cause severe mucosal damage. Lactose absorption depends on lactase enzymes in the mucosal brush border, which recover slowly after mucosal damage. Until recovery, lactose in milk and other dairy foods cannot be absorbed, so it ferments in the bowel lumen, causing abdominal discomfort, flatulence and acid diarrhoea soon after feeding. Other nutrient absorption systems are rarely affected by intestinal infection.

Infants with severe lactose intolerance can be maintained on lactose-free feeds, such as Galactomin® Formula 17, until gaining weight satisfactorily, when normal feeding is gradually introduced. Older children, including weaning infants, can be given non-dairy solids to replace milk feeds.

⚠️ Secondary acquired lactose intolerance must be distinguished from primary acquired alactasia, a permanent loss of lactase, particularly affecting adult oriental and Caribbean people.

Drug treatment of vomiting and diarrhoea

Antiemetic drugs

Antiemetic drugs can help in reducing fluid loss, which aids oral rehydration. Antiemetic drugs include: antihistamines, such as cyclizine and promethazine, phenothiazines such as prochlorperazine, metoclopramide, whose action is similar to phenothiazines, domperidone, which acts at the chemoreceptor trigger in the brainstem, and the $5HT_3$ receptor antagonists (e.g. ondansetron), which are more often used for treating post-operative or chemotherapy-induced vomiting. Antihistamines are effective, and promethazine hydrochloride can be given orally to infants from age 2 years upwards. Phenothiazines and metoclopramide are dopaminergic drugs, and are slightly sedative at antiemetic doses. Both may cause dystonic reactions, especially in children and teenagers. Metoclopramide is available for injection, and can be used in infants and children of all ages. Domperidone also has dopaminergic actions, but does not cross the blood–brain barrier, so causes fewer central effects.

Antidiarrhoeal drugs

Antidiarrhoeal drugs are rarely successful in severe diarrhoea. They reduce gut motility, allowing fluid faeces to accumulate, but symptoms resume as soon as the expanded bowel volume is filled. Atropine-like side-effects include dry mouth and the risk of urinary retention. Severe atropine-like side-effects are a risk in small children. Antidiarrhoeal drugs are not therefore recommended for this age group. In general it is better to restore hydration than to resort to antidiarrhoeal drugs (this is the recommendation of the World Health Organization for treating cholera and childhood diarrhoea).

Laboratory diagnosis of diarrhoea

The diagnosis of infective diarrhoea depends on the identification of the pathogen from faeces by electron microscopy, culture or demonstration of antigens. For viruses, electron microscopy, culture and PCR are usually used. As most bacterial bowel pathogens are closely related, diagnosis is usually made by culture and formal identification of individual pathogens. Toxins of organisms such as *Clostridium difficile* can be identified in faeces by antigen detection.

At least three faecal specimens should be examined. A faecal suspension is inoculated on to a range of selective media. The more selective the medium, the easier it is to identify the suspected pathogen, but selective media also inhibit the pathogen to some extent, so that low numbers of pathogens may not be detected. A relatively non-selective medium such as MacConkey agar is therefore also inoculated (Fig. 8.2).

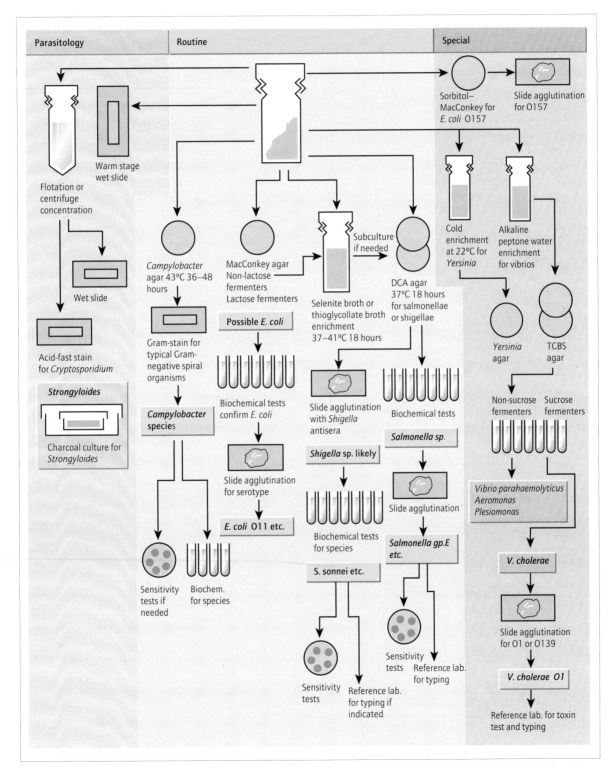

Figure 8.2 Scheme for the bacteriological investigation of stool specimens. TCBS, thiosulphate–citrate–bile-salt–sucrose.

Multiplex real-time PCR techniques are now becoming available, which use sets of reactions to detect bacterial, viral and protozoal pathogens. This permits aetiological diagnosis within a few hours.

Managing food-borne disease in the community

In low-income countries without safe water supplies, most gastrointestinal infections are water-borne. In high-income countries, food is a more important source of infec-tion. More than 50% of infectious gastrointestinal cases in the UK are thought to be food-related (Fig. 8.3).

Ideally, all food for human consumption should be free from pathogens. In practice, food is frequently contami-nated and measures are required to ensure its safety at the point of consumption. Measures that reduce bacterial contamination at source include hygienic husbandry and slaughtering of livestock. These measures are normally the responsibility of central government. Further down the supply chain, a complex legislative framework exists that controls processing, storage and preparation of food by

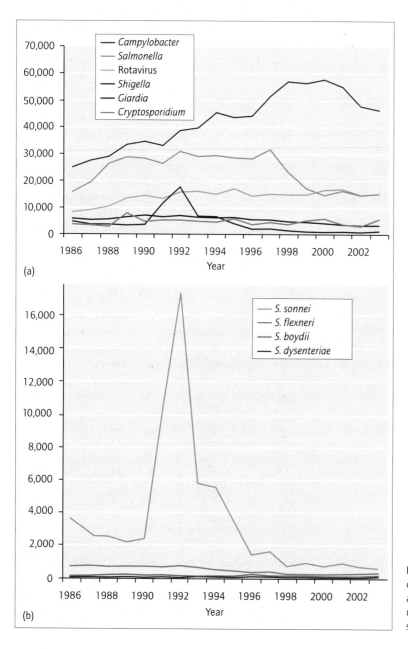

Figure 8.3 (a) Trends in the incidence of common food-borne infections in England and Wales since 1980. (b) Changes in the reported incidence of the four *Shigella* species. Source: Health Protection Agency.

retailers. The enforcement of these regulations is the responsibility of local government.

At consumer level, prevention of food poisoning depends on safe cooking, storage and food preparation, and on personal hygiene. Food should be cooked thoroughly and eaten immediately. Cooked and raw food should be stored and prepared separately. Perishable items should be kept in properly maintained refrigerators. Food not in a refrigerator should be covered to protect it from flies, rodents and other animals. Food-handlers should wash hands before and after food preparation. Cuts, abrasions or other skin lesions should be covered before preparing food. A food-handler who is ill (particularly with diarrhoea or vomiting) should not prepare food until fully recovered.

Prevention of food poisoning

1 Safe food production: healthy flocks and herds, avoiding use of sewage to fertilize food crops.
2 Food-manufacturing processes: hygienic slaughtering and meat packing, rodent-free storage of crops, storage at chill or refrigeration temperatures, hygienic packaging, cold chain during distribution.
3 Domestic and commercial food hygiene: adequate refrigeration, avoidance of cross-contamination, usage within spoilage dates, adequate decontamination of food by washing and/or cooking, personal and kitchen hygiene.

Food poisoning is a notifiable disease in the UK and many other countries. Outbreaks should be promptly investigated to determine the source and implement any necessary control measures (see Chapter 26).

Viral infections of the intestinal tract

Introduction

Recent large surveys have shown that viral infections are the commonest cause of symptomatic intestinal infections in the Western world. They tend to be highly infectious, causing family and community outbreaks. Although rarely severe or fatal, they cause many days of debility and work absence.

Organism list

- Rotavirus
- Adenoviruses (high serotype numbers)
- Caliciviruses
- Noroviruses
- Small round viruses
- Astroviruses
- Coronaviruses

(Hepatitis A and E viruses; see Chapter 9)

Rotavirus gastroenteritis

Epidemiology

Rotaviruses are the commonest viral cause of severe gastroenteritis in young children. Up to a million children worldwide die each year from rotavirus diarrhoea, mostly in developing countries. The disease causes important morbidity in developed countries. In the UK about half, and in the USA around a third of hospital admissions for diarrhoea in children are due to rotavirus.

Over 10 000 laboratory-confirmed cases are reported each year in the UK, with a peak incidence between ages 6 and 24 months. This probably represents only a small fraction of all infections. Illness is rare in children over 5 years of age, although subclinical infection is probably common in children and adults. The disease has a seasonal pattern, with most infections occurring during winter. Transmission takes place between humans by the faecal–oral route and by environmental contamination.

Virology and pathogenesis of rotavirus infection

Rotaviruses are members of the Reoviridae: icosahedral non-enveloped viruses with a segmented double-stranded RNA genome. They have a distinctive appearance on electron microscopy, with two shells surrounding a central protein core. They are classified into seven serogroups (A–G). The different serogroups appear stable and distinct, with no genetic exchange between them. They are further subdivided by serotyping and can also be distinguished by genotype. Genome sequencing has identified great variation between rotavirus strains, caused by both genetic drift and by re-assortment of the segmented genome between human viruses and, occasionally, with animal viruses.

Serogroups A to G are defined by VP2 and VP6 antigens but only strains in groups A–C are associated with human infection. Group A is associated with epidemic gastroenteritis of children. Group B is associated with adult diarrhoea and group C with a milder infections in children and adults. Other serogroups are mainly animal pathogens. Serotypes are defined in group A strains on the basis of the VP4 antigen ('P' serotype) and VP7 antigen ('G' serotype). The majority of infections are caused by just four serotypes.

The genome is divided into 11 segments. It codes for proteins VP1–4, VP6 and VP7, and six non-structural proteins (NSP1–6). The viral structure is constructed from the VP2 (innermost shell), VP4 (outer protruding spikes), VP6 (middle shell) and VP7 (outer glycoprotein) proteins. The outermost proteins VP4 and VP7 are the target of protective antibodies. The virus is naturally robust and resists inactivation by many disinfecting processes including the concentrations of chlorine used to treat drinking water.

The mechanism of viral attachment to enterocytes is unknown. Direct penetration of the cell membrane is mediated by VP5, and the virus can also enter cells, via vesicles and lysosomes, by receptor-mediated endocytosis. Mature enterocytes at the distal part of the villi are preferentially infected. They become cuboidal, their microvilli become shortened and the cells soon die and are shed, leaving stunted and denuded intestinal villi. Diarrhoea is thought to be caused mainly by loss of absorptive mucosal cells, and their replacement by secretory cells. However, the non-structural protein NSP4 acts as a viral enterotoxin (the first viral enterotoxin to be described). It is thought to produce diarrhoea early in infection by increasing enterocyte membrane permeability, causing chloride secretion, leading to secretory diarrhoea.

Clinical features

After about 48 hours' incubation, there is an abrupt onset of diarrhoea and vomiting. Moderate fever is common at the onset, but rarely persists. Vomiting is not severe, and rarely prevents oral feeding. Prolonged vomiting is rare; if it continues for more than 48 hours the diagnosis should be reviewed. The diarrhoea varies in severity but occasionally causes gross dehydration or even shock. Stools may be bloody, especially in infants under the age of 1. This probably reflects the large-scale destruction of mucosal cells, leaving denuded areas in the duodenum and upper ileum. Even severely affected cases tend to recover after 2–4 days.

Most adults are immune to group A rotaviruses, but a few are susceptible, often suffering transient vomiting as their major symptom. The illness is highly infectious while vomiting lasts, and susceptible children may be infected by an affected nurse or mother. Group B and C rotaviruses are uncommon, and adults are more often susceptible to them. They have caused large outbreaks of adult gastroenteritis.

Diagnosis

Electron microscopy, if available, often demonstrates typical rotavirus particles in diarrhoea stools (Fig. 8.4). Human rotaviruses do not grow in cell culture. Enzyme-linked immunosorbent assay (ELISA) tests for rotavirus antigen are available. Latex agglutination is less sensitive but is simple to perform. Real-time polymerase chain reaction (PCR) techniques are now routinely used. They provide both rapid diagnosis and the ability to determine serotypes.

Management

Most cases recover rapidly with oral rehydration treatment. A few develop secondary acquired lactose intoler-

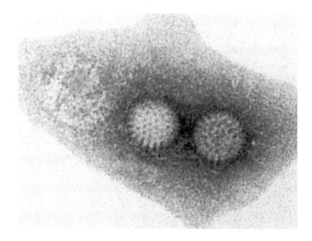

Figure 8.4 Rotavirus particles demonstrated by negative-stained electron microscopy preparation of a diarrhoea stool. Courtesy of Professor P. D. Griffiths and Ms G. Clewley, Department of Virology, Royal Free Hospital School of Medicine.

ance, and require a lactose-free diet until lactase systems are restored.

A few infants have exacerbations of diarrhoea after each feed of milk formula or oral rehydration fluid, suggesting that there are too few functioning mucosal cells to absorb the glucose in the fluid. This transient problem can be managed by giving intravenous electrolyte solutions, or glucose- and lactose-free oral preparations such as Galactomin® Formula 19, or Pregestemil®.

Rare effects of rotavirus infection

Some 1–2% of children admitted with rotavirus infection present with fever and shock. The aetiology of the condition is often in doubt until fever declines and diarrhoea occurs, suggesting an intestinal infection. A few reports exist of rotavirus particles being detectable in the cerebrospinal fluid in such cases.

Rotavirus can cause persisting diarrhoea in immunosuppressed patients, including those with human immunodeficiency virus infection.

Rotavirus vaccines

Orally administered live rotavirus vaccines are currently being developed, using both bovine and human virus strains. Their efficacy is variable. A vaccine containing reassortant rhesus monkey and human rotavirus genes was briefly used in infants in the USA, but was withdrawn following reports of intussusception in vaccine recipients.

Other viral infections of the gastrointestinal tract

Epidemiology

Noroviruses, adenoviruses, astroviruses, coronaviruses

and other small round viruses have all been associated with outbreaks of gastroenteritis. All age groups are affected. Transmission is usually from person to person by the faecal–oral route. Large school and cruise-ship outbreaks are well-documented. Outbreaks affecting hospitals and elderly care facilities are highly disruptive, often forcing the closure of wards. Food-borne incidents have also been reported, often due to shellfish harvested from sewage-polluted estuaries being eaten raw or without sufficient cooking. Aerosol spread may be important in some outbreaks, particularly in winter vomiting disease (see below). Norovirus is the most common cause of infectious gastroenteritis in England and Wales, causing between 600 000 and a million cases, and between 130 and 250 reported outbreaks, annually.

Norovirus virology
Noroviruses (previously known as Norwalk viruses) are small, round viruses, members of the Calicivirus group (a name that is derived from the Latin word for cup). They are single-stranded positive sense RNA viruses 26–35 nm in diameter. Genetic analysis of the Caliciviruses has permitted the establishment of four genera of which two, Norovirus and Sapovirus, are associated with gastrointestinal disease in humans. Noroviruses can be divided into at least three genogroups and numerous genotypes. Diverse genotypes circulate in the human community but occasionally a single genotype will emerge and become responsible for the majority of infections.

Clinical features
The incubation period is short, 6–36 hours, with more variation between individuals with the same infection than between infections. Illness lasts for 1–8 days.

Adenoviruses tend to affect children, causing mainly diarrhoea. Noroviruses affect children and adults, causing vomiting and diarrhoea. Adenoviruses have often been identified during surveys of asymptomatic children, but noroviruses usually cause symptomatic infection.

Winter vomiting disease
Noroviruses are the common cause of this disease. The incubation period of about 1 day is followed by several hours of profuse vomiting. There is then a period of 24–36 h of fatigue before recovery is complete. This is an intensely infectious disease, which often causes school and family outbreaks, affecting both adults and children.

Diagnosis of viral gastroenteritis

An aetiological diagnosis in viral gastroenteritis is of most importance in outbreak investigations, especially in hospitals and elderly care facilities. For individual cases diagnosis rarely affects management, and its expense is therefore

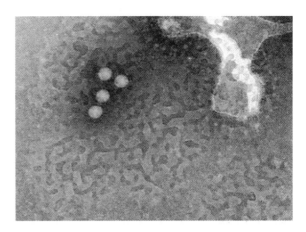

Figure 8.5 Norovirus particles demonstrated in negative-stained electron microscopy preparation of stool from an elderly patient infected in a hospital outbreak of diarrhoea.

rarely justified. The principal method of diagnosis is electron microscopy, which demonstrates characteristic viral morphology (Fig. 8.5). ELISA tests are available for the diagnosis of rotavirus and norovirus. RT-PCR diagnosis is more sensitive than electron microscopy and EIA but antigenic variations can render the test falsely negative. The investigation of viral gastroenteritis is approached on a syndromic basis.

Bacterial diseases of the gastrointestinal tract

Organism list

Common infections
- *Campylobacter* spp.
- *Salmonella* spp.
- *Shigella* spp.
- *Escherichia coli*
- *Clostridium perfringens*
- *Helicobacter pylori*

Uncommon or rare infections
- *Aeromonas* spp.
- *Vibrio parahaemolyticus*
- *Plesiomonas* spp.
- *Yersinia* spp.
- *Clostridium difficile*
- *Vibrio cholerae*

Bacteria that elaborate preformed toxins
- *Staphylococcus aureus*
- *Bacillus cereus*
- *Clostridium botulinum*

Escherichia coli intestinal infections

Introduction

E. coli strains are the major facultative anaerobic inhabitants of the human gut. Their role in human disease has been increasingly appreciated as it has become possible to distinguish commensal types from types behaving as pathogens.

In the 1960s certain serotypes of *E. coli* caused large epidemics of gastroenteritis in infants. Since then, more *E. coli*-mediated diseases have been recognized, including toxin-mediated traveller's diarrhoea (see Chapter 24), and haemorrhagic colitis associated with certain serotypes of verocytotoxin-producing organisms.

E. coli are facultative anaerobes that ferment a wide range of sugars, including lactose, producing acid and gas. They are readily isolated from human material on simple laboratory media. They can be identified by biochemical tests such as their characteristic ability to ferment lactose. They possess flagella and are actively motile.

Microbiology and pathogenesis of *Escherichia coli* infections (Fig. 8.6)

The capsular K antigens of *E. coli* facilitate adherence to host cells. Strains possessing a K1 capsule are the most frequent *E. coli* isolated from cases of neonatal meningitis. The mechanism for this is not understood, although it is known that colonization of neonatal rat gut by K1 strains often leads to bacteraemia and meningitis. K88 antigen-bearing strains readily cause scour (diarrhoea) in piglets.

Type I fimbriae are possessed by all strains and mediate attachment to mucosal surfaces (first recognized in the contribution of P fimbriae to the pathogenesis of urinary tract infection). In addition *E. coli* express many different fimbrial types. Fimbriae play a role in the pathogenesis of enterotoxigenic *E. coli* (ETEC) infections, as toxigenic *E. coli* that cannot adhere to gut mucosa are non-pathogenic. Some fimbrial colonizing factor antigens (CFAs) have been identified, and these include CFA/I, 6–7 nm rigid fimbriae, and the CFA/II family of antigens, 2–3 nm fibrillar fimbriae that are analogous to the K88 antigens of porcine strains. Other similar CFA families have been described in other strains.

Like other Gram-negative bacteria, *E. coli* has a lipopolysaccharide (LPS) antigen with a central lipid A core, an oligosaccharide moiety and polysaccharide chain. The lipid A portion is an endotoxin (see Chapter 1). The polysaccharide chain protects the organism from serum lysis and is the main (O) antigen of the organism. More than 150 O serotypes have been described, many of which are linked to other *E. coli* pathogenicity factors. For example, the most common enteropathogenic *E. coli* (EPEC) are serotypes O26, O55, O111, O114, O119, O125–9 and O142. Strains responsible for the haemolytic–uraemic syndrome (HUS) are usually serotype O157.

Enterotoxigenic *E. coli* (ETEC)

E. coli toxins are important in the pathogenesis of diarrhoeal and systemic disease. Two main enterotoxins are produced by ETEC – heat-labile (LT) and heat-stable (ST) toxins.

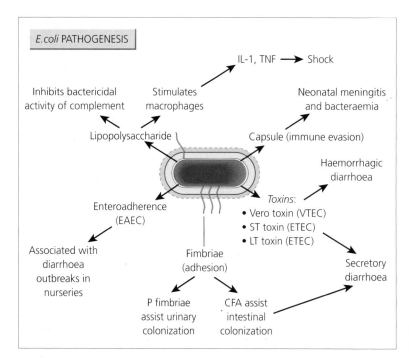

Figure 8.6 The pathogenesis of *E. coli* infections.

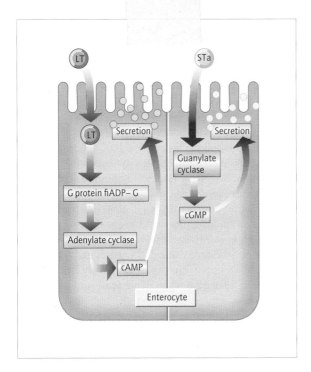

Figure 8.7 Actions of *E. coli* heat-labile (LT) and cholera toxin and of *E. Coli* heat-stable (ST) toxin: LT and cholera toxin stimulate adenylate cyclase via a specific G-protein-coupled membrane receptor, while ST directly stimulates its cell-surface receptor to generate guanylate cyclase activity. cAMP and cGMP both cause phosphorylation of membrane ion channels, leading to chloride, sodium and water secretion.

The LT toxin consists of polypeptide subunits, A and B. In the native toxin, five B subunits mediate attachment to cells via the GD1 receptor, and one A subunit then enters the cell and activates adenylate cyclase. This toxin is biologically and immunologically closely related to cholera toxin (Fig. 8.7).

The ST toxin exists in several forms. It induces hypersecretion by stimulating guanylate cyclase synthesis.

Enteroinvasive *E. coli* (EIEC)
These strains closely resemble *Shigella* spp., and often cannot ferment lactose. They are capable of invading and replicating within enterocytes and of spreading from cell to cell. This property depends on possession of a plasmid that encodes specific outer membrane proteins. EIEC and shigellae also produce toxic substances that are not related to LT or ST.

Verocytotoxic *E. coli* (VTEC) (or enterohaemorrhagic *E. coli*, EHEC)
Some *E. coli* strains elaborate verocytotoxin (VT) or shiga-like toxin (SLT), which destroys vero-cell monolayers. This bacteriophage-encoded toxin exists in two forms, VT1 and VT2. VTEC strains that also possess the attach-

ing and effacing property found in EPEC are known as enterohaemorrhagic *E. coli*. VT1, but not VT2, is neutralized by antibodies to Shiga toxin. Both toxins are important in the pathogenesis of haemorrhagic colitis and HUS.

The infective dose of VTEC is small, probably between 10 and 100 organisms. Large outbreaks can therefore occur when low numbers of organisms exist in food, milk or the environment.

Enteropathogenic *E. coli* (EPEC)
EPEC cause diarrhoea outbreaks in institutions, and infant diarrhoea in developing countries. These strains lack ST, LT, and shiga-like toxin and are not enteroinvasive. Their pathogenetic mechanism is not fully understood. The genes encoding the attachment-effacing properties exist on a pathogenicity island known as the locus of enterocyte effacement (LEE). They include *EspA*, which encodes a filamentous protein found between the bacterium and the enterocyte. Many EPEC strains have an enteroadherence factor (EAF), a plasmid-encoded type IV pilus, whose absence is associated with reduced pathogenicity. Electron microscopic studies show that EPEC cause destruction of microvilli without evidence of invasion (attachment-effacing lesions). Unusually for *E. coli*, they adhere to Hep-2 cells via an EPEC adherence factor (EAF), a 94-kDa plasmid-encoded protein. EPEC serotypes without this gene cannot cause disease. Infection with EAF-positive strains stimulates production of EAF antibody, which may be protective.

Enteroaggregative *E. coli* (EAEC)
These are not classic EPEC serotypes and do not possess the EAF plasmid, or produce recognized toxins. Through a plasmid-encoded adherence factor, they adhere to Hep-2 cells in either an aggregative or a diffuse pattern, depending on different factors. It is not yet clear whether the pathogenicity of this organism depends on its enteroadherence property. EAEC strains cause mucosal damage with loss of microvilli and cell death. They produce dispersin, an immunogenic surface protein. Some strains also express a toxin similar to heat-stable enterotoxin.

Pathogenicity factors for *Escherichia coli*

1 Capsular K antigens (K1).
2 Fimbriae (colonizing factor antigens – CFA/I, CFA/II and others).
3 Lipopolysaccharide O antigens (endotoxin).
4 Enterotoxins (heat-labile and heat-stable).
5 Verocytotoxin (shiga-like toxin).
6 Enteropathogenic *Escherichia coli* adherence factor (EAF).
7 Enteroadherence (mechanism unknown).

Epidemiology
The routes of transmission and epidemiological features of *E. coli* vary between different pathogenic types and in different geographical locations. Where hygiene is poor,

ETEC is the commonest bacterial cause of childhood diarrhoea. The disease is uncommon in western Europe and the USA, although ETEC are an important cause of traveller's diarrhoea. Humans are the main source of infection, which is transmitted by contaminated food and water.

EPEC usually spread from person to person by the faecal–oral route. Contaminated baby food has also caused outbreaks. Most infections occur as outbreaks of infantile enteritis. The incidence of EPEC infections is decreasing in many countries.

Enteroinvasive *E. coli* (EIEC) infections are also rare in developed countries. Infection is usually food-borne but direct person-to-person spread may also occur. All age groups are affected.

VTEC includes several serogroups of *E. coli*, though only serogroup O157 is common. During the 1980s the recognition of VTEC infection and its clinical sequelae (haemorrhagic colitis and HUS) increased rapidly in the USA and Canada. VTEC is now the commonest recognized cause of acute renal failure in children in Western countries (Fig. 8.8). Infection is mainly food-borne, particularly via hamburger meat or unpasteurized milk. Large outbreaks have affected the community, nursing homes for the elderly and children's day-care centres. Outbreaks have also followed contact with young animals (e.g. during farm visits). The largest UK outbreak, involving 500 cases, with 18 deaths, was associated with contaminated meat sold from a butcher's shop.

Clinical features

The incubation period is 1–5 days. The onset is abrupt, with vomiting and diarrhoea for the first 6–24 h, followed by watery diarrhoea alone. There is often moderate fever at the onset. There is little abdominal pain in uncomplicated gastroenteritis, which is clinically similar to viral gastroenteritis or salmonellosis. A minority of cases exhibit severe disease with rapid dehydration and collapse, or prolonged illness with persisting diarrhoea and sometimes lactose intolerance.

Haemorrhagic colitis

Haemorrhagic colitis, caused by EHEC, affects both children and adults. It begins as an unremarkable diarrhoeal illness, but quickly progresses to a syndrome of bloody diarrhoea with abdominal pain. Fever is not an important feature. Sigmoidoscopy reveals acutely inflamed colonic mucosa and the condition can be mistaken for acute inflammatory bowel disease. Most cases are self-limiting, recovering in 5–10 days. A handful of cases need prolonged symptomatic support. In outbreaks, about 10% of children develop haemolytic–uraemic syndrome (HUS, see below), but almost all of the fatalities are in patients over age 65–70.

Haemolytic–uraemic syndrome

Haemolytic–uraemic syndrome is mainly a disease of children, but can also affect adults. The adult form overlaps clinically with the syndrome of thrombotic thrombocytopenic purpura (TTP), but in TTP thrombosis and platelet depletion are related to functional abnormality of uncleaved von Willebrand's factor (see Chapter 21), while the pathogenesis of HUS is based on toxin-mediated mucosal damage, renal vascular damage and microangiopathic anaemia. The features of HUS are a rising blood urea and creatinine, microangiopathic haemolytic anaemia and thrombocytopenia (Fig. 8.9). Clinical warn-

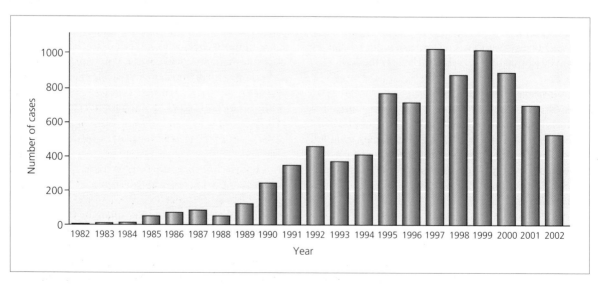

Figure 8.8 Increasing reports of verocytotoxin-producing *Escherichia coli* infection in England and Wales.

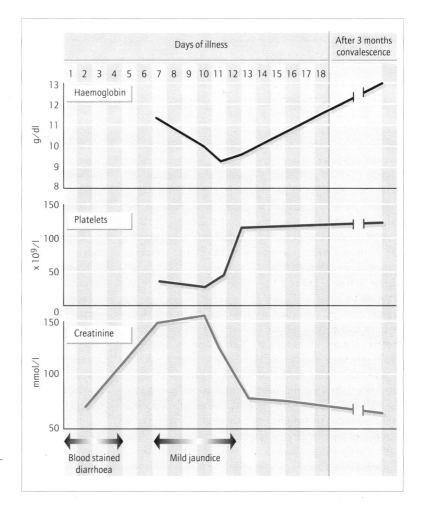

Figure 8.9 The evolution of haemolytic–uraemic syndrome after an attack of bloody diarrhoea.

ing signs are a raised blood pressure, persistent vomiting or fits. Although most cases of HUS follow a gastrointestinal illness, diarrhoea may be mild without bleeding or abdominal pain.

Antidiarrhoeal agents may exacerbate HUS by delaying clearance of the pathogen from the bowel. Antibiotic treatment may increase the risk of HUS in VTEC infection. Evidence suggests that some antibiotics stimulate verocytotoxin production, but that quinolones inhibit it.

Over 70% of all HUS cases recover completely. However, about 25% have persisting hypertension or impaired renal function. The overall mortality for HUS is about 5%.

Diagnosis

Most gastrointestinal illnesses produced by *E. coli* are non-specific. Epidemiological evidence such as recent travel or HUS will alert clinicians and public health specialists.

In HUS, diarrhoea may have ceased by the time systemic illness is apparent. Stool cultures are often negative after the fifth or sixth day of illness. The VT gene may be detectable in stools by PCR. Serum antibodies to VT often become detectable after 5–7 days.

Laboratory diagnosis

E. coli pose significant problems for diagnosis in the microbiological laboratory, as the pathogen and the normal flora are the same species. Conventional methods of selection cannot therefore be employed.

EPEC are limited to a relatively small number of serotypes. In children under the age of 3 years in whom the diagnosis is suspected, an additional blood or nutrient agar plate is inoculated and several colonies of *E. coli* are 'screened' with polyvalent antisera to the EPEC strains.

VTEC strains are found in several O serotypes, among which O157 predominates. This strain is unusual in not fermenting sorbitol. A modified MacConkey's agar, incorporating sorbitol in place of lactose, can be used to identify the majority of these strains. In addition to this, toxin production must be confirmed by immunological or gene-detection techniques. Unselected stool cultures can be tested by DNA hybridization for the presence of the verocytotoxin-producing gene.

Management

In most cases symptomatic treatment is adequate. Specific chemotherapy is indicated in severe or prolonged disease, or in elderly and debilitated patients, who are at high risk of complications. Pathogenic *E. coli* are resistant to many antimicrobial agents, including broad-spectrum penicillins and cephalosporins, trimethoprim, sulphonamides, chloramphenicol and aminoglycosides. Specific treatment should therefore be guided by the results of culture and sensitivity tests. The agents most likely to be effective are quinolones; oral ciprofloxacin 500 mg twice daily for 3–5 days often succeeds, intravenous ceftriaxone is an alternative. Antibiotic therapy should be avoided, when possible, in the management of HUS.

Prevention

Prevention of *E. coli* intestinal infections is by adequate sanitary and hand-washing facilities. Scrupulous attention to hygiene is important, particularly in nurseries where infection is common. Hamburger meat should be cooked thoroughly before eating. Travellers to tropical countries should avoid drinking untreated water and eating high-risk foods such as raw vegetables, salad, peeled fruit and undercooked meat (see Chapter 24). A recently introduced cholera vaccine (Dukoral) contains modified cholera toxin antigen, and may cross-protect against traveller's diarrhoea caused by LT-producing *E .coli*. However, it is unlicensed for this use in the UK and data on its efficacy in travellers is limited.

Salmonella infections

Introduction

The food-poisoning serotypes of *Salmonella* infect both humans and animals. They are biochemically and clinically distinct from the serotypes that cause enteric fevers, which are exclusively human pathogens. Enteric fevers are mainly associated with travel, and will be discussed in Chapter 24.

Epidemiology

Around 2200 different serotypes of *Salmonella* exist in animals, and most are capable of causing salmonellosis in humans. About 200 serotypes are reported in any one year in the UK. Currently the most common are *S. typhimurium*, *S. enteritidis* and *S. virchow*.

Salmonellae commonly infect flocks and herds of food animals. Most human infections are therefore food borne. Poultry meat is the commonest source of human infections. The shell and occasionally the white of eggs can be contaminated. Other meats such as beef and pork are also well-recognized 'sources' and outbreaks have been caused by contamination of seafood and vegetables. Some *Salmonella* serotypes are particularly associated with one type of food, e.g. *S. enteritidis* phage types 4 and 8 with chicken.

Food-handlers do not usually transmit infection unless they remain at work with diarrhoea. Faecal–oral transmission sometimes occurs, usually in institutions such as psychogeriatric hospitals and old people's homes.

Salmonellosis typically occurs sporadically or in small outbreaks, often within households. It affects all age groups, though morbidity is greatest in children and the elderly. Large outbreaks are less common, often affecting large, easily identified functions such as weddings, or occurring in institutions. Most cases occur during late summer and early autumn, especially in hot weather.

Many countries experienced an unprecedented rise in salmonellosis during the late 1980s. The commonest epidemic serotype varies between countries, and over a period of years within the same country. In the UK, beef-associated *S. typhimurium* was replaced by poultry- and egg-associated *S. enteritidis* phage type 4 in the 1980s. In mainland Europe the predominant type in the 1990s was *S. enteritidis* phage type 8. More recently, the epidemiology of *S. enteritidis* infections in England and Wales has changed. The incidence of *S. enteritidis* phage type 4 infection has declined from around 10 000 cases in 1998 to around 2500 cases in 2003, while non-phage type 4 infections increased from 3500 cases in 2000 to 7000 cases in 2003. The number of reported food-borne outbreaks of *S. enteritidis* non-phage type 4 increased, mostly associated with use, in the catering industry, of contaminated eggs imported from Spain.

The 10 commonest *Salmonella* serotypes in the UK

1 *S. enteritidis.*
2 *S. typhimurium.*
3 *S. virchow.*
4 *S. dublin.*
5 *S. bovis-morbificans.*
6 *S. hadar.*
7 *S. newport.*
8 *S. braenderup.*
9 *S. heidelberg.*
10 *S. montevideo.*

Microbiology and pathogenesis of salmonella infections

Salmonellae belong to the family Enterobacteriaceae. *Salmonella* is a non-lactose fermenter, which produces acid and gas from glucose, metabolizes citrate and produces hydrogen sulphide. These and other biochemical characteristics are used to identify salmonella growth in selective media. *S. typhi, S. cholerae-suis, S. paratyphi* and *S. arizona* can be differentiated from other salmonellae on the basis of their biochemical reactions.

Salmonellae possess lipopolysaccharide (LPS), which is their somatic O antigen, and exists in many types. These, together with the flagellar H antigens, define the serotype, which is often given a name, e.g. *S. enteritidis, S. typhimurium, S. virchow*, etc. The LPS protects the bacterial cell from the bactericidal activity of serum, influences macrophage interactions, decreases susceptibility to host cationic proteins and functions as endotoxin.

DNA hybridization, protein isozyme analysis and DNA sequence suggest that *Salmonella* serovars belong to a single species, but modern taxonomists classify them as two species: *Salmonella enterica* and *S. bongori*. *S. enterica* includes six subspecies differentiated by biochemical variations (*enterica, salamae, arizonae, diarizonae, houtenae* and *indica*). Thus, the designation of *S. enteritidis* phage type 4 (PT4) is *S. enterica* subsp. *enterica* serotype enteritidis PT4, usually abbreviated to *S. enteritidis*. Serovars of subspecies other than subsp. *enterica* are designated by their antigenic formulae, for example, *S. enterica* subsp. *arizonae* 61:k:1,5,7.

Salmonellae cause localized infection of the intestinal mucosa but can also invade macrophages, to multiply in the reticuloendothelial system, occasionally leading to severe systemic infection.

Non-*Salmonella typhi* serotypes

Salmonellae traverse the M cells of the intestinal Peyer's patches, destroying the invaded M cells and adjacent epithelial cells. The *Salmonella* genome contains five pathogenicity islands. Most of the genes controlling entry into the M cells reside on *Salmonella* pathogenicity island (SPI)-1. They encode a type III secretion system (TTSS), which mediates protein export from the bacterium to the host cell cytosol. Four proteins encoded on SPI-1 (SopE, SipA, SipC and StpP) enter the host cell in this way and initiate cytoskeletal rearrangements, membrane ruffling and, ultimately, bacterial internalization via large vesicles (macropinosomes). Within macrophages, *Salmonella* remains inside phagosomes (*Salmonella*-containing vacuoles; SCVs). SPI-2 encodes proteins such as SpiC, which inhibit fusion of the SCV with host cell lysosomes and endosomes, which contain the microbicidal enzymes of the cell. Other SPI-2 proteins may disrupt host-cell intracellular trafficking, further enhancing the survival of *Salmonella* within the phagosome. For example, SPI-2 proteins may reduce intra-macrophage concentrations of NADPH oxidase, which is essential for the production of microbicidal compounds, such as reactive oxygen and nitrogen intermediates.

Virulence plasmids

The virulence of *S. typhimurium* depends on possession of a cryptic plasmid, which carries, among others, the *rck* gene. This encodes a protein that prevents the formation and insertion of the fully polymerized attack complex of complement into the bacterial outer membrane.

Non-typhoid salmonellae, which frequently cause extra-intestinal invasion (*S. dublin, S. cholerasuis, S. gallinarum-pullorum* and *S. virchow*), possess plasmids of differing sizes that contain genes important for invasive infection. All these plasmids possess *spv* (*Salmonella* plasmid virulence) genes contained on an 8 kb regulon. The regulon carries the positive regulator (*spvR*) and four structural genes (*spvA, spvB, spvC* and *spvD*). Efficient transcription of *spvR* depends on a chromosomally encoded alternative s factor (RpoS). The activity of RpoS increases in the post-exponential phase of growth. The plasmid-encoded *spv* genes are not expressed during early exponential growth, but the synthesis of SpvA begins in the transition period between the late logarithmic and early post-exponential phases of growth. The other proteins are expressed later. It appears that nutrient starvation, iron limitation and pH (but not cell density) stimulates *spv* expression, which enhances *Salmonella* growth within the macrophages and other phagocytic cells.

The virulence of *S. typhi* (see Chapter 25)
Virulence-related genes
S. typhi has some gene sequences not found in non-typhoid salmonellae, e.g. SPI-7, a pathogenicity island that contains the genes for Vi antigen production. Almost all virulent *S. typhi* possess a Vi antigen – a polysaccharide composed of *N*-acetylglucosamine uronic acid that inhibits phagocytosis and masks the O antigen, reducing the infective dose of the organism. Organisms lacking the antigen require a very high dose to achieve infection. Antibodies to Vi develop during recovery from typhoid fever, and immunization with Vi antigen confers protective immunity against typhoid. The *S. typhi* genome also contains over 200 pseudogenes, stretches of DNA that encode gene-like sequences but are inactive, usually due to single base mutations. One hundred and forty-five *S. typhi* pseudogenes exist in active form in *S. typhimurium*, a *Salmonella* serovar with a wide host range that causes enteritis.

Lipopolysaccharide
Like other Enterobacteriaceae, *S. typhi* and other salmonellas possess a lipopolysaccharide with a lipid A core, which activates the release of tumour necrosis factor α (TNF-α) and other cytokines from macrophages. Injection of small amounts of *S. typhi* LPS in human volunteers can reproduce typhoid-like symptoms. *S. typhi* (and

also *S. paratyphi* C and some *Citrobacter* spp.) possess the Vi antigen, a polysaccharide capsule consisting of alpha-1,4,2-deoxy-*N*-acetylgalacturonic acid. As in *S. typhi*, it prevents phagocytosis and masks the O antigen, reducing the minimum infective dose for organisms possessing this antigen. Injection of small amounts of *S. typhi* LPS in human volunteers can reproduce typhoid-like symptoms.

Lipopolysaccharide also protects the organism from the effects of complement. The long 'O' side chain on the molecule prevents the membrane attack complex from disrupting the bacterial outer membrane.

Clinical features of *salmonella* intestinal infections

After 12–36 hours' incubation illness begins with malaise, nausea and vomiting, often with fever. Diarrhoea becomes the main feature within 24 hours: it is watery and brown, often becoming greenish if it persists. Fever usually resolves by the first or second day and intestinal symptoms improve soon afterwards. In severe cases illness may persist for many days with variable fever, continuing diarrhoea and progressive debility. Abdominal pain is rarely an important feature.

In elderly patients, sudden fluid loss during acute diarrhoea may lead to low-output cardiac failure, myocardial infarction or stroke. Frail patients tolerate rehydration poorly and easily develop pulmonary oedema. Confusion, hypostatic chest infections and the risk of deep-vein thrombosis during immobility are longer-term hazards.

Patients with achlorhydria are especially susceptible to salmonellosis. A small infective dose of organisms can cause severe illness with high fever, intense watery diarrhoea and a high blood urea. These patients are at increased risk of salmonella bacteraemia. Recrudescences of bacteraemic infection are a marker of immunodeficiency in HIV infection.

Salmonella colitis

Salmonella colitis, with colic and bloody stools, occurs in up to 10% of patients. Sigmoidoscopy often confirms the presence of colonic inflammation.

Salmonella bacteraemia

Salmonella bacteraemia and metastatic infections are uncommon. They can occur with or without preceding bowel symptoms. Some salmonellae are more likely than others to produce bacteraemia; *S. dublin* and *S. cholerae-suis* do so in 30–40% of reported cases, but fortunately these are uncommon serotypes.

Susceptible tissues may become metastatically infected by salmonellae, even when bacteraemia has not been apparent. Commonly affected sites are bones and joints, including those of sickle-cell sufferers (Fig. 8.10), and

arterial aneurysms. Occasionally a patient presents with a soft-tissue abscess (Fig. 8.11), and laboratory investigation reveals salmonellae in the pus.

Diagnosis of salmonellosis

The clinical features of salmonellosis are non-specific. It is, however, the commonest cause of persisting diarrhoea without abdominal pain. Recovery of a salmonella from the diarrhoea stool confirms the specific diagnosis. Blood cultures are indicated if fever is high or persists for more than 48 h. A high blood urea, significant dehydration, shock or a history of achlorhydria makes blood cultures mandatory.

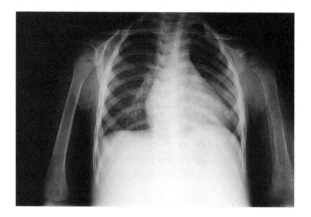

Figure 8.10 Salmonella osteomyelitis in the left humerus of a child with sickle-cell disease; there is periosteal elevation and new bone formation, with patchy irregularity in the medulla of the bone.

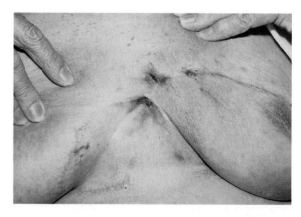

Figure 8.11 *Salmonella typhimurium* breast abscess: this abscess was unresponsive to treatment with anti-staphylococcal agents, but resolved after prolonged co-trimoxazole therapy (there had been no preceding diarrhoeal illness).

Stool cultures should always be carried out in patients presenting with extra-intestinal *Salmonella* infections, as the bowel may be the reservoir of infection.

Isolation and identification of salmonellae and shigellae

The selective procedures to identify salmonellae and shigellae are largely the same.

A range of selective media are available for the isolation of salmonellae and shigellae. All contain compounds inhibitory to the normal faecal flora, but which also inhibit the pathogens to some extent. There is therefore a balance between selection and diagnostic yield. Bile salts select for organisms that inhabit the bowel: sodium desoxycholate, found in xylose lysine desoxycholate (XLD) agar, or desoxycholate citrate agar (DCA), which is more strongly selective. Bismuth sulphate, used in Wilson and Blair's medium, is even more selective and inhibitory.

Most microbiological laboratories use a combination of selective media; one less inhibitory medium (MacConkey or XLD) for the diagnosis of the more fastidious shigellae, and a more inhibitory (DCA or Wilson and Blair) to select for salmonellae. These media contain indicator systems to distinguish colony types for further study. On media containing lactose and neutral red, lactose-fermenting organisms generate acid, producing pink/red colonies (Fig. 8.12). Neither salmonellae nor shigellae ferment lactose so these organisms form clear colonies (easily distinguished from the pink colonies of *E. coli* and klebsiellae). In XLD agar, coliforms and salmonellae both ferment xylose and produce acid, but salmonellae also decarboxylate lysine, causing a counterbalancing rise in pH. Shigellae neither ferment xylose nor decarboxylate lysine. Coliform organisms only produce acid, and have yellow opaque colonies.

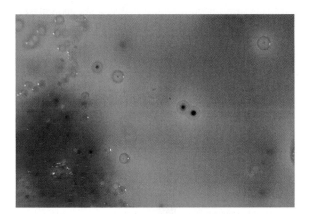

Figure 8.12 Colonies of enteric bacteria growing on MacConkey agar: lactose-fermenting Enterobacteriaceae (with pink centres) and transparent colonies of non-lactose-fermenters are both seen.

Salmonellae and shigellae produce neutral colonies, for different reasons, and appear red (neutral). Ferric ammonium citrate in this medium, and in DCA, indicates colonies of hydrogen sulphide producers (salmonellae), which produce a central black dot of iron sulphide.

The next step is the use of biochemical screening tests. Commercially produced kits (e.g. API ZYM) also detect the characteristic range of enzymes produced by salmonellae, indicating which organisms require full biochemical and serological identification. Serological identification of salmonellae and shigellae is unreliable without biochemical confirmation, because of extensive antigenic cross-reactions with commensal gut flora.

Molecular techniques can provide rapid detection of *Salmonella* spp. Commercial kits based on real-time PCR are available.

For routine diagnosis and infection control, serotyping and phage typing of salmonellae is sufficient. Further typing is only indicated for the identification and investigation of outbreaks. Available typing methods include antibiotic sensitivity profiles, phage typing, plasmid typing, ribotyping, insertion sequence typing or comparison of organisms by pulsed-field gel electrophoresis.

Salmonellae that can be phage-typed

1 *S. enteritidis.*
2 *S. typhimurium.*
3 *S. virchow.*
4 *S. typhi.*
5 *S. paratyphi.*

Management of salmonellosis

Most cases require only oral or intravenous rehydration.

Indications for antibiotic therapy of salmonellosis

1 No spontaneous improvement is evident after 3 days' supportive therapy.
2 Hypotension and tachycardia persisting after adequate rehydration.
3 For patients at special risk: sickle-cell patients, immunosuppressed patients and the frail elderly.
4 All patients with positive blood cultures.

Salmonellae have acquired increasing antibiotic resistance in recent years. Most are zoonotic organisms, and have been widely exposed to antibiotics used in both veterinary and medical practice. They are usually resistant to tetracyclines, sulphonamides, aminoglycosides, broad-spectrum penicillins and many cephalosporins. A substantial proportion are resistant to trimethoprim and chloramphenicol. Quinolone agents are often effective against salmonellae contracted in Western countries, but quinolone resistance is increasingly common in Asian countries where salmonellosis is common and quinolones are widely used.

Recommended regimens include oral ciprofloxacin or ofloxacin. Infections contracted in the Indian subcontinent may respond to trimethoprim, co-trimoxazole or ceftriaxone. Chloramphenicol is still available for use in cases resistant to other drugs. Short courses of 4 or 5 days' treatment are usually sufficient.

Patients with shock or bacteraemia should be treated with intravenous antibiotics for 10 to 14 days.

Treatment of salmonellosis

1 First choice: ciprofloxacin orally 500 mg twice daily for 3–5 days, *or* norfloxacin or ofloxacin orally 400 mg in the same regimen.

2 Alternatives: trimethoprim orally 200 mg twice daily for 5–7 days, *or* co-trimoxazole orally 960 mg in the same regimen.

3 Invasive salmonellosis: ciprofloxacin or ofloxacin i.v. 400 mg twice daily.

4 Alternative: ceftriaxone 2.0 g daily.

5 Chloramphenicol i.v. 500 mg 6-hourly if other agents are unsuitable.

Sequelae of *Salmonella* infections

Continuing excretion of salmonellae

Prolonged or permanent excretion is rare. Predisposing factors include gut disorders, such as diverticulosis, inflammatory bowel disease or ischaemia, and immunological disorders, including acquired immunodeficiency syndrome. Treatment with inappropriate antibiotics, particularly aminoglycosides or ampicillin, may also prolong excretion. Excretion is rare after treatment with ciprofloxacin (but not after norfloxacin).

Salmonella excretion does not compromise the health of otherwise fit patients, and is only a minimal infection hazard once diarrhoea has ceased. Excretors who are food, water or dairy workers may be excluded from working until excretion has ceased. Three consecutively negative, twice-weekly, stool tests are adequate evidence of this.

Metastatic *Salmonella* infections

Metastatic infections should be treated with an antibiotic to which the salmonella is sensitive. Prolonged treatment is often needed. Abscesses should be drained if possible, and may need 3 or more weeks' treatment. Bone and joint infections require at least 6 weeks. Clinical progress and the results of follow-up imaging, erythrocyte sedimentation rate or C-reactive protein measurements may help to indicate when treatment can be stopped. Resistance to ciprofloxacin has been shown to develop during prolonged treatment; ceftriaxone may be a useful alternative.

Post-infectious disorders

Salmonella bowel infections are occasionally followed by reactive arthritis, which can persist for many months (see Chapter 21). This is usually monoarticular, affecting a large joint such as the knee. Symptomatic treatment with non-steroidal anti-inflammatory drugs is helpful, but severe cases may need step-down courses of corticosteroids.

Prevention and control

General principles for the prevention and control of bacterial food poisoning apply to salmonellosis. In the UK salmonella infections in farm animals are notifiable under the Zoonoses Order, and are investigated by the State Veterinary Service.

Shigellosis (bacillary dysentery)

Introduction

Bacillary dysentery is an important disease worldwide. In Western countries the endemic *Shigella* spp. cause self-limiting illnesses that are generally mild. Tropical shigelloses tend to be both more severe and persistent, causing serious morbidity, especially in children. Malnutrition is exacerbated by the prolonged diarrhoea and fever of shigellosis.

Epidemiology

Shigellosis spreads from person to person by the faecal–oral route, by direct contact with faecally contaminated hands, or indirectly from contaminated food, milk or water. Water-borne shigellosis is important in rural tropical areas. Secondary spread within households is common. Outbreaks occur under conditions of crowding and poor sanitation or personal hygiene, for example in prisons, psychiatric institutions and nursery schools. Food-borne outbreaks are uncommon. An incident due to contaminated iceberg lettuce affected several European countries during 1994.

About 1000 cases are reported annually in the UK, of which two-thirds are due to *S. sonnei*. The peak incidence is in children under 5 years of age. In recent years there have also been a number of outbreaks in men who have sex with men.

Microbiology and pathogenesis of *Shigella* infections

There are four species of *Shigella*: *S. sonnei*, *S. flexneri*, *S. boydii* and *S. dysenteriae*. They are members of the Enterobacteriaceae, closely related to *E. coli*. They are relatively inert biochemically and are, for example, non-lactose fermenters, and non-motile. *S. sonnei* and *S. dysenteriae* are biochemically distinct, but *S. boydii* and *S. flexneri* are very similar. Serological characterization is therefore important for identification and typing. *S. dysenteriae* is divided into 10 serotypes on the basis of O antigens, *S. flexneri* into 6 types and *S. boydii* into 15. *S. sonnei* is serologically homogeneous, so colicine typing is used. A phage-typing system for *S. sonnei* has recently been developed.

Shigellae mainly cause mucosal infection of the distal ileum and colon. *S. flexneri* gains access to the basal epithelium via M-cells, the specialized antigen-presenting cells of the lymphoid follicles. Antigen presentation leads to the release of cytokines, which recruit neutrophils. These break down the tight junctions between cells and provide a path for the shigellae to invade the sub-mucosa, where they make contact with the basolateral membrane of the epithelial cells. Shigellae possess a virulence plasmid that encodes the *ipa* and the *mxi-spa* loci, necessary for invasion. The genes IpaB, IpaC and IpgD encode factors that combine to insert a pore into the host-cell membrane. Bacterial proteins are then secreted into the host cell and polymerize actin, leading to the development of cell surface extensions that surround the bacteria and take them up into vacuoles. *IpaB*, and possibly *IpaC*, lyse the vacuoles, releasing bacteria into the cytoplasm, where they replicate. The epithelial cells eventually lyse, probably due to the immune response of the host. Shigella exploits the host cell actin, Arp 2/3 complex, to migrate into adjacent epithelial cells via the outer membrane protein. It does this by expressing actin filaments on one pole of the bacterium, which creates a polymeric actin tail to propel the *S. flexneri* through the cytoplasm until it presses on the plasma membrane. This creates a protrusion into the neighbouring epithelial cell. Both membranes are lysed, releasing *S. flexneri* into the neighbouring cell.

Shigella dysenteriae type 1 produces a potent exotoxin (Shiga toxin) that potentiates local vascular damage. It is structurally related to the Shiga-like toxins (VTs) of *E. coli*. It produces fluid accumulation in the rabbit ileal loop model but does not appear to be necessary for destruction of mucosal cells. Systemic distribution of the toxin results in microangiopathic renal damage, leading to haemolytic uraemic syndrome. Shiga toxin is a bipartite molecule. The multimeric B subunit binds to a host cell surface glycolipid (Gb3, which possesses a Gal-α-1,4-Gal carbohydrate moiety). The enzymatic A subunit can then enter the cell. It is activated by proteolytic cleavage and, in turn, cleaves a specific adenosine residue of the 28S rRNA, inhibiting the host cell 60S ribosomal subunit.

Clinical features

After an incubation period of 3 or 4 days there is a feverish prodromal illness lasting 12–24 h. Non-specific features of this prodrome often occur in children and include meningism, convulsions and confusion. The white cell count is usually high ($12\text{–}16 \times 10^9$/l). However in most cases the

fever resolves suddenly, and diarrhoea with colic appears, indicating the true diagnosis.

S. sonnei and *S. boydii* tend to cause brown watery diarrhoea, containing varying amounts of mucus. Blood may be seen in the stool, but the amount is rarely large. The diarrhoea usually subsides spontaneously after 3–5 days.

S. flexneri and *S. dysenteriae* often cause more severe disease in which irregular fever continues and increasing amounts of mucus and blood appear in successive stools. Distressing colic is exacerbated by attempts to eat or drink. Untreated disease of this kind can lead to dehydration and progressive malnutrition.

Both types of illness can be followed by asymptomatic excretion of the pathogen lasting from days to weeks. Because of the low infective dose of shigellae, excretors can be hazardous, especially in nurseries and children's institutions.

Diagnosis

The clinical picture usually suggests the diagnosis. Bacillary dysentery must be distinguished from amoebic dysentery and inflammatory bowel disease. The mainstay of differential diagnosis is laboratory identification of the pathogen (see p. 183).

Management

Mild shigellosis requires only symptomatic treatment. Antispasmodic agents are helpful in relieving colic. In more severe cases antibiotic treatment can terminate the illness, though it often takes 2 or 3 days for diarrhoea to cease completely. Ciprofloxacin is often successful, and tends to eliminate excretion as well as curing disease.

Prevention and control

The infection can be prevented by safe disposal of faeces, the availability of safe drinking water and adequate handwashing. *Shigella* dysentery is notifiable. Patients should avoid handling food until their stools are normal.

Campylobacter infections

Introduction

Campylobacter infections occur as commonly as salmonella infections in most parts of the world, and add significantly to the burden of childhood gastroenteritis in at-risk communities. In high-income countries they cause less morbidity than salmonella infections because they rarely produce metastatic or bacteraemic disease. Asymptomatic excretion is uncommon and campylobacters are not transmitted from person to person; however, large food- and water-borne outbreaks have been documented. *C. jejuni* is rapidly killed in the stomach, and the chance of infection is increased if the organism is consumed in milk, which may protect it from gastric acid. The infective dose may be as low as 500 organisms.

Microbiology and pathogenesis of *Campylobacter* infections

Campylobacters and the related species *Arcobacter* form part of a group of spiral organisms found in human and animal intestinal tract and in certain environments, such as the sediments of salt marshes. Molecular analysis distinguishes them from *Helicobacter* spp. The cells are of variable length and wavelength. Four species exist only as straight rods. All campylobacters grow under microaerobic conditions, although some also grow aerobically or anaerobically. Growth is optimal between 30–37 °C. Campylobacteraceae is divided into three groups.
• The first group is thermophilic, including *C. jejuni* (with two subspecies), *C. coli*, *C. lari*, *C. upsaliensis* and *C. helveticus*.
• The second group includes *Campylobacter fetus* (with two subspecies), *C. hyointestinalis* (with two subspecies), *C. sputorum* (with three biovars) and *C. mucosalis*.
• A third group, found in human periodontal disease, contains *C. concisus*, *C. curvas*, *C. rectus*, *C. showae* and *C. gracilis*.

Campylobacters possess a low molecular weight cell-wall LPS, containing a lipid A moiety, similar to that of *Haemophilus* or *Neisseria* sp. It is antigenically diverse, possessing many serotypes. The similarity between *Campylobacter* LPS and human gangliosides is thought to be important in the pathogenesis of Guillain–Barré syndrome. Campylobacters possess flagella and are motile, which may be important in establishing infection, as non-motile strains are not pathogenic. Convalescent serum contains antibody to the surface-exposed flagellar flagellin. Flagellins are predominantly strain-specific but species-specific antigens are present.

C. jejuni produces an enterotoxin that causes fluid accumulation in ileal loops, and a cytopathic effect in Chinese hamster ovary cells and Y-1 mouse adrenal cells. Its mode of action is similar to that of cholera toxin and *E. coli* LT. Other toxins include a cytotoxin, which may contribute to the pathogenesis of enteritis. However, strains lacking toxins are reported to be fully virulent.

Pathogenicity factors for *Campylobacter jejuni*

1 Lipid A-containing lipopolysaccharide.
2 Flagellar antigens.
3 Toxins: enterotoxin and cytotoxin.

Epidemiology

Up to 45 000 cases of *Campylobacter* enteritis are reported each year in the UK. Following increasing incidence during the 1990s, numbers are now decreasing.

Most cases are sporadic and associated with consumption of raw or undercooked meat (particularly poultry), unpasteurized milk, untreated water, owning domestic pets with diarrhoea or occasionally with drinking milk contaminated by birds that peck through the tops of doorstep milk bottles. Person to person transmission occurs rarely, in conditions of poor hygiene. The infective dose is relatively low but *Campylobacter* does not multiply in food and food-borne outbreaks are rare. Large outbreaks are usually due to consumption of unpasteurized milk or untreated water.

The peak incidence of *Campylobacter* enteritis is during the summer months, about 8 weeks earlier than that of salmonellosis (Fig. 8.13). There is a bimodal age distribution with the greatest incidence in infants and young adults. Infection occurs more frequently in rural areas than in cities.

Clinical features

The incubation period is usually 3 or 4 days, but can be up to 8 or 9 days. This is followed by approximately 24 h prodromal illness with fever, headache and prostration. The main illness consists of diarrhoea, which is watery and may be bloody. Vomiting may occur at the onset. Abdominal pain is common, and is constant rather than colicky. Abdominal examination often reveals rebound tenderness that may be severe.

Campylobacter infection occasionally causes very severe abdominal pain with little diarrhoea, or only tiny stools containing mostly mucus and blood. Distinction from acute surgical conditions such as acute appendicitis, salpingitis and ectopic pregnancy may therefore be difficult. The abdominal X-ray shows multiple fluid levels in acute bowel infections, and is often unhelpful but it may show other diagnostic features such as free gas in the perito-

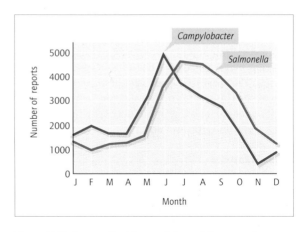

Figure 8.13 Seasonal incidence of *Campylobacter* and *Salmonella* infections.

neum or, in children, the C-sign of intussusception. Early ultrasound examination may assist diagnosis.

Rare cases of systemic *Campylobacter* infection occur, often affecting children with diseases causing iron overload. The patient is feverish and blood cultures are positive. Prolonged infection can mimic inflammatory bowel disease, with persistent bloody diarrhoea, abdominal pain and low-grade fever, but stool cultures will indicate the infectious aetiology.

Diagnosis

The diagnosis can rarely be made on clinical findings alone.

Laboratory diagnosis

There are several selective media that permit detection of campylobacters in heavily contaminated specimens such as faeces. The selective agents employed include a mixture of antibiotics including vancomycin, polymyxin B, trimethoprim, amphotericin B and cephalothin. Selection is improved if cultures are incubated at 43 °C. Media that provide effective selection at 37 °C are increasingly used, as they can also identify the small number of pathogenic campylobacters that are not thermotolerant.

Campylobacters are microaerophilic organisms that grow optimally at reduced oxygen tension and with an increased concentration of carbon dioxide. For some species the isolation rate is improved by the addition of hydrogen. Commercially available gas-generating systems are available for this.

Identification can be confirmed by Gram staining to demonstrate the characteristic 'gull wing' morphology, and by obtaining a positive oxidase reaction. Biochemical testing can be used to differentiate the species, and disc testing of antimicrobial sensitivities can be performed. No single DNA amplification technique can reliably identify campylobacters but a combination of 16S rRNA, 23S rRNA, *ceuE*, *flaA*, *mapA*, and *glyA* genes can be used.

Microbiological characteristics of campylobacters

1 Optimum growth at 37 °C but capable of growth at higher temperatures.
2 Characteristic spiral morphology.
3 Microaerophilic.
4 Oxidase-positive.
5 Resistant to vancomycin, polymyxin B, cephalothin, trimethoprim.
6 Sensitive to erythromycin and ciprofloxacin.

Clinical management

Many mild cases are self-limiting. Specific treatment is indicated in severe or prolonged illness (and a good response will obviate concerns about surgical or inflammatory conditions). The treatment of choice is a 3- or 4-day course of oral erythromycin. Ciprofloxacin is often effective, but resistance rapidly develops. These drugs can be used parenterally in severe or systemic infections.

Treatment of *Campylobacter* infection

1 First choice: erythromycin orally 250 mg 6-hourly for 3 or 4 days.
2 Alternative: ciprofloxacin orally 500 mg twice daily for 3 days (resistance easily develops).

Prevention and control

Campylobacter enteritis, like salmonellosis, may be prevented by good hygienic practice. Milk-borne spread can be avoided by pasteurization. The most effective long-term measure would be to control infection in poultry, but the existence of many environmental sources of infection makes this difficult.

Infection with 'food-poisoning' vibrios

These members of the family Vibrionaceae, like *V. cholerae*, are all natural residents of water. The range of Vibrionaceae is wide, some living in salt water, some in brackish or fresh estuaries and ponds. None of them produce cholera toxin, but in *V. parahaemolyticus* at least, pathogenicity is probably related to the production of a haemolysin.

Vibrio parahaemolyticus is a natural resident of brackish water, which may be concentrated in the gut of filter-feeding and scavenger shellfish. The most common vehicle of infection is a meal of shrimps or prawns. Most cases occur sporadically, but outbreaks have been reported. The disease is rare in the UK but common in south-east Asia and the USA. Other non-cholera vibrios including non-toxigenic and non-O1 *V. cholerae*, *V. alginolyticus*, *V. fluvialis* and *V. mimicus* are occasionally reported. The main symptoms are watery diarrhoea and colicky abdominal pain.

Aeromonas spp. are members of the Vibrionaceae that are natural inhabitants of water. They can be isolated from diarrhoea stools of some cases of 'food poisoning', and may also be contracted directly from water by fishermen or boatmen. *Aeromonas* infection is relatively common in Japan where freshwater fish is often eaten raw. The number of infections reported in the UK is small (less than 50 per annum), but increasing. Half of all reported infections are in young adults.

Plesiomonas shigelloides is related to *Aeromonas* and is less certainly associated with food poisoning. Between 20 and 40 laboratory reports are received each year in England and Wales.

Diagnosis

If the clinician suspects *Vibrio* infection on epidemiological grounds, a specific request should be made for the organisms to be sought (see below). Diagnosis is usually of epidemiological rather than therapeutic value, as the associated illnesses are rarely severe, and require only symptomatic treatment.

Laboratory diagnosis

Vibrionaceae cannot be identified by techniques used to detect Enterobacteriaceae. Instead, a high-pH selection-indicator medium containing bile salts is used, which inhibits many other bowel flora. The most used medium is thiosulphate–citrate–bile-salt–sucrose (TCBS) medium containing a bromothymol blue indicator. Food-poisoning vibrios grow well on this medium. Most are non-sucrose fermenters, producing blue-green colonies, which distinguishes them from *V. cholerae*, a sucrose fermenter that produces yellow colonies.

A suitable enrichment medium for vibrios is alkaline peptone water.

Yersinia infections

Yersinia enterocolitica and *Y. pseudotuberculosis* are capable of causing bowel disease. In some countries, for example Belgium, the infection is as common as salmonellosis and *Campylobacter* infections. It is rarely diagnosed in the UK. The number of reported cases increased more than 10-fold during the 1980s to 726 *Yersinia* infections in 1989 but has subsequently declined again with only 33 in 2002. This mainly reflects changes in laboratory practices in identifying the organism.

Transmission occurs from contaminated food, milk and water. Most infections are sporadic, often following consumption of raw or undercooked pork, although raw vegetables have also been implicated. A few milk-borne outbreaks have been reported in the USA.

Most infections occur in children under 5 years of age. There is a seasonal pattern, with more infections during autumn and winter.

Yersinia enterocolitica causes a mild to moderate gastroenteritis, often with aching abdominal pain. Post-infectious problems, particularly arthritis and erythema nodosum, are common, and can persist for some weeks (Fig. 8.14).

Diagnosis requires special investigations. Culture of *Yersinia* sp. from stool is possible (see below), but diagnosis is often considered late, when post-infectious symptoms occur. Serological diagnosis must then be sought by demonstrating rising antibody concentrations in paired serum samples.

Y. enterocolitica produces only pinpoint colonies on MacConkey's agar after 48 h incubation. Isolation is simpler using specialized *Yersinia* media. Selection is obtained by inclusion of sodium desoxycholate, crystal violet and an antibiotic combination (Irgasan, novobiocin and cefsulodin). *Y. enterocolitica* has translucent colonies with

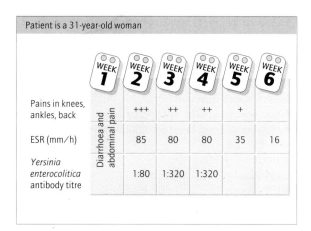

Figure 8.14 Progress of *Yersinia* food poisoning associated with arthritis. ESR, erythrocyte sedimentation rate.

a dark pink centre, which may be surrounded by a pale opalescent zone of precipitated bile salts.

Y. pseudotuberculosis more often causes mesenteric adenitis than diarrhoea. Fever and right lower quadrant abdominal pain must be differentiated from appendicitis, salpingitis or the onset of Crohn's disease. Ultrasound examination will often reveal oedematous terminal ileitis and groups of enlarged mesenteric lymph nodes. Similar findings, even in the absence of diarrhoea, can occur in *Salmonella* and *Campylobacter* infections, demonstrating the overlapping symptom complexes of intestinal infections. It is therefore worthwhile requesting stool cultures in patients with terminal ileitis and mesenteric adenitis, in case specific treatment can be offered.

Yersiniosis can be treated with oxytetracycline 250–500 mg four times daily or doxycycline 100 mg daily, both for 7–10 days.

Helicobacter pylori and acute gastritis

Microbiology

H. pylori is a spiral, Gram-negative obligately microaerophilic bacterium, motile by means of 4–8 sheathed flagella. Analysis of 16S rRNA sequences of helicobacters reveals two large groups of species. The first contains *H. pylori*, *H. heilmanni* and other *Helicobacter* species that colonize gastric mucosa. The second, more diverse group contains the enteric, biliary and blood-associated *Helicobacter* species.

H. pylori is found in the gastric mucus and surrounding the apices of gastric parietal cells in patients suffering from acute hypochlorhydric gastritis and chronic gastritis. It probably causes about 95% of duodenal ulcers, and a small minority of gastric ulcers. It is also found in association with tumours of mucosa-associated lymphoid tissue (MALT), which will often regress if the *Helicobacter* infection is eradicated.

Diseases associated with *Helicobacter pylori*

1 Acute hypochlorhydric gastritis.
2 Chronic gastritis.
3 Duodenal ulcer.
4 Mucosa-associated lymphoid tissue lymphoma.

Pathogenicity of *H. pylori*

H. pylori causes acute and chronic inflammation in the gastric antrum. The organism elaborates a potent urease, which produces a local alkaline environment. This enables it to overcome the inhibitory effect of gastric acid and to establish itself in the mucosa. This is thought to stimulate the feedback mechanism for gastric acid, resulting in a high gastrin level and hyperacidity in the rest of the stomach. The phenotype associated with peptic ulcer disease possesses the *cagA* gene, associated with a pathogenicity island that encodes a type IV secretion system, and also proteins that stimulate pro-inflammatory cytokines. The cagA product may deregulate cell growth and induce proliferation of gastric epithelial cells, ultimately leading to malignant transformation.

The *vacA* gene encodes a toxin that produces vacuolation in tissue culture cells. Evidence from animal studies indicates that VacA has an important role early in infection. The gene has a mosaic structure, and variations cause differences in the VacA level and the degree of damage in human tissue.

Most strains of *H. pylori* produce an LPS, antigenically similar to either Lewis^x or Lewis^y blood group antigens; some strains express both. Antigenic mimicry between *H. pylori* LPS epithelial cell surface glycoforms of human gastric cells can lead to immune tolerance of *H. pylori* antigens. In chronic infection, the resulting autoantibodies may promote chronic active gastritis.

Epidemiology

Humans are the probable reservoir of infection, but *H. pylori* has been recovered from untreated water. The origin of the organisms colonizing the stomach is uncertain, but possibilities include person-to-person or food-borne spread. In high-income countries the prevalence of *H. pylori* infection is 20 to 40%, but up to 95% of duodenal ulcer patients are colonized. *H. hominis* has recently been identified in the human stomach, and may be another cause of peptic ulcer disease.

Diagnosis

A non-invasive diagnosis can be made using the urea breath test in which the patient swallows ^{13}C or ^{14}C urea orally and samples of exhaled breath are collected over 30 min. Urease in the stomach breaks down the radiolabelled urea and this activity is detected by the presence of radiolabelled CO_2. Biopsy specimens can be examined using rapid urease tests, by histology and by cultivation. A stool antigen test provides a simple non-invasive diagnosis with excellent sensitivity and specificity. Bacterial isolation and sensitivity testing is becoming increasingly important, as antibiotic resistance increases.

Treatment

The most successful treatment regimen is a combination of amoxicillin, plus clarithromycin or metronidazole, with a proton pump inhibitor such as omeprazole, continued for 1 month. Combined bismuth/clarithromycin preparations may also be used with omeprazole. Resistance to antibiotics is an increasing problem.

Diseases caused by bacterial toxins

Organism list

- *Staphylococcus aureus*
- *Bacillus cereus* and other *Bacillus* spp.
- *Clostridium perfringens*, *C. difficile*, *C. botulinum*

Staphylococcal food poisoning

Introduction

Staphylococcus aureus is able to produce up to 15 enterotoxins, among which are enterotoxins (A–E), which are preformed toxins, produced during replication of the staphylococci in prepared food. Food poisoning occurs when food becomes contaminated by an infected individual, usually from a skin lesion or nasopharyngeal secretions. The organism subsequently multiplies and produces toxin, which is absorbed and travels in the circulation to the vomiting centre, causing vomiting by a central effect. It may also cause bowel irritation, leading to mild, transient diarrhoea.

In some countries, including the USA, *S. aureus* is a leading cause of food poisoning. In the UK it is comparatively rare.

Clinical features

The incubation period varies from 30 min to 6 h. Malaise and nausea are quickly followed by vomiting, which may be severe and repeated, causing dehydration or even shock. Occasional deaths are reported. The vomiting lasts from 2 to 6 h and is followed by several hours' exhaustion before recovery is complete.

Diagnosis

The illness is similar to winter vomiting disease and other viral infections. Single cases are rarely extensively investigated. The absence of diarrhoea limits the availability of specimens, though staphylococci may be recovered from vomitus if this is available, and can then be shown to produce enterotoxin.

Family and catering-associated outbreaks are relatively common and may indicate a food source. Remaining food can then be examined for the presence of enterotoxin and/or enterotoxin-producing staphylococci.

Diagnosis of staphylococcal food poisoning

1 Demonstration of toxin-producing *Staphylococcus aureus* in suspected food.
2 Demonstration of toxin-producing *S. aureus* in vomitus.
3 Demonstration of an enterotoxin in suspected food.

Management
Management is symptomatic, including vigorous rehydration when indicated.

Prevention
Prevention is by hygienic food preparation, particularly covering skin lesions with a waterproof dressing. Food-handlers with extensive staphylococcal skin infection should be excluded from work.

Bacillus cereus food poisoning
Bacillus cereus is an aerobic, spore-forming Gram-positive rod. It is found in soil and dust, and easily contaminates foods such as cereals and beans. It may also exist in animal faeces and occasionally contaminates meat and vegetables. The disease is relatively uncommon in the UK.

Two toxins are clinically important: a heat-stable (10 kDa) emetic toxin, which is usually associated with rice-based meals, and a diarrhoeal toxin, which is a mixture of two heat-labile proteins, and mostly associated with meats and vegetables.

Food-poisoning serotypes produce spores that resist boiling. For the emetic form, boiled foods, particularly rice, may be stored at ambient temperatures for later re-warming (or frying). The spores germinate in the stored food and highly heat-resistant preformed toxins accumulate, which are not inactivated by flash frying. For the diarrhoeal form, it is thought that sporulation in the stomach leads to the elaboration or release of toxin.

Some other *Bacillus* spp., e.g. *B. licheniformis*, also occasionally cause food poisoning.

Botulism
Introduction
Classical botulism is caused by preformed toxins of *Clostridium botulinum*, elaborated during the germination of spores in anaerobic conditions. Different strains of *C. botulinum* produce one of seven toxins (A–G). Human disease is usually caused by A or B, derived from soil; oc-casional cases are caused by type E, derived from estuarine or marine mud. Botulinum neurotoxin (BoNT) causes disease by preventing the release of acetylcholine in neuromuscular junctions.

Home-canned or home-bottled vegetables and salads are the commonest sources of illness, usually causing family outbreaks of type A or B. Home-smoked or preserved meats and fish have also been implicated. Commercial canning is highly controlled, but rare community outbreaks have been caused by commercial products. An outbreak of 27 cases occurred in the north-west of England in 1989, caused by canned hazelnut purée contained in yoghurt (Fig. 8.15). Four cases of type E intoxication were caused by a faulty can of salmon in the 1970s.

Other types of exposure are wound botulism, in which toxin is elaborated by *C. botulinum* spores germinating in a closed wound or abscess (see page 107), and infant botulism, in which the infant's bowel is colonized by replicating *C. botulinum*, and toxin is produced locally.

Nowadays, case fatality rates are around 10% for food-borne botulism and 15% for wound botulism.

Clinical features
The toxin prevents fusion of acetylcholine-containing vesicles with the terminal membrane of neurones, thus blocking neuromuscular transmission and causing paralysis, which appears 24–48 hours or more after toxin ingestion, depending on the dose consumed. This may be preceded by malaise and mild gastrointestinal symptoms. The onset is insidious and often initially dismissed as 'hysteria'. Dry mouth and blurred vision are followed by difficulty with

Figure 8.15 A 'blown' can of hazelnut purée that had not been properly heat-treated, resulting in the formation of botulinum toxin type b; 27 people developed botulism after eating hazelnut yoghurt prepared from the contents of such a can. Courtesy of Dr Richard Gilbert, Central Public Health Laboratory.

swallowing and speech, producing the so-called 'four Ds' presentation: *diplopia, dysarthria, dysphonia, dysphagia.* Ptosis is usually also present. Paralysis develops from the head downwards, aiding differentiation from the opposite process in Guillain–Barré syndrome. Increasing generalized weakness and respiratory paralysis soon follow. Once established, this can last for many weeks, gradually recovering as the muscles are re-innervated. Autonomic dysfunction and smooth-muscle paralysis mean that the blood pressure may be unstable and bowel function will be disturbed.

Mild cases also occur, and may not progress to generalized weakness or respiratory paralysis.

Diagnosis

Initial diagnosis depends on recognizing the clinical presentation and obtaining a history of recent consumption of suspect food. The cerebrospinal fluid is usually normal. Electromyography shows features of developing denervation (but in the earliest stages may suggest an axonal lesion). *C. botulinum* is not usually found in the stool, but toxin may be demonstrable in serum or stool. Both the organism and its toxin are demonstrable in left-over food, food containers or utensils. The toxins are destroyed by heating and cooking.

Laboratory diagnosis of botulism

1 Demonstration of *Clostridium botulinum* in suspected food.
2 Recovery of *C. botulinum* from unwashed food containers or utensils.
3 Demonstration of BoNT in food, food containers or, occasionally, patient's serum.

Management

Without intensive care and ventilatory support, about 60% of cases are fatal. Ventilatory support is usually needed for 2 to 8 weeks, sometimes longer, and near-normal muscle strength may take several months to return.

Trivalent, A, C and E antitoxin neutralizes circulating toxin, limits its permanent binding to neuromuscular junctions, and reduces the severity and duration of paralysis. The sooner it is given, the greater its effect. It is rapidly available from the Health Protection Agency's Centre for Infections and other designated centres. Antitoxin infusion can be repeated at 12–24-hour intervals, if indicated by a response. It is also used prophylactically for others who ate the suspect food.

A minority of cases are mild, not requiring ventilation, but respiratory function should be closely monitored, and measures taken to protect the airway, to avoid aspiration of saliva or stomach contents.

Antitoxin treatment for botulism

Botulism antitoxin is derived from animal plasma, and contains 0.45% phenol. Tests for hypersensitivity should be performed, according to the instructions in the pack, before full dosage. Different formulations are available:

Undiluted BoNT antitoxin
Prophylaxis
20 ml, i.m. as soon as possible after exposure.

Treatment
1 20 ml, diluted to 100 ml with sodium chloride 0.9%, by slow i.v. infusion.
2 10 ml, diluted to 100 ml with sodium chloride 0.9%, by slow i.v. infusion 2–4 hours later.
3 Repeat doses of 10 ml may be given at 12–24-hour intervals, if they still give improvement.

Botulism antitoxin (Behring)
This is dispensed in 250 ml containers, for intravenous infusion.

Treatment
1 Infuse 250 ml slowly, while observing the patient closely (this dose may be used for prophylaxis).
2 If there is no adverse reaction, a further 250 ml is infused rapidly.
3 A further 250 ml infusion may be given 4–6 hours later, depending on the patient's clinical condition.

 Adults and children should receive the same dose.

Infant botulism

This is a rare condition with worldwide occurrence caused by rapid replication of *C. botulinum* in the infant bowel. Six cases were reported in the UK between 1975 and 2001. Bacteria, and high concentrations of toxin, are found in the stool of affected infants. The source of the *C. botulinum* is unknown, but may be direct from soil or derived from food. Some affected infants have consumed honey before their illness.

Affected infants are aged between 6 weeks and 3 months. The onset is insidious, with weak suckling, constipation, reduced muscle tone and shallow respiration. Antitoxin is not reliably effective, and is rarely given. Many infants recover with supportive measures only, including postural drainage, careful feeding or tube feeding and close observation of respiratory function. Less than half require ventilation. No secondary case has been reported following exposure to the stool of an affected infant.

Clostridium perfringens food poisoning

Clostridium perfringens is an anaerobic Gram-positive rod that inhabits the bowel of many animals, and can contaminate meat during butchery and storage. Its spores survive boiling and will germinate in the anaerobic conditions in stored stews, soups, gravies and large joints of meat. Infection occurs when contaminated cooked food is allowed to stand at room temperature for long periods, typically in large-scale catering situations with inadequate facilities for cooling and storing food.

Ingested bacteria multiply in the large bowel, elaborating toxin locally. After 18–36 hours incubation the toxin causes watery diarrhoea with colicky abdominal pain, which may last for 3–5 days.

C. perfringens can be recovered from the diarrhoeal stool and identified as a 'food-poisoning' serotype. It is also possible to demonstrate toxin gene in the stool by PCR.

Clostridium difficile and pseudomembranous colitis

Pseudomembranous colitis is almost always antibiotic-associated, though rare cases occur spontaneously. A milder version of the disease is recognized as antibiotic-associated diarrhoea. Commonly associated antibiotics are those that strongly inhibit normal gut flora, especially enterococci. These include clindamycin, ureidopenicillins and third-generation cephalosporins. However, ampicillin, rifampicin and ciprofloxacin have occasionally been associated with cases.

Microbiology and pathogenicity of *C. difficile* infections

C. difficile is a Gram-positive, sporing, obligate anaerobe with variable morphology. The colonies produce a very distinctive odour, similar to elephant or horse manure. Colonies grown on blood-based selective media fluoresce yellow-green under long wave (365 nm) ultraviolet illumination, but this is medium dependent and not unique to *C. difficile*. *C. difficile* does not readily sporulate on agars containing selective agents. Colonies therefore become non-viable if plates are left in air for prolonged periods. However, after 72 hours incubation on non-selective agars colonies usually sporulate heavily, and will then survive prolonged exposure to air.

C. difficile establishes itself where the 'colonization resistance' of the normal anaerobic flora is disrupted. *C. difficile* usually produces two toxins, toxin A and B, often referred to as an enterotoxin and cytotoxin, respectively. Both have a role in pathogenesis but toxin A–, B+ isolates can cause disease. The toxin genes and other, minor, toxin genes are found on a chromosomal pathogenicity island. Some strains hyper-produce toxin, causing more severe disease. A binary toxin (an actin-specific ADP-ribosyltransferase) similar to *Clostridium perfringens* iota enterotoxin has also been described in *C. difficile* but its significance is unknown.

Clinical features

Watery diarrhoea and abdominal pain occur abruptly, usually 3 or 4 days after beginning antibiotic treatment. Low-grade fever is common. *C. difficile* and its toxin can both be demonstrated in stool specimens. Biopsies show typical changes, with focal inflammation in lymphoid tissue, extending through the muscularis mucosae to involve the submucosa. Mucosal debris and inflammatory exudate produce a thick pseudomembrane. Slow improvement may follow discontinuation of antibiotics, but progressive inflammation and perforation can occur in untreated cases.

Diagnosis

The isolation of *C. difficile* from stools is not diagnostic, as up to 30% of hospital patients are asymptomatic carriers. The diagnosis is made by detecting toxin in patients' stools. Toxins can be detected by their cytopathic effect on cell cultures, but an ELISA method is nowadays more often used. Nucleic acid amplification techniques are not used, as culture and toxin detection are sensitive and inexpensive. The typical histological changes in colonic biopsies are also diagnostic (Fig. 8.16).

Treatment

Oral metronidazole 400 mg 8-hourly for 5 to 7 days is often effective. Parenteral treatment can be given if oral treatment is impossible. Relapse sometimes occurs and can be treated by repeating this course. An alternative is oral vancomycin 250 mg three times daily for a week. Some evidence indicates that probiotic therapy may be helpful in patients with recurrent disease.

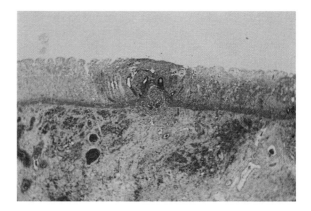

Figure 8.16 Pseudomembranous colitis: intense inflammation erupts from beneath the muscularis mucosae, producing an accumulation of necrotic debris in place of the damaged mucosa.

Other toxins and food poisoning

Toxin list

- Bean haemagglutinins
- Scombrotoxin
- Shellfish toxins (diarrhoetic and paralytic)
- Ciguatera toxin
- Cyanobacterial toxins (blue-green algae)

Bean haemagglutinins

These haemagglutinins are found in many species of beans. Red kidney beans contain large amounts and are the commonest cause of illness. The toxin is destroyed by vigorous boiling; when the beans are soft they are also safe to eat. Incomplete cooking is particularly likely if beans are insufficiently soaked, or cooked as a casserole ingredient. Canned beans are safe. Illness, consisting of severe vomiting followed by profuse diarrhoea, develops within minutes to an hour after consumption of as few as five or six beans.

Scombrotoxin

Scombrotoxin is a substance that develops during spoilage of scombroid fish, such as mackerel, tuna and bonito. Sardines have also been implicated in cases of food poisoning. Even mildly spoiled fish can be toxic. The toxin is probably either histamine or a closely related substance, and is highly heat-resistant. It is not destroyed by cooking or canning.

Symptoms develop 1–4 hours after eating affected fish. They resemble the symptoms of histamine toxicity, with headache, flushing, urticarial rash and swelling or tingling of the lips and mouth. Spontaneous recovery takes 4–6 hours.

Diarrhoetic shellfish poisoning

Diarrhoetic shellfish poisoning is caused by a toxin produced by shellfish. Some shellfish are well known to be toxic (for instance, the red whelk) but are occasionally mistaken for edible species. The resulting illness is a short-incubation acute diarrhoea whose severity is proportional to the dose of toxin.

Paralytic shellfish poisoning

Neurotoxins are elaborated by dinoflagellates, which are low in the food chain of many sea creatures. The dinoflagellates multiply rapidly when the sea is warm and nu-

trients are plentiful, often forming a visible red or brown 'bloom' (the so-called red tide). Shellfish and other filter feeders consume large quantities of this and the toxin accumulates in their bodies. Humans become poisoned by eating affected clams or crustacea. In tropical waters fish such as groupers can accumulate very large amounts of ciguatera toxin, becoming more poisonous with increasing age and size.

Symptoms of poisoning are caused by interference with neuromuscular junctions. Bradycardia, tingling of the lips and fingers, and muscle weakness begin soon after eating affected fish, and can continue for many hours, or days. In severe cases profound muscle weakness can affect respiratory muscles. Bradycardia and hypotension can require intensive care support, and deaths have been reported in patients who consumed large quantities of affected fish. Full recovery of neuromuscular transmission is slow; for many weeks afterwards even a meal of 'safe', or canned, fish can precipitate weakness.

Ciguatera poisoning is mainly seen in the Caribbean, where it is well recognized. Local people do not eat large fish.

Cyanobacterial toxins

Cyanobacterial toxins are produced by the blue-green algae that thrive in freshwater when the weather is warm and nitrates or other nutrients are plentiful. An algal 'bloom' is visible in the water. The toxins are extremely irritant, and cause erythema, burning and even blistering of exposed skin and mucosae. Ingestion affects mainly animals, but occasional human cases of diarrhoea, vomiting and abdominal pain are reported. Abnormal liver function tests and sometimes jaundice also occur. Species of cyanobacteria can readily be identified by direct microscopy of affected water.

Poisoning incidents have been reported from many countries, but only rarely in the UK. In 1989, 10 young army recruits were affected following canoeing and swimming exercises in a lake containing algal bloom.

Parasitic infections of the gastrointestinal tract

Organism list

Protozoa
- *Giardia intestinalis*
- *Cryptosporidium parvum*
- *Cyclospora catayensis*
- *Entamoeba histolytica*

Helminths
- *Enterobius vermicularis*
- *Trichuris trichiura*
- *Ascaris lumbricoides*
- Hookworms
- *Strongyloides stercoralis*

Tapeworms

Laboratory diagnosis of intestinal parasite infections

The simplest investigation is to examine a saline suspension of stool microscopically under the ×10 and ×40 objective. To improve the diagnostic yield, concentration techniques are employed. These include flotation methods such as the zinc sulphate, magnesium sulphate and sucrose flotation techniques or the formol ether centrifugation technique. Flotation methods vary in their ability to concentrate different species, whereas the formol ether technique effectively concentrates helminth ova and protozoal cysts. Individual parasites are identified on the basis of their size (measured by a microscope eyepiece graticule calibrated against a stage micrometer), and their charac-teristic morphology. Some examples of protozoan and helminthic parasites are demonstrated in Fig. 8.17.

In patients with acute amoebic dysentery the transit time through the large bowel may be so rapid that motile amoebae may be identified by stool microscopy. The organisms are susceptible to cooling and desiccation, so specimens must be processed without delay. Microscopic examination is performed on a heated stage to preserve trophozoite motility for as long as possible. Scrapings from the base of ulcers demonstrated on proctosigmoidoscopy may also show motile trophozoites.

The larvae of *Strongyloides stercoralis* are infrequently seen in the stools. The diagnostic yield is increased by incubating the stool, mixed with activated charcoal, in a Petri dish kept inside a larger dish, at room temperature for 5 days. Larvae differentiate and migrate into the second dish, where they can be seen using a plate microscope.

A string test may aid diagnosis of strongyloidiasis and giardiasis, although there are differing reports of its efficiency. The patient swallows a weighted string, which passes into the upper small intestine where it remains for approximately 2 h. The string is then retrieved. Attached

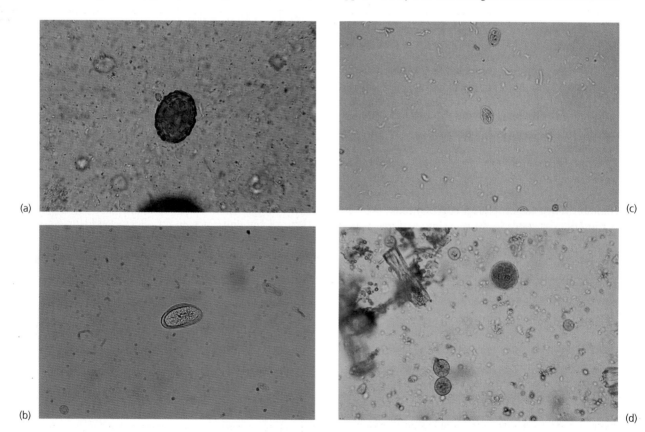

(a)   (b)   (c)   (d)

Figure 8.17 Examples of parasitic infections demonstrated by microscopy of unstained faeces: (a) ovum of *Ascaris lumbricoides*; (b) ovum of *Enterobius vermicularis*; (c) cysts of *Giardia intestinalis*; (d) cyst of *Entamoeba histolytica* containing six nuclei (smaller cyst of non-pathogenic *Endolimax nana* is also seen). (See also Figure 3.1.)

material is concentrated by washing and centrifugation before examination using the ×10 and ×40 objectives. Enzyme immunoassay (EIA) and PCR methods have been described for most of the common protozoan infections, but in practice few are routinely used as most laboratories in developed countries have a low throughput of specimens and these methods are not therefore cost-effective. Facilities and resources for their use are often lacking in high burden countries.

Giardiasis

G. intestinalis, a flagellate protozoan, can cause acute diarrhoea, persistent grumbling intestinal symptoms or asymptomatic colonization of the bowel. Infectious cysts are passed in the faeces of both sick patients and asymptomatic carriers. Cysts survive for many days in sewage-contaminated water and infect a new host when the water is consumed. Large outbreaks can occur when water treatment procedures are defective or when breakage of water pipes allows contamination. Faecal–oral transmission is uncommon. Dogs can carry the organism and also excrete infectious cysts.

The disease is common in areas of poor sanitation, where children under 5 years are most frequently affected. Travellers are at risk when visiting countries with inadequately treated water supplies. The number of reports increased in the UK up to a peak of just over 6000 cases per year in the mid-1990s, probably due to more widespread examination of faecal samples for cysts. Reports have subsequently declined again to around 3000 per year.

Giardia adheres to the mucosa of the jejunum and upper ileum, using a ventral 'sucker', and disrupts the enterocyte brush border. In heavy infections the whole areas of mucosa may be affected, leading to malabsorption and steatorrhoea with rapid weight loss. Chronic cases are common; flatulence and passage of greasy, loose stools are particularly noticeable in the mornings and may persist for weeks or months.

Treatment of giardiasis

1 Metronidazole 400 mg three times daily for 5 days *or* 500 mg twice daily for 7–10 days *or* 2 g daily for 3 days.
2 Single 2 g dose of tinidazole.
3 If other drugs fail: mepacrine (unlicensed) 100 mg 8-hourly for 5–7 days.

Cryptosporidiosis

Cryptosporidium parvum is a member of the phylum Sporozoa. Serotype A has its reservoir in humans. Serotype B is excreted by cattle but also infects other animals and is the commonest serotype found in human disease. The in-fectious dose is low, probably under 50 organisms. Person-to-person transmission often causes family outbreaks.

Surface water containing infectious oocysts can contaminate water reservoirs and mains, causing large outbreaks (Fig. 8.18) and sporadic infection. Outbreaks have been reported in children's nurseries, probably due to faecal–oral spread. Transmission also occurs between animals and humans by the faecal–oral route. Agricultural and veterinary workers, farmers and their families are at risk. The organism was first identified in the UK in 1983 and reports have risen steadily as testing for the organism has increased. Between 3000 and 6000 laboratory confirmed cases per year are currently reported in England and Wales. Seasonal peaks occur in spring and late autumn.

The parasite attaches to the small-bowel mucosa, damaging the brush border of the enterocytes (Fig. 8.19). After 3–6 days incubation there is an acute onset of diarrhoea, often with abdominal pain and colic. The average duration of illness is about 3 weeks. Immunosuppressed patients may fail to recover completely.

Diagnosis depends on demonstrating oocysts in the faeces, by acid-fast stains such as modified Ziehl–Nielsen or auramine techniques. No effective specific treatment exists but azithromycin has been reported to ameliorate the symptoms.

Figure 8.18 Leakage from this drainage system contaminated a public swimming pool, causing a large outbreak of cryptosporidiosis. Courtesy of Dr David Casemore.

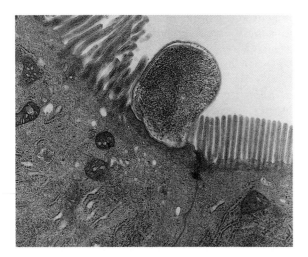

Figure 8.19 Electron micrograph of a developing trophozoite of *Cryptosporidium parvum*, attached to mucosal cells in the small bowel. The parasite is within a vacuole formed from the host-cell brush border membrane. Courtesy of Dr David Casemore.

Cyclosporiasis

Cyclospora catayensis is a common cause of springtime diarrhoea, discovered in children in Central America. Several countries reported outbreaks caused by soft fruit, particularly raspberries, imported from endemic areas. Protection of fruit crops from faecal contamination during irrigation and fertilization is important.

Diagnosis is by demonstration of characteristic cysts in the stool.

Amoebiasis

Introduction

Entamoeba histolytica causes infection and disease in many parts of the world where sanitation is poor. The clinical effects include asymptomatic excretion of cysts, chronic intestinal infection, intestinal granuloma formation, and acute amoebic dysentery. Extra-intestinal abscesses may also occur, and can threaten life if untreated.

Manifestations of amoebiasis
1 Asymptomatic cyst passage.
2 Chronic intestinal symptoms.
3 Caecal granuloma (amoeboma).
4 Amoebic dysentery.
5 Amoebic liver abscess.
6 Rarely, other extra-intestinal disease.

Epidemiology

Amoebiasis spreads mainly via contaminated water and food, especially raw vegetables. Infection is common in areas of poor sanitation and personal hygiene. The disease occurs principally in adults and is rare in children below 5 years of age.

Up to 200 cases per year are reported in England and Wales though these numbers have declined in recent years. At least half of cases are believed to have been acquired overseas, principally in the Indian subcontinent. *E. histolytica* has been differentiated from the non-pathogenic *E. dispar*, which is morphologically identical. *E. histolytica* infection makes up approximately 10% of those cases with *Entamoeba* cysts present in the stool.

Pathogenesis

Cysts of *E. histolytica* are passed in the stools of affected individuals. After ingestion the trophozoites excyst and attach to the mucosa of the large bowel. Binding is mediated by an amoebic surface membrane glycoprotein, galactose N-acetyl-D-galactosamine, which acts as a lectin. Amoebae secrete three closely related types (a, b and c) of a pore-forming enzyme, amoebapore. Amoebapores each contain six cystine residues in a similar structure to that found in saponins and pulmonary surfactant. *E. histolytica* also expresses several proteolytic enzymes, which may mediate the characteristic tissue destruction found in amoebic abscesses.

Clinical features

After a variable incubation period, from a few days to 4–6 weeks, illness begins with swinging fever, and watery diarrhoea, which rapidly becomes bloody. Colicky abdominal pain increases and may be replaced by constant pain as disease progresses. The stool is eventually replaced by mucus mixed with large quantities of blood. Severe, untreated cases can progress to acute colonic dilatation, with risk of perforation.

Diagnosis

Acute amoebiasis must be distinguished from bacillary dysentery. Other causes of bloody diarrhoea include VTEC infection, pseudomembranous colitis, ulcerative colitis and Crohn's colitis. Inflammatory bowel disease should be suspected if repeated stool examinations reveal no pathogens. Sigmoidoscopic biopsy is helpful in demonstrating typical changes of inflammatory bowel disease or pseudomembranous colitis. Biopsies from ulcer margins may also reveal undermined (flask-shaped) amoebic ulcers with amoebae packed into their eroded borders. It may be difficult to demonstrate amoebae or cysts in the stools.

Laboratory diagnosis

Intestinal amoebiasis can be confirmed by detection, in fluid diarrhoea, of amoebic trophozoites containing red cells. The specimen must be examined with minimum delay, and kept warm, so that motile amoebic trophozoites can be visualized. In semiformed or formed stools, or stools examined after a delay, amoebic cysts should be sought by the formol ether technique. *E. histolytica* can only be differentiated from *E. dispar* by specialized tests such as EIA, or culture followed by zymodeme analysis.

Less than half of patients with amoebic dysentery produce antibodies to *E. histolytica* but these antibodies are almost invariably present in patients with invasive amoebiasis. Immunofluorescence antibody and ELISA tests have been described. These tend to remain positive after an attack of systemic amoebiasis. A rapid technique using cellulose acetate precipitation is the first to become positive in amoebic liver abscesses and should be used for urgent diagnosis. This positivity is rapidly lost after successful treatment. Species-specific diagnosis has been described, based on detection of antibody to the surface-binding lectin.

Management

The treatment of choice is metronidazole 500 mg three times daily intravenously, switching to 800 mg three times daily by mouth when oral treatment is possible. Treatment should be continued for 5–7 days, depending on the speed of response.

Once cured of the acute disease, many patients continue to excrete infectious cysts. This can be terminated by treatment with the intestinal amoebicide diloxanide furoate; the dose is 500 mg every 8 h for 1 week (this may be omitted if the patient is soon to return to an endemic area, as reinfection is inevitable).

Prevention

The disease can be prevented by filtration of drinking water, adequate disposal of human faeces and good personal hygiene. Travellers to tropical countries should avoid inadequately treated drinking water, unpeeled fruit and raw vegetables.

Chronic intestinal amoebiasis

This is common in endemic areas, causing fluctuating abdominal symptoms and recurrent diarrhoea. *E. histolytica* preferentially affects the caecum, causing right iliac fossa pain and tenderness, which must be differentiated from appendicitis. Persisting mucosal inflammation stimulates granuloma (amoeboma) formation. Large amoebomas may produce the same symptoms and imaging appearances as caecal carcinoma. Stool examination in such cases will reveal amoebic cysts. The condition responds to treatment with intestinal amoebicides, but surgery may still be needed if the bowel is obstructed.

Amoebic abscess

This may follow either apparent or subclinical bowel infection. The liver is by far the commonest site affected; others include pleura, pericardium and occasionally excoriated perineal or abdominal skin. The localized infection causes intense inflammation and tissue damage, which is visible as dark necrosis in advancing skin lesions.

Clinical features

The patient presents with high fever, leucocytosis and a raised C-reactive protein. Some patients have palpable enlargement and tenderness of the liver, but many have a paucity of physical signs. An elevated right diaphragm, small right-sided pleural effusion or slight collapse of the lower lung segment should direct attention to the liver in a returned traveller with pyrexia of unknown origin.

Diagnosis

Useful investigations include ultrasound examination and computed tomographic (CT) scanning of the liver. The abscess is usually a single lesion in the right lobe, accessible to diagnostic aspiration and drainage. Left-sided abscesses are rare and difficult to localize on physical examination or ultrasound scan. CT scanning is the investigation of choice for this and other unusual sites.

Amoebae are rarely demonstrable in aspirate, as they are closely adherent to the abscess wall. However, the pus has a typical pinkish (anchovy-sauce) appearance that is diagnostically helpful.

Serological tests are usually positive and permit rapid diagnosis. Culture of the pus should be performed to exclude super-added bacterial infection.

Management

Metronidazole is the treatment of choice. It readily penetrates the abscess wall and can allow rapid healing, often avoiding the need for further aspiration. The dose is 500 mg intravenously or 800 mg orally three times daily for 10–14 days (see Chapter 9).

Helminth infections of the bowel

Introduction

Helminth infections of the bowel are endemic wherever there is environmental contamination with human or animal faeces. Some are directly transmitted by the faecal–oral route.

Helminths with direct person-to-person transmission

Some species, *Enterobius vermicularis* (threadworms) and *Trichuris trichiura* (whipworms), inhabit the large bowel and rectum, deriving nutrients from the faeces. They release ova that are detectable by microscopic examination of unstained faeces.

Whipworms fix themselves to the bowel wall by their narrow head; their wider body may be seen on proctoscopy, dangling from the mucosa.

Threadworms are freely motile and are often seen in the lumen of the appendix in surgical specimens. They emerge from the anus, particularly at night, to deposit ova on the perineal skin. This causes irritation, which may spread into the vagina, especially in children. Worms can be seen macroscopically as tiny moving threads in the faeces or on the skin. Adults or their ova can be trapped by pressing the sticky side of transparent adhesive tape on to the anal margin at night or just after waking. Microscopy of the tape, stuck to a glass slide, can then be performed to make the diagnosis.

Mebendazole, which is given as a single oral dose of 100 mg, kills the worms and terminates the infection. Threadworms can be paralysed by treatment with piperazine. This is usually given as a mixture of piperazine phosphate and senna in powder form. The senna ensures that worms are passed in the faeces before regaining motility.

Treatment of threadworm infection

1 First choice: mebendazole single oral 100 mg dose (not recommended in pregnancy or for children under 2).
2 Second choice: pripsen (piperazine 4 mg and sennoside 15.3 mg per sachet). Adult and child over 6 years: one sachet stirred into milk or water and repeated after 14 days (age 3 months–1 year, one-third of a sachet; 1–6 years, two-thirds of a sachet); each sachet contains 7.5 ml of powder.

 Re-infection or recrudescence is common: treatment may be repeated after 2–3 weeks.

Ascaris lumbricoides

Epidemiology

Ascariasis is transmitted by ingestion of soil containing infective eggs, or vegetables contaminated by soil or sewage. In tropical countries, up to 50% of children may be infected. Children who eat soil are most at risk. The peak incidence of infection is between 3 and 8 years of age. There have been around 100 cases reported in England and Wales each year since 1999, a third of which are known to have been acquired abroad.

Clinical features

Ingested ova germinate in the upper gastrointestinal tract and the larvae migrate through the tissues to the lungs. Intense larval invasion can cause clinically evident lung irritation, with cough, wheezing and mild fever, associated with an eosinophilia (the syndrome of pulmonary eosinophilia). Larvae ascend via the trachea to the pharynx, and are swallowed to complete their lifecycle in the bowel.

Uncomplicated infection is asymptomatic, except for the occasional passage of a worm *per rectum*. Worms may also be expelled orally if vomiting occurs. Adult worms are incidental findings in contrast X-ray studies of the bowel, when they are outlined and their gut is filled by radiopaque medium.

Very heavy worm loads can cause complications, by entering and obstructing small ducts, such as the bile and pancreatic ducts. Tangled masses of worms can obstruct the small bowel (especially when the worms are irritated in the early stages of piperazine intoxication). These conditions may resolve spontaneously, but without treatment there is a risk of recurrence.

Diagnosis

Clinical diagnosis is based on seeing worms in stools, vomitus or contrast X-rays. Eosinophilia without other cause may raise suspicion, especially in the presence of respiratory symptoms.

A laboratory diagnosis is made by demonstrating ova in wet preparations or concentrates of faeces.

Management

The treatment of choice is single-dose mebendazole, which kills the worms without causing hyperactivity. Single-dose combinations of piperazine and purgatives are also effective (see text note above), but are less pleasant to take, as piperazine causes significant nausea at effective doses, and may also provoke hyperactivity of the worms. The most effective and well-tolerated treatment is levamisole (available on a named patient basis) in a single dose of 120–150 mg for an adult. In endemic areas treatment can significantly reduce the worm load, lessening the likelihood of complications and reducing the 'stealing' of nutrients from the bowel (which may be important in an undernourished child).

Prevention

The infection can be prevented by education of children in hand-washing after defecation and by safe disposal of human faeces.

Hookworms

Introduction

Hookworm infection is not endemic in Westernized countries, but is often present in recently immigrated individuals. Hookworms feed on blood, and heavy worm loads can cause significant anaemia, especially in children and others whose diet is low in iron.

Epidemiology

The infection is transmitted from faecally contaminated soil via infective larvae, which penetrate the skin, especially of the feet. All ages are affected, although infection is commonest in children.

The disease is common in the tropics, and rare in temperate climates. Between 1990 and 2003 there were just over 70 laboratory reports of hookworms in England and Wales. The majority of cases are likely to have been in people migrating to the UK from endemic countries.

Pathogenesis

Hookworm ova are passed in faeces and infective larvae develop in an extracorporeal soil life cycle. The larvae can attach to and penetrate human skin, from where they migrate through the tissues to the lungs. They then pass to the pharynx and are swallowed, producing intestinal infection. In the intestine they attach to the bowel wall by their grinding mouth parts, damaging the mucosa and releasing blood, which is their source of nutrition.

Clinical features

Larval migration in the early stages of infection causes both local skin irritation (ground itch) and pulmonary eosinophilia. Once the intestinal infection is established there are no specific features other than those of iron-deficiency anaemia.

Cutaneous larva migrans

Larva migrans is a condition of abortive human infection by animal hookworms, usually those of dogs. The larvae penetrate human skin, but remain locally in the tissues until they die. Persisting skin irritation results, often with serpiginous tracks, transient urticarial rashes and respiratory symptoms (see Case 24.1, p. 500). A variant, caused by ingestion of dog hookworm eggs, has been described, where the pathogen causes inflammation in the terminal ileum resembling acute appendicitis.

Management

The treatment of hookworm infection is oral mebendazole, 100 mg twice daily for 3 days. Dietary iron supplements may be needed in anaemic children. In rare cases of severe anaemia, transfusion may be given. Cutaneous larva migrans can be treated topically with a paste containing thiabendazole, with oral albendazole 400 mg twice daily for 3 days, or with ivermectin 200 μg/kg as a single dose (both available on a named patient basis). Irritation persists for a while after treatment.

Prevention

Prevention of hookworm infection is by wearing shoes, and by safe disposal of human faeces.

Strongyloidiasis

Strongyloides stercoralis has a lifecycle similar to that of hookworms, except that rhabditiform larvae rather than ova are excreted in the faeces. Many larvae undergo a developmental stage to form infectious filariform larvae in soil before penetrating the skin of a subsequent host. However, some rhabditiform larvae develop to infectious filariform larvae before being excreted and reinvade the original host by penetrating the bowel wall or perianal skin. Strongyloidiasis can therefore persist lifelong, after the host has left the endemic area.

The disease is common throughout tropical countries, but also occurs in temperate zones. Fewer than 50 cases are reported annually in the UK.

Larva currens: intermittent larval migration causes a local urticarial rash with a moving, linear component (Fig. 8.20). This appears in various sites, usually on the trunk or thighs. Eosinophilia may be present, and larvae may be demonstrated in faeces, duodenal aspirate or string-test material.

Immunosuppressed patients may suffer massive tissue invasion by larvae, without any eosinophilia. Sometimes immunosuppression is due to increasing doses of prednisolone, given to control 'asthma', caused by pulmonary eosinophilia. The migrating larvae carry bowel bacteria with them, and may enter the lungs (see Fig. CS.2, p. 165) or the meninges, causing secondary coliform infections. The diagnosis must be considered in immunosuppressed patients who originate from the Far East, Africa, Asia or other endemic areas. Transplanted organs may also contain larvae, and transmit infection to the recipient if the donor had strongyloidiasis.

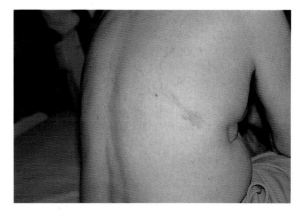

Figure 8.20 Larva currens in strongyloidiasis: recurrent urticarial rash, containing meandering tracks; this patient suffered repeated episodes after being infected as a prisoner of war in Burma in the 1940s; attacks ceased after albendazole treatment in 1981.

The treatment of choice is ivermectin, orally (available on a named patient basis) 200 µg/kg daily for 2 days. Alternatives are albendazole orally, 400 mg twice daily for 3 days, with a second course after 3 weeks if indicated, or thiabendazole orally 25 mg/kg (maximum 1.5 g) 12-hourly for 3 days; this drug has a significant incidence of gastrointestinal side-effects. Immunosuppressed patients may need longer treatment, depending on their response.

Treatment of strongyloidiasis

1 First choice: ivermectin 200 µg/kg daily for 2 days *or* albendazole orally 400 mg twice daily for 3 days (may be repeated after 3 weeks.)

2 Alternative: thiabendazole orally 25 mg/kg (maximum 1.5 g) 12-hourly for 3 days.

3 In immunosuppression: albendazole (may be given in doses of 400 mg twice daily for up to 4 weeks).

 In immunosuppressed patients these courses may be repeated up to three times with intervals of 14 days.

Toxocariasis

Introduction

Toxocariasis is usually a mild disease, predominantly of children, caused by *Toxocara canis*, whose embryonated eggs are excreted in the faeces of young dogs. After 1–3 weeks' maturation in the soil, the eggs become infectious. When a dog ingests infectious eggs, larvae are released and migrate via the bowel wall, through the tissues and lungs, to ascend to the pharynx from where they re-enter the gut and mature to adult worms. The full cycle occurs in puppies up to 6 months old. In adult dogs, the migrating larvae do not fully mature but persist, inactive, in the tissues. When a bitch becomes pregnant, the larvae activate and migrate, infecting her puppies by trans-placental spread, and through breast milk.

Children (and non-canine animals) become infected by eating contaminated soil, or raw unwashed vegetables. After ingestion, embryonated eggs hatch in the intestine, larvae penetrate the wall and migrate to the liver and lungs, but do not fully mature. From the lungs, organisms spread to the abdominal organs (visceral larva migrans) or the eyes (ocular larva migrans). Larvae do not mature to enter the gut, so ova are not excreted by non-canine hosts.

Clinical features

Toxocara larvae excrete highly pro-inflammatory antigens, *Toxocara* excretory-secretory antigens (TES). The inflammatory response to these can cause abdominal pain, failure to thrive, or the classical visceral larva migrans syndrome of fever, cough, wheeze, hepatosplenomegaly and eosinophilia, which has a course of many weeks.

In the eye, uveitis, endophthalmitis or retinal granulomata may occur; lesions affecting the macula can threaten sight.

Diagnosis and treatment

Diagnostic tests include an ELISA test for the presence of serum antibodies to TES. *Toxocara* larvae can be found in liver biopsies, if the investigation is indicated.

Treatment of visceral larva migrans is with albendazole orally, 400 mg daily for 7 days. Ivermectin may be an effective alternative. For ocular larva migrans, it is uncertain whether antihelminthic drugs are useful. Success has been reported with a combination of albendazole and corticosteroids.

Trichinellosis

Trichinellosis is a disease caused by a systemically invasive roundworm, *Trichinella spiralis*, contracted from eating raw or undercooked pork, horse or bear meat, containing larval cysts. It is rare in the UK, with around 20 cases reported each year.

Symptoms are variable; conjunctivitis, oedema of the eyelids and retinal haemorrhages are characteristic early features, often followed by muscle pains, thirst, sweating, chills and remitting fever. Blood levels of muscle enzymes may be elevated. Cardiac and neurological complications may occur after 3–6 weeks. There is a marked eosinophilia, which may assist in diagnosis. Serological tests are available. Antihelminthic drugs are not effective in treatment but corticosteroids may be effective in limiting inflammatory damage in severe cases.

Case study 8.1: Difficult recovery from diarrhoea

History
A previously healthy 6-year-old boy became ill with watery diarrhoea and severe abdominal cramps 2 days after a school outing to a 'petting zoo', where he had contact with rabbits, piglets, lambs and kids. Of the 20 other children on the outing, six also developed diarrhoea. Despite plentiful electrolyte drinks, he became fatigued and the diarrhoea became bloody. On the third day of illness he was admitted to hospital.

Investigations
Haemoglobin 12.2 g/dl, white cell count 11.4×10^9/l (7.8×10^9/l neutrophils), platelets 108×10^9/l, blood urea 4.0 mmol/l, creatinine 65 mmol/l, electrolytes normal.

Management and progress
Rehydration was achieved with intravenous dextrose–saline. After overnight treatment he was able to take fluids orally and the diarrhoea had abated. The following day, he appeared pale and listless, and became unwilling to eat and drink, though he had only three small, loose stools. On the sixth day of illness he vomited once, and appeared minimally jaundiced. Two of the other affected children had now been admitted to hospital with bloody diarrhoea.

Repeated investigations showed haemoglobin 11.2 g/dl, white cell count 10.9×10^9/l and platelets 40×10^9/l; the blood film report described microcytic and fragmented red cells (schistocytes), with no platelets seen on the film. Blood urea was 7.8 mmol/l, creatinine 152 mmol/l, sodium 130 mmol/l, potassium 4.9 mmol/l, serum bilirubin 27 mmol/l, aspartate transaminase 50 IU/l, alkaline phosphatase 230 IU/l (normal for this age group).

Questions
- What blood disorder does the blood count and film indicate?
- What is the likely diagnosis?
- What infectious organism is probably responsible?
- What pathogenicity factor(s) are important in the aetiology?
- Is antimicrobial therapy indicated?
- What is the prognosis?

Further management and progress
Based on the blood film evidence of microangiopathic haemolytic anaemia and the renal failure, a diagnosis of haemolytic uraemic syndrome (HUS) was made. This syndrome is usually caused by infection with a verocytotoxin-producing *Escherichia coli* (VTEC), which also causes haemorrhagic colitis. Most VTEC isolated from cases of haemorrhagic colitis and HUS are of serotype O157. *E. coli* O157 does not metabolize sorbitol. It can be recognized in the laboratory by identifying its sugar metabolism profile and by agglutination testing with specific O157 antiserum. Rare cases of colitis and HUS have been related to infections with other *E. coli* serotypes. Two phage-encoded types of verocytotoxin, VT1 and VT2 (which exists in two forms VT2a and VT2b), can be identified serologically, by DNA hybridization tests and by variations in their biological effects. Because of their close homology with the toxin of *Shigella dysenteriae*, the toxins are also called shiga-like toxins (SLT1 and SLT2). The toxins cause death and shedding of the colonic mucosal cells. Toxin-mediated endothelial cell damage is responsible for HUS. In outbreaks of *E. coli* O157 haemorrhagic colitis, approximately 5% of patients develop HUS.

Stool cultures in this case revealed a heavy growth of *E. coli* O157, which was later shown to produce VT2.

There is no evidence that antibiotic therapy ameliorates the course of HUS; there is some evidence that it may worsen the prognosis. Some antibiotics, such as co-trimoxazole, appear to stimulate increased VT production, though others, such as ciprofloxacin, may reduce it. The causative *E. coli* becomes undetectable in stools by the fourth or fifth day of illness. Supportive therapy includes control of fluid and electrolyte balance, diuretic therapy for oedema, and blood and platelet infusions as indicated. Although renal support (haemoperfusion or haemodialysis) is required in up to half of cases, 70% of children recover from the renal insult (see Fig. 8.9). The 5% overall case fatality in HUS is mainly due to cerebral complications related to oedema, haemorrhage and vasculitis.

Questions
- Could it be shown that the ill children were part of an outbreak?
- Could the animals have been the source of infection?

Epidemiological investigation
E. coli O157 strains were recovered from four of the seven affected children. All produced VT2a. Phage typing showed them all to be phage type 8. Pulsed-field gel electrophoresis patterns were identical in the four isolates.

Faeces was collected from animal pens, and rectal swabs from individual animals at the zoo. Both the lamb and the kid pens produced positive results, and two of the three kids sampled were positive. The lamb pen samples and two of the three isolates from kids were of the same type as the isolates from the children.

Haemorrhagic colitis and HUS outbreaks in children are commonly related to animal contact. Lambs and goats are the most often implicated, but calves and adult cattle may also be involved. Consumption of meat, meat products and unpasteurized milk are also sources of infection in children, as in adults.

Infections of the Liver

9

Introduction

The liver is a complex organ. It receives blood from the intestinal tract via the hepatic portal system, and is also sustained by 'systemic' blood via the hepatic artery. The hepatocytes are the site of detoxification and metabolism of intrinsic and extrinsic substances, including ammonia, complex amines, endotoxin, steroids and many drugs. They are the main site where glycogen is manufactured and are the only important source of blood glucose. They are the only site of synthesis of urea, and of several proteins, including albumin and the clotting factors III, VII, IX and X. C-reactive protein is also synthesized by the liver.

The biliary system is the route of excretion of bilirubin, derived from the metabolism of haem substances. It is also important in the enterohepatic circulation of bile salts. Several drugs are excreted and concentrated in the bile; these include penicillins, cephalosporins and rifampin. Bile also contains cholesterol, which easily crystallizes and may form gallstones. Patients with a high turnover of haem excrete high bilirubin loads, and are at risk of pigment stones.

Infections of the liver tend to affect either the cells, producing typical hepatocellular disorder, or the biliary tract, producing cholestasis and features of infection in a hollow organ. Space-occupying lesions in the liver (e.g. abscesses or granulomas) tend to produce the biochemical changes of cholestasis, but rarely cause jaundice.

Viral infections of the liver

Organism list

- Hepatitis A virus (HAV)
- Hepatitis E virus (HEV)
- Hepatitis B virus (HBV)
- Delta virus (hepatitis D)
- Hepatitis C virus (HCV)
- Hepatitis G virus
- Transfusion-transmitted virus (TTV)
- Epstein–Barr virus (EBV)
- Cytomegalovirus (CMV)

Hepatitis A

Introduction

Hepatitis A is a highly infectious disease with worldwide distribution, which therefore spreads easily among children. Many cases are subclinical; others cause moderate morbidity. Overall mortality is less than 0.1%, but about 1% of hospital admissions with hepatitis A suffer severe or life-threatening disease.

Epidemiology

In conditions of poor hygiene and sanitation, most infections occur in childhood, and are mild or asymptomatic. Immunity is lifelong and outbreaks in adults are uncommon. In high-income countries, infection in children is less usual and infection occurs more commonly in adults, in whom it is more likely to be symptomatic. Three patterns of disease are then seen:

- sporadic infections among adult travellers to endemic countries;
- epidemics affecting mainly school-age children; and
- explosive common-source outbreaks (Fig. 9.1).

Epidemics of hepatitis A tend to evolve slowly, last for several months and affect large geographical areas. Spread is from person to person, mainly by the faecal–oral route,

occurring within households, nurseries, schools and other institutions.

Common-source outbreaks are usually associated with food contaminated by an infected handler, or with undercooked shellfish harvested from contaminated waters. Soft fruits have been associated with outbreaks, and water-borne outbreaks occur occasionally.

The incidence of hepatitis A infection in the UK has fallen progressively since 1990, particularly in the age-group 5–14. There were around 6700 laboratory reports of hepatitis A infection in 1992, declining to around 1300 by 2003, with the highest proportion of cases affecting the age group 15–34. In parallel with the overall decline in cases, imported hepatitis A infections have apparently declined since 1990. However, this may be partly a reporting

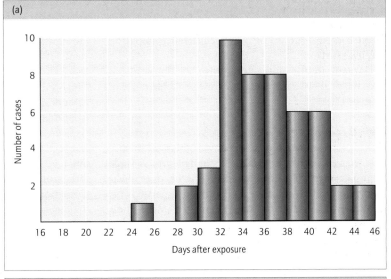

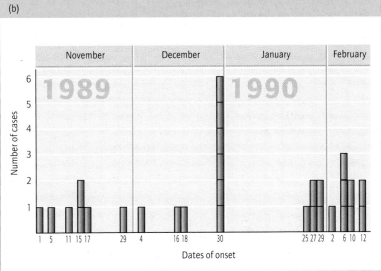

Figure 9.1 (a) Rapidly evolving or explosive epidemic of hepatitis A, originating from a common source – in this case raspberries that were contaminated with human excreta before packaging. (b) Slowly evolving community outbreak of hepatitis A, spreading by the person-to-person route.

artefact, since an increasing proportion of reports over this time have not included information about travel history. The Indian sub-continent was the single most reported region of acquisition during the period 1990 to 2002.

Virology and pathogenesis of hepatitis A

Hepatitis A virus (HAV) is a heparnavirus in the family Picornaviridae. It is a small RNA virus, 27 nm in diameter with a single-stranded, approximately 7.5 kb genome of positive-sense RNA. A single polypeptide is synthesized and cleaved by viral proteases. The virus particle is more stable to heat, detergents and proteases than the other picornaviruses and this may explain the ease with which it is transmitted. The virus particle consists of four capsid polypeptides VP1–4. There is only one major serotype and four subtypes defined on the basis of the sequence of the VP1/2 junction.

The virus appears not to be cytopathic. Hepatic damage occurs after the feverish stage of illness, and is probably immunologically mediated. Neutralizing antibodies are directed against groups of closely clustered epitopes on the virus surface.

Humans are the only natural host, but the virus has been adapted by serial passage to grow in a wide range of primate cells.

Clinical features

The incubation period of 15–45 days is followed by malaise, nausea, lassitude, myalgia, arthralgia and variable fever. The intestinal mucosa is inflamed at this stage: sugar absorption tests are abnormal and children, particularly, may have loose stools. Viruses are excreted in the stools and urine. Features of hepatitis gradually appear after 2–7 days. The urine darkens as bilirubinuria increases, and the stools may be noticeably pale. Jaundice is first seen in the sclerae and later in the skin (Fig. 9.2).

Fever resolves and virus excretion ceases as jaundice develops. Patients are now no longer infectious, and most patients feel better. After 7–10 days the appetite returns and the jaundice begins to resolve.

During the prodrome the white cell count is normal, with a few atypical mononuclear cells. As the prodrome ends, rising plasma transaminases indicate hepatocellular damage, and conjugated bilirubin appears in blood and urine. The early transaminase levels are often 2000–7000 IU but these fall rapidly in most cases, and are often below 100 IU after 7–10 days.

Cholestasis, with a rise in alkaline phosphatase to 250–400 IU, is common during convalescence. Abnormality of bile-salt excretion may then cause itching. Occasionally cholestasis is severe and prolonged, with deepening jaundice and severe pruritus, persisting for months if untreated. This is a post-infectious event, not associated with viral replication.

Diagnosis

The history of malaise followed by jaundice, and the absence of risk factors for blood-borne hepatitides, is highly suggestive. Prolonged vomiting in an otherwise well child can occur in anicteric hepatitis. Laboratory investigations should include tests to exclude hepatitis B and C, and E where appropriate (see pp. 207–214).

Glucose-6-phosphate dehydrogenase deficiency can cause haemolysis and clinical jaundice, often following a viral infection, ingestion of proprietary antipyretic drugs or eating beans (favism). Spherocytosis, sickling conditions, and ovalocytosis may also cause haemolytic episodes. In travellers, malaria is a common cause of haemolysis, usually accompanied by significant fever. Haemolytic jaundice is acholuric (not accompanied by bilirubinuria); urinalysis therefore shows normal bilirubin levels and raised urobilinogen.

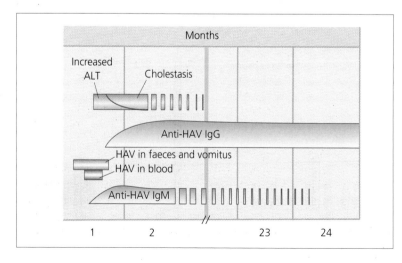

Figure 9.2 The clinical and immunological evolution of hepatitis A infection.

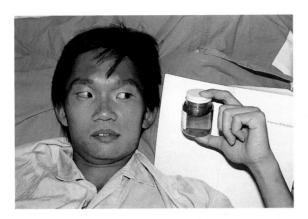

Figure 9.3 Hepatitis A: typical appearance of jaundice and dark urine in viral hepatitis; the patient is no longer suffering from fever or malaise.

Persisting fever in viral hepatitis is unusual once jaundice has developed. In febrile hepatitis, Epstein–Barr virus (EBV) infection, or leptospirosis, may need to be excluded. Malaria must be excluded in tropical travellers.

Acute cholecystitis presents with fever, right hypochondrial pain and often jaundice (see below). Gallstones may cause pain and obstructive jaundice. Painless cholestatic jaundice is a common feature of biliary or pancreatic carcinoma, which should be considered in all middle-aged or elderly patients. An elevated alkaline phosphatase is a useful warning sign of obstructive or cholestatic pathology.

Some drugs cause hepatocellular damage. Alcohol is the most common, but paracetamol toxicity is important. A history of paracetamol ingestion should be sought in jaundiced patients.

Differential diagnosis of viral hepatitis

1 Other infections: Epstein–Barr virus infections, cytomegalovirus, leptospirosis.
2 Infections of the biliary system: acute cholecystitis, acute cholangitis, acute pancreatitis.
3 Obstructive jaundice: gallstones, tumours of the pancreaticobiliary system.
4 Haemolysis, e.g. glucose-6-phosphate dehydrogenase deficiency, malaria.
5 Drug jaundice, e.g. alcohol and paracetamol (hepatocellular), phenothiazines (cholestatic).

Laboratory diagnosis

Specific diagnosis depends on demonstration of specific immunoglobulin M (IgM) antibodies by antibody-capture enzyme immunoassay (EIA). These become detectable when jaundice develops and persist for 3 months to 2 years. IgG antibodies persist, and confer immunity lifelong. An EIA to detect specific IgG may be used to determine immune status.

Treatment

There is no specific antiviral treatment. Symptomatic treatment with mild analgesics and antiemetics improves prodromal symptoms. Antipruritic drugs may be helpful in convalescence, but cholestyramine is more effective for severe pruritus, as it interrupts enterohepatic recirculation of bile salts. There is little evidence that bed rest shortens the illness or prevents complications.

Prolonged severe cholestasis after hepatitis A responds to corticosteroid treatment. Initial dosage with prednisolone 30–40 mg/day can be reduced over 2–4 weeks depending on the clinical and biochemical response.

Complications
Liver failure

Liver failure is the most common complication. It may be fulminant and early, presenting with altered consciousness before jaundice develops, or subacute and progressive, with inexorable deterioration in liver function and deepening of jaundice. The prognosis is best in patients suffering rapid-onset liver failure and worst in those with insidious onsets.

Clinical indicators of liver failure are persistent vomiting, disturbed behaviour, flapping tremor of the outstretched hands and increasing drowsiness. Poor cerebral function is classically demonstrated by showing the patient's inability to copy a drawing of a five-pointed star, although able to copy a square (constructional apraxia).

Biochemical warning signs are the disappearance of blood constituents synthesized by the liver: glucose, clotting factors and urea (albumin has a longer half-life and declines later). Transaminase levels may not accurately reflect hepatocellular damage as they decline with extensive cell death; initial levels above 7000 IU/l may predict severe hepatocellular damage (Fig. 9.4). In hepatic coma the electroencephalogram shows characteristic 3/s slow-wave patterns. A blood pH of <7.3, or prothrombin time of >100 s is an indication for urgent referral to consider early liver transplantation.

Management of liver failure

⚠ Note that about 55% of all liver failure cases in the UK are due to various types of viral hepatitis.

1 Maintain the blood glucose to avoid hypoglycaemic convulsions or cerebral damage; this often requires constant infusion of 10% dextrose solution, with occasional further supplementation.
2 Minimize amine production by avoiding excess protein loading (the patient should receive sufficient protein to fulfil daily nitrogen requirements). Dietary advice and enteral nutrition should be considered early. Empty the bowel by giving oral lactulose. Non-absorbable antibiotics, such as neomycin

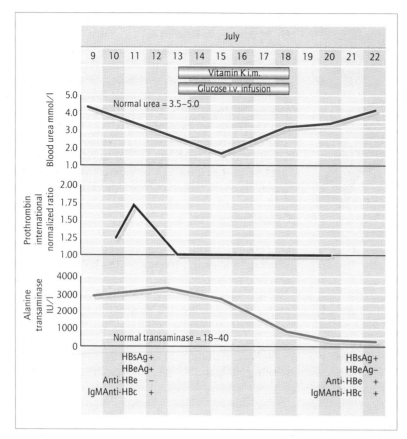

Figure 9.4 Development of liver failure in a patient with a severe attack of hepatitis B; the patient made a spontaneous recovery with supportive treatment.

(up to 4 g daily, in divided doses), by mouth or nasogastric tube, may reduce ammonia production from bacterial proteins in the bowel.

3 Minimize the risk of bleeding while clotting factors are deficient. Give a protein pump inhibitor to protect against gastric erosions, avoid vigorous or traumatic procedures.

4 Avoid drugs such as loop diuretics, opiates, major tranquillizers, caution with drugs with hepatotoxic effects and hepatically metabolized drugs.

5 Refer early to a specialist unit, where liver transplantation can be planned; HAV replication is transient in hepatitis, so transplant infection does not occur.

6 Other therapeutic considerations: MARS (molecular adsorbent recirculating systems) may be available to maintain patients with liver failure, pending recovery or transplant.

⚠ The prothrombin time (PT), and factor V levels are useful markers of progress.

⚠ Short-acting benzodiazepines may be given for sedation. Paracetamol is a useful analgesic, as its toxic metabolite is not produced during liver failure. Blood products constitute a protein load, and should be reserved for treating haemorrhage. Clotting factors are quickly consumed and do

not prevent continuous oozing. Transfusions may generate alloantibodies, with risk of rejection if liver transplantation is performed. Small doses of vitamin K may avoid deficiency during a catabolic state, but frequent dosing should be avoided.

Prevention and control

Hepatitis A is preventable by good sanitation and personal hygiene, important in nurseries and schools where spread is likely to occur. Clean facilities for hand washing should be available in all settings and communities with higher risks of infection. The risk of common-source outbreaks can be minimized by proper education of food-handlers and adequate cooking of shellfish. Cases of hepatitis should be isolated until 1–2 days after the onset of jaundice. The disease is notifiable.

Non-replicating vaccines against hepatitis A are highly effective, producing lasting immunity from 8–10 days after a single dose. They are recommended for pre-exposure prophylaxis for travellers to endemic areas and for certain high-risk groups, e.g. individuals with special needs (and their carers) who may have difficulty in maintaining good hygiene standards; laboratory workers working with HAV or HAV-infected non-human primates; and sewage workers who frequently have direct contact with raw sewage. In addition, people with haemophilia, hepatitis B or C infection, or liver cirrhosis of any cause, should be offered HAV vaccination Several outbreaks of hepatitis A have occurred in injecting drug users and men who have sex with men; these groups should also be considered for preventive vaccination. (Subcutaneous, rather than i.m. injection, is recommended for haemophilia patients, to avoid causing large haematomas.)

Post-exposure prophylaxis (PEP) should be considered for close (household and sexual) contacts of a confirmed case. The available measures for PEP are HAV vaccine, human normal immunoglobulin (HNIG, a blood product), or both together (though administering HAV vaccine and HNIG simultaneously may reduce the long-term efficacy of the vaccine). Both vaccine and HNIG have effectively controlled outbreaks in defined communities and in the general population.

HAV vaccine is usually the PEP of choice because it does not require the use of human blood products. Vaccination must be given within one week of exposure.

HNIG has well-established efficacy in preventing disease when given within two weeks of exposure, and can modify disease if given within 28 days. HNIG is preferred where the exposed person is at particular risk of an adverse outcome of HAV infection or has been identified too late to be protected by vaccine.

> **Prevention of hepatitis A**
> 1 Safe water supplies.
> 2 Effective sanitation.
> 3 Personal hygiene.
> 4 Safe food-handling.
> 5 Pre-exposure prophylaxis: inactivated vaccine (immunoglobulin for temporary protection of immunosuppressed individuals).
> 6 Post-exposure prophylaxis: vaccine is effective if given in the first week after exposure; human normal immunoglobulin can be used if prophylaxis is delayed.

Hepatitis E

Introduction and epidemiology
Hepatitis E is an acute hepatitis, transmitted by the faecal–oral route, particularly in north India, Nepal, Bangladesh, North Africa and parts of tropical South America. Epidemics have occurred in several countries, particularly in South-east Asia and the Indian subcontinent, predominantly affecting young adults, especially males. It is unaccountably severe in pregnancy, causing a mortality of 10–40%, compared with under 1% in non-pregnant patients. Jaundice is often prolonged, lasting 4 or 5 weeks. Infection is usually thought to originate from contaminated water but, recently, HEV has been identified in pigs,

raising the possibility that hepatitis E could be a zoonosis.

Virology
Hepatitis E (HEV) is a small (32–24 nm) non-enveloped RNA virus, classified in its own genus *Hepevirus*. There is one serotype, differentiated by genetic heterogeneity into five genotypes. It possesses a single-stranded, positive-sense, polyadenylated RNA of approximately 7.5 kb with three open reading frames. Two putative proteins are predicted, one of which is an RNA polymerase. Structural proteins are coded on the second of the three open reading frames. The virus is unstable in storage and has not yet been cultivated artificially.

Diagnosis
Diagnostic tests for hepatitis E are in continuing development. An antibody-capture EIA, based on recombinant proteins, is used for demonstrating IgM anti-hepatitis E virus (HEV) antibodies. IgM becomes detectable once jaundice has developed. Virus-like particles may be demonstrated by immunoelectron microscopy examination of stool. PCR methods have been developed to detect hepatitis E virus in stools, and are useful in studying transmission.

Clinical features
After an incubation period averaging 40 days, typical viral hepatitis develops. The duration of illness is more prolonged than hepatitis A: the average duration of clinical jaundice is 5 or 6 weeks. HEV is excreted in the stools for about three weeks after the onset of clinical illness. Liver failure can occur, and is more common in pregnant women, who have a high mortality (up to 40%). Persisting cholestasis with high alkaline phosphatase levels has been observed, but is rare.

Prevention and control
General sanitary measures and personal hygiene are effective in limiting spread. No vaccine or immunoglobulin is available.

Hepatitis B

Introduction
The importance of hepatitis B lies in its ability to cause chronic infection with liver inflammation, leading eventually to cirrhosis and malignant change in the liver. Vertical and horizontal infection from mother to child can produce successive generations of chronically infected, infectious carriers.

Virology and pathogenesis of hepatitis B diseases
Hepatitis B is a small enveloped virus containing partially

double-stranded DNA. The DNA contains four major genes that each encode more than one protein: The surface protein gene encodes three polypeptides (including hepatitis B surface antigen, HBsAg), which make up the viral envelope:

- the C gene encodes the core or capsid protein (hepatitis B core antigen, HBcAg) and the precore protein;
- the P gene encodes the reverse transcriptase and RNAase; and
- the X gene encodes transactivator proteins.

HBsAg is the major protein of hepatitis B virus (HBV). The major antigenic determinant of HBsAg is the a antigen, antibodies to which confer protection after infection or vaccination. There are also two pairs of mutually exclusive subdeterminants – d or y, and w or r – giving four phenotypes – adw, adr, ayw and ayr. The frequency of these phenotypes varies in different parts of the world. Eight genotypes of HBV are recognized (A–H), of which types A to D are common; genotype A is the predominant type in the UK.

HBcAg is the major nucleocapsid antigen. Hepatitis B e antigen (HBeAg) is a cleavage product of HBcAg, and is a marker in blood for HBV replication. Mutation in the precore region of the HBV genome causes faulty HBeAg transcription. Around half of chronically infected HBsAg-positive, HBeAg-negative patients possess this mutation, and may have high blood levels of HBV DNA, despite being e-antigen negative.

The mechanism of HBV binding to liver cells is not yet known. HBV infection typically causes acute hepatitis, characterized by centrilobular and multifocal hepatocyte necrosis, with periportal histiocytic infiltration. Rarely, massive immune-mediated necrosis occurs, associated with major disruption of the reticulin framework of the liver. The mechanism of liver cell damage is uncertain, although most studies suggest that the virus is not directly cytotoxic. In up to 10% of patients, persistent infection causes chronic hepatitis, progressing to cirrhosis after an average of about 30 years, Persistent infection is thought to result from failure of the initial immune response.

Liver carcinoma is an important complication of hepatitis B-associated cirrhosis, with a 100-fold excess risk compared with the general population. The molecular basis of carcinogenicity is not understood, but the carcinoma is derived from the clonal expansion of a single infected cell. Hepatitis B virus is incorporated into the host chromosome but most of the viral genes are truncated or inactivated. They are inserted at random positions and genetic analysis of tumours has not revealed any consistent pattern to indicate a tumour-triggering integration site.

Epidemiology

There are around 350 million chronic carriers of the hepatitis B virus worldwide. The incidence of acute disease and prevalence of carriage varies considerably from country to country. In parts of south-east Asia, 10–20% of the population may be carriers, whereas most countries in Europe and North America have carriage rates below 2%.

Where carriage rates are high, acute infection occurs mainly in infants and young children, mostly via intrapartum and horizontal transmission within households. Skin disease and biting arthropods may facilitate the transfer of body fluids from person to person.

In low-prevalence countries most infections are sporadic and arise in adults through needlestick injuries, shared syringes, bites and scratches, or by sexual contact. Those most at risk include intravenous drug abusers, homosexual men, residents and staff of institutions for the mentally handicapped, surgeons, dentists, laboratory workers, morticians, renal dialysis patients and recipients of unscreened blood and blood products.

Risk groups for hepatitis B in developed countries

1 Intravenous drug abusers.
2 Homosexual men.
3 Sexual contacts of antigen-positive persons.
4 Residents in long-stay homes for mentally handicapped people.
5 Renal dialysis patients.
6 Recipients of multiple blood products (e.g. haemophiliacs).
7 Surgeons, dentists and morticians.
8 Infants of infectious HBsAg-positive mothers.

Acute HBV infection is relatively uncommon in the UK, with 600 to 800 confirmed cases each year, mostly in high-risk groups. A large outbreak occurred in 1984, associated with intravenous drug abuse. Small common-source outbreaks have been reported in association with tattoo parlours and other skin-piercing activities. A large outbreak occurred in 1998, associated with an alternative therapy centre. The UK prevalence of chronic HBV infection (HBsAg carriage) is less than 0.5% overall. Certain ethnic subgroups have a much higher chronic infection rate, and many chronically infected cases affect individuals who acquired their infection overseas, often in infancy. Up to half of these individuals may possess the pre-core mutant virus, which emerges as their immune system destroys HBeAg-producing viruses. HBsAg carriage is often increased in long-stay institutions for mentally handicapped people.

Because they are both transmitted by similar routes, chronic co-infection with HBV and HIV can occur, particularly affecting homosexual men or intravenous drug users. HBV and HCV co-infection and triple infections are also seen, and are more difficult to treat than single infections.

Clinical and laboratory features

The incubation period is from 3 to 6 months. Up to 90%

of cases, particularly those acquired in infancy, are sub-clinical. In clinical illness, prodromal symptoms are followed by afebrile jaundice similar to that of hepatitis A. The prodrome may resemble serum-sickness, with mild arthritis and a faint rash. Clinical distinction of acute hepatitis B from other types of viral hepatitis is rarely possible but cholestasis, common in convalescent hepatitis A or E, is rare in hepatitis B.

During HBV replication, a large excess of HBsAg is produced, and is detectable in the serum 2–8 weeks before clinical illness. As transaminases rise, other viral products become detectable, including viral DNA, DNA polymerase and HBeAg. During recovery and convalescence, antibodies to these proteins appear in the blood, and the proteins gradually disappear. Anti-HBc IgM is detectable at the onset of illness. Anti-HBe usually appears in the first 1–3 weeks and HBeAg then becomes undetectable. Anti-HBs appears after 6 weeks–6 months, and HBsAg then disappears.

This sequence can cease at any stage, leaving the patient with persisting antigenaemia. Subclinical infections are thought more likely to progress to chronicity. If anti-HBe antibodies fail to develop, the patient's blood remains significantly infectious, even if HBeAg disappears. Patients with only HBs antigenaemia are less infectious. However, in chronic infection, virus with the pre-core mutation may emerge and HBeAg may disappear, but the patients remain highly infectious. Patients with serum HBV DNA levels above 10^3 copies per ml are considered infectious by the blood-borne route.

If antigens remain detectable after 6 months the patient is considered to be chronically infected. This occurs in 5–10% of all infections, but is more common in infants and in patients with defective immunity. Many chronically infected patients have normal liver function and are well, but those with persisting viral replication, indicated by positive HBV DNA and/or abnormal transaminases, are at risk of chronic persistent hepatitis, chronic active hepatitis, cirrhosis and eventual hepatocellular carcinoma. HBeAg- and HBV DNA-positive women are at risk of infecting their infant during or soon after its birth.

Diagnosis
Both acutely and chronically infected individuals have HBs antigenaemia. The diagnosis of acute disease is confirmed by demonstrating IgM anti-HBc in the serum. This appears 2 weeks after HBsAg (Fig. 9.5) and disappears a few months after uncomplicated infection. IgG anti HBc persists probably lifelong, and is a marker of previous infection. The stage of evolution of antigenaemia and antibody production is determined by EIA tests. Viral persistence can be confirmed by PCR-based detection of HBV DNA in serum. Detection of HBe is still used as a marker of enhanced infectivity and risk of chronic liver disease.

Treatment and complications of acute hepatitis B
Acute hepatitis B
Treatment of acute hepatitis B is symptomatic. The patient should be followed up to ensure that HBeAg and HBsAg are cleared from the serum, and that anti-HBs becomes detectable. Patients with antigenaemia persisting after 6 months should be considered chronically infected, and assessed for antiviral treatment.

Liver failure
Liver failure is the most important early complication of acute hepatitis B. Its presentation and management are the same as for hepatitis A. However, liver transplantation may not result in permanent cure because replicating virus, persisting in the pancreas and elsewhere, can infect the transplanted liver. This is suppressed by intra-operative and post-operative treatment with hepatitis B immunoglobulin (HBIG) and antiviral drugs, to prevent rapidly progressive disease in the immunosuppressed patient.

Rare complications
Rare complications include renal failure due to nephritis, and rash (more common in children, affecting the peripheries early in the illness).

Management of chronic hepatitis B
The aim of treatment is the eradication of e-antigenaemia. Failing that, there is a health gain from inhibiting HBV replication to low blood levels of viral DNA. Successful treatment normalizes transaminases, improves liver histology and reduces the future likelihood of carcinoma. Most experts would offer treatment to patients with HBe antigenaemia, and patients with HBs antigenaemia who have abnormal transaminases or detectable HBV DNA in the blood. Patients should be evaluated by liver function tests, serology, liver scan and liver biopsy to exclude other causes of chronic liver disorder (including hepatitis C), and to determine the severity of existing inflammation and fibrosis of the liver. Already existing carcinoma should also be excluded by confirming normal alpha-fetoprotein levels.

Drugs active against HBV
1 Lamivudine, 100 mg daily, orally (also used for HIV).
2 Adefovir dipivoxil, 10 mg daily, orally.
3 Tenofovir (used for HBV/HIV co-infected patients).
4 Alternative: interferon alpha, 5–10 MIU three times weekly, subcutaneously for 6 months.

⚠ If an antiviral drug effective against HBV is also being used to treat HIV co-infection, the HIV-treatment dose should be given (this is often higher than the dose for HBV).

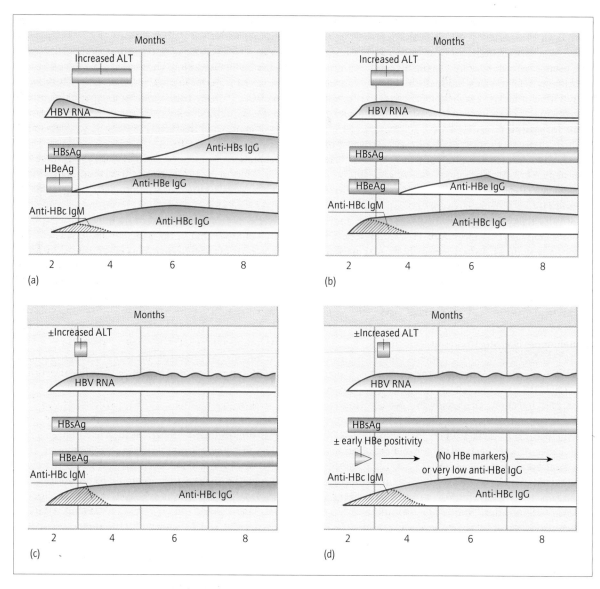

Figure 9.5 The evolution of hepatitis B infection: occurrence of hepatitis B virus markers and antibodies in the blood of infected patients. (a) Acute hepatitis B with recovery and immunity: immunity is confirmed by presence of anti-HBs IgG. (b) Hepatitis B infection with persistence of HBsAg (anti-HBs IgG does not develop), a low or sub-detectable HBV viraemia persists. (c) Hepatitis B infection with persistence of HBsAg and HBeAg, with sustained HBV viraemia: this is common after perinatal infection. (d) Hepatitis B infection with persistence of HBsAg but no HBe markers: some patients have high persisting HBV RNA due to active replication of HBV pre-core mutant.

Lamivudine, a nucleoside reverse transcriptase inhibitor, can achieve HBe seroconversion in up to 35% of patients. Treatment may be discontinued when this is confirmed. It can suppress HBV DNA levels for several years in patients with chronic hepatitis B acquired in infancy. It can be used in decompensated liver disease. After one year up to 20% of patients have a population of lamivudine-resistant viruses (usually with a well-characterized mutation in the reverse transcriptase gene), and this proportion increases each year. Treatment should be continued until effectiveness is lost; as discontinuing effective treatment results in a rebound of viral replication. Adverse events at this dosage include nausea, vomiting, malaise, headache and insomnia and, occasionally, rash and fever.

Adefovir dipivoxil is a prodrug that provides high bioavailability of adefovir. Adefovir inhibits HBV DNA polymerase, and is at least as effective as lamivudine. It is active against lamivudine-resistant viruses, and the emergence of adefovir-resistant viruses is rare. Adverse effects include nausea, anorexia and weight loss. Adefovir causes dose-dependent nephrotoxicity, and is excreted by the kidney. The dosing interval must be extended if the creatinine clearance is below 30 mmol/dl.

Interferon alpha (5×10^6 IU per day or 10×10^6 IU three times weekly, s.c. or i.m. for 16 weeks) can achieve seroconversion in 25–40% (33% in one meta-analysis) of cases. If there is no improvement in viral DNA levels after three months, the risk of continued therapy outweighs the likelihood of success. Seroconversion is heralded by a short illness with raised transaminases, which usually occurs during treatment, but is delayed up to 6 months after completion of treatment in about 15% of patients.

Problems and adverse effects with interferon alpha

1 It is contraindicated in immunosuppressed patients.
2 It is toxic in patients with decompensated liver disease.
3 It is not effective in patients who were infected perinatally.
4 Influenza-like side-effects may be reduced by paracetamol therapy before dosing, but may need dose reduction or discontinuation.
5 Cytopenias, especially neutropenia and thrombocytopenia.
6 Precipitation of depression and other neuropsychiatric events; contraindicated in the presence of depression and should be given with great caution if there is a history of depression.
7 Thyroiditis, and exacerbation or unmasking of autoimmune conditions.
8 Disorders of hepatic and renal function.

⚠ Trials of combined interferon plus lamivudine have not produced unequivocal evidence of benefit: other combinations, some including pegylated interferon, are under investigation.

Prevention and control
Immunization

Hepatitis B vaccines contain HBsAg generated through recombinant DNA technology. A course of three injections provides protection, which is probably lifelong in immunocompetent individuals who seroconvert. Approximately 10% of recipients fail to achieve an effective response after three doses; post-vaccination screening can

be carried out and booster doses given if necessary. The risk of an inadequate response increases with age, obesity, female sex, and in those who receive the vaccine intradermally or in the buttock.

A common cause of failure to produce an anti-HBs response to HBV vaccination is that the patient is already chronically infected with HBV, has antigenaemia, and cannot generate anti-HBS: all persistent non-responders should be tested for HBsAg.

In the UK, vaccination is indicated for pre-exposure prophylaxis of high-risk groups (see above). In higher-incidence countries universal immunization is offered in infancy or adolescence.

HBIG (hepatitis B immunoglobulin), prepared from the plasma of selected donors, provides passive protection against hepatitis B. It is normally used in combination with vaccine to provide passive–active immunity: (i) for infants born to mothers who are HBsAg-positive, and have detectable HBe or HBV DNA; (ii) following percutaneous or mucosal exposure of non-immune individuals to infected body fluids; and (iii) for sexual contacts of acute cases. Routine antenatal screening for HbsAg is now recommended for all pregnancies in the UK, in order to identify mothers whose infants should receive neonatal prophylaxis.

Immunoglobulin must be given within 1 week of exposure, and preferably within 48 h. A course of vaccine should be started at the same time.

Other measures

Hepatitis B can also be prevented by ensuring that all blood and blood products are screened and that blood is not collected from donors at risk of infection. All syringes, needles and skin-piercing equipment should be disposable. Infectious carriers of HBsAg may be restricted from certain occupations. The disease is notifiable.

Prevention of hepatitis B

1 Pre-exposure prophylaxis with non-replicating vaccine.
2 Post-exposure prophylaxis with specific immunoglobulin and vaccine.
3 Universal antenatal screening.
4 Universal testing of donated blood products.
5 Use of condoms.
6 Use of disposable needles and other skin-piercing equipment.
7 Safe disposal of hospital sharps.
8 Use of disposable haemodialysis equipment.

Delta hepatitis

Introduction

Delta hepatitis, or hepatitis D, is caused by a satellite virus that uses HBsAg as its own surface antigen. It therefore only infects antigen-positive patients with either acute hepatitis B, or long-term HBsAg carriage. Co-infection with hepatitis B and D causes more severe and rapidly progressive disease than hepatitis B alone.

Virology and pathogenesis of hepatitis D infection

Hepatitis D virus, the delta agent, is an enveloped RNA virus of similar size to hepatitis B virus (38–41 nm compared to 42 nm for hepatitis B). The envelope consists of the same three surface proteins as hepatitis B virus, although the relative amounts are different. The virus core contains only one known protein, the delta antigen. Replication occurs in the nucleus of the host cell via a host RNA polymerase. Two complementary strands of RNA are synthesized from the circular genome. One, the antigenome, is an exact complement of the genome. The second, a smaller fragment, is polyadenylated and acts as the messenger RNA for the delta antigen. Infected liver cells become packed with several hundred thousand copies of hepatitis D virus genome.

The delta agent is probably directly cytotoxic, although *in vitro* studies are not conclusive. There is evidence that asymptomatic infection is common. The presence of hepatitis B virus is required to provide the envelope proteins, enabling hepatitis D virus to spread from cell to cell, and to express its pathogenic potential. The coexistence of hepatitis D with chronic hepatitis B antigenaemia is associated with an accelerated progression to carcinoma. Delta antigen and IgM anti-delta antibody are detectable in the blood during infection.

Epidemiology

Hepatitis D always coexists with hepatitis B infection, sharing its routes of transmission and its epidemiology. Infection occurs endemically among infants and children in areas of high HBsAg prevalence and sporadically among adults exposed to blood and body fluids in low HBsAg prevalence areas.

Diagnosis

The most sensitive method of making a diagnosis of hepatitis D is demonstration of viral RNA by RT-PCR. Additionally, IgG and IgM assay for antibodies to the delta antigen should be performed.

Prevention and control

No specific immunoglobulin or vaccine exists. Control of hepatitis D depends on controlling hepatitis B. Treatment of hepatitis B reduces HBV replication and deprives the delta virus of its surface antigen: HDV RNA levels in the blood are therefore reduced in parallel with HBV DNA.

Hepatitis C

Introduction

Hepatitis C virus (HCV) is an RNA virus related to the flaviviruses and animal pestiviruses. It is a common cause of blood-borne hepatitis, most cases of which are subclinical. Its importance is that chronic infection develops in up to 85% of cases, with persistently abnormal transaminases in many, and a long-term risk of cirrhosis and hepatic carcinoma.

Virology and pathogenicity of hepatitis C infection

The existence of HCV was proven, and antigens for immunoassay were developed, when viral messenger RNA was purified from the serum of infected patients. Complementary DNA, prepared by reverse transcriptase, was inserted into the phage, lambda gt11. Bacteria into which the phage was inserted produced HCV proteins for screening against sera from cases of 'non-A–non-B hepatitis'.

Hepatitis C is a single-stranded RNA virus, with a 9.4-kb positive-sense genome, encoding a single polypeptide. Although related to flaviviruses (such as yellow fever or dengue) it is sufficiently distinct to be classified in its own group, Hepacivirus. The genome encodes a single large open reading frame, which encodes the core and envelope proteins, flanked by highly conserved untranslated regions. The virus exists in six major genotypes, which possess varying potential for causing persisting viraemia, and for responding to antiviral treatment. The genome shows considerable sequence heterogeneity due to low-fidelity replication, with only 65% homology between the major genotypes. Within the genotypes there are sub-types with 75–85% homogeneity. Virus clades in an individual patient are 91–99% homologous. The virus replicates within hepatocytes and some suggest that it may also replicate in T and B cells in chronic infection.

Innate immunity is important in limiting the initial infection. A strong antibody response develops but viral clearance and cure depends on a strong, broadly based cellular immune response. The mechanism of viral persistence is poorly understood. HCV appears to block interferon induction and also impairs the efficiency of liver dendritic cells. The genetic diversity of the virus is likely to play a role in permitting escape from effective immunity.

Epidemiology

Hepatitis C has been found in every country where it has been sought. Between 4000 and 6000 cases are confirmed in England and Wales each year. Transmission is by exposure to blood and body fluids. Blood donations are screened for HCV, so transmission occurs mainly via contaminated needles and syringes. Intravenous drug use accounts for about 75% of infections in the UK, and about 37% of drug-users are HCV-positive. HIV/HCV co-infection is common, affecting 7–15% of HIV cases in different communities. Chronic HCV infection progresses more rapidly in these patients, up to half of whom may die of end-stage liver disease.

Sexual transmission occurs much less readily than with hepatitis B. Vertical transmission from an infected mother is uncommon. The prevalence of anti-HCV antibody in serum is 1% to 2% in most developed countries. The prevalence of chronic HCV infection in England is around 0.4% and in the UK, 1 in 1400–2000 of volunteer blood donors possesses antibodies. Most of these are HCV RNA-positive. Around 20–30% of chronically infected patients develop cirrhosis within 20 years, of whom 1–4% may suffer from hepatocellular carcinoma.

Clinical features

The incubation period of blood-borne hepatitis C is 14–60 days (mean 50 days). Many cases are subclinical; typical illness with jaundice is unusual. Liver failure is extremely rare in immunocompetent individuals, but can occur rapidly in patients with agammaglobulinaemia or cell-mediated immunodeficiency.

Following acute infection, serum transaminases may fluctuate for many weeks. About one-third of patients recover, with clearance of viraemia, another third have persistent viraemia and appear well, but may develop hepatocellular disorder later; the remainder have persisting liver inflammation and are at high risk of developing cirrhosis and later carcinoma. Patients infected at a younger age seem to have more slowly progressive liver disease.

Diagnosis

Panels of HCV antigens can be detected in blood by antibody capture and antibody competition EIA. A positive test in acute infections can be confirmed by an alternative method, of which the best is viral genome detection by reverse transcriptase PCR. Commercial methods are available to quantify hepatitis C viral load and to monitor therapy. Genotyping, achieved by sequencing sections of viral RNA, is useful in predicting the response to therapy.

Treatment

Drugs used for the treatment of hepatitis C

1 Ribavirin, orally 1000–1200 mg daily (at least 11 mg/kg daily) in 2 divided doses (not effective unless used in combination with 2 or 3 below).
2 Interferon alpha, s.c. 3 MIU 3 times weekly for 48 weeks.
3 Peginterferon alpha (PEG IFN alpha) s.c. up to 1.5 μg/kg weekly for 48 weeks (peginterferon is IFN conjugated with polyethylene glycol units, to obtain a longer serum half-life, without repeated dips in serum IFN levels).

Other options: amantadine in combination with IFN, or with IFN plus ribavirin, is under investigation.

The aim of treatment is to achieve a sustained viral response (SVR) in which HCV RNA is suppressed below detectable levels, and remains undetectable at 6 months after the end of treatment. Patients who are likely to respond achieve viral suppression by 12 weeks of therapy; these patients should complete the full 48-week course, to maximize the chance of a SVR. The response to dual therapy is high with genotypes 2 and 3, so interim measurement of viral response is not always performed.

The proportion of patients who respond to treatment depends on the ability to tolerate optimum therapy, the genotype of HCV with which they are infected, and on other prognostic factors (Table 9.1).

The optimum treatment is combined ribavirin plus peginterferon. The adverse effects of peginterferon are the same as those for non-pegylated interferon. The commonest adverse effect of ribavirin is a dose-related haemolytic anaemia, often causing a drop of about 1 g/dl in haemoglobin levels during dosing. Nausea, asthenia or rare pancreatic or hepatic enzyme elevations also occur. When ribavirin is contraindicated or not tolerated, interferon monotherapy may be offered, with a lower chance of success. Patients co-infected with HIV/HCV can be treated if their cell-mediated immunity is not significantly

Table 9.1 Sustained virological responses to different treatment regimens for chronic hepatitis C

Infected population	IFN-alpha	PEG IFN-alpha	IFN-alpha plus RBV	PEG-IFN alpha plus RBV
All HCV-infected, treated patients.	3–12%	23–36%		
Genotype 1	0–6%	12–34%	33–43%	41–56%
Genotypes 2 and 3	25–36%	41–68%	58–79%	74–81%
Genotypes 4, 5 and 6	0–3%	60%	38%	50%

Higher percentage response rates are achieved in patients with low viral loads (HCV RNA <2 million copies/ml).
Other factors associated with better responses are age less than 40 years, female sex and absence of bridging fibrosis or cirrhosis on liver biopsy.

depressed, or has been reconstituted after antiretroviral therapy.

Complications
Mixed cryoglobulinaemia
About half of all cases of mixed (type 2) cryoglobulinaemia are associated with chronic HCV infection. Clinical features are proteinuria, renal vasculitis and vasculitic skin rashes. Successful treatment of the HCV infection causes regression of the cryoglobulinaemia.

Aplastic anaemia
This occasionally follows acute hepatitis C, usually in the convalescent stage. The aplasia is profound and permanent. Bone marrow transplant offers the best hope of cure. Blood and platelet transfusions should be avoided as far as possible, so that anti-haemocyte antibodies are not induced. Expert haematological advice should be sought immediately.

Prevention and control
General measures for the prevention of hepatitis B (see above) apply also to hepatitis C. Needle exchange schemes help to reduce new cases among intravenous drug users. There is no vaccine or specific immunoglobulin available. Screening of donated blood has been carried out since 1991.

Hepatitis G virus and transfusion-transmitted virus

Hepatitis G virus (HGV) is a single-stranded RNA virus of the family Flaviviridae. It is found in the serum of 1–2% of American blood donors, and up to 15% of west Africans, who have an increased likelihood of raised transaminases. To date, there is no evidence linking positive tests for HGV RNA with chronic hepatitis or cirrhosis.

Transfusion-transmitted virus (TTV), like HGV, has been identified in the blood of those receiving multiple donations of blood products. Its role as a pathogen is not yet clear.

Epstein–Barr virus

Approximately 15% of clinically recognized cases of EBV infection have clinical hepatitis. This is usually accompanied by fever, atypical lymphocytosis and positive heterophile antibody tests. The liver function tests initially show moderately elevated transaminases. Cholestasis is common, with alkaline phosphatase levels reaching 400–1000 IU/l, but with few or no clinical consequences. Liver failure is extremely rare (see Chapter 6).

Cytomegalovirus

In high-income countries 40% of adults have evidence of immunity to cytomegalovirus; this figure may reach 100% in low-income countries. Infection is almost always asymptomatic in the immunocompetent. On rare occasions fever, mild jaundice, low-grade atypical lymphocytosis and a negative heterophile antibody test are accompanied by seroconversion to cytomegalovirus. About 1800 (3%) of the 600 000 children born each year in England and Wales are congenitally infected with cytomegalovirus and about 10% of these have some permanent handicap (see Chapter 17).

Bacterial infections of the liver parenchyma

Organism list

- *Leptospira* spp.
- *Coxiella burnetii*
- *Brucella* spp.
- Mycobacteria

It can be seen from this list that bacterial hepatitis is rare. These infections are multisystem diseases in which liver disorder or jaundice is an important or predominant part of the clinical picture. They are important in the differential diagnosis of infectious liver disorders.

Leptospirosis
Introduction
Leptospiroses are zoonotic diseases caused by infection with a variety of *Leptospira* spp., often transmitted from small rodents, cattle, pigs or occasionally dogs. The organisms are excreted by their animal host and survive in water. Contaminated water infects humans. The disease is important economically among farming communities.

Epidemiology
Transmission occurs when skin or mucosa is exposed to freshwater contaminated by the urine of infected animals or by direct contact with animals. In the UK, the principal animal reservoirs are rats (*L. interrogans* var. *icterohaemorrhagiae*) and cattle (*L. hebdomadis* var. *hardjo*). Approximately 50 cases are reported each year, most of which are due to these two serotypes. Two groups are affected: those exposed occupationally (mainly farmers, vets and sewer workers), and those infected during recreational exposure.

In both groups the disease occurs mostly in men. Recreations associated with infection are swimming, canoeing, windsurfing and fishing. A number of imported cases are identified each year, especially from south east Asia and Australia, where the serotypes may be different from those commonly encountered in the UK.

People at risk of leptospirosis
1 Farmers.
2 Sewage workers.
3 Veterinarians.
4 Water-sport enthusiasts.
5 River fishermen and water bailiffs.

Microbiology
Leptospires are tightly coiled spirochaetes, usually $0.1\ \mu m \times 6$–$20\ \mu m$ with Gram-negative cell walls. They are obligate aerobes and grow slowly in simple enriched media. Originally, two species were recognized: *L. interrogans*, comprising all pathogenic strains; and *L. biflexa*, containing saprophytic strains isolated from the environment. Nowadays, a phenotypic classification, based on DNA reassociation, distinguishes 17 species. Pathogenic and non-pathogenic serovars occur within the same species and some traditional serovars have been split into more than one species. Serological typing remains the principal method of identifying isolates as pathogens. The organisms usually have a preferred mammalian host but can infect a wide range of species, including humans. Thus, *L. icterohaemorrhagiae*, the causative agent of Weil's disease, has its reservoir in the rat, and *L. hebdomadis* occurs in cattle. *L. canicola*'s preferred host is the dog, while in humans it causes canicola fever.

Clinical features
The clinical picture is a variable combination of fever, hepatitis, nephritis, meningitis and rash. In severe cases there may be a marked bleeding tendency. Mild and subclinical infection is common with all types of leptospirosis. While the full severe picture of Weil's disease is often associated with *L. icterohaemorrhagiae* infections, it can also occur with other types.

The incubation period is from 1 to 3 weeks, occasionally shorter or longer, and is followed by an illness which may evolve through two or three phases.

• First, a phase of acute infection lasting up to a week with fever, malaise and (unusually in a bacterial infection) widespread myalgia; at this stage there is bacteraemia and bacteriuria, and a polymorph leucocytosis gradually develops.

• Second, the appearance of conjunctival suffusion. This may overlap both the preceding and subsequent phase.

• Finally, the phase of immunopathology with liver, kidney and central nervous system involvement, and

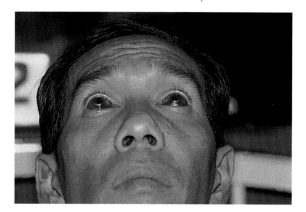

Figure 9.6 Leptospirosis: jaundice and haemorrhagic conjunctivitis in a feverish farmer. Courtesy of Dr D. Lewis.

the appearance of antibodies in the blood; fever persists throughout the illness (Fig. 9.6).

The meningitis is associated with moderate elevation of the cerebrospinal fluid protein level and lymphocyte count; the glucose level is not altered. Meningitis is more associated with canicola fever, while hepatorenal disturbance is marked in *L. icterohaemorrhagiae* infection.

Thrombocytopenia is quite common. Severe disseminated intravascular coagulation is rare, and often associated with widespread vasculitis, which is occasionally very severe, causing bleeding, thromboses, multiple necrotic lesions, loss of digits or loss of limbs. An increasingly recognized feature of severe disease is widespread nodular pneumonitis.

Many patients are thought to have viral hepatitis or meningitis at presentation. Warning signs are the association of red eyes and fever with hepatocellular liver disturbance, or of hepatocellular disorder with meningitis. A maculopapular rash can be a helpful sign. Most cases have significant proteinuria and many have microscopic or macroscopic haematuria with or without casts. Renal failure can develop suddenly in untreated cases.

Important clinical features of leptospirosis
1 Early myalgia.
2 Hepatitis with fever.
3 Renal impairment.
4 Lymphocytic meningitis.
5 Conjunctivitis.
6 Rash, sometimes haemorrhagic.
7 Thrombocytopenia.
8 Blood, protein and/or bilirubin in the urine.
9 Rare, nodular pneumonitis.

There is a 5–10% mortality in severe cases. Fatalities are due to cardiovascular collapse, often with haemorrhagic myocarditis, and usually associated with vasculitis, renal failure and haemorrhage.

Diagnosis

The diagnosis can be suspected on the basis of epidemiological and clinical features. A useful pointer in flu-like presentations such as fever and myalgia, is the finding of protein, blood and/or bilirubin on urine testing.

Laboratory diagnosis

Leptospira can be visualized in the blood of patients who present during the bacteraemic phase, but false-positive results are common. Freshly voided urine can be examined by dark-ground microscopy but this must be done quickly, as the acidic environment kills the organisms.

Leptospira can be cultivated from the blood during the first week of illness and from the urine later in the course. The diagnostic yield is low, as shedding may be intermittent. A specialized medium such as the semisolid agar of Ellinghausen and McCulloch is required for successful culture. *Leptospira* are microaerophilic and grow beneath the surface of the medium. Suspensions of cultures are examined by dark-ground microscopy to confirm the presence of the organisms, which are then identified by slide agglutination, using panels of antisera. Organisms can also be typed by pulsed-field gel electrophoresis.

Diagnosis is usually made by serology. In the *Leptospira* agglutination technique, *Leptospira* of various serogroups (including groups of closely related serovars) are exposed to the patients' serum, to determine which serovars are agglutinated. More recently, an enzyme-linked immunosorbent assay (ELISA) technique has been described, which can confirm the diagnosis by demonstrating specific IgG and IgM using a single serum specimen.

DNA amplification methods are useful in confirming the diagnosis in patients with early severe disease, before antibodies are produced.

Diagnostic methods for leptospirosis

1 Dark-ground microscopy of blood or urine.
2 Cultures of blood or urine.
3 Serum *Leptospira* agglutination tests.
4 Serum immunoglobulin M enzyme-linked immunosorbent assay.
5 DNA amplification.

Treatment

Many cases recover in 10–14 days. Even after the bacteraemic phase, antibiotic treatment may modify the course of severe leptospirosis. The treatment of choice is benzylpenicillin. The organisms are also sensitive to tetracycline, sulphonamides and erythromycin, but the effectiveness of cephalosporins is unpredictable.

Treatment of leptospirosis

Benzylpenicillin i.v. 1.2–2.4 g 6-hourly *or* oxytetracycline orally 500 mg 6-hourly, *or* doxycycline orally 100 mg 12-hourly (treat for 10 days).

Complications

Recovery from leptospirosis is usually complete. Some 1–2% of cases develop late uveitis some weeks after recovery. This can be treated with standard ophthalmological anti-inflammatory preparations.

Prevention and control

No effective human vaccine is available in the UK. For short periods of high risk, especially through occupational exposure, prophylaxis with doxycycline (200 mg weekly) may be effective.

The most important preventive measure is the use of appropriate protective clothing during occupational and recreational exposure and avoidance of skin exposure in water that might be contaminated with animal urine. Cuts and abrasions should be covered with waterproof dressings. Rodent control measures may further reduce the risk of infection on farms and in recreational areas. Vaccines are available for use in animals, and provide some protection against common serotypes.

Granulomatous hepatitis

Granulomatous hepatitis may be caused by a number of infections, most of which are chronic or subacute. They have to be distinguished from non-infectious causes of parenchymal granuloma, including sarcoidosis, vasculitides, and occasionally drugs such as quinine.

Organism list

- *Coxiella burnetii*
- *Brucella* spp.
- *Mycobacterium* spp.
- *Histoplasma capsulatum*

Chronic Q fever (*Coxiella burnetii*)

Chronic Q fever can follow acute Q fever, or can commence without previous illness. It usually presents with swinging fever, abnormal liver function tests with a cholestatic picture, and moderate jaundice. Half of patients with chronic Q fever also have *Coxiella burnetii* endocarditis (see Chapter 12).

Acute Q fever is a zoonosis that occurs in those exposed to sheep and sheep products. It often presents as atypical pneumonia (see Chapter 7).

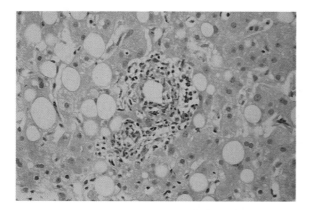

Figure 9.7 Q fever: liver biopsy showing small granulomata with a peripheral zone of scattered eosinophils.

Chronic Q fever can be confirmed by demonstrating phase I antibodies to *C. burnetii*. Liver biopsy shows a characteristic histological picture, with many small, non-caseating granulomata, which have a peripheral zone of eosinophils (Fig. 9.7).

Tetracycline is the treatment of choice. Two or three weeks' dosage is sufficient in uncomplicated cases. If endocarditis is present, long-term treatment is advisable (see Chapter 12).

Brucellosis

Brucellosis is a zoonosis. It is a multisystem disease often with low-grade bacteraemia. There is probably some granulomatous change in the liver of most infected patients. Liver disorder is rarely the main feature of brucellosis, but liver granulomata may be discovered during the investigation of pyrexia of unknown origin and should prompt consideration of the diagnosis.

In acute brucellosis blood cultures are often positive, though prolonged culture may be required. Serodiagnostic tests can demonstrate IgG or IgM antibodies to *Brucella* spp.

The treatment of choice is 6 weeks' treatment with tetracycline (doxycycline is useful in a once-daily dose, but some experts favour twice-daily demeclocycline). The addition of rifampicin or gentamicin is recommended to ensure a bactericidal effect (see Chapter 25).

Tuberculous granulomata of the liver

Granulomata are always present in miliary tuberculosis. They are often small and non-caseating, and must be distinguished from the granulomata of sarcoidosis, autoimmune diseases, brucellosis and rare fungal infections. Diagnosis is aided by ensuring that biopsy material is cul-

tured for mycobacteria, fungi and pyogenic organisms. A chest X-ray may show typical features of tuberculosis or sarcoidosis.

Serum should also be obtained to permit diagnosis of brucellosis, yersiniosis or autoantibody-positive conditions. The serum angiotensin-converting enzyme level may be elevated if pulmonary sarcoidosis coexists with liver disease, as is often the case.

Abscesses of the liver

Introduction

Liver abscesses are a relatively common cause of pyrexia of unknown origin, and sometimes present without localizing signs. Before the advent of accurate imaging techniques, half of all abscesses were diagnosed at postmortem examination. Nowadays they can be readily demonstrated, and modern antibiotics permit effective treatment.

Pyogenic liver abscess

Organism list
- *Escherichia coli*
- *Klebsiella* spp.
- *Serratia* spp.
- Other Enterobacteriaceae
- Enterococci
- *Streptococcus* 'milleri'
- *Staphylococcus aureus*
- *Bacteroides* spp.
- Anaerobic streptococci
- Other bowel anaerobes

Many abscesses contain a mixture of organisms. Pyogenic abscesses may be a result of bacteraemia (when other abscesses may also exist, for example, in the brain). They may originate from portal system bacteraemia, from abdominal sepsis (including appendicitis or perforated peptic ulcer), or they may be apparently spontaneous. They vary from single large lesions to multiple moderate or small ones, and may even be widespread and microscopic.

Clinical features
Patients usually present with high, swinging fever, marked neutrophilia and elevated C-reactive protein (CRP). Large abscesses may produce tender enlargement of the liver, or even fluctuant lesions pointing between the ribs. Even without abdominal signs, inflammation below the diaphragm sometimes produces a small right-sided pleural effusion (Fig. 9.8a).

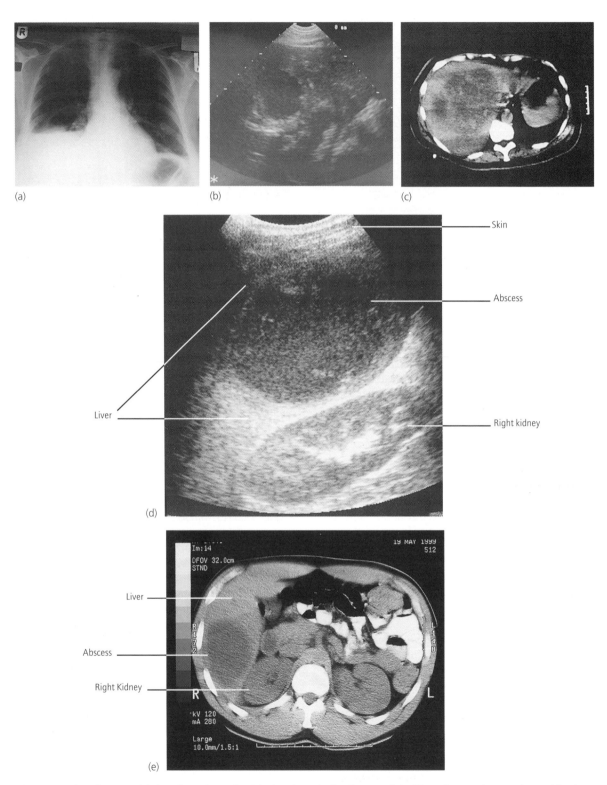

Figure 9.8 Liver abscesses: (a) pleural reaction at the right lung base; in the absence of significant liver tenderness, the combination of this opacity with neutrophilia and a raised erythrocyte sedimentation rate can lead to the erroneous diagnosis of atypical pneumonia; (b) ultrasound; and (c) computed tomographic scan, showing the large, multilocular, pyogenic abscess causing the changes in (a); (d) ultrasound and (e) CT scans demonstrating an amoebic abscess; in contrast to the pyogenic abscess, this is unilocular and relatively smooth-walled.

The liver function tests may be abnormal, showing a moderate elevation of alkaline phosphatase. Clinical jaundice is rare.

Diagnosis

The investigation of choice is imaging (Fig. 9.8b,c). Ultrasound examination will demonstrate most large or moderate-sized lesions. Small lesions, or those whose sonic density is near to that of liver, are best demonstrated by computed tomographic scan. Gallium or magnetic resonance scanning will show moderate to small lesions. Liver biopsy may be necessary to demonstrate multiple small or microscopic lesions.

Microbiological information is obtained from aspiration of the abscess and culture of pus. When the liver lesions complicate bacteraemia, blood cultures are often positive. In bacteraemic cases caused by *Streptococcus 'milleri'*, the likelihood of abscess formation in the brain must be remembered. Patients who have visited endemic tropical areas may have amoebic liver abscesses (see below). Immunocompromised patients may have abscesses caused by unusual organisms, e.g. *Nocardia* spp. (see Chapter 22). DNA amplification techniques using 16S and 18S rRNA genes as targets are often successful in identifying causative organisms when culture is insufficiently sensitive (see Chapter 3).

Management

Empirical treatment should be commenced as soon as blood cultures and pus have been obtained. It should be effective against the likely range of causative organisms, and can be adjusted later in the light of cultural evidence.

A third-generation cephalosporin or broad-spectrum penicillin will be effective against Enterobacteriaceae. The addition of an aminoglycoside can provide a synergistic action against enterococci. Metronidazole may be added to account for anaerobic organisms. If blood-borne infection is suspected, it may be worth including an antistaphylococcal drug such as flucloxacillin in the initial treatment regimen.

Drainage of abscesses

Drainage of pyogenic abscesses is advisable for moderate or large lesions. Diagnostic aspiration may be sufficient if it is followed by a good response to antimicrobial therapy. Repeated image-guided needle aspiration may speed the healing of larger lesions. Persistent re-accumulation of pus requires the insertion of an indwelling drainage catheter and prolongation of chemotherapy, until satisfactory resolution is demonstrated by temperature response and imaging. Some abscesses are inaccessible to needle aspiration, and require surgical evacuation.

> **Empirical treatment for pyogenic liver abscess**
> Co-amoxiclav i.v. 1 g 6-hourly *or* ceftriaxone i.v. 1–2 g 12-hourly; *plus* gentamicin i.v. or i.m. 2–5 mg/kg daily in three divided doses (providing peak blood levels up to 10 mg/l, trough levels less than 2 mg/l); *plus* metronidazole i.v. or rectally 500 mg 8-hourly.
> *Alternative*: meropenem 500 mg–1.0 g 8-hourly.

An adequate response to treatment is indicated by reduction in size of the lesion or lesions, reduction in fever and a falling CRP.

Complications

Rupture of a liver abscess into the peritoneal or pleural cavity causes a sudden deterioration in the patient's condition, and may lead to endotoxaemia and collapse. Early treatment should prevent this by controlling the infection and draining abscesses at risk.

Amoebic liver abscess

Introduction

Exposure in endemic areas may lead to amoebic abscess, the commonest complication of *Entamoeba histolytica* infection of the bowel (see Chapter 8). Amoebic abscesses tend to be large and single, usually in the right lobe of the liver (see Fig. 9.8d,e).

Diagnosis

The presentation is identical to that of pyogenic liver abscess. Over half of patients have no bowel symptoms at any time. Diagnostic aspiration is useful, as amoebic abscesses contain brownish-pink, thick (anchovy-sauce) pus rather than greenish or yellow-grey pyogenic pus. Amoebae are rarely seen in the pus, as they are fixed in the wall of the abscess. Bacterial culture should also be carried out, as pyogenic and amoebic infections occasionally coexist.

Definitive diagnosis is confirmed by serology. Patients with current or previous systemic amoebiasis have positive fluorescent antibody tests (FAT). Residents of Western countries are unlikely to have positive FATs from previous amoebiasis. Residents of endemic areas may have pre-existing FAT responses, but a gel precipitin test is also positive in acute systemic amoebiasis, and quickly becomes negative when the condition is cured. A positive gel precipitin test is strongly suggestive of amoebic abscess. Where facilities exist, DNA amplification can be used to identify the presence of *Entamoeba*.

Treatment

The treatment of choice is metronidazole 500 mg intravenously 8-hourly for 10 days. Many small and moderate-sized abscesses will heal with medical treatment alone. A few large lesions, or ones that continue to enlarge despite treatment, may require aspiration or drainage.

It is unusual for metronidazole to fail, but the addition of chloroquine 600 mg daily may induce a response. Chloroquine readily penetrates the abscess wall and has an additive effect to that of metronidazole. If the abscess still fails to resolve, the possibility of coexisting bacterial infection must be considered, and broad-spectrum antibiotic treatment may be indicated.

Complications

Rupture of an amoebic abscess

Rupture of an amoebic abscess may release pus into the peritoneum, the pleura or the pericardium, causing severe inflammation, shock and collapse. Secondary abscesses themselves may require drainage. Rupture is unusual in right lobe abscesses, which are accessible to percutaneous drainage. Abscesses in the left lobe of the liver are more difficult to approach. Surgical drainage should be considered if medical treatment and/or needle aspiration are unsuccessful.

Secondary pyogenic infection

Secondary pyogenic infection is uncommon, but well recognized. It should be considered in all cases of poor response to metronidazole treatment. Serology usually suggests acute amoebic infection, but the pus may not have the typical appearance. Gram-negative rods, enterococci and/or anaerobes may be recovered on culture. Presumptive treatment should be commenced if there is doubt about the aetiology of the abscess (see above). Metronidazole must be given to cover both amoebae and anaerobes.

Prevention and control

See Chapter 8.

Cholangitis

Introduction and epidemiology

Cholangitis is infection of the biliary tree, often associated with surgical disorders of the gallbladder or bile ducts. Predisposing conditions include gallstones, chronic cholecystitis, benign and malignant strictures of the bile duct or of larger hepatic ducts and, occasionally, pancreatitis. Cholangitis can complicate instrumentation of the common bile duct, for instance endoscopic retrograde cholangiopancreatography or choledochojejunostomy, for which antibiotic prophylaxis should always be given. Biliary stents are also an important predisposition to bacterial cholangitis.

Pathology

The causative organisms colonize the biliary tree by ascending from the duodenum. *Enterococcus*, *Escherichia coli*, *Klebsiella pneumoniae* and *Serratia* spp. are common pathogens. Anaerobes, notably *Bacteroides fragilis*, are also found. *Clostridium perfringens* is an occasional cause, which may produce gas in the biliary tree. Debilitated patients and those who have received broad-spectrum antibiotics occasionally develop candidal cholangitis.

Clinical features

The common presentation is with right hypochondrial pain, fever, features of sepsis, and a variable degree of jaundice. Laboratory tests show a neutrophil leucocytosis, a raised alkaline phosphatase level and a raised bilirubin.

Many patients have a past history suggestive of biliary or pancreatic disease.

Diagnosis

Contrast imaging of the biliary tree is difficult when the biliary pressure is raised by infection, but ultrasound, computed tomographic or magnetic resonance imaging may show gallstones and/or a dilated duct system.

Blood cultures should be obtained, but are not always positive. Bile may be obtained by cannulation of the bile duct or image-directed percutaneous aspiration from dilated intrahepatic ducts. Bile duct cannulation may carry a risk of introducing further organisms, but this should be weighed against the therapeutic advantage of draining the infected bile.

Management

Initial treatment is usually empirical. Gentamicin or other aminoglycoside, or a broad-spectrum cephalosporin, is appropriate for coliform infections. Ampicillin or amoxicillin, plus an aminoglycoside, will provide effective treatment for enterococcal infection. Metronidazole is the treatment of choice for anaerobes. Meropenem provides broad-spectrum action with an antianaerobe effect. The antibiotic regimen can be modified if the response is unsatisfactory or the bacteriological results suggest a need for change.

Surgery or drainage is indicated if purulent bile is trapped or loculated. An early surgical opinion is of value, as cholecystectomy is curative, and avoids the risk of recurrences.

Complications

The most important of these are Gram-negative bacteraemia, shock, abscess formation and erosion of the gallbladder. Clinical vigilance must be maintained, and surgical assistance should be sought if improvement is not maintained.

Colonization of the biliary tree

A patient with one or more strictures in the biliary tree can acquire colonization of the bile without symptoms of infection. At varying intervals there are transient bacteraemias, with sudden fever, rigors and then sweating and defervescence. The episodes may last anything from 15 min to several hours.

It is often difficult to obtain blood cultures at the correct time for a positive result. Some patients become debilitated and lose weight because of frequent fevers, some have clinical jaundice or laboratory signs of cholestasis. The diagnosis depends on suspicion, and imaging should be performed to identify the cause of the condition. Single strictures may be amenable to correction. Occasionally, a condition such as sclerosing cholangitis is responsible. There is then a choice of treating bacteraemias as they arise, or of attempting chemoprophylaxis. Culture of the colonized bile will identify the organism and its sensitivities.

Parasites of the liver

Organism list

- Schistosome species (*Schistosoma mansoni* is most important)
- Hydatids, especially *Echinococcus granulosus*
- *Fasciola hepatica* and other flukes

Schistosomiasis

Introduction

Schistosomes are trematodes that derive their name from the large genital cleft in the body of the male, in which the longer, slimmer female lies throughout its adult life. The adult forms live in the visceral veins: *Schistosoma mansoni* in the veins of the small bowel and the portal veins of the liver, *S. japonicum* in the veins of the small bowel and *S. haematobium* in the vesical plexus. The paired schistosomes produce many fertile ova, some of which lodge in the tissues, causing an intense inflammatory response, and some of which penetrate the mucosae and are released in faeces or urine. They 'hatch' on contact with freshwater.

Free ova release motile miracidia, which invade freshwater snails. They are eventually released from the snails as cercariae, which use their motile tails to swim in the water. The cercariae die in a few hours if they are unable to attach to a host's skin and penetrate it, when they shed their tails and become invasive schistosomules. These migrate through the tissues causing an inflammatory reaction and

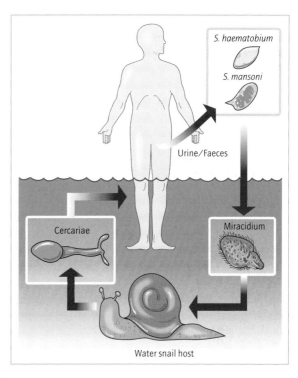

Figure 9.9 Schistosomiasis: the life cycle of the parasite.

eosinophilia, until they mature and settle as mating pairs in their favoured venous site, producing up to 2000 eggs per day (Fig. 9.9).

Many eggs are excreted from the gut or bladder to complete the lifecycle. Some pass through the portal veins, to cause liver inflammation leading to fibrosis and portal hypertension. In the lungs (usually in *S. japonicum* infections) they can cause lung fibrosis and cor pulmonale. Ectopic eggs in the spinal cord or central nervous system occasionally cause transverse myelitis or epilepsy. Eggs in the bladder wall can result in epithelial metaplasia and bladder wall fibrosis. Chronic heavy infection is associated with bladder malignancy.

Epidemiology

It is estimated that 200 million people are infected worldwide. Transmission occurs wherever there is exposure to water inhabited by infected snails, for instance in rice-growing areas or where freshwater rivers are used for washing. The peak prevalence of infection is in children aged 5–15 years. Transmission is greatest during the rainy season.

The reservoir of infection for *S. haematobium* and *S. mansoni* is humans, although domestic animals are also a reservoir for *S. japonicum*. The geographical distribution of disease follows that of the infected snails, which are species-specific. The vector snail for *S. haematobium* occurs throughout the Middle East and Africa, particularly along the Nile valley. *S. mansoni* has a wider distribution in Africa, the Middle East, Brazil, Surinam and the West Indies. *S. japonicum* occurs in the Far East and South-east Asia.

Two geographic patterns of disease are seen: focal (associated with small water sources) and non-focal (associated with large marshy areas, mainly in South-east Asia). In some areas the disease incidence has recently increased, due to the construction of reservoirs.

Clinical features and diagnosis

The severity of clinical disease depends on the numbers of invading parasites. Slight infections are rarely symptomatic. More intense invasion may cause a sequence of symptoms and signs, while massive parasite loads can cause severe or fatal disease. Heavy or repeated infections are an important cause of chronic ill health in endemic areas.

There are several stages of schistosomal infection, as detailed below.

Swimmer's itch

Swimmer's itch is an irritation, sometimes with a rash, caused by cercarial invasion of the skin. It begins within a few hours of exposure and clears up in 24–48 h.

Invasive stage

The invasive stage begins after about 24 h as schistosomules migrate through veins and lymphatics to the lungs and viscera. The patient may complain of dry cough and abdominal discomfort. The spleen may be palpable, and eosinophilia is common.

Toxaemic phase

A toxaemic phase, often called Katayama fever, follows significant infection after about 15–20 days. Clinical features are fever, lymphadenopathy, splenomegaly, diarrhoea and eosinophilia. Severe cases can be life-threatening, especially in non-immune travellers.

Acute disease

Acute bowel disease begins after 6–8 weeks as ova are deposited in the bowel wall. The main effects are dysentery-like diarrhoea and weight loss, which sometimes persist for several months.

Chronic disease

The bowel may show fibrotic, polypoid or granulomatous lesions. The liver may become firm and large. Severe cases develop portal fibrosis and hypertension, with splenomegaly, hypersplenism and risk of variceal bleeding (hepatocellular function is usually well preserved).

Rare, severe effects include nephrosis, and granulomatous transverse myelitis following ovum deposition in the spinal cord.

S. japonicum tends to cause more intense invasion and more widespread disease. Chronic features include cardiac involvement with cor pulmonale and alterations in the electrocardiogram. In all types of schistosomiasis, adventitious ova may be deposited in many tissues, including the brain, but overt tissue damage is rare.

Diagnosis

Only clinical diagnosis is possible up to, and sometimes for 2 or 3 weeks after, the stage of Katayama fever. From then on, eggs are increasingly produced, and antibodies to soluble egg antigen begin to appear, permitting serodiagnosis. Alternatively, detection of circulating 'cathodic' or 'anodic' antigen can be detected.

Ova appear in the stool from 6–8 weeks after infection. If ova are scanty and not detected in stool, they can often be seen in rectal snips (fragments of mucosa teased off the rectal wall with a needle or biopsy forceps, and squashed on a microscope slide).

Management

Praziquantel is the treatment of choice. It is given as a single dose of 40 mg/kg, preferably after an evening meal, to minimize abdominal discomfort or vomiting. For *S. japonicum* 60 mg/kg is given in two divided doses on the same day. Treatment must be given without delay in suspected neurological cases.

Treatment is not fully effective unless given *after* the deposition of ova has begun. Katayama fever and neurological disease respond poorly. In these cases, treatment with step-down dosage of prednisolone, starting with 40–60 mg/day, may be helpful in limiting inflammation and tissue damage.

The faeces can be examined 1 month after treatment, when ova will still be seen, but should by then all be dead. Death is demonstrated by diluting the stool with distilled water. This stimulates any live ova to hatch rapidly, releasing active miracidia.

Prevention and control

Prevention in infected areas depends on provision of safe water and sanitation for communities, and education on avoiding contact with potentially infected water. Snail control is expensive and only practical in small focal areas of infection. An alternative approach is to reduce egg excretion from the human reservoir, using chemotherapy. In high-prevalence areas, it may be necessary to treat the whole population, whereas in low-prevalence areas it is more cost-effective to treat children (who are high egg excretors) and clinically affected adults. Tourists to infected areas should be advised of the risks of swimming in fresh-water.

Hydatid disease

Introduction and epidemiology

Hydatid disease is caused by the cystic phase of the small tapeworm *Echinococcus granulosus*, whose primary host is the dog (particularly sheepdogs), and whose cysts affect sheep. The rarer *E. multilocularis* tapeworm affects foxes (but can infect dogs), and cysts are found in small rodents on which the foxes predate.

E. granulosus is the commonest cause of human disease. In the natural cycle, infected dogs excrete tapeworm eggs, contaminating fields, where they are consumed by their natural secondary host, the sheep. The eggs migrate from the gut to the liver and occasionally the lung or other organs, where they develop into cysts containing several tapeworm scolices. If the cysts are eaten by a dog, the scolices are released and attach to the bowel wall, where they develop into mature tapeworms.

Human disease occurs when tapeworm ova are ingested by humans, often as a result of close contact with an infected working or pet dog. Cysts then develop in the human host, and may enlarge for many years before becoming clinically apparent.

Clinical features

E. granulosus infection usually causes a solitary cyst in the liver, or occasionally the lung. Eventually, the enlarging cyst causes abdominal pain or swelling, biliary obstruction, or respiratory embarrassment. Lung cysts can cause partial bronchial obstruction, with repeated chest infections. Rupture of cysts is rare, and often related to attempted surgery. Eosinophilia is not seen, unless leakage of cyst contents causes an allergic reaction.

E. multilocularis is more invasive than *E. granulosus*. Its cyst wall cells can invade locally, and also metastasize to distant sites, including the brain.

Diagnosis

Cysts of *E. granulosus* may be visible as thin-walled, fluid-filled structures in plain radiographs, especially of the lungs, and on liver imaging. Computed tomographic scans may reveal daughter cysts, and on magnetic resonance scans magnified images may demonstrate scolices within them.

ELISA tests, used to detect cyst-associated hydatid antibodies, are highly sensitive. Antigen detection tests can be useful in making an acute diagnosis and, importantly, in monitoring the response to chemotherapy.

Management

Albendazole is effective in treating *E. granulosus*. It is given in a dose of 400 mg twice daily for 4 weeks, repeated up to four times, with intervals of 1 or 2 weeks between courses. Liver and renal function should be checked during treatment. As the cysts die, they may become painful and inflamed for a time. Analgesics are then helpful. Corticosteroids are not usually required.

E. multilocularis does not respond fully to albendazole; lifelong suppressive therapy is often necessary.

Surgery without prior chemotherapy is undesirable, as rupture of the cyst may cause an anaphylactic reaction, and can also seed the tissues with daughter cysts (Fig. 9.10). A common approach is to prepare patients for operation with albendazole for at least 2 weeks. Praziquantel is given preoperatively and for 2 weeks following to act against the protoscolices and reduce the risk of secondary seeding.

It is impractical to excise multiple cysts, or some of those in the brain. Albendazole offers a good chance of disease regression in such cases, and courses may be alternated

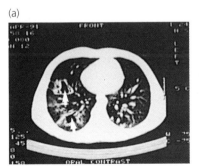
(a)

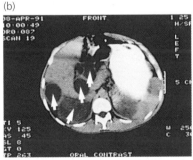
(b)

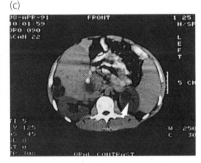
(c)

Figure 9.10 Hydatid disease: computed tomographic body scans showing: (a) two thin-walled cysts in the right lung (arrows); (b) several cysts in the liver (arrows); and (c) a large cyst in the spleen (arrow). The patient's abdominal pain and the size of the cysts improved dramatically with albendazole therapy.

with praziquantel 20 mg/kg daily for 1 week. Recent trials suggest that aspiration of cysts can safely be performed after chemotherapy, avoiding the need for major surgery in frail patients or those with multiple cysts.

Liver flukes

Introduction

Liver flukes are trematodes whose adult forms inhabit the intrahepatic bile ducts. Their lifecycle includes multiple hosts. *Fasciola hepatica* is the only liver fluke endemic in the UK. Its second host is a freshwater snail. Cercariae migrate on leaving the snail's body, and become attached to vegetation, existing as non-motile metacercariae until eaten by humans or sheep. Wild watercress is the commonest source of human infection. In oriental countries *Clonorchis (Opisthorchis) sinensis* is common. This fluke affects humans, dogs, pigs, cats and even rodents. Its next host is the snail; cercariae then invade fish (especially freshwater carp) or shrimps, which are again eaten by humans and other animals.

Ingested metacercariae excyst in the duodenum, penetrate the bowel wall, cross the peritoneum and eventually migrate through the capsule and parenchyma of the liver to reach the bile ducts.

Clinical features

Acute

The acute presentation is with fever, abdominal pain, liver tenderness and eosinophilia, caused by parasite migration and liver invasion. Symptoms vary in severity, depending on the parasite load, and can last for 3 or 4 months. Episodes of obstructive jaundice may occur as flukes invade the biliary tree.

Chronic

Chronic fascioliasis reflects persisting irritation of bile ducts. Upper quadrant pains, fevers, cholangitis and persisting eosinophilia may occur. Localized stasis and infection can produce liver abscesses. Over a period of years the flukes die, being passed in the faeces or becoming calcified in the liver.

Diagnosis

Adult flukes release eggs, which are passed in the faeces. Eggs are often scanty and are best demonstrated by concentration techniques. Serological tests are positive early in *Fasciola* infections, but are not helpful in early *Clonorchis* cases.

Management

The treatment of choice is praziquantel 20 mg/kg daily for 3 days, preferably taken at night after food to avoid abdominal discomfort and nausea in waking hours. Treatment may need repeating after an interval if egg excretion continues.

Prevention and control

Fascioliasis can be prevented by avoiding consumption of wild watercress, especially if grown near land grazed by sheep. The use of sheep faeces for fertilization of water plants should be avoided. Snail control is technically feasible, but not economically justified.

Case study 9.1: Hepatitis B outbreak in an alternative medicine clinic

History and diagnosis
A case of acute hepatitis B was notified in a middle-aged woman whose only risk factor was having received injections at an alternative medicine clinic 4 months previously. The patient's daughter, who had visited the clinic at the same time as her mother for treatment of acne, also developed acute hepatitis B.

Question
How could the consultant in communicable disease control determine whether transmission of hepatitis had occurred in the clinic?

Further enquiries and investigation
The clinic offered a variety of alternative therapies, including autohaemotherapy, in which patients receive an injection of their own blood after it has been mixed with saline solution.

Several possible routes of transmission were identified at the clinic, the most likely being the use of multidose bottles, into which the needles of successive syringes were introduced, for several of the procedures. The immediate control measures were to prevent the use of multidose bottles in the clinic, to give advice on sharps and clinical waste disposal, and to arrange for all clinic staff to be tested for hepatitis B virus (HBV) and HCV.

Question
How can it be determined whether other patients have been at risk?

Further investigation
A 'look-back' investigation was conducted (Fig. CS.3), in which all patients were identified and followed up who had attended the clinic in the 6 months before the first case developed symptoms (i.e. the maximum incubation period for hepatitis B). The look-back period was extended to 1 year when an earlier case of hepatitis B was identified in a clinic attender.

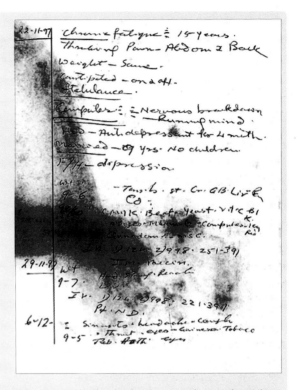

Figure CS.3 Blood-stained patient records at an alternative therapy centre associated with an outbreak of hepatitis B; a look-back exercise identified 31 cases and one HBe Ag-positive staff member.

Two of the practitioners in the clinic had evidence of HBV infection and one was HBeAg positive, i.e. highly infectious. Both were required to cease practice by the Health and Safety Executive.

The look-back exercise traced 353 patients who had received autohaemotherapy, of whom 33 subsequently developed acute hepatitis B. Thirty were infected with the same virus strain, suggesting a common exposure at the clinic. This is the largest reported iatrogenic outbreak of hepatitis B in the UK.

Infections of the Urinary Tract 10

Introduction

Urinary tract infections are common in most communities. Both the collecting system and the kidneys themselves may become infected.

The collecting system is usually bacteriologically sterile, but the urethra opens to the exterior at the perineum, a site rich in potentially pathogenic flora. Protection from ascending infection is derived from: the regular flow of urine, diluting and expelling pathogens; the mucosal defence mechanisms of the urinary tract; the antibacterial properties of the urine itself; and the integrity of the sphincters separating the urethra from the bladder and upper tract. Disturbance of any of these mechanisms predisposes to ascending infection. Examples include the following.

1 Mechanical abnormalities of the urinary tract, which may cause stagnation of urine, with bacterial replication in residual pools of urine.

2 Disruption of the urothelium (e.g. by polyps), allowing the establishment of epithelial infection.

3 Abnormal chemical constituents in urine, e.g. glucose, which may encourage the survival and replication of organisms.

4 Foreign bodies, such as stones, tumours, schistosome eggs and associated granulomata, which may become colonized with pathogens and act as a reservoir of infection.

5 Loss of sphincter function (including that due to indwelling catheters), which destroys an important barrier to ascending colonization of the bladder.

6 Pregnancy causes dilatation of the collecting system, reduced motility and a large volume of stagnant urine, predisposing to ascending infection.

The kidneys have a complex glandular structure of tubules and blood vessels. They receive about one-third of the cardiac output and are therefore susceptible to blood-borne pathogens, as well as to ascending infections. Previously healthy kidneys may show only a temporary decline in excretory function during bacteraemic disease, but localized infection in the form of renal abscesses can also occur. This is more likely if the kidney has a pre-existing abnormality such as cysts, scarring or infarcts.

Pathogenesis of urinary tract infections

Uropathic strains of *Escherichia coli* possess pili (fimbriae:

see also Chapter 2). Type I, mannose-sensitive pili appear to be important in strains colonizing the bladder. Type P pili favour colonization of the kidney. These pili are encoded by *pap* genes (pyelonephritis-associated pili). Although P pili are divided into many antigenic types they all bind to α-D-Gal-(1,4)-β-D-Gal (globobiose). Expression of the *pap* product changes in response to various stimuli, including temperature and glucose concentration.

Uropathogenic strains also adhere by non-fimbrial mechanisms. Examples include afimbrial adhesin 1 (AFA 1), and the Dr adhesin. Both of these recognize the Dr blood group antigen.

Bacteria sometimes undergo phase variation after attachment to the host's epithelium, changing their expressed antigens to those less readily recognized by phagocytes.

Endotoxin-mediated inflammation (initiated by bacterial lipopolysaccharide) appears to act synergistically with P pili. There is also evidence that uropathogenic strains produce exotoxins, designated RTX haemolysins.

Symptoms

These are classically described as symptoms of urethral irritation, bladder irritation or upper tract symptoms originating in the pelvicaliceal system and/or the kidney. Obstruction or severe inflammation of a ureter may also produce typical, colicky ureteric pain, felt in the flank and radiating to the lower abdomen and perineum.

Urethral symptoms
Urethral symptoms are burning or stinging at the meatus and in the perineum, with a continuous desire to micturate, causing marked frequency. There is often severe dysuria.

Bladder symptoms
Bladder symptoms include suprapubic aching, often relieved by micturition. Suprapubic tenderness is common. Urethral symptoms often coexist.

Upper tract symptoms
Pyelonephritis is represented by loin pain, with tenderness in the renal angle on the affected side. Rigors are common. An obstructed or hydronephrotic kidney may produce a tender palpable or ballottable mass in the upper abdomen.

The patient's symptoms correlate poorly with the true site and extent of infection. Any or all of the classical symptoms may be present in infections of various parts of the urinary tract, and many urinary infections occur without specific symptoms. Spurious symptoms are common, particularly in children. These include failure to thrive, episodic nausea and vomiting, and symptomless

fever. Without a high index of suspicion, persisting infection can cause impaired health or, in infants, long-term renal damage. Acute urinary infections may present with watery diarrhoea, spurious meningism or febrile convulsions. Urinary infection can exist in the absence of fever or significant neutrophilia.

Non-specific symptoms of urinary tract infections
1 Failure to thrive.
2 Episodic nausea.
3 Diarrhoea.
4 Meningism.
5 Pyrexia of unknown origin.
6 Febrile convulsions.

Diagnosis

The urinary tract above the urethra is normally bacteriologically sterile, but the urethral meatus and surrounding perineum are colonized with a mixture of skin and bowel flora. In women vaginal flora or pathogens may contaminate the urethra.

Three-glass test

In the three-glass test, urination is commenced into the first container (voiding bottle 1; VB1); when the flow is established, the mid-stream urine (MSU) is directed into the second container; the last millilitres of the urine flow (terminal urine) are passed into the third container, making an effort to empty the bladder and urethra.

The first urine passed contains debris, cells and organisms from the urethra and often contains strands of mucus if urethritis is present. The MSU contains bladder urine. A normal specimen appears clear and transparent, while cloudiness indicates the presence of cells, bacteria or crystals. The final sample contains matter from the clefts of the trigone, from the prostate, or from glands adjacent to the pelvic urethra. Mucus strands are often seen in cases with prostatitis. Schistosome ova from the bladder wall may be recovered from terminal urine (Fig. 10.1).

Other useful investigations

A number of screening techniques are available for the rapid detection of urinary tract infection. These techniques usually use 'dip-strip' tests. Glucose can usually be detected in overnight fasting urine in normal subjects, but is undetectable in urinary tract infection, when glucose is metabolized by bacteria. In urothelial inflammation, blood cells and protein may be released, and are detectable by dip-strip tests for blood and protein. Many uri-

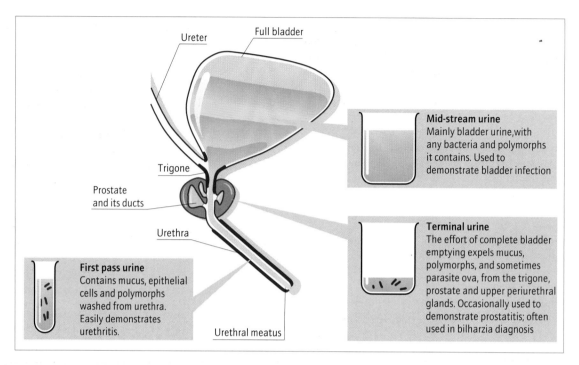

Figure 10.1 The three-glass procedure for examining urine.

nary pathogens catalyse the reduction of nitrate to nitrite, which can also be detected by a dip test.

Large numbers of white cells may indicate urinary tract infection: but white cells are a normal part of the urine. Disease is indicated only when they are present in an excess ($>10/mm^3$). Excess white cells may appear in urine as a result of fever, exercise or contamination from a vaginal discharge. Absence of white cells does not exclude infection as they rapidly lyse in acid urine. Excess white cells in urine can be indicated by the presence of leucocyte esterase activity. Positive dip-test results for nitrite and leucocyte esterase can provide a rapid presumptive diagnosis of infection. Clinical trials show this to be both sensitive and specific; many laboratories now use the tests to detect those samples suitable for culture.

Bacteriuria and its relationship to infection

Although bladder urine is usually sterile, bacterial colonization of the bladder is common in girls and women, and infrequently occurs in boys and men. Because both colonization and asymptomatic infection are common, criteria are required to decide whether infection is truly present.

The most helpful test is semiquantitative bacterial culture. Infection is likely to be present if colony counts from a MSU exceed $10^8/l$ in women. A concentration of less than $10^7/l$ suggests that infection is unlikely, while inter-

mediate concentrations require further investigation. The different anatomy of the lower urinary tract in men affects this principle. Infection is considered likely in men and boys when the bacterial count in the MSU is $10^6/l$ or greater. The purity of the culture is also important, as a mixed culture is more likely to indicate a contaminated specimen.

When the significance of bacteriuria is in doubt, urine can be obtained from the bladder by catheterization using an aseptic technique. Suprapubic aspiration of urine is possible in infants, in whom the full bladder is an abdominal organ. Numerical criteria are not applied to these specimens, in which any bacteria are taken to indicate infection. In patients who have long-term indwelling urinary catheters bacterial colonization is always acquired, so that a urine colony count greater than $10^8/l$ may not indicate infection. In catheterized patients specimens should only be sent when there is clinical evidence of infection. Even then identifying the pathogen may be difficult as more than one bacterial species may be present.

Microbiological diagnosis of urinary tract infection

Handling the specimen
After collection the specimen should be processed rapidly or refrigerated to prevent multiplication of any contaminating organisms. The results of semiquantitative culture

cannot be interpreted too rigidly; the specimen may have been taken later in the day, so that organisms have not had the opportunity to multiply in the bladder. Early morning urine is more concentrated than specimens taken later. Also, in specimens delayed in transit, organisms that were present in non-significant numbers may have multiplied sufficiently to give a positive result. Despite the potential pitfalls, examination of MSU is the most useful method for investigating urinary tract infection.

The laboratory diagnosis of urinary tract infection is divided into three stages:

1 microscopy;
2 semiquantitative culture; and
3 sensitivity testing.

Each of these has a contribution to make to clinical diagnosis.

Microscopy

The presence of pyuria (more than 10/mm³ neutrophils in the urine) is useful evidence of infection when accompanied by bacteriuria. In most laboratories, an unspun specimen of urine is examined either in a counting chamber (disposable chambers are commercially available) or by using a microtitration tray and an inverted microscope. Commercially available flow cytometry techniques can also be used to quantify bacteria and white cells.

A number of non-infectious conditions also cause pyuria; these include urinary stones, tumours of the urinary tract and reactions to drugs and chemicals such as cyclophosphamide. Pyuria in apparently sterile urine can also be caused by chlamydial infection, tuberculosis, rare conditions such as brucellosis and when bacterial growth has been suppressed by antibiotic therapy.

Causes of sterile pyuria

1 Urinary stones.
2 Urinary tract tumours.
3 Drug reactions.
4 Chlamydial infection.
5 Brucellosis.
6 Tuberculosis.
7 True urinary tract infection partly suppressed by antibiotics.

Bacteria can be seen on microscopy of unstained uncentrifuged preparations of urine, if at least 10⁴ CFU/ml are present. This is similar to the significance level for confirming infection, and some laboratories will inoculate a primary sensitivity plate if bacteria are seen (see below).

Microscopy also enables the quality of the specimen to be evaluated. The presence of epithelial cells indicates skin contamination, suggesting that the specimen should be repeated, or the results interpreted with caution. White cell casts may indicate renal infection, and red cells or red cell casts may be seen in glomerulonephritis or endocarditis.

Culture

There are many methods for semiquantitative culture of urine. The simplest is to inoculate a standard loop of 1 µl. The growth of more than 100 colonies will be the equivalent of >10⁵ CFU/ml (10⁸/l). It is usual to report cultures with 10⁴ CFU/ml as indicating significant infection (see above). A method using filter paper strips to deliver a standard inoculum has been described (Fig. 10.2).

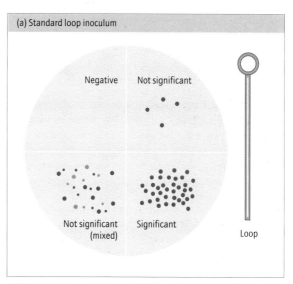

(a) Standard loop inoculum

Negative   Not significant

Not significant (mixed)   Significant

Loop

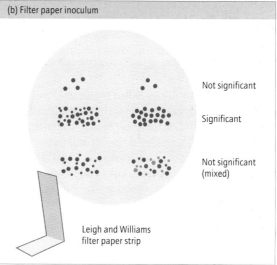

(b) Filter paper inoculum

Not significant

Significant

Not significant (mixed)

Leigh and Williams filter paper strip

Figure 10.2 Methods of semiquantitative urine culture. (a) Standard loop inoculum; (b) filter paper inoculum: the filter paper is dipped into the urine and then pressed on to the agar plate.

Figure 10.3 *Proteus* spp. swarming on a blood agar plate after overnight incubation; the same inoculum on a MacConkey's agar plate forms discrete colonies.

In all culture systems it is important to inhibit the motility of *Proteus* spp. Itself a common pathogen of urine, its swarming growth may obscure other organisms (Fig. 10.3). The bile salts incorporated into MacConkey agar inhibit swarming, and most urinary pathogens will grow in their presence. However, some fastidious organisms are inhibited. Electrolyte-deficient media, such as cysteine lactose electrolyte-deficient (CLED) medium, prevent swarming and support the growth of more fastidious organisms.

Most laboratories use a number of criteria to determine the clinical significance of a positive culture result. The criteria for the significance of bacteriuria were originally determined in a study of early morning urine specimens presented by pregnant women. As not all patients present such concentrated specimens, results should be reported in a way that avoids mistakenly interpreting lower colony counts as negative.

The adequacy of the specimen is indicated by the absence of epithelial cells on microscopy – significant numbers of epithelial cells indicate meatal contamination. Excess neutrophils support a diagnosis of infection.

The number of organisms (see above) and the purity of culture are very important. Lower numbers of organisms may be significant if present in pure culture, whereas larger numbers of mixed enterococci and coliforms probably indicate a contaminated specimen. A further specimen should be processed.

The identity of the organism is important because some organisms usually present on the skin, such as coagulase-negative staphylococci, are also recognized as common urinary pathogens. If there is doubt about the origin of positive growth, a further specimen should be tested.

Some patients with persistent symptoms have repeatedly negative cultures. This may arise because the pathogen is a species that does not grow on media used routinely for urine cultures (e.g. rare cases of *H. influenzae*

or *Mycoplasma* spp.) Molecular diagnostic methods may, in future, help in these cases.

> **Indicators that bacteriuria is significant**
> **1** Adequate specimen (no epithelial cells).
> **2** Excess neutrophils.
> **3** Purity of culture.
> **4** Number of organisms (appropriate for type of urine specimen).
> **5** Identity of the organism.

Automated methods

Several automated methods have been described that rapidly identify specimens containing a significant number of organisms. Growth detection methods are based on changes in optical density, or electrical impedance. Subculture is then performed using conventional techniques.

A semiautomated system uses a series of microtitration trays containing different identification and sensitivity test media. Specimens are inoculated by multipoint inoculator and the plates incubated at 37 °C for 18 h. The changes in the media are read by a computer-driven image analyser. Contaminated cultures are detected by visual inspection of the reactions of test media included in the trays. Isolation, identification and sensitivity testing occur in parallel rather than sequentially, so a final result is usually available within 24 h.

Bacterial identification and sensitivity

For most clinical situations, species identification is not necessary, and exhaustive laboratory testing need not be performed. Thus, a lactose-fermenting organism that produces indole may be labelled a coliform. Urease-positive organisms will be reported as *Proteus* sp., and enterococci diagnosed on the basis of their characteristic colonial morphology and hydrolysis of aesculin. Automated and semiautomated methods produce a species or genus diagnosis by use of a computer database.

Sensitivity tests should be performed on all potentially significant isolates. Antibiotics used in the treatment of urinary tract infection include trimethoprim, co-amoxiclav, cephalexin, nalidixic acid, nitrofurantoin, and 4-fluoroquinolones such as ciprofloxacin. In-patient specimens should also be tested against injectable penicillins (e.g. azlocillin), cephalosporins (e.g. cefotaxime) and aminoglycosides.

Urethritis and the urethral syndrome

Introduction

Infection of the urethra often occurs as part of a more

extensive urinary tract infection. However, the urethra is also often involved in sexually transmitted diseases. In women, it is also often affected by genital infections such as candidiasis and non-specific vaginitis.

Organism list

- Herpes simplex
- *Escherichia coli*
- *Staphylococcus saprophyticus* (in young women)
- Other Enterobacteriaceae
- *Neisseria gonorrhoeae*
- *Chlamydia trachomatis*
- *Gardnerella vaginale*
- (Lactobacilli)
- (Diphtheroids)
- *Candida albicans*
- *Mycoplasma hominis*
- *Ureaplasma urealyticum*

Epidemiology

This varies according to the infecting organism. Urethritis occurring as part of a general urinary tract infection is commoner in females than males and increases with age. More than two-thirds of urinary tract infections are due to *E. coli*.

Urethritis associated with sexually transmitted disease is usually due to *Chlamydia trachomatis* or *N. gonorrhoeae*, less commonly herpes simplex or *Trichomonas vaginalis* (see also Chapter 15). Males are clinically affected more commonly than females because of the higher preponderance of asymptomatic infections among females and the higher incidence of infection in male homosexuals. The incidence of gonorrhoea and trichomoniasis declined throughout the 1980s, especially among older patients who modified their sexual behaviour in response to the HIV epidemic, but infection rates are now increasing, especially in younger age groups. The incidence of both genital herpes and *Chlamydia* infection has been steadily increasing since the early 1980s in both sexes but especially among females for genital herpes.

Microbiology and pathogenesis (see also Chapter 15)

Chlamydia

The family Chlamydiaceae contains two genera: *Chlamydia* spp. and *Chlamydophila* spp. There are 13 serotypes: A–C are associated with trachoma and D–K associated with acute urethritis, neonatal conjunctivitis and pelvic inflammatory disease. Serotypes L1–L3 are associated with lymphogranuloma venereum. The lifecycle of chlamydiae has two phases:

- the intracellular metabolically active form, called the reticulate body; and
- the extracellular form, adapted for transmission and infection, the elementary body.

Neisseria gonorrhoeae

N. gonorrhoeae is a small, fastidious Gram-negative organism that infects the urogenital tract of humans but can spread to cause pelvic inflammatory disease or bacteraemia. Attachment to the urethral epithelium depends on pili, which are essential for infectivity, but the lipo-oligosaccharides (LOS), outer membrane proteins and porins are also involved. The Opa proteins mediate attachment through heparin sulphate; carcinoembryonic antigens CD66 mediating intracellular invasion. Bacteria traverse the epithelial cells to infect the subepithelial layer, where cytokines stimulate intense inflammation. Phagocytosis by macrophages is thought to be essential in controlling the infection. Systemic IgG and local IgA antibodies may promote this. Constant antigenic variation in the pilus, Opa and LOS means that repeated infection is possible whenever the host encounters a new strain.

Clinical features

The main features are frequency and urgency of micturition, with burning dysuria. In men, particularly those with sexually transmitted diseases, there may be a mucoid or mucopurulent urethral discharge, often most noticeable first thing in the morning.

Just over half of cases are associated with pyuria and a diagnostic growth of bacteria from the MSU. In most cases the pathogen is a coliform, but in women it may be *Staphylococcus saprophyticus*.

Both chlamydial infection and gonorrhoea affect the urethra, and they occasionally coexist. In women with sexually transmitted urethritis there is often an associated vaginal infection and there may be a discharge. The meatus of both the male and female urethra may be affected by genital herpes simplex, with typical vesicular lesions, and often with painful inguinal lymphadenopathy.

Urethral syndrome

The urethral syndrome in women is the presence of symptoms without significant bacterial growth or demonstrable vulvovaginal infection. About half of sufferers have midstream bacteriuria of less than 10^7/l. Surveys show that these women have positive growth in catheter specimens of bladder urine, and probably have low-grade urinary infection that can be treated conventionally. About half of the remainder have positive investigations for *Chlamydia*.

Among the rest, fastidious organisms such as lactobacilli or *Gardnerella* are sometimes found. The clinical significance of these findings is uncertain, but appropriate treatment is sometimes helpful. Finally, *Candida albicans* should be treated if vulval swabs are positive.

There are some non-infectious causes of urethral irritation. Post-menopausal women may have dryness and mild atrophy of the vulval mucosa. In middle-aged women urethral caruncle produces a small, red swelling at the meatus, which responds to surgical treatment. Irritants in the urine may include drugs such as rifampicin, warfarin and cyclophosphamide. Dietary irritants such as cayenne may also cause urinary symptoms when excreted.

Diagnosis (see also Chapter 15)

Diagnosis depends on a combination of adequate history and appropriate investigations. Herpes virus should be sought by culture or PCR. A presumptive diagnosis of gonococcal infection can be made from a Gram smear of urethral or cervical material (around 95% sensitive in male urethritis but only 60% in cervical specimens). Ideally specimens should be inoculated at the clinic, otherwise a transport medium is essential. Gonococci can be isolated on nutrient-rich selective media such as New York City or Thayer Martin, which contain a selection of antibiotics to inhibit the normal flora. Rapid and sensitive nucleic acid amplification tests are now entering clinical practice but culture remains necessary to confirm the antibiotic susceptibility of the organism. The diagnosis of *Chlamydia* is nowadays based on nucleic acid amplification techniques, which are more sensitive and specific than ELISA or culture.

Management

Many patients can be managed with simple antimicrobial chemotherapy, as for urinary tract infection. *Staphylococcus saprophyticus* usually responds to a 5–7-day course of flucloxacillin. Chlamydial infection responds to tetracyclines or erythromycin given in conventional doses for 2 weeks. Other organisms may be treated according to laboratory sensitivity tests.

Relapse or failure to respond should prompt a search for predisposing factors such as vaginal prolapse or atrophic vaginitis, surgical conditions such as caruncle, or carriage of a pathogen by a sexual partner.

Reasons that urethral symptoms may fail to respond to antimicrobial therapy

1 There is a predisposing factor (atrophic vaginitis, urethral caruncle, small anterior prolapse).
2 There is chemical irritation from drugs or strong spices excreted in the urine.
3 There is herpes simplex or candidal infection.

Cystitis and ascending urinary infections

Organism list

- Adenoviruses (haemorrhagic cystitis of children)
- *Escherichia coli*
- *Staphylococcus saprophyticus* in young women
- *Klebsiella pneumoniae*
- Other coliforms
- *Proteus mirabilis*
- Other Proteaceae
- *Candida albicans*
- *Staphylococcus aureus* (often post-operative)

Introduction

Cystitis is infection of the bladder. It often also involves the upper urinary tract. It affects the sexes and age groups differently, depending on the prevalence of bacteriuria in each group. Thus, among infants, boys are more commonly affected, particularly those who are uncircumcised. Among children and young adults, females outnumber males by 10:1. In the older age groups the occurrence of infection is favoured by prostatism in men, by prolapse or vaginal atrophy in women, and by incontinence in the frail or disabled of both sexes.

Clinical features

It is impossible to distinguish clinically between simple bladder infection and that which involves the ureters and renal calyces. Symptoms of frequency, urgency, dysuria and suprapubic discomfort are common. The urine is often cloudy, pale pink or frankly blood-stained. Proteinuria and microscopic haematuria are the rule. Dip tests for nitrite and leucocyte esterase are positive. Spurious or non-specific symptoms are common, especially in young patients.

Symptoms often resolve spontaneously, but low-grade infection may persist without symptoms. Fever may be slight or absent; neutrophilia is not the rule in simple urinary infections. Continuing infection may be harmful in patients at risk of renal damage or of severe exacerbations of infection (see below).

When infection affects the renal parenchyma there is often pain and tenderness in the renal angle, upper abdomen or loin. Nausea is common; vomiting or loose stools may occur. Abscesses, either renal or perinephric, may form; warning signs are increasing pain, fever and neutrophilia.

Some urinary tract infections are associated with bacteraemia. High fever and rigors are warning signs, and septic shock may follow. Bacteraemia is made more likely by foreign bodies in the urinary tract, and often complicates instrumentation or catheterization performed while untreated infection exists.

Diagnosis

A high index of suspicion is needed to detect less obvious cases. Laboratory examination of the urine should be requested whenever infection is a possibility. Renal function should be checked if recurrent infections occur. Feverish patients in hospital should always have two blood cultures before chemotherapy is commenced.

Urine examination is sometimes negative when a patient has strongly suggestive signs of urinary infection, or when infection has failed to respond to initial therapy. Microbiological reasons for this have already been discussed. Additionally, pus and organisms from an infected kidney may be trapped behind an obstruction in the ureter. This prevents both drainage of the loculated infection and microbiological diagnosis. A renal ultrasound scan may reveal hydronephrosis, or an excretion scan may show reduced excretion from the affected kidney.

Management

Most infections are caused by Gram-negative rods. A suitable choice of antibiotic for initial treatment would be nitrofurantoin, trimethoprim, cephalexin or co-amoxiclav. All of these agents are concentrated in the urine, and reach adequate therapeutic levels when given orally. Flucloxacillin is usually effective against *S. saprophyticus*. *Klebsiella* and some other unusual urinary pathogens are uniformly resistant to amoxicillin. Very broad-spectrum agents are rarely required, and should only be given when indicated by laboratory test results.

> ### Oral treatment of urinary tract infection
> **1** First choices: cephalexin 250–500 mg 6-hourly (child 25–50 mg/kg daily in three divided doses); *or* trimethoprim 200 mg 12-hourly (child 2–5 months, 25 mg; 6 months–5 years, 50 mg; 6–12 years, 100 mg, all twice daily; contraindicated in pregnancy and in neonates).
> **2** Second choices: co-amoxiclav (as amoxicillin) 250 mg 8-hourly (child up to 10 years, 125 mg 8-hourly); dose may be doubled in severe infection *or* amoxicillin 250 mg 6-hourly (if organism sensitive); *or* nitrofurantoin 50 mg 6-hourly with food (child over 3 months, 3 mg/kg daily in four divided doses). Nitrofurantoin has many side-effects, including occasional severe nausea. Co-trimoxazole carries a risk of severe skin reaction or neutropenia, especially in the elderly. It should only be used if there is no satisfactory alternative. All of the above regimens for 7 days.

Severe infections with fever, neutrophilia or shock should be treated initially with parenteral antibiotics. Broad-spectrum cephalosporins such as cefotaxime or ceftriaxone are useful and have low toxicity. When fever is controlled, treatment may be continued orally, with any effective antibiotic.

The duration of treatment is controversial. Although single-dose amoxicillin treatment is possible, almost half of infecting organisms are now resistant to this drug.

Infection behind an obstruction usually demands drainage. Debris or a ureteric stone can often be removed via a ureteric catheter, but if the obstruction is impassable the kidney must be drained by nephrostomy until the infection is controlled.

Prophylaxis of urinary tract infection

Some individuals suffer repeated urinary infections and often have continuing bacteriuria between attacks. Low-dose antimicrobial chemoprophylaxis will often suppress bacteriuria, improving both health and quality of life, and preventing deterioration in renal function. Temporary prophylaxis is indicated while awaiting treatment of a predisposing condition.

> ### Prophylaxis of urinary tract infection
> **1** First choice: trimethoprim 100 mg at night (child 1–2 mg/kg); or nitrofurantoin 50–100 mg at night (child over 3 months, 1 mg/kg).
> **2** Alternatives: cephalexin 250 mg at night (child 125 mg at night); *or* co-amoxiclav 250 mg at night (child 125 mg at night).

Urinary tract infections in children

Urinary infections in children are often associated with renal tract abnormalities. Examples include urethral valves in boys, bladder outflow obstruction, duplex drainage systems and stones (which may themselves be associated with metabolic disorders such as renal tubular acidosis). Many children have ureteric reflux, which allows infection to ascend to the kidneys. This is thought to cause renal scarring. As 20% of all cases of chronic renal failure have evidence of scarring, it is important to detect and treat childhood infections. A consensus view of experts in childhood infections may be summarized as follows.

1 Detect the infection. Urine should be collected and cultured from all feverish children, and those who fail to thrive, unless there is an obvious alternative cause.

2 Treat promptly. Antibiotic treatment should be commenced as soon as specimens have been obtained (it can be modified later, if necessary). The UK Cochrane database collaborative finds that 14 days oral therapy with cefixime, or 2–4 days with i.v. antibiotics followed by 10 days oral therapy, is effective. Once daily gentamicin is an

effective i.v. regimen in Gram-negative infections. At this stage all children should have a renal ultrasound examination to indicate renal size and detect hydronephrosis, and a plain abdominal radiograph to detect stones.

3 Maintain prophylactic chemotherapy (see above) and investigate further. Once infection is controlled, an excretion scan should be performed to demonstrate excretory function and to outline the collecting system (children over 1 year of age may have an intravenous urogram instead).

4a All infants, and older children with abnormalities of the other tests, should have a voiding cystourethrogram (VCUG) to search for ureteric reflux. Cystourethrograms may be direct (contrast or isotope is introduced into the bladder) or indirect (the bladder and urethra are imaged using contrast or isotope that has been excreted via the kidneys). Direct studies are necessary to demonstrate minor reflux confined to the lower urethra. Indirect studies are useful for follow-up of major reflux, and avoid the need for catheterization. Up to half of children will have reflux. Antibiotic prophylaxis should be continued in these children, and VCUG repeated after 2 (for mild reflux) to 3 years. If reflux has resolved, antibiotic prophylaxis can be discontinued.

4b Alternatively, if all investigations are normal and bacteriuria has ceased, prophylactic antibiotics may be discontinued. Repeat urine cultures should be performed every 3 or 4 months until the age of 2, following which culture need only be performed if infection is suspected.

Urinary tract infections in men

Urinary infections are relatively common in infant boys, who are four times more likely than girls to have bacteriuria, and in elderly men, when prostatism predisposes to stagnation of residual urine in the bladder. Recurrent urinary symptoms may be related to intravesical malignancy. In later childhood and adult life, males are only one-tenth as likely to have bacteriuria as are females.

Men with urinary infections should therefore be investigated, as the likelihood of urinary tract abnormality is high. Imaging of the renal pelvis, calyces and ureters should be performed, to demonstrate deformities or reduplications of the collecting system. If there is evidence of back pressure, the possibility of partial urethral obstruction should be considered. This may be due to bladder exit obstruction, urethral stricture, or to urethral valves persisting into adulthood.

If no pyogenic organisms are found on urine culture, the patient should be investigated for sexually transmitted or other genitourinary infection, even if urethral discharge is not evident.

Urinary tract infections in pregnancy

This is one of the commonest complications of pregnancy. Nearly half of all women who have bacteriuria detected at the first antenatal assessment will develop overt infection. Even asymptomatic bacteriuria should therefore be treated. A further discussion of infections in pregnancy is found in Chapter 17.

Reflux nephropathy (chronic pyelonephritis)

Reflux nephropathy is thought to be the result of damage to the growing kidney from ascending urinary tract infections in childhood. As in chronic bronchitis, a mechanical or anatomical abnormality predisposes the normally sterile organ to bacterial colonization. Recurrent infections then cause further damage and impair function.

Reflux nephropathy, with scarring of the kidney, is associated with a 10% risk of renal failure in later adult life and with a 20% risk of hypertension. Each successive urinary infection causes a dip in renal function, which never quite returns to its previous level. About half of all renal scars are present when the first urinary tract infection is diagnosed.

Treatment is directed at controlling infections and preventing recurrences. Recurrent infections are often with the same organism, having the same antimicrobial sensitivities. Prophylactic antimicrobial chemotherapy may have a place. Nitrofurantoin is a useful prophylactic drug, as it is concentrated in the urine while attaining only very low levels in the blood and other systems. Its general side-effects are therefore few (confined to nausea or mild gastrointestinal symptoms). Trimethoprim also produces few side-effects, and its spectrum is such that it causes little significant disturbance of the bowel flora.

Surgery is important in relieving obstruction and removing stones. Its place in correcting ureteric reflux is more controversial, but many would advocate surgery for gross reflux through a patulous ureteric orifice.

In established reflux nephropathy the treatment of hypertension is important, as this can damage the kidneys further, as well as causing hypertensive vascular disease.

Bilharzia

In endemic areas, bilharzia is a common cause of chronic urinary tract disease. The disease occurs throughout Africa and the Middle East, but especially along the Nile valley (see Chapter 9). The adult *Schistosoma haematobium* resides in the venous plexus of the bladder wall. The spiked ova penetrate the mucosa to be excreted in the urine. Haematuria is common in heavy infections, which cause severe inflammation. Inflammatory granulomata, fibrosis and calcification distort the bladder and lower ureters, causing repeated infections and progressive reflux nephropathy.

Haematuria and acute urinary symptoms follow invasion of bladder wall venules by schistosomes. Following this, typical ova are excreted. These are best demonstrated in the pellets from centrifuged early morning or terminal urine specimens. Serology is usually positive by this stage.

The treatment of choice for bilharzia is praziquantel. For prevention and control, see Chapter 9.

Acute bacterial prostatitis

Acute prostatitis is caused by urinary tract pathogens.

Organism list

- *Escherichia coli*, 80%
- (*Pseudomonas aeruginosa*)
- (*Serratia* spp., 10–15%)
- (*Klebsiella* spp.)
- (*Proteus* spp.)
- Enterococci, 5–10%
- Staphylococci, occasionally

Clinical features

This is an acute, severe systemic illness, usually accompanied by fever and sometimes with a sepsis syndrome. There is deep-seated aching pain in the perineum, the lower back and sometimes the penis or rectum. There are usually signs of urinary tract infection with frequency of micturition and dysuria. Oedema of the prostate may compress the urethra, causing acute urinary retention.

Gentle rectal examination reveals a smooth, swollen, hot and very tender prostate gland.

Diagnosis

Immediate investigations include a mid-stream urine examination and blood cultures.

Prostatic massage should not be attempted for diagnisis of acute prostatitis; it is extremely painful, and may precipitate bacteraemia. The causative organism is almost always identified in the MSU specimen.

If urinary tract infection coexists, VB2 (the MSU) contains a high concentration of bacteria, usually the same organism as that which causes the prostatitis.

Treatment

General measures include rest, hydration and adequate analgesia.

In urinary retention, urethral catheterization should be avoided, as it may damage the prostate and precipitate bacteraemia. Suprapubic bladder drainage should be carried out.

Empirical antibiotic treatment should be commenced immediately. It is facilitated by the intense inflammatory response, which allows antibiotics to penetrate the normally impervious prostate.

> ### Antibiotic treatment of acute prostatitis
> **1** Oral treatment: Ciprofloxacin 500 mg, 12-hourly for 28 days *or* ofloxacin 200 mg, 12-hourly for 28 days.
> **2** Alternative oral regimens: co-trimoxazole 960 mg, daily for 28 days *or* trimethoprim 200 mg, 12-hourly for 28 days.
> **3** Intravenous treatment: cefuroxime 750 mg to 1.5 g, 6–8 hourly *or* ciprofloxacin 400 mg, 12-hourly (switch to oral therapy when patient is well enough and sensitivities of the pathogen are known).

After treatment, the urinary tract should be investigated for a causative mechanical disorder, as for acute urinary infection in adult men.

Complications

Failure to respond to therapy may be caused by development of a prostatic abscess. This can be demonstrated by transrectal ultrasound, or CT scan. It is treated by perineal or transurethral drainage.

Chronic prostatitis

This is a poorly understood condition, which is nowadays divided into three syndromes:
1 chronic bacterial prostatitis (CBP);
2 chronic abacterial prostatitis/chronic pelvic pain syndrome-inflammatory (CAP/CPPS-inflammatory);
3 chronic abacterial prostatitis/chronic pelvic pain syndrome-non-inflammatory (CAP/CPPS-non-inflammatory).

Chronic bacterial prostatitis may be caused by *Staphylococcus aureus*, *Streptococcus faecalis* or enterococci. There is controversy over the possible pathogenic role of coagulase-negative staphylococci, 'diphtheroids' or non-group D streptococci. *Chlamydia trachomatis* is not thought to be a significant cause.

Diagnosis of CBP is made by comparative quantitative culture of urine collections and expressed prostatic secretions (EPS; obtained by prostatic massage). The patient should have been treated for any other urogenital infection, and have remained off antibiotics for one month. He should not have ejaculated for two days. He should have a full, but not distended, bladder. The foreskin should be fully retracted, and the glans of the penis should be carefully cleaned.

Specimen collection for diagnosis of CBP

1 The patient collects a first-glass urine (VB1).
2 After discarding a further 1–200 ml urine, the patient collects a second-glass (MSU) sample.
3 The prostate is vigorously massaged from laterally to the midline for one minute, via digital rectal examination, while a sterile container is held over the glans to collect any expressed secretions (EPS).
4 Immediately after this, the patient passes a post-massage urine specimen (VB3).

To assign an organism to the prostate, the colony count in EPS and VB3 must be at least ten times greater than in VB1 and MSU (if there is significant bacteriuria in VB1 and MSU, this can be treated with nitrofurantoin, which does not penetrate the prostate; and the procedure is then repeated).

The pH of EPS rises in CBP: a level of 8 or higher makes CBP likely. Clumps of neutrophils (>5 cells), and the presence of lipid-laden macrophages suggests prostatitis. Neither of these tests is diagnostic.

Treatment is with an oral antibiotic that penetrates the prostate. The antibiotics of choice are ciprofloxacin or ofloxacin (as for acute prostatitis), or other quinolones. Alternatives are doxycycline 100 mg, 12-hourly, trimethoprim 200 mg, 12-hourly, or co-trimoxazole 960 mg, 12-hourly. All regimens are given for 28 days (some studies have evaluated 90 days' treatment, but there is no evidence of greater effect).

Some cases of abacterial chronic prostatitis also gain benefit from antibiotic treatment.

Acute epididymo-orchitis

Organism list

- Mumps virus
- Coxsackievirus
- *Neisseria gonorrhoeae*
- *Chlamydia trachomatis*

- *Escherichia coli*
- Other coliforms

In men under 35 years of age, sexually transmitted pathogens are the commonest cause of acute epididymo-orchitis. In older men, Gram-negative urinary pathogens are more likely. Instrumentation or catheterization of the urethra predispose to the condition.

Acute epididymo-orchitis usually has an acute onset. There is pain, swelling and redness of one, or occasionally both, testicles with or without symptoms of cystitis or urethritis. A hydrocoele may be found on the affected side. Fever and other systemic symptoms are common.

There may be obvious features of the underlying systemic disease, e.g. parotitis in mumps or severe muscle pain in coxsackie B infections. In sexually transmitted cases, a urethral discharge may be evident.

Torsion of the testis, sudden and common in fit young men, must be excluded. Testes that undergo torsion hang horizontally, below the epididymis instead of lying anterior to it. Ultrasound examination is the investigation of choice to detect torsion. Surgical correction must be performed without delay, to rescue the ischaemic testis. Chronic or persisting epididymo-orchitis, with indurated swelling, is a rare feature of tuberculosis (particularly of the renal tract), and of brucellosis. Up to 20% of men with Behçet's disease may suffer epididymo-orchitis.

Bacterial pathogens may be demonstrable on MSU or terminal urine examination or by urethral swab. A first-pass urine specimen may contain *Neisseria gonorrhoeae* or give a positive PCR-test for *Chlamydia trachomatis* (see Chapter 15). Viral infections are diagnosed initially on clinical grounds and ultimately by cultural and serological methods applicable to the individual viruses.

Treatment of epididymo-orchitis is that of the underlying condition

1 For gonorrhoeal infection: single-dose ceftriaxone i.m. 250 mg *or* ciprofloxacin 500 mg, orally, single dose *or* doxycycline 100 mg, 12-hourly, orally for 10–14 days.
2 For probable chlamydial infection: doxycycline, as above.
3 For suspected Gram-negative infection: ciprofloxacin, orally, 500 mg, 12-hourly for 10 days *or* as indicated by culture and sensitivity.

⚠ In viral infections a short course of prednisolone treatment will alleviate oedema and inflammation, and will not prolong the natural course of the disease.

A well-fitting scrotal support or underpants will reduce discomfort. Anti-inflammatory analgesics are helpful. Stronger analgesics may be required early in the disease.

Bartholin's abscess

Bartholin's glands open into the posterior part of the vulval vestibule via a duct on each side. Infection is usually unilateral, causing intense pain, swelling and tenderness deep to the labia on the affected side. Common causative organisms include *Staphylococcus aureus* and *E. coli*, which probably ascend from the perineal skin. Broad-spectrum antibiotics may be effective, but surgical drainage is often required.

On rare occasions the abscess contains chlamydiae or mycobacteria. Culture of the pus obtained at surgery is therefore advisable, to confirm the aetiology.

Childhood Infections

11

Introduction

Many infections are common in children, who have not yet accumulated a repertoire of antibodies to protect against common pathogens. Examples include acute respiratory virus infections (see Chapters 6 and 7), primary herpes simplex, hand, foot and mouth disease and meningococcal disease. The more highly infectious the disease, the earlier in childhood it tends to occur. The epidemic diseases of children discussed in this chapter are almost all highly infectious by the airborne or droplet route, have a high rate of clinical morbidity, often have a rash as one of their clinical features, and affect up to 90% of all susceptible individuals by the end of childhood.

Many childhood infections are nowadays controlled by immunization programmes (Fig. 11.1), but a high rate of immunization must be maintained to prevent their resurgence. In epidemics, the few non-immune adults are at risk of infection, and often suffer more severe disease, with more complications, than those infected in childhood.

A few diseases mentioned here are not highly infectious, but are important differential diagnoses of childhood diseases with rashes. It is convenient to present them in this context.

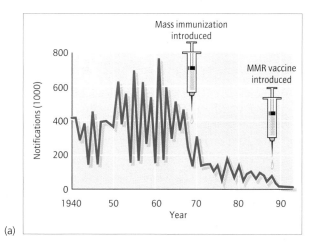
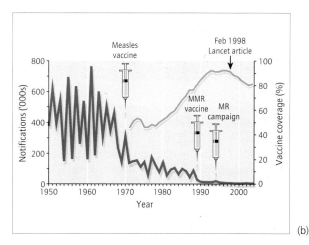

Figure 11.1 (a) Progressive control of measles with immunization in England and Wales. (b) Fluctuations in measles reporting in England and Wales, due to periods of anxiety about the adverse effects of MMR vaccine. Source: Health Protection Agency.

Measles

Introduction

Measles is a systemic viral infection whose main features are respiratory disease and rash. It is highly infectious and usually produces clinical disease in those infected. In unprotected populations it tends to occur in large epidemics, mainly affecting children. It can cause individual fatalities and devastating outbreaks in refugee camps and other populations lacking organized healthcare and vaccination. The widespread use of effective immunization has made it uncommon in many parts of the world.

The important features of measles are:
- it can be a severe and debilitating illness;
- secondary bacterial respiratory disease is common and may be severe;
- post-measles encephalitis is life-threatening and can leave severe sequelae;
- it will emerge and spread in populations with immunization rates below 95%.

Measles virus RNA, and raised measles IgG levels in perilymph, have been reported in otosclerosis. The incidence of otosclerosis has decreased, and the age of onset has risen, since measles vaccination has been introduced.

Viral RNA has been demonstrated in osteocytes by some workers, but not found by others, seeking to confirm a viral aetiology for Paget's disease. Early reports of virus detection in Crohn's disease were disproved by extensive further research.

Epidemiology

The disease is highly infectious with a reproduction rate (see Chapter 1) of 10–18. The attack rate in a susceptible population is usually 95% or greater and the average age at infection is 4 years. An unvaccinated person is unlikely to avoid infection. Transmission occurs mainly by droplet spread or direct person-to-person contact, less commonly by airborne spread or by fomites. Maternal antibody provides protection in young infants, lasting until about 6 months of age

The maximum period of communicability is during the prodromal period and in the first 2 days of the rash. The incubation period is 10 days (range 8–13 days) from exposure to the onset of fever, and 14 days to the onset of rash. Recovery is followed by lasting immunity. Humans are the only reservoir of infection.

In England and Wales, large epidemics occurred at regular 2-yearly intervals until the introduction in 1968 of a mass vaccination programme with a single measles vaccine (see Fig. 11.1). Initially, vaccine coverage was low (about 50%), and epidemics still occurred every 2–3 years, although the number of cases dropped by 80%. Following the introduction of MMR vaccine in October 1988 and coverage levels in excess of 90%, measles notifications fell progressively to the lowest levels since records began. The epidemic cycle has been broken, though a suggested but unproven association between the vaccine and autism recently led to reduced uptake and outbreaks of measles in infants and schoolchildren. Cases also occur in young adults who have not received a full course of two vaccinations. The peak incidence is now in spring and early summer. The case fatality ratio has declined from 1 per 100 in 1940 to 0.02 per 100 in 1989. The last two deaths from acute measles in England and Wales were in children who died in 1992 and 2006.

Virology and pathogenesis of measles

Measles virus is a paramyxovirus of a single serological

type, which only infects primates. It is related to canine distemper and rinderpest viruses. Its helical nucleocapsid contains a single strand of RNA, coated with a protein and associated with an RNA-dependent RNA polymerase. The virus is enveloped and varies in diameter from 120 to 200 nm.

The genome encodes six structural proteins. The envelope contains three major antigens:
• the matrix or M protein;
• H protein, a glycoprotein responsible for haemagglutination and adsorption of the virus to host cell receptors; and
• F protein, a glycoprotein that mediates fusion with the host cell membrane and haemolysis.

Unlike other paramyxoviruses, the measles virus does not contain neuraminidase. Among the non-envelope proteins, the large protein (L) interacts with the pyrophosphate protein (P) to form an RNA-dependent RNA polymerase complex. The H antigens are genetically most variable and many genotypes have been described.

Like other enveloped viruses, measles is sensitive to ether and is readily inactivated. Despite its infectiousness for humans, it is difficult to culture artificially. It can be grown in primary human or simian cells, producing a cytopathic effect (CPE) characterized by multinucleate giant cells. In contrast, vaccine strains produce a spindle-cell CPE.

Viral replication in the respiratory mucosa is followed by a primary viraemia during which the virus invades reticuloendothelial cells, leading to the secondary viraemia. The virus can invade most cell types, and affects most organs. Viral damage to the respiratory epithelium leaves the submucosa susceptible to secondary bacterial infections. Koplik's spots represent buccal mucosal infection, and virus is found in the lesions of the skin rash. Multinucleate giant cells become detectable in the skin and respiratory tract as the first antibodies develop.

Measles encephalitis is thought to be caused by cell-mediated hypersensitivity to the virus, leading to demyelination, gliosis and infiltration of fat-laden macrophages around cerebral blood vessels. In immunocompromised individuals giant cell pneumonia occurs, often without a rash.

Clinical features

After 8–13 days' incubation, illness begins with fever and respiratory symptoms. Extreme irritability and febrile convulsions are common at this stage. There is conjunctival inflammation, running eyes and nose, persistent croupy cough, mucoid sputum and often coarse crepitations on auscultation of the chest. Diarrhoea is common, especially in children.

The buccal mucosa is inflamed and Koplik's spots appear. These are raised white lesions resembling bread-crumbs or grains of salt on an inflamed background (Fig. 11.2). They occur during the prodromal period, on the mucosa of the cheek, adjacent to the upper premolars and molars, but can be much more extensive in some cases. They fade as the skin rash evolves.

The rash starts in the hairline and behind the ears on the third or fourth day of fever. It reaches the hips on the next day and then the lower legs. On the face it can produce large, swollen blotches (Fig. 11.3) but elsewhere it is maculopapular with elements of 1–2 cm in the greatest

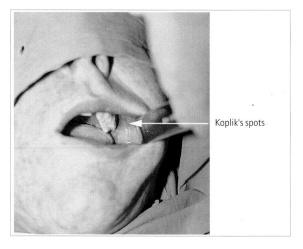

Koplik's spots

Figure 11.2 Measles: Koplik's spots during the prodrome.

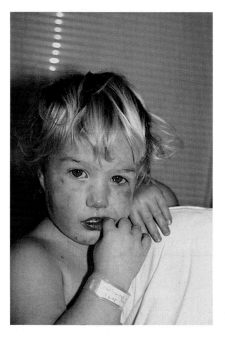

Figure 11.3 Measles: facies on development of the rash.

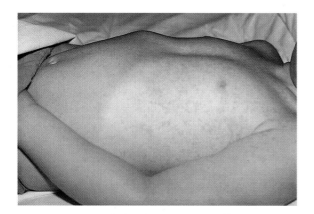

Figure 11.4 Measles: evolution of the maculopapular rash. By the second day the rash has reached the hips.

diameter (Fig. 11.4). It is not pruritic but the skin feels hot and uncomfortable.

When the rash reaches the lower legs the fever improves. In the next few days the rash fades to a café-au-lait colour ('staining'); the cough should then subside and the fever resolve. The staining rash fades completely within 7 to 10 days.

Differential diagnosis

The evolution of the illness is characteristic, even in mild or immunized cases, often permitting clinical diagnosis. The white cell count is low or normal unless secondary bacterial infection develops. The absence of a prodromal illness or Koplik's spots makes the diagnosis unlikely. The rash of rubella does not spread from the head downwards, and its maculopapular elements are much smaller. Allergic rashes are usually pruritic, and occur without respiratory features.

Laboratory diagnosis

Laboratory diagnosis is increasingly important; as measles is now rare, its clinical features are unfamiliar. It is necessary in immunodeficient patients, who may lack typical clinical features.

Measles is confirmed by a positive IgM ELISA, which can be detected in saliva or blood between 1 and 6 weeks after the onset of symptoms. IgM antibodies are lost after 3 months. Measles virus RNA can be demonstrated by reverse transcriptase polymerase chain reaction (RT-PCR) methods. Patients with subacute sclerosing panencephalitis (SSPE) can have persistently elevated IgM concentrations.

Smears of material from Koplik's spots or aspirated nasopharyngeal secretions may contain multinucleate giant cells. Immunofluorescent stains can demonstrate

viral antigen. Blood, nasopharyngeal secretions, conjunctival secretions and urine may be cultured to reveal a haemadsorbing agent, which is further identified by immunofluorescence of the infected cells. Specimens should be transported to the laboratory at 4 °C with minimum delay, due to the lability of virus infectivity. Specimens may be inoculated into primary human cells, or continuous-line Vero cells. Isolated virus is identified by the giant-cell CPE, and intranuclear inclusions.

Diagnosis of measles

1 Demonstration of measles immunoglobulin M by enzyme-linked immunosorbent assay (ELISA).
2 Immunofluorescent staining of cells to demonstrate measles antigen (a rapid diagnostic method).
3 Culture if specimens rapidly transported to the laboratory.
4 Cytology of Koplik's spots or respiratory mucosal cells (shows multinucleate giant cells).
5 Rising titres of various antibodies in paired sera.
6 RT-PCR to amplify measles virus.

Treatment

There is no specific treatment for acute measles. Severe secondary bacterial disease is as important as the viral infection itself, and is responsible for most of the 5–20% fatality rate in third-world outbreaks. Bacterial bronchitis should be easily suspected, sputum and blood cultures should be obtained, and antimicrobial therapy should include an antistaphylococcal agent.

Problems and complications

Unusually severe measles

Danger signs include early, severe respiratory disease, widespread petechial or haemorrhagic components in the rash and/or features of sepsis. Giant-cell viral pneumonitis, trivial in most cases, can cause severe or fatal respiratory failure, particularly in the immunosuppressed (see Chapter 22).

Nebulized ribavirin has been used in complicated measles cases, but the evidence for its effectiveness is not strong enough to make firm recommendations.

Keratitis

This is a rare sequel of measles, usually presenting as blurred vision in one or both eyes. Cloudy oedema of the cornea is apparent on slit-lamp examination. It is usual to offer antibiotic drops or ointment to prevent secondary bacterial infection. The disorder is often self-limiting over a period of 3 or 4 weeks.

Secondary bacterial infections

Acute otitis media and lower respiratory infection are the

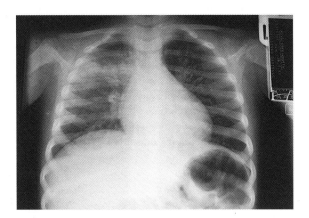

Figure 11.5 Secondary bacterial bronchopneumonia in measles: this 2-year-old girl developed respiratory distress and high fever as the rash began to stain. The chest X-ray shows extensive nodular pneumonitis as well as a distinct area of consolidation. The patient responded rapidly to intravenous ampicillin plus flucloxacillin, and *Staphylococcus aureus* was recovered from sputum cultures.

commonest complications of measles. Otitis media occurs during the catarrhal phase of the disease, while bronchitis or bronchopneumonia are indicated by persisting cough and fever when the rash is staining (Fig. 11.5). *Staphylococcus aureus* is a common secondary pathogen, and can cause severe disease. *Streptococcus pneumoniae* and *Haemophilus influenzae* are also common.

Mucocutaneous infections occur in debilitated or undernourished children, starting as impetiginous lesions at the angles of the mouth or around the nares. In conditions of poor hygiene, necrotizing, synergistic infection may develop. This is called cancrum oris and is nowadays rare except in the poorest rural communities.

Temporary immunosuppression and vitamin A

Measles produces significant immunosuppression with temporary loss of response to skin-test antigens. This predisposes to secondary infections, and in susceptible communities may precipitate or exacerbate tuberculosis. Vitamin A deficiency is common in such communities and exacerbates the immune paresis. A single dose of 10 000 IU of vitamin A in acute measles can reduce mortality by up to 70% in affected populations.

Post-measles encephalitis

Post-measles encephalitis complicates about 1 in 4000 infections, developing in the second week as the acute illness is resolving. The encephalitis is severe and no specific treatment is available. There is no evidence that corticosteroids or other immunosuppressive drugs influence the course, though dexamethasone may be used to treat raised

intracranial pressure. Mortality is about 15% and about 50% of survivors have permanent neurological sequelae.

Measles inclusion body encephalitis (MIME)

Measles inclusion body encephalitis is a severe, progressive brain disorder that is the result of measles infection in immunosuppressed individuals (see Chapter 22).

Subacute sclerosing panencephalitis

Subacute sclerosing panencephalitis is rare, affecting about 1 in 1 000 000 cases, almost always patients who had measles before the age of 2. Neurological disease begins 7–10 years after the acute measles. Clumsiness and poor school performance are followed by progressive spasticity. Myoclonic episodes and salaam-like seizures are common and there is a typical abnormality of the electroencephalogram. The outcome is uniformly fatal.

Measles antigen is plentiful in the brains of SSPE patients, with high antibody titres in cerebrospinal fluid. It appears that virus lacking surface protein spreads within the syncytium of the brain, without escaping from the cells. Viral antigen can also be demonstrated in peripheral blood mononuclear cells.

Complications of measles

1 Severe, haemorrhagic disease.
2 Measles keratitis.
3 Secondary acute suppurative otitis media.
4 Secondary bacterial bronchopneumonia.
5 Cancrum oris (rare).
6 Post-infectious encephalitis.
7 Subacute sclerosing panencephalitis
8 MIME, in the immunosuppressed.

Prevention and control

Live-attenuated measles vaccines are offered to all children at 12–15 months of age, as part of a combined measles/mumps/rubella (MMR) preparation. Before this age there is a poor immune response, possibly because of interference by maternal antibody. The vaccine is contraindicated in children with immune deficiency disorders. However, children with human immunodeficiency virus (HIV) infection should be vaccinated, as the risks from measles are greater than those from the vaccine.

A single injection of measles vaccine induces protective antibody in over 95% of recipients. Minor side-effects occur in 5–10% of vaccine recipients. They include fever, loss of appetite and sometimes a measles-like rash (mini-measles) appearing six to ten days after vaccination. Febrile convulsions may occur. Simple measures such as tepid sponging, removal of warm clothing and antipyretic therapy reduce the likelihood of this. Very rarely (approximately 1 per 10 million doses) SSPE occurs following vaccination. The risk of neurological complications is 10–100 times greater after the disease than after the vaccine.

More than 95% of the population must be immune to prevent the spread of measles. Not all children seroconvert following vaccination, so even in countries such as Sweden, with vaccination rates close to 100%, the disease has not been completely eliminated. Most countries recommend a second dose of vaccine at either 4–6 or 11–12 years of age, to increase the rate of seroconversion, and to prevent the decline of vaccine immunity over time.

Measles is a notifiable disease, and should be urgently reported to local health protection services, so that preventive measures may be taken. Children with measles should not attend school until 5 days after the appearance of the rash. It is particularly important to avoid contact with immunosuppressed children.

Unvaccinated children who have been in contact with a case of measles can be protected by post-exposure immunization, given within 3 days of exposure. Immunosuppressed children, for whom vaccination is contraindicated, may be given human normal immunoglobulin, which contains sufficient measles antibody to provide temporary passive protection.

Mumps

Introduction

Mumps is a systemic viral infection commonly regarded as epidemic parotitis. It also has many systemic effects capable of causing significant morbidity. Fatalities, however, are very rare.

Epidemiology

The disease is moderately infectious, with a reproduction rate of 4–7. Transmission is through droplet spread and direct contact with saliva of an infected person. The maximum period of communicability is in the 2 days before onset of illness. However, virus may be recovered from saliva from 6 days before the onset of parotitis to 9 days after, and from urine up to 14 days after onset. Mild or subclinical infections are common, and these can also be infectious. The incubation period is 2–3 weeks (average 18 days). Humans are the only reservoir of infection.

In unimmunized populations, epidemics occur at 3-yearly intervals, during winter and spring. The maximum incidence is in children aged 5–9 years. Deaths from mumps are extremely rare; however, approximately 1500 children were admitted to hospital in the UK each year before vaccination was introduced in 1988, as a result of mumps complications. Mumps meningitis was the commonest viral cause of meningitis in children. Immunity following natural mumps infection is generally lifelong.

Virology and pathogenesis of mumps infection

Mumps virus is a paramyxovirus of one serological type. It contains a single linear strand of RNA, which is associated with an RNA-dependent RNA polymerase to form a helical nucleocapsid. The virus is enveloped and is of variable size, between 120 and 200 nm. It is ether sensitive, and is destroyed by heating to 56 °C for 20 min, and by ultraviolet irradiation. The envelope is approximately 10 nm thick and consists of three layers: the outer layer contains glycoproteins with haemagglutinin, neuraminidase and cell fusion activity. The middle layer is made up of host cell membrane, and the inner layer contains nonglycosylated viral structural proteins.

Six major proteins are encoded: the nucleocapsid protein (NP), the phosphoprotein (P) and the large protein (L) are associated with the ribonucleoprotein complex. The envelope consists of matrix protein (M), and two glycoproteins that mediate fusion (F) and haemagglutination (HN).

Complement-fixing antibodies recognize the NP antigen. Antibodies to S antigen appear soon after infection and decline over the next few months. Anti-V antibodies rise more slowly, peaking 2–4 weeks after the beginning of the infection, and persist for years. Neutralizing antibodies develop during convalescence and persist for years.

When pathological specimens of parotid glands have been examined the tissue is oedematous with neutrophil invasion and areas of necrosis. Multinucleate cells are not seen. A similar picture is seen in the pancreas or testis although haemorrhage is more common in orchitis. The pathology of encephalitis is similar to other cases of post-infectious encephalitis with demyelination and perivascular cuffing. In cases of primary encephalitis demyelination is not seen but there is widespread neuronolysis.

Clinical features

The features of mumps can appear in any order. Parotitis is the most common, occurring in over 70% of cases. It is usually bilateral but may be asymmetrical and other salivary glands may be involved. Parotid tenderness and pain on salivation precede swelling by 2–4 days. After increasing for about 3 days, parotitis subsides over 7–10 days.

Orchitis is common in adult men but rare before puberty. It varies greatly in severity, but can cause extreme swelling and pain of the testicle, and takes up to a month to resolve completely. Severe inflammation can result in testicular atrophy; but bilateral atrophy is very uncommon, so infertility rarely results from mumps.

Meningitis and meningoencephalitis affect around 15% of cases, but are often mild, and overlooked. Clinically significant cases are often admitted to hospital.

Rarer manifestations of mumps include mastitis (which can affect both sexes and all age groups), cochlear infection with hearing impairment, oophoritis in women, and arthritis (which often appears in the second week). Myocarditis can occur and may be involved in the rare fatalities in adult cases.

Diagnostic problems arise when mumps presents as isolated meningitis, pancreatitis (occasionally with acute diabetes) or orchitis. It should always figure in the differential diagnosis of these conditions, but not necessarily be first on the list.

Clinical features of mumps

1 Parotitis.
2 Meningitis and meningoencephalitis.
3 Orchitis.
4 Pancreatitis.
5 Oophoritis.
6 Cochlear inflammation.
7 Arthritis.
8 Mastitis.
9 Myocarditis.

Differential diagnosis

Parotitis must be distinguished from cervical lymphadenitis, which is common in children. The angle of the jaw is enclosed in the swollen parotid gland, whereas it is superficial to cervical lymph nodes. The serum amylase is often raised in parotitis. The white blood cell count is variable, as the severe inflammatory effects of mumps can induce neutrophilia. Antibiotic treatment (to cover *Staphylococcus aureus* and *Streptococcus pyogenes*) may be necessary while awaiting laboratory diagnosis. Pyogenic parotitis, or parotid abscess may also cause a raised amylase. This is usually unilateral; the gland may be fluctuant and any abscess will tend to point below the lobe of the ear, where the cartilage joins the external meatus. Surgical drainage is often required. Sarcoidosis can present with persisting bilateral granulomatous parotitis.

Isolated orchitis must be distinguished from testicular torsion or pyogenic epididymo-orchitis. Ultrasound imaging is the best investigation for distinguishing these.

In meningitis, orchitis or pancreatitis, a neutrophil leucocytosis is not uncommon.

Diagnosis

Saliva, cerebrospinal fluid or urine may be collected for culture. Specimens are cultured in primary monkey kidney cell or human embryonic kidney cells. Cultured virus may be identified by haemadsorption, neutralization or fluorescent antibody techniques. An RT-PCR technique is also available for diagnosis.

Paired serum samples should be taken for serological investigation. A number of techniques have been described, including haemagglutinin inhibition (HI), virus neutralization, complement fixation (CF) and ELISA. Virus neutralization is the best technique for establishing immunity to mumps but it is too cumbersome for routine diagnostic use.

HI and CF tests are simple to perform but lack sensitivity. A fourfold rise in CF antibodies would confirm a diagnosis of mumps. An elevated titre to S antigen with a low or high titre to V antigen in an acute-phase specimen would indicate recent mumps infection. IgM antibodies may be detected by ELISA and titres reach a peak 1 week after the onset of symptoms, remaining elevated for approximately 6 weeks. An IgG ELISA may be used to detect local production of antibodies in the cerebrospinal fluid of patients with mumps encephalitis or meningitis.

Treatment

No specific treatment is available. Analgesics and bed rest are helpful in severe cases. A short course of corticosteroids may reduce pain and swelling in severe orchitis.

Prevention and control

Mumps vaccine is a live, attenuated preparation, usually administered as part of combined MMR vaccine. Two mumps vaccine strains (Urabe Am9 and Jeryl Lynn) have been widely used in immunization programmes. MMR vaccination of children aged 12–15 months was introduced in the UK in 1988 and greatly reduced the incidence of the disease. The vaccine is highly effective. More than 95% of recipients develop immunity, which is usually lifelong, after a single dose. A sustained coverage level of 85% or greater is required to prevent transmission of mumps and to eliminate the disease.

If coverage levels fall the incidence of mumps may continue to decline, but the average age of cases will increase. This factor, together with the presence of an unimmunized young cohort (those who were older than 15 months in 1988), has led to outbreaks among students and young adults in the UK. Complications occur more frequently in these older age groups. Campaigns to vaccinate university students have recently been carried out.

As with other live vaccines, immunodeficiency is a contraindication (see Chapter 26). Minor side-reactions include fever and parotitis (mini-mumps). These occur in up to 5% of recipients, usually in the third week after vaccination. A mild, self-limiting meningoencephalitis occurs after approximately 1 per 11 000 vaccinations with the Urabe Am9 strain, which is not therefore used in the UK. The incidence of meningitis following the Jeryl Lynn strain is very low. These complications must be compared to the risk of clinical meningoencephalitis following natural infection, estimated to occur in 1 per 200 cases.

Mumps is a notifiable disease. Children with the disease should be kept out of school until 5 days after the onset of parotitis. Vaccination of exposed contacts is of no value, although it may be used during outbreaks to protect those who have not previously been vaccinated.

Rubella

Introduction

Rubella is a systemic viral infection with many features, including a rash. Although highly infectious, it often produces subclinical or trivial disease. It is important because even subclinical viraemia can infect the developing fetus, causing severe tissue damage and progressive developmental defects (see Chapter 17).

Epidemiology

The disease is moderately infectious, although slightly less so than measles or varicella. The reproduction rate is 7–8. Transmission is by droplet spread or direct person-to-person contact. The period of communicability lasts from about 1 week before the onset of rash to at least 4 days after. The incubation period from exposure to onset of fever is 2–3 weeks. Humans are the only reservoir of infection.

In the absence of mass immunization, rubella epidemics occur approximately every 6 years. Children are predominantly affected (the average age at infection is 8 years); however, cases may also occur in adolescents and adults. During rubella epidemics, up to 5% of susceptible pregnant women may catch the disease, resulting in subsequent epidemics of congenital rubella syndrome and rubella-associated terminations of pregnancy. Women who immigrate to the UK in adolescence or adulthood may be unvaccinated and susceptible to rubella. Most incidents of congenital disease occur in this group.

Virology and pathogenesis of rubella

Rubella is caused by rubivirus, an alphavirus member of the family Togaviridae. Unlike many other members, it does not need an arthropod vector for transmission. It is immunologically distinct and differs serologically from the other alphaviruses. It is an icosahedral enveloped virus

60 nm in diameter. The central core containing the nucleocapsid is 30 nm in diameter. It consists of a single-strand RNA and a nucleocapsid protein (protein C) in a helical form. Two viral glycoproteins, E1 and E2, are associated with the envelope. Haemagglutinin activity is associated with the viral envelope. The virus is readily inactivated by ether, trypsin, ultraviolet light, heat and extremes of pH. It will grow in Vero cells, without cytopathic effect.

IgG and IgM antibodies begin to rise as the rash appears, reaching a peak after 7 to 14 days illness. IgM antibody concentrations fall to low levels within 1 month. HI (haemagglutination inhibition) and CF antibodies rise in the first week and remain elevated for a prolonged period. Passive haemagglutinating antibody begins to rise after 1 month and remains elevated for years.

The pathological course of rubella is similar to measles, with primary and secondary viraemia. The rash is likely to be caused by immune damage because it appears as viraemia ceases.

Clinical features

The average incubation of 17–18 days is followed by a mild sore throat and conjunctivitis, often just a gritty feeling in the eyes. Fever is rarely high and the rash appears on the second or third day. It consists of fine macules; papules are unusual, petechiae rare. The macules coalesce to a generalized 'blush' in 1 or 2 days, and this fades without desquamation in 3–5 days. Lymphadenopathy commonly affects the neck, and suboccipital nodes may be large and painful.

Arthralgia is common in young adults. It affects the small joints of the hands and feet and occasionally large joints. It can last some weeks, so non-steroidal anti-inflammatory agents may be needed until the discomfort gradually subsides.

Differential diagnosis

Clinical diagnosis is unreliable, because many patients lack the rash. Rubelliform rashes are common in parvovirus infection (in which arthralgia is also common), enterovirus infections, mild allergic rashes and sometimes mild scarlet fever or toxic shock syndrome. A severe rubella rash may be confused with measles.

Rubella in pregnancy, or exposure of a susceptible pregnant woman, can have very serious consequences. Readiness to suspect rubella and prompt laboratory diagnosis are extremely important in this context. It is important to respond to the threat, even if the exposed pregnant woman has rubella antibodies, as rare re-infections in 'immune' individuals have caused fetal damage (see Chapter 17).

Laboratory diagnosis

The diagnosis of rubella is usually made by serological techniques. Viral culture should be attempted when strain characterization is required, when vaccine-related infection is suspected, or in complicated neonatal cases. In the absence of a cytopathic effect, viral growth is detected by challenging infected cells with enterovirus. Control monolayers are destroyed but infected cells resist enteroviral infection and remain intact. Cultured virus is identified by neutralization of virus infectivity with polyclonal rabbit immunoglobulin. Virus shedding from the pharynx may be scanty, but urine is a useful specimen for viral culture.

Haemagglutination inhibition (HI) was once widely used to detect both IgM and IgG antibodies by their ability to inhibit the agglutination of chick red cells by rubella antigen. Titres of antibody correlate well with the degree of immunity. The diagnosis of acute rubella can be established if a fourfold rise in HI titre is detected. ELISA based methods are now the simplest methods for detecting specific rubella IgM. Congenital rubella can be diagnosed by detecting the presence of IgM antibody in the neonate, but this may not become detectable for many weeks. Viral culture or RT-PCR detection of viral RNA can also be used. Serial estimations of IgG concentrations show falling levels (of maternal antibody) in an uninfected infant, but rising levels if the infant is responding to neonatal infection.

Diagnosis of rubella

1 Immunoglobulin M detection by enzyme-linked immunosorbent assay, radioimmunoassay or particle agglutination.
2 Fourfold rise in haemagglutinin inhibition antibodies in paired sera.
3 Viral culture.
4 Demonstration of viral RNA by RT-PCR.

In most countries, all pregnant women are tested to confirm immunity to rubella. Most laboratories use the detection of adequate antibody levels by IgG ELISA tests. Older tests remain reliable and inexpensive, and are still used where the simple methods and reagents fit in with laboratory practice. The passive haemagglutination test utilizes red cells, coated with rubella antigens, which are agglutinated in the presence of rubella antibodies. The results obtained correlate well with the more complex HI test. In the single radial haemolysis test the presence of rubella antibody is indicated by complement-mediated lysis of sensitized red cells suspended in an agar gel (Fig. 11.6). It is simple and rapid, facilitating the screening of large numbers of specimens.

Detection of immunity to rubella

1 Rubella IgG by ELISA.
2 Passive haemagglutination test.
3 Single radial haemolysis test.

Complications

Immune thrombocytopenic purpura

Rubella is one of the commonest infectious precursors of thrombocytopenic purpura. The thrombocytopenia is transient, lasting from 1 to 3 weeks. If mild it needs no treatment, but it can be treated if necessary with a brief course of prednisolone or, if severe, with intravenous immunoglobulin.

Encephalitis

This is clinically evident in about 1 in 5000 cases, occurring soon after the rash. It is variable in severity, and can only be treated symptomatically.

Prevention and control

Rubella vaccine is a live attenuated vaccine, given as part of the combined measles/mumps/rubella (MMR) vaccine. The aim of the programme is to interrupt rubella transmission and thereby avoid infection in pregnancy (Fig. 11.7; see also Chapter 17). Rubella is a notifiable disease. Children should be excluded from school for 5 days after the onset of the rash.

Contact with pregnant women should be avoided. It is particularly important that all healthcare workers who are likely to be in contact with pregnant women are vaccinated against rubella. Pregnant women who have been in contact with a case, particularly during the first trimester, should be tested serologically for susceptibility (if not already tested), and for evidence of early infection (IgM antibody) and advised accordingly. Laboratory diagnosis of the original case should be sought, as the clinical diagnosis of rubella is unreliable.

Chickenpox (varicella)

Introduction

Chickenpox is a systemic viral infection with a characteristic vesicular rash. Infection is almost always symptomatic but not usually severe in immunocompetent children. Recovery from the infection is followed by latency of the virus in the dorsal root ganglia of the central nervous sys-

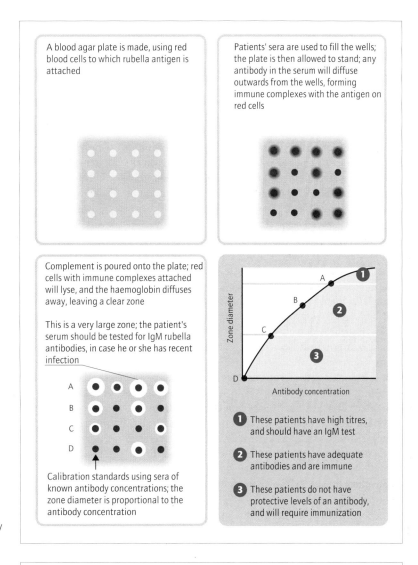

Figure 11.6 The principle of the radial haemolysis test in screening for immunity to rubella.

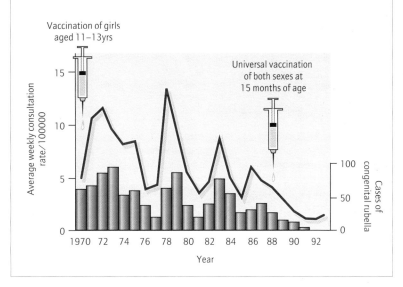

Figure 11.7 The effect of selective, followed by universal, childhood immunization on the incidence of rubella consultations in general practice and congenital rubella. Sources: Health Protection Agency and Royal College of General Practitioners.

tem. This can cause the reactivation disease, herpes zoster, which causes high morbidity in the elderly. Although zoster occurs mainly in adults and the elderly, children, especially those with immune deficiency, may also develop typical lesions. Patients on moderate and high doses of corticosteroids are at particular risk of severe illness. Infection with VZV (varicella zoster virus) is dangerous in individuals with cell-mediated immunodeficiency. Rare cases of fetal injury have occurred after infection in pregnancy (see Chapter 17). Active immunization is available, using a live-attenuated vaccine. Passive immunization, using specific VZV immunoglobulin, can prevent or ameliorate clinical disease in those at risk.

Epidemiology

The disease is highly infectious. The most important sources of infection are cases of chickenpox. Patients with herpes zoster are somewhat less infectious, but their susceptible contacts can develop chickenpox. Transmission occurs directly by person-to-person contact, by air-borne spread of respiratory secretions or vesicular fluid, and indirectly through articles recently contaminated by discharge from vesicles and mucous membranes. The period of communicability is usually from 1–2 days before, to 6 days after, the appearance of the first crop of vesicles, but may be longer, particularly in patients with immune deficiency. The incubation period averages 15–18 days (range 10–25 days). Humans are the only reservoir of infection.

In Western countries, chickenpox occurs predominantly in young children, with epidemics every 1–2 years, usually in winter and early spring. Over 90% of adults have naturally acquired immunity, but in the UK and some other Western countries the incidence of chickenpox in adults increased in the 1980s and 1990s (Fig. 11.8), though case-reports have now stabilized. In tropical countries the incidence of chickenpox is low in young children; older children and adults therefore remain susceptible.

Virology of varicella zoster virus

Varicella zoster virus (VZV) is a member of the Herpesviridae, and shares morphological and pathological characteristics with the other members. Antigenic variation has not been shown, although strain variation can be demonstrated by restriction endonuclease digestion. The virion consists of a central nucleocapsid core 90–95 nm in diameter and an outer membrane envelope 150–220 nm, from which glycoprotein spikes project. The genome consists of double-stranded DNA of mass 80 MDa, coding for 75 protein antigens. There are five major glycoprotein antigen families that have been detected: gp I–V. Viral infectivity may be neutralized by monoclonal antibodies to gp I, II and III.

VZV can be isolated in continuous cell culture using many simian and human cell lines, but the virus is strongly cell associated and therefore difficult to culture from serum and even from cell-free vesicle fluid. The VZV CPE is characterized by discrete foci of rounded, enlarged cells, and destruction of the monolayer.

Like other herpesviruses, VZV produces lifelong latency, and may reactivate in later life. This reactivation is usually confined to one or two dermatomes served by the dorsal root ganglia in which the latent virus exists. The mechanism of VZV reactivation remains unknown.

Clinical features

Children rarely have a prodromal illness but adults may suffer an initial few days of fever, headache and myalgia. Skin lesions first appear as small round papules, which are rarely noticed (Fig. 11.9). They progress rapidly to vesi-

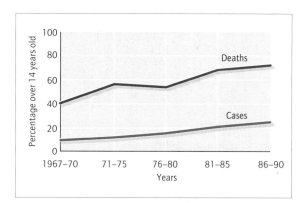

Figure 11.8 Chickenpox: the increasing incidence in older children and adults in the UK. Courtesy of Dr Elizabeth Miller, Communicable Disease Surveillance Centre.

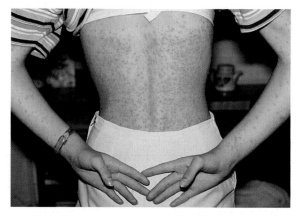

Figure 11.9 Chickenpox: the very early rash. Many papules are seen, which will all become vesicles in the next few hours.

cles, many of which are oval with their long axis along the creases of the skin, and are sometimes intensely itchy and inflamed. The appearance of clear vesicles on the head or trunk is often the first recognized sign of disease. These evolve to opaque pustules, which become umbilicated as they dry to crusts. For three to five days, new lesions appear as the older ones evolve (a process called cropping) but newer lesions are smaller and eventually fail to develop (Fig. 11.10). Lesions appear earliest and most densely on the trunk and face. The hands and feet are relatively spared (Fig. 11.11).

Superficial mucosal lesions can affect the conjunctiva, mouth and perineum, and usually heal without scarring. Oesophageal, gastric and intestinal lesions occasionally cause significant abdominal pain, and synovial lesions can cause joint pain and swelling.

A focal pneumonitis occurs in parallel with the rash. It is rarely clinically evident, but occasionally causes severe

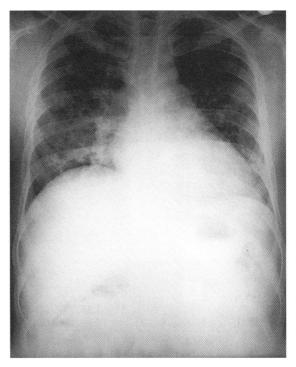

Figure 11.12 Chickenpox pneumonitis: extensive nodular pneumonitis. Respiratory failure required 5 weeks of assisted ventilation.

respiratory disease with blood-stained expectoration and widespread nodular opacities on chest X-ray (Fig. 11.12). Smokers are at greatly increased risk of severe pneumonitis. Lung lesions occasionally calcify on healing, producing a typical X-ray appearance (chickenpox lung).

Differential diagnosis

The diagnosis of chickenpox and herpes zoster (see Chapter 5) is usually clinically evident, because of the typical rash. Disseminated herpes simplex, hand, foot and mouth disease, guttate psoriasis and, occasionally, vesiculating allergic eruptions can cause confusion. Rickettsial pox, an arthropod-transmitted disease, is occasionally seen in deprived inner cities in the Americas, and can cause a chickenpox-like rash.

Laboratory diagnosis

The optimum specimens for viral diagnosis are smears of scrapings from the base of skin lesions, tissue biopsy or post-mortem material. Multinucleate giant cells can be found in the cellular debris from the base of vesicles and smears of this material can also be examined by electron

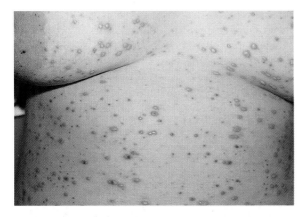

Figure 11.10 Chickenpox: typical rash showing non-coalescing lesions at all stages of development.

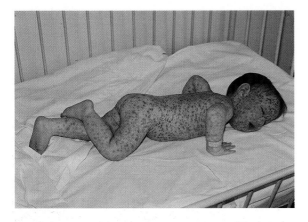

Figure 11.11 Chickenpox rash: the centripetal distribution spares the hands and feet.

microscopy, direct immunofluorescence and culture. Immunofluorescence (and electron microscopy) are more sensitive than viral culture for the diagnosis of VZV infection. VZV DNA can be detected by PCR. Virus growing in cell cultures is identified by electron microscopy or by immunofluorescence.

Serological diagnosis can be achieved using IgM and IgG ELISA to detect the primary antibody response to infection, to detect seroconversion, or to demonstrate established immunity.

Diagnosis of chickenpox

1 Varicella zoster virus DNA detection by PCR.
2 Immunofluorescent staining of vesicle scrapings.
3 Culture of vesicle scrapings.
4 Demonstration of IgM antibodies by ELISA.
5 Electron microscopy of vesicle scrapings.

Treatment

Specific treatment is rarely required in children. Adolescents and adults may suffer severe disease and should be treated, ideally starting on the first day of the rash. Treatment with oral aciclovir reduces the duration of fever and active rash by an average of 1 day, and may prevent serious complications. Immunosuppressed patients should be referred urgently for intravenous therapy. Pruritus may be ameliorated by antihistamines. Fewer lesions will develop, and irritation will be less if the skin is kept cool, with light clothing and frequent cool washes. Mild analgesia may be helpful for painful lesions.

Problems and complications

Severe forms of chickenpox

Widespread tissue damage occurs, with pneumonitis, often a variable degree of intravascular coagulation and, rarely, multi-organ failure.

Warning signs of severe chickenpox

1 Dense and/or haemorrhagic rash.
2 Numerous mucosal lesions.
3 Severe substernal and epigastric pain (probably indicating many mucosal lesions).
4 Reduced arterial oxygen saturation.

By the time that abnormal physical signs appear in the chest, respiratory failure is often already established. In the immunocompromised, pneumonitis may occur with little or no rash.

Treatment with aciclovir (10 mg/kg, 8-hourly by intravenous infusion) may ameliorate the disease, but must be given early for the best effect, as the advancing lesions are not halted until 24–48 h after commencing therapy. Aciclovir must be given by slow infusion over 1 h to avoid nephrotoxicity, which is related to peak blood levels. The blood urea and creatinine levels should be monitored during therapy. Renal failure demands modification of the dose of aciclovir, which is excreted via the kidneys.

Secondary bacterial bronchopneumonia is common (see below), and appropriate antibiotic therapy should always be added.

Secondary bacterial infections

Super-infection of skin lesions by *Staphylococcus aureus* or *Streptococcus pyogenes* is common, particularly in children. Children lack antibodies to the toxins produced by these organisms, and are at risk of erysipelas, scarlet fever or toxic shock syndrome (Fig. 11.13). They should be promptly treated with appropriate antibiotics. If skin lesions become severely inflamed, painful, abscess-like or necrotic, swabs should be taken and parenteral antibiotics commenced.

Secondary staphylococcal pneumonia is particularly common in adults, especially smokers. Any patient with respiratory symptoms should be treated with an antistaphylococcal antibiotic as mixed viral and staphylococcal pathology is common. Admission to hospital should be considered. Other secondary bacterial chest infections may occur, and should be treated with co-amoxiclav or a broad-spectrum cephalosporin.

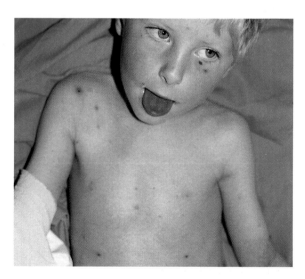

Figure 11.13 Chickenpox with staphylococcal secondary infection. An abscess has developed in an affected lesion, and the child has an erythematous rash of toxic shock syndrome.

Treatment of chickenpox

1 Symptomatic – keep the skin cool and give antipruritics.

2 Aciclovir orally; adult 800 mg five times daily for 7 days; child 20 mg/kg (maximum 800 mg 6-hourly).

3 *For immunosuppressed or severely ill patients*: aciclovir i.v. (by infusion over 1 h) 10 mg/kg 8-hourly for 5–7 days.

4 Antibiotic treatment of secondary skin or chest infection – include an antistaphylococcal drug.

⚠ Aciclovir causes blood-level dependent renal impairment, and is renally excreted; the dose should be modified in patients with creatinine clearance <60 ml/minute.

Post-chickenpox encephalitis and other neurological diseases

Post-chickenpox encephalitis is uncommon. While many cases are self-limiting, severe disease can also occur. Cerebellar disturbance is rare and usually seen in boys or young men, who present with ataxia and nystagmus as the rash heals. Cases with other neurological signs or altered consciousness have a more unpredictable outcome. Specific treatment is indicated for post-infectious conditions.

Rare cases of meningitis, necrotizing retinitis, transverse myelitis or ascending myeloencephalitis are seen, and can occur before, during or after the rash. All are more common in the immunosuppressed. The diagnosis should be sought, as aciclovir is effective in many cases.

Thrombocytopenia and other purpuras

This is not uncommon. It may make the rash appear haemorrhagic, but the concurrent appearance of dependent purpura and microscopic or macroscopic haematuria will indicate the diagnosis (Fig. 11.14). The condition is transient and can be managed by a brief course of corticosteroids, with or without platelet transfusion. Intravenous immunoglobulin is also effective.

There are rare reports of severe purpura, caused by transient production of antibodies to protein S or protein C.

VZV-associated vasculitis

Vasculitic events, including strokes and retinal vasculitis, occurring during or after VZV infection, have been described in both adults and children. Some investigators have found PCR evidence of VZV in affected vascular tissue, while others could not. Intrathecal production of VZV antibodies has also been described. The hypothesis that VZV activity can precipitate medium and small vessel vasculitis has not been proven.

Prevention and control

Live attenuated varicella vaccine is highly effective. In

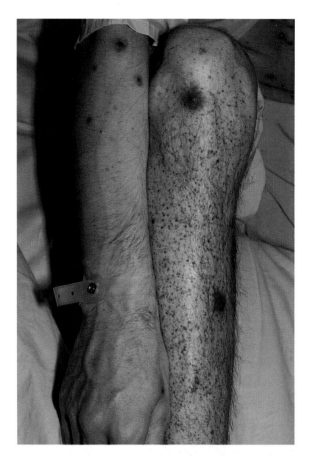

Figure 11.14 Thrombocytopenia in chickenpox. There has been bleeding into the vesicular lesions and there is a purpuric rash on the leg. The patient had gross haematuria.

the UK it is available for the immunization of healthcare workers seronegative for VZV, who are at high risk of occupational infection. This protects both the worker, and the patients to whom they could transmit VZV. The vaccine can also protect children with leukaemias and lymphomas, when it is given during chemotherapy-induced remission. A minority of vaccinees develop a rash from the vaccine. This is infectious, and healthcare workers should remain away from work until the lesions have healed. Herpes zoster can also be caused by the vaccine virus (more often in immunosuppressed recipients).

Universal vaccination in early childhood has been introduced in the USA, where a combined measles/mumps/rubella/varicella (MMRV) vaccine is available.

Varicella zoster immunoglobulin (VZIG) prepared from human plasma containing high titres of specific antibody is available for passive immunization. It can be requested from the Health Protection Agency Centre for Infections and other designated centres. It is indicated for those without antibodies to the virus who have had a significant exposure to chickenpox or herpes zoster and who have a clinical condition that increases the risk of severe varicella. There is no evidence that children with asthma or eczema suffer unusually severe or complicated chickenpox.

Children with chickenpox should be excluded from school for 5 days after the appearance of the first crop of vesicles. Patients are no longer infectious when all of the lesions are dry and scabbed. In hospital, cases of chickenpox and shingles should be isolated, because of the risk to immunosuppressed patients.

Prevention of chickenpox

1 Isolation of patients with fluid in their vesicles.
2 Post-exposure varicella zoster immunoglobulin (VZIG) for individuals who fulfil all the following criteria:
  • a clinical condition that increases the risk of severe varicella, e.g. immunosuppressed patients, neonates, pregnant women;
  • no antibodies to varicella zoster virus;
  • significant exposure to chickenpox or herpes zoster.
3 Live attenuated vaccine for healthcare workers and other at-risk groups.

Human parvovirus B19 and erythema infectiosum (slapped cheek syndrome)

Introduction

Human parvovirus B19 (HPV-19) causes a viraemic disease with a rash (erythema infectiosum) in which the virus particularly attacks rapidly dividing cells. Epidemics of feverish illness with rash and arthralgia are recognized. Infection of red cell precursors occurs and can cause a transient aplastic crisis in children with inherited haemolytic anaemias. Infection during the second trimester of pregnancy carries a risk of fetal anaemia and hydrops (see Chapter 17). Persistent infection in the immunocompromised can cause prolonged aplastic anaemia (see Chapter 22).

Epidemiology

The disease appears to be highly infectious. Transmission is person to person, by respiratory droplet infection or, occasionally, through contaminated blood products. Infection is commonest in children between 5 and 14 years. Outbreaks in schools are common, and usually occur in late winter or spring.

Virology and pathogenesis of human parvovirus B19 infection

Parvoviruses are small DNA viruses that infect a wide range of animal species. A single serotype, B19, infects humans, and replicates only in cells that are in the S phase of mitosis. It is a member of the Parvovirinae subfamily, which includes Parvovirus (which can replicate autonomously), Densovirus (which requires a helper virus) and Erythrovirus, which replicates in erythroid progenitor cells. As parvovirus B19 replication only occurs in erythrocyte precursors, this virus is an Erythrovirus.

Parvovirus is a small, dense, icosahedral virus that is not enveloped. The genome is a single-stranded linear DNA, which may be of positive or negative polarity. Three non-structural proteins are produced, which participate in replication and encapsidation, as well as in regulation of transcription. The two capsid structural proteins are VP1 and VP2, of which VP2 is the major component. Antigenic variation is limited, usually less than 2%. The virus may be cultivated in tissue culture but this is too insensitive for routine clinical use. The virus is stable to heating at 56 °C for more than 1 h. It is resistant to ether and chloroform and survives at room temperature for a prolonged period.

The tissue tropism of parvovirus B19 is limited to the erythroid progenitor cells because it depends on a globoside P antigen, which is the receptor for the virus. This receptor exists only on red cell progenitors. Individuals who lack this antigen cannot be infected with parvovirus. Infection results in an interruption of red cell production. Since the life span of red cells is approximately 120 days this deficit is not of importance to healthy individuals. Those who have a condition with high red cell turnover, such as a haemolytic anaemia, or persisting blood loss, cannot tolerate this interruption and an aplastic crisis results. The characteristic rash is thought to be an effect of immune complexes.

Clinical features

The average incubation of 14 days (range 1–3 weeks) is followed by viraemia and often fever. In 2 or 3 days the fever falls and the rash appears. In children the cheeks may be bright red (slapped cheek syndrome). In all age groups the rash affects the limbs, and less often the body. It may be rubelliform but is often reticulate. It fades quickly, recurring transiently if the skin is warm. Bizarre petechial or haemorrhagic rashes can occur, particularly affecting the peripheries or perineum, but the patient is rarely severely ill. Painful generalized arthralgia particularly affects young women, and can last for several weeks, improving with a fluctuating course.

Infection of red cell precursors in the bone marrow causes a pause in erythropoiesis coinciding with the rash. Reticulocytes become undetectable in the blood film and there may be transient anaemia, often profound in children with inherited haemolytic anaemias. Erythropoiesis resumes as the infection resolves. Immunosuppressed patients may be unable to clear their infection, leading to continued hypoplastic anaemia. Aplasia can occur in patients who have no rash, causing diagnostic confusion.

Rare presentations of parvovirus infection include encephalitis and hepatitis.

Differential diagnosis

The acute illness must be distinguished from rubella. Clinical distinction is difficult, so laboratory tests for both rubella and parvovirus should be performed (both infections can adversely affect pregnancy: see Chapter 17).

Other rash diseases such as scarlet fever and toxic shock syndrome should be suspected if the white cell count is raised. Wide fluctuations in joint symptoms may lead to a suspicion of rheumatic fever.

Laboratory diagnosis

The diagnosis is usually made on the basis of the characteristic clinical features. The difficulties of *in vitro* cultivation of parvovirus mean that serology and genome detection by PCR are the principal methods of diagnosis. Specific IgM and IgG ELISA tests can be used to detect rising antibody titres. Antibodies develop at the end of the transient viraemic phase, and are usually present when the rash appears.

Parvovirus can be demonstrated by electron microscopy in the serum of infected patients. Antibodies to parvovirus can be found in up to 50% of the normal population. Investigation of possible fetal infection requires sampling from the amniotic fluid, fetal cord blood or post-mortem tissue.

Treatment

There is no specific treatment. Non-steroidal anti-inflammatory agents help the arthralgia. Transfusion may be required for aplastic crises. Normal human immunoglobulin contains antibodies to parvovirus, and intravenous immunoglobulin (IVIG) treatment may terminate or ameliorate bone marrow infection in the immunosuppressed. Anecdotal reports exist of improvement in chronic aplasia after IVIG treatment.

Prevention and control

Exclusion of cases from school is of no value, as most susceptible children will already have been exposed to the viraemic patient by the time the characteristic rash appears. If a midwife or healthcare worker has HPV-19 infection, susceptible contacts such as pregnant women and immunosuppressed individuals, or those with haematological disorders, should be identified, as they can be offered human normal immunoglobulin, which may give a degree of protection from infection.

Human herpesvirus type 6

Human herpesvirus type 6 (HHV-6) was originally thought to be a B-lymphotrophic virus, but is now known to infect many types of cells. Serosurveys suggest that most people are infected early in childhood, subsequently developing antibodies and harbouring latent virus.

HHV6 shares the characteristics of herpes viruses. The genus can be divided into two groups, A and B, that are genetically homogeneous. Transmission is thought to occur via saliva. The virus replicates in lymphocytes and persists in the mononuclear cell population throughout the body. The virus persists by down-regulating host immune responses. There is evidence that HHV-6 and HIV mutually up-regulate one another's expression.

The diagnosis of infection can be made by ELISA, but since almost everyone possesses antibodies after the age of two, acute infection is indicated by seroconversion or a significant rise in titre. Although HHV-6 can be cultured from peripheral blood lymphocytes, PCR represents a more straightforward approach to diagnosis, and quantification of the viral load permits the presumptive diagnosis of active disease.

Roseola infantum (erythema subitum) is the clinical manifestation of HHV-6 seroconversion. It is an illness of infants in which 2 or 3 days of fever are followed by a widespread morbilliform rash that starts on the trunk. As in other viral infections, leucopenia is common. Small and large outbreaks have been described, with an incubation period of 10–15 days. HHV-6 may also be a rare cause of severe hepatitis, especially in the immunosuppressed.

Human herpesvirus type 7

Human herpesvirus type 7 (HHV-7) has been demonstrated in the lesions of pityriasis rosea, and seroconversion to HHV-7 has been associated with this disease. However, almost everyone is infected by the age of five and there is no certainty of the role this virus plays in human disease.

Kawasaki disease (mucocutaneous lymph-node syndrome)

Introduction

Kawasaki disease is a vasculitic disease of children whose main features are fever, rash, mucocutaneous inflammation and lymphadenopathy. Its cause is unknown, but is likely to represent the response to an infection because moderate to large epidemics sometimes occur. Although uncommon, the disease has a significant morbidity and mortality. Cardiac ischaemia due to coronary aneurysms makes Kawasaki disease the commonest reason for heart transplant in children.

Clinical features

The patient is usually a toddler, but rare cases occur in older children and adults. Illness begins with irregular fever, which is often severe. Swelling of the hands and feet is common and a rash soon appears. There is usually an intense 'glove-and-stocking' erythema; adjacent areas of the limbs develop round, slightly raised lesions of various sizes while the rest of the body may be covered with a faint erythema. A shiny or scaly rash may affect the perineum (Fig. 11.15). Other early features include oral inflammation, fissuring of the lips, angular cheilitis and conjunctival injection. There is moderate or gross enlargement of cervical lymph nodes.

After 3–6 days the peripheral rash begins to desquamate. The skin from the ends of the digits often breaks away, characteristically in one piece (Fig. 11.16). Elsewhere the rash may persist for many days.

Even after desquamation, the fever persists, and systemic features become increasingly important. Arteritis predominates and, like polyarteritis nodosa of adults, can cause aneurysm formation. Coronary artery aneurysms can be large and multiple, leading to myocardial ischaemia or infarction.

Mucocoele of the gallbladder, which is often present, varies in severity from simply an ultrasound finding to an acute surgical emergency.

Mild, self-limiting forms of the disease probably occur. Cardiac events without a preceding illness have not, however, been recognized.

Diagnosis

There is no specific diagnostic test, but the evolution of the skin and mucous membrane lesions combined with lymph node enlargement is quite characteristic. The

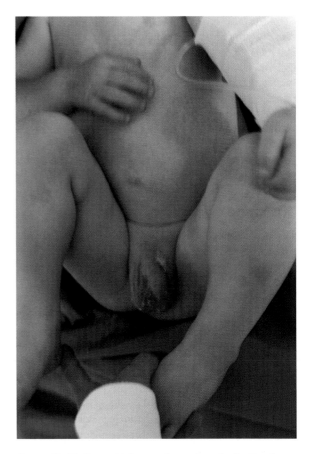

Figure 11.15 Kawasaki disease: the acute rash affecting the peripheries and the perineum.

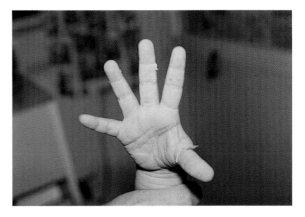

Figure 11.16 Kawasaki disease: typical desquamation of the fingertips.

British Paediatric Surveillance Unit uses a clinical definition.

Clinical definition of Kawasaki syndrome

1 Fever, otherwise unexplained, lasting 5 days or more.
2 Bilateral conjunctival congestion.
3 Generalized erythema of buccal and pharyngeal mucosae.
4 Localized cervical lymph node enlargement.
5 Polymorphous rash with changes in extremities.
6 Desquamation of hands and feet.

As the illness progresses, the erythrocyte sedimentation rate and the platelet count both rise dramatically. There is a moderate leucocytosis. Abnormal liver enzyme levels are common, particularly a rising alkaline phosphatase. Over half of patients have positive anti-neutrophil cytoplasmic antibodies (ANCA).

Ultrasound imaging often shows distension of the gallbladder. It is essential to perform echocardiography, both at presentation to detect any existing aneurysm and later to check that coronary artery dilatation is not occurring.

Treatment

Treatment with intravenous immunoglobulin inhibits the development of coronary artery aneurysms, and also contributes to reduced fever and distress (Fig. 11.17). An infusion of 400 mg/kg should be commenced as soon as the diagnosis is evident. This may be repeated once or twice more if fever continues.

Kawasaki disease is an indication for aspirin treatment, even in children. Full doses of 50 mg/kg daily should be given while the platelet count is raised. Smaller doses, e.g. 10–15 mg/kg daily, may then be used. Most experts would continue aspirin for at least 3 months, longer if the erythrocyte sedimentation rate and platelet count remain raised. There is no evidence of benefit from the addition of dipyridamole to this regimen.

Complications

Myocardial ischaemia or infarction

This is the most important complication. Repeated echocardiography will give warning of developing aneurysms. Large aneurysms of more than 0.9 mm diameter are dangerous, and paediatric cardiologists may consider surgical intervention if these are seen. Myocardial ischaemia is accompanied by pain, as in adults, and electrocardiogram and cardiac enzyme estimations must be performed. Rarely, infarction has occurred after cessation of active treatment. Severe heart failure following infarction may be an indication for heart transplant.

Mucocoele of the gallbladder

This is a rare cause of abdominal emergency in Kawasaki disease. If peritonism or gross pain and enlargement of the gallbladder occur, surgery may be indicated, but the risk of anaesthesia must be taken into account if the heart is compromised.

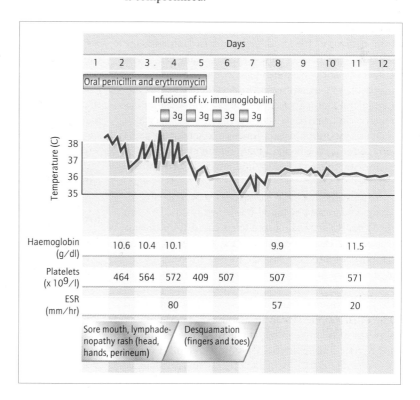

Figure 11.17 The course of Kawasaki disease and response to intravenous immunoglobulin. ESR, erythrocyte sedimentation rate.

Infections of the Cardiovascular System

12

Introduction

The cardiovascular system has three main structural components.

1 The intimal lining of the blood vessels, and the endocardium, are composed of endothelium and are in contact with the blood. Damage to the endothelium or endocardium disrupts its smoothness and also initiates the mechanisms of platelet adhesiveness, causing the development of a sticky plaque of platelet thrombus. This can trap circulating organisms and protect them from the immunological mechanisms of the blood. Endothelium has numerous functions, mediated by chemical factors, which initiate or modulate platelet function, coagulation pathways, leucocyte chemotaxis, margination and migration, inflammatory responses and fluid and electrolyte movement.

2 The muscular media of the arteries is occasionally invaded and damaged by blood-borne organisms. This can cause aneurysms, leading to altered circulation, thrombosis and occlusive disease. Infected aneurysms can act as a source of infectious emboli, further affecting the circulation distal to it. The myocardium is also susceptible to infection; organisms can penetrate from infected heart valves and cause myocardial abscesses; many viruses and some other pathogens can invade the myocardium and cause infective myocarditis, sometimes compromising cardiac function. Toxic shock syndrome toxins, *Corynebacterium diphtheriae* toxin and other toxins can damage myocardial cells.

3 The pericardium is a mesothelial structure, with the same embryonic origin as the pleura. The visceral and parietal pericardial layers have a potential space between them. The pericardium can be invaded directly, causing isolated pericarditis, or it can be involved by extending pleural or myocardial infection. Exudate in the pericardial space can immobilize the heart, reducing its ability to fill and sometimes leading to dangerous tamponade.

Some infecting organisms target a particular part of the cardiovascular system. Rickettsiae attack endothelial cells and cause endovasculitis, predisposing to thrombosis, and haemorrhage. Late syphilis causes aortitis, which damages the elastic layer of the ascending aorta, with aneurysm formation.

Pericarditis

Introduction

Pericarditis is inflammation of the pericardium. While it is often infective in origin, other causes include reactive pericarditis after myocardial infarct, autoimmune diseases, hypothyroidism, malignant invasion and involvement in the pancarditis of rheumatic fever. Non-infectious pericarditis should therefore be considered in the initial assessment of the patient.

A variety of pathogens can cause pericarditis, either alone or accompanied by myocarditis. Acute pericarditis can be a focal complication of bacteraemic disease, such as staphylococcal or streptococcal septicaemia. Contiguous spread of pus from an empyema or from a liver abscess can cause pericarditis with organisms such as pneumococci, enterococci or *Entamoeba histolytica*. Tuberculosis can cause subacute pericarditis, sometimes with thickening and stiffening of the pericardium (constrictive pericarditis).

Organism list

- Enteroviruses, especially coxsackieviruses
- Influenza viruses
- *Mycoplasma pneumoniae*
- *Chlamydophila pneumoniae*
- *Streptococcus pneumoniae*
- Other Gram-positive cocci
- *Mycobacterium tuberculosis*
- *Coxiella burnetii*

Clinical features

Viral pericarditis

The commonest type of pericarditis is a self-limiting illness with fever, normal or neutropenic white cell count and typical chest pain, accompanied by cardiographic evidence of pericarditis. It is often preceded or accompanied by symptoms of malaise, myalgia or arthralgia, or sore throat, suggesting a viral aetiology. Precordial pain then develops. Usually pleuritic, it often varies with different postures (it is frequently exacerbated by lying flat), with swallowing or with the heart beat, and sometimes radiates to the neck, shoulders or back. Auscultation may reveal a pericardial rub, which is often transient and variable, disappearing if effusion separates the pericardial layers.

Bacterial pericarditis

This can occur in isolation or as part of a condition, such as atypical pneumonia or empyema. Pain is often severe, and tamponade may occur, with a falling blood pressure and pulse pressure, and a paradoxical rise in the jugular venous pressure on inspiration. In pyogenic conditions there is swinging fever, neutrophilia, and sometimes features of sepsis.

Diagnosis

Electrocardiography (ECG) often shows upward-curved elevated S-T segments in the anterior chest and sometimes other leads. Echocardiography demonstrates pericardial thickening or effusion, and permits an assessment of cardiac function. On chest X-ray, a pericardial effusion shows as an enlarged, globular heart shadow, which may have a double outline: one for the heart itself and one for the border of the distended pericardium. CT scanning also clearly demonstrates pericardial effusions (Fig. 12.1).

In viral-type pericarditis, nose swabs, throat swabs, urine and stool should be submitted for virus culture. These specimens, or blood, can also be examined by PCR techniques. An early serum sample should be taken. Immunoglobulin M (IgM) antibodies to enteroviral antigens may be demonstrable in viral pericarditis. Detection of *Mycoplasma*-specific IgM or detection of rising IgG titres indicates a diagnosis of *Mycoplasma pneumoniae* infec-

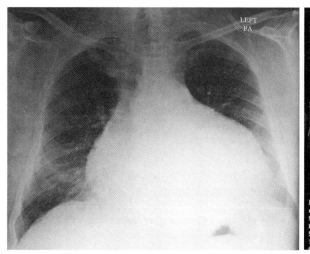
(a)

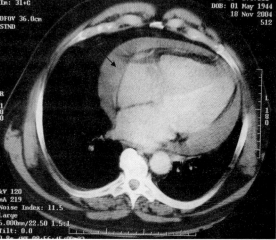
(b)

Figure 12.1 Pericarditis with effusion. This large effusion probably developed quite slowly, as there was no evidence of tamponade. (a) Chest X-ray showing enlarged, globular heart shadow; (b) computed tomographic scan showing a rim of pale material surrounding the ventricles (arrow).

tion, and serological test results can also reveal rarer conditions such as *C. pneumoniae* or Q fever.

In suspected pyogenic infections, blood and sputum cultures should be performed. The chest X-ray should be examined for evidence of pneumonia, abscess or tuberculosis. If pericardial fluid is obtained, specimens should be submitted for PCR examinations, cytology, bacterial culture, acid-fast staining and mycobacterial culture, as indicated.

> ⚠ Tapping the pericardium carries a risk of damage to the atrial wall or coronary vessels. It should only be performed if clearly indicated for diagnostic or therapeutic reasons, and should preferably be carried out by an experienced operator with imaging control (such as computed tomographic (CT) scanning).

Management

Viral pericarditis can be treated symptomatically with bed rest and analgesia. Improvement in fever, pain and ECG signs usually occurs in 2 or 3 days and gradual mobilization can then begin. Dysrhythmia or heart failure is rare, as are ECG abnormalities other than S-T elevation; if present, these may warn of accompanying myocarditis.

Pyogenic pericarditis must be treated urgently with an appropriate choice of empirical intravenous antibiotics. If there is localized pneumonia, suggesting pneumococcal aetiology, high doses of a broad-spectrum cephalosporin are indicated. Flucloxacillin or another antistaphylococcal drug should also be given if the aetiology of the infection is unknown or a lung abscess suggests possible staphylococcal bacteraemia. It is also advisable to add it when treating an intravenous drug abuser. In a severely ill patient with few clinical signs, *Streptococcus pyogenes* must be considered; high-dose penicillin may be added to the regimen in this case.

If tuberculosis is suspected, sputum, bronchoalveolar lavage specimens and urine should be examined. A tuberculin test should be 'planted' and triple or quadruple therapy then considered. If a decision is made to give antituberculosis treatment, corticosteroids should be given at the same time, to avoid a sudden increase in pericardial effusion or oedema (see Chapter 18).

Complications

Cardiac tamponade is caused by effusion or thickening of the pericardium that interferes with the normal filling of the heart. Warning signs are low systolic or pulse pressure, shortness of breath first appearing on exertion, and raised jugular venous pressure, often with a paradoxical rise on inspiration (Kussmaul's sign). Tamponade can develop suddenly, and occasionally causes severe heart failure, which can be relieved by drainage (see cautions above).

In all cases of severe pericarditis, there is a risk of pericardial fibrosis, with slow development of tamponade (constrictive pericarditis). Such cases often present with congestive cardiac failure, and signs of tamponade must be carefully sought, both clinically and by imaging or cardiac angiography. Surgical treatment (by forming a large window in the pericardium, or stripping adherent tissue) may be curative.

Myocarditis

Introduction

Myocarditis is inflammation of the myocardium. Many cases are probably of infectious aetiology, and the majority of these are viral. Many viral cases are purely myocarditis, but myocarditis can also complicate multisystem infections such as influenza, infectious mononucleosis, mumps, adenovirus or *Mycoplasma* infections. The incidence varies with the seasons and from year to year, as would be expected for a viral infection. A few cases occur as part of a systemic disease such as a bacteraemia, brucellosis or a rickettsial infection. Diphtheria toxin causes destruction of myocardial cells (see Chapter 6).

Borrelial (Lyme) myocarditis is a feature of early disseminated borreliosis, particularly common in *B. burgdorferi* infections, acquired in the USA. It rarely causes clinically apparent heart failure, but commonly presents as a self-limiting episode of complete heart block.

Non-infectious causes of cardiomyopathy include inflammatory conditions such as sarcoidosis, reactions to some drugs (e.g. doxorubicin), thyrotoxicosis, vitamin B1 deficiency (in alcoholism and starvation), infiltrations (particularly amyloid) and some muscular dystrophies.

Organism list

Viruses
- Coxsackie B (causes about 50% of enteroviral myocarditis)
- Coxsackie A
- Echovirus
- Influenza A and B viruses
- Rubella
- Epstein–Barr virus
- Cytomegalovirus
- Adenovirus
- Mumps
- Rarities: rabies, hepatitis viruses, vaccinia vaccination

Bacteria
- *Coxiella burnetii*
- *Leptospira* spp.
- *Borrelia* spp. (particularly *B. burgdorferi*)
- *Mycoplasma pneumoniae*
- *Neisseria meningitidis* and other pyogenic organisms

Pathology and epidemiology

The pathogenesis of myocarditis is probably different in different types of infection. In systemic diseases, the myocarditis is often concurrent with the acute infection. Examples include influenza and mumps infections, *Mycoplasma pneumoniae* and *Coxiella burnetii* infections.

In some viral infections, particularly coxsackie B infections, the myocarditis is delayed and occurs when viruses can no longer be isolated from the respiratory tract or bowel. In animals, this type of myocarditis can be prevented by disabling cell-mediated immunity. Nevertheless, coxsackievirus RNA has been demonstrated by polymerase chain reaction (PCR) in the myocardium of a high proportion of patients with this type of myocarditis (and also in patients with dilated cardiomyopathy of unknown aetiology).

Clinical features

The clinical presentation is variable: acute cases may complain of fever and prostration; mild cases may comprise only self-limiting symptoms of fatigue. There is often a history of feverish symptoms in the previous 2 weeks, sometimes more recently. Fatigue, exertional dyspnoea and palpitations are common symptoms, sometimes with dull precordial pain. Severe cases may present with intractable cardiac failure. Physical examination may reveal fever, hypotension, tachycardia, dysrhythmia or features of right and/or left ventricular failure. Conduction abnormalities with heart block, extrasystoles or tachycardias may occur: heart block is common in borelliosis and diphtheria. Symptoms and signs of pericarditis may coexist.

The ECG shows non-specific changes in the T waves, often inversion. There may be an abnormal axis, small complexes, prolongation of the P-R, QRS or QT interval, extrasystoles or heart block. S-T depression is also seen, or elevation, especially if pericarditis coexists. Laboratory tests may reveal elevation of the cardiac enzymes, which can persist for many days, but is not always present. The chest X-ray may show a widened heart shadow. In bacterial disease, such as Q fever or leptospirosis, other systemic signs of the disease are usually present (see Chapters 7 and 9).

Myocarditis is rare in infants, but is often severe or fulminating, with significant cardiac failure accompanying signs of active viral infection. Older children tend to have relatively mild disease.

Diagnosis

The diagnosis is suggested by the relationship of fever, viral symptoms or typical features of Lyme borreliosis, to the development of cardiological abnormalities. Examination of the patient may reveal features of infection in other organ systems. A chest X-ray may reveal atypical pneumonia. Cardiac function should be assessed clinically and by echocardiography. A coexisting pericardial effusion should be sought, and also investigated if found.

Enteroviruses may be recovered in culture from the throat or stools. PCR-based techniques are becoming available for rapid diagnosis, and can also be applied to blood samples. Respiratory viruses can be detected by direct immunofluorescence in nasopharyngeal or throat specimens.

Rickettsial or mycoplasma infections, or borreliosis, may be diagnosed by serology.

Endomyocardial biopsy may allow demonstration of a pathogen by culture, immunofluorescence or PCR. Diffuse myocardial inflammation can be demonstrated by magnetic resonance scanning.

Non-infectious causes of myocardial disease include thyroid disease, vasculitic and autoimmune diseases, sarcoidosis, thiamine deficiency, drugs such as ethanol or anticancer drugs, and infiltrations such as amyloidosis.

Management

Specific antiviral treatment is not yet available for most viral causes of myocarditis, and treatment must therefore be symptomatic. Bacterial pericarditis should be managed with appropriate antibiotic treatment. Supportive treatment is important. Myocardial damage in diphtheria may be limited by early antitoxin treatment (see Chapter 6). Bed rest in the early stages may limit both inflammation and the development of heart failure. If heart failure or severe cardiomyopathy develops, increasing anti-failure medication, inotropic support or ventricular assistance may be required, and should be supervised by a cardiologist. In severe disease, once the acute viral infection is over, cardiac transplant may be indicated. Many viral cases will recover to a greater or lesser extent; slow improvement is possible even after prolonged disability.

Infective endocarditis (IE)

Introduction

Infective endocarditis (IE) is infection of the lining of the heart, particularly the cusps of valves. It can also affect other intracardiac sites where turbulent blood flow and/or endothelial damage encourages deposition of platelets or bacteria. Platelet thrombi form on the affected site, offering a protected nidus where pathogens can adhere and replicate. Many thrombi are friable, producing emboli. Endocarditis is important because it can be difficult to diagnose, may require prolonged and closely supervised treatment and is uniformly fatal if untreated.

Epidemiology

Up to the 1950s, endocarditis was closely linked to rheumatic heart disease, and affected the damaged mitral or sometimes aortic valve. Bacteraemia of oral or dental origin was the usual source of infection. The high prevalence of dental caries and gingival disease, and the resulting high rate of dental interventions, increased this risk.

Rheumatic heart disease and poor oral health are now uncommon, but the detectable rate of endocarditis has remained at 1000–1500 cases per year. Predisposing factors are now different. Congenital valve conditions, particularly bicuspid aortic valve and floppy mitral valve, are common precursors of native valve endocarditis. Small ventricular septal defects are at risk, because of the large pressure gradient and intense turbulent flow that they cause; vegetations readily form on the downstream side of the orifice. Elderly patients may have fibrosis or calcification on valves, on chordae tendinae or on the myocardial wall, or occasionally a mural thrombus may become infected. Younger patients remain at risk from mouth organisms. Older patients may have bacteraemias of genital or bowel origin, and enterococci are relatively common in native valve endocarditis of the elderly.

Intravenous drug abusers suffer repeated bacteraemias, often with skin organisms such as *Staphylococcus aureus*, but also with pathogens that contaminate their drug preparations. These may include organisms such as *Candida albicans*, derived from lemon juice used as a diluent. Because the source of the contamination is venous, the right side of the heart is often affected, and gives rise to numerous infected pulmonary emboli. Right-sided endocarditis is otherwise rare.

The advent of high-dependency care and transplant surgery has created a new population at risk of nosocomial infective endocarditis, mainly related to intravascular devices. Cardiac valve replacement generates a risk either from early, intrasurgical infection, predominantly with skin staphylococci, or from later haematogenous infection of the implanted valve. This affects particularly children treated for congenital disorders, or the middle-aged and elderly with acquired valvular damage. Long-term anticoagulation, to protect mechanical valves from thrombus formation, may reduce the risk of vegetations and endocarditis.

> **Predispositions to endocarditis**
> **1** Congenital valve disease.
> **2** Septal defects (usually ventricular).
> **3** Degenerative valve disease.
> **4** Rheumatic heart disease.
> **5** Mural thrombus.
> **6** Intravenous drug abuse (right-sided infections).
> **7** Cardiac surgery, including artificial or biological valve implants.

Pathology

Organism list
Native valve infection: Young patients
- Viridans streptococci
- *Streptococcus sanguis*
- *Streptococcus mitior*
- *Streptococcus mutans*
- *Streptococcus mitis*
- *Streptococcus 'milleri'*
- Enterococci
- Staphylococci
- *Coxiella burnetii**

Native valve infection: Elderly patients
- Enterococci
- *Enterococcus faecalis*
- *Streptococcus bovis*
- *Enterococcus faecium*
- *Streptococcus durans*
- Viridans streptococci
- Staphylococci
- *Coxiella burnetii**

Prosthetic valve infection
- *Staphylococcus aureus*
- *Staphylococcus epidermidis*
- Viridans streptococci
- Enterococci
- *Coxiella burnetii**

Rare organisms
- Small Gram-negative rods (*Haemophilus, Actinobacter, Cardiobacter, Eikenella, Kingella* 'HACEK organisms'*; cause about 3% of all IE cases)
- Enterobacteriaceae
- *Chlamydia psittaci**
- *Brucella* spp. (in endemic areas)
- Fungi (e.g. *Candida* spp.)*
- Mycobacteria (often 'atypical')*
- *Bartonella henselae**

Nosocomial infective endocarditis
- *Staphylococcus aureus* (including methicillin-resistant organisms)

- Enterococci
- *Aspergillus* spp.*
* These organisms are causes of culture-negative endocarditis

Clinical features

Clinical features can be divided into:
1 early manifestations of infection;
2 embolic events; and
3 late complications of sepsis and inflammation.

The earliest features are usually fever and heart murmur, with or without malaise and fatigue. Infection with aggressive organisms, such as *Staphylococcus aureus*, can present with high fever or sepsis. However, the onset is often subtle, the fever slight and the murmur easily dismissed as a flow murmur. Characteristically, the murmur changes as vegetations develop and valvular patency alters, but this may not occur for days or sometimes weeks. At this stage there is often a mild leucocytosis, and the C-reactive protein level is elevated. Tissue damage affects the valves of the heart, which may rupture or fragment, causing sudden heart failure. Pus may form in the valve rings or cardiac septum, and occasionally ruptures into the pericardial sac.

Features of embolization appear after days or weeks. Early emboli are seen in more aggressive endocarditis, for instance *Staphylococcus aureus* endocarditis. Fungal endocarditis can produce frequent and massive emboli, which occasionally manifest as a stroke, or other acute event, before endocarditis is suspected. Endocarditis usually affects the left side of the heart, so emboli are usually systemic. Showers of small emboli may cause petechial skin lesions, episodes of haematuria or splinter haemorrhages in the nails. Large emboli occasionally cause strokes, infarcts of the kidneys or acute arterial occlusions. In rare cases of right-sided endocarditis (often in intravenous drug abusers), showers of pulmonary emboli or pulmonary infarcts are typical. These can occur, paradoxically, in left-sided endocarditis with a septal defect and left-to-right shunt. Emboli may be infected and produce local infective vasculitis (Janeway lesions; Fig. 12.2), infective (marantic) aneurysms or lung abscesses.

The long-term effects of endocarditis, now rarely seen, are immunologically mediated. Immunological effects include splenomegaly, nephritis, vasculitic rashes or lesions of the eyes and skin. Single vasculitic lesions occur in the finger pulp or nail margin (Osler's nodes) and in the retina (Roth's spots). Clubbing of the fingers (Fig. 12.3) is caused by expansion of the capillary loops of the nail beds. In some cases of endocarditis clubbing occurs early, and responds quickly to effective therapy. Amyloidosis can complicate longstanding disease.

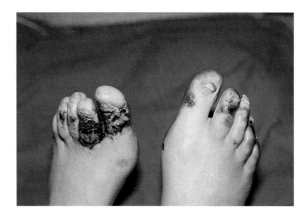

Figure 12.2 Septic embolic lesions in the feet of a man with acute endocarditis and a huge, friable aortic vegetation.

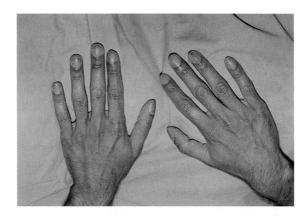

Figure 12.3 Clubbing of the fingers in endocarditis.

The 'Duke' criteria

These are clinical and investigational criteria, originally developed at Duke University, USA, that are often used to support diagnostic decision-making in suspected cases of infective endocarditis (Table 12.1). They have been modified since their original development, to take account of improved cardiac imaging. Most proven cases have clinical diagnostic features based on these criteria.

Nosocomial endocarditis

This is a complication of nosocomial bacteraemia, and is usually Gram-positive or fungal in aetiology. It becomes increasingly likely (affecting up to half of patients) in bacteraemias persisting for more than 36 to 48 hours, and occurs in over 25% of all patients with nosocomial staphylococcal bacteraemia. It can be difficult to diagnose in intensive care patients, as fever may be ascribed to other nosocomial causes, emboli may be considered sepsis-re-

Table 12.1 Duke University criteria for the diagnosis of infective endocarditis.

Type of criterion	Description of criterion
Major	**1** Positive blood culture, either with an organism recognized as a cause of endocarditis or more than one positive from cultures taken more than 12 hours apart, or three positives from three or four cultures drawn over a period of 1 hour **2** Evidence of endocardial damage. Echocardiographic (preferably trans-oesophageal echocardiography, which is much more sensitive than trans-thoracic) evidence of vegetations on valves or supporting structures, on an implant or in the turbulent path of a high-flow jet (as in damaged valve, septal defect or patent ductus); *or* new valvular regurgitation (not just a changed murmur)
Minor	Known predisposition to endocarditis Temperature of 38 °C or higher Evidence of focal vasculitis or emboli Immunological features (Osler's nodes, nephrosis, etc.) Single positive blood culture or blood culture result of uncertain significance Serological evidence of infection known to cause endocarditis (such as chlamydial or *Coxiella* infection) Echocardiographic finding of uncertain significance

Diagnostic groups of criteria: two major criteria, *or* one major and three minor criteria *or* five minor criteria.

lated, and imaging procedures such as trans-oesophageal echocardiography are difficult to perform in ventilated patients.

Aggressive endocarditis

This term describes the damaging and rapidly progressive infective endocarditis caused by highly pathogenic organisms. *Staphylococcus aureus*, group B streptococci, *Pseudomonas aeruginosa*, *Streptococcus pyogenes* and *Aspergillus* spp. are recognized causes. In contrast to less aggressive infections there is high fever, large vegetations develop quickly, catastrophic valve failure occurs early, and septic emboli or marantic aneurysms are more common. These factors result in a higher case fatality rate.

Diagnosis

The diagnosis must be considered in all patients with fever and a heart murmur. A known predisposition to endocarditis should heighten suspicion. However, in the elderly, endocarditis can present with heart failure, malaise or confusion but very little fever.

Positive blood cultures, in conjunction with a compatible history of predisposition and clinical features (see Table 12.1), will strongly indicate the diagnosis and provide essential information on the bacterial cause and antibiotic sensitivity. Three sets of cultures should be taken (a total of 30–60 ml blood) although there is little additional benefit in taking more than three sets. It is essential that blood cultures are taken before any antibiotics are given. Non-aggressive endocarditis rarely progresses sufficiently quickly to justify introducing antibiotic therapy before waiting for cultures, which should be incubated for 7–10

days, though positive results are usually available within 2 or 3 days.

Care must be taken in interpreting culture results, especially with organisms that are normally resident on the skin. Multiple isolates with identical antimicrobial susceptibility are required to confirm the significance of, for example, a *S. epidermidis* isolate. Further tests, or nucleic-acid based typing, can help to confirm that isolates are identical.

About 10% of patients with endocarditis have negative blood cultures. Blood cultures should then be performed in media containing nutritional supplements, and serological tests performed for *Coxiella burnetii*, *Chlamydia psittaci* and *Borrelia* spp.

After surgery, tissue specimens become available for culture and microscopy. Organisms may then be recovered from vegetations, or rarities such as mycobacteria demonstrated, by staining and culture. Cardiac tissue from culture-negative endocarditis should be subjected to 16S DNA amplification techniques, which may increase the diagnostic yield and enable antibiotic regimens to be redirected.

Echocardiography

Echocardiography is of critical importance in demonstrating vegetations on affected valves, identifying structural defects, showing abnormal blood flow patterns and sometimes detecting intracardiac abscesses (Figs 12.4 and 12.5). Trans-thoracic echocardiography (TTE) will often demonstrate significant mitral or tricuspid valve vegetations, but a negative result does not exclude endocarditis. Small aortic valve lesions are particularly likely to be missed. Trans-oesophageal echocardiography (TOE) will

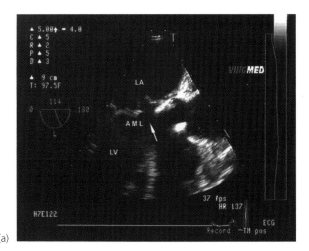
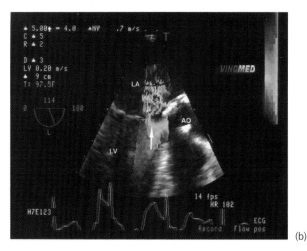

(a)   (b)

Figure 12.4 (a) Trans-oesophageal echocardiogram showing perforation of the base of the anterior leaflet of the mitral valve due to septic erosion of the valve (the arrow indicates the hole in the valve); small vegetations are seen, protruding into the left atrium at both margins of the perforation. (b) Trans-oesophageal Doppler echocardiogram: same view of the same valve lesion showing regurgitation of blood through the leaking valve: confused colours are generated by the turbulent regurgitant jet streaming into the left atrium during systole. Courtesy of Dr Joseph Davar, Royal Free Hospital. LA, left atrium; LV, left ventricle; AML, anterior mitral leaflet; AO, aorta.

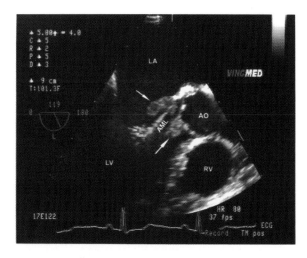

Figure 12.5 Trans-oesophageal echocardiogram showing two mobile vegetations (arrows) on the aortic valve, one of which prolapses into the aorta during systole while the other is forced through a defect in the thickened aortic valve ring, to prolapse into the left atrium; the thickened and perforated valve ring indicates the presence of an aortic root abscess. Courtesy of Dr Joseph Davar, Royal Free Hospital. LA, left atrium; LV, left ventricle; RV, right ventricle; AML, anterior mitral leaflet; AO, aorta.

also demonstrate intracardiac abscesses (see Fig. CS.4, in Case study 12.1). Contrast ventriculography and angiography may be necessary to define connections between abscesses and the bloodstream, or confirm the patency of the coronary ostia in aortic valve disease.

Management

The mainstay of management is adequate bactericidal antibiotic treatment. Many alpha-haemolytic streptococci are sensitive to benzylpenicillin, but enterococci respond more reliably if gentamicin is added. Other organisms may need different regimens, depending on the results of sensitivity testing. Table 12.2 gives examples of typical regimens recommended by British and American specialist groups.

Resolving fever and stabilization of cardiac function are signs that infection is resolving. The C-reactive protein level is a useful monitor of progress, falling to normal as infection is controlled.

During therapy, cardiac function, renal function and C-reactive protein should be monitored twice weekly. The resolution of vegetations may be followed by echocardiogram.

Complications

Continuing fever

Continuing fever and elevated C-reactive protein are rarely due to inappropriate antibiotic treatment. The usual cause is abscess formation in the valve ring or the cardiac

reveal vegetations in 25% or more of cases with negative TTE, and also has a much better specificity than TTE. Computed tomography or magnetic resonance scans can

Table 12.2 Summary of recommendations for the treatment of infective endocarditis

Organism	Preferred treatment	Alternatives
Empirical treatment	*Aggressive IE:* flucloxacillin i.v. 8–12 g daily in 4–6 divided doses *plus* gentamicin i.v. 1 mg/kg 8-hourly* *Indolent IE:* amoxicillin i.v. 2 g 6–hourly *plus* gentamicin i.v. 1 mg/kg 8-hourly	*For penicillin allergy, intracardiac prosthesis or suspected MRSA:* vancomycin i.v. 1 g 12-hourly* *plus* gentamicin i.v. 1 mg/kg 8-hourly* *plus* rifampin orally 300–600 mg 12-hourly
'Viridans' streptococci (strains fully sensitive to penicillin) Also suitable treatment for sensitive *S. bovis* (NB in *S. bovis* IE, the risk of associated bowel malignancy is high: the bowel should therefore always be investigated when the IE is controlled)	Penicillin i.v. 1.2–2.4 g 4-hourly **for 4 weeks** *plus* gentamicin i.v. 1 mg/kg 8-hourly for 1st 2 weeks (2 weeks of penicillin *plus* gentamicin is adequate if the vegetation is 0.5 cm or less, there is no intracardiac abscess or septic embolization, and fever is abolished in the first week)	*If risk of gentamicin toxicity is high:* penicillin i.v. 2.4 g 4-hourly **for 4 weeks** *Penicillin allergy:* vancomycin i.v. 1 g 12-hourly* **for 4 weeks**, *or* (US recommendation, if allergy is not anaphylaxis:) ceftriaxone i.v. 2 g daily for 4 weeks
'Viridans' streptococci and *S. bovis* (less sensitive or resistant to penicillin) and enterococci	Ampicillin or amoxicillin i.v. 2 g 4-hourly *plus* gentamicin i.v. 1 mg/kg 8-hourly* **for 4 weeks**	*For penicillin allergy, or ampicillin-resistant organisms:* vancomycin i.v. 1 g 12-hourly* *plus* gentamicin i.v. 1 mg/kg 8-hourly*, *or* teicoplanin* i.v. 10 mg/kg daily *plus* gentamicin i.v. 1 mg/kg 8-hourly **for 4 weeks**
Staphylococcal endocarditis (coagulase-positive or -negative organisms)	*Methicillin sensitive:* flucloxacillin i.v. 2 g 6-hourly (4-hourly for patients >85 kg) **for 4–6 weeks** (US recommendation for tricuspid valve infections: flucloxacillin *plus* gentamicin, for 2 weeks)	*For penicillin allergy or methicillin resistance:* vancomycin i.v. 1 g 12-hourly (US recommendation for non-anaphylaxis penicillin allergy: cefazolin i.v. 2 g 8-hourly for **4–6 weeks** *plus* gentamicin i.v. 1 mg/kg 8-hourly for 1st 3–5 days)
HACEK organisms	Ceftriaxone i.v. 2 g daily for **4 weeks**	*For ampicillin-sensitive organisms:* ampicillin *or* amoxicillin i.v. 2 g 4–6 hourly **for 4 weeks** *plus* gentamicin 1 mg/kg 8-hourly for 1st 2 weeks
Bartonella henselae	Ampicillin *or* amoxicillin i.v. 2 g 4–6 hourly *plus* gentamicin 1 mg/kg 8-hourly for **4–6 weeks**	*For penicillin allergy:* doxycycline orally 100 mg daily **for 4–6 weeks**
Coxiella burnetii	Doxycycline orally 100 mg daily *plus* ciprofloxacin orally 500 mg 8-hourly **for at least 2–3 years**: monitor antibody titres 6-monthly on treatment, and 3-monthly after stopping	Chloroquine may be a useful addition if the response is poor
Fungal endocarditis	Antimicrobial treatment is not satisfactory – surgery is often required	Amphotericin B and caspofungin are fungicidal to *Candida* spp. Amphotericin or voriconazole may be useful adjuncts to management of *Aspergillus* spp. infections

*When a prosthetic valve is involved, treatment should always be for at least 6 weeks.

septum. Large vegetations, or the presence of necrotic tissue may also inhibit clearance of infection. In these circumstances, surgery is indicated, to terminate sepsis, permit healing and prevent deterioration in valve function. Accurate preoperative imaging of the infected site, and of the rest of the heart (to assess ventricular function, the extent of any shunts and the integrity of the coronary circulation), greatly assists the surgeon in operating effectively and safely. Damaged and infected valves may be replaced, or partially excised, refashioned and repaired. Intracardiac abscesses can be drained and, where possible, septal defects closed.

In endocarditis due to drug injecting, or nosocomial bacteraemia, it is important to rule out a coexisting focus of infection, such as a lung, soft tissue, liver or cerebral abscess, which may cause continuing fever. These lesions may also require surgery or drainage.

Persisting production of emboli

Persisting production of emboli can lead to disabling strokes, myocardial infarction or limb ischaemia. This is most common with very large or mobile vegetations, or vegetations with frond-like projections. Large vegetations are common in infections with *S. aureus*, *S. bovis*, *Pseudomonas* spp. (sometimes seen in renal transplant patients) and *Aspergillus* spp. Anticoagulation is considered unwise in uncomplicated endocarditis, due to the risk of bleeding from sites of intra-arterial infection. However, it may be justified when embolic complications threaten life or function.

Thrombolysis after an embolic stroke or myocardial infarct carries a high risk of haemorrhage in endocarditis, and should be avoided wherever possible. Angioplasty and stenting of coronary arteries are possible, but are best carried out when the infection is controlled (when fever and bacteraemia have been documented absent for at least 2 weeks).

In some cases, especially of fungal endocarditis, the best chance of arresting the problem is surgery to remove the affected valve and vegetations.

Severe valve dysfunction

Severe valve dysfunction is unpredictable, and more likely if infection is uncontrolled. Friability and dehiscence of devitalized tissue can cause sudden disruption or fenestration of a valve, with dramatic haemodynamic changes. In left-sided endocarditis this usually causes acute heart failure, with signs of aortic or mitral regurgitation. Emergency surgery and valve replacement or repair is the treatment of choice. This is partly curative, but should be followed by 2 weeks of antibiotic treatment. In tricuspid endocarditis, valve failure is less catastrophic, and often does not require intervention.

Prevention and control

Two-thirds of endocarditis cases affect patients with known cardiac abnormalities, usually with a positive medical history or physical signs of a susceptible lesion. Antibiotic prophylaxis should be offered to at-risk patients undergoing procedures likely to cause bacteraemia. Although there is no firm evidence that prophylaxis prevents cases, the costs of prophylaxis are low, and the benefit of saving even one case of infective endocarditis is relatively enormous.

The British Society for Antimicrobial Chemotherapy recommendations for the prophylaxis of infective endocarditis

1 For standard risk patients (heart valve lesion, septal defect, patent ductus arteriosus or prosthetic valve) for dental extractions, scaling, periodontal surgery without general anaesthesia; for surgery or instrumentation of the upper respiratory tract; for urogenital examination or instrumentation:

- amoxicillin 3 g orally 1 h before the procedure; or
- for penicillin-allergic patients, or those who have received more than a single dose of penicillin in the previous month) clindamycin 600 mg orally 1 h before the procedure.

For both of these regimens, children under 5 receive one-quarter of the adult dose, and children aged 5–10 years receive half of the adult dose.

2 For patients with a previous history of endocarditis:

- amoxicillin plus gentamicin, as for procedures performed under anaesthesia.

3 For the above procedures performed under general anaesthesia in standard-risk patients without prosthetic valves:

- amoxicillin 1 g intravenously 1 h before induction and 0.5 g orally 6 h later; or
- amoxicillin 3 g orally 4 h before and 3 g orally as soon as possible postoperatively; or
- amoxicillin 3 g orally plus probenecid 1 g orally 4 h before induction (child doses not recommended).

4 For these procedures under general anaesthesia in patients with prosthetic valves or a previous history of endocarditis:

- intravenous amoxicillin 1.0 g plus intravenous gentamicin 120 mg at induction, then 0.5 g amoxicillin orally or intravenously 6 hours later (child under 5, amoxicillin one-quarter of the adult dose plus gentamicin 2 mg/kg; child 5–10 years, amoxicillin half of the adult dose plus gentamicin 5 mg/kg).

5 For all penicillin-allergic patients having these procedures under general anaesthetic:

- vancomycin 1.0 g intravenously over at least 100 min, followed by gentamicin 120 mg intravenously at induction or 15 min before the procedure (child under 10 years, vancomycin 20 mg/kg plus gentamicin 2 mg/kg); or
- teicoplanin intravenously 400 mg plus gentamicin 120 mg at induction or 15 min before the procedure (child under 14 years, teicoplanin 6 mg/kg plus gentamicin 2 mg/kg); or
- clindamycin intravenously 300 mg over at least 10 min at induction or 15 min before the procedure (child under 5 years, a quarter of the adult dose; child aged 5–10 years, half of the adult dose).

6 For genitourinary procedures:

- As for 4 and 5 above, except that clindamycin is not given.

⚠️ NB: for genitourinary procedures in the presence of colonized or infected urine, the prophylaxis chosen should include an antibiotic effective against the urinary organism.

7 For obstetric, gynaecological and gastrointestinal procedures:

- prophylaxis is indicated only for patients with a previous history of endocarditis or with a prosthetic valve: the regimens described for genitourinary procedures are appropriate.

Infective endarteritis

Infective endarteritis can arise by haematogenous spread in any bacteraemic condition. Adherence of organisms is more likely in the presence of aneurysm, intravascular thrombus or vascular injury. The atheromatous lower aorta is particularly at risk, because of its susceptibility to endothelial ulceration and other risk factors. In the affected age groups, bacteraemias are often of bowel or urinary tract origin, so *Escherichia coli* or enterococci are common pathogens. Aneurysms in other vessels, such as the popliteal or femoral arteries, may also become infected. These are often traumatic or autoimmune in origin, affecting younger people. They may be infected with staphylococci, endocarditis organisms or rarities such as salmonellae. Infected emboli from endocarditis itself may damage or occlude arteries, causing a metastatic or marantic arteritis that can become aneurysmal. Direct inoculation of organisms, such as staphylococci, can occur in intravenous drug abusers, often affecting the femoral arteries.

Clinical warning signs are not always obvious, but include fever, local pain or enlargement of any aneurysm, raised erythrocyte sedimentation rate and C-reactive protein and episodes of embolization or occlusion in the circulation affected by the disease. Post-operative graft occlusion or leakage of the anastomosis site may be indicators of post-operative infection. Imaging of the affected site is helpful in demonstrating vegetations, exuberant thrombus, dissection or leakage from the vessel.

Syphilitic aortitis (see Chapter 15) is a special case in which late infection of the ascending aorta destroys the elastic tissue, causing the development of a potentially massive and fragile aneurysm. This must be treated with care, as the Jarisch–Herxheimer reaction can increase inflammation and friability of the infected tissue. When infection is eradicated, replacement of the ascending aorta can be carried out, to remove the risk of rupture.

Case study 12.1: PUO in a fit young man

History
A 20-year-old manual worker complained of 2 or 3 weeks' history of fevers and fatigue with no other symptoms. He had recently been working at a bridge restoration where rats and nesting pigeons were often found. Six weeks before, he had stayed on his parents' farm in rural Wales, but denied consuming any unpasteurized dairy products. He had always been extremely fit, and had never travelled overseas.

Physical findings
Temperature 37.8 °C, pulse 88 bpm, blood pressure 115/70 mmHg, a faint aortic systolic murmur (possibly a 'flow' murmur), some traumatic scratches on the hands, a blister on the right great toe (from his new work boots). Physical examination was otherwise normal, including neurological testing and fundoscopy.

Questions
- What immediate investigations and laboratory tests would you request?
- Which infectious diagnoses would you consider at this stage?

Laboratory investigations
Haemoglobin 12.3 g/dl, white cell count 11.2×10^9/l (8.7×10^9/l neutrophils), ESR (erythrocyte sedimentation rate) 45 mm/hour, CRP (C-reactive protein) 30 IU/l. Urine dip testing showed + blood but no neutrophil reaction, renal and liver function tests were all normal. The chest X-ray was normal. Blood was drawn for cultures, and serological tests for leptospirosis, brucellosis and *Coxiella burnetii* infection were requested. Serum was saved for possible histoplasmosis investigation, but this was deferred in the absence of skin or chest abnormalities.

Management and progress
After 2 days' observation and bed rest in hospital, his temperature had fallen to 37 °C and urine and blood cultures had produced no evidence of growth. He was discharged with early outpatient follow-up arranged. Thirty-six hours later, he was recalled to the ward because four of four blood culture bottles had signalled positive, and Gram-staining showed small Gram-negative rods.

Re-examination showed temperature 38.2 °C, pulse 92 bpm, blood pressure 120/55 mmHg, and a $\frac{1}{4}$ ejection systolic murmur radiating to the left neck with a faint early diastolic murmur at the left sternal edge. The haemoglobin was 11.2 g/dl, white cell count 13.4×10^9/l. Urine dip test showed ++ blood and + protein.

Questions
- What is the likely diagnosis?
- Does it fulfil accepted criteria?
- What immediate investigation is indicated?
- Should antimicrobial therapy be commenced, what regimen would you use?

Further management and progress
The diagnosis of infective endocarditis was made, and fulfilment of the 'Duke University criteria' was complete, with two major criteria documented (see Table 12.1). Treatment was commenced immediately with intravenous amoxicillin plus gentamicin. A Doppler echocardiogram was urgently arranged (presence of diagnostic abnormalities would comprise another diagnostic criterion, but information was also required about cardiac function and the degree of valvular damage at this stage). The result confirmed the presence of vegetations on both leaflets of a bicuspid aortic valve, with restricted outflow and a pressure gradient of 12 mmHg across the valve. Mild regurgitation was demonstrated. Left ventricular volumes and function were normal.

After 4 days' therapy, fever persisted at 38 °C, the pulse was 120 bpm and the blood pressure 130/30 mmHg.

The patient complained of shortness of breath and rapid palpitation.

Questions
- What is the most likely reason for his failure to respond to therapy?
- What urgent investigation is indicated?
- What mode of treatment is likely to be needed?

Further management and progress
Laboratory results identified the Gram-negative rod as *Cardiobacterium hominis*, sensitive to amoxycillin plus gentamicin and also to broad-spectrum cephalosporins. Inappropriate antimicrobial chemotherapy was therefore not the cause of the patient's deterioration. In almost every case, the problem is caused by infection inaccessible to the antimicrobial agent, either because of gross vegetations or, more often, to intracardiac abscess formation. Urgent cardiac imaging (in this case, dynamic magnetic resonance imaging, though trans-oesophageal echocardiography or angiography is often employed) showed a large abscess surrounding the origin of the ascending aorta (Fig. CS.4). Surgery is the only curative option in such cases; urgent abscess clearance and aortic valve replacement was followed by 2 weeks' therapy with intravenous ceftriaxone, and resulted in complete resolution of fever and cardiac abnormalities.

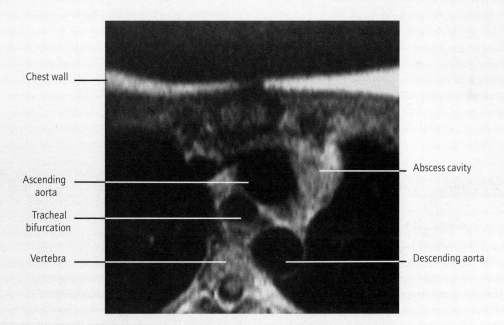

Figure CS.4 Trans-axial NMR image of the upper mediastinum at the level of the tracheal bifurcation, showing a large abscess surrounding the origin of the ascending aorta.

Infections of the Central Nervous System

13

Introduction

Structural and functional considerations

The brain and spinal cord form a special compartment of the body. The cerebrospinal fluid (CSF) is different from other body fluids, in particular having a low protein content, including very low immunoglobulin levels. The central nervous system (CNS) has no lymphatics, but depends on migration of lymphocytes to and from the blood circulation.

The layers of the meninges support and protect the brain and spinal cord, and contain the CSF. Together with their blood vessels, choroid plexuses and arachnoid processes, the meninges constitute the blood–brain barrier, a structural and functional barrier between the blood and the CSF. It permits the brain to maintain its own homeostatic environment, with its own acid–base equilibrium. It inhibits the entry of organisms, and of many drugs and toxins.

Inflammation of the meninges increases the permeability of the blood–brain barrier. During an inflammatory process, acute-phase proteins appear in the CSF, mostly by diffusion from the blood, with the low molecular weight, more diffusible proteins appearing in highest concentrations. Immunoglobulins diffuse into the CSF but can also be locally manufactured once lymphocytes have migrated from the circulation. Both lymphocytes and neutrophils are able to enter the CSF during infection and inflammation.

It is possible to demonstrate local production of a specific antibody, rather than simple diffusion into the CSF, by comparing the ratio of its CSF and blood concentrations with the ratio for an antibody unlikely to be involved with acute CNS disease. For instance, if *Borrelia afzelii* antibody is manufactured locally in a case of *Borellia* encephalitis, its CSF:blood concentration ratio will be higher than that of, say, rubella antibodies. The relationship between the ratios is an antibody index.

An antibody index

$$\frac{\text{CSF : serum ratio of the antibody in question}}{\text{CSF : serum ratio of reference (e.g. rubella) antibody}}$$

If this index is greater than 2, local production of the antibody in question may be assumed. This may be useful when culture or PCR tests for cerebral or meningeal infections are not reliable.

When the blood–brain barrier is disrupted by inflammation, proteins and cells readily enter the CSF. This may permit more effective immune responses, and also allows the rapid entry of therapeutic drugs such as penicillins or vancomycin that do not readily penetrate the intact blood–brain barrier. The disadvantages of inflammation and exudation are: elevation of intracranial pressure; reduction in cerebral perfusion pressure (the difference between blood pressure and CSF pressure); and reduced cerebral blood flow. Intense inflammation causes protein accumulation and fibrin deposition. This can obstruct the aqueduct, or the foramina of the brain, causing local or general hydrocephalus.

Inflammatory changes can be inhibited by the use of corticosteroids, especially if these are given early, and this may improve the outcome in some diseases.

Rising CSF pressure is also related to changes in the water content of the brain. Increased brain water is not necessarily equivalent to cerebral oedema, as the water may be in other compartments than the interstitium. In meningitis particularly, it is no longer the custom to limit fluid intake as, rather than preventing a rise in intracranial pressure, it reduces blood volume and impairs cerebral perfusion. In encephalitis there may be true cerebral oedema, often demonstrable by computed tomographic (CT) or magnetic resonance (MR) scanning. It may then be possible to reduce oedema with dexamethasone treatment.

Effects of meningeal inflammation
1 Increasing CSF protein and cell count.
2 Increased entry of water-soluble antibiotics.
3 Increased brain water.
4 Increased CSF (= intracranial) pressure.
5 Reduced cerebral perfusion pressure.
6 Risk of CSF obstruction leading to hydrocephalus.

The outcome of some types of acute meningitis, including the incidence of deafness, is improved by early dexamethasone treatment. This may be due to reduction of oedema or inflammation, or possibly protection of the cochlear hair cells from the action of endotoxin.

Pathogenesis of CNS infections

Many infections of the CNS are blood-borne. The causative organism colonizes or infects other sites of the body before causing CNS disease. Thus, the viruses of meningitis are often found in the throat and bowel as well as in the CSF; a neonate with cutaneous herpes simplex lesions is at risk of developing encephalitis; in bacterial meningitis, the causative organisms appear first in the nasopharynx, then in the blood and lastly in the CSF.

It is relatively uncommon for the CNS to be infected by the spread of organisms from adjacent structures. However, pneumococcal meningitis may arise from chronic infection in the paranasal sinuses or the middle ear. This is more likely if there is a defect in the dura mater following head injury, surgery or inflammatory damage (in this situation recurrent meningitis may occur). It is thought that amoebic meningitis arises by migration of amoebae through the cribriform plate after nasal exposure to infected mud or water.

Rarely, pathogens advance along nerve trunks or between segments of the spinal cord. Rabies viruses enter peripheral nerves and migrate to the CNS where they then cause meningoencephalitis. The incubation period is proportional to the length of nerve that must be traversed. Transection of the nerve trunk or amputation of the injured limb will prevent rabies in animals. Herpes zoster viruses can enter the spinal cord via a dorsal root and spread segment by segment, causing an ascending myeloencephalitis.

Modes of pathogenesis of central nervous system infections
1 Infection during viraemic phase of viral infections.
2 Blood-borne spread from local or bacteraemic bacterial infection.
3 Contiguous spread from intracranial infective focus.
4 Entry of bacteria through a defect in the dura.
5 Rare spread through cribriform plate.
6 Rare spread along nerve fibres and connections.

Meningitis and meningism

Introduction

Meningitis is inflammation of the meninges. Meningism is the group of symptoms and signs that accompany the inflammation. The symptoms are headache, neck stiffness, nausea and vomiting, and photophobia. The headache is global and usually described as the worst ever experienced. Photophobia is the most variable of the symptoms; in severe cases it may be very marked, but it can occur in a number of other conditions such as tension headache or migraine.

Main features of meningism

1 Headache.
2 Neck and back stiffness.
3 Nausea and vomiting.
4 Photophobia.

Typical meningism is uncommon in infants and small children, who rarely exhibit neck and back stiffness. They usually simply appear hypotonic, though later they may develop opisthotonus. A bulging fontanelle is a useful sign of raised intracranial pressure in an infant. Fever, weak, high-pitched crying, convulsions and persistent vomiting are important warnings of meningitis in this age group.

Warning signs of meningism in infants

1 Bulging fontanelle.
2 Vomiting.
3 Strange high-pitched cry.
4 Convulsions.
5 Opisthotonus.

Meningism can occur without meningitis. It can accompany upper lobe pneumonia, urinary tract infection and high fevers such as the prodromal fever of dysentery. In these conditions CSF examination is normal, and the subsequent investigation and evolution of the disease reveal the true diagnosis. The mechanism of meningism without meningitis is unknown.

Meningism can also occur without significant fever in non-infectious conditions. In subarachnoid haemorrhage, headache and nausea have a very abrupt ('thunderclap') onset and may be accompanied by collapse, loss of consciousness and/or abnormal neurological signs. Malignant infiltration is a cause of meningism in leukaemia and metastatic melanoma. Meningism is a rare adverse reaction to non-steroidal anti-inflammatory drugs, described most often after ibuprofen.

Conditions where meningism can occur without meningitis

1 Small children with high fevers.
2 Upper lobe pneumonias.
3 Acute urinary tract infections.
4 Subarachnoid haemorrhage.
5 Meningeal malignancies.

Physical signs of meningism

Traditional tests for meningism are:
1 to demonstrate an inability to flex the neck and touch the chin to the chest; and
2 to elicit Kernig's sign (Fig. 13.1).

These will easily demonstrate moderate to severe meningism, but mild meningism is sometimes missed. More subtle signs are: the tripod sign, in which the patient is unable to sit up from a supine position without making a tripod with the hands resting on the bed behind him or her; and the inability to curl forwards enough to touch the nose to the knees (Fig. 13.2).

The detection of mild meningism is important: it permits early clinical suspicion of meningitis, including suspicion of diagnoses such as tuberculous meningitis, which may cause only mild meningism before major neurological deficit occurs.

Figure 13.1 Kernig's sign of meningism: the patient's leg cannot be straightened because of hamstring spasm.

(a)

(b)

Figure 13.2 Meningism can be easily demonstrated: this patient with meningococcal meningitis could not oppose his nose and knees (a), but after recovery (b), the manoeuvre was easily performed.

Lumbar puncture

This is the most rapid diagnostic test for meningitis. It permits distinction between bacterial meningitis, viral meningitis or meningoencephalitis and non-meningitis meningism. It often allows a rapid aetiological diagnosis of infection by Gram stain, polymerase chain reaction (PCR) or slide agglutination test on CSF.

Nevertheless, herniation of the brain through the foramen magnum can occur after fluid is removed from the spinal theca. This is rare if the duration of inflammation before presentation is short, and has not yet interfered with normal CSF flow. With communicating hydrocephalus, removing CSF from the lumbar theca simply reduces intracranial pressure. Since fluid is incompressible, only a tiny amount need be removed to reduce intracranial pressure and increase cerebral perfusion pressure and, often, to reduce headache (Fig. 13.3).

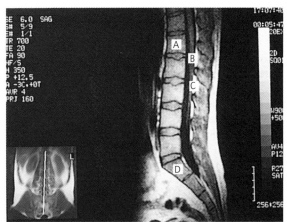

Figure 13.3 Lateral magnetic resonance image of normal lumbar spine, showing the termination of the spinal cord at the L1/L2 level, leaving a space below filled with cerebrospinal fluid and nerve roots; the optimum track for lumbar puncture is shown by the dotted line. A, body of 1st lumbar vertebra; B, lower extreme of the spinal cord; C, track of the lumbar puncture between spines of the third and fourth lumbar vertebra; D, the sacrum.

The risk of herniation is greater when there is a non-communicating raised CSF pressure (because lumbar puncture then reduces the pressure in the spinal theca but the intracranial pressure remains high, and forces the hindbrain through the foramen magnum).

Anxiety about causing herniation results in many patients having CT scans before lumbar puncture; indeed, some practitioners always perform a scan. However, CT of the head delivers a substantial radiation dose to the brain, in the form of X-rays (therapeutic investigations are the greatest component of ionizing radiation exposure nowadays in the UK). There is some evidence that long-term effects may result from this. CT scan in acute meningism reveals a contraindication to lumbar puncture in less than 0.5% of cases. Also, the delay involved in obtaining CT evidence means that many patients receive at least one dose of broad-spectrum antibiotic, at least half of whom have viral or non-infectious conditions. Although this is less likely than irradiation to produce adverse effects, it is undesirable from the microbiological and environmental point of view. For these reasons, most experts advise that, for adults, lumbar puncture should be performed without waiting for CT scan, provided that the meningism is of recent onset, and there are no contraindications.

The British Infection Society/Meningitis Research Foundation recommend that lumbar puncture should be avoided when the clinical history and examination reveal the factors listed in the box below. Imaging should also be performed before lumbar puncture when the history of headache is longer than 3 or 4 days or when cerebellar tonsillar herniation is known to exist. The presence of papilloedema also indicates increased risk. Papilloedema, however, develops slowly, and is not always present when the CSF pressure is acutely raised.

Lumbar puncture may be avoided if it is possible to deduce the diagnosis of intracerebral infection and give empirical treatment without performing a lumbar puncture. This is usually so with childhood meningitis, particularly in the presence of a typical meningococcal rash. If lumbar puncture is unavoidable, an emergency CT or MR scan can be obtained, and usually shows no contraindication. CSF can sometimes be obtained by ventricular puncture, but there is a risk of brain herniation through a burr hole if brain swelling is severe.

Indicators of increased risk of lumbar puncture

1 Reduced cerebral function (Glasgow Coma Scale < 13).
2 Sluggish or dilated pupils.
3 Prolonged or focal seizures.
4 Abnormal or decerebrate posturing.
5 Papilloedema.
6 Arterial hypertension, bradycardia or abnormal respiratory pattern.
7 Severe shock.
8 Coagulopathy.

Unrelieved intracranial hypertension will 'incarcerate' the brain inside the skull, occluding the cerebral circulation and leading to cerebral vein thrombosis or cerebral ischaemia. Anaesthesia, artificial hyperventilation or insertion of a CSF drain may occasionally be needed to avoid this.

Laboratory diagnosis

Urgent diagnostic and prognostic information can be obtained from CSF examination. However, it is essential that a simultaneous specimen of blood be obtained for glucose estimation so that the result of CSF glucose may be interpreted. Other important investigations include blood cultures, nasopharyngeal swabs and stool specimens (for viral culture). A blood sample should also be sent for PCR since this can permit a positive diagnosis of meningococcal infection when cultures are negative. Serum should be obtained, so that it can be examined for serological evidence of infection, and compared with a second serology investigation on a specimen obtained two to three weeks later.

CSF examination

Three specimens of CSF are collected into numbered universal containers, and a further one is collected into a blood sugar estimation bottle. From infants, only 10–12 drops of fluid should be taken for each specimen, while from older children and adults 1–2 ml can be collected.

The CSF cell count is performed by examining an unspun, unstained specimen in a counting chamber. The cell types are identified by examination of a Giemsa-stained centrifuged deposit. The presence of neutrophils is associated with bacterial infections such as *Neisseria meningitidis*, *Streptococcus pneumoniae* or *Listeria monocytogenes*. The presence of lymphocytes is associated with infection with viruses, mycobacteria or *Leptospira*. A lymphocytic or mixed pleiocytosis may be seen in cases of partially treated pyogenic infection or in brain abscess.

Further information can be obtained from biochemical tests. The CSF protein level reflects the degree of meningeal inflammation. It is often two or three times the normal level in viral infections of the CNS, and up to five or 10 times the normal level in bacterial infections. In abscesses and chronic infections, such as tuberculous meningitis, it may reach very high levels. The CSF glucose level is usually normal in viral infections, significantly low in bacterial meningitis, and negligible in advanced tuberculous meningitis. As acute intracranial disorders can raise the blood glucose (often to 8–12 mmol/l), it is essential to compare the CSF glucose with a simultaneous blood glucose estimation.

Normal CSF composition includes:
1 Cells: <6 white blood cells, all lymphocytes; no red blood cells (in an atraumatic tap).
2 Protein: 0.1–0.4 g/l (lumbar theca); 0.07–0.25 g/l (ventricular).
3 Glucose: up to 1.7 mmol/l below blood glucose.

The centrifuged deposit should be stained by Gram's method to demonstrate bacteria. The sensitivity of Gram stain is about 10^4 CFU/ml. This can be improved upon by using acridine orange. Ziehl–Nielsen stain should be employed if tuberculosis is suspected, or indicated by the cell count or biochemical examination. If cryptococcosis is suspected in an immunosuppressed patient, the CSF should be examined by an India ink method, which outlines the fungal capsule in sharp relief.

Specimens from the centrifuged deposit are inoculated on to a range of media capable of growing the main pathogens. This usually includes blood and chocolate agar, together with an enrichment broth (such as tryptose broth, intended to amplify the sometimes scanty organisms). All colonies growing after 24 h are identified and sensitivity testing performed. If there is no bacterial growth after 24 h the enrichment broth is subcultured. It is important that all isolates are sent to reference laboratories for confirmation and typing (Fig. 13.4). In England

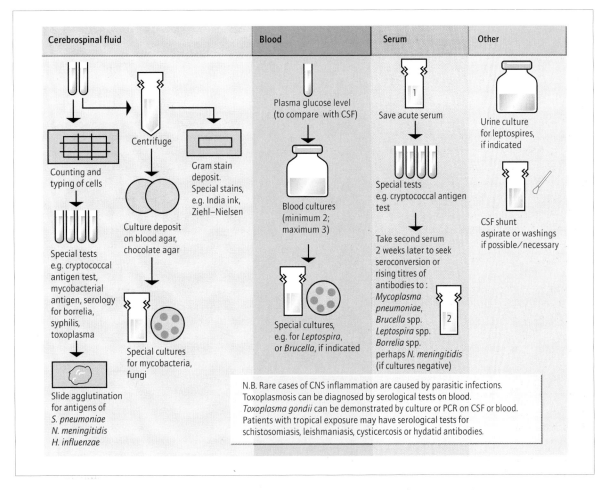

Figure 13.4 Scheme for the investigation of bacterial meningitis. CNS, central nervous system; CSF, cerebrospinal fluid; PCR, polymerase chain reaction.

and Wales, specimens and isolates from cases of meningococcal disease should be referred to the Meningococcus Reference Laboratory for detailed identification and typing. The results are reported to the Health Protection Agency Centre for Infections, to update national records of current types and antimicrobial sensitivities of important pathogens.

The aetiological diagnosis of viral meningitis by CSF culture is often unrewarding. A better diagnostic yield, in enteroviral infections, is obtained from stool culture. Mumps or measles virus may be recovered from urine cultures. In suspected herpetic encephalitis, molecular diagnostic methods are preferred (see below); other methods include the measurement of anti-HSV or anti-VZV antibodies in the CSF, and the CSF:serum antibody ratio compared with that of rubella or measles (Fig. 13.5). Positive identification of viral agents in CNS disease should be reported to the National Database for Health Protection. Uncommon infections, such as tick-borne encephalitis or West Nile meningoencephalitis, can be diagnosed by serological methods, using a single serum specimen.

Rapid diagnostic techniques

The common bacteria causing meningitis possess polysaccharide capsules, whose antigens can be detected by rapid methods. These methods include latex agglutination, coagglutination, counterimmunoelectrophoresis and enzyme immunoassay (EIA). Positive results can be obtained after antibiotics have been commenced and cul-

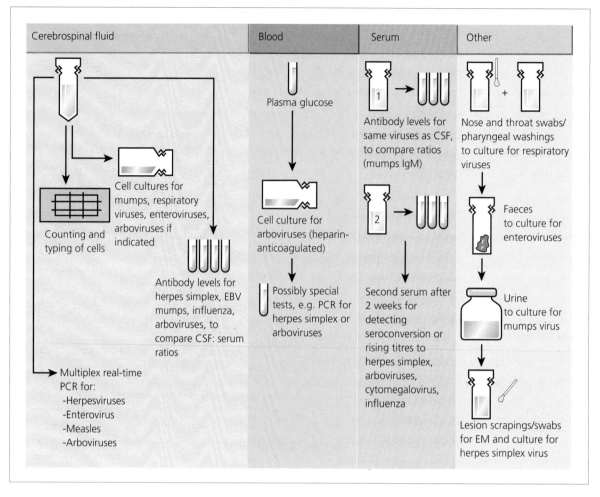

Figure 13.5 Scheme for virological investigation of central nervous system disease. CSF, cerebrospinal fluid; EBV, Epstein–Barr virus; EM, electron microscopy; IgM, immunoglobulin M; PCR, polymerase chain reaction.

tures become negative. Effective tests are available for *S. pneumoniae*, *N. meningitidis* serogroups A, C and W135 and *H. influenzae*. Satisfactory results are more difficult to obtain with the less immunogenic *N. meningitidis* serogroup B. The concentration of antigen indicates prognosis in *S. pneumoniae* infection, but studies for other pathogens are awaited.

PCR-based detection methods have had a significant impact on the diagnosis of pyogenic meningitis and are proving especially valuable in the diagnosis of meningococcal disease. PCR has also improved the ability of laboratories to diagnose tuberculosis meningitis and has also found an important role in herpes simplex encephalitis and varicella zoster virus meningitis. Increasingly, laboratories are developing multiplex DNA amplification assays especially for viral causes. Methods that amplify a wide range of pathogen nucleic acid have been linked to high-density DNA array analysis, which determines the identity of the amplimers.

Rapid diagnostic techniques that can be performed on cerebrospinal fluid

1 Gram stain of centrifuged deposit.
2 Special stains:
   • Ziehl–Nielsen (for acid-fast organisms).
   • India ink (for capsulated yeasts).
3 Slide agglutination tests:
   • *Neisseria meningitidis* A, C, W135, Y.
   • *Streptococcus pneumoniae*.
   • *Haemophilus influenzae*.
4 Antigen test: *Cryptococcus neoformans*.
5 Examples of polymerase chain reaction nucleic acid amplification tests:
   • *Mycobacterium tuberculosis*.
   • Herpes simplex.
   • Varicella zoster virus.
   • Enteroviruses.
   • *Neisseria meningitidis*.
   • *Streptococcus pneumoniae*.

Examination for other pathogens

CSF examination contributes to the investigation of other infective conditions such as syphilis, Lyme disease and trypanosomiasis. In syphilis and Lyme disease serological investigations are important. In suspected cerebral trypanosomiasis, CSF should only be examined after the blood has been cleared of parasites with suramin. Trypanosomes can be concentrated by an anion exchange chromatography. The presence of parasites or of morula cells indicates CSF infection and the need for treatment with arsenical preparations.

Viral meningitis

Introduction

Viral meningitis is common worldwide. Approximately 1000 laboratory-confirmed cases are reported each year in the UK, but this is only a fraction of the total incidence as most infections are mild or inapparent. The illness is usually self-limiting, rarely lasting for more than a week. Its importance lies in the differential diagnosis from tuberculosis (see Chapter 18) and acute bacterial meningitis, and from rare diseases such as listeriosis, borreliosis and cryptococcosis. Despite the World Health Organization campaign for polio eradication, rare outbreaks of acute disease still occur in places where immunization programmes are interrupted, and may be followed by viral meningitis and paralytic disease.

Organism list

- Echovirus
- Coxsackievirus
- Poliovirus
- Mumps virus
- Herpes simplex type 2
- Varicella zoster virus
- Influenza virus type A or B
- Arboviruses (usually meningoencephalitis) e.g.:
  - Tick-borne encephalitis
  - St Louis encephalitis
  - West Nile meningoencephalitis

> ⚠ Meningitis or meningoencephalitis is a rare effect of many virus infections, such as rubella, varicella, influenza and Epstein–Barr virus infection.

Enterovirus meningitis

Introduction and epidemiology

Young children are the usual source of enterovirus infection. Faecal virus shedding may persist for several weeks and transmission occurs through environmental contamination, particularly under conditions of crowding and poor hygiene. Enterovirus meningitis is commonest in children aged 5–14 years, but also occurs in other age groups. Intra-familial spread is common, and outbreaks have been described in hospital nurseries, boarding schools and other residential institutions.

Echovirus infection is by far the commonest cause of enteroviral meningitis. The predominant serotypes are 6, 9, 11, 19 and 30. There are annual epidemics in the late summer (Fig. 13.6), presenting as so-called 'summer flu', with fever, sore throat and often headache. A proportion of infected individuals develop meningitis, with or without a preceding or accompanying sore throat.

Enteroviruses types 70 and 71 are also important causes of outbreaks. Coxsackievirus meningitis is caused predominantly by serotypes A9, B4 and B5. Coxsackie A in-

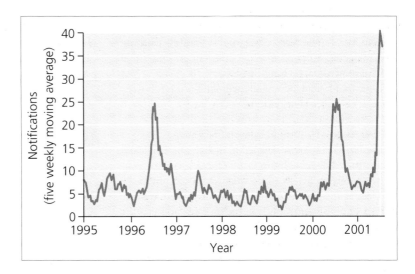

Figure 13.6 Viral meningitis long-term trends: this common infection dominates the trends for all meningitis notifications.

fections affect all age groups; type B disease occurs mainly in infants and pre-school children.

Enteroviruses can cause a rare, generalized chronic brain infection in individuals with agammaglobulinaemia.

Clinical features

There may be a personal or family history of sore throat in the preceding 5–7 days. The patient then develops increasing headache over 12–36 h, usually with nausea and vomiting. There is no alteration in consciousness or neurological function. Meningism varies from minimal to severe and is not always in proportion to the severity of the headache.

Other clinical features are few. There may be cervical lymphadenopathy and, rarely, pharyngitis or rash. The white cell count and differential are almost always normal, as is blood biochemistry.

Diagnosis

The typical clinical presentation during the epidemic period of the year should suggest the diagnosis. In Europe the differential diagnosis of tick-borne encephalitis or neuroborreliosis must be considered in spring and summer; in the USA, St Louis encephalitis or West Nile virus infection can present with meningitis, while in the Far East, Japanese encephalitis must be considered. In the UK, most differential diagnoses are rarities; they include tuberculous meningitis, neuroborreliosis and, when the CSF cell count includes neutrophils, intracerebral abscess and partly treated bacterial meningitis.

Rarely, exacerbations of multiple sclerosis or an autoimmune transverse myelitis cause CSF pleiocytosis, usually accompanied by typical neurological abnormalities. Meningism with mild lymphocytosis is a rare adverse event associated with non-steroidal anti-inflammatory drugs.

CSF examination is useful in patients with other than mild meningism. This is to exclude bacterial meningitis or other diagnoses such as subarachnoid haemorrhage. Cases with minimal meningism often improve rapidly with rest and analgesia. A period of observation often shows that lumbar puncture is no longer necessary.

Typical CSF changes in viral meningitis are as follows:
1 Cells: 40–250, all lymphocytes.
2 Protein: 0.45–1.2 g/l.
3 Glucose: 1.0–1.7 mmol/l below blood glucose.

The blood glucose is elevated in diabetic patients, and can be raised in acute brain insult (though viral meningitis is rarely sufficient to do this). It should always be checked at the time of lumbar puncture. There may be a proportion of neutrophils in the CSF cell count if the lumbar puncture is performed early in acute viral meningitis. The presence of 50% or more of lymphocytes,

however, is reassuring evidence against acute bacterial meningitis.

A lymphocytosis, or predominantly lymphocytic pleiocytosis, can occur in the CSF in:
1 Tuberculous meningitis.
2 Partly treated bacterial meningitis.
3 Intracranial abscess.
4 Leptospirosis.
5 Lyme borreliosis.
6 Viral encephalitis.
7 Lymphocytic leukaemias.
8 Rare reaction to non-steroidal anti-inflammatory drugs.

These conditions should be considered if the presentation or course of viral meningitis is unusual. CSF glucose levels are usually depressed in tuberculous meningitis (see Chapter 18).

Enteroviruses are transiently present in the pharynx and the CSF, but are excreted for up to three weeks in the faeces. Enteroviruses and their detection are discussed more fully in Chapter 6. PCR examination of CSF is a rapid diagnostic test. Test systems exist for detecting immunoglobulin M (IgM) antibodies to enteroviruses, but the multiplicity of serotypes makes specific serological diagnosis impractical.

Management

There is no licensed antiviral treatment for enteroviral diseases. Symptomatic treatment with analgesics, antiemetics and bed rest is effective in reducing pain and malaise. Fever and meningism often improve within 48–72 h. Occasionally severe tension headache or cervical spondylosis delays recovery. A non-steroidal anti-inflammatory analgesic will often help, as will a small dose of promethazine for its mild sedative and muscle-relaxing effect. If there is no improvement in 5–7 days, the differential diagnosis should be reconsidered.

An antiviral drug, pleconaril, which inhibits the attachment and entry of enteroviruses into host cells, is available on a named patient basis for the treatment of enteroviral encephalopathy in agammaglobulinaemia.

Poliomyelitis

Introduction

Poliomyelitis is an enteroviral infection that causes meningitis, but also damages anterior horn cells. This is exceptionally rare in other enteroviral infections, but is seen in some West Nile virus infections. Polio is now rare, owing to a World Health Organization campaign to eradicate the disease, using live-attenuated vaccine. Until the 1970s, it was the commonest worldwide cause of paralysis and limb-wasting in young age groups.

Epidemiology

Humans are the only known reservoir of infection. In acute infection, virus excretion persists for a week in oropharyngeal secretions and up to 6 weeks in faeces. Transmission is predominantly faecal–oral in conditions of poor hygiene, and pharyngeal in better conditions.

Polio occurs in three patterns: endemic, epidemic and sporadic. As over 99% of polio infections in young children are asymptomatic, paralysis is relatively rare when the disease is endemic. Outbreaks of endemic-type disease have recently occurred in Nigeria, and spread to parts of West and Central Africa and Yemen where paralytic cases affected infants and young children who were not recently vaccinated (see Chapter 1).

Epidemic polio occurs in improved living conditions. Virus circulation in young children diminishes and infection then occurs in teenagers and young adults, in whom around 10% of infections are paralytic. Large outbreaks of paralytic polio occurred in western countries during the 1940s and 1950s.

Since the 1950s, vaccination has rapidly brought polio under control. The last UK outbreak, in 1977, affected unvaccinated 'travellers'. No cases of polio have been reported in England and Wales since 1998. An unusual outbreak occurred in Finland in 1984, due to an antigenically altered type 3 poliovirus against which vaccination with inactivated vaccine (routinely used in Finland) was ineffective.

Rarely, live polio vaccines mutate to a virulent form, and can cause paralytic disease in vaccine recipients or their contacts. Only inactivated vaccines are now used in regions where polio no longer circulates.

Virology and pathogenesis of poliomyelitis

Poliovirus is an enterovirus and shares the virological characteristics of this group (see Chapter 6). There are three serotypes, 1, 2 and 3, and immunity is type specific.

Poliovirus enters the body through the alimentary tract and the oropharynx. This is followed by a viraemic phase, during which the virus enters the meninges and spinal cord. The virus infects and kills anterior horn cells, leading to lower motor neurone degeneration.

Clinical features

Average incubation is 1 to 3 weeks (range 3 to 35 days). Around 90% of cases are inapparent or mild. Pharyngitis or mild diarrhoeal disease may occasionally be recognized as caused by poliovirus. Otherwise the disease may be clinically classified as non-paralytic or paralytic.

Non-paralytic poliomyelitis

Non-paralytic poliomyelitis is viral meningitis caused by poliovirus. It is indistinguishable from other enteroviral meningitides, other than by laboratory diagnosis. It is epidemiologically important, as patients excrete the virus in faeces, and are infectious, for up to 6 weeks.

Paralytic poliomyelitis

Paralytic poliomyelitis begins with fever, often with myalgia and/or meningism, followed after 2–5 days by asymmetrical paralysis developing over about 48 h. Affected muscles are often painful in the day or two before paralysis is evident. Fasciculation appears first, then reflexes and movement are lost. Upper motor neurone or sensory features are absent. If the upper arms are affected, the diaphragm is at risk; if bulbar weakness occurs, the airway must be guarded.

Trauma, including exercise, intramuscular injection, vaccination or tonsillectomy, makes paralysis of the local muscles more likely. Paralysis is more likely and more extensive in older age groups.

Clinical spectrum of poliovirus infection

1 Inapparent seroconversion.
2 Non-neurological disease (usually pharyngeal or diarrhoeal).
3 Non-paralytic (viral meningitis only).
4 Paralytic (anterior horn cell death).

Examination of the CSF reveals a lymphocyte count of up to 500 cells/mm^3, with other changes consistent with viral meningitis. The peripheral white cell count may be raised during the myalgic early stages, but is otherwise normal.

Diagnosis

Specimens from suspected polio cases should be handled with great care, to prevent inadvertent spread of the virus during the last stages of disease eradication. Laboratories receiving specimens should be warned beforehand. Rapid diagnosis is important, and can be made by PCR-based testing of CSF or other specimens. Culture of throat swabs, CSF and faeces permits identification and typing of the causative virus. In later cases, rising titres of antibodies may be detectable in serum, and local synthesis of antibody may be demonstrable by comparing serum and CSF titres.

Virus typing is important in identifying outbreaks and in distinguishing between wild and vaccine viruses. Poliovirus can be serotyped into types 1, 2 and 3, and further characterized by RNA studies. The attenuated vaccine viruses can be distinguished from wild virus by RNA sequencing or PCR.

Diagnosis of poliomyelitis

1 PCR-based detection of poliovirus RNA in CSF or other specimens.

2 Cell culture of stool, throat swab or cerebrospinal fluid (CSF).

3 Serotyping of virus isolates by neutralization tests.

4 RNA studies if vaccine-associated polio is suspected.

5 Demonstration of local antibody production in CSF.

6 Demonstration of rising titres of serum antibodies.

Management

There is no licensed antiviral therapy. Bed rest is recommended until the fever has resolved, as this may limit the severity of paralysis.

Once the patient is afebrile and stable, physiotherapy should be commenced. The rate and extent of recovery of power are variable, but improvement can continue for up to 2 years. Children are more likely than adults to have a good recovery.

Complications

Secondary bacterial infections commonly affect the chest when the glottis or respiratory muscles are weak. Appropriate cultures and antibiotic treatment are important in maintaining the patient's health during recovery.

Muscle paralysis may lead to contracture or deformity, especially in growing limbs. Expert orthopaedic advice may be needed to optimize posture and give the best functional result.

Prevention and control

Inactivated polio vaccine (IPV) was introduced in 1956, and live oral polio vaccine (OPV) in 1962. Since 2004, only IPV has been used in the UK. It is given in three doses, as a component of the pentavalent childhood vaccine, at monthly intervals starting at 2 months of age, with booster doses before entry to school and at 15 years of age (see Chapter 25).

Vaccination protects against poliovirus types 1, 2 and 3. OPV is orally administered and provides intestinal as well as humoral immunity, whereas immunity from IPV is largely humoral. OPV strains protect from mucosal invasion, and are excreted in the faeces for up to 6 weeks. They therefore infect and immunize close contacts of vac-

cinees, producing good population immunity where vaccine uptake is low. They compete with wild polioviruses in the body. OPV therefore remains the vaccine of choice for eradicating disease, and controlling epidemics.

The main disadvantage of OPV is that it causes paralysis in approximately 1 per 2 million recipients, with a similar risk for non-immune contacts. When polio is eradicated, the risk of OPV-related paralytic disease is unacceptable, and IPV is then the vaccine of choice, until polio is declared extinct, when vaccination programmes may cease.

Advantages of oral polio vaccine

1 Can be administered orally.

2 Confers mucosal as well as systemic immunity.

3 The vaccine infects and immunizes close contacts.

4 Vaccine of choice for eradication and epidemic control.

Disadvantages of oral polio vaccine

1 Vaccine must be kept refrigerated.

2 Three doses are needed.

3 There is a small risk of mutation to virulence, and paralytic disease.

4 The vaccine is infectious within the family and the community.

5 Should not be used where polio virus is eradicated.

Polio is a notifiable disease. Cases should be isolated in hospital. In an outbreak, extensive virus circulation usually precedes the first case; a single case of indigenously acquired polio requires vaccination of a wide network of contacts.

Herpes simplex type 2 meningitis

This is almost always associated with primary genital herpes simplex, of which it is a complication. Occasional cases occur in the absence of detectable genital lesions. The meningitis is benign, behaving like an enteroviral infection, but may be treated with aciclovir if the diagnosis is known before resolution occurs.

Mollaret's recurrent meningitis is caused by relapsing attacks of herpes simplex meningitis.

It is sometimes accompanied by sacral myeloradiculitis with pain in the perineum, buttock or leg, associated with paraesthesiae, and occasionally difficulty of micturition or defaecation. Although the condition is usually self-limiting, it can persist for weeks. It is therefore treated with aciclovir, which should be given intravenously in a dose of 10 mg/kg 8-hourly (see Chapter 15).

Bacterial meningitis

Introduction

This section deals with diseases of children and adults. Neonates have particular types of bacterial meningitis associated with the birth process and early bacterial colonization. Neonatal meningitis is discussed in Chapter 17.

Bacterial meningitis is a medical emergency because of the high mortality of untreated or late-treated cases. Different types of meningitis cause potentially preventable morbidity in various age groups. Prompt treatment can minimize the mortality to 4–8% in childhood types, and to 8–25% in adults. Early treatment can also minimize neurological deficit. Empirical antibiotic treatment should be given before microbiological diagnosis is available, and before lumbar puncture if the diagnosis is strongly suspected and lumbar puncture is not indicated, or is delayed.

Empirical treatment of bacterial meningitis

1 Cefotaxime i.v. 2 g immediately, then 12-hourly or ceftriaxone i.v. 2 g immediately, then daily. Treat for 5–14 days, depending on microbiological information and clinical progress.
2 Dexamethasone 0.15 mg/kg 6-hourly for 4 days commenced just before, or with, antibiotics (omit if unsure that antibiotic treatment is optimal).
3 For patients over age of 55 years, who may have listerial meningitis, ampicillin or amoxicillin i.v. 2 g 6-hourly should be added.

Organism list

- *Neisseria meningitidis*
- *Streptococcus pneumoniae*
- *Listeria monocytogenes*
- *Haemophilus influenzae*
- *Mycobacterium tuberculosis*
- *Streptococcus 'milleri'*
- Other Gram-positive cocci
- *Escherichia coli*
- *Mycoplasma pneumoniae*
- Rarities:
  - *Leptospira* spp.
  - *Borrelia* spp.
  - *Brucella* spp.

⚠️ Rarely, after neurosurgery or head injury, *Acinetobacter* spp., *Pseudomonas* spp., staphylococci, anaerobic cocci or bacilli, enterococci, or *Candida* spp.

Between 15 and 25% of bacterial meningitis cases have negative blood and CSF cultures. This is often due to sampling problems or previous antibiotic treatment rather than unusual bacterial aetiology.

Lumbar puncture and CSF changes in purulent meningitis

With the exception of tuberculosis, leptospirosis, *M. pneumoniae* and borreliosis, bacterial meningitis causes neutrophil pleiocytosis in the CSF. The presence of neutrophils in the CSF is always abnormal; in meningitis, numbers range from $10/mm^3$ to $2000/mm^3$ or more. The protein level is usually higher in bacterial than in viral meningitis; levels greater than 1.0 g/l are common. A low glucose level is rarely found in viral meningitis, but is usual in bacterial infections. Endotoxin and lactate levels rise, but are usually only measured for research purposes.

In bacterial meningitis, the degree of CSF abnormality is closely related to the prognosis. High neutrophil, protein, lactate, endotoxin and bacterial antigen concentrations are all associated with a poor prognosis. Since these levels tend to rise in parallel, the protein level and neutrophil count often give a reasonable indication of prognosis. However, some severe cases of pneumococcal meningitis with a high bacterial load have a poor neutrophil response.

Meningococcal meningitis

Introduction

This disease, caused by *Neisseria meningitidis*, is an important cause of morbidity and mortality in children and young adults. It is a severe bacteraemic disease with a rapid onset. About 80% of cases have signs of meningitis, which aid early diagnosis. Those with bacteraemia alone sometimes present with more advanced disease, and suffer a higher mortality than those with meningitis.

Epidemiology

N. meningitidis is the commonest cause of bacterial meningitis in the UK (Fig. 13.7). Approximately 2000 laboratory-confirmed cases of meningococcal disease are reported annually. However, this underestimates the true incidence of disease since only around 50% of cases are confirmed microbiologically. Around 40% of cases occur between January and March. In the UK, the incidence of meningococcal disease has been rising since the mid-1990s, but has declined again since childhood immunization against serogroup C was introduced (Fig. 13.8).

This is in contrast to the 'meningitis belt' of sub-Saharan Africa, where the predominant strains, group A, and sometimes group C, cause explosive epidemics during the dry season from December to March.

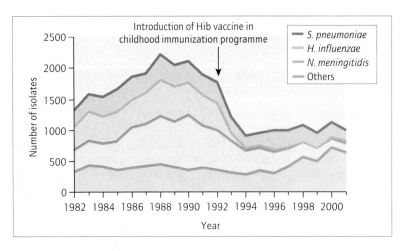

Figure 13.7 Epidemiology of bacterial meningitis in the UK. Source: Health Protection Agency.

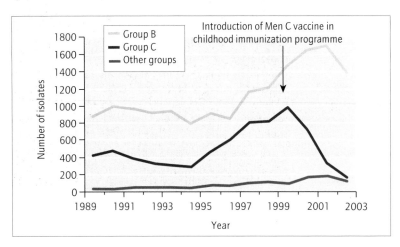

Figure 13.8 Recent trends in the epidemiology of meningococcal disease in the UK. Source: HPA, Meningococcal Reference Unit and Statutory Notifications of Disease (From 1996 data includes PCR confirmed reports in addition to culture confirmed isolates).

The peak incidence of meningococcal meningitis is at 6 months of age, coinciding with loss of maternal antibody. A second, smaller peak occurs in teenagers and young adults. Many infections are sporadic, but outbreaks sometimes occur in households, schools and military establishments. Group C strains caused outbreaks in school pupils until vaccination was included in the UK national childhood vaccination programme. During outbreaks, the age distribution shifts towards older children.

Certain conditions predispose to meningococcal infection, notably functional or anatomical asplenia, complement deficiency and exposure to cigarette smoke. There is evidence that recent upper respiratory infection (especially influenza) temporarily increases the risk of meningococcal disease.

Bacteriology and pathogenesis of meningococcal disease

N. meningitidis is a small Gram-negative diplococcus. It is carried in the nasopharynx by up to 40% of individuals in closed communities, and causes septicaemia and meningitis in a minority of hosts. The organism is aerobic, catalase- and oxidase-positive and non-motile. It possesses a polysaccharide capsule, which is the main antigen and determines the serogroup of the species. The organism has at least 13 serogroups of which the most commonly implicated in disease are A, B, C and W135. The prevalence of a serogroup varies with time and geographical location. Meningococcal serogroups can be subdivided into serotypes, based on the outer membrane protein (OMP) 2 antigen. This is one of three OMPs that act as porins, controlling the influx of water-soluble molecules through the lipophilic outer membrane. Some OMP types are associated with more serious infections.

Neisseria spp. express either of two types of pili, which are important in adherence to host epithelium. Class I pili are similar to those produced by *N. gonorrhoeae*. Pilus production can be switched on and off and this may give the organism advantages in colonization and transmission. Attachment to host cells is also mediated through the outer membrane proteins OPa and OPc. Capsule production must be switched off during intracellular invasion and this is achieved by a molecular switch mechanism.

Up to 50% of the bacterial outer membrane is a lipo-oligosaccharide (LOS), analogous to the lipopolysaccharide of Gram-negative bacteria, in that it contains a lipid A subcomponent. This antigen causes activation of macrophages and release of tumour necrosis factor, a mediator of shock in severe meningococcal septicaemia. The LOS may also assist in invasion of the CNS, by altering the permeability of the blood–brain barrier.

Antibodies to the capsular polysaccharide are bactericidal and protective, but the response to serogroup B antigen is weak because it cross-reacts with human host tissues. There is a strong immune response to meningococcal iron regulation proteins, which may assist in controlling infection by denying the organism iron. Like many other upper respiratory tract colonists, *N. meningitidis* produces an IgA protease that cleaves secretory IgA.

Complement is important in protection against meningococcal disease. Defects of the alternative complement pathway cause enhanced susceptibility to infection. Rare defects in properdin predispose to catastrophic infection. Immunization permits recruitment of the classical complement pathway and abolishes this predisposition.

> **Pathogenicity factors in *Neisseria meningitidis***
> 1 Capsular polysaccharide.
> 2 Outer membrane lipo-oligosaccharide (endotoxin-like).
> 3 Outer membrane proteins.
> 4 Pili.
> 5 Immunoglobulin A protease.

Clinical features

The incubation period can be very short. Most secondary cases present 2–5 days after the primary case, although some occur up to 4 weeks later. Illness usually develops over less than 24 h, but a slower or stuttering onset may occur, and cause diagnostic uncertainty. Fever, malaise and increasing headache are accompanied by nausea and often vomiting. Photophobia may be extreme.

A petechial and purpuric rash appears in most patients (Fig. 13.9), sometimes preceded by a measles-like rash or generalized erythema that lasts for only a few hours. The rash, a feature of disseminated intravascular coagulation (DIC), is an important clue to diagnosis but varies widely in severity and extent. Some cases have very scanty lesions and a few have petechiae only in the conjunctiva or other mucosal surface (Fig. 13.10). A careful search for petechiae is warranted, as their presence supports the clinical diagnosis. A few cases have vasculitic lesions (Fig.

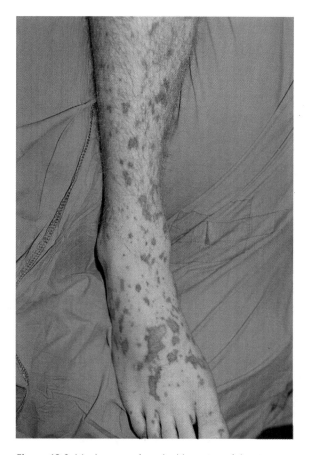

Figure 13.9 Meningococcal meningitis: acute rash in a teenage boy; the legs, buttocks and elbows are often most affected.

13.11), or extensive, necrotic lesions (purpura fulminans; Fig. 13.12).

Patients with rapidly advancing disease quickly develop features of endotoxaemia such as hypotension, pulmonary oedema and sluggish peripheral circulation, with bluish, mottled skin. Drowsiness and confusion are common and coma sometimes develops.

The white cell count is usually raised to 15–$20 \times 10^9/l$, and the C-reactive protein is elevated. The erythrocyte sedimentation rate is low, consistent with DIC, and fibrin degradation products are increased. The prothrombin time or international normalized ratio (INR) is often slightly prolonged, but serious bleeding or very low platelet counts are unusual.

Lumbar puncture reveals purulent CSF with a neutrophilia, raised protein and low sugar levels. Gram-negative diplococci may be seen inside and outside the neutrophils (Fig. 13.13), though they are often undetectable after antibiotic treatment. Blood and CSF cultures are usually positive.

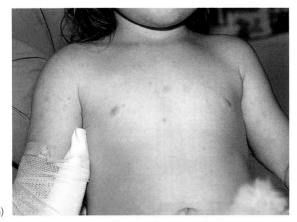

(a)

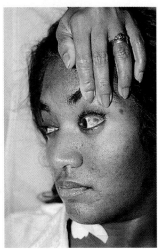

(b)

Figure 13.10 Acute meningococcal meningitis: (a) this 3-year-old girl had only a handful of odd-shaped spots on the trunk and shoulders; (b) this 20-year-old woman had a conjunctival lesion.

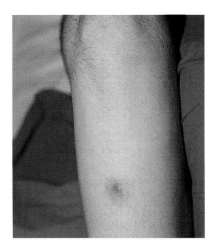

Figure 13.11 Acute meningococcal meningitis: a vasculitic lesion on the shin; there were three others on the hand and arms.

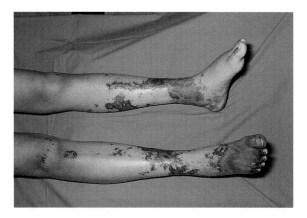

Figure 13.12 Acute meningococcal meningitis: purpura fulminans.

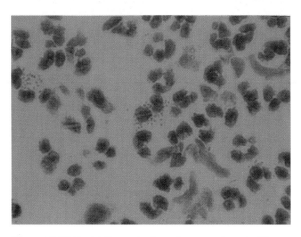

Figure 13.13 Rapid diagnosis of meningococcal meningitis. Cerebrospinal fluid smear shows neutrophils, and intracellular and extracellular Gram-negative diplococci.

Unusual presentations

Meningitis in an infant sometimes has a fluctuating onset, with intermittent fever, lassitude, high-pitched crying and vomiting, separated by periods of apparent normality. Persisting fevers, bulging fontanelle or a seizure eventually prompts the performance of diagnostic tests.

Occasionally the disease progresses so fast, typically in a toddler, that the patient presents with profound sepsis and shock, but no rash. The case fatality rate with this presentation approaches 40%.

Older children and adults may have chronic meningococcal bacteraemia, with fluctuating fever, rash and arthralgias. The initially flat rash becomes progressively more papular and vasculitic (Fig. 13.14), and then closely resembles Henoch–Schönlein disease, which is a common misdiagnosis. Joint swelling and small effusions may ap-

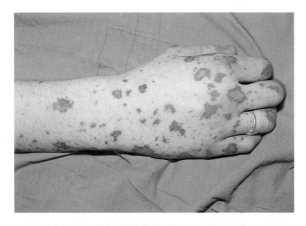

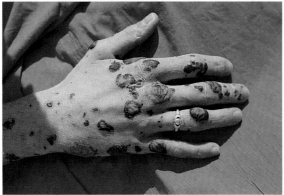

Figure 13.14 (a) Early meningococcal rash; (b) the same rash after 1 week's evolution.

pear in the hands, knees or ankles. The white cell count is usually elevated. The erythrocyte sedimentation rate rises, further suggesting autoimmune or collagen disease. Blood cultures are positive during feverish episodes, though several may be taken before the true diagnosis is confirmed. This fluctuating course can persist for weeks, with the risk of sudden progression to acute meningitis or bacteraemia.

Diagnosis

Typical clinical features often allow immediate treatment without the need for a confirmatory lumbar puncture. However, the rash of DIC can occur with other bacteraemic conditions, and some meningococcal cases have no rash. Where there is doubt, lumbar puncture can provide a rapid diagnosis. Blood cultures should always be taken.

The diagnosis can be confirmed by culture of blood and/or CSF. PCR-based DNA detection allows rapid identification and typing of organisms even when cultures

are rendered sterile by pre-admission antibiotic therapy. Latex agglutination tests can detect group A or C antigen in CSF. IgM antibodies can be detected in acute serum. For epidemiological reasons, all *N. meningitidis* should be typed by referring cultures and PCR specimens to the reference laboratory. Gram-staining of material from incision and scraping of petechial lesions may permit rapid detection of Gram-negative diplococci. Scrapings can also be cultured.

Management

Antibiotic therapy

Over many years, meningococci have become less sensitive to benzylpenicillin. Meningococci with reduced sensitivity will initially respond to high-dose treatment but penicillin will later fail, as the blood–brain barrier recovers and penicillin cannot penetrate the CSF in effective concentrations.

Attending general practitioners should give an intravenous dose of benzylpenicillin (or intramuscular if i.v. access is impossible) pending hospital referral. Many patients claim to be allergic to penicillin, often describing transient rash, diarrhoea or vomiting in infancy. The treatment of meningococcal disease is so urgent that only a history of anaphylaxis or angioneurotic oedema should be taken as a contraindication to penicillin. This policy saves lives, as the disease can progress alarmingly during the journey to hospital.

In hospital, the treatment of choice is a third-generation cephalosporin, such as cefotaxime or ceftriaxone. These are effective against resistant organisms, and also against *S. pneumoniae* and rarer organisms that may cause meningitis.

Chloramphenicol is effective against *N. meningitidis*, is well absorbed and penetrates the CSF well, even in oral dosage. It is little used because of an association with agranulocytosis. In developing countries it is inexpensive, and is available for use as a single- or two-dose course of 'oily injection' (which gives cure rates of over 80%). It is widely used for treating meningitis, though reports of resistant organisms are now emerging.

Other management issues: corticosteroids

Uncomplicated cases become afebrile after 6–12 h of treatment. The headache progressively resolves. Skin lesions fade, heal by scabbing, or ulcerate and re-epithelialize, depending on their size. Neurological sequelae are rare, but include cranial nerve lesions, visual impairment, deafness and hydrocephalus. Treatment with corticosteroids shortens the illness and reduces neurological sequelae only if it is initiated just before, or together with, antibiotic therapy.

Treatment of meningococcal meningitis

1 General practitioner treatment: benzylpenicillin i.v.: infant below 1 year, 300 mg; child 1–6 years, 600 mg; older child or adult, 1.2 g; given immediately.

2 Alternative: cefotaxime i.v. 2.0 g 8-hourly (child 150–200 mg/kg daily) for 5 days; or ceftriaxone i.v. 2–4 g daily (child 50–80 mg/kg daily) as a single dose.

3 Hospital treatment: benzylpenicillin i.v. 1.2 g 2-hourly for 24–48 h (child 1–12 years, 200 mg/kg daily), reducing to 1.2 g 6-hourly to complete 5 days' course (child reduced to 100 mg/kg daily).

4 In severe allergy to beta-lactam antibiotics: chloramphenicol i.v. or orally, 2–3 g daily in three or four divided doses (child 50–100 mg/kg daily).

5 For inexpensive emergency short-course therapy: chloramphenicol oily injection, one or two doses within 48 h.

 PLUS: Dexamethasone, 0.15 mg/kg 6 hourly for 4 days initiated just before, or with, antibiotic treatment.

Managing seizures

A single dose of diazepam, intravenously (250 µg/kg) or rectally (500 µg/kg), may be sufficient to control convulsions occurring during the acute fever at presentation. If fits persist, phenytoin 15 mg/kg may be given by slow intravenous infusion not exceeding 50 mg/minute (neonate 15–20 mg/kg at 1–3 mg/kg per minute), followed by maintenance dosage of 100 mg 6–8-hourly (child 2.5 mg/kg 12-hourly). Alternative: Phenobarbital by i.v. infusion, 10 mg/kg at a rate not exceeding 100 mg/minute; maximum 1 g, followed by adult maintenance dose of 60–180 mg at night, or child dose of 5–8 mg/kg at night. As fever and inflammation are controlled, fits often cease, and weaning from the anticonvulsant may be attempted.

Managing sepsis

Profound shock, pulmonary oedema and severe purpura at presentation are adverse prognostic signs, indicating a 25–40% risk of fatality. Early, vigorous resuscitation contributes significantly to a favourable outcome. Patients with severe presentations usually become afebrile after 36–48 hours of treatment. However, the pathology of the sepsis and accompanying DIC resolves more slowly. There are platelet plugs containing trapped organisms in many small blood vessels. The activated platelets promote inflammation, complement activation and coagulation. Haemodynamic status and cerebral function remain unstable in the first 24–36 h, requiring management and support as for Gram-negative sepsis (see Chapter 19). There is no evidence that heparin benefits the DIC once effective antibiotic treatment has been commenced.

Plasmapheresis and extracorporeal membrane oxygenation do not appear to improve the prognosis. Activated protein C is not indicated in meningitis, because of the risk of intracerebral haemorrhage. Bacterial permeability-increasing factor, which binds endotoxin, appears to confer some benefit in limited trials. Many who die have bilateral haemorrhagic necrosis of the adrenal glands (Waterhouse–Friderichsen syndrome), but the plasma cortisol levels in such cases are usually high.

Complications

Other manifestations of meningococcal infection

Other manifestations of meningococcal infection may precede or complicate the meningitis, or occur alone. Many trivial meningococcal infections occur, including conjunctivitis, pharyngitis or otitis media, which are often inadvertently treated with empirical antibiotics. Severe non-meningeal presentations include endophthalmitis, pericarditis, endocarditis, myocarditis and, rarely, septic arthritis. Bacteraemic disease can accompany any of these cases.

Necrotic skin lesions

Necrosis of purpuric and ecchymotic lesions can cause severe pain. Adequate analgesia is important. Some lesions ulcerate and a few leave full-thickness skin defects that later require grafting. Large peripheral lesions can cause gangrene, with loss of digits or extremities. This risk is increased if inotropic support is required, as this favours central circulation, to the disadvantage of peripheral perfusion.

Reactive arthritis

Reactive arthritis is a common late complication, affecting more than 50% of adolescents and young adults after 5–7 days. The fingers or knees are most often affected, usually asymmetrically (Fig. 13.15). Aspiration reveals neutrophils but no organisms. Immune complexes containing meningococcal antigen are demonstrable in the synovium. The condition resolves spontaneously and symptomatic relief is afforded by non-steroidal anti-inflammatory agents.

Serositis

Serositis with effusion is less common than arthritis, but is probably also an immune-complex disorder. A pleural or pericardial effusion may be detected on X-ray, and occasionally requires aspiration. A rapidly reducing course of prednisolone will often terminate persistent or recurrent effusions.

Neurological complications

Neurological sequelae may occur in severe cases. Deafness or other cranial nerve lesions are the most likely, and some improve or recover during convalescence. Occasionally an apparently recovering patient will collapse with a

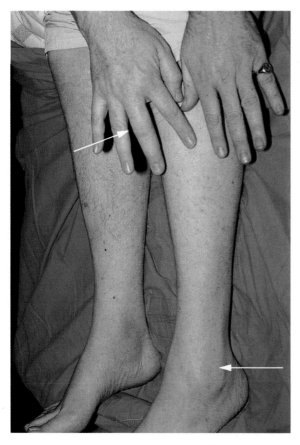

Figure 13.15 Post-meningococcal synovitis affecting the finger and ankle (arrows).

major cerebral event, usually secondary to cerebral artery damage caused by meningeal or intracerebral purpuric lesions.

Abscess formation

Abscesses can affect sites such as the middle ear or a paranasal sinus. They may respond to treatment of the meningitis, but surgical decompression and drainage should be considered, to clear persisting infection and speed recovery from the meningitis.

Prevention and control

Meningococcal vaccines are available against groups A, C, Y and W135 strains. No group B vaccine is currently available in the UK, though several promising candidates are in clinical trials. A group BH vaccine has been developed for local use in Cuba.

Divalent type A and C polysaccharide vaccine contains purified capsular antigen. This is poorly effective in young children, providing only temporary T-cell-independent

immunity. Its use is indicated for those at particular risk from meningococcal disease, including military recruits, household contacts of cases due to vaccine-preventable strains, asplenic patients, travellers to the African meningitis belt and those with certain complement deficiencies. Vaccination has been successfully used to control outbreaks of group C disease in schools and military training camps.

Conjugate serogroup C meningococcal vaccines produce lasting immunity in infants from 2 months of age. A conjugate serogroup C vaccine was introduced into the UK immunization schedule in the autumn of 1999.

Hajj pilgrims visiting Mecca are at high risk of meningococcal disease, including outbreaks of type W135. They should also be vaccinated before travel. A tetravalent A, C, Y, W135 vaccine is available for this use, on a named patient basis.

Chemoprophylaxis is indicated for household and other contacts with mouth-kissing exposure to cases in the 10 days preceding their illness. During outbreaks, classroom, nursery or other institutional contacts may be offered chemoprophylaxis (see Chapter 25). Ciprofloxacin is effective and well tolerated. The alternative rifampicin often causes gastrointestinal side-effects that limit compliance. It also interferes with oral contraception, and causes red coloration of urine, sputum and tears, staining of contact lenses, skin rashes and gastrointestinal reactions. Ceftriaxone may be used in pregnancy. The index case is considered clear of meningococci after 24 hours of ceftriaxone treatment.

Prophylaxis of meningococcal disease

1 Ciprofloxacin orally, 500 mg, single dose.
2 Alternative: Rifampicin orally, 600 mg 12-hourly for 2 days (infant 6–12 months, 5 mg/kg 12-hourly; child 1–12 years, 10 mg/kg 12-hourly). This differs from the regimen for Hib prophylaxis (see p. 288).
3 In pregnancy: ceftriaxone i.m. 1 g, single dose.
4 *Meningococcus* types A plus C polysaccharide vaccine, for school and military outbreaks with these types, 0.5 ml i.m. or deep s.c. (not effective below age of 18 months).
5 Conjugate meningococcus C vaccine for infants.

 Swabbing of contacts and cases after chemoprophylaxis is not necessary.

A case of meningococcal meningitis, particularly in a school, often causes alarm and may attract media attention. Good communication and public education can minimize this. It is advisable to inform parents when a case occurs in a school, nursery or college. Meningococcal meningitis and septicaemia are both notifiable diseases (see Chapter 25).

Haemophilus influenzae meningitis and bacteraemic diseases

Introduction and epidemiology

Haemophilus influenzae meningitis is a rare disease caused by an encapsulated *H. influenzae* of capsular type b (Hib). Childhood immunization is universal in industrialized countries, and antibodies to the organism are naturally acquired by age 4 or 5 in unimmunized populations. Other rare childhood bacteraemic diseases caused by Hib include pneumonia, acute epiglottitis, facial cellulitis and some bone and joint infections. Immunocompromised adults occasionally suffer *H. influenzae* bacteraemias. Rare cases of *H. influenzae* meningitis occur in adults and older children, often complicating chronic CSF fistulae or other predispositions. Infections with non-type b capsulated *H.* *influenzae* are nowadays more common than Hib in these settings (Fig. 13.16).

Microbiology and pathogenesis of Hib infections

H. influenzae is a small Gram-negative coccobacillus that is catalase- and oxidase-positive. As it requires nicotinamide adenine dinucleotide phosphate (NADP) and haematin for active growth, it produces little or no growth on unsupplemented blood agar, but it grows well on chocolated blood agar. The organism is aerobic but isolation and growth are favoured by an atmosphere with enhanced carbon dioxide.

The main pathogenicity determinant of *H. influenzae* is the capsule. Although the organism is capable of producing one of six different capsular polysaccharides (or none), almost all isolates from clinical disease are serotype b. Sus-

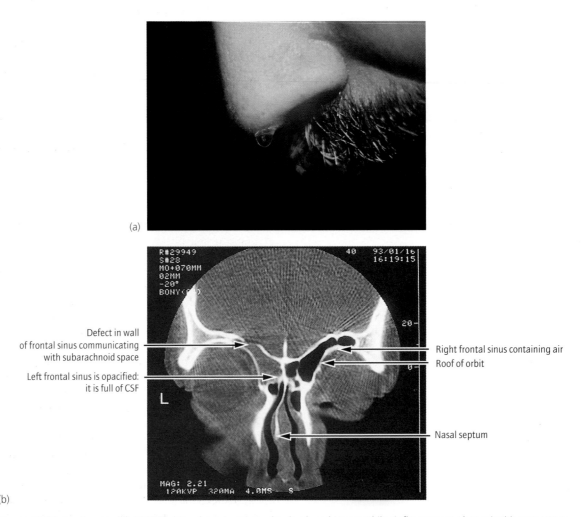

Figure 13.16 Cerebrospinal fluid (CSF) rhinorrhoea in a man who developed *Haemophilus influenzae* type b meningitis many years after a head injury. (a) Clear CSF dripping from the nose; (b) contrast computed tomographic scan showing CSF leaking via a skull defect into the frontal sinus (arrow). The sinus is opaque compared with the air-filled sinus on the right. Courtesy of Dr W. R. C. Weir and the *Journal of Infection*.

ceptibility to Hib infection correlates with the absence of antibodies to the type b capsule. Molecular studies have confirmed the importance of the polyribitol ribosyl phosphate (PRRP) capsule in experimental meningitis. Other pathogenicity determinants include the lipopolysaccharide, outer membrane protein, pilus proteins and the IgA protease.

Pathogenicity factors of *Haemophilus influenzae*

1 Polyribitol ribosyl phosphate (PRRP) capsule, especially capsular type b.
2 Cell-wall lipopolysaccharide.
3 Outer membrane protein.
4 Pilus proteins.
5 Immunoglobulin A protease.

The capsule is not essential for pathogenicity in the respiratory tract, where most infections are caused by noncapsulate *Haemophilus influenzae*. These infections tend to occur in individuals with bronchiectasis or chronic obstructive airways disease, where the organisms are partly protected from the host immune system.

Clinical features

Hib meningitis often has a slower onset than meningococcal meningitis, developing over 3 or 4 days. Clinical features may fluctuate initially, improving when the temperature is lower. A 'febrile' convulsion may warn of severe underlying pathology.

There is usually a neutrophilia, with a white cell count of $15–20 \times 10^9$/l. The blood sugar may also be raised to 8–10 mmol/l in the face of acute cerebral insult.

Diagnosis

CSF examination shows neutrophils, and often small, Gram-negative rods, but the bacteria are not easy to identify and are rarely seen if antibiotic treatment has already been given. A positive latex agglutination test on the CSF, using the Hib PRRP antigen provides evidence of Hib infection. Blood cultures are usually positive. DNA amplification tests enhance the sensitivity of diagnosis and may remain positive after antibiotics have been commenced.

Diagnosis of *Haemophilus influenzae* type b (Hib) meningitis

1 Demonstration of small, Gram-negative rods in cerebrospinal fluid (CSF) deposit.
2 Demonstration of Hib antigen in CSF by latex agglutination.
3 Positive CSF culture.
4 Positive blood culture.

Management

The treatment of choice is a broad-spectrum cepha-losporin, such as cefotaxime or ceftriaxone, which penetrates excellently into the CSF and is highly bactericidal to Hib. Cefuroxime does not maintain good CSF concentrations once the blood–brain barrier recovers, and carries a risk of relapse. Antibiotic treatment should be continued for at least 1 week, and probably for 10 days in severe cases.

Dexamethasone makes only a few hours' difference to the duration of fever, and does not alter mortality, but it reduces the incidence of deafness after recovery in some trials. The child's dose is 8 mg twice daily during the first 3 days of antibiotic treatment.

Treatment of *Haemophilus influenzae* type b meningitis

1 Cefotaxime i.v. 150–200 mg/kg daily in three or four divided doses; *alternative*, ceftriaxone i.v. or i.m. 50 mg/kg daily in a single dose (up to 80 mg/kg can be given i.v.); both for 7 days.
2 Second choice: chloramphenicol i.v. or orally 50–100 mg/kg daily in four divided doses for 10–14 days.
Plus dexamethasone orally, i.m. or i.v. 8 mg twice daily for the first 3 days only.

Other priorities include adequate rehydration and haemodynamic support, control of convulsions (see p. 284) and, on rare occasions, specific measures to control raised intracranial pressure.

Complications

Although the fatality rate is lower (around 4%) adverse sequelae are more common than with meningococcal disease, probably because Hib meningitis occurs in such young infants.

• Recurrence of fever, sometimes with neurological signs, is often due to a small subdural area of infection, or cerebral vein thrombosis. This can be demonstrated by CT or MR scan. Continued or higher-dose antibiotic treatment is usually curative, but drainage of subdural collections is occasionally indicated.

• Convulsions are rarely caused by frank abscess formation, but may be due to cerebral venous thrombosis, subdural collection, rising intracranial pressure or local gliotic scarring. A CT or MR scan is indicated to exclude these causes.

• Hydrocephalus is rare but can develop at any stage. It may present with seizures or deteriorating cerebral function. Imaging is required to make the diagnosis, and temporary or permanent CSF drainage may be necessary.

• Neurological impairment, most commonly reduced hearing, occurs in 9–15% of survivors. It should always be sought, as it affects subsequent schooling and social development. Other sequelae include squint, ptosis, reduced visual field or acuity, and impaired balance. These may improve considerably with time. Monoparesis, spasticity, learning difficulties and epilepsy are all rare sequelae.

Complications of *Haemophilus influenzae* type b meningitis
1 Impaired hearing in 9–15% of survivors.
2 Convulsions (rarely, persisting epilepsy).
3 Subdural collections of fluid (rarely, subdural empyema).
4 Cerebral vein thrombosis.
5 Rare visual, motor or sensory deficit.
6 Very rare hydrocephalus.

Prevention and control
Conjugated polysaccharide Hib vaccines are highly effective in infants from 2 months of age, and were introduced in the UK immunization schedule in 1992. Three doses are given at 2, 3 and 4 months of age (see Chapter 25). Children and adults with asplenia are also recommended to be immunized. There are no specific contraindications to Hib vaccine. A mild to moderate local reaction occurs in up to 10% of vaccine recipients.

All household contacts of patients with Hib should be given chemoprophylaxis if there is a child under 4 years in the household (and also vaccine, if unvaccinated and below the age of 10). The case should also receive prophylaxis before discharge from hospital. Rarely, recurrent infections have been documented, so the index case should also receive vaccine if under 10. Nursery contacts should be given chemoprophylaxis (and vaccine, if unvaccinated) where two or more cases occur within 120 days of each other.

Prophylaxis of *Haemophilus influenzae* type b (Hib) meningitis
Rifampicin orally 20 mg/kg daily (maximum 600 mg daily) for 4 days, to *all* household or nursery contacts, when other children below age four are present; *plus* (in unvaccinated children) Hib vaccine 0.5 ml i.m. or deep s.c. as a single dose (infants below 13 months require three injections at intervals of 1 month; the vaccine is not indicated above the age of 10 years except for asplenic individuals).

Haemophilus meningitis is a notifiable disease in the UK (see Chapter 25).

Pneumococcal meningitis

Introduction and epidemiology
Pneumococcal meningitis is the commonest bacterial meningitis in the middle-aged and elderly, but can affect all age groups. While it can be apparently spontaneous, certain susceptible patients are at increased risk. Pneumococcal meningitis is an important disease because of these increased susceptibilities. It has an average fatality rate of 20%, the highest of the common bacterial meningitides.

Conditions causing susceptibility to *S. pneumoniae* infections
1 Defects of the dura mater, for instance after head injury or surgery (see Fig. 13.16).
2 Chronic infections in the skull, for instance chronic suppurative otitis media or sinusitis. Patients with large nasal polyps are susceptible, especially after polypectomy.
3 Functional or anatomical asplenia, for instance in inherited haemolytic diseases, or after splenectomy for any reason.
4 Alcoholics are at increased risk of all types of pneumococcal disease.
5 Age over 65 years.

Microbiology and pathogenesis of pneumococcal infections
These are discussed in detail in Chapter 7.

Clinical features
Most cases are intrinsically derived, caused by pneumococci that have colonized the upper respiratory tract. The onset is often rapid, over 1 or 2 days, but some cases evolve gradually, with meningism developing during the course of an ear or sinus infection.

Most patients have marked meningism by the time of presentation, and impairment of consciousness is common. Watery diarrhoea and confusion are frequent, nonspecific presenting features; the patient may be unable to complain spontaneously of headache. Signs of meningism must be actively sought if such cases are not to be missed.

Other features of severe pneumococcal disease, such as pneumonia or peritonitis, may coexist, and should be actively excluded.

The white cell count is usually $15–25 \times 10^9/l$. A low count indicates a poor prognosis. The plasma creatinine is often slightly elevated. Lumbar puncture shows a marked neutrophil pleiocytosis, and Gram-positive diplococci are relatively easily demonstrated in most cases. Blood cultures are positive.

Management
A significant proportion of *S. pneumoniae* have reduced penicillin sensitivity, and a few are highly penicillin-resistant (PRSP). Penicillins act by binding and inhibiting proteins (known as penicillin binding proteins) that are important in cross-linking bacterial peptidoglycan. *S. pneumoniae* can become resistant to penicillin by acquisition of small segments of DNA from genetically similar organisms, such as *S. oralis* and *S. mitis*, living in the same environment. These fragments are incorporated into the

penicillin binding protein gene, producing a new (mosaic) gene, which encodes a protein with reduced affinity for penicillin (see Fig. 7.6). Moderate resistance is caused by changes in PBP2b. Changes in PBP2x produce resistance to cefotaxime. Changes in both these proteins and PBP1A produce high-level beta-lactam resistance, which is very rare in the UK.

Cefotaxime or ceftriaxone are the initial treatments of choice, as they are usually effective, and penetrate well into the CSF.

In Spain, Portugal, South Africa, parts of the Pacific and increasingly in the Americas, there are increasing reports of treatment failure in cases infected with organisms having cefotaxime minimum inhibitory concentrations (MIC) above 2 mg/l. In patients from areas with known resistance, the addition of teicoplanin or vancomycin should be considered. Experimental (*in vitro*) evidence suggests that adding teicoplanin or other agents provides useful synergy, but there is only anecdotal evidence of their clinical effectiveness. Vancomycin-tolerant strains of *S. pneumoniae* have been reported.

Patients with PRSP meningitis require careful monitoring, and may require repeat lumbar puncture to document CSF sterility.

Treatment of pneumococcal meningitis

1 Cefotaxime 2 g 8-hourly for 10–14 days; *or* ceftriaxone i.v. 2–4 g daily single dose for 10–14 days.
2 *For penicillin-sensitive pneumococci:* benzylpenicillin i.v. 1.2 g 2-hourly, then 2.4 g 4–6-hourly for 10–14 days.
3 *For multiply resistant pneumococci (PRSP):* cefotaxime as above *plus* teicoplanin i.v. 400 mg 12-hourly for three doses, then 400 mg daily; *or* vancomycin i.v. 500 mg 6-hourly initially, but then to maintain plasma levels at peak not more than 30 mg/l and trough not more than 10 mg/l.

Severe sepsis and multiorgan failure should be treated as for any patient with sepsis (see Chapter 19).

Complications

Abscess formation

An abscess may be the origin of the meningitis, or may form during its course. Persistent fever, deteriorating cerebral function, seizures or the appearance of focal signs indicate the need for brain imaging to exclude sepsis in the meninges or adjacent intracranial structures. Surgical drainage is urgently indicated in such cases.

Recurrence

Recurrence is a real possibility in a patient at increased risk. It is particularly likely in those with dural defects, who may have many attacks. CSF rhinorrhoea or otorrhoea can be confirmed by the presence of tau protein in the fluid, and the dural defect may be demonstrable by contrast injection and CT scan (see p. 286 and Fig. 13.16). Recurrences can be prevented by repair of the predisposing condition, by immunization and prophylaxis. Susceptible patients may also suffer meningitis attacks caused by *Haemophilus influenzae* or other organisms.

Other complications

Other complications are those of meningitis in general. They include cranial nerve lesions, early or late hydrocephalus, visual defects or pareses of varying severity.

Prevention and control

Pneumococcal polysaccharide vaccine contains antigen from 23 capsular types of pneumococci, which account for 90% of the invasive infections in the UK. The efficacy is approximately 70% in adults with normal immunity. In 2003, pneumococcal polysaccharide vaccination was recommended in the UK for all people over 65 years of age. This is being introduced over three years. For those with immunological impairment and children under the age of five years a 7-valent conjugate vaccine is recommended for those at risk (see Chapter 7), and will soon be offered to children. Revaccination is not generally recommended but antibody levels decline rapidly in individuals with splenic dysfunction or chronic renal disease. Re-immunization with polysaccharide vaccine is therefore recommended every 5 years in these groups.

Splenectomized patients should be advised that they are at increased risk of pneumococcal disease. In addition to vaccine they should receive long-term penicillin prophylaxis (see Chapter 22).

Listerial meningitis

Introduction and epidemiology

Listeriosis is best known as an infection of pregnant women and neonates (see Chapter 17). However, 35–40% of reported cases occur in older children and adults. Approximately half of non-neonatal infections affect immunosuppressed individuals. Among the immunosuppressed, those at most risk are those on corticosteroid or anticancer therapy, and those with chronic uraemia or transplant-related immunosuppression. Listeriosis rarely affects patients with HIV-related immunodeficiency. Most cases of listerial meningitis occur in the summer months.

Microbiology and pathogenesis of listerial meningitis

Listeria monocytogenes is a small, Gram-positive rod that is common in the environment. It is found in human and animal faeces, sewage slurry and land on which it is spread, vegetables, soft cheeses, pâtés and some prepared meat meals that have been kept chilled. The organism is robust, multiplies readily between 4 and 40 °C, and can survive temperatures of up to 60 °C. Minor pasteurization failures have been followed by large, milk-borne outbreaks. Its ability to replicate at refrigeration temperatures allows it to contaminate chilled food while other bacteria are inhibited. *L. monocytogenes* is one of six *Listeria* species, but the others, *L. seegleri*, *L. welshimeri*, *L. innocua*, *L. ivanovii* and *L. grayi*, are rarely associated with human infection. *L. monocytogenes* can be typed using phages, but more modern methods such as pulsed-field gel electrophoresis and multi-locus sequence typing are now more important.

The organism grows well on blood agar and exhibits a narrow band of beta-haemolysis. It is catalase-positive, is motile by virtue of its flagella, and exhibits a characteristic, end-over-end 'tumbling' motion at 22 °C. *Listeria* survives inside macrophages and has frequently been used as a model of intracellular parasitism.

Listeriolysin is now being recognized as the major pathogenicity determinant. This is one of the family of cysteine-based toxins homologous to streptolysin O and pneumolysin (see Chapters 6 and 7). It appears to act by allowing the organism to escape from phagolysosomes into the cytoplasm of the host cell, where it can then multiply. Multiplication is followed by cell to cell transfer by means of 'filopods', which protect the organisms from exposure to the extracellular environment. Iron enhances the virulence of *Listeria*; listerial infection is more common and severe in patients with haemochromatosis and other forms of iron overload.

Clinical features
Subclinical or mild gastroenteritis-like illness is common, and may occur alone or preceding more invasive disease. The commonest severe manifestation of non-neonatal listeriosis is purulent meningitis. In the immunocompetent it often presents as acute meningitis, and must be differentiated from the commoner types. In about a third of patients it causes either bacteraemia alone or an unusual type of encephalitis with a predominance of brainstem abnormalities (listerial rhombencephalitis).

In the immunosuppressed listeriosis occurs as meningitis, bacteraemia or occasionally peritonitis; but not as brainstem encephalitis.

In pregnant women, infection results in colonization of the bowel or genital tract, or trivial feverish illness. The only sign of maternal infection may be the birth of an affected infant.

Manifestations of listeriosis
1 In the immunocompetent: acute meningitis; rhombencephalitis; or, less commonly, bacteraemia.
2 In the immunosuppressed: acute or subacute meningitis; bacteraemia; peritonitis.
3 In pregnant women: mild feverish illness; mild 'gastroenteritis'; asymptomatic colonization of bowel or genital tract.
4 In the neonate: early bacteraemic multisystem disease or later meningitis (see Chapter 17).

Diagnosis
Diagnosis depends on clinical suspicion, the use of appropriate imaging and the results of blood and CSF culture. In meningitis the CSF microscopy and culture are usually positive while blood cultures can remain negative, even without prior antibiotic treatment. It is therefore important to obtain a CSF culture if listerial meningitis is suspected. However, CSF culture is rarely positive in encephalitic listeriosis, and positive blood cultures are the usual confirmatory finding. DNA amplification techniques are sometimes available as part of a multi-organism testing system, and can be used to make a rapid diagnosis.

Clinical suspicion is important in cases of brainstem encephalitis, where there may be little or no meningism, and the CSF may show no abnormalities. Magnetic resonance scanning is the investigation of choice, and much more sensitive than CT scanning, for showing focal brainstem inflammation.

Management
L. monocytogenes is sensitive to penicillin, ampicillin, tetracycline and many other agents, but cephalosporins are poorly active and should not be used. High-dose amoxicillin is often effective. Some would advocate additional gentamicin, but there is no definite evidence of benefit. Meropenem has been successful in treating penicillin-allergic patients.

Treatment of listeriosis
1 Ampicillin or amoxicillin i.v. 2 g, 6-hourly for 10–14 days (child up to 1 month, 50 mg/kg, 6-hourly; 1 to 3 months, 50–100 mg/kg 6-hourly; 3 months to 12 years, 100 mg/kg 6-hourly, maximum 12 g daily) *plus or minus* gentamicin 2–5 mg/kg daily in divided doses (avoid peak concentration above 10 mg/l and trough above 2 mg/l).
2 Alternative: meropenem i.v. 2 g, 8-hourly (child <50 kg, 40 mg/kg 8-hourly: *not licensed for children under 3 months*).

Prevention and control
Prevention and control in the community depend largely

on food hygiene. Adequate reheating of cook–chill and cook–freeze meals substantially reduces the risk of infection among immunocompromised patients. High-risk foods (especially pâté and soft cheeses) should not be served to such patients. Pregnant women should also avoid these foods. Affected mothers and babies should be nursed in isolation in delivery units.

Streptococcus 'milleri' meningitis and ventriculitis

Streptococcus 'milleri' is the name used for a group of related streptococci including the species *S. anginosis*, *S. constellatus* and *S. intermedius*. They are uncommon causes of bacteraemia and meningitis, *S. intermedius* particularly tending to produce widespread infection with collections of thick pus. They are among the few causes of simultaneous abscess formation in the brain and the liver. *S. 'milleri'* meningitis often leads to purulent obstruction of CSF pathways, resulting in loculated 'ventriculitis', as well as severe meningitis.

The organisms may be alpha-, beta- or non-haemolytic, and of Lancefield group A, C, F or G. The commonest type is beta-haemolytic group F. Group A forms typically produce microcolonies on culture. Diagnosis depends on careful assessment of streptococci isolated from blood, CSF or pus cultures.

The importance of the organism is its reduced sensitivity to penicillin, which is, however, more effective than ampicillin. The treatment of choice is either a broad-spectrum cephalosporin or high-dose benzylpenicillin plus standard doses of gentamicin. A search should be made for distant infection and abscesses, particularly in the liver, as drainage may remove an additional source of sepsis.

Encephalitis and meningoencephalitis

Introduction

Encephalitis is inflammation of the brain. It can occur independently of meningitis or the two can coexist, when the condition is called meningoencephalitis.

The clinical features are those of cerebral irritation and dysfunction. Irritation often appears first, initially resembling bad temper, restlessness or personality change, with distractibility or mild confusion. Some patients complain of ataxia or generalized weakness. If the temporal lobes are affected there may be aphasia or loss of short-term memory. Seizures may occur at this stage. On examina-

tion the reflexes are often brisk and the plantar responses may be up-going.

As cerebral dysfunction increases, brain swelling may entrap cranial nerves, causing focal signs such as ophthalmoplegia or ptosis. Seizures become more likely and drowsiness or coma develops. In severe cases the pupil reactions are sluggish or absent, bulbar function is impaired, intermittent breathing patterns may develop, followed by brainstem death.

General features of encephalitis
1 Irritability.
2 Altered personality.
3 Drowsiness.
4 Ataxia.
5 Excessively brisk tendon reflexes.
6 Up-going plantar responses.
7 Signs of cerebral or brainstem failure (sluggish or absent pupil reflexes, intermittent breathing patterns).
8 Signs of brain swelling (focal neurological signs, papilloedema).

 Fever is usually present; signs of meningitis may coexist.

There are many causes of cerebral dysfunction, which can mimic encephalitis. Acute confusional states complicate fever, especially in the old and the very young. Certain infections commonly cause confusion or encephalopathy; legionnaires' disease, typhoid fever and typhus are examples. Non-cerebral infections should therefore be excluded in cases of global cerebral disturbance. Acute disseminated encephalomyelitis (ADEM), is considered to be a reactive condition; it can complicate many infections, and is characterized by widespread central neurological deficit, with altered consciousness, sometimes seizures, and characteristic high signal bilaterally in the cerebral white matter (or also areas of grey matter in severe cases). Drugs may also cause encephalopathy. Alcohol, alcohol withdrawal, opiates, high doses of phenytoin, indometacin, cycloserine, amantadine and some antidepressants are examples.

Organism list

- Herpes simplex virus type 1 (HSV 1)
- Arboviruses
  - Tick-borne encephalitis virus
  - West Nile virus
  - St Louis encephalitis virus
  - Japanese B encephalitis virus
  - Equine encephalitis viruses
- Rabies virus

> ⚠ An outbreak of encephalitis in Malaysia in 1999 was found to have been caused by a newly identified paramyxovirus, designated Nipah virus, which was transmitted from sick pigs to their keepers by droplet spread (see Chapter 26).

Encephalitis can be part of many other viral syndromes, or post-viral conditions. Acute encephalitis is common and transient in mumps, and is rare but well recognized in influenza A and B. Acute encephalitis can follow natural measles infection, and cerebellar or generalized encephalitis is seen after chickenpox (see Chapter 11). Yellow fever vaccination and vaccinia vaccination are rarely complicated by encephalitis. These types of encephalitis are mentioned in the relevant chapters.

Other infections causing fever and encephalopathy

1 Legionnaires' disease.
2 Typhoid and paratyphoid fever.
3 Typhus fevers.
4 Rare neurobrucellosis.
5 Rare neuroborreliosis.

Herpes simplex encephalitis

Introduction
Herpes simplex is a common infection of humans, usually affecting the skin or mucosae, and followed by lifelong latency of the virus in nerve cells (see Chapter 6). Herpes simplex encephalitis is rare. It is important because it attacks previously healthy individuals, and can cause high mortality or prolonged disability. If recognized early it responds to antiviral therapy.

Virology and pathogenesis of herpes simplex infections
These are discussed in detail in Chapter 6.

Epidemiology
Fewer than 50 laboratory-confirmed cases are reported in the UK each year. Encephalitis is usually caused by HSV type 1 virus, whereas meningitis is usually due to type 2 virus. The disease is sporadic, affecting mainly young and middle-aged adults. Infection in children is rare. Both sexes are equally affected. The case fatality rate approaches 70% in untreated cases.

Clinical features
As the disease is intrinsic, originating from latent or colonizing herpes simplex virus, the incubation period is uncertain. The onset may be rapid, with fever and 'influenzal' symptoms, quickly followed by neurological abnormalities, fits and coma. Other cases have several days of slowly declining cerebral function. The temporal or frontotemporal regions of the cerebral cortex are usually affected earliest and most severely. Focal signs may reflect this, with memory loss, dysphasia and unsteady gait in a drowsy or irritable patient. Headache is not a common complaint.

Up to half of cases have a herpetic lesion of the lip, skin or eye. There is rarely any other physical sign. The white cell count is normal, except when cerebral necrosis occurs, when there may be a neutrophilia. If there is significant cerebral destruction, aspartate transaminase levels may be raised.

The CSF almost always contains an excess of lymphocytes, varying from a few dozen to more than 200. Red blood cells may also present in the absence of a traumatic tap. The protein level is moderately elevated, and the glucose is normal.

Diagnosis
When present, focal neurological signs help to distinguish encephalitis from other confusional conditions. Imaging is helpful in showing cerebral inflammation, often localized in the temporal lobes (see Fig. CS.5). MR is probably most sensitive, but CT is also useful. There are typical electroencephalographic abnormalities in many cases. However, not all herpes simplex encephalitis shows localization, and temporal lobe involvement is not pathognomonic of herpes simplex.

Lumbar puncture is carried out as soon as possible after imaging. PCR-based HSV DNA detection is the most sensitive diagnostic test. It is unusual to isolate herpes simplex virus by culture. A wide variety of antigen detection systems have been tried, but none are reliable. Differential herpes simplex and rubella antibody levels may be estimated in blood and CSF, and may suggest intrathecal HSV antibody production. It is rarely justifiable to perform brain biopsy, as neither a positive nor a negative result would influence the necessity to attempt specific antiviral therapy.

Diagnosis of herpes simplex encephalitis

1 Demonstration of temporal lobe oedema on brain imaging.
2 Demonstration of herpes simplex virus (HSV) DNA in cerebrospinal fluid by polymerase chain reaction.
3 Demonstration of intrathecal anti-HSV antibody production.
4 Demonstration of encephalitic electroencephalographic changes in the temporal cortex.
5 Demonstration of HSV immunoglobulin M, seroconversion or rising immunoglobulin G titres in serum.

Management
Antiviral therapy should be commenced urgently. Since

herpes simplex is the commonest cause of infective encephalitis in Britain, it is reasonable to treat all encephalitis cases until the diagnosis is confirmed or refuted.

The treatment of choice is intravenous aciclovir 10 mg/kg 8-hourly. There is no evidence that the addition of other antiviral agents or of corticosteroids offers any benefit, though dexamethasone can be used if imaging demonstrates generalized cerebral oedema.

The response to treatment is gradual. Fever usually subsides in 2 or 3 days, but cerebral abnormalities improve slowly, and recovery is often incomplete. Treatment is continued for 2 weeks, sometimes longer, but improvement can continue for many weeks after cessation of therapy. Physiotherapy and speech therapy may help the patient regain social and motor skills.

Arbovirus encephalitides

Although not endemic in the UK, arboviruses are a common cause of encephalitis in both the New and the Old World. The arboviruses are a heterogeneous group of organisms found in the families of Alphaviridae, Flaviviridae and Bunyaviridae. They replicate in the cells of both vertebrate and invertebrate hosts and some cause encephalitic diseases. They are transmitted by many different vectors, including mosquitoes, ticks and sandflies (Table 13.1). In the forests of Germany, Austria, Scandinavia and eastern Europe, tick-borne encephalitis is endemic. It is transmitted to humans from its natural rodent hosts in the summer months when ticks are active. Human cases are common.

In north India, Pakistan, parts of Indonesia and the Far East, including Japan, Japanese B encephalitis is transmitted by mosquitoes, and causes outbreaks and epidemics with high morbidity.

In the USA there are five main types of encephalitis. St Louis encephalitis has been recognized for many years,

West Nile virus emerged in 1999, and now affects all mainland states. Eastern equine, western equine and Californian encephalitis also cause significant human disease. All are transmitted from birds and rodents, by mosquitoes, to horses and humans.

Clinical spectrum

These infections have incubation periods of 7–12 days. The onset of illness is rapid, with features of viral-type meningitis. Cerebral disturbance develops over the next 1–2 days. The diseases vary in severity and there are many inapparent infections, especially in children and young adults. However, Japanese B encephalitis can have a mortality rate of 7–20%, with up to 30% incidence of intellectual or psychological problems in survivors.

Tick-borne encephalitis causes less severe sequelae, such as poor concentration and disorders of balance, occurring in around 20% of cases.

West Nile virus infections are associated with self-limiting uveitis and retinitis. Parkinsonian dyskinesias are seen, but tend to recover. Rare flaccid paralysis, identical to that of poliomyelitis (and now called WNV poliomyelitis), is caused by anterior horn cell infection, and is usually permanent.

There is no specific treatment for arboviral encephalitides. Trials of prednisolone and dexamethasone have failed to show benefit.

Diagnosis, management and control

Arbovirus infections can be diagnosed by PCR-based rapid tests, by culture or serology. Culture diagnosis requires specialized high-containment laboratory facilities and is rarely attempted in clinical practice. Immunofluorescence, EIA and PCR may be used to diagnose specific viral infections. The detection of virus-specific IgM in the blood indicates recent infection.

Table 13.1 Principal human neurological diseases caused by arboviruses

Type of virus	Name	Vector	Distribution
Alphaviruses	Eastern equine encephalitis	Mosquito	Americas
	Venezuelan equine encephalomyelitis	Mosquito	Americas
	Western equine encephalomyelitis	Mosquito	Americas
Flaviviruses	Tick-borne encephalitis	Tick	Europe, Asia
	West Nile virus	Mosquito	Mediterranean, S. Europe, N. America
	St Louis encephalitis	Mosquito	N. America
	Japanese encephalitis	Mosquito	Asia, Pacific islands, N. Australia
	Murray Valley encephalitis	Mosquito	Australia, New Guinea
	Powassan virus encephalitis	Tick	North America, Russia
	Kyanasur forest disease	Tick	India
	Louping ill	Tick	UK
Bunyaviruses	Rift Valley fever	Mosquito	Africa
	California encephalitis	Mosquito	USA

Mosquito control is an important factor in prevention of the mosquito-borne encephalitides. Surveillance of disease in horses or seroconversion in sentinel bird flocks are used to guide the need for enhanced control measures. People exposed in endemic areas can use mosquito-repellents and other avoidance methods to protect themselves. Safe and effective vaccines exist for tick-borne encephalitis (for which an immune globulin is also available for post-exposure prophylaxis) and for Japanese B encephalitis.

Rabies

Pathology and epidemiology

Rabies is caused by neurotropic lyssavirus infection, transmitted from infected animals to humans by salivary contamination of a bite or open skin lesion. Human infections can result from contact with bats, dogs, cats, squirrels, skunks, mongooses, wolves, and occasionally horses or other animals in endemic areas of the world. Rarely, iatrogenic transmission has occurred to recipients of corneal grafts or solid organs inadvertently harvested from patients who have died with unrecognized rabies.

Virology and pathogenesis of rabies

Rhabdoviridae are bullet-shaped single-stranded negative-sense RNA viruses. Of the three genera that infect animals, rabiesvirus is the type species of the lyssavirus genus. The remaining lyssaviruses rarely cause human infection. The rabies virus is an enzootic virus that can occasionally become epizootic. The genome encodes five proteins. Of these the G protein is the antigen that induces neutralizing antibodies and is involved in binding to host cells via the acetylcholine receptor. Variations in this gene are responsible for the differences in serotype. Mutations affecting a particular site disrupt the virulence of the virus.

Viruses enter peripheral nerves in the infecting lesion and migrate antidromically to the CNS, where they initiate an encephalomyelitis. Viruses then pass down the nerves to the salivary and lachrymal glands and the other body tissues. The incubation period varies with the length of nerve that the virus must traverse. It is often 3 or 4 weeks for a bite on the face or head, and 60 days or more for a bite on the foot.

Clinical features

Illness often begins with aching or paraesthesia at the site of the originating lesion. Furious rabies, more common in humans, is a basal encephalitis, with fever, altered personality and periodic extreme agitation. Stimulation of the face and mouth, by attempts to drink or by draughts of air, can precipitate pharyngeal and facial spasms with regurgitation of saliva, typical of 'hydrophobia'. Electrocardiographic signs of myocarditis are common, and death often occurs in 5–7 days, with cardiac failure or arrest. If the history of exposure is not identified, the agitation of the encephalitis can be mistaken for hysteria or 'rabies phobia'.

Paralytic (dumb) rabies accounts for around 30% of human cases. It presents as a spreading, often ascending, paralysis or myelitis. It must be distinguished from Guillain–Barré syndrome, ascending myelitis or poliomyelitis. It has a course of 2 or 3 weeks, with death from respiratory complications or progressive encephalopathy.

Diagnosis and management

The initial diagnosis of rabies is clinical, based on the clinical and epidemiological evidence. Laboratory diagnosis is impossible until viruses are released from the brain in established disease. Rabies virus is a hazard category 4 organism. All virological tests are performed in specialist laboratories (in the UK, the Central Veterinary Laboratories). RT-PCR techniques will identify rabies genome in CSF, saliva corneal scrapings or urine. Rabies antigen can be demonstrated by immunofluorescence in the nerve endings of nuchal skin biopsies, or in post-mortem brain samples. Virus can be cultured in cell cultures, and antibodies demonstrated in serum and CSF. Histological examination of brain tissue shows perivascular inflammation. Characteristic, eosinophilic intraneuronal inclusion bodies known as (Negri bodies) are absent in up to a third of patients, particularly in persons who have received vaccination.

Rabies is virtually always fatal in unimmunized individuals, though one or two reports of recovery exist. Intensive support, high-dose immunoglobulins, antiviral drugs and interferon have all failed to influence the course of established disease.

Prevention and control

Pre-exposure immunization is recommended for those occupationally exposed and for some travellers.

Control of stray animals is important in limiting epizootics, and preventing human exposure to affected animals. In endemic countries, many family pets are vaccinated with live-attenuated vaccines, and certificates are issued. In Europe, a pet travel scheme permits identified and vaccinated animals to travel between rabies-free countries (e.g. the UK) and other countries, provided that they comply with simple animal health procedures.

Post-exposure prophylaxis is highly effective, and consists of wound treatment and risk assessment for appropriate immunization. However, post-exposure prophylaxis is only sought if the exposure is recognized. Education of travellers is important in alerting them to the risks of contact with unsupervised or wild animals.

Management of rabies-prone wounds

1 Washing the wound with any disinfectant will reduce the viral load and greatly reduce the risk of infection. Suturing and debridement may increase the entry of virus into nerves, and should be deferred.

2 Risk assessment: each case requires an expert risk assessment in conjunction with the relevant national health protection organization. This will take into account the country and circumstances of exposure, the immune status of the person bitten, and the site and severity of the wound.

3 Appropriate post-exposure prophylaxis, based on the risk assessment: this may include post-exposure immunization with high-potency cell-culture-derived vaccine (two or five doses depending on immune status) and human rabies immunoglobulin (HRIG) in a dose of 20 IU/kg given as soon as possible; infiltrated into the site of the wound/s with any remaining given intramuscularly at a different site. (If wounds are extensive, the HRIG should be diluted to permit treatment of all injured sites).

Neuroborreliosis

Among the manifestations of early disseminated Lyme disease is a meningoradiculitis (Bannwarth's syndrome), which can occur 2–6 months after the initial infection. Persistent or recurrent lymphocytic meningitis or, rarely, non-specific encephalopathy may also occur. A history of tick bite and/or erythema chronicum migrans, and/or facial or other nerve palsy (see Chapters 5 and 25) is present in approximately 70% of cases. Late manifestations of neuroborreliosis include peripheral neuropathy, which may be accompanied by a thinning, reddening or discoloration of the skin of the extremities (chronic atrophic acrodermatitis).

Microbiology

The infecting organism, *B. burgdorferi*, is maintained in and transmitted by ticks of the *Ixodes* genus. In Europe, three genospecies of the *B. burgdorferi* sensu lato complex are pathogenic, including *B. burgdorferi* sensu stricto, *Borrelia garinii*, *Borrelia afzelii* and the newly described *B. vientiana*. *Borrelia burgdorferi* is the only pathogenic species in North America. Borrelias are helical bacteria with 7 to 11 periplasmic flagella.

Clinical features

These are headache, fever, varying degrees of neck and back stiffness, with localized pain, weakness and reduced tendon reflexes in the affected nerve root territory, often in the lumbosacral area.

The CSF contains excess lymphocytes, and often a few red cells and/or plasmacytes. The protein is raised and the sugar may be normal or slightly low.

Diagnosis

In early infection, culture is very specific and more sensitive than serology. It is often performed using biopsy material from ECM lesions. Later in the infection, PCR testing is greatly superior to culture (for instance in detecting *B. burgdorferi* in joint fluid). Culture and PCR of CSF is rarely positive. Serology, usually by ELISA, is the most important routine diagnostic method. IgM is detected only during the early stages of infection, with IgG concentrations developing slowly and remaining positive even after antibiotic therapy. IgM antibodies are present in 50–70% of cases at presentation with early cutaneous disease, and in most cases by 3 weeks after the onset of infection. Positive ELISA tests can be confirmed by immunoblotting.

IgM antibodies to *Borrelia burgdorferi* may have disappeared by the time neurological features develop, but IgG antibodies should be present both in serum and CSF.

Treatment

Treatment is with intravenous cefotaxime or ceftriaxone. However, the optimum duration is uncertain; a significant proportion of patients are not cured by 2 weeks' therapy. Doxycycline orally 200 mg daily may also be effective. Effective treatment at this stage prevents further manifestations of infection, such as late arthritis, neuropathy or acrodermatitis.

Spongiform encephalopathies

Spongiform encephalopathies are progressive brain diseases characterized by dementia, movement disorder and terminal paralysis and coma. Typical 'spongiform' histological changes are seen in affected areas of the brain, with vacuolation of damaged nerve cells. Their aetiology is not fully understood, but no nucleic acid is demonstrable in infectious material. Infectivity is associated with an abnormal variant (PrP^{sc}) of a cell surface glycoprotein called prion protein. This agent is extremely resistant to heat and chemical agents, including formalin disinfection; it requires prolonged autoclaving (up to 1 h at 137 °C) to sterilize contaminated materials.

Genes for normal prion protein (PrP) are found in all animals and susceptibility to infection seems to have a strong genetic basis. It is suspected that infectious PrP^{sc} initiates irreversible conversion of the host's PrP to the abnormal form.

Spongiform encephalopathies of humans include:

1 Creutzfeldt–Jakob disease (CJD): sporadic or iatrogenic; onset age 55–75; mainly dementia/ataxia leading to death after 4–6 months.

2 Gerstmann–Straussler–Scheinker syndrome: inherited form of CJD; genetic abnormality of PrP; ataxia is marked; survival is longer than with CJD.

3 Fatal familial insomnia: another inherited disorder of PrP, with malignant insomnia and autonomic dysfunction.

4 Kuru: now disappearing; ataxic disease with tremor and incoordination; transmitted by cannibalism, only in Papua New Guinea.

5 Variant CJD (vCJD): first described in 1995, now accepted as human infection with agent of bovine spongiform encephalopathy, probably diet-acquired; course of about 2 years with paraesthesiae, depression, paralysis, death.

Natural infection is probably acquired via the diet. Iatrogenic transmission has followed the use of reusable stereotactic instruments for neurosurgery, human dura mater grafts, and (in the past) human growth hormone or gonadotrophin manufactured from pituitary glands harvested post-mortem. The incubation period for natural infection is probably in the region of a decade or more. PrP^{sc} can be found in lymphoid tissue of the appendix, terminal ileum and tonsil for many months before the onset of clinical disease. Infection may also be transmitted via lymphocytes in whole blood from infected donors.

Creutzfeldt–Jakob disease is a presenile dementia. Rare before the age of 40, its peak onset is in the 60s. Illness begins with clumsiness, ataxia and tremor, and progresses to intellectual and motor impairment, leading to death in 4–24 months. No treatment alters this course.

A new variant of Creutzfeldt–Jakob disease (vCJD) was first diagnosed in the UK in 1995, about 10 years after bovine spongiform encephalopathy (BSE) had appeared in British cattle. vCJD is characterized by a distinct, rapid clinical course and unique neuropathology.

Diagnostic criteria for variant Creutzfeldt-Jakob disease (UK Department of Health)

I   A) Progressive neuropsychiatric disorder.
   B) Duration of illness > 6 months.
   C) Routine investigations do not suggest an alternative diagnosis.
   D) No history of potential iatrogenic exposure.
II A) Early psychiatric symptoms.
   B) Persistent painful sensory symptoms.
   C) Ataxia.
   D) Myoclonus *or* chorea *or* dystonia.
   E) Dementia.

III A) EEG does not show the typical appearance of sporadic CJD (or no EEG performed).
   B) Bilateral pulvinar high signal on MRI scan.
IV A) Positive cerebellar tonsil biopsy.

 Neuropsychiatric symptoms include depression, anxiety, apathy, withdrawal or delusions.

Painful sensory symptoms include frank pain and/or unpleasant dysaesthesiae. EEG changes comprise triphasic periodic complexes at approximately one per second. Positive biopsy findings include spongiform change, extensive PrP deposition and florid plaques.

There is now adequate evidence to implicate BSE as the source of vCJD (see Case study 26.1). The BSE epidemic in the UK was controlled by 2000. The vCJD epidemic appeared to peak with 28 cases in 2000, and numbers fell to 9 in 2004. Strict control measures have been instituted, to prevent transmission through surgical instruments, endoscopes and the use of potentially contaminated blood products.

Transverse myelitis and longitudinal myelopathies

Transverse myelitis is an inflammatory condition that affects a restricted area of the spinal cord, usually defined as causing both motor and sensory disorder affecting both sides of the cord. It may affect one or several contiguous spinal levels (Fig. 13.17). The aetiology is not fully understood, but many cases are probably related to the generation of anti-phospholipid antibodies, which damage nerve sheaths or nerve cell membranes. Rarely, direct inflammatory damage may result from active viral infection (such as VZV virus) or the deposition of parasites such as schistosome ova.

In case-series, a minority of 20–40% of cases are associated with various previous infections, such as *Mycoplasma pneumoniae*, HSV 1, HSV 2, syphilis, borreliosis, cytomegalovirus, Epstein–Barr virus, tuberculosis, HHV 6, hepatitis C, other hepatitis viruses, HIV and HTLV 1.

Non-infectious aetiologies include other inflammatory conditions, particularly sarcoidosis, Sjögren's syndrome, systemic lupus erythematosus and mixed connective tissue disorders.

Multiple sclerosis can initially present with a spinal cord lesion, and 20–50% of cases are shown to have multiple sclerosis, on the basis of multiple central nervous system lesions, oligoclonal bands in CSF protein electrophoresis and disease progression.

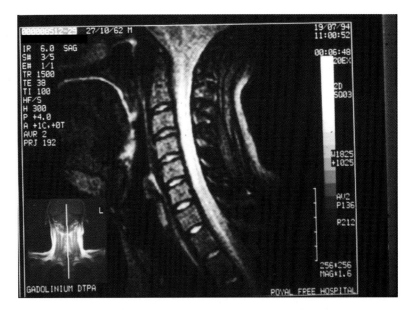

Figure 13.17 CT scan of the cervical and upper thoracic spinal cord, showing a patchy area of irregular pale oedema, typical of transverse myelitis involving several segments.

Clinical features

The onset is usually rapid, with maximum signs developing in 4 hours to 21 days. Pain at the level of affected segments is common. Motor, sensory and sphincter functions are affected. A sensory level is usually demonstrable, though it may appear variable and asymmetrical.

The investigation of choice is MR imaging of the spinal cord, which shows an altered signal in over 95% of cases. There is usually a lymphocytic pleiocytosis of the cerebrospinal fluid, with a proportionate elevation of the protein level (in contrast to the dissociation between cell and protein elevations seen in Guillain–Barré syndrome). CSF eosinophilia may be present in parasitic conditions. Electromyography shows evidence of radiculitis or peripheral nerve lesions (which are not usually seen in multiple sclerosis myelopathies) in half or more of cases.

Treatment

Non-infectious cases may respond to treatment of the underlying cause. Direct infection can be treated (for instance with anti-herpesvirus drugs), and parasite-related inflammation may respond to step-down courses of corticosteroids. 'Para-infectious' cases have a variable course, and do not always improve significantly.

Human T-lymphotrophic virus I and HTLV-associated myelopathy (HAM)

Human T-lymphotrophic virus I (HTLV-I) is an oncovirus member of the family Retroviridae that infects T lymphocytes. Transmission occurs through blood transfusion, during breast-feeding (with an 18–30% risk of transmission), and, uncommonly, by the sexual and perinatal routes.

HTLV-I is most common in southern Japan, where the seroprevalence is up to 16%, and in the Caribbean, where it ranges from about 2.5 to 6%. The seroprevalence in native Europeans is probably below 0.5%.

Seroconversion is probably symptomless, but two serious diseases can follow: (i) HTLV-associated myelopathy (HAM, or tropical spastic paraparesis), after an average interval of about 4 years (lifetime risk approximately 0.25%); and (ii) acute T-lymphoblastic leukaemia, after an average interval of about 30 years (lifetime risk 2–5%). Infected children may suffer repeated attacks of infective dermatitis, which respond only temporarily to antibiotics. Uveitis is relatively common in infected individuals.

Diagnosis is with serology using EIA, confirmed by Western blotting. Reverse transcriptase PCR assays are available.

The results of antiretroviral drug treatment have been disappointing, with no sustained effect on viraemia. However, treatment with alpha-interferon has been reported to produce symptomatic improvement in some patients with HAM, probably through an immunomodulatory effect.

Prevention depends on safe blood transfusions, screening and advice for the families of infected individuals, avoidance of breast-feeding by infected mothers and, to a lesser extent, on the use of condoms. For donated blood, the best available screening tests are utilized in various countries.

Human T-lymphotrophic virus II

This is a retrovirus closely homologous to HTLV-I, which

has been identified in T lymphocytes less commonly than HTLV-I. It has been recovered from the lymphocytes in a case of hairy-cell leukaemia, but an aetiological association has not been established. It has never been associated with neurological disease.

Cerebral and intracranial abscesses

There are many ways in which space-occupying infections of the CNS can occur. Most are derived from bacteraemia, either overt or subclinical. Others follow CNS invasion by contiguous spread from adjacent bone or intracranial sinus. Extradural or subdural collections of pus may compress the brain or spinal cord, or occlude an important blood supply, producing both local and long tract signs without actual CNS invasion.

Infectious lesions of this type often, but not always, produce fever, and are otherwise difficult to distinguish clinically from tumours or localized vascular disease. Imaging is important in making an early diagnosis, and allowing biopsy or aspiration under imaging control. A CT scan will define lesions and their position, and will demonstrate enhancement of the oedema surrounding inflammatory lesions. An MR scan will readily demonstrate oedema, and define fluid in an abscess cavity; enhancement can be used to demonstrate inflammation in the abscess wall.

The importance of early suspicion and diagnosis is paramount as the prognosis is relatively good before impaired consciousness and severe neurological deficit occur, but is definitely bad once these changes are established.

Cerebral abscess

Epidemiology

Brain abscesses are often associated with suppurative disease of the middle ear and mastoid cavity, and less often with ethmoid or frontal sinus disease. In children, an almost equal number of abscesses are of bacteraemic origin, often associated with cyanotic congenital heart disease.

Rarer sources of brain abscesses are bacteraemias in patients with bronchiectasis, or with necrotic and ischaemic bowel lesions. Several reports exist of brain abscess as a rare complication of injection sclerotherapy for oesophageal varices. Staphylococcal, S. 'milleri' or pneumococcal bacteraemias may be complicated by brain abscess, and listeriosis is occasionally accompanied by brainstem abscess.

Organism list

- Anaerobic Gram-positive cocci
- *Prevotella melaninogenicus*
- *Bacteroides fragilis*
- *Fusobacterium* spp.
- *Actinomyces* spp.
- Aerobic staphylococci and streptococci
- *Haemophilus aphrophilus*
- Other *Haemophilus* spp.
- *Listeria monocytogenes*
- Facultative Gram-negative rods

The bacteriology of abscesses is often mixed, with more than one anaerobe identified, or a mixture of aerobes and anaerobes.

Clinical features

These are a mixture of headache, reduced consciousness, features of raised intracranial pressure, localizing signs and, rarely, fits (which are more common when the abscess is subdural or cortical). Up to half of patients have papilloedema at presentation, and a similar number have vomiting and altered consciousness. Only about half of patients have fever. Features of raised intracranial pressure are global headache (worse on lying down and often waking the patient at night), increasing somnolence, rising blood pressure and falling pulse rate.

The headache may develop against the background of pain in the ear or the sinus, and this offers the chance to make an early diagnosis. If the abscess affects the meninges, or ruptures into the CSF, meningism or obvious meningitis are seen.

Localizing signs depend on the site of the abscess. Frontal abscesses cause subtle signs of altered personality, paucity of conversation and apparent depression; parietal lesions cause classic features of hemiparesis and sometimes dysphasia; temporal lesions cause memory defects, sometimes aphasia and altered personality; visual field defects may be a sign of occipital lesions, or ataxia and tremor of cerebellar abscess.

Approximately equal numbers affect the frontal, parietal or temporal lobes. Occipital lobe abscesses are less common and brainstem abscesses are rare. Multiple lesions are uncommon but may complicate bacteraemic disease, actinomycosis and nocardiosis. Subdural empyema occurs less often than intracerebral abscess, but the organisms involved and the clinical presentations are similar.

Diagnosis

Early imaging is essential to make the diagnosis (Fig. 13.18). In the presence of evolving localizing signs or raised intracranial pressure, it is unsafe to perform lumbar puncture because deformity of the brain may block the flow of CSF and cause herniation of the cerebellum or brainstem through the foramen magnum (coning) when the theca is opened. Lumbar puncture should be deferred until after imaging, even when meningism is present.

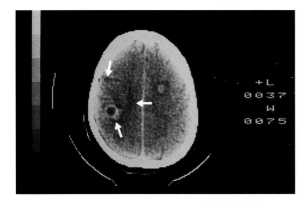

Figure 13.18 Computed tomographic brain scan, showing abscesses surrounded by oedema (arrows) and zones of bright enhancement (contrast study).

Blood cultures should always be set up to identify any coexisting bacteraemia.

It is a neurosurgical choice whether a brain abscess can be addressed by needle drainage alone or whether indwelling drainage is required. Craniotomy and formal evacuation may be indicated for large compressing lesions. Pus should be obtained for aerobic and anaerobic culture when drainage is performed.

Amplification of the 16S rRNA gene, and comparison of its sequence with a database of organisms nowadays allows rapid aetiological diagnosis. This technique can detect the presence of bacteria even when cultures are rendered negative by antibiotic therapy. As mixed infection is common in this situation it is often necessary to clone the amplification products to generate a sequence from each of the infecting organisms (see Fig. 3.17).

Management

Broad-spectrum antibiotics should be given at diagnosis, to cover the likely range of pathogens. Agents that penetrate the brain well and are effective include cefotaxime, ceftriaxone and metronidazole. High doses of the cephalosporin should be given to achieve high tissue levels as early as possible. Some specialists still use penicillin, chloramphenicol and metronidazole, a mixture that has stood the test of time. However, penicillin penetrates the blood–brain barrier less well than the broad-spectrum cephalosporins. Treatment can be modified if necessary when the results of blood and pus culture are known.

Neurosurgery is an important part of treatment. Single abscesses can be aspirated or drained via burr holes or at craniotomy. In a few cases the abscess cavity may require excision. Small abscesses, especially if multiple, may be treatable with antibiotics alone, but require close follow-up by imaging to ensure adequate resolution.

Treatment of cerebral abscess

1 Cefotaxime i.v. 2–4 g 8–hourly; *or* ceftriaxone i.v. 2–4 g daily; *plus* metronidazole i.v. or rectally 500 mg 8-hourly.
2 Alternatives: meropenem 2 g i.v. 8-hourly; *or* benzylpenicillin i.v. 2.4 g 4–6-hourly *plus* chloramphenicol i.v. or oral 2–3 g daily in three or four divided doses; *plus* metronidazole i.v. or rectally 500 mg 8-hourly.

 Drainage by needle or burr hole approach is often indicated.

It is important to examine the patient clinically, and by imaging, to detect any precipitating lesion. Sinusitis, otitis or, in the spinal column, osteomyelitis are all common and easy to miss if not deliberately sought. The liver, bowel and the lung should also be examined for evidence of coexisting abscesses.

Other space-occupying lesions of the CNS

Apart from tumours, tuberculomas and cysticercal lesions can cause chronic space-occupying lesions in the brain. Cerebral toxoplasmosis causes a similar picture in AIDS sufferers.

Tuberculomas appear similar to abscesses on CT and MR scanning. They may be associated with tuberculosis of the meninges, or of another body system. A positive biopsy result or positive PCR on CSF may confirm the diagnosis. Treatment is with quadruple antituberculosis therapy. Corticosteroids are also indicated to avoid early increases in inflammation (see Chapter 18).

Cysticercosis is the condition in which tissue cysts of *Taenia solium* develop in the brain and/or other tissues. It is rare in Western countries where tapeworm infection is extremely rare, but is the commonest cause of adult-onset epilepsy in individuals from endemic areas. Imaging of the brain shows cystic lesions, often with surrounding oedema (Fig. 13.19). Cysts with surrounding oedema will often die as a result of the inflammatory immune reaction; calcified cysts are usually dead. Scolices are sometimes visible in living cysts.

The commonest manifestation is epilepsy, but raised intracranial pressure, nausea or vomiting are also often seen. Single cysts may block the flow of CSF, or affect the spinal cord. ELISA tests for cysticercal antibody and antigen are useful for making the diagnosis. Stool examination for eggs and proglottides is advisable, but is often negative. Treatment with corticosteroid is given to alleviate the inflammatory reaction and swelling. Treatment with anti-epileptic agents may be required until the inflammatory reaction has subsided; it can then usually be discontinued. Symptomatic cysts are usually dead or

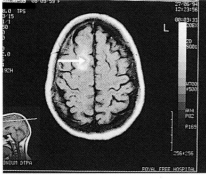
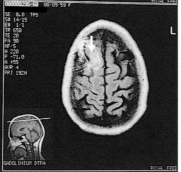

Figure 13.19 Cerebral cysticercosis: this 44-year-old woman from Kenya had recent-onset epilepsy. The computed tomographic brain scan shows a cystic lesion with enhancing halo in the right frontal lobe (arrows). Serology was positive.

dying, but albendazole or praziquantel can be given in cases of doubt.

Treatment of cerebral cysticercosis

1 Albendazole orally 400 mg 12-hourly for 8–10 days.
2 Alternative: praziquantel orally 20 mg/kg 8-hourly for 1 day.

 Consider simultaneous prednisolone 40–60 mg, continued for 3–5 days.

Tetanus

Epidemiology

Tetanus is a neurological disease caused by tetanospasmin, the toxin of the anaerobic, spore-forming, Gram-positive rod, *Clostridium tetani*. Horses, other farm animals, dogs, cats and guinea-pigs excrete *C. tetani* spores in their faeces, and soil is often contaminated. Spores can germinate, producing toxin, in faeces- or soil-contaminated deep, necrotic, penetrating or puncture wounds, in which an anaerobic environment exists. Neonatal tetanus follows contamination of the unhealed umbilical stump. Cephalic tetanus rarely complicates *C. tetani* chronic ear infection. Intravenous drug abusers are occasionally infected by contaminated 'street' heroin.

Pathology

Tetanospasmin's heavy chain binds to the GT1 ganglioside of neurons at neuromuscular junctions in the infected wound, and distally by bloodborne spread. The light chain then enters the cytoplasm, passes antidromically along nerve axons to the anterior horn cells, and trans-neuronally within the central nervous system. Toxin inhibits the release of neurotransmitters, particularly GABA and glycine from inhibitory neurones and alpha motor neurons. This disinhibits motor reflexes and autonomic responses.

Clinical features

After 3 to 21 days' incubation (average 8 days) disease develops over 1 to 5 days. Shorter incubation, and clinical evolution within 48 hours, predict more severe disease.

Generalised tetanus has four stages: 1, increasing jaw stiffness and dysphagia; 2, muscular spasms of the jaw, and sometimes brief, mild spasms elsewhere; 3, generalized spasms which can be severe, causing lip retraction (so called risus sardonicus) and opisthotonus, with pain, respiratory embarrassment, vertebral and other fractures; 4, autonomic dysfunction with cardiovascular and haemodynamic disruption.

Local tetanus produces stiffness near the infecting wound, but there is a high risk of developing (usually mild) generalized disease. Cephalic tetanus may cause spasms on one side of the face with lower motor neurone paralysis (from 7th nerve damage in the middle ear) on the other.

Diagnosis and management

Diagnosis depends on clinical vigilance: there are no pathognomonic changes in electromyography or in cerebrospinal fluid. Culture of *C. tetani* from wounds is helpful, if clinical disease coexists. Treatment is based on the following.

1 Eradication of *C. tetani*, using metronidazole (first choice) or intravenous benzyl penicillin: necrotic or contaminated wounds should be debrided and cleaned.

2 Neutralization of toxin using human tetanus immunoglobulin (HTIG), 150u/kg i.m. at diagnosis (intrathecal antitoxin treatment is of unproven benefit, and easily stimulates spasms).

3 Control of spasms and autonomic instability: Stage 1 cases can be managed with sedation and muscle relaxation using infused i.v. diazepam or midazolam. For more severe disease, intramuscular (or intrathecal) baclofen, or intravenous magnesium sulphate infusion have proved effective in managing spasms; magnesium also helps to stabilize autonomic reflexes.

Airway management, adequate nutrition and pain control are important. Severe cases may need a period of paralysis and ventilatory support. With intensive care support, the mortality rate can be as low as 1%. In low-income countries 20–30% of patients may die (up to 40–50% with neonatal tetanus).

Prevention and control

An attack of tetanus does not confer immunity. The disease is preventable by immunization with tetanus toxoid.

The five doses provided in the UK vaccination schedule provide adequate protection: a booster dose (adult diphtheria/tetanus/inactivated polio vaccine; Da/T/IPV) may be given at any time in the event of a contaminated wound.

Prenatal vaccination of pregnant women protects the infant from tetanus.

Unvaccinated individuals who sustain at-risk wounds should be offered immunoglobulin (250 units – or 500 units if >24h after injury) and a full course of immunization (3 doses of adult Da/T/IPV at monthly intervals).

Case study 13.1: Feverish and lost for words

History

A 56-year-old taxi driver was accompanied to hospital by his wife, who gave a history on his behalf. Four days previously he had developed a fever and malaise, interpreted as a viral infection. Two days previously, he had developed difficulty in recalling the names of people and objects, and since then he had become increasingly drowsy and lost his memory of recent events.

Physical examination

This showed a drowsy man, with a Glasgow Coma Score of 13/15, and temperature 38.2 °C. He could recall his address and the names of famous individuals, but had no recall of the events of the day. He had a severe nominal aphasia. There was no meningism and no focal neurological deficit. The examination was otherwise normal.

Laboratory tests

The haemoglobin was 15.7 g/dl, white cell count 7.8×10^9/l with a normal differential. Renal function, liver function and chest X-ray were normal.

Questions

■ What investigations would you like to perform next?
■ What diagnoses would you consider?

Further management and progress

An urgent MR scan was performed, as the neurological features suggested pathology in the left temporoparietal region. This showed no evidence of infarct, space-occupying lesion or abscess. A pale area consistent with localized oedema was demonstrated, affecting a major portion of the left temporal lobe (Fig. CS.5).

Question

■ Does this suggest an aetiology for his illness?

Diagnosis and management

A feverish illness with central nervous system signs and temporal lobe inflammation is strongly suggestive (but not diagnostic) of herpes simplex encephalitis. Lumbar puncture was performed, and showed a CSF white count of 350 lymphocytes, protein of 0.7 g/l and glucose of 4.5 mmol/l (blood glucose 6.7 mmol/l).

Intravenous aciclovir 10 mg/kg 8-hourly was commenced. Two days later his temperature was normal, and his nominal aphasia had markedly improved. Polymerase chain reaction tests showed the presence of herpes simplex DNA in the CSF. Intravenous aciclovir treatment was continued for a further week. Three months later his only remaining deficit was an unreliable short-term memory, which he overcame by keeping a diary and daily workbook.

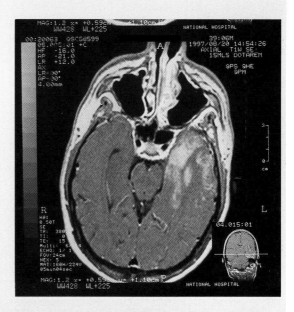

Figure CS.5 T2-weighted MR scan of the brain, in which water molecules appear white, showing extensive water accumulation, suggestive of oedema, in the left temporal lobe.

Bone and Joint Infections

14

Introduction: structural considerations

Structure of bone

Bone is a complex connective tissue formed of osteoid material hardened by the calcium salt hydroxyapatite. Its basic structure, best seen in the cortex of long bones, is the Haversian system, composed of concentric lamellae of bone surrounding a small blood vessel that runs in a central canal, or Haversian canal. The bone is laid down by osteocytes, cells that surround the blood vessel and also form a layer beneath the highly vascular periosteum. The osteocytes separate perivascular tissue fluid from bone tissue fluid (Fig. 14.1), which fills the Haversian system and the transverse connections across the bone lamellae (Volkmann's system). Each Haversian system is bounded by a 'cement line', which separates each set of lamellae, and across which blood vessels do not pass (Fig. 14.2). Each bone is surrounded by tough, vascular periosteum, rich in sensory nerves and having a lymphatic supply, both of which are absent from the interior of the bone.

Loss or damage of the blood supply of a Haversian system causes death of the bone lamellae within the cement line. The dead tissue may demineralize, fibrose or occasionally be replaced by cartilage. Healing and remodelling occur if the disease process is controlled. Remodelling is achieved by small, advancing points of bone resorption.

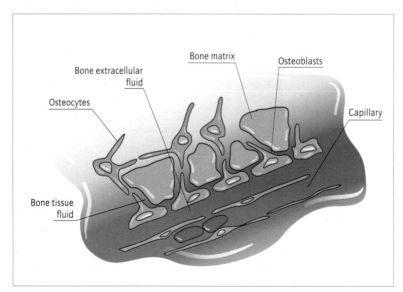

Figure 14.1 The cells and fluid compartments of bone; antibiotics must enter the bone tissue fluid to treat intraosseous infection effectively.

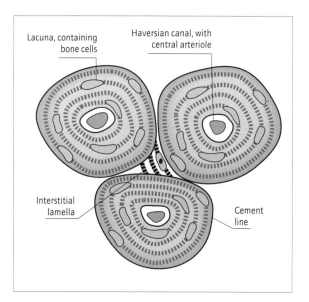

Figure 14.2 The Haversian system of bone structure.

This is followed by osteogenesis, producing a new Haversian system, usually in a different orientation to the first, fragments of which are left supporting the new system.

Bone growth in childhood

Bone growth in childhood takes place by the extension of the central, or diaphyseal, part of a bone at a cartilaginous plate, the epiphyseal plate. The growing part of the bone is called the metaphysis (Fig. 14.3). On the other side of the epiphyseal plate is the epiphysis, which is initially entirely cartilaginous but develops a centre of ossification as the child grows. When the rate of ossification overtakes that of cartilage growth, the epiphysis becomes fully ossified.

Blood vessels do not cross the epiphyseal plate; the epiphysis has its own blood supply, often derived from a single afferent vessel, the epiphyseal artery, which usually enters the epiphysis directly. In the head of the femur it must traverse the joint capsule, as the epiphysis is entirely intracapsular. This artery is more at risk than others of damage caused by pressure or deformity affecting the joint.

The epiphyseal plate possesses a proliferation zone, in which osteoid is manufactured, and a deeper maturation zone. Mineralization, or ossification, occurs in the deepest layer. These areas are served by many capillary loops, in which blood flow slows as it passes from the arterial to the venous side. This slowing may predispose to the deposition of bacteria, and may explain the susceptibility of the growing end of the metaphysis to haematogenous osteomyelitis.

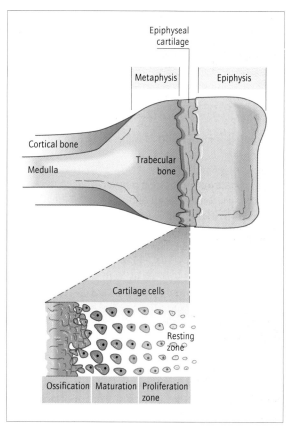

Figure 14.3 General structure of a growing bone.

Structure of joints

Most large joints are synovial joints, in which the adjacent bones are covered with articular cartilage. The bones are held together by a tough, fibrous joint capsule, lined with vascular synovial membrane that secretes synovial fluid. Where the capsule joins with the articular cartilage there is a fibrocartilaginous zone overlapped by the vascular edge of the synovial membrane.

The articular cartilage ceases to grow in adulthood. It is not bound to the underlying bone, but is attached by irregular-shaped, interlocked surfaces of bone and cartilage. The cartilage receives no blood supply from the underlying bone. Its nutrition depends on diffusion from the synovial fluid. As cartilage has an open, water-saturated structure, this process is enhanced by joint movement. Damaged cartilage is not replaced, but may repair by forming a fibrous scar.

The joint capsule is usually attached at the end of the articulating bone. The metaphysis is then extracapsular. In the femur, the capsule of the hip joint is attached far down the neck, and the femoral neck is therefore intracapsular. This means that infection of the neck of the femur can extend directly into the joint space, or that joint infection may directly invade the bone.

Osteomyelitis

Osteomyelitis is infection of bone. It may be initiated in several ways:
1 haematogenous spread;
2 extension from an adjacent infected joint;
3 direct invasion as a result of trauma;
4 extension of infection from an overlying soft-tissue infection or deep ulcer;
5 iatrogenic infection following surgery or instrumentation.

Osteomyelitis may be acute, with a rapid onset and typical signs of acute infection, such as fever and acute inflammation. Chronic osteomyelitis occurs when bone necrosis, vascular damage, or the nature of the infecting organism prevent the clearance of infection and, in turn, increase the tissue damage. Foreign material, such as pins, plates, screws, intramedullary nails and bone cement, offer a nidus for infection, protected from the immune defences of living tissues. They make it difficult or impossible to eradicate infections in which they are involved.

Suppurative arthritis

Suppurative arthritis can arise in a number of ways.
1 Haematogenous infection following inapparent bacteraemia or complicating bacteraemic disease. In children the infection often arises apparently spontaneously, while in adults pre-existing disease such as rheumatoid arthritis, gout or pseudogout may predispose to septic joint disease, and complicate its diagnosis.
2 Extension from adjacent infected bone (most often affecting the hip joint, especially in children).
3 Introduction of pathogens by penetrating trauma or, rarely, following aspiration or arthroscopy.

Suppurative arthritis in a prosthetic joint is a special situation. Not only is infection more likely in the presence of foreign material, but biofilms may form on the artificial surfaces. Neutrophil responses are down-regulated by contact with biofilm material.

Acute bone and joint infections

Epidemiology

Acute haematogenous osteomyelitis is often a childhood disease, affecting growing bones. The causative bacteraemia originates from the skin, mouth or respiratory system. The infection usually affects a single long bone (Fig.

14.4), and arises in the metaphysis, possibly because of the rich supply of capillary loops (see above). Salmonella osteomyelitis affects the bones of individuals with sickle-cell disease, possibly because of stagnant bone circulation following sickling crises (Fig. 14.5). Osteomyelitis in adults less often affects metaphyses, and more often occurs in sites of trauma, affecting bones of the wrist, knee or ankle.

Vertebral osteomyelitis usually affects adults and the elderly. Predispositions include a combination of degenerative spinal disease, increased risk of bacteraemias arising from the abdomen or pelvis, and the presence of valveless venous plexuses of the pelvis and lower spine. In typical cases, infection or bacteraemia arising from the lower bowel or urogenital tract is accompanied or followed by

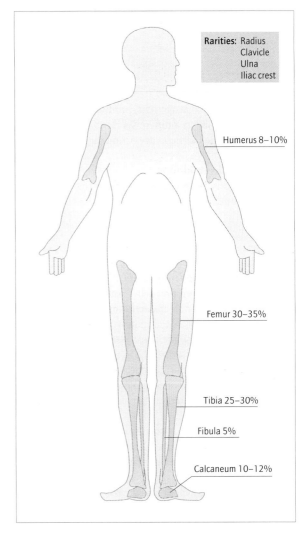

Figure 14.4 Bones most often affected by childhood osteomyelitis. The spine is rarely affected in children, but is more commonly involved in adults.

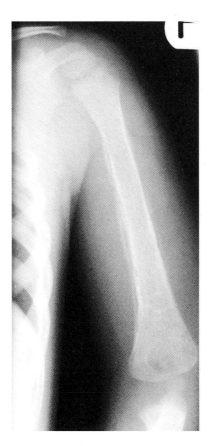

Figure 14.5 *Salmonella typhimurium* osteomyelitis of the humerus in a child with sickle-cell disease. Note the elevated periosteum.

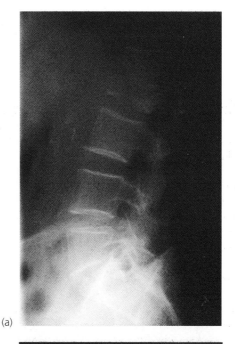

(a)

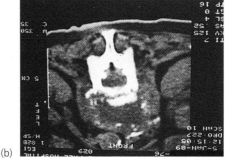

(b)

Figure 14.6 Osteomyelitis of the lumbar spine complicating *Salmonella enteritidis* colitis. (a) X-ray; (b) computed tomographic scan showing destruction of the body of the second lumbar vertebra.

Gram-negative (commonly *Escherichia coli*) spinal osteomyelitis. Salmonella infection of a vertebra occasionally occurs after bowel infection (Fig. 14.6). Other bones, such as the pelvis, damaged by degenerative or malignant disease, may become infected. Some intravenous drug users develop pseudomonal osteomyelitis of the spine or pelvis, possibly after injecting into inguinal veins.

Finally, bone may be infected by extension from an overlying site of soft-tissue infection. An example of this is osteomyelitis underlying a deep sacral pressure sore or a diabetic foot ulcer (Fig. 14.7).

Acute suppurative arthritis in young people is usually of haematogenous origin, caused by a similar list of organisms to that of osteomyelitis. A small number of pathogens are especially likely to invade joints; these include *Neisseria gonorrhoeae* (which has a predilection for the knee, ankle and wrist), and *Brucella* spp. (which can affect many joints, but seems often to select the cervical spine, where it also causes chronic osteomyelitis). In elderly people and others with pre-existing joint disease, affected joints are at risk of haematogenous infection, and

this is more likely if immunosuppressive drugs are used to modify the disease process.

High-risk endemic areas are recognized, where some infections are more common than elsewhere. Lyme arthritis is relatively common in the eastern states of the USA; tuberculosis of bones and joints is common in sub-Saharan Africa and brucellosis is common in eastern parts of Europe and the Mediterranean.

Pathology

In osteomyelitis, inflammatory oedema and pus form in the infected bone and track through the Haversian and Volkmann's canals. At the site of infection, bone ischaemia and necrosis occur. When the exudate reaches the perio-

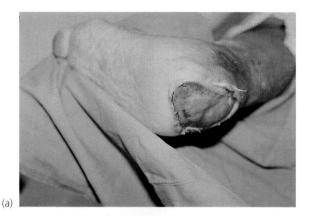

(a)

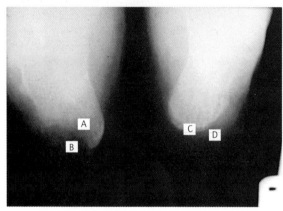

(b)

Figure 14.7 Osteomyelitis of the calcaneum: this immobile, elderly man had a deep heel ulcer colonized with methicillin-resistant *Staphylococcus aureus*; this infection extended to the underlying periosteum and bone. (a) The ulcer; (b) X-ray showing loss of structure of the calcaneum. A, defect in both cortical layer and bone structure; B, absence of soft tissue; C, normal outline of continuous layer of bone cortex; D, soft tissue of the heel.

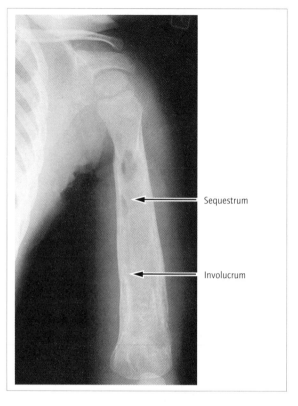

Figure 14.8 Osteomyelitis in a long bone, showing formation of a sequestrum and involucrum.

steal surface of the bone, the periosteum is elevated by the fluid. Osteocytes underlying the periosteum then begin to generate new bone.

Formation of sequestra and involucra

If effective treatment is delayed, an infected area of bone may undergo necrosis. An accumulation of new bone eventually encloses it. The old, dead bone is called a sequestrum, and the surrounding live bone is termed an involucrum. On X-ray, or MR imaging, the sclerotic sequestrum is surrounded by the more normal-appearing involucrum, from which it is separated by a lucent area (Fig. 14.8). The sequestrum has no blood supply, and is inaccessible to antibiotics and immunological processes. As with implanted foreign materials, it can become a site of persisting infection that prevents the healing of the osteomyelitis.

Brodie's abscess

Sometimes infected bone is replaced by pus. An intraosseous abscess forms, enclosed in a fibrous membrane, and surrounded by a sclerotic bony reaction. Infection may become quiescent when it is contained in this way, but there is a high risk of future recrudescence and extension.

Tracking of pus, and sinus formation

Accumulating pus may track through tissues, causing a local abscess or, in the skin, producing a draining sinus. Pus forming near to the joint surface of an intracapsular bone may penetrate the synovial membrane, to cause pyogenic arthritis.

Organism list

Osteomyelitis: long bones
• *Staphylococcus aureus* (90%)

- *Streptococcus pyogenes*
- *Streptococcus pneumoniae*
- *Haemophilus influenzae* (rare)
- *Salmonella enterica* (especially in sickle disease)
- *Escherichia coli*
- *Neisseria meningitidis*

Vertebral osteomyelitis
- *Staphylococcus aureus*
- *Escherichia coli*
- Anaerobic bacteria (anaerobic streptococci, *Bacteroides* spp., *Fusobacterium* spp., etc.)
- Mycobacteria
- *Proteus* spp.
- *Klebsiella* spp.
- *Pseudomonas* spp.
- *Brucella* spp.
- *Salmonella enterica*

Suppurative arthritis
- *Staphylococcus aureus* (70% in adults, 50% in children)
- *Streptococcus pyogenes* (16% in children, 7% in adults)
- Gram-negative rods (15% in adults)
- *Streptococcus pneumoniae*
- *Neisseria gonorrhoeae* (3%)
- *Borrelia burgdorferi*

Rarities
- *Haemophilus influenzae*
- *Salmonella enterica*
- *Neisseria meningitidis*
- Mycobacteria
- *Brucella* spp.
- *Pasteurella multilocularis* (on cat and dog contacts)
- Anaerobic bacteria (anaerobic streptococci, *Bacteroides* spp., *Fusobacterium* spp., etc.)

Clinical features

The onset of osteomyelitis is often insidious, with a feverish illness and poorly localized pain, which progressively worsens. The white cell count is often normal, and early illness may be passed off as 'flu-like'. Infants and toddlers cannot describe the pain, but often stop using the affected limb, displaying a 'pseudoparalysis'.

Later changes, such as overlying redness and swelling, the development of draining sinuses, deformity or pathological fractures, are rare consequences of delayed diagnosis or ineffective treatment.

An infected joint may be acutely swollen and red (see Fig. 15.3), sometimes with a detectable effusion. However, a less distinct onset can occur in an already inflamed rheumatic or gouty joint. There may be doubt whether increased inflammation is due to infection or exacerbation of underlying pathology.

Vertebral discitis has a similar onset to that of vertebral osteomyelitis.

> **Misleading clinical signs in childhood hip infections**
> 1 Pseudoparalysis of the leg.
> 2 Abdominal pain.
> 3 Abdominal distension.
> 4 Constipation.

Diagnosis

Clinical suspicion is important in making a diagnosis. For joint infections, it may be easy to detect redness, pain, immobility or effusion but if these are already present because of pre-existing inflammatory disease, specific investigation is required to detect the new infection. Osteomyelitis is only diagnosed as a result of performing specific investigations. Suspicion of infective bone and joint disease should be confirmed by imaging of bones and joints, by demonstrating the presence of acute inflammation and by identification of a causative organism.

For osteomyelitis, other than following trauma or surgery, adults are more susceptible than children to infection of the vertebrae or small bones. For synovitis, the joints most commonly infected are similar in children and adults. The knee is involved approximately twice as commonly as any other joint. The hip, ankle, elbow and wrist are progressively less often affected. In adults the shoulder rivals the hip or ankle in frequency; in children it is about as commonly affected as the wrist.

Imaging
Plain X-ray does not usually show bone changes of osteomyelitis at presentation; the first signs of bone loss or periosteal reactions appear only when infection has existed for at least 7–10 days. Effusion in an infected joint can be seen on X-ray, and increased soft tissue shadows may help to confirm the presence of surrounding inflammation. Magnetic resonance (MR) scanning is the investigation of choice, as it clearly shows bone oedema, bone loss, effusions, damage to joint surfaces, and fluid collections in soft tissues. Bone oedema is a reliable early sign of osteomyelitis, and oedema persists until the infection is cured (Fig. 14.9). MR scanning also demonstrates the anatomy, in case a surgical procedure is indicated.

Ultrasound scanning can often demonstrate early periosteal elevation in children. Technetium or gallium isotope scans are of limited value, as they cannot distinguish between infection, inflammation of other origins, or tumour circulation.

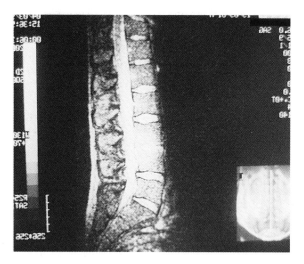

Figure 14.9 Magnetic resonance scan showing infection of the L4/5 intervertebral disc (discitis): the patient 'strained' his back while suffering from a severe sore throat. Persisting back pain and fever led to imaging and needle aspiration, revealing a *Fusobacterium necrophorum* infection (see Chapter 7).

Bacteriological diagnosis

Blood cultures are positive in about half of cases with acute bone and joint infections. Aspirated joint fluid can be Gram-stained and cultured. Imaging-directed aspiration of fluid collections, pus, or biopsies of bone, periosteum or synovium can be performed in difficult cases. Direct aspiration of material from infected bone is possible in infants and small children.

Aspiration of fluid from inflamed joints confirms the presence of inflammatory cells, and Gram-staining may reveal bacteria. Examination of the fluid also provides a rapid diagnosis of haemarthrosis, gout or pseudogout. However, it cannot always rapidly exclude infection co-existing with these conditions, as they all produce neutrophilic joint inflammation, and bacteria may not be demonstrable by direct staining. Image-guided or open biopsy may be needed to obtain pus or infected material from deep infections such as vertebral or pelvic osteomyelitis.

For a large collection of pus, a sequestrum or a Brodie's abscess, operative drainage and debridement may contribute to treatment, and also provide diagnostic specimens.

A search should be made for a focus of infection elsewhere in the body, including skin, chest, throat or ear sepsis, and the stools should be cultured. Specimens may be obtained from the genital tract, or the pustular rash, in suspected gonococcal arthritis (see Chapter 15). Especially in the elderly, the urine should be cultured. In Gram-negative bone and joint infections, bowel, or genitourinary disorders and malignancy should be excluded by imaging and/or endoscopy.

Management

Antimicrobial treatment

The aim of treatment in acute infections is to eradicate infection before any damage occurs to bone, periosteum, cartilage or other tissues. As soon as there is loss of bone or cartilage, or tissue necrosis, infection becomes more difficult to treat and the functional outlook is compromised. Empirical antimicrobial treatment should be commenced as soon as specimens have been obtained, and should be effective against *Staphylococcus aureus*, *Streptococcus pyogenes* and (for adults, or where abdominopelvic disease exists) Gram-negative rods. Initial treatment should always be given intravenously. In special circumstances, such as infection following an animal bite, pelvic osteomyelitis in an intravenous drug-user or when there is risk of infection with methicillin-resistant *S. aureus*, empirical treatment should include relevant cover. Antibiotic therapy may be modified when culture results are known.

Clindamycin is particularly useful in osteomyelitis, being effective against staphylococci, streptococci and anaerobes, and having excellent bone penetration, even when given orally. However, this drug should be discontinued immediately if the patient develops antibiotic-associated diarrhoea or colitis.

Fusidic acid is highly bioavailable by mouth and penetrates bone well. It is only effective against staphylococci, and should never be used as monotherapy.

Rifampin is an effective antistaphylococcal drug that is effective against slow-replicating organisms in biofilms and oedematous tissues. It should never be used as monotherapy because of the likely emergence of resistance. It is effective when given orally.

Ciprofloxacin is effective against many Gram-negative pathogens, including salmonellae, and is often also active against staphylococci. It penetrates bone well, is well absorbed when given by mouth and reaches good concentrations in bone. It is also active against some MRSA strains. Newer quinolones may also be effective.

Second- and third-generation cephalosporins are effective against Gram-negative rods. They are effective against *Hemophilus influenzae* of capsular type b (Hib), which can occur in unimmunized pre-school children (see Chapter 6). They penetrate well into bone, but must be given intravenously.

Suggested antibiotic treatments for acute bone and joint infections

For childhood, or adult staphylococcal osteomyelitis

1 Flucloxacillin i.v. 2 g 6-hourly (child under 2 years, 250–500 mg 6-hourly; 2–10 years, 500 mg–1 g 6-hourly) *plus* (for an adult) fusidic acid orally or i.v:. adult over 50 kg, 580 mg 8-hourly; adult under 50 kg and child, 6–7 mg/kg 8-hourly; *or* orally 750 mg 8-hourly, *or* (when Gram-negative infection is possible) ciprofloxacin by i.v. infusion 400 mg 12-hourly (child 5–17 years: up to 10 mg/kg 8-hourly, maximum 1.2 g daily).

2 Rifampin orally or by i.v. infusion, 450–600 mg 12-hourly (child 10–15 mg/kg 12-hourly) *plus* fusidic acid orally or i.v; adult over 50 kg, 580 mg 8-hourly; adult under 50 kg and child, 6–7 mg/kg 8-hourly; *or* orally 750 mg 8-hourly (child up to 1 year, 50 mg/kg daily in three divided doses; 1–5 years, 250 mg 8-hourly; 5–12 years, 500 mg 8-hourly).

⚠️ Rifampin is an enzyme-inducer; it will reduce the effectiveness of warfarin, sulphonylureas, corticosteroids and a number of other drugs including oral contraceptives.

For child under 5 years

3 Cefuroxime i.v. 60–100 mg/kg daily in three or four divided doses.

For elderly people and for vertebral osteomyelitis

4 Cefuroxime i.v. 750 mg–1.5 g 8-hourly *or* cefotaxime i.v. 2 g 8-hourly plus an antistaphylococcal drug; *plus* or *minus* metronidazole 400–500 mg i.v. or rectally, 8-hourly.

For any bone or joint infection, including vertebral osteomyelitis

5 Clindamycin 600–900 mg by i.v. infusion 6-hourly (child over 1 month, *of any weight*: 300 mg daily in 3 or 4 divided doses) *plus* ciprofloxacin (dosage as in 1).

For methicillin-resistant *Staphylococcus aureus* or enterococci

6 Teicoplanin i.v. 400 mg 12-hourly for three doses, then 400 mg daily (child 10 mg/kg daily, reducing to 6 mg/kg daily after first 2–5 days) *or* vancomycin by slow i.v. infusion, 1 g 12-hourly (500 mg 12-hourly or 1 g daily for patient over 65 years) (neonate up to 1 week: 15 mg/kg initially then 10 mg/kg 12-hourly; 1–4 weeks: 15 mg/kg, then 10 mg/kg 8-hourly; child over 1 month: 10 mg/kg 6-hourly).

Treatment should be continued until the fever has subsided, the C-reactive protein level has fallen to the normal range, and healing of the bone is established. In acute, uncomplicated osteomyelitis in children, a 2–3-week course may be sufficient, while 4–6 weeks is more appropriate for adults, and older patients with Gram-negative or mixed infections. The disappearance of bone-marrow oedema can be demonstrated by MR scanning.

Patients with 'difficult' organisms, such as methicillin-resistant *Staphylococcus aureus* (MRSA), enterococci or mixed facultative and anaerobic infections, are often long-term patients, frail, demented or otherwise debilitated. It is sometimes impossible to maintain continued parenteral treatment in such patients. As long as a steady response is obtained, treatment with combinations of oral drugs such as fusidic acid, rifampicin, ciprofloxacin, trimethoprim, clindamycin and/or metronidazole can be successful.

Joint washout

There is no firm evidence for the effectiveness of this process, but many expert units undertake arthroscopic washout of infected joints that contain purulent effusions. The aim is to prevent damage to synovial cartilage, caused by the inflammatory process, and to remove pus, which may become loculated. The procedure is most often performed on the shoulder, which is an extensive joint, having bursa-like extensions of the capsule.

Surgical treatment

Surgical treatment is usually reserved for those cases requiring release of clinically or radiologically apparent pus. It is rarely justified to operate empirically on a suspected site, as it may not be the exact site of infection. Exceptions are the calcaneum and the sacroiliac area, which are accessible to aspiration.

Evidence of pus formation should be sought by daily examination for fluctuance or localization of inflammation. The C-reactive protein level can be used to monitor the inflammatory reaction during treatment. When pus is present, drainage will alleviate the infection, enhance penetration by antibiotics, reduce fever and allow healing of the bone.

Surgery may also be indicated when a sequestrum has formed. The orthopaedic specialist will decide whether or when there is adequate new bone formation to compensate for the defect caused by sequestrum removal.

Problems and complications

Failure to respond

Care should be taken that an adequate antibiotic dosage is being used, and that the route of administration is optimal. The causative organism may be unusual (for instance, *Brucella* sp.), and not covered by the chosen antibiotic regimen. Unexpected antibiotic-resistance may be the problem, as with MRSA.

Problems of prolonged antibiotic treatment

Problems may result from the need to use antibiotics for

several weeks in severe infections, though many patients tolerate this surprisingly well. Candidal infections of mucosae or skin folds may be troublesome if broad-spectrum agents must be used; intermittent or concurrent anti-*Candida* medication will often help, if necessary using oral itraconazole or fluconazole (though fluconazole-resistant *Candida* species can emerge, especially in hospital environments).

Prolonged ciprofloxacin treatment can cause anorexia and weight loss. Long-term metronidazole can cause peripheral neuropathy. Prolonged high-dose beta-lactam medication can cause granulomatous hepatitis or, rarely, idiosyncratic agranulocytosis, which quickly resolves if treatment is stopped, but which precludes further use of any beta-lactam because of complete cross-reactogenicity.

Mistaken diagnosis

Bone tumours and cysts, or the effects of trauma, can all cause a painful, warm and immobile limb. Failure to respond to antibiotics should prompt a reconsideration of the diagnosis and a review of imaging studies.

Damage to epiphyseal cartilage

Irreversible damage to epiphyseal cartilage occasionally follows extensive osteomyelitis. This causes progressive distortion and, in children, failure of the bone to elongate at the affected site. The least favourable sites for this to occur are the lower femur and the upper tibia, as gait, stature and knee function are all severely affected.

Adjacent suppurative arthritis

Invasion and suppurative arthritis of a neighbouring joint can occur if a bone metaphysis is intracapsular, or partly so. The hip and the shoulder are therefore the joints most affected by this problem.

Chronic osteomyelitis

Epidemiology

The epidemiology of chronic osteomyelitis has changed with the development of powerful antibiotics. Only half of cases, or fewer, are nowadays the result of persisting infection after acute blood-borne osteomyelitis. The remainder of cases follow fractures (sometimes associated with non-union), or complicated surgery.

Bones

The femur and the tibia together account for over 70% of cases in most case series. Any bone may be involved, however, including vertebrae, the bones of the foot (especially in diabetics) or the skull (complicating severe otitis externa, for instance).

Bacteria

The bacteria are similar to those of acute disease. Over half of cases are caused by *S. aureus*. Gram-negative infections are slightly less common, and are more often caused by *Pseudomonas* spp. or *Proteus* spp. than by *Escherichia coli*. Mixed Gram-positive and Gram-negative infections occur. Between 10 and 20% of cases are due to anaerobic infection.

Causes of chronic osteomyelitis

1 *Staphylococcus aureus* (>50%).
2 Anaerobic infections (10–20%).
3 *Pseudomonas* spp.
4 *Proteus* spp.
5 *Escherichia coli*.
6 Mixed Gram-positive and Gram-negative infections.

Clinical features

Clinical features include pain, swelling, deformity or (following fracture or surgery) defective healing, or loosening of a prosthesis or pin. Some cases have intermittently or continuously discharging sinuses. A pathological fracture at the site of a metal insert or a prosthesis may be due to chronic infection.

Diagnosis

Diagnosis is made on clinical and radiological grounds, and by identifying a causative organism. Pus or swabs from sinuses are often contaminated by skin commensals or saprophytes, so specimens obtained directly from infected bone or periosteum are preferred.

In patients at risk through known exposure or overseas residence, cultures for mycobacteria should be set up. About 80% of patients with tuberculous arthritis have a strongly positive tuberculin test. Brucellosis is a recognized cause of infective arthritis in African, Middle Eastern and east European countries. It often affects the knee or the spine. A positive IgM ELISA test may indicate recent infection, and is useful if recent antibiotic therapy makes culture unreliable.

A syphilitic gumma of bone can mimic chronic osteomyelitis. In patients who originate from high-risk communities, serological tests for syphilis should be considered.

Differential diagnosis

Neuropathic joints (joints subjected to repeated, gradually destructive trauma due to reduced sensation), are still seen in patients with peripheral neuropathy, including that of diabetes. Rare cases of syphilitic tabes are still seen in elderly patients, especially those previously resident in at-risk communities. Acute inflammation accompanying exacerbations of trauma or osteoarthropathy may cause the appearance of infection in these deformed joints. An appropriate clinical history and examination should help to make the diagnosis, though infection must still be excluded.

Management

Appropriate chemotherapy is important in chronic osteomyelitis, but surgery often also has a major role. The condition should not be managed without the support of surgical colleagues.

The principles of antibiotic therapy, and the agents used, are the same as for acute bone and joint infection, but courses may need to be prolonged, and continued, or modified during periods of surgery. Chemoprophylaxis during bone surgery should be provided, using antibiotics other than those employed for ongoing management of the chronic infection.

Appropriate surgical interventions include removal of affected pins, screws and plates, or of sequestra and other devitalized tissue. Plastic surgery and vascular surgery procedures may be used to fill tissue defects and optimize blood supply, and bone grafting can restore anatomy and function when infection has been controlled. Spacers, or temporary external fixation, may be required to preserve anatomy while infection is controlled. The resolution of inflammation on imaging, and normalization of inflammatory indices, will help to indicate when the infection is no longer active. Definitive reconstruction or re-insertion of a prosthesis can then be carried out.

In the rare case of extensive joint destruction, arthroplasty or arthrodesis may be the best option, once the infection has been fully controlled (Fig. 14.10).

In some cases, especially when an elderly or frail patient has a low-grade infection, it may not be prudent to aim for complete cure or reconstruction. A usable, and relatively pain-free bone or joint can sometimes be maintained by continued oral antibiotic therapy (such as rifampin plus fusidic acid for staphylococci, or clindamycin plus ciprofloxacin for low-grade osteomyelitis in a diabetic patient). Other oral combinations can be tailored to the patient's need and the infecting organism(s). Useful antibiotics include doxycycline, trimethoprim, co-amoxiclav and the oral quinolones.

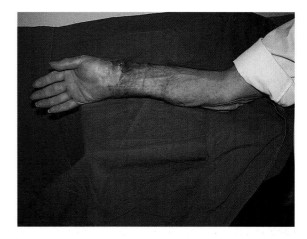

Figure 14.10 Destructive tuberculous infection of the wrist, with draining sinus. Concurrent pulmonary tuberculosis responded well to triple therapy, but the wrist only healed after debridement and arthrodesis.

Infection in prosthetic joints

The management of prosthetic joint infections is largely surgical, though advice on effective antibiotic therapy may be sought. The usual aim of management is to maintain healthy bone and joint capsule tissue while controlling the infection. Plastic, and to a lesser extent metal, implants readily support the formation of biofilms. Rifampin and quinolone antibiotics are among the best available for penetrating these. The surgeon will wish to control spreading infection by giving antibiotic treatment, and then remove the prosthesis that is the nidus of residual organisms. Any pus, devitalized or infected tissue are completely removed. If the adjacent remaining tissue is healthy, a one-step replacement of the prosthesis may be performed, usually using gentamicin-containing bone cement to discourage continuing infection. If this cannot be achieved, a period without the prosthesis may be necessary, allowing tissue healing to take place before a new prosthesis is installed, or arthroplasty or arthrodesis is carried out. A temporary 'spacer' prosthesis may be inserted in the meantime, to preserve anatomy and minimize contracture of soft tissues (gentamicin-impregnated spacers are available for this use).

As for chronic osteomyelitis, there is an option for controlling low-grade infection in frail patients, by continuing oral antibiotic therapy.

Part 3: Genital, Sexually Transmitted and Birth-Related Infections

Genital and Sexually Transmitted Diseases

Introduction

The structures of the male and female genital tract differ greatly, but they have their pelvic position and perineal connections in common. Both the vagina and the urethra are lined with squamous epithelium and, in both sexes, the upper genital tract has unique structures that are both germinal and secretory.

In men the genital tract flora reside on the glans of the penis and the urethral meatus. Colonizing and infecting organisms may exist in the normally sterile upper urethra, prostate and epididymis.

In women the vulva and vagina have a complex flora partly derived from the perineal skin, and partly dictated by the acidic environment of the adult vagina, which is maintained by colonizing bacilli. The normal vaginal flora help to inhibit the establishment of *Candida* infections, to which this site is especially vulnerable. The female urethra and cervix are vulnerable to invasion by sexually transmitted pathogens, which may then ascend to cause endometrial and tubal infections, and to affect intra-abdominal organs.

In both sexes the genital tract can act as the portal of entry for systemic or bacteraemic disease. Fastidious organisms are carried and protected in genital secretions, and may be deposited in the consort's genital epithelium, mouth, eye or rectum, depending on the type of sexual activity. It is important to remember these associations when investigating possible sexually transmitted infections.

Trends in incidence of the commonest sexually transmitted diseases are shown in Fig. 15.1. There is a difference in incidence between different ages and sexes. This is because women often become sexually active at an earlier age than heterosexual men, and homosexual men at a later age (Fig. 15.2). The incidence of gonorrhoea and trichomoniasis declined throughout the 1980s, especially among older patients who modified their sexual behaviour in response to the acquired immunodeficiency syndrome (AIDS) epidemic. Since 1989, reported levels of gonorrhoea have increased, and several outbreaks of syphilis have been detected. In contrast, the incidence of both genital herpes and *Chlamydia* infection has been increasing since the early 1980s in both sexes, but especially among females for genital herpes.

> ⚠ Any patient with a proven sexually transmitted condition has an increased risk of other sexually transmitted infections. In specialist clinics, a full range of screening and diagnostic tests is performed, to ensure that no additional diagnosis is overlooked.

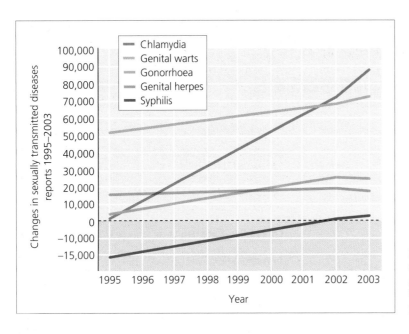

Figure 15.1 Number of new diagnoses of selected sexually transmitted infections; genitourinary medicine clinics, England, Wales and Northern Ireland: 2003. Source: Health Protection Agency.

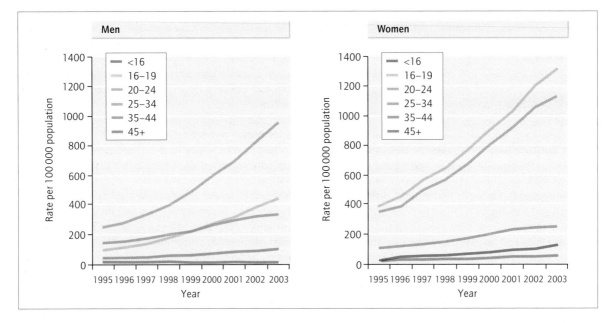

Figure 15.2 Rates of diagnosis of uncomplicated genital chlamydial infection by sex and age group, genitourinary medicine clinics, United Kingdom 1995–2003. Source: Health Protection Agency

Viral infections of the genital tract

Organism list

- Papillomaviruses
- Herpes simplex viruses

Human papillomaviruses, warts and intraepithelial neoplasia

Virology

Human papillomaviruses (HPVs) are small viruses with a circular genome of double-stranded DNA. They are non-enveloped viruses with an icosahedral capsid consisting of 72 capsomeres. The viral genome is just under 8000 base pairs. Nine open reading frames are transcribed, producing seven 'early' and two 'late' proteins. The 'early' proteins are involved in establishing infection. They include E5, which prevents acidification of the endosome, and the products of E6 and E7, which are oncogenic. The L1 and L2 genes encode the major and minor capsid proteins. Papillomavirus is typed by determining the genome sequence. There are around 100 types of papillomavirus, loosely associated with types of cutaneous, genital and laryngeal warts. Some types (2, 3, 5, 8, 9, 10, 12, 14, 15, 17) are only found in the skin lesions of epidermodysplasia verruciformis.

Pathogenesis

HPV replication occurs in the nuclei of epithelial cells as they mature and progress to the epithelial or mucosal surface. In squamous epithelium, the largest concentrations of virus are found in shed squames (or skin keratinocytes). Viral replication causes excessive proliferation of all dermal layers except the basal layer, leading to acanthosis and hyperkeratosis. The histologically characteristic cell is the koilocyte, a large polygonal squamous cell with a smaller nucleus found within a vacuole. Keratohyalin inclusion bodies are seen in the cytoplasm.

Malignant change is accompanied by integration of HPV DNA into the chromosome. Cytological changes include excessive proliferation of basal cells with a high nuclear/cytoplasmic ratio and many mitoses. In benign lesions HPV DNA is located in the nucleus, but not integrated into the chromosome. An effective immune response, which is not currently well understood, terminates HPV infection. Patients with severe primary and acquired immunodeficiency suffer severe or recurrent HPV disease.

Epidemiology

Papillomaviruses are usually transmitted to the genital tract by sexual contact, though perinatal infection and transfer via digital contact can occur, especially in children. Most infections are caused by HPV types 6 and 11, and are benign. It is likely that the great majority of infections are subclinical and unnoticed.

Cervical intraepithelial neoplasia and high-risk papillomavirus types

Hybridization and polymerase chain amplification have identified papillomavirus genome in neoplastic cells of the cervix. Types 16 and 18 predominate in neoplastic cells; types 31, 33, 35, 39, 45, 51, 52, 56, 58, 59, 66 and 68 are less often detected, and other papillomavirus types have similar distributions in malignant and non-malignant cells. Thus, carcinoma of the cervix is associated with sexual activity at a young age, and with multiple sexual partners.

There is no firm evidence to connect papillomavirus infections with neoplasias other than cervical carcinoma.

Clinical features of HPV infection

The clinical lesions of HPV infection are warts. They can occur anywhere on the anogenital skin, often at sites of trauma related to sexual intercourse. Single or scanty lesions often occur on the shaft of the penis, glans, vulva, perineum or perianal skin. Occult lesions can affect the urethral meatus, vagina or cervix and, associated with anal receptive sexual intercourse, the anal canal. The mouth is occasionally affected. On moist, hairless skin the lesions are often flat and soft.

Genital warts cause distress, sometimes itching and irritation and, rarely, cause disrupted urine flow or bleeding from the urethra or anal canal. Large areas of moist, shaggy lesions occasionally develop, and can be a problem in pregnancy. If very extensive and hypertrophic, they can interfere with vaginal delivery.

The diagnosis is clinically obvious. Doubtful or pigmented lesions may be biopsied for histological diagnosis. Detection of HPV DNA can clarify the diagnosis. Commercial methods are based on hybridization of suspensions of cytological specimens with HPV RNA probes. Hybridization techniques are rapid, and do not involve nucleic acid amplification, so avoiding the major errors that can result from cross-contamination in PCR techniques.

Treatment and prevention

Podophyllotoxin, podophyllin or trichloroacetic acid topically are effective for soft, non-keratinized warts. Large areas (>4 cm^2) should be treated under the supervision of a physician, as the first two agents are cytotoxic and the third is caustic. The surrounding skin should be protected with soft paraffin ointment and the solution, or cream should be washed off after 4 hours. Podophyllotoxin and podophyllin are contraindicated in pregnancy, and should not be used for treating the cervix or the anal canal.

Cryotherapy, electrocautery or excision are more effective for keratinized lesions. A combination of podophyllin and cryotherapy is widely used.

Imiquimod topical cream induces a cytokine response that facilitates immunological resolution of the lesions. It is useful for both soft and keratinized lesions. It is contraindicated in pregnancy, and should not be used internally.

Extensive warts can be treated in successive small areas, but after delivery the warts shrink somewhat, and can be treated conventionally.

Current sexual partners may be offered examination and treatment. Condom use is effective in interrupting transmission.

Genital herpes simplex infections

Pathology and epidemiology

Genital herpes simplex is a common infection that spreads by direct contact, usually but not always sexual. Many infections are caused by herpes simplex virus type 2 (HSV2), but in recent years type 1 has become equally common. Both types of infection can be associated with HSV infection in other sites, particularly cold sores and other cutaneous infections. The incidence of both genital herpes and genital warts has increased in recent years with around 18 000 infections reported in 2003, two-thirds of which were in females. Some of this is related to infection with human immunodeficiency virus (HIV). Attendance at genitourinary clinics, and ascertainment of infection has also recently increased.

As with other herpesvirus infections, there is a primary and post-primary type of disease, and periods of asymptomatic excretion of virus from the previously affected genital tract. Virus has been demonstrated in the urethra and the vas deferens of asymptomatic men. Antiviral treatment of acute disease does not prevent relapses or asymptomatic excretion. Contact tracing is therefore ineffective in controlling genital herpesvirus infections.

 Virology, pathogenesis and diagnosis of herpes simplex infection are discussed in more detail in Chapter 5.

Clinical features

Primary infection

The incubation period is variable, averaging 4 or 5 days. Mild fever and malaise, with tender, usually unilateral, inguinal lymphadenopathy is followed after a day or two by tense superficial vesicles, which quickly evolve into painful ulcers. The commonest affected site is the coronal sulcus or glans of the penis, the vulva or the anal margin. In severe infections pain may inhibit micturition and cause acute urinary retention. Rare cases of faecal impaction are seen when the perianal area is involved.

An effective cell-mediated immune response is important in controlling genital herpes simplex infections. Patients with immunodeficiency often suffer severe, persistent or complicated disease.

Associated neurological problems

Viral meningitis is not uncommon, and may be missed if mild. Lumbar puncture reveals typical cerebrospinal fluid (CSF) changes with lymphocytosis, and HSV2 can be demonstrated in CSF. The meningitis is usually benign, and resolves in a few days.

Radiculitis of the pelvic nerve roots causes pain and stiffness of the lower back, sometimes with associated meningism. Asymmetrical neurological lesions may cause paraesthesia or anaesthesia of the buttock, thigh or perineum, with associated disturbance of micturition and defecation. Occasionally, there is loss of the sartorius or quadriceps tendon reflexes. The course of the disorder is variable, but is often 5–10 days or more. Relapses may occur over several weeks. Recovery is almost always complete.

Clinical presentations of primary genital herpes

1 Painful inguinal lymphadenopathy.
2 Painful genital ulcers.
3 Lymphocytic meningitis (herpes simplex virus type 2).
4 Pelvic radiculitis (HSV 2).

Post-primary genital herpes

Post-primary genital herpes is like a cold sore. It may affect the buttock, thigh, perineum or genitalia. The appearance of vesicles is heralded by malaise, with local pain or burning. The vesicles last for 2–5 days before healing by drying and epithelialization. Abortive attacks, with the transient appearance of a few papules, also occur. Recurrences can be frequent, and in women may occur premenstrually. The great majority are caused by HSV2. Although there is little fever or systemic change, patients feel fatigued and suffer radiation of pain to the thighs and back. Recurrences often slowly decline in frequency, but some patients suffer relentless and frequent attacks.

Diagnosis

This is often clinically obvious. Herpesvirus particles can be demonstrated by electron microscopy of vesicle fluid or scrapings. Both HSV1 and HSV2 grow rapidly in cell cultures, producing typical cytopathic effects in 48–72 h. Seroconversion is demonstrable in primary attacks. Polymerase chain reaction (PCR)-based diagnosis is useful in confirming the diagnosis of neurological presentations.

Management

Primary attacks can be shortened by oral treatment started within 5 days of onset, using famciclovir, aciclovir or valaciclovir, usually in courses of 5 days. Topical treatment may have a weak local effect, but is ineffective against virus in the lymph nodes and nervous system.

Severe primary infections are best treated with intravenous aciclovir 5 mg/kg 8-hourly, and usually require courses of at least 5 days. Meningoradiculitis may respond better to 10 mg/kg 8-hourly.

Most recurrent attacks are brief and self-limiting. They can be managed with topical emollients or lidocaine gel. Frequent recurrences can be prevented by suppressive therapy with oral aciclovir, valaciclovir or famciclovir. When recurrences are confined to the premenstrual days, treatment need only be taken at this time.

Treatment of genital herpes

1 Acute attacks: aciclovir orally 200 mg five times daily for 7 days; *or* valaciclovir 500 mg twice daily; *or* famciclovir 125 mg twice daily.
2 Meningoradiculitis: aciclovir i.v. 10 mg/kg 8-hourly for 5–10 days.
3 Suppression of recurrences: aciclovir orally, 400 mg twice daily (or 200 mg 6-hourly) *or* valaciclovir 250 or 500 mg twice daily *or* famciclovir 250 mg twice daily, given continuously, or in the week before menstruation every month.

Complications and cautions

True complications are rare; staphylococcal secondary infection is marked by increasing inflammation and exudation, often of yellowish pus. It responds to treatment with an antistaphylococcal antibiotic such as flucloxacillin orally for 5 or 6 days.

Herpes simplex infection is severe in the neonate and in immunosuppressed patients. In pregnancy it can be transmitted during delivery and (rarely) trans-placentally, causing extensive, life-threatening disease in the neonate (see Chapter 17).

HSV infection is often persistent, extensive and invasive in AIDS; extensive lesions lasting more than 1 month comprise an AIDS-defining opportunistic infection.

Aciclovir-resistant organisms sometimes emerge during prolonged suppressive therapy; they lack, or have altered, viral thymidine kinase, and are of low pathogenicity, but can cause important disease in highly immunodeficient patients. Valganciclovir or foscarnet are options for treatment of these organisms.

Erythema multiforme accompanies herpes simplex recurrences in a small group of people, among whom tissue types DQW3 and/or DRW53 predominate. HSV DNA can be demonstrated in erythematous skin but not in normal skin. Attacks diminish in severity over a period of 24–36 months, and eventually cease. Suppression of the herpes simplex recurrences prevents the erythema multiforme.

Prevention

Patients should avoid sexual contact during primary and recurrent attacks. Condom use may prevent transmission, but is unproven. Antiviral treatment and suppression of attacks reduces infectiousness. Individuals known to shed viruses frequently can take suppressive treatment. Sexual partners may have unrecognized attacks or shedding, and should therefore be examined, and treated and advised where appropriate.

Pregnant women may be treated as indicated. Suppressive treatment in the last 4 weeks of pregnancy will prevent recurrent genital lesions, and avoid the need for other interventions (see Chapter 17).

Bacterial infections of the genital tract

Organism list

- *Chlamydia trachomatis* (serogroups D–K)
- *Neisseria gonorrhoeae*
- *Treponema pallidum*
- *Gardnerella vaginale*
- *Actinomyces* spp.
- *Haemophilus ducreyi*
- Lymphogranuloma venereum (*Chlamydia trachomatis* serogroups L1, L2, L3)
- *Mycoplasma hominis* and *M. genitalis*
- *Ureaplasma urealyticum*

Chlamydial genital infections

Introduction

Chlamydia trachomatis is the commonest cause of non-specific or non-gonococcal genital infections (causing up to 50% of cases in the USA), and is the commonest genital pathogen worldwide. Diagnosis of chlamydial infections has increased steadily over the past 20 years with 90 000 infections reported in England, Wales and Northern Ireland in 2003. This is partly as a result of increased testing in the National Chlamydia Screening Programme, which commenced in England in 2002. Highest diagnosed case numbers occur amongst females and people below the age of 25. Chlamydial genital infections are important not only because of the high transmission rates and high morbidity that they cause, but because of their contribution to infertility due to chronic pelvic infection and their ability to cause significant intrapartum infection of neonates.

Pathology and microbiology of chlamydial infections

C. trachomatis is an obligately intracellular bacterium that lacks a cell wall. It is a member of the genus *Chlamydia*, which contains one other species, *C. psittaci*, a zoonotic pathogen. The closely related *Chlamydophila pneumoniae* is a respiratory pathogen of humans (see Chapter 7).

C. trachomatis exists in several serotypes: A, B, Ba and C, which are associated with ocular trachoma (see Chapter 5); D–K, associated with oculogenital infections and neonatal infections (see Chapter 17); and L1, L2 and L3, the causes of lymphogranuloma venereum.

The infectious form of *C. trachomatis* is the elementary body (EB). This is a metabolically inert extracellular infectious transport particle, 200–300 nm in diameter. This form alternates with a larger intracellular reproductive particle of 1000 nm diameter, called the reticulate body

(RB). Phagocytes take up the EB and form a phagosome within which the EB survives by inhibiting lysosomal fusion. The EB rapidly develops into an RB (within about 6 hours) and the RB undergoes numerous binary fissions. After 18 to 20 hours some of the RBs begin to condense, via an intermediate form, into EBs. After around 48 hours, programmed rupture of the host cell membrane culminates in release of the EBs into the surrounding environment. Arrest of this life cycle can lead to latency, which may be an important process in chronic genital infection.

Clinical features

In men around 50% of infections are asymptomatic. The commonest presentation is urethritis, causing dysuria, urethral and meatal soreness, and early morning urethral discharge (before micturition). Infection may occur alone or cause persisting symptoms after treatment for gonorrhoea. Ascending infection can cause epididymitis, but is a rare cause of acute or chronic prostatitis. Chlamydial proctitis, often mild or asymptomatic, is increasingly recognized.

In women around 80% of infections are asymptomatic. Urethritis and cervicitis are common, causing soreness, dysuria, purulent vaginal discharge, and sometimes postcoital bleeding, without significant systemic features. Ascending infection typically causes acute salpingitis, with fever, neutrophilia and lower abdominal pain. It must be distinguished from appendicitis. Wider invasion can lead to perihepatitis (Curtis–Fitz-Hugh syndrome), with high fever, upper abdominal pain and guarding, abnormal liver function tests and perihepatic enhancement on imaging. Rare cases of perihepatitis are seen in men, though the route of infection is uncertain. Pelvic or abdominal disease can occur without detectable preceding genital symptoms.

Persisting conjunctivitis, and asymptomatic pharyngeal infection can affect both sexes.

 The occurrence of chlamydial infection in a neonate suggests asymptomatic infection of the parents.

Clinical presentations of chlamydial infections
1 Urethritis.
2 Cervicitis.
3 Proctitis.
4 Conjunctivitis.
5 Salpingitis.
6 Prostatitis.
7 Perihepatitis.
8 Infected neonate.

Diagnosis

Urethritis and cervicitis may be clinically apparent, but asymptomatic infections and proctitis are detectable only by screening. Gonorrhoea is an important differential diagnosis and may coexist with chlamydial disease.

Acute salpingitis does not have the distinctive evolution of appendicitis, and may be accompanied by suprapubic pain and guarding, and vaginal discharge, which can be examined microbiologically. Swelling, induration and tenderness is often detectable in one or both fornices, on vaginal examination.

Perihepatitis must be differentiated from gallbladder disease and liver abscess. Imaging studies may show a healthy gallbladder and/or oedema of the liver capsule.

Laboratory diagnosis

The most useful rapid tests for *C. trachomatis* are nucleic acid amplification techniques, which are clearly superior to ELISA tests, and should be the investigation of choice where there is sufficient expertise in the laboratory to support it. Direct fluorescent antibody tests are much less sensitive, time-consuming and depend on skilled performance. A major advantage of NAATs is that urine specimens can be used for diagnosis making it possible to perform large-scale screening. The most reliable diagnostic specimen for men is the first-voided ('first glass') urine specimen, passed after holding the urine for at least 1 hour.

C. trachomatis can be cultured from swab specimens in cultures of cycloheximide- or idoxuridine-pretreated McCoy cells. HeLa cells are also suitable for *Chlamydia* cultures. Positive results are demonstrated by the presence of chlamydial inclusion bodies on Giemsa- or iodine-stained cells, or by antigen demonstration using direct IF or ELISA techniques. Culture is the 'gold standard', and may be used for cases with legal implications, but its poor sensitivity means that few laboratories now perform this technique.

LGV can be identified when material from patients positive in the *Chlamydia* NAAT is typed by molecular methods.

Serology is not helpful in the diagnosis of chlamydial genital infections.

Management

The treatment of choice is a 1-week course of doxycycline. Azithromycin in a single dose is at least as effective, and is effective against *N. gonorrhoeae*, but more commonly causes nausea. Erythromycin is an alternative. The simplest regimen possible is chosen to encourage compliance.

Systemic infections require parenteral treatment. Erythromycin is then the treatment of choice.

Treatment of chlamydial infections

1 First choice: doxycycline orally 100 mg twice daily for 7 days (contraindicated in pregnancy and lactation) *or* azithromycin, orally, 1 g single dose.

2 Alternatives: erythromycin orally 500 mg 6-hourly for 1 week (or 500 mg 12-hourly for 2 weeks) *or* Deteclo, orally 300 mg 12-hourly for 1 week *or* ofloxacin, orally 200 mg 12-hourly for 1 week (or 400 mg daily for 1 week) *or* tetracycline, orally 500 mg 6-hourly for one week (re-examine after 3 weeks to confirm cure).

3 Systemic infections: erythromycin i.v. 500 mg–1 g 6-hourly for 5–7 days.

⚠️ Sexual partners should be examined and offered treatment; they should also be offered screening for other sexually transmitted infections.

Surgery for acute salpingitis is avoided if possible, as salpingectomy reduces fertility. Fibrosis of the lumen can adversely affect tubal function if healing is slow or incomplete.

Role of *Ureaplasma urealyticum* and *Mycoplasma genitalium* in non-specific genital infection

Some non-specific genital infections are not associated with *C. trachomatis*. *U. urealyticum* or *M. genitalium* can be detected by cultural techniques (using media suitable for mycoplasmas; see Chapter 7) in a substantial proportion of these. The importance of this is that *U. urealyticum* is often resistant to tetracycline. For this reason, nongonococcal urethritis not known to be chlamydial should be treated with azithromycin (or erythromycin).

Complications

The most important complication of non-specific genital infection is Reiter's syndrome. This is a post-infectious condition affecting mainly men, and associated with the human leucocyte antigen (HLA) B27 tissue type. It causes prolonged synovitis and connective tissue inflammation (see Chapter 21).

Lymphogranuloma venereum

Lymphogranuloma venereum is a systemic sexually transmitted disease usually caused by serotypes L1, L2 and L3 of *C. trachomatis*. It has an incubation period of 2–30 days. It is endemic in east, west and south Africa, south east Asia, Papua New Guinea and some Caribbean islands, and causes 1–5% (rarely up to 10%) of genital ulcers in these areas.

A primary lesion: a small, shallow ulcer or ulcers may appear at the site of inoculation in the genitalia or, rarely, the mouth. Most patients present with secondary lesions:

slowly enlarging, usually bilateral, inguinal or femoral lymph nodes (buboes). Suppuration is common, with one or more draining sinuses and slow healing. Tertiary features, appearing in a minority of cases, include fever, meningism, pericarditis, keratitis and skin rashes, all of which slowly resolve. Years later, perineal and inguinal fibrosis may occur, causing rectal strictures in women and genital lymphoedema in men.

Diagnosis requires a high index of suspicion. It can be confirmed by positive nucleic acid amplification-based detection of chlamydial genome (see above). Chlamydial culture is an alternative diagnostic test. Other methods include serological testing for *C. trachomatis*, or demonstration of chlamydiae on histology of the primary lesion or lymph nodes.

The treatment of choice is doxycycline 100 mg 12-hourly or tetracycline, which must be given for at least 3 weeks. Erythromycin is an alternative. Rifampin may also be added. Repeated courses may be needed for late or persistent disease. Treatment of early disease prevents late fibrosis, which is not amenable to antibiotic therapy. Sexual partners should be examined.

Gonorrhoea

Introduction and epidemiology

Although less common than non-specific genital infection, gonorrhoea still causes millions of infections worldwide. As well as local infection, it can also cause bacteraemic disease, which is often recognized and treated late.

In contrast to viral infections of the genital tract, the incidence of gonorrhoea fell in many developed countries in the late 1980s. This decline was mainly among patients over 25 years of age. In recent years the incidence has been climbing again particularly in those under 35. Between 1999 and 2003 the number of cases in England, Wales and Northern Ireland increased by 50%, two-thirds of the cases being males. This trend suggests a poor response to prevention initiatives among young adults.

Clinical features

Men

- Urethral discharge (80%)
- Dysuria (50%)
- Rectal infection with anal discharge or pain (80% asymptomatic)
- Pharyngeal infection (>90% asymptomatic)
- Asymptomatic infection (<10% overall)
- Ascending infection: epididymitis, rare prostatitis
- Up to 20% have *C. trachomatis* co-infection

Women

- Increased or altered vaginal discharge (+/– 50%)
- Lower abdominal pain (approx. 25%)
- Dysuria (12%), but not frequency

- Pharyngeal infection (>90% asymptomatic)
- Rarely, intermenstrual bleeding or menorrhagia
- Ascending infection: salpingitis, perihepatitis
- Up to 40% have *C. trachomatis* co-infection

> ⚠ Salpingitis usually presents surgically, as a differential diagnosis of appendicitis or 'acute abdomen'. It is clinically identical to chlamydial salpingitis. Suprapubic tenderness and guarding, the absence of the usual evolution of appendicitis, adnexal swelling, tenderness and vaginal discharge point to the diagnosis.

Gonococcal bacteraemia

Gonococcal bacteraemia tends to present with local inflammatory lesions, rather than a sepsis syndrome. The accompanying genital infection may be mild or subclinical. Arthritis is common, frequently affecting both knees, but the ankle and the wrist are also vulnerable (Fig. 15.3). The patient is feverish; the joint or joints are painful and swollen, and may be surrounded by considerable erythema. There is usually an effusion, which can be large in the knees.

Many patients have a sparse, vasculitic rash, of individual, painful vesicular lesions, with an intensely inflamed halo (Fig. 15.4). These lesions look very like herpes simplex vesicles but are found on extensor surfaces, particularly of the fingers, the elbow or the foot. A few patients have a bruising or haemorrhagic component to the rash, making it similar to a meningococcal rash.

Diagnosis

Urethral and/or cervical swabs are the specimens of choice. Other specimens that should be taken include pharyngeal swabs, rectal swabs in individuals who practise receptive

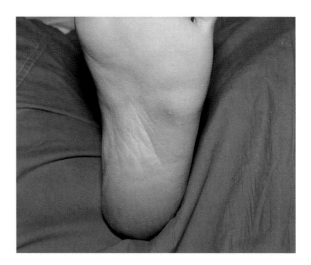

Figure 15.4 Gonococcal skin lesion: an intensely inflamed, rather vasculitic lesion; the same patient as in Figure 15.3.

anal intercourse, and material obtained by prostatic massage.

Gram staining of swab specimens is a useful rapid test. A positive test demonstrates neutrophils, with intracellular and extracellular Gram-negative diplococci. This test is more sensitive in men (90–95% of symptomatic and 50–75% in asymptomatic men) than in women (50% overall).

Culture on selective media is the most sensitive diagnostic method, being positive in over 95% of cases.

Gonococcal bacteraemia should be suspected in any young patient with a fever and arthritis. Aspiration of the affected joint shows many neutrophils, but demonstration of gonococci is difficult until late in the illness. Blood cultures are the main diagnostic test. Gonococci are present in the skin lesions, and can be recovered from vesicle fluid.

Neisseria gonorrhoeae is a delicate organism that does not survive for long outside the body. Specimens from the male urethra may be obtained using a platinum loop as many swab materials impair the survival of this organism. Ideally, specimens should be examined in a side room by microscopy and inoculated on to culture medium and incubated locally, or transported with the minimum delay to the microbiology laboratory.

Specimens in which *N. gonorrhoeae* is sought are usually heavily contaminated with other bacteria. Antibiotics must therefore be added to the medium to aid selection. The medium must be very nutritious, containing lysed blood and additional growth factors such as yeast extract. The usual medium inoculated is a modified New

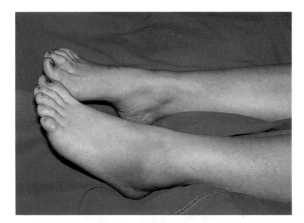

Figure 15.3 Gonococcal arthritis affecting the ankle 1 week after contact with a new sexual partner.

York City medium or Thayer–Martin medium in parallel with a non-selective medium such as chocolate agar. The plates should be incubated in 5–10% carbon dioxide and increased humidity, and inspected after 24 and 48 h. Suspect colonies are identified by the oxidase test and sugar oxidation tests. Confirmation of identification can be made by serological agglutination tests. The presence of a beta-lactamase enzyme can be detected rapidly using the commercial kits based on a cephalosporin that changes colour when the beta-lactam bond is broken. Susceptibility to penicillin, tetracycline, spectinomycin, ciprofloxacin and a third-generation cephalosporin should be tested.

More recently NAATs have been introduced, which detect *N. gonorrhoeae* often in parallel with *Chlamydia* diagnostic tests. These provide a rapid and sensitive approach to diagnosis. Swab or pus samples are first screened with the NAAT and those positive for *N. gonorrhoeae* are cultured so that susceptibility testing can be performed.

Gonococci can be typed, using multi-locus sequence typing (see Chapter 3).

Management

Treatment may be offered immediately to patients with positive rapid diagnostic tests, or those with known exposure to an infectious case. Patients with positive cultures should be treated without delay. Penicillin-allergic patients can be given cefuroxime 1.5 g intramuscularly (half the dose in each of two sites) or oral ciprofloxacin 500 mg as a single dose.

Salpingitis, perihepatitis and bacteraemia require admission to hospital and treatment with intravenous antibiotics. A week's course is usually sufficient. Care must be taken that complications of bacteraemia, such as endocarditis, are not overlooked.

Treatment of gonococcal infections

1 Ciprofloxacin orally, 500 mg single dose *or* ofloxacin orally, 400 mg single dose (contraindicated in pregnancy and lactation).

An increasing proportion of *N. gonorrhoeae* isolates are resistant to ciprofloxacin, especially if the patient was exposed in Asia or the Far East (Fig. 15.5).

2 Alternative: cefuroxime i.m. 1.5 g single dose (half the dose in each of two sites).
3 Alternative, when penicillin-resistant organisms unlikely (<5% local prevalence): ampicillin orally, 2 or 3 g single dose plus probenecid 1 g orally (not marketed in the UK: special purchase necessary).
4 Alternatives, when infection has been contracted in an area of high antimicrobial resistance: ceftriaxone i.m. 250 mg single dose *or* cefotaxime, i.m. 500 mg single dose *or* spectinomycin, 2 g i.m. single dose.
5 Systemic disease: benzylpenicillin i.v. 1.2–2.4 g 4–6-hourly for 7 days *or* ceftriaxone i.v. 1 g daily for 7 days.

Partners should be examined for gonorrhoea and other infections (all partners up to 2 weeks before onset, or last partner if longer): re-testing should be performed at least 72 hours after treatment, to ensure cure.

Complications

The commonest complication of gonorrhoea is re-infection. The extreme antigenic variability of gonococcal pilus proteins, the main immunogenic proteins of the organism, means that even repeated infections do not provide protection from other strains. Repeated infections, severe infections or those in which effective treatment is delayed

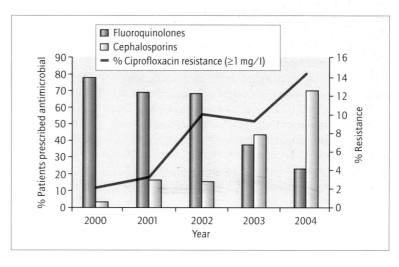

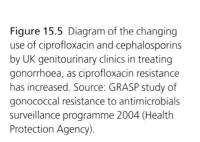

Figure 15.5 Diagram of the changing use of ciprofloxacin and cephalosporins by UK genitourinary clinics in treating gonorrhoea, as ciprofloxacin resistance has increased. Source: GRASP study of gonococcal resistance to antimicrobials surveillance programme 2004 (Health Protection Agency).

can lead to tissue damage. In men urethral strictures, prostatic damage and chronic prostatitis may occur. In women tubal damage and obstruction can cause infertility and predispose to chronic pelvic inflammation.

Syphilis

Introduction and epidemiology

Syphilis, caused by the spirochaete *Treponema pallidum*, is now relatively uncommon in industrialized countries. It remains important, however, because it is prevalent in some places overseas and in travellers, and remains more common in homosexual men than in other social groups. Early cases must be detected and treated to avoid the problems of congenital infection and of late manifestations, which cause great morbidity and dependence.

The incidence of syphilis has started to rise again in some countries, notably inner cities in the USA, and more recently Russia. Several outbreaks have occurred in the UK in recent years and the incidence has been steadily increasing here. The number of cases diagnosed in England, Wales and Northern Ireland increased seven-fold between 1999 and 2003, with more than half the cases occurring in homosexual men.

Classification of syphilis

Acquired syphilis

- Early syphilis: primary, secondary and early latent syphilis, manifested in the first 2 years of infection.
- Late syphilis: late latent syphilis, gummatous (tertiary) disease, cardiovascular and neurological disorders (sometimes called quaternary syphilis), manifested from 2 years after infection.

Congenital syphilis

- Early congenital syphilis: manifested in the first 2 years of life.
- Late congenital syphilis: manifested after age 2 years, including the stigmata of congenital disease.

Pathogenesis of treponemal infections

The pathogenesis of treponemal infection is poorly understood. The infective dose is thought to be as few as about 10 organisms, which can enter through intact skin. The primary lesion results from local multiplication but dissemination also occurs during the primary stage, possibly within minutes of infection. The pathological changes are the result of the host immune response. *T. pallidum* has limited toxigenic capabilities and its survival in the host may be as a result of its poor antigenicity.

Clinical features

Primary syphilis

Appearing about 10 to 70 days after infection, the typical primary lesion, or chancre, is usually a single, painless, indurated ulcer. Commonly affected sites are the foreskin, coronal sulcus, vulva, fourchette, uterine cervix or adjacent structures such as the urethra or penile shaft. There is painless enlargement of local lymph nodes. About 5% of chancres are extragenital, affecting the lips, mouth or nipple. Atypical chancres are common, and may be multiple or painful, easily mistaken for herpes simplex, chancroid or small malignant lesions.

Secondary syphilis

Secondary syphilis develops 6–8 weeks after the primary manifestations, though some patients have no history of chancre. It is a spirochaetaemic disease with fever, lesions of the skin and mucosae, and generalized lymphadenopathy.

The rash is usually generalized, maculopapular or papular, extending to the palms and soles. It is notoriously variable; serpiginous or discoid lesions make differential diagnosis difficult. Other cutaneous and mucous membrane lesions are common, including: superficial oral erosions covered with greyish exudate (mucous patches), serpiginous mouth ulcers (snail-track ulcers) and flat, moist, warty lesions (condylomata lata) on the perineum, especially around the anus. Systemic manifestations include meningitis, arthritis, arthralgia, mild nephrotic syndrome, patchy alopecia and, in about 5% of cases, iritis or retinitis.

Latent syphilis

Latent syphilis is an asymptomatic state that may persist for years following untreated infection. Tissue damage progresses very slowly, and a few patients have elevated CSF protein levels or mild pleiocytosis. Many patients with latent syphilis will eventually develop late manifestations of the disease.

Late syphilis

Late syphilis is characterized by the gumma, an indolent, granulomatous lesion that may undergo central mucoid degeneration. Gummata can occur in any body tissue, causing chronic osteomyelitis or periostitis, nodular liver enlargement or skin lesions, which may ulcerate, producing a sticky discharge.

Cardiovascular syphilis: in the cardiovascular system aortitis occurs 10 years or more after primary infection. There is intimal thickening and loss of elastic tissue from the root and ascending part of the aorta. The aortic valve ring is dilated, the coronary arteries may be occluded and an aortic aneurysm commonly develops.

Neurosyphilis: meningovascular syphilis causes vasculitis and leptomeningitis, particularly at the base of the brain and the upper spinal cord. It can present as meningitis, often with papilloedema, but more often causes focal disorders, such as cranial nerve palsy, weakness and wasting of the hands, transverse myelitis or ataxia. The vascular disease can produce strokes or epilepsy. There is a moderate lymphocytosis and a raised protein level in the CSF.

Tabes, now exceptionally rare, occurs 15 years or more after primary infection. It is caused by degeneration and demyelination in the posterior columns of the spinal cord and the dorsal nerve roots. It presents with dysaesthesia and anaesthesia, pains and ataxia. Pains are typically shooting girdle or limb pain, sometimes with prolonged or repetitive abdominal pain and vomiting (gastric crisis). Ophthalmoplegia is common. The pupils may be small and unreactive to light, while remaining reactive to accommodation (Argyll Robertson pupil). Hypoaesthesia leads to trophic arthropathy of the legs and feet (Charcot joints).

Syphilitic paresis is a progressive dementia, with or without tabetic features. The CSF contains many lymphocytes and the protein content is high, unlike Alzheimer's disease or idiopathic psychosis.

Diagnosis

Lumbar puncture is only necessary when neurological involvement is evident, when it is important in differential diagnosis. A chest X-ray or CT scan should be performed if aortitis is a possibility. Ophthalmological examination by slit-lamp may show typical changes of uveitis or keratitis.

Laboratory diagnosis

Syphilis is a rare but important part of many differential diagnoses, especially of neurological conditions. *T. pallidum* has not been cultivated in artificial medium, so diagnosis is based on dark-ground microscopy, antigen tests, PCR-based tests and serology.

Rapid diagnosis: in early syphilis a rapid diagnosis can be made by dark-ground microscopy of exudate from lesions or of aspirate from enlarged lymph nodes. The spirochaetes are seen as tightly coiled, 'watch-spring' structures. They also have typical motility, rotating about their long axis, or bending at an angle. A direct fluorescence antibody test (DFA) is also available. PCR tests are offered by the reference laboratory, and are particularly useful in immunosuppressed patients.

Enzyme immunoassay (EIA) for IgM anti-treponemal antibodies becomes positive at the end of the second week in primary infection. The VDRL (venereal disease reference laboratory test) or RPR (rapid reagin test) are tests for total antibodies against cardiolipin antigens, which cross-react with *Treponema* species, and are usually positive in early syphilis; false-negatives in secondary disease may be due to a 'prozone' phenomenon (see Chapter 3). Their titre diminishes in the first 6 months after successful treatment, and they eventually become negative (or stabilize at a low titre). Specific EIA tests for immunoglobulin G are suitable for diagnosis at all stages of post-primary syphilis, and can be automated, allowing large numbers of screening samples to be processed efficiently. Specific treponemal tests use a cultivatable treponeme, the Reiter's treponeme, as the antigen and are therefore less subject to false reactions. Positive results appear later in the course of infection than cardiolipin-based tests, but remain positive for life.

The most specific test available for routine laboratories is the fluorescent treponemal antibody absorption (FTA-abs) test. *T. pallidum* is bound to a glass slide and patient serum is first absorbed to remove group-reactive antibody, then placed on the slide and incubated. After washing, the binding of specific anti-*T. pallidum* antibody is detected using a fluorescent antihuman immunoglobulin specific for either IgG or IgM. The IgM test is especially useful for detecting acute and congenital infection (Table 15.1).

The diagnosis of late syphilis: this requires a high index of suspicion. It depends on the demonstration of positive specific treponemal antibodies, with or without positive reagin tests, in patients who have no clinical features of early syphilis, and who may have a history of early syphilis, contact with syphilis or genitourinary symptoms. A search should also be made for clinical features of congenital syphilis (see Chapter 17). Patients with late syphilis and neurological disorder have abnormal cerebrospinal fluid, which contains excess white cells, and gives positive results on reagin tests.

Table 15.1 Results of serological tests for syphilis at different disease stages

	VDRL/RPR	TPHA	FTA-abs	IgM (ELISA or FTA)
Congenital infection	+	+	+	+
Primary infection	+	–/+	–/+	+
Untreated secondary infection	+	+	+	+
Treated or late disease	–	+	+/–	–

VDRL, Venereal Disease Reference Laboratory; RPR, rapid plasma reagin; TPHA, *Treponema pallidum* haemagglutination assay; FTA-abs, fluorescent treponemal antibody absorption; IgM, immunoglobulin M; ELISA, enzyme-linked immunosorbent assay.

Management

Penicillin remains the treatment of choice for all forms of syphilis. Early syphilis is treated with doses that reach treponemicidal levels for 10 days, while late syphilis and early syphilis with neurological involvement (when the treponemal division time may be prolonged beyond the usual 30–33 hours), is treated for 17 days. Long-acting penicillins are no longer routinely available in the UK, but a range of preparations can be obtained by special purchase. A range of other drugs is also effective, but high-quality evidence for dose and duration of therapy is lacking.

Treatment of syphilis

1 Early syphilis: procaine penicillin G, i.m. 750 mg daily for 10 days or benzathine penicillin, i.m. 2.4 MU, single dose or two doses (in days 1 and 8).

2 In penicillin allergy: doxycycline, orally 100 mg 12-hourly for 14 days (contraindicated in pregnancy and lactation) or erythromycin, orally, 500 mg 6-hourly for 14 days or azithromycin 500 mg daily for 10 days or ceftriaxone i.m. 500 mg daily for 10 days.

3 If parenteral treatment is refused: amoxicillin, orally, 500 mg 6-hourly plus probenecid orally 500 mg 6-hourly for 14 days or an oral regimen from point 2 (above).

4 Late syphilis (without neurological disease): procaine penicillin, i.m. 750 mg daily for 17 days or benzathine penicillin, i.m. 2.4 g single dose on days 0, 14 and 21 or doxycycline, orally, 200 mg 12-hourly for 4 weeks or amoxicillin, orally, 2 g 8-hourly plus probenecid, orally, 500 mg 6-hourly for 4 weeks.

5 Late syphilis with neurological involvement, or when neurosyphilis has not been ruled out: procaine penicillin, i.m. 2 g daily plus probenecid 500 mg, orally, 6-hourly for 17 days or benzylpenicillin 1.8 to 2.4 g 4- to 6-hourly for 17 days.

6 Alternatives for neurosyphilis: doxycycline, orally, 200 mg 12-hourly for 4 weeks or amoxicillin, orally, 2 g 8-hourly plus probenecid, orally, 500 mg 6-hourly for 28 days.

Follow-up

The VDRL titre should show significant decline after 6 months and become undetectable, or unchanging ('serofast') after 18–24 months. Patients are discharged after remaining seronegative or serofast for one year. A persisting positive test is an indication for re-treatment. More specific tests remain positive indefinitely. After neurosyphilis lumbar puncture should be performed 6-monthly until the cell count is normal. Any appearance of new neurological signs, or an increase in cellularity or a fourfold or greater rise in antibody levels, is an indication for re-treatment.

Jarisch–Herxheimer reaction

The Jarisch–Herxheimer reaction is an inflammatory response to spirochaetal antigens. It is much commoner in early than in late disease, and occurs within hours of the first antibiotic dose, with fever and exacerbation of swelling and inflammation. This can be a serious problem if, for instance, a coronary ostium is further occluded, or if epilepsy or stroke is precipitated. The reaction can be minimized by starting with low doses of penicillin and/or by adding corticosteroids such as prednisolone 10–20 mg 8-hourly during the first few days of treatment.

Complications

The most important complications of syphilis are related to infection in pregnancy. Trans-placental infection of the fetus is likely, causing abortion, or neonatal and long-term disease (see Chapter 17).

Prevention and control

All pregnant women, new genitourinary clinic attenders, and all sexual contacts of syphilis cases are offered serological testing. About 40% of contacts of infectious cases (most early cases) will be infected. They may be offered 'epidemiological' treatment, with a single i.m. dose of 2.4 MU benzathine penicillin, or oral doxycycline 100 mg 12-hourly for 14 days. Alternatively, they are serologically tested at first attendance and again at 6 weeks and 3 months, with treatment if results become positive.

Non-sexually transmitted (endemic) treponemal diseases

These are nowadays rare. They are caused by treponemes that are distinguishable from *T. pallidum* only by one or two altered base-pairs on DNA sequencing. The commonest example is yaws, usually a childhood disease, affecting children in rural Africa and parts of the Caribbean. Its causative agent, *T. pertenue*, is indistinguishable from *T. pallidum*. It is spread by direct skin and mucosal contact, and produces a disease that follows a course very similar to that of syphilis. The primary nodule (mother yaw), at the site of inoculation is often itchy and may reach up to 5 cm diameter. It heals spontaneously in 3–6 months. Post-primary lesions appear weeks to two years later. They are nodular, reddish, and soft, with a wide distribution on the body, and may be accompanied by lymphadenopathy, and uncomfortable periostitis of finger and long bones. The skin lesions resemble raspberries, giving the condition its common name (framboesia). The diagnostic tests for syphilis are applicable to yaws, and the treatment of yaws is the same as for syphilis. Late yaws, including the development of gummas, is occasionally seen in patients whose early disease was untreated. Characteristic lesions are destruction and secondary infection of the palate, nasal cartilages and pharyngeal soft tissues (gangosa), tibial periostitis and hyperkeratosis of the palms and soles.

Pinta, caused by *T. carateum*, is a mild disease confined to some areas of the New World. A scaly, red primary lesion at the site of inoculation is followed after some months by a secondary stage of many, similar lesions, which may be discoloured, reddish or brown, and are usually itchy. Late, atrophic, depigmented skin lesions may last for years, and continue to be infectious.

Endemic syphilis tends to affect children. The primary lesion is usually inapparent, and is followed by skin and mucosal lesions similar to those of syphilis or yaws. Fissured papules at the angles of the mouth are typical. A latent phase may be followed by manifestations similar to late yaws.

Chancroid (soft sore)

This is a local, ulcerating genital infection caused by *Haemophilus ducreyi*, a Gram-negative rod uncommon in the UK, but widely prevalent in the West Indies, south-east USA, north Africa, the Middle East, China and parts of the Mediterranean. Up to the 1990s, it caused 20–60% of genital ulcer disease in these areas, but this proportion has dropped to well below 10% as herpes simplex virus type 2 has become more prevalent. In western countries, it is an uncommon imported disease. Its importance is that it is often mistaken for syphilis, which may coexist. It can be transmitted by asymptomatic carriers. Small abrasions are susceptible to infection with chancroid, which, in turn, increases the chance of infection with other sexually transmitted pathogens, including HIV.

After an incubation period of 3–10 days, the typical, soft sore lesion quickly develops. This is a large, irregular, painful ulcer with a necrotic base and undermined edges, almost always on the genitalia. Unilateral, painful lymph node enlargement is common. Suppuration and purulent discharge can occur, producing a large ulcer crater.

The diagnosis is usually made by Gram stain and culture of scrapings from lesions or discharge from lymph nodes. The lesion should be thoroughly cleansed, and a cotton wool swab applied vigorously. The swab should be plated on to isolation medium with the minimum of delay. Mueller–Hinton or enriched gonococcal medium with the addition of charcoal has been shown to improve the isolation rate. A PCR-based diagnostic test is also available. There is considerable research interest in the rapid diagnosis of chancroid by EIA and other techniques but no method has yet gained acceptance in routine use. The ulcers have a characteristic histological appearance, so biopsy is useful in cases of doubt. Other causes of genital ulcer disease, and HIV, should always be excluded, as up to 10% of patients have co-infections.

Treatment recommendations are: azithromycin, orally 1 g single dose *or* ceftriaxone, i.m. 250 mg single dose *or* ciprofloxacin, orally, 500 mg single dose *or* (WHO-recommendation) erythromycin 500 mg to 1 g 6-hourly for 7 days. Some resistance to erythromycin and ciprofloxacin has been reported. Resistance to co-trimoxazole has forced the discontinuation of its use. Sexual partners should always be treated, as asymptomatic infection is common.

Buboes may be treated by needle aspiration from adjacent, healthy skin (or by incision and drainage where surgical facilities are available).

Granuloma inguinale (donovanosis)

Granuloma inguinale is a syndrome of one or more papular and nodular lesions, often developing to friable ulcers on the genitalia, cervix or sometimes the mouth. It is a *Klebsiella granulomatis* infection distributed in India, Vietnam, the south-west Pacific, tropical south America and central Africa. It can lead to scarring and deformity of the genital area, dissemination to distant tissues and bone, and metastatic abscess formation.

Diagnosis is by demonstration of capsulated Gram-negative organisms forming aggregations (Donovan bodies) within large macrophages. This can be done by examining smears or crush preparations of tissue pinched off or biopsied from lesions. They can be seen in histological preparations stained with Giemsa, Wright or Leishman stains. While it may spread by sexual contact, granuloma inguinale is only slightly infectious, and is possibly mainly an autoinfection of faecal origin.

Effective antibiotics include: azithromycin, orally, 1 g weekly or 500 mg daily; ceftriaxone i.m or i.v 1 g daily; co-trimoxazole, orally, 160/800 mg 12-hourly; doxycycline, orally, 100 mg 12-hourly; gentamicin i.m or i.v. 1 mg/kg 8-hourly. Treatment should be continued until all lesions have healed (and for a minimum of 3 weeks). Sexual contacts should be examined and treated if indicated.

Empirical treatment of genital ulcer disease

In tropical and developing countries, genital ulcer disease is common, and usually due to granuloma inguinale or chancroid. Genital ulcer disease is an important risk factor in the acquisition of HIV infection. The World Health Organization therefore encourages empirical treatment of genital ulceration whenever possible. Ciprofloxacin is an effective treatment that is readily available and inexpensive in many countries, and a 7-day course is usually the treatment of choice. Azithromycin is also effective, but is expensive. Co-trimoxazole is a less effective alternative.

Bacterial vaginosis

Bacterial vaginosis is an inflammation of the vagina in

women of childbearing age, associated with overgrowth of predominantly anaerobic organisms, including *Gardnerella vaginalis*, *Prevotella* spp., *Mycoplasma hominis* and *Mobiluncus* spp. These organisms replace the usually present lactobacilli, causing a rise in vaginal pH from its normal acidic level. The infection can arise and remit spontaneously, and is not considered to be sexually transmitted. Bacterial vaginosis is highly prevalent in women with pelvic inflammatory disease. In pregnant women, it is associated with an increased risk of pre-term delivery.

Infection is asymptomatic in about half of cases. In others it produces a slight, whitish discharge with an offensive, fishy odour, which is not associated with soreness or irritation.

Diagnosis is based on the presence of three of four diagnostic criteria (the Amsel criteria: see text box).

The Amsel criteria for diagnosis of bacterial vaginosis

1 Thin, white discharge, coating the vagina and vestibule.
2 Clue cells (epithelial cells heavily coated with bacteria) present in the discharge.
3 Release of a fishy odour on adding 10% potassium hydroxide to the discharge.
4 pH of vaginal fluid greater than 4.5.

 Three of these criteria must be present to confirm the diagnosis.

Isolation of *G. vaginalis* is not diagnostic, as more than half of normal women carry the organism. Treatment options are: metronidazole, orally, either 400–500 mg twice daily for 7 days, or as a single dose of 2 g *or* clindamycin, orally 300 mg 12-hourly for 7 days. Alternatives are: topical metronidazole gel 0.75% daily for 5 days or clindamycin cream 2% daily for 7 days. Screening of sexual partners is not indicated. Treatment may be offered to affected women before termination of pregnancy, or vaginal hysterectomy. It may reduce the likelihood of further preterm births in affected women with a history of this.

Pelvic inflammatory disease (PID)

Introduction

This is a condition of chronic gynaecological symptomatology that causes morbidity and infertility throughout the world. There is evidence of inflammation in intrapelvic organs, and often in vaginal and cervical swab specimens. Occasionally there is extensive oedema, fibrosis or abscess formation, with much distortion of pelvic structures.

Although *Neisseria gonorrhoeae* or *Chlamydia trachomatis* may be responsible, *Mycoplasma hominis*, *M. genitalium*, *Gardnerella vaginalis* or other genital tract organisms may also play a part. A single causative organism is often not identified.

Clinical features

These are abdominal pain, deep dyspareunia, abnormal vaginal bleeding and vaginal discharge. The pain may be suprapubic, radiating to the thighs or referred to the lower back. Many cases are asymptomatic, but may present with infertility. In severe cases there is chronic fever and/or weight loss.

Examination may show lower abdominal tenderness, tenderness, swelling or induration in the vaginal fornices, pain on motion or palpation of the cervix and, sometimes, fever.

Rare cases present with acute perihepatitis (Curtis–Fitz-Hugh syndrome; see above).

Diagnosis

Material from vaginal and endocervical swabs often contains moderate numbers of neutrophils, though this is not diagnostic; absence of pus cells is strong evidence against PID. Elevated inflammatory indices support the diagnosis. Positive tests for gonorrhoea or chlamydial infection support the diagnosis. *Actinomyces* spp. may be associated with intrauterine devices. Ectopic pregnancy, appendicitis, endometriosis and malignancy should be excluded. Urinary tract infection should also be sought.

Pelvic ultrasound scan may show fluid in the pouch of Douglas, distortion of the fallopian tubes or inflammatory masses in the pelvis. In difficult cases, cervical cytology and uterine curettage are useful tests, with histological examination and microbiological tests for bacteria, including mycobacteria (see Chapter 18). Laparoscopy permits direct examination of the pelvic organs, and small biopsies or other sampling may be possible. Actinomycosis, rarely, can produce granulomatous infiltration of the pelvis, with loss of tissue planes.

Management

Early treatment should be given, to prevent complications and reduce the risk of infertility. This is usually empirical, using broad-spectrum antibiotics, based on the sensitivities of likely causative organisms. For mild to moderate disease, outpatient treatment is as effective as inpatient management, which is indicated for severe or complicated cases, or when a surgical emergency is suspected.

Antibiotic treatment of PID
Outpatient treatment
1 Ceftriaxone, i.m. 250 mg single dose *or* cefoxitin, i.m. 2 g single dose; *followed by* doxycycline, orally, 100 mg 12-hourly *plus* metronidazole 400 mg 12-hourly, for 14 days.
2 Ofloxacin, orally, 400 mg 12-hourly *plus* metronidazole, orally, 400 mg 12-hourly for 14 days.

Inpatient treatment
1 Initiate with cefoxitin, i.v. 2 g 8-hourly *plus* doxycycline, orally or i.v. 100 mg 12-hourly; continue when appropriate with: doxycycline, orally 100 mg 12-hourly *plus* metronidazole, orally, 400 mg 12-hourly, to complete 14 days' therapy.
2 Alternative: ofloxacin, i.v. 400 mg 12-hourly *plus* metronidazole, i.v. 500 mg 12-hourly; continued with ofloxacin, orally, 400mg 12-hourly *plus* metronidazole, orally, 400 mg 12-hourly, to complete 14 days' therapy.

⚠ Removal of any intrauterine device should be considered, as this may contribute to cure, but the risk of pregnancy should be balanced against this.

In pelvic actinomycosis, antimicrobial chemotherapy will often reduce the bulk of the infection and restore tissue planes. Prolonged treatment may be needed before surgery can safely be undertaken.

Candida albicans genital infections

Candida albicans is a resident of moist skin and mucosae that readily infects the squamous epithelium of the vagina and vulva or the glans of the penis. Inflammation and maceration of the affected surface are often accompanied by a soft, white exudate, which tends to form plaques or spots. Predispositions include diabetes, antibiotic treatment, hypercalcaemia and disorders of cell-mediated immunity. The genitalia of adult women are protected from candidal infection by an acid pH, associated with colonization by *Lactobacillus acidophilus*. This colonization is lost in pregnancy and after antimicrobial chemotherapy.

The inflammatory symptoms of vulvovaginal candidiasis are often accompanied by a voluminous, cheesy vaginal discharge with little odour. Itching and oedema of the mucosa are common.

The diagnosis is often clinically apparent. Microscopy of the exudate shows many budding yeasts, and extensive formation of pseudohyphae, a mark of aggressive candidal growth. Positive cultures are readily obtained.

Vulval infections respond to topical treatment with nitroimidazoles such as clotrimazole or econazole creams. Vaginal candidosis is better treated with pessaries or vaginal cream. Clotrimazole or econazole pessaries are well tolerated in courses of 1–5 days, depending on the preparation used. Single-dose clotrimazole vaginal cream is also effective. Nystatin pessaries are an alternative, but must be used daily for 2–4 weeks.

Persistent or recurrent infections can be treated with oral triazole drugs. Itraconazole or fluconazole can be given as a 1-day course.

Treatment of vaginal candidiasis
1 Clotrimazole vaginal tablets 500 mg one at night as a single dose, *or* econazole pessaries 150 mg one at night as a single dose, *or* clotrimazole 10% vaginal cream 5 g at night as a single dose.
2 Alternative: nystatin pessaries one or two at night for 14–28 days.
3 Systemic treatment: itraconazole orally 200 mg twice daily for 1 day *or* fluconazole, single 150 mg dose.
4 Additional treatment for vulval involvement: clotrimazole 1% cream *or* econazole 1% cream twice daily; *or* nystatin cream to anogenital area three or four times daily.

Trichomonas vaginalis

Trichomonas vaginalis is a flagellate protozoan that thrives in the low oxygen tension of the vagina in women, and in the urethra and paraurethral glands of both sexes. In women it produces symptoms of irritating vulvitis or vaginitis, serous or mucoid vaginal discharge, dysuria and, sometimes, an offensive odour. Men are often asymptomatic, and present as partners of affected women, but they may suffer from urethritis with variable discharge. Prostatitis is a rare complication. There is accumulating evidence that infection in women is associated with pre-term delivery and low birth weight.

Diagnosis in women can often be made by examination of direct, or acridine-orange-stained wet preparations of material from the posterior fornix. Culture of material in *Trichomonas* medium will diagnose about 95% of cases. *T. vaginalis* can be seen on cervical smears, but should be confirmed by other methods, as false-positives are relatively common.

In men, wet preparations give poor diagnostic results. Cultures from urethral swabs and/or first-voided urine will confirm about 80% of cases.

Treatment is with oral metronidazole 400 mg 8-hourly for 1 week *or* metronidazole 2 g single dose *or* tinidazole 2 g single dose.

Other infections transmissible during sexual contact

Sexual contact usually includes prolonged skin-to-skin contact, as well as significant exchange of body fluids, particularly saliva and genital secretions. Minor trauma or small skin lesions can permit exchange of blood, or inoculation of blood on to a partner's mucosae. Extensive anal and rectal contact, or contact with urine droplets increase the likelihood of transmission of several infections not usually spread by sexual contact.

Infections spread by skin contact

- Scabies
- Pubic lice
- Tinea cruris
- Impetigo

Infections spread by body fluids

- Hepatitis B
- Cytomegalovirus
- Epstein–Barr virus

Infections spread by blood

- Hepatitis B
- HIV (see Chapter 16).

Infections spread by faeces

- Shigellosis
- Giardiasis
- Amoebiasis

HIV infection and AIDS

Introduction

The human immunodeficiency viruses HIV-1 and HIV-2 cause lifelong infection of many tissues, but particularly of CD4 helper lymphocytes. Untreated individuals develop profound helper-cell deficiency and associated immunodeficiency over a period of several years (acquired immunodeficiency syndrome, AIDS) and die of uncontrolled opportunistic or HIV disease. HIV-1 is distributed worldwide and produces more severe and rapidly progressive disease than the less aggressive HIV-2, which is mostly found in West Africa. HIV is spread by sexual, blood-borne and vertical transmission, and is capable of affecting whole communities, profoundly disrupting social structure and economic activity. HIV viraemia can be controlled, and the development of AIDS prevented, by multidrug antiviral therapy (highly active antiretroviral treatment, or HAART) but this requires a substantial investment in drugs and medical services, which cannot be sustained by some poorer countries.

Epidemiology

AIDS was first recognized in the early 1980s among homosexual men in the USA. The number of cases has risen exponentially, and in 2004 there were 39 million people living with HIV worldwide (Fig. 16.1). Five million of these were newly infected that year, in which 3 million people died from AIDS. Around two-thirds of all HIV infections have occurred in Africa (three-quarters of global HIV infections in women), but in recent years the steepest increases in case numbers have been in East Asia, eastern Europe and Central Asia.

Two major patterns of transmission are recognized. In Africa and some other low-income nations, transmission is mainly through heterosexual intercourse and by verti-cal transmission to infants from infected mothers. Genital ulcer diseases (see Chapter 15) increase the risk of transmission of HIV. In high-income countries, most cases in the early years of the pandemic were acquired through sexual intercourse between men, and sharing of contaminated drug-injecting equipment. In the UK, however, since 1999 the number of new HIV diagnoses in heterosexuals has exceeded the number for homosexual men, though homosexual men remain the group with the highest percentage risk of acquiring infection. A similar trend is seen in other countries. Most of the heterosexually transmitted infections in 2003 were probably acquired in Africa, but indigenous heterosexual transmission is increasing. Receipt of contaminated blood and blood products was an important route of spread, until most countries began to screen blood donations.

Virology and pathogenesis

HIV-1 is an enveloped RNA virus belonging to the lentivirus subfamily of Retroviridae. HIV-2 is a closely related strain that has been identified from patients in West Africa. The simian immunodeficiency virus of chimpanzees is the precursor of HIV-1. In common with other RNA viruses, HIV has a relatively unstable genome, giving rise to numbers of closely related strains or 'clades' of virus, designated a to f. These strains form groups that have different distributions in the world, and slightly different characteristics of infectiousness.

The genetic information of HIV-1 and HIV-2 is found on a single strand of RNA. The replication strategy is to transcribe infectious RNA into linear double-stranded DNA, which is integrated into the host-cell genome. The virus life cycle therefore has two phases: the DNA provirus, within the host-cell, and the infectious RNA virion. Incorporation into the host chromosome, resulting in long-term latency, provides a major survival advantage.

The genome is typically 10 kb, with three structural genes encoding a number of proteins essential for viral

GLOBAL ESTIMATES FOR ADULTS AND CHILDREN, END **2004**

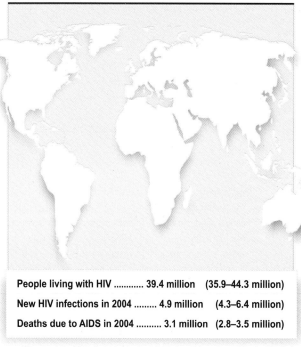

People living with HIV .... 39.4 million   (35.9–44.3 million)

New HIV infections in 2004 .... 4.9 million   (4.3–6.4 million)

Deaths due to AIDS in 2004 .... 3.1 million   (2.8–3.5 million)

The ranges around the estimates in this table define the boundaries within which the actual numbers lie, based on the best available information.

ADULTS AND CHILDREN ESTIMATED TO BE LIVING WITH **HIV** AS OF END **2004**

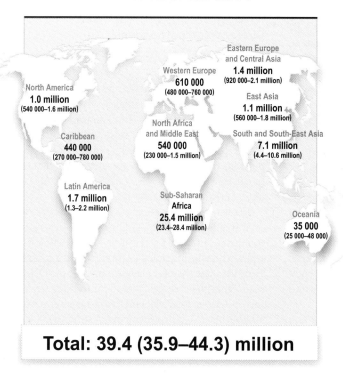

Western Europe
610 000
(480 000–760 000)

Eastern Europe
and Central Asia
1.4 million
(920 000–2.1 million)

North America
1.0 million
(540 000–1.6 million)

East Asia
1.1 million
(560 000–1.8 million)

Caribbean
440 000
(270 000–780 000)

North Africa
and Middle East
540 000
(230 000–1.5 million)

South and South-East Asia
7.1 million
(4.4–10.6 million)

Latin America
1.7 million
(1.3–2.2 million)

Sub-Saharan
Africa
25.4 million
(23.4–28.4 million)

Oceania
35 000
(25 000–48 000)

Total: 39.4 (35.9–44.3) million

Figure 16.1 Maps showing the global impact of HIV infections and acquired immunodeficiency syndrome (AIDS), as of 2004. Reproduced by kind permission of UNAIDS (www.unaids.org).

Estimated number of adults and children newly infected with HIV during 2004

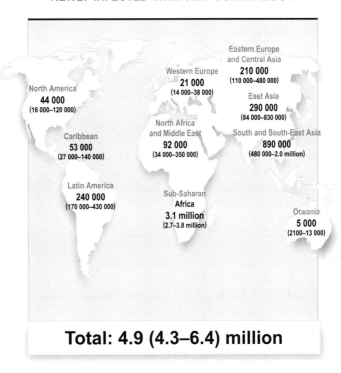

Total: 4.9 (4.3–6.4) million

Estimated adult and child deaths from AIDS during 2004

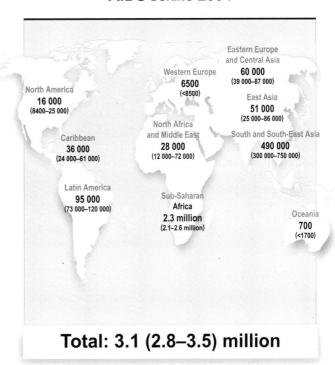

Total: 3.1 (2.8–3.5) million

Figure 16.1 (*Continued*).

replication. These include, among others, *pol*, which encodes reverse transcriptase, the enzyme that generates complementary DNA from viral RNA. The *gag* gene encodes a protein precursor that is cleaved to form the p24 core protein, detectable in early infection. The *env* gene codes for a 160-kĎa precursor of the envelope glycoprotein that permits HIV attachment to the CD4 receptor of target cells.

To enter cells HIV binds to CD4 receptors and a second receptor: CCR5 in macrophages and CXCR4 for T cells. A mutation in CCR5 protects against HIV infection. Reverse transcription occurs in the cytoplasm and double-stranded DNA then enters the nucleus. Viral integration is not site-specific but occurs preferentially near to active genes. Synthesis of new viral RNA is accomplished by host enzymes operating in a tightly regulated fashion. Virus is found in CD4-positive T cells and in tissues rich in CD4-positive T cells, macrophages and follicular dendritic cells.

The half-life of virus-producing T cells is just less than a day and the generation time of HIV-1 is approximately 2 days. HIV-infected macrophages are less able to respond to stimuli and do not proliferate as efficiently. Damage to immune responses results in an increase in viral turnover and immune cell depletion and eventual immunodeficiency. HIV infection of various cell-lines leads to impaired haemopoiesis, impaired thymopoiesis, and apoptosis of uninfected immune cells.

Viral enzymes are important targets for antiretroviral drugs. Protease is necessary for the assembly of infectious virus particles. Reverse transcriptase is an RNA-dependent DNA polymerase that can synthesise DNA copies from RNA or DNA. Frequent mis-translations of the *pol* gene produce multiple mutants within each patient. These numerous 'quasi-species' allow the rapid emergence of resistance to antiretroviral drugs, and also make vaccine design difficult. Integrase mediates integration of viral double-stranded DNA into the host genome (Fig. 16.2).

Clinical features

HIV infection evolves through three phases:
• seroconversion, during which HIV viraemia disperses the virus to its target cells and serum antibodies to HIV develop;
• latent (asymptomatic) infection, during which virus replication destroys CD4 and other cells until immunity deteriorates;
• symptomatic infection, when cell-mediated immune responses have declined to a level at which increasingly serious opportunistic diseases occur, fulfilling the clinical definition of AIDS.

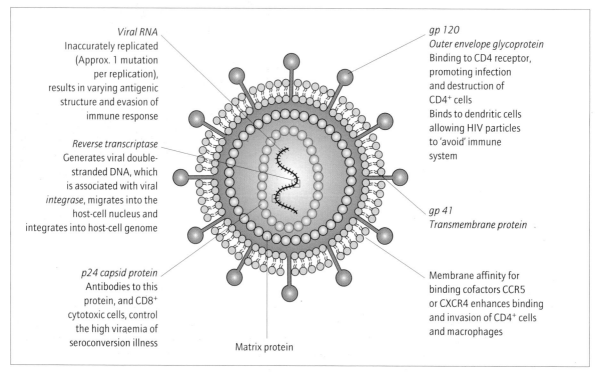

Figure 16.2 Structure and pathogenicity of HIV virus.

Seroconversion

Seroconversion usually occurs 6–8 weeks after infection, but can range from 4 to 12 weeks. At least 75–80% of infected patients are feverish during seroconversion, and about 60% have a rash. Clinical examination reveals generalized, moderate enlargement of lymph nodes, which are firm, non-tender and mobile. Atypical mononuclear cells are seen in the blood, suggesting acute viral infection. Rashes may be macular, acne-like or display ovoid and papular lesions, characteristically on the upper chest and back (Fig. 16.3). Lesions may appear rough, but are not painful or itchy. Oral lesions and genital ulcers are commonly seen. A few patients present with lymphocytic meningitis. Most illnesses are mild, but all tend to be prolonged, lasting from 2 to 6 weeks. They resolve without specific treatment.

During seroconversion there is a high viraemia, with a profound fall in the CD4 lymphocyte count and significant immunosuppression. Some patients present with an opportunistic infection, such as candidal oesophagitis or *Pneumocystis jiroveci* pneumonia (PCP). These respond to specific treatment and do not recur until after the latent phase. Pre-existing tuberculosis may advance, or recrudesce, during this stage, acting as an early marker of HIV disease.

The low CD4 count is accompanied by a raised cytotoxic CD8 T lymphocyte count, which contributes to termination of the viraemia. However, the CD4 count never quite returns to its original level, and a variable, low viraemia (< 1000 genomes/ml; the 'set point') may persist throughout the following latent phase. A severe seroconversion illness and a high set point predict early progression to symptomatic disease.

Latent infection

Untreated, the latent phase of HIV infection lasts from 18 months to 15 years or more, averaging about 8 years. For much of this time the individual is well, not unduly susceptible to infections and recovers apparently normally from common and seasonal infections. The total CD4 helper-cell population slowly declines, and CD4 helper function is increasingly impaired. Intercurrent infections probably speed this decline, as activated lymphocytes are more readily invaded by HIV virus.

The viral load in the blood remains relatively constant at around the 'set point', usually below 1000 genome copies per ml. The lower the set point, the longer the asymptomatic period is likely to last. Patients with prolonged latent periods (long-term non-progressors), have been studied to identify good prognostic factors or factors that inhibit progression. They tend to have sustained, strong antiviral CD8 lymphocyte responses, and some lack the co-ligand chemokine CCR5, which aids the entry of HIV

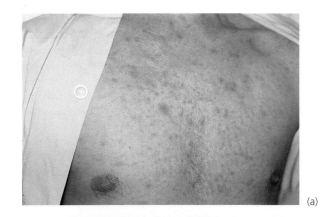

(a)

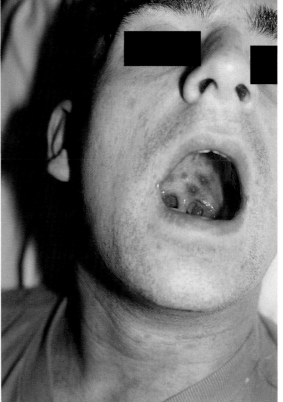

(b)

Figure 16.3 (a) Rash of large erythematous lesions and (b) oral lesions in human immunodeficiency virus (HIV) seroconversion illness.

into CD4-positive cells. Their virus does not evolve from non-syncytium-inducing (NSI) to syncytium-inducing (SI) forms, typical of isolates from late, symptomatic cases. However, even these patients progress slowly until increasing viraemia and immunosuppression lead to symptomatic disease.

Symptomatic HIV infection and acquired immunodeficiency syndrome (AIDS)

An increasing decline of CD4 cell numbers and increasing viraemia herald the end of the latent phase. Before they become truly symptomatic many patients develop generalized lymphadenopathy. Rubbery, mobile nodes from 1 cm upwards in diameter are easily palpable. These have a characteristic histological appearance of reactive histiocytosis. The lymphadenopathy persists through the symptomatic stages.

Infections that are common in debilitated and mildly immunosuppressed individuals tend to occur at about this time. Herpes zoster is common and can be severe, recurrent or multidermatomal, sometimes leading to scarring of the skin or eye. Prompt antiviral treatment is therefore indicated. Bacterial infections also occur, including pneumococcal and *Salmonella* infections. Pneumococcal pneumonia is often bacteraemic, but with a misleadingly low neutrophil and total white cell count. It responds to vigorous antimicrobial treatment (see Chapter 7). *Salmonella* infections may also be bacteraemic, and repeated recurrences of both diarrhoea and bacteraemia may occur despite adequate treatment. Tuberculosis may become apparent, or advance at this stage. Skin and mucosal infections become increasingly common; these include seborrhoeic dermatitis, molluscum contagiosum, relapsing herpes simplex infections, and oral and genital candidiasis.

As the CD4 cell count falls, and the circulating viral load rises, opportunistic conditions become increasingly common, and are caused by organisms that are less and less pathogenic in normal hosts (Fig. 16.4). These include malignancies, such as Kaposi's sarcoma, lymphomas or invasive carcinoma of the cervix, which have an infectious aetiology – for instance human herpesvirus type 8 (HHV-8) in Kaposi's sarcoma, human papillomaviruses (HPV 16 and 18) in carcinoma of the cervix, and Epstein–Barr virus (EBV) in some lymphomas. They represent a defect of immune control of latent, potentially oncogenic infections.

Many opportunistic conditions are characteristic of HIV-related cell-mediated immunodeficiency, so that the disease can be recognized, even without virological confirmation (Table 16.1). This is the clinical condition of acquired immunodeficiency syndrome (AIDS). With high-quality diagnosis, and the availability of antiretroviral therapy, the need for clinical diagnosis is now rare, and the stage of progression is usually deduced from the CD4 count and viral load. There is, however, no firm relationship between CD4 counts and viral load; both are independent predictors of the development of such conditions.

HIV disease in infants and children

There are several differences between adults and children in the presentation of HIV disease and the apparent degree of CD4 cell depletion. The CD4 cell count is higher in infants and young children than in adults. Adult reference values are not reliable indicators either of CD4 cell loss or of immunosuppression in children. Adults possess

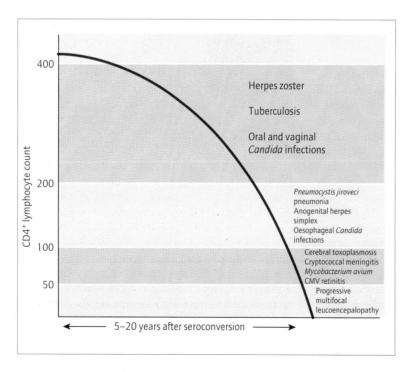

Figure 16.4 The association of opportunistic infections with approximate CD4 count.

Table 16.1 Conditions defining a clinical diagnosis of acquired immunodeficiency syndrome (AIDS). In the USA a total CD4 lymphocyte count below $0.2 \times 10^9/l$ is also considered diagnostic

Disease
Bacterial infections; multiple or recurrent (child < 13 years)
Candidiasis (trachea, bronchi or lungs)
Candidiasis (oesophagus)
Cervical carcinoma (invasive)
Coccidioidomycosis (disseminated or extrapulmonary)
Cryptococcosis (extrapulmonary)
Cytomegalovirus retinitis
Encephalopathy (milestone loss with no other cause in children)
Herpes simplex; ulcers for >1 month, or bronchial, lung or oesophageal lesions
Histoplasmosis (disseminated or extrapulmonary)
Isosporiasis (diarrhoea >1 month)
Kaposi's sarcoma
Lymphoid interstitial pneumonia (child <13 years)
Lymphoma (Burkitt's, immunoblastic, cerebral)
Mycobacteriosis (disseminated, including extrapulmonary TB)
Mycobacteriosis: (pulmonary TB)
Pneumocystis jiroveci pneumonia
Pneumonia (recurrent within 1 year)
Progressive multifocal encephalopathy
Salmonella (non-typhoid) septicaemia, or recurrent
Toxoplasmosis (cerebral: onset >1 month old)
Wasting syndrome: weight loss (>10% of baseline with >30 days' fever or diarrhoea; no other cause

TB, tuberculosis.

humoral immunity to a wide range of pathogens encountered before the onset of HIV infection. Children infected with HIV perinatally, or in early childhood, do not develop effective antibody responses, due to reduced helper-cell function. Furthermore, in those under 2 years of age, T-cell-independent antigens are ineffective immunogens. HIV-infected children suffer repeated bacterial infections such as pneumonias, gastroenteritis, and skin and upper respiratory infections. Some viral infections cannot be terminated or controlled in a latent state. Measles, chickenpox and especially EBV infection can cause severe disease and persisting fever. EBV infection causes progressive pulmonary lymphoid hyperplasia (PLH) or lymphoid interstitial pneumonitis (LIP).

These problems can be reduced by immunization, immunoglobulin prophylaxis and early treatment of feverish illnesses. Measles vaccine should be given to HIV-infected children, whether they have HIV-related symptoms or not, because the risk from measles is much greater than that from the live attenuated vaccine.

Diagnosis and staging of HIV infection

A diagnosis of HIV commits the patient and health services to long-term healthcare investment. Diagnostic tests must therefore be both sensitive and highly specific. The main tests in current use are antibody tests, and PCR-based detection of viral genome.

A range of formats can be used for antibody detection, including antibody capture enzyme immunosorbent assay (EIA), competitive EIA and rapid agglutination tests. These may incorporate HIV-1 and HIV-2 peptides or recombinant antigens. All indeterminate and positive results must be verified by alternative methods. Discrepant results and suspected HIV-2 infections should be confirmed by a reference laboratory. For a short period (the window period) immediately after infection, patients may test HIV antibody-negative. Therefore, after a recent exposure, a negative result cannot be confirmed by antibody testing until a further negative result is obtained after a 3-month interval. PCR-based tests for viral RNA can provide earlier confirmation of infection in this situation.

Polymerase chain reaction (PCR) techniques are used to demonstrate viral RNA in the blood in adults and in neonates possessing antibody that may be of maternal origin. Quantitative PCR tests permit viral load measurement, used in follow-up of cases. Enzyme-linked immunosorbent assay (ELISA) tests for antibodies to HIV antigens are very sensitive and become positive soon after PCR-based tests, with antibody to the p24 antigen appearing first.

In the Western blot test, HIV antigens are separated by sodium dodecyl sulphate–polyacrylamide gel electrophoresis (SDS-PAGE) and blotted on to nitrocellulose. Antibodies in the patient's serum bind to the different antigens. A positive result is indicated if two or more of the bands p24, gp41 and gp120/160 are positive. During seroconversion, antibody bands appear sequentially on successive Western blots, allowing distinction between early and late seroconversion illness. This is useful in research on therapy.

Diagnosis of HIV infection

1 Screening test for HIV gp41 antibodies.
2 Confirmatory test using a different technique.
3 PCR-based test for HIV viral RNA (useful in suspected perinatal transmission when the infant may have passive antibody in the blood, and for early detection of recent infection).
4 Test for p24 antibodies in suspected early disease.

The progress of HIV disease can now be closely monitored by a combination of viral load measurement, CD4 lymphocyte counts and the patient's clinical status (Fig. 16.5). These are used to guide the initiation of treatment, monitor the effectiveness of the regimen chosen and as a warning of developing drug resistance. Most patients have these tests approximately 6-monthly when their disease is stable and more often after initiation or change of treatment.

Management

In communities where it is available, the effectiveness of HAART has dramatically reduced the progression of HIV infection to AIDS. Opportunistic infections are uncommon except in some cases during seroconversion, late-presenting cases, or immigrants from areas without access to antiretroviral drugs. Many HIV-infected individuals can hope to lead a normal life. However, HAART is a life-long commitment; adverse effects of various drugs can be severe, and drug interactions between antiretrovirals and other commonly used drugs can cause significant problems. Resistance to individual drugs, or classes of drugs, inevitably accumulates, particularly if adherence to effective therapy is unreliable. The search therefore continues for drugs with novel modes of action, and for both prophylactic and therapeutic vaccines.

Antiretroviral drugs
Nucleoside-related reverse transcriptase inhibitors (NRTIs)
These include zidovudine (AZT, ZDV), lamivudine (3TC), stavudine (D4T), abacavir (ABC), zalcitabine (DDC) and emtricitabine. Useful combinations of NRTIs are Combivir (zidovudine plus lamivudine, for twice-daily dosage) and Trizivir (abacavir plus zidovudine plus lamivudine for twice-daily dosage). Tenofovir disoproxil is a nucleotide (i.e. phosphorylated nucleoside) reverse transcriptase inhibitor (see also Chapter 4).

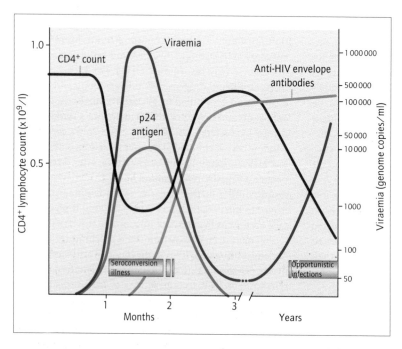

Figure 16.5 The progression of antigen (Ag) and antibody appearance in the blood of a person infected with human immunodeficiency virus (HIV).

Adverse effects include nausea (often diminishing with continued therapy), vomiting, diarrhoea, and a varying incidence of peripheral neuropathy, pancreatitis, hepatic disorder and reduced red or white blood cell counts. Hypersensitivity reactions are common with abacavir, and can be life-threatening. Lactic acidosis is a rare, late reaction to NRTIs, caused by damage to mitochondrial DNA synthesis.

Viral protease inhibitors (PIs)

These include atazanavir, amprenavir, saquinavir (best tolerated), ritonavir, indinavir (most potent) and nelfinavir. All interact with the cytochrome CYP450, particularly the 3A4 system (of which ritonavir is a powerful inhibitor), causing adverse reactions with many drugs, including each other. They are often given in combination to exploit these pharmacokinetic interactions, providing optimum blood levels, while limiting the toxicity of one or both drugs by minimizing dosages. The fixed-dose combination of lopinavir (not suitable for sole use) plus ritonavir (Kaletra) is often used, or saquinavir may be given with low-dose ritonavir. Rifampin, an enzyme inducer, significantly reduces blood levels of PIs, and should be avoided. Rifabutin dosage must be halved if given with most PIs; its ocular toxicity is particularly enhanced by ritonavir. Adverse interactions may also occur with macrolides, azole antifungal drugs, some anti-arrhythmic drugs and some antihistamines. Saquinavir enhances the risk of muscle disorder when given with simvastatin, but other statins may be given. Atazanavir should not be combined with indinavir or nevirapine. Indinavir is relatively insoluble, so that extra fluid intake is necessary to prevent crystalluria and renal tubular damage.

Long-term use of PIs is associated with raised cholesterol and triglyceride levels, decreased glucose tolerance, hyperuricaemia and lipodystrophy with typical 'buffalo hump'. Statin drugs can ameliorate some of these problems.

Non-nucleoside reverse transcriptase inhibitors (NNRTIs)

Nevirapine is the longest-used NNRTI. Rash, including Stevens–Johnson syndrome, limits its use in about 20% of patients. Efavirenz has a low toxicity profile, but reduces the plasma levels of a number of drugs, particularly atazanavir and saquinavir. Its liver toxicity is enhanced by ritonavir. The dose of rifabutin must be increased when the two drugs are co-administered. Delavirdine is licensed for use in some countries. NNRTIs are effective in combination with NRTIs, and efavirenz is often included in initial therapy.

Other drugs

Enfuvirtide inhibits the fusion of HIV to target cells. It is licensed for use in patients who cannot tolerate other drugs, or whose virus is resistant to other drugs.

Principles of antiretroviral therapy (highly active antiretroviral therapy, HAART)

The aims of antiretroviral therapy are: to minimize viral replication (and therefore the rate of viral mutation to resistance); to reduce infectiousness; to prevent the progress of immunosuppression; and where possible to allow reconstitution of lymphocyte counts and immunological function.

Aims of HAART

1 To minimize viral replication.
2 To minimize the emergence of resistance to antivirals.
3 To reduce infectiousness.
4 To halt the progress of immunological damage.
5 To permit reconstitution of immune function.

Commencing antiretroviral therapy

When to commence treatment and prophylaxis for HIV

For seroconversion illness

The high viraemia and infectiousness at this stage might make treatment theoretically desirable, even though the condition is self-limiting. It is thought that about 70% of sexually transmitted HIV infections result from contact with a seroconverter. However, the risks of toxicity or the emergence of resistance could outweigh any potential advantage in limiting viral replication. A rebound of viral activity always follows discontinuation, but may settle at a low level. Adverse effects and difficult compliance make it unlikely that lifelong treatment will be practicable. Hopes that viral eradication could be achieved by early treatment have not been fulfilled.

For established HIV disease

Opinions differ slightly over the balance between viral suppression and the long-term problems of therapy. Most experts would begin HAART in asymptomatic patients at viral loads between 50 000 and 100 000 copies per ml or at CD4 cell counts of $0.3–0.5 \times 10^9/l$. Patients developing HIV-related symptoms or opportunistic infections should always be treated.

HAART regimens

Previously untreated patients usually respond to most types of drugs, but all will possess small populations of virus with resistance to one or more drugs or drug classes. These must be reduced, rather than selected for, by antiviral therapy. In some countries, especially where treatment modalities and options are limited, up to 10% of patients have primary resistance to one type of drug. Multidrug therapy with good treatment compliance and close follow-up of response is important in treating the pandemic, as well as the individual.

Multidrug therapy with at least three drugs is the ideal. Two of the drugs are usually NRTIs; an effective and well-tolerated combination for initial therapy is zidovudine plus lamivudine. The third drug may be a PI (often lopinavir with ritonavir), or an NNRTI such as efavirenz. A fall to undetectable HIV plasma levels should be achieved in 4 to 6 months in previously untreated patients. With continued treatment viral levels may fall in less accessible 'sentinel' sites such as the genital tract and central nervous system. The longest-surviving infected cells are the 'memory' T cells, with a life span of many months.

Continuing therapy then allows gradual recovery of CD4 cell counts, with reconstitution of immune function, though, particularly after profound immune depletion, this may never fully recover. Reactivation of immune responses may cause inflammatory responses to existing opportunistic infections. 'Immune reconstitution diseases' then occur, such as increased fevers, lymphadenopathy in mycobacterial infections, increased respiratory symptoms in pulmonary tuberculosis, or increased ocular inflammation in retinal cytomegalovirus disease. These conditions can be treated specifically, and the inflammatory response controlled with step-down corticosteroid therapy.

While primary prophylaxis may be discontinued soon after commencing HAART, secondary prophylaxis and treatment for episodic opportunistic infections must be continued until CD4 counts are in the range of 0.4–0.5×10^9/l. Abolition of cytomegalovirus viraemia and *Mycobacterium avium-intracellulare* (MAI) infection usually occurs at this stage, and the risk of *Pneumocystis jiroveci* pneumonia is negligible. The use of HAART resulted in a five-fold fall in the incidence of clinical AIDS in Europe in 1998, compared with numbers in 1993.

Emergence of resistance to therapy

RNA viruses have an unstable genome because they lack an RNA 'proof-reading' mechanism during replication. Thus, many genetic variants ('quasi-species', or 'clades') exist in a virus population. Variants may be as successful as the parent virus, or less 'fit' if they replicate or infect less well. Some mutations yield a replicative advantage in the presence of immunological or therapeutic pressure. Around 10 billion HIV virus particles are produced daily, undergoing one mutation in 9200 nucleotides. Thus, every possible single drug-resistant mutant is likely to be generated every day, but multiple changes are increasingly less likely. Resistant viruses can be detected in patients not previously exposed to treatment, but almost all will be suppressed by multidrug therapy. A single mutation can confer resistance to lamivudine and certain NNRTIs. When these drugs are given in partially suppressive regimens, resistant quasi-species will emerge within weeks. For zidovudine and certain protease inhibitors, three or more mutations

must occur to achieve resistance. Agents that require multiple mutations to resistance should therefore be included in multidrug regimens.

Even with multidrug therapy, effectiveness is eventually limited by emerging viral resistance, and a change (switch) to alternative drugs is indicated, but a long duration of symptom-free life and avoidance of opportunistic disease is possible. Prolonged therapy is exacting for the patient because of demanding dosing regimens, the need to schedule drugs relative to meals, to drink extra fluids and to avoid interactions with other drugs. Multidrug formulations are increasingly available to minimize these problems.

Examples of antiretroviral regimens

For initiating treatment

Zidovudine *plus* lamivudine *plus either* lopinavir/ritonavir *or* efavirenz.

Alternative

Abacavir *plus* zidovudine *plus* lamivudine (combined tablet after 6–8 weeks stabilization on individual drugs) *plus either* saquinavir (with or without low-dose ritonavir) *or* efavirenz.

Switch therapy

Tenofovir *plus* zalcitabine *or* abacavir *plus either* amprenavir *or* atazanavir.

Salvage of late increase in viral load

Regimen based on sensitivity testing and previous drug experience, possibly including envufirtide.

⚠ **UK and USA recommendation for prophylaxis after percutaneous exposure**

Zidovudine *plus* lamivudine *plus* indinavir for 4 weeks.

Switch of therapy

Switch of therapy is needed if primary therapy is not effective, for drug intolerance or adverse effect, and for relapse of previously controlled infection. Many experts would consider switching therapy if viral load increases by 1–1.5 logs or rises from undetectable levels to 5–10 000 genomes per ml. Compliance must always be reviewed before a decision to switch therapy.

Prophylaxis after percutaneous exposure

The risk of HIV infection after percutaneous exposure to infected blood is about 0.3%, increased five- to 16-fold in deep injuries, with hollow needles and with high 'donor' viral loads. Zidovudine monotherapy offers about 80% protection (as does nevirapine in chimpanzees exposed to HIV-1). Individuals with recognized percutaneous exposure are nowadays offered triple therapy with zidovudine plus lamivudine plus indinavir for a period of 4 weeks. This should commence as soon as possible after the event, preferably within an hour or two. Pre-treatment blood samples should be taken for HIV testing. Consent should be obtained when possible for testing of the 'donor' for HIV and other blood-borne infections. Further tests should be performed at 6 weeks, 3 months and 6 months post-exposure. Readily accessible occupational advice, consultation 'hotlines' and pre-prepared treatment packs are extremely useful. Support is required to minimize anxiety and to manage nausea and other adverse events.

Prophylaxis for pregnancy in an HIV-positive woman
Perinatal transmission from an HIV-positive woman can be minimized by safe delivery techniques, antiretroviral therapy for the woman and her newborn, and avoidance of breast-feeding (see Chapter 17).

Prophylaxis after sexual exposure to HIV
There is no consensus on the efficacy or practicality of prophylaxis after sexual exposure to HIV, but it is offered to victims of sexual assault.

Vaccines against HIV
A number of potential vaccines have been developed, which induce both humoral and cellular immune responses against HIV virus or its antigens. No significant protective efficacy has so far been demonstrated. Large trials of therapeutic vaccination have been aimed at inhibiting the replication of virus, or enhancing the natural immune response and sparing the use of antivirals. So far the results have been disappointing.

Prevention and treatment of opportunistic diseases
Treatment of opportunistic conditions is important when immune competence is reduced during seroconversion or late HIV disease. Primary and secondary prophylaxis are important as immunocompetence declines in end-stage HIV disease. Immunodeficient patients should be rigorously investigated and treated for opportunistic infections, to maintain their health, and to prevent accelerated damage to the immune system (for instance, by CMV infection). Impaired inflammatory responses make the presentation of infections insidious and atypical; investigations such as bronchoalveolar lavage or brush biopsy are necessary to test for *Pneumocystis*, mycobacterial, herpesvirus and fungal infections, all of which may present similarly, or coexist (Fig. 16.6). Skin or lymph-node biopsy is also useful in the diagnosis of poxvirus, mycobacterial and fungal infections, and in distinguishing between lymphoma, Kaposi's sarcoma and bacillary angiomatosis (due to *Bartonella henselae* infection). Syphilis may present with alopecia, neurological features or indolent rashes, but can often be diagnosed by standard methods. If syphilis is locally prevalent, 3-monthly serological testing may be advisable in sexually active individuals.

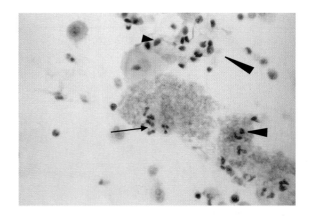

Figure 16.6 Material from bronchoalveolar lavage, examined with a Papanicolaou-type stain, showing dark blue, oval bodies of *Pneumocystis jiroveci* (arrowheads) and a pale blue foamy macrophage; a group of small, Gram-positive diplococci is also present (arrow), probably *S. pneumoniae.*

Prophylaxis is usually commenced after a first episode of an opportunistic infection (secondary prophylaxis). Primary prophylaxis is rarely offered as it might encourage infection with resistant organisms in, for instance, mucosal candidiasis. However, PCP is so common and serious that patients are offered primary prophylaxis when their CD4 count is consistently below 0.3×10^9/l. In countries where BCG immunization is not used, skin test-positive individuals may be offered isoniazid prophylaxis when their CD4 count falls below 0.4×10^9/l (as there is a 10% per year incidence of tuberculosis in such cases).

Common regimens of prophylaxis include co-trimoxazole or nebulized pentamidine for PCP, aciclovir for herpes simplex and valganciclovir to suppress cytomegalovirus retinitis (Table 16.2). Children often receive regular intravenous immunoglobulin to prevent recurrent pyogenic infections. Rare infections have occurred with aciclovir-resistant herpes simplex and fluconazole-resistant fungi.

Prevention and control

The most effective control measures are the avoidance of high-risk sexual activity and the use of condoms. Despite extensive educational campaigns, the spread of HIV infection continues, particularly among heterosexuals who either do not perceive themselves at risk, or do not use preventive measures for complex social or psychological reasons. Treatment of genital ulcer disease may reduce the risk of HIV acquisition. Other preventive measures include screening of blood donations, needle exchange schemes for intravenous drug abusers and antenatal screening, which is now routinely recommended for pregnant women in the UK. Individuals in high-risk categories

Table 16.2 Prophylaxis and treatment of some opportunistic conditions in acquired immunodeficiency syndrome

Opportunistic infection	Prophylaxis	Treatment
Pneumocystis jiroveci pneumonia	**1** Co-trimoxazole orally 960 mg 12-hourly (rash may compel use of alternative) **2** Pentamidine isethionate by nebulized inhaler: 300 mg every 2 weeks or 600 mg every 4 weeks	**1** Co-trimoxazole 120 mg/kg daily in divided doses for 14 days (may be given i.v. 1.44 g 12-hourly) **2** Atovaquone orally 750 mg 8-hourly for 21 days **3** Pentamidine isethionate i.v. 4 mg/kg daily for 14 days **4** Pentamidine isethionate by nebulizer 150 mg daily for 21 days
Cytomegalovirus retinitis (suppression)	**1** Ganciclovir by i.v. infusion 5 mg/kg daily or 6 mg/kg on 5 days per week (check blood count frequently) **2** Oral ganciclovir 1 g 8-hourly (2 g 8-hourly after a recrudescence) **3** Oral valganciclovir 900 mg daily	**1** Ganciclovir by i.v. infusion (over 1 h) 5 mg/kg 12-hourly for 14–21 days **2** Foscarnet by i.v. infusion 20 mg/kg over 30 min, then 20–200 mg/kg daily (according to renal function) for 14–21 days (check blood count, liver and renal function)
Cryptococcal meningitis	Fluconazole orally 100–200 mg daily	**1** Amphotericin up to 1 mg/kg daily by i.v. infusion (depending on renal function, plasma potassium, blood count, febrile reaction) for 8–12 weeks **2** Liposomal amphotericin up to 6 mg/kg daily if amphotericin fails or is intolerable **3** Fluconazole i.v. or orally 400 mg, then 200–400 mg daily for 8–12 weeks
Toxoplasmic encephalitis	Pyrimethamine-sulfadoxine, one tablet weekly	Sulfadiazine 1.0 g three times daily plus pyrimethamine 50 mg daily for several weeks with repeated CT scans to confirm response
Mucocutaneous *Candida*	Not usually given due to risk of resistant yeasts emerging	**1** Amphotericin orally (suspension) 200 mg 6-hourly **2** Fluconazole 50 mg daily orally for 7–30 days **3** Itraconazole 200 mg daily orally for 15 days
Mycobacterium avium-intracellulare	**1** Rifabutin 300 mg/day orally **2** Azithromycin 1200 mg weekly	See Chapters 18 and 22
Herpes simplex (mucocutaneous)	Aciclovir 200 mg 6-hourly orally	Aciclovir orally 200–400 mg five times daily; i.v. 5 mg/kg 8-hourly (reduced in renal impairment: check blood urea)
Molluscum contagiosum		Cidofovir 10% ointment topically

should be encouraged to be tested for HIV, so that they know their diagnosis, obtain treatment and participate in prevention and control measures.

Strategies for the prevention of human immunodeficiency virus (HIV) infection

1 Safe sex: condom use.
2 Screening of blood products.
3 Needle exchange schemes.
4 Antenatal screening.
5 Voluntary testing of those in high-risk categories.
6 HAART treatment of known cases.
7 Treatment and control of genital ulcer diseases.

Other retroviruses

Animal retroviruses

A number of retroviruses have been identified in animals. Some, such as simian immunodeficiency virus (SIV) and feline immunodeficiency virus (FIV), can cause immunodeficiency in their infected hosts. SIV has caused asymptomatic seroconversion in a few humans who handle monkeys.

Other animal retroviruses appear to exist only as inserts into the host genome, with no pathological action, others are capable of expression in some circumstances, but their potential to cause disease is not known. An example is bovine immunodeficiency virus, a retrovirus found in cattle that has no association with known cattle disease (and that has not been identified in humans who have contact with cattle, or who drink cows' milk).

Three types of endogenous retroviruses have been identified in pigs (porcine endogenous retroviruses, PERVs; types A, B and C). Type C can infect human cells, and this has been a barrier to the development of xenotransplantation (transplantation of animal organs into humans).

PERVs have not so far been identified in human recipients of pig islet cells or other limited 'transplants'.

Human endogenous retroviruses (HERVs)

In common with other animals, humans have genetic evidence of endogenous retroviruses. It is likely that all animal endogenous retroviruses have been acquired over millennia, possibly modified or inactivated by being altered during mutation, insertions or deletions in the genome of the animal in which they exist, and passed from generation to generation in the germ line.

Introduction

The fetus and neonate can be exposed to infection in a variety of ways (Fig. 17.1). Some maternal infections cause viraemia, bacteraemia or parasitaemia during pregnancy, with the risk that organisms will cross from the maternal to the fetal circulation, causing trans-placental infection. The fetus may die of the infection, may recover *in utero* and be born with or without long-term sequelae, or may have a continuing infection at birth (so-called congenital infection).

Even if a blood-borne pathogen does not cross the placenta, the baby may be infected by contact with the mother's blood during birth. Alternatively, a pregnant woman's genital tract may be colonized or infected by a transmissible pathogen at the time of delivery. The neonate may become infected during birth, by contact with infectious genital secretions. The signs of these infections usually appear in the first 2–6 weeks of life (or, rarely, after several years as in the case of syphilis or toxoplasmosis). If the membranes rupture prematurely, organisms can ascend from the maternal genital tract, colonizing the amniotic sac and sometimes causing amnionitis. The infant is then exposed to a kind of intrapartum infection a few days before delivery.

Some blood-borne pathogens, such as *Plasmodium falciparum* or *Brucella* spp., cause severe placental infection, leading to abortion or premature delivery.

In neonatal life, the infant has close contact with its mother, ingesting her breast milk and having intimate face-to-face contact. Older children may eat food from their mother's spoon or have food chewed for them by her; she may clean a dropped comforter or pacifier by licking it. Infants occasionally ingest maternal blood if breast-feeding from cracked nipples. The extent of such contacts varies in different cultures. Infection can thus be transmitted from mother to child via saliva, blood or milk in early infancy. Occasionally, infections are transmitted via intimate contact with other family members.

General effects of trans-placental and intrapartum infection

Trans-placental infection

Trans-placental infection causes intrauterine infection of the fetus, but not necessarily any long-term effect. After intrauterine infection with varicella or parvovirus, most fetuses have recovered completely by the time of delivery, and few either die *in utero* or have permanent sequelae.

Congenital infections

Children born with congenital infections often have a multisystem disease, as is the case with congenital rubella or listeriosis. Hepatitis, pneumonitis, meningoencephalitis and blood disorders are common, thrombocytopenic purpura is often seen, and the infant often excretes large

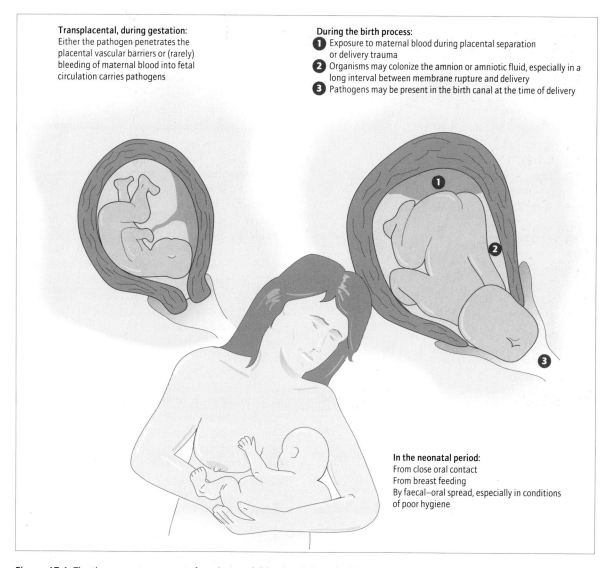

Transplacental, during gestation:
Either the pathogen penetrates the placental vascular barriers or (rarely) bleeding of maternal blood into fetal circulation carries pathogens

During the birth process:
1. Exposure to maternal blood during placental separation or delivery trauma
2. Organisms may colonize the amnion or amniotic fluid, especially in a long interval between membrane rupture and delivery
3. Pathogens may be present in the birth canal at the time of delivery

In the neonatal period:
From close oral contact
From breast feeding
By faecal–oral spread, especially in conditions of poor hygiene

Figure 17.1 The three common routes of mother-to-child transmission of infections.

amounts of the causative pathogen. Many of these acute problems will resolve if the infant survives, but permanent tissue damage may also be present.

The brain, the heart and the inner ear are the tissues most often permanently affected, causing microcephaly, epilepsy, heart murmurs and deafness, which can be progressive. Retinopathy is also common, but may not affect sight severely. Isolated nerve deafness is probably the commonest result of intrauterine infection.

Intrapartum and perinatal infections

Intrapartum and perinatal infections are often inapparent. However, exposure of the infant to commensal maternal organisms, such as *Escherichia coli*, group B streptococci or, rarely, *Listeria monocytogenes*, can lead to bacteraemia or meningitis in the first week or month of life. Because they take time to develop (i.e. have an incubation period) there is an opportunity for prophylaxis or early treatment of these infections.

> **Outcomes of intrauterine and intrapartum infection**
> 1 Abortion or premature delivery.
> 2 Intrauterine infection and recovery.
> 3 Intrauterine infection and fetal death or stillbirth.
> 4 Born infected but no disease develops.
> 5 Born infected, disease develops later.
> 6 Born infected with active disease.
> 7 Born with permanent or progressive tissue damage.

Trans-placental, intrapartum and post-natal infections

Organism list

Trans-placental

- Rubella virus
- Cytomegalovirus
- Human parvovirus B19
- Herpes simplex virus
- Human immunodeficiency virus
- Varicella zoster virus
- Vaccinia virus
- *Listeria monocytogenes*
- *Treponema pallidum*
- *Toxoplasma gondii*

Intrapartum

- Cytomegalovirus
- Herpes simplex virus
- Hepatitis B virus
- Human immunodeficiency virus
- *Escherichia coli*
- Group B *Streptococcus*
- *Chlamydia trachomatis*
- *Neisseria gonorrhoeae*
- *Listeria monocytogenes*

Post-partum

- Cytomegalovirus
- Hepatitis B virus
- Varicella zoster virus
- Human T lymphotropic virus I
- Human immunodeficiency virus
- Herpes simplex virus
- Enteroviruses (e.g. echovirus type 11)

⚠️ Many of these pathogens can infect the fetus and infant by more than one route. Therefore, for each organism mentioned, the routes of transmission and the different clinical effects will be discussed.

Congenital rubella

Introduction and epidemiology

The risk of congenital rubella depends on the stage of pregnancy at which maternal infection occurs (Table 17.1).

Before the introduction of rubella vaccine, outbreaks of congenital rubella occurred every 5–7 years, following the natural epidemic cycle of rubella. In a typical epidem-

Table 17.1 Risk of congenital rubella after exposure during different stages of pregnancy

Stage of pregnancy	Risk
<11 weeks	90%
11–16 weeks	20%
16–20 weeks	Minimal risk of deafness only
>20 weeks	No increased risk

ic year, nearly 1000 infections in pregnant women were reported in the UK. Approximately 90% of these pregnancies were terminated; however, where termination of pregnancy is not readily available, the impact of these epidemics is much greater.

Rubella immunization was introduced in the UK in 1970 for pre-pubertal girls and non-immune women of childbearing age, and this resulted in a reduction in the number of congenital rubella infections. In 1988, universal immunization with MMR was introduced as part of the immunization programme in the UK. This augmented the previous selective immunization policy and congenital rubella (Fig. 17.2) has now almost been eliminated in the UK, as in many countries. Some women may not have received rubella vaccination, usually because they were born in countries where it was not available. They remain at risk of infection through contact with young unimmunized men or after exposure abroad.

Permanent or progressive effects of congenital rubella infection

1 Cardiac: patent ductus arteriosus with or without pulmonary stenosis.
2 Cataracts.
3 Dysplasias of the retina or uveal tract.
4 Delayed motor and sensory development.
5 Nerve deafness.

Clinical features

Rubella in pregnancy approximately doubles the risk of fetal death. Among survivors the severity and nature of the clinical outcome depend on the stage of pregnancy at which infection occurred. Early infection leads to severe, multisystem damage and/or large cataracts; later infection causes isolated defects, low birth weight or an apparently normal infected neonate.

Reversible effects

Reversible effects of active infection are often present at birth, causing hepatitis and jaundice, haemolysis and thrombocytopenic purpura, and often low-grade meningoencephalitis. Some cases have dysplasia with patchy

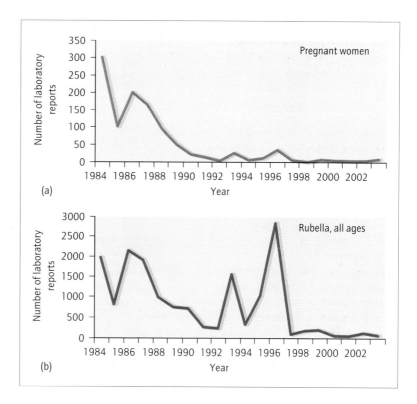

Figure 17.2 The impact of the MMR vaccine on rubella infections, especially in pregnancy. Source: Health Protection Agency.

mineralization of the metaphyses of the long bones. Most affected infants have low birth weight and fail to attain their expected developmental milestones. Mortality is high in severely affected infants, but in survivors these reversible effects resolve in the first 2–6 months of life.

Permanent effects

After infections in the first month of pregnancy, over half of those born will have multiple defects. Patent ductus arteriosus with or without pulmonary stenosis is the commonest. Deafness may be profound, or only detectable by audiometry. Dense cataracts are common after infection in the first month; smaller or faint opacities may occur after second-month infections. Retinal pigment dysplasia is often evident on ophthalmoscopy. Most infants have a severe brain syndrome inhibiting motor, sensory and intellectual development. This can result in a severely disabled child, further impaired by defective vision and hearing.

Hearing defects are the commonest effect of infection later in pregnancy, and may worsen during the early years of life.

Diagnosis

Suspected exposure during pregnancy should be con-firmed by serological testing of the contact, as clinical diagnosis is extremely unreliable (see Chapter 11). A positive immunoglobulin M (IgM) antibody test confirms recent, infectious, rubella in the contact. The immune status of the pregnant woman should be known from her routine testing, but serological 'immunity' to rubella does not absolutely protect from re-infection. Just under 10% of congenital rubella cases in England are shown to follow maternal re-infection during pregnancy.

The pregnant woman who was seronegative at booking, or whose immune status is unknown, should have an immediate test for IgM rubella antibodies. If negative, the test should be repeated after 2 weeks, or sooner if a feverish illness develops. A final test may be performed after a further week or 10 days. A woman who previously possessed antibodies may not develop an IgM response, but should be followed up to detect a secondary rise in IgG antibody levels. The aim is early detection of rubella infection, so that a risk assessment can be made, and termination of pregnancy can be offered if appropriate.

A rubella-infected neonate will have a positive IgM rubella antibody test, which persists until the third month of life. The absence of IgG antibodies excludes the diagnosis of congenital rubella, as neither mother nor child can have been infected.

Laboratory diagnosis of rubella

1 In pregnant woman: serum immunoglobulin G (IgG) at booking, to check immune status; IgM after suspected exposure or illness (repeat after 2 weeks).

2 In an exposed woman who possessed IgG antibodies at booking, a fourfold or greater rise in IgG antibodies, or the appearance of IgM antibodies, indicates probable infection (prenatal diagnosis may then be attempted).

3 In fetus: culture of rubella virus from amniotic fluid; fetal blood sampling to detect IgM antibodies (for suspected infection after 20 weeks gestation).

4 In neonate: positive IgM in the first three months of life (absence of IgG excludes the diagnosis).

Management

Pregnant women with possible rubella infection must be isolated from other antenatal patients.

Many women infected during the first two to three months choose to terminate the pregnancy; in later pregnancy there is a balance between the likely fetal damage and the desirability of termination.

There is no evidence that human normal immunoglobulin offers reliable post-exposure prophylaxis, though it has sometimes been given where termination of pregnancy is not acceptable to an infected woman.

The rubella-affected neonate remains highly infectious for several months, and should be isolated from pregnant women. Supportive treatment during the active infectious phase often includes blood or platelet transfusions, sometimes with treatment for immune thrombocytopenia.

The extent of permanent defects should be ascertained by detailed examination. Hearing tests are difficult to perform in infants; observational tests can be complemented with audiometry when the child is old enough. Cataracts can be treated surgically during the first year of life, allowing normal visual development. Cardiac abnormalities often require correction in the early months. Those with severe physical and intellectual disabilities need lifelong support.

Prevention

Rubella vaccine is a live attenuated vaccine produced from the RA 27/3 strain of virus. A single dose elicits protective antibodies in over 95% of recipients. The duration of protection is not yet known, as the vaccine was first produced less than 30 years ago; however, long-term follow-up shows that vaccine-induced antibodies wane at a similar rate compared with those acquired from natural infection. This suggests that, for most individuals, protection will be lifelong.

Immunization of all infants with combined measles/mumps/rubella (MMR) vaccine aims to provide indirect protection, by interruption of rubella transmission among children. Achieving high vaccine coverage (greater than 90%) has led to the success of this strategy in many countries.

Prevention of congenital rubella

1 Universal immunization in childhood.
2 Antenatal screening for immunity.
3 Post-partum immunization of susceptibles (to protect subsequent pregnancies).
4 Isolation of cases in antenatal units.
5 Active attention to diagnosis in pregnancy.
6 Counselling in pregnancy.

Live rubella vaccine is contraindicated in immune suppression, and in early pregnancy because of theoretical damage to the fetus. However, follow-up of several hundred babies whose mothers were accidentally vaccinated in early pregnancy has shown no increased risk of congenital abnormalities. Termination of pregnancy following inadvertent rubella vaccination is not therefore indicated.

Congenital and neonatal cytomegalovirus infections

Epidemiology

Cytomegalovirus (CMV) is the commonest cause of congenital infection in the UK. It almost always complicates primary CMV infection of a pregnant woman, though there are rare reports of congenital infection following recrudescence of latent infection. Although most babies have no symptoms at birth, about 10% are symptomatic, and a further 10% will subsequently develop deafness and neurological impairment. In the USA, primary CMV infection affects between 1% and 3% of all pregnancies and around 2500 infants a year are born with symptoms.

Clinical features

Congenital infection

When symptoms are present at birth, they may include: prematurity, low birth weight, hepatomegaly, splenomegaly, thrombocytopenia and prolonged jaundice. About 25% of clinically affected neonates have cerebral irritability, fits or abnormal muscle tone or movement. Pneumonitis is uncommon, and ventilation is rarely required.

Permanent defects occur in about half of all symptomatically affected infants. Microcephaly and sensorineural deafness are the commonest problems, and often coexist. Microcephaly improves or disappears with growth in a third to a half of those affected. Deafness is a solitary finding in about 10% of cases. Other problems include cerebral calcification, hemiplegia, diplegia or quadriplegia, and psychomotor retardation. Choroidoretinitis and myopathy have also been reported.

Permanent effects of congenital cytomegalovirus infection

1 Microcephaly.
2 Nerve deafness.
3 Cerebral calcification.
4 Upper motor neurone disorders.
5 Psychomotor retardation.
6 Choroidoretinitis (rare).
7 Myopathy (rare).

The prognosis for later childhood development is poorest in those who have neurological defects at birth. Microcephaly alone does not confer such a poor prognosis as hard neurological signs. Children without neurological defects at birth have a good prognosis, but it is not known whether deafness or other defects may become apparent in late childhood or adulthood.

Intrapartum and perinatally acquired infection

Intrapartum and perinatally acquired infection is often inapparent, although fever, poor growth, pneumonia or late-onset jaundice occasionally occur. It is common for an infant to contract cytomegalovirus infection from its seropositive mother; the majority of all seroconversions occur before school age.

Diagnosis

Infection in pregnancy

Cytomegalovirus infection in pregnancy is rarely apparent, and is often diagnosed by finding congenital infection in the neonate. In the rare situation of symptomatic primary infection in pregnancy, seroconversion would be demonstrable. Demonstration of IgM antibodies is not completely reliable, as these can also appear in post-primary infections, which rarely affect the fetus. If the fetus is infected, cytomegalovirus DNA is often detectable in amniotic fluid and fetal blood samples.

Diagnosis in the neonate

Blood, urine and throat swabs are the best specimens for diagnosis. Cytomegalovirus DNA can be demonstrated in these specimens by polymerase chain reaction (PCR) techniques. IgM antibodies to cytomegalovirus can be demonstrated in serum. Congenital infection is considered confirmed when these tests are positive in the first 20 days of life.

Intrapartum and perinatal infection

Intrapartum and perinatal infection produces positive IgM antibodies, PCR tests and cultures occurring after 20 days of age, while the presence of IgM and cytomegalovirus excretion up to 20 days of age indicates intrauterine infection. Infants who are not tested before 20 days of age cannot have a certain microbiological diagnosis of congenital infection because infection in early infancy is extremely common.

Management

In pregnancy

If new infection is diagnosed during pregnancy, counselling is based on the knowledge that only 10% of children have any detectable disorder at birth, and that a proportion of these will recover. There is no specific treatment.

Neonatal

There is no high-quality evidence that foscarnet or cidofovir treatment of an infected neonate can produce long-term benefit, and both drugs are relatively toxic. There are individual reports of the arrest of infection in severe neonatal disease treated with foscarnet. Trials of ganciclovir, given intravenously or orally to symptomatic neonates, indicate that progression of sensorineural deafness can be prevented, or even improved in some cases. For other features of congenital cytomegalovirus infection, supportive treatment is given. Long-term follow-up to detect sensorineural deafness is important. Those who present with neurological disorders require additional follow-up with regular hearing and neurological assessments, and physiotherapy, rehabilitation and/or educational support as appropriate.

Prevention

Opportunities for prevention are limited. Blood transfusion and organ donation from cytomegalovirus-seropositive donors to seronegative recipients should be avoided. Since cytomegalovirus infection in adults is often asymptomatic, screening in pregnancy would require repeated blood sampling. It has been suggested that susceptible pregnant women can reduce their risk by avoiding contact with the urine and saliva of young children, although this is of unproven benefit. Vaccines are being developed, but are not near to clinical use.

Occupational exposure to young children has not been shown to carry additional risk of cytomegalovirus infection.

Congenital parvovirus infection

Congenital parvovirus infection is probably quite common during parvovirus B19 epidemics, when children and adults may develop 'slapped cheek syndrome' or arthralgia and rash (see Chapter 11). It is important to undertake diagnostic tests (detecting IgM antibodies, or parvovirus B19 DNA) in a woman with these symptoms, or with fever during a parvovirus epidemic, and to exclude rubella, which parvovirus infection may closely resemble. Women who possess IgG antibodies to parvovirus B19 are immune, and not at risk of infection.

Fetal infection occurs in about half of maternal infections during pregnancy. Usually, both mother and fetus recover uneventfully and a normal infant is born. Termination of pregnancy is not indicated for maternal parvovirus B19 infection.

Infection in the first 20 weeks of pregnancy carries a risk of fetal anaemia and hydrops. Rapid fetal growth during these weeks must be paralleled by equally rapid production of red blood cells if the fetus is not to become anaemic. Human parvovirus B19 can only infect cells undergoing the S phase of mitosis, and infection of the rapidly dividing red cell precursors causes temporary aplasia, which can be severe enough to produce significant anaemia. Fetal death occurs in about 9% of pregnancies affected in the first 20 weeks, and hydrops affects around 3% of those infected between the 9th and 20th weeks. Hydrops is detectable by serial fetal ultrasound, and severely affected cases may be offered intrauterine blood transfusion.

Congenital and intrapartum herpes simplex infections

Primary herpes simplex infections may be accompanied by viraemia and, if this occurs in pregnancy, trans-placental infection can result.

Congenital infection

Infants born with congenital infection tend to have severe disease with a high mortality. The commonest features are pneumonitis, meningoencephalitis with fits and neurological signs, hepatosplenomegaly and cytopenias. A minority of infants have herpetic lesions of the skin or mucosae.

Diagnosis of congenital infection is important, as treatment reduces mortality from 80–90% to 10–15%. Virus particles can be seen on electron microscopy of throat swabs, bronchial secretions and scrapings from lesions. The virus grows well in cell culture, producing early and diagnostic cytopathic effects. Herpes simplex DNA is detectable by PCR in these specimens and in cerebrospinal fluid. Intravenous aciclovir, 10 mg/kg 8-hourly, is effective and well tolerated.

Intrapartum infection

Intrapartum infection occurs when a mother is excreting herpes simplex virus at the time of delivery. In this case the infant develops skin, conjunctival, oral or genital lesions within a few days of birth. The condition should be treated vigorously, as half of cases will develop disseminated disease. Intravenous aciclovir is the treatment of choice.

Prevention of neonatal herpes simplex infection

Although aciclovir is not licensed for use in pregnancy, extensive experience has shown no adverse effects. A pregnant woman may be treated for primary herpes simplex infection. Caesarean section is often performed if a woman has active herpes simplex lesions at the time of delivery. However, the risk of severe disease in the infant appears to be small, and the risks of a surgical delivery may be weighed against the risk of the infant needing treatment for perinatally acquired lesions.

Varicella embryopathy and neonatal varicella

Varicella embryopathy

Varicella embryopathy is a rare deformity following varicella infection, usually during the 13th to 20th week of pregnancy. About 12% of adults are susceptible to varicella, so maternal infection during pregnancy is not uncommon. Large surveys suggest that varicella embryopathy occurs in about 2% of infected pregnancies.

The usual deformity is contracture of a limb, with hypoplasia and zonal scars suggestive of old, zoster-like lesions. If the head is affected, microcephaly and unilateral microphthalmia may occur. A woman who develops varicella during pregnancy may be treated with aciclovir. There is no evidence that aciclovir or immunoglobulin treatment prevents embryopathy after the mother has developed chickenpox, though some studies suggest that aciclovir treatment may reduce the risk.

If a woman with no history of chickenpox is exposed to infection during pregnancy, she should immediately have her anti-varicella antibody titres estimated. If IgG antibodies are present, no further action need be taken. If she is not immune, the woman may be offered post-exposure prophylaxis with varicella zoster immunoglobulin (VZIG). This can prevent or ameliorate infection if given within 10 days of exposure.

Neonatal varicella

Neonatal varicella is a life-threatening, disseminated disease, with a mortality of up to 40%. It occurs in neonates born to women with no immunity to varicella, as the neonate has no protective maternal antibodies. It can acquire severe infection if its mother develops chickenpox within 1 week before or after delivery. In maternal infection occurring soon before delivery, the mother cannot develop antibodies in time to confer trans-placental protection to the infant. If a woman without antibodies develops varicella soon after delivery, the infant has no maternal antibodies to protect it, and is highly susceptible to severe disease.

Immunoglobulin given to the pregnant mother does not cross the placenta to protect the fetus. The neonate should therefore be given post-exposure prophylaxis with VZIG. If the mother has had chickenpox for less than 8 days prior to delivery, the infant should receive VZIG immediately after birth (the VZIG can be provided in time for the delivery by the Health Protection Agency, Centre for Infections or Regional Centre). If the mother's disease appears within 7 days after birth, VZIG should be given to the infant as soon as possible, preferably within 48 h. The infant should be followed up, and treated with aciclovir if varicella develops despite VZIG prophylaxis.

> ⚠ Herpes zoster in pregnancy does not require any protective action, as it affects women already immune to varicella and cannot therefore cause fetal varicella infection.

Maternal varicella

Clinical varicella in a pregnant woman is often no more severe than in the non-pregnant woman, though reports exist of severe disease occurring in late pregnancy. As in other adults there is a risk of pneumonitis, which is greater in smokers. Severe feverish illness can occasionally precipitate early or premature labour. Many experts therefore recommend offering oral aciclovir treatment if a pregnant woman presents in the first 48 hours of the illness. In severe or complicated disease, intravenous aciclovir and appropriate antibiotics should be used as indicated (see Chapter 11). Aciclovir is not licensed for use in pregnancy, but increasing experience so far shows no evidence of adverse effects.

Women with no history of varicella may be tested for immunity before conception. Varicella vaccine (a live attenuated vaccine) may be given to those who are not immune, to protect from infection during future pregnancies.

> ⚠ Pregnancy should be avoided for 3 months after varicella vaccination; vaccination should be avoided during breast-feeding.

Congenital HIV infection

Epidemiology

Congenital human immunodeficiency virus (HIV) infection is rare in the UK, where the prevalence of infection in the antenatal population is low (0.16% in England as a whole in 2003, but 0.5% in Inner London). Antenatal detection of HIV is actively carried out, and measures may be recommended to reduce vertical transmission in pregnant women with uncontrolled viraemia (see below). Since the prevalence of HIV infection in the UK is continuing to rise in the antenatal population these measures remain very important. In many African countries, however, the prevalence of infection in pregnant women can reach 25%, with a risk of vertical transmission approaching 30%. HIV-1 is the commonest congenital infection in these countries.

These figures are well established for HIV-1 infection. They differ for the less common HIV-2 infection, which appears to be less easily transmissible. There are insufficient data from which to deduce the probable risk from HIV-2.

Clinical features

A variable percentage of infants born to HIV-positive mothers are premature or of low birth weight. This may be related to the mother's health and does not inevitably mean that the child is infected.

Some infected infants fail to thrive, and develop severe infections within the first 12–18 months of life, while others remain antibody-positive for several years without any adverse effect. The reasons for these different presentations are uncertain.

Diagnosis

All infants of infected mothers have HIV antibodies in the blood at birth. Most antibodies are trans-placentally acquired. They decline in titre, and disappear between 6 weeks and 6 months of age. Demonstration of HIV RNA in the infant's blood is reliably diagnostic of congenital HIV infection. HIV antigen is diagnostic if present, but may be absent for the initial weeks or years of life.

Management

In pregnancy

HIV infection during pregnancy should be treated as indicated with multidrug therapy; the aim is to provide treatment for the mother that will prevent progression of HIV disease, control HIV viral replication and reduce maternal viraemia, with minimum toxicity to the fetus. There is no information on toxicity of antiretroviral drugs in pregnancy, so they may be used if the benefits outweigh the risks. A rare syndrome of mitochondrial dysfunction, possibly related to nucleoside reverse transcriptase inhibitors, has affected a small number of neonates. Some drugs used in treatment or prophylaxis of HIV-associated infections, including sulphonamides, some antifungal drugs and systemic pentamidine, are contraindicated in pregnancy; others, such as rifabutin, are not recommended but are not known to produce adverse effects. Care should therefore be taken with drugs given prophylactically or electively, though life-saving treatment should be given as in the non-pregnant.

Before and after delivery

Although multidrug treatment is often indicated for the mother's condition, some women conceive at a stage in their HIV disease when they are not receiving therapy. Zidovudine monotherapy of the mother in late pregnancy, and the infant for the early weeks of life, reduces the risk of vertical transmission of HIV-1 by around 70%. Single-dose nevirapine during labour is an alternative. Multidrug therapy is nowadays preferred and probably reduces the risk even further. With multidrug therapy and/or optimum management of birth (e.g. Caesarean section in which a bloodless field is obtained before the fetal sac is opened) and avoidance of breast-feeding, the risk of vertical transmission can be reduced to 1% or less.

If a woman is already taking multidrug therapy during pregnancy, this should be continued, and the viral load monitored. A switch of therapy may be used for the last trimester, to aim for a maximum antiviral effect. In this case, the infant should receive zidovudine, unless the mother's virus is known to be resistant to it, when an alternative may be offered (or treatment omitted if the mother has an undetectable viral load).

> ⚠ The management of HIV in pregnancy, and prevention of vertical transmission of infection, require expert supervision and should be carried out at a centre of expertise.

The World Health Organization continues to recommend breast-feeding by untreated HIV-positive mothers in developing countries, where the protective effect of breast milk against opportunistic infections outweighs the small additional risk of HIV transmission.

The HIV-infected infant

The HIV-infected infant may need no specific treatment for months or years. The normal programme of infant immunization should be offered, including measles/mumps/rubella vaccine as, even in symptomatic children, the risk from childhood infection is far greater than that from the vaccines. However, inactivated polio vaccine (IPV) should be substituted for live, oral vaccine (OPV) in countries where this is still given.

There is no reason to exclude HIV-positive children from nursery or school. As is good practice in all situations, cuts and grazes should be cleaned with water, soap or mild disinfectant and covered with a simple dressing. Spillages of blood or body fluids should be absorbed with absorbent wipes or granules, and the area cleaned with soapy solution or a mild household disinfectant/cleanser; the wipes or granules may be placed in a plastic bag for disposal (commercial spillage kits are convenient for this purpose).

Neonatal ophthalmia

Neonatal ophthalmia is a severe conjunctivitis occurring within the first 48 h of life. Most cases are due to *Chlamydia trachomatis* infection; *Neisseria gonorrhoeae* is an uncommon cause. Chlamydial infection can be confirmed by a positive PCR-based test for chlamydial DNA. In gonococcal infection, Gram stain of the purulent conjunctival discharge shows many Gram-negative diplococci.

Chlamydial infection is usually accompanied or followed by chlamydial pneumonitis, and should be treated with oral or intravenous erythromycin (see below).

Gonococcal infection responds to chloramphenicol eye ointment, applied three times daily for 7 days. If a mother is known to be infected with *N. gonorrhoeae* at the time of delivery, gonococcal ophthalmia can be prevented by treating the infant with an effective antibiotic. Benzylpenicillin is effective for sensitive gonococci; cefuroxime or cefotaxime is recommended for penicillin-resistant gonococci.

Both parents should be offered investigation and treatment for chlamydial infection, gonorrhoea and other sexually transmitted diseases.

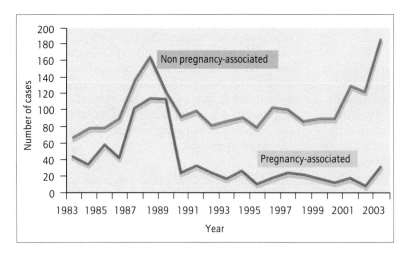

Figure 17.3 Recent trends in listeriosis in England and Wales. Source: Health Protection Agency.

Congenital and neonatal listeriosis

Introduction

Both congenital and neonatal listeriosis are rare in the UK. Fewer than 50 cases a year are reported, particularly since the 1990s following health education and food hygiene measures to limit pregnant women's consumption of contaminated foods such as soft cheese or pâté (Fig. 17.3). A discussion of the microbiology and pathogenicity of *Listeria monocytogenes* can be found in Chapter 13.

Clinical features

Congenital listeriosis

Congenital listeriosis is the result of trans-placental infection with *Listeria monocytogenes* during maternal bacteraemia. The bacteraemic infection in the mother is usually inapparent. Features such as transient fever, diarrhoea, backache and pruritus are described, but these are common in pregnancy and rarely lead to specific diagnosis. The first sign of a problem is usually recognition of the disease in the offspring. An infected mother–baby pair may become the focus of nosocomial transmission in a neonatal unit.

Listeriosis in early pregnancy often causes fetal death. Otherwise, affected infants may be born prematurely, or at term, with severe disease. Characteristically this is bacteraemic multisystem disease with microabscesses affecting many parenchymal organs. There is often an erythematous rash containing many pale nodules on the infant's body and legs (granuloma infantisepticum). Hepatosplenomegaly is common, and other features include meningoencephalitis, thrombocytopenia and variable pneumonitis.

Neonatal listeriosis

Neonatal listeriosis follows intrapartum exposure to maternal birth passages colonized by *L. monocytogenes*. The infant becomes ill within the first 2–4 weeks of life, usually with meningitis and bacteraemia.

Diagnosis

In congenital listeriosis the unusual rash may suggest the diagnosis. Neonatal listeriosis must be distinguished from the more common *Escherichia coli* or group B streptococcal meningitis and bacteraemia. The recent occurrence of a case or cases in a delivery unit should increase the index of suspicion.

Blood and/or cerebrospinal fluid cultures will provide the definitive diagnosis. In neonates with fever or sepsis, meconium, urine and gastric aspirate should also be submitted for culture at the time of birth. Products of conception may yield positive culture after abortion. Positive cultures may also be obtained from placental tissue or lochia. If the mother is feverish, blood cultures should also be obtained from her.

Serological tests may demonstrate anti-listeria antibodies, but these often originate from remote infections or infections by strains of low pathogenicity. Serology is unreliable in making a diagnosis of recent or current listeriosis.

Diagnosis of neonatal and congenital listeriosis

1 Culture of meconium, urine and gastric aspirate at birth.
2 Culture of blood and cerebrospinal fluid at birth or in neonatal fever or meningitis.
3 Culture of products of conception (placental tissue or lochia).
4 Culture of a feverish mother's blood.

Management

Affected mothers and their infants should be managed in isolation, as outbreaks of listeriosis in neonatal units have been well-described.

L. monocytogenes is sensitive *in vitro* to many antibiotics, including penicillin, ampicillin, tetracyclines, aminoglycosides, imipenem, trimethoprim and co-trimoxazole. It is also sensitive to rifampicin, ciprofloxacin and vancomycin, which are either contraindicated in pregnancy or the neonate, or penetrate the cerebrospinal fluid poorly. Most cephalosporins are ineffective or only weakly effective.

The response to treatment is not always as satisfactory as expected from laboratory tests. Improvement may be transient, and relapses often occur. High-dose ampicillin or amoxicillin are currently the treatment of choice; there is evidence from animal studies that the addition of gentamicin improves the outcome, but there is little evidence for this in human neonatal disease. Penicillin plus gentamicin, used as empirical treatment for neonatal meningitis or septicaemia, is also likely to be effective. If a good response is obtained, 2–3 weeks' treatment may be sufficient, but 4 or even 6 weeks may be wise to avoid relapse after initial slow response.

When the response to other treatments is suboptimal, chloramphenicol has been found effective, but is probably less so than ampicillin.

Treatment of congenital listeriosis

1 First choice: ampicillin *or* amoxicillin i.v. 50–100 mg/kg daily in four divided doses, for 2–6 weeks, depending on response.
2 Alternative: benzylpenicillin i.v. 50 mg/kg daily 8-hourly; *plus* gentamicin i.v. – up to 2 weeks, 3 mg/kg 12-hourly; over 2 weeks, 2 mg/kg 8-hourly (may be given intrathecally 1–2 mg daily). Plasma concentrations should be measured to obtain peak (1 h) concentration not above 10 mg/l and trough not above 2 mg/l.
3 Second choice: chloramphenicol i.v.; infant under 2 weeks: 25 mg/kg daily in four divided doses; age 2 weeks to 1 year: 50 mg/kg daily in four divided doses.

> ⚠ Plasma concentration monitoring is required to avoid 'grey syndrome' (pallor, cyanosis, abdominal distension and collapse, due to inefficient metabolism of chloramphenicol in infants): pre-dose level should not exceed 15 mg/l, post-dose (1 hour after intravenous dosing) should be 15–25 mg/l.

Prevention and control

Pregnant women should be discouraged from eating foods likely to contain high counts of *L. monocytogenes*. These include pâtés and soft, ripened cheese such as Brie and Camembert. Other food that may contain significant bacterial counts include cook–chilled meals and ready-to-eat poultry: pregnant women should reheat these types of food thoroughly before consumption. They should also avoid contact with potentially contaminated material such as aborted animal fetuses on farms.

Congenital syphilis

Congenital syphilis is now extremely rare. Untreated syphilis acquired during pregnancy often causes fetal death and early abortion. However, if the infection remains untreated succeeding pregnancies are more likely to survive to term. The infant may have signs of disease at or soon after birth. If untreated, later manifestations can occur, appearing in childhood, the teens and even in adulthood.

Early manifestations (can appear up to 2 years postnatally)

The affected baby is often feverish and has features similar to those of secondary syphilis: rash, condylomata, and mucosal fissures and inflammation. Osteochondritis may cause pain. Half of those affected even mildly have persistent rhinitis ('snuffles'). More severe manifestations include lymphadenopathy, hepatosplenomegaly, non-immunological hydrops, glomerulonephritis, eye or neurological disorders and thrombocytopenia.

Early diagnosis is best made by demonstrating treponemes on dark-ground microscopy of material from mucosal or skin lesions. Serology results may be misleading because of the presence of maternal antibodies; only antibodies persisting after 3–6 months of age, or the demonstration of seroconversion on repeated testing, indicate true infection. The fluorescent treponemal antibody-absorption (FTA-ABS) test and IgM enzyme-linked immunosorbent assay (ELISA) are unreliable in the neonate. A VDRL titre more than two dilutions above the mother's titre is strong evidence of infection in the neonate.

Late manifestations

Late manifestations tend to appear between the ages of 12 and 20 years. Neurological disorders such as nerve deafness, optic atrophy or paretic neurosyphilis respond poorly to treatment. Interstitial keratitis often causes damaging corneal opacity. Synovitis of the knees (Clutton's joints) commonly accompanies this protracted inflammatory condition. Other features include bossing of the frontal bones, saddle nose, chronic periostitis of the tibias (sabre tibia), notching of the incisors (Hutchinson's teeth), 'mulberry' deformity of the first permanent molar and a high arched palate. Cold haematuria may occur. Gummas may rarely develop in any site.

Treatment

The treatment of choice for infants is benzylpenicillin. Neonates can be treated with 300 mg 6-hourly for 17 days. Older children may be treated with intramuscular procaine penicillin 150 000 U/kg daily. Neurological disease, interstitial keratitis, Clutton's joints and other periostitis respond only slowly to treatment, and courses of several weeks may be justified. There seems to be a hypersensitivity component to the keratitis, which can be suppressed by topical corticosteroids; this treatment should be continued until remission occurs.

Congenital toxoplasmosis

In the UK toxoplasmosis is rare in pregnancy, affecting between 1 in 1000 and 1 in 2000 pregnancies. It is much more common in France and other continental countries, where 2–6% of pregnancies are affected. The great majority of maternal infections are subclinical.

Infection occurring late in pregnancy is more likely to affect the fetus than early infections (60–65% affected in the last 3 months, versus 15–25% in the first 3 months). Overall, trans-placental infection occurs in about a third of affected pregnancies, and has only been reported in primary infection of the mother.

Early infections are more likely to cause severe disease in the infant. Clinical disorders are evident at delivery in 20–30% of infants infected between the second and sixth month of gestation. Later infections often cause only seropositivity or isolated choroidoretinitis. Overall, about 10% of affected infants have severe disease at birth.

Severely affected neonates may be stillborn or die soon after birth. Others have the typical features of hydrocephalus, cerebral calcification and choroidoretinitis, which lead to developmental delay, epilepsy, and often cerebral palsy. Choroidoretinitis may not be evident until some months after birth. Other, non-specific features can include anaemia, thrombocytopenia, jaundice and pneumonitis.

Clinical features of congenital toxoplasmosis

1 Intrauterine death or stillbirth.
2 Cerebral calcification.
3 Cerebral palsy.
4 Epilepsy.
5 Early or late choroidoretinitis.

Diagnosis

In the pregnant woman, this is confirmed by the presence of IgM antibodies or by seroconversion.

In affected neonates IgM antibodies may also be demonstrated. High titres of specific IgM in the infant's cerebrospinal fluid are strongly suggestive of congenital infection. *Toxoplasma gondii* can be detected in products of conception or in the infant's cerebrospinal fluid by PCR, or by culture in a continuous cell line.

Diagnosis of *Toxoplasma* infection in mother–infant pairs

- *In the pregnant woman*: immunoglobulin M (IgM) antibodies or seroconversion in other antibodies (these are indications to perform fetal sampling).
- *In the fetus*: PCR or culture of amniotic fluid or fetal blood samples; serodiagnosis on fetal blood samples (from 20 weeks).
- *In the neonate*: IgM antibody in blood; positive PCR or culture of blood, cerebrospinal fluid, placenta or products of conception.

Management

Termination may be considered for infections in the second to sixth month of pregnancy. Expectant management is usual for later infections. When the decision is made to continue an affected pregnancy, antiparasitic treatment can be offered. Spiramycin, a macrolide antibiotic, is thought to reduce the incidence of placental infection by reducing maternal parasitaemia. However, fetal infection can occur before, or despite, spiramycin treatment. Fetal blood spiramycin concentrations are only 50% of maternal blood levels, and are ineffective in treating fetal infection. It is not known whether newer macrolides and azolides are more effective.

The most effective regimen for treating the fetus is a combination of pyrimethamine plus sulfadiazine. Antifolate agents may have teratogenic effects in early pregnancy, but treatment is justified because of the risk of severe damage from known toxoplasmosis in the fetus at this stage of gestation. Folinic acid supplements and combined antiparasitic therapy are continued throughout the pregnancy.

Treatment of mothers and infants with toxoplasmosis

1 For the infected pregnant woman: spiramycin, orally, 1 g 8-hourly for 3 weeks (perform fetal sampling afterwards).
2 If fetal infection is confirmed, and termination is not contemplated: sulfadiazine, orally, 1 g 6-hourly *plus* pyrimethamine, orally, 50 mg daily *plus* folinic acid 15 mg daily until delivery.
3 For the infected neonate: sulfadiazine 50–100 mg/kg daily in two divided doses *plus* pyrimethamine, single doses 1 mg/kg daily *plus* folinic acid 5 mg daily for 6–12 months.

Prevention

Women of childbearing age who suffer toxoplasmosis are advised to wait until the IgM antibodies have disappeared before attempting to conceive. A small minority have IgM antibodies persisting for many months; after a year it is unlikely that this is related to persisting parasitaemia, and further delay is not usually recommended.

Infections transmitted only by the intrapartum and perinatal routes

Neonatal and perinatal hepatitis B

Epidemiology

The risk of vertically transmitted hepatitis B (HBV) infection is between 10 and 25% for infants born to mothers who are hepatitis B e antigen (HBeAg)-negative, but rises to 90% for those born to e antigen-positive mothers, or those with significant HBV DNA levels in blood. Women who are HBsAg-positive and HBeAg-negative, but who lack anti-HBe antibody (i.e. who have viraemia, but no HBe markers) are also likely to infect their newborn.

In the UK, the prevalence of hepatitis B surface antigen (HBsAg) carriage in the general population is less than 0.5%. Both neonatal and perinatal infection are therefore uncommon. However, in communities originating in some parts of the world, notably Asia and some Mediterranean areas, the prevalence of infection in the antenatal population exceeds 10%, and neonatal and perinatal infections are very common.

Clinical features

Infants who acquire HBV infection neonatally rarely suffer significant illness; they simply develop the antigen and antibody sequence characteristic of acute hepatitis B infection. In the great majority (70–90%) healing is arrested at the stage of hepatitis B e and s antigenaemia (HBeAg and HBsAg), leaving the child at risk in later life of cirrhotic or malignant complications. Female children will also be at high risk of passing infection to their own infants.

Children in whom infection at birth is avoided or prevented may still be exposed during intimate contact with an antigen-positive mother or other family member during nursing and early childhood. Childhood hepatitis B infection is usually mild but may be icteric and in exceptional cases can be life-threatening or fatal. The risk of persisting e-antigen positivity after the acute infection is greatest when seroconversion occurs below 6 months of age; after the age of 9 months the risk is no higher than in older age groups.

Diagnosis
All pregnant women are screened to identify those with hepatitis B antigenaemia. The populations most at risk are Mediterranean, Far Eastern, Asian and Caribbean. Effective post-exposure prophylaxis for the newborn is possible because the infant's infection is contracted at the time of birth, and almost never *in utero*.

Management
The labour and delivery of a hepatitis B-antigenaemic mother should take place in a single-occupancy suite, as her blood and body fluids present an infection hazard to other mothers and infants.

As soon as possible after birth the neonate should receive appropriate immunization (Table 17.2). In infants at high risk, hepatitis B immune globulin (HBIG) prolongs the incubation period of the infection and allows time for active immunization with hepatitis B vaccine. The first dose of vaccine should be given at the same time as HBIG, but in a different intramuscular site. Neonates and infants mount a good immune response to the vaccine. Passive–active immunization reduces the incidence of permanent antigenaemia to about 5% in infants exposed to perinatal infection.

Human T-lymphotropic virus type I infection in infancy

Human T-lymphotropic virus type I (HTLV-I) is a retro-virus that infects human T lymphocytes. Seropositivity for this virus is associated with repeated skin infections and, later, the occurrence of T-cell leukaemia and with a rare acquired neurological disease called tropical spastic paraparesis or HTLV-associated myelopathy (HAM: see Chapter 13). The populations in which HTLV-I is prevalent include people of African and Caribbean origin, and Far Eastern populations, particularly from southern Japan.

The main route of transmission from mother to infant is via breast milk (which contains significant numbers of infected lymphocytes). Infants of infected mothers are many times more likely to acquire infection if breast-fed than if they are given artificial feeds. Avoidance of breast-feeding, if possible, offers important protection to the child.

Serological tests for HTLV-I are currently unreliable, as there are many false positives for each true positive result. Only Western blotting or tests dependent on identifying seropositivity to recombinant *env* and *gag* gene products offer dependable results, and these are too cumbersome and expensive to use as screening tests. Screening is recommended for the families of patients known to be infected with HTLV-I, as advice about blood donation and prevention of neonatal transmission of disease can be given.

Neonatal and infant chlamydial infections

Introduction
Chlamydial infections are among the commonest sexually transmitted diseases now encountered throughout the world. They have superseded gonorrhoea and syphilis, which are more readily detected and treated before and during pregnancy. Many pregnant women shed increasing numbers of chlamydiae from the genital mucosa as term approaches, so many neonates are exposed to intrapartum infection.

Clinical features
Chlamydial neonatal ophthalmia
Chlamydial ophthalmia is a severe conjunctivitis that appears within 3 or 4 days of birth. It may appear first in one

Table 17.2 Recommendations for prophylaxis of hepatitis B in infants of antigen-positive mothers (see Chapter 9)

Hepatitis B status of mother	Baby receives HBIG?	Baby receives vaccine?
HBsAg-positive and HBeAg-positive	Yes: 200 IU immediately	Yes: three doses plus booster
HBsAg-positive, no e-markers (or not determined)	Yes: 200 IU immediately	Yes: three doses plus booster
HBsAg-positive, HBeAg-negative, but HBV DNA-positive	Yes: 200 IU immediately	Yes: three doses plus booster
Mother had acute hepatitis B in pregnancy	Yes: 200 IU immediately	Yes: three doses plus booster
HBsAg-positive and anti-HBe-positive	No	Yes: three doses plus booster

(Modified from: UK National Screening Committee, 1998.) Neonate's dose of hepatitis B vaccine: half of adult dose given at birth, 1 month and 2 months of age; a booster is given at 1 year.

eye, but usually becomes bilateral. There is oedema of the lids, closing of the eye and a purulent exudate, which runs from the palpebral fissure. Parting the lids reveals a swollen, bulging conjunctiva.

Neonatal chlamydial pneumonitis
Neonatal chlamydial pneumonitis is a moderately severe, persistent infection that is often, but not always, preceded by ophthalmia. It presents at 3–6 weeks of age with tachypnoea and a repetitive, staccato cough. The chest X-ray is often abnormal, showing streaky or radiating perihilar opacities with or without areas of consolidation and air bronchograms. Untreated, it can last for many weeks, producing debility, failure to thrive and significant respiratory impairment.

Diagnosis
Differential diagnosis
Differential diagnosis of neonatal ophthalmia is important, as both parents may be colonized by the causative organism, and should be offered appropriate examination and treatment. An identical ophthalmia is caused less often by *Neisseria gonorrhoeae*. Staphylococcal or pneumococcal infections also occasionally cause purulent neonatal conjunctivitis.

Specimens should be obtained for bacteriological examination, and should include swabs in appropriate transport medium for isolation of gonococci. Scrapings from the inflamed conjunctiva can be examined by NAAT. Serology is unhelpful in the diagnosis of mucosal infections. However, chlamydial pneumonitis can be confirmed by the diagnostic presence of IgM antibodies to *C. trachomatis* in the infant's blood.

Management
Chlamydial neonatal ophthalmia
For chlamydial ophthalmia the treatment of choice is erythromycin, orally, 125 mg 6-hourly for 10 days, as pneumonitis often coexists with or follows eye infection. Tetracycline eye ointment applied three times daily for 10–14 days is an alternative, or may be added. Chloramphenicol drops or ointment are not recommended as relapse can follow apparently effective treatment.

Mild 'sticky eye' without conjunctival redness or positive cultures is often non-infectious, caused by sticky secretions from an engorged lacrimal sac, and relieved by twice-daily massage of the sac to empty it.

Chlamydial pneumonitis
For chlamydial pneumonitis the treatment of choice is erythromycin orally in a dose of 125 mg 6-hourly for 10–14 days.

Prevention
Prevention depends on detection and treatment of infection in one or both parents before the birth of the infant. Screening of urine samples is effective in detecting chlamydial infection (see Chapter 15).

Neonatal bacteraemias

Epidemiology
The epidemiology of neonatal bacteraemia is markedly different from that in other age groups. Approximately 60% of infections are caused by *Escherichia coli* and group B streptococci. The remainder are due to other Gram-negative bacteria, staphylococci, *Listeria monocytogenes*, *Neisseria meningitidis*, *Haemophilus influenzae* and *Streptococcus pneumoniae* (Fig. 17.4).

Neonatal infection is closely related to colonization of the maternal genital tract. Bacteraemia tends to occur in the first week of life. Meningitis often occurs later, at 2–3 weeks. It is more likely in cases where there is a long interval between rupture of the membranes and delivery, which allows colonization of the amniotic sac with maternal genital flora.

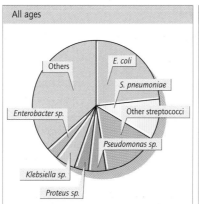
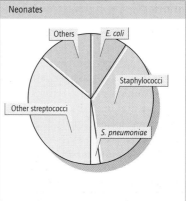

Figure 17.4 Pie charts comparing the aetiology of neonatal bacteraemia with bacteraemia in other age groups.

Clinical features

In the first few days of life there are few specific clinical features of bacteraemia. There is often a fever. The infant is listless, with poor or absent muscle tone (floppy baby) and is uninterested in feeding. The neutrophil count may rise, though in very small babies this is not always a reliable finding.

Meningitis also presents non-specifically, with features of bacteraemia. There may also be vomiting, drowsiness, convulsions or a strange, high-pitched cry.

These presentations should be taken seriously, especially if convulsions occur in the absence of a biochemical cause. All cases of doubt should be investigated. Blood cultures, urine cultures and cerebrospinal fluid examination should all be performed. Gastric aspirate, meconium and urine may also be cultured in the first days of life. Treatment for bacteraemia and meningitis should be commenced while results are awaited.

Management

Intravenous antimicrobial chemotherapy should be given, with a spectrum broad enough to cover both *E. coli* and group B streptococci.

Broad-spectrum cephalosporins penetrate the cerebrospinal fluid readily, have low toxicity and do not require blood-level monitoring. Drugs such as cefotaxime or ceftazidime are convenient and effective. Treatment should be continued for about a week, with high doses for the first 2–3 days. The choice of antibiotic can be modified if indicated when the results of cultures are available.

Prevention of neonatal sepsis

The prevention of neonatal sepsis depends largely on hygienic and expeditious delivery of an infant. Delayed delivery (longer than 48 h after rupture of the membranes) should be avoided. Attempts to screen mothers for group B streptococcal colonization immediately before delivery have not proved reliable in identifying all cases at risk.

Neonatal staphylococcal infections

The commonest neonatal infection with *Staphylococcus aureus* is umbilical infection. The cut end of the umbilical cord easily accepts colonization and infection, leading to an impetiginous or weeping lesion. Spread to surrounding skin and subcutaneous tissues can occur.

Any area of broken skin, or the conjunctival sac, may also become infected.

Bullous impetigo (Lyell's syndrome)

Although *S. aureus* bacteraemia is rare, the effects of staphylococcal toxins are more common. Staphylococcal pyrogenic exotoxins damage the layers of the epidermis allowing the surface layers to separate and form fragile blisters and bullae. The surface of erythematous, non-blistered skin can be rubbed off with gentle shearing stress (Nikolsky's sign). The resulting bare patches closely resemble scalds, and the condition is often called the scalded skin syndrome.

The lesions are heavily colonized with staphylococci, which can be recovered by culture from swabs. Staphylococcal skin lesions are highly infectious. The infant and mother should be separated from others in the neonatal nursery. Most cases will recover readily on treatment with oral flucloxacillin; severe infections need parenteral treatment. Lyell's syndrome can present with extensive exfoliation. This needs vigorous treatment with antibiotics; the infant requires warmth, humidity, ample fluid replacement and strict hygiene to avoid super-infection of the bare areas.

Control of infection measures should always include hand-washing after each contact with affected individuals. Chlorhexidine is often applied topically to the umbilical stump in the first day or two of life, but this does not preclude the need for strict hygiene generally in the nursery.

Maternal infections related to childbirth

Introduction

Around the time of delivery a woman may have increased susceptibility to conditions such as urinary tract infections because of her changed anatomy and physiology. She may also be more severely affected than others by certain infections such as genital warts or candidiasis because of altered physiology during pregnancy. A severe, feverish illness can cause premature labour, particularly in the third, and sometimes the first trimester.

There are other susceptibilities in pregnancy and the puerperium because of the presence of tissues or body functions that do not occur at any other time. The placenta is unique to pregnancy; it can be invaded and damaged by some pathogens causing severe maternal morbidity and putting the pregnancy at risk. This is important in some unusual infections including brucellosis, enzootic ovine abortion and, in exposed populations, falciparum malaria, which is dangerous to both mother and fetus. After parturition the placental bed provides a portal of entry for pathogens, which can invade the blood, causing puerperal fever. Finally the breasts become highly vascular and filled with secreted milk. Engorgement and stasis are common, and occasionally lead to the formation of breast abscesses.

Puerperal fever

Organism list
- *Escherichia coli*
- *Streptococcus pyogenes*
- *S. pneumoniae*
- *Staphylococcus aureus*
- Other coliforms
- *Mycoplasma hominis*
- *Clostridium perfringens*
- *Bacteroides fragilis*

Introduction and epidemiology
Puerperal fever is defined as any significant feverish illness occurring within 14 days of childbirth, miscarriage or termination of pregnancy. It is a severe, usually bacteraemic, infection caused by entry of pathogens either through the bare placental bed or through traumatic lesions of the cervix, vagina or perineum.

Clinical features
Most cases of puerperal fever begin within 4–7 days of delivery. Fever may be the only sign, but there may be back pain, offensive lochia, faintness or signs of sepsis with or without shock. Disseminated intravascular coagulation occurs early, especially if there are retained products of conception.

There may be clinical features particular to the causative organism. Erysipelas-like lesions on the trunk or arms or a scarlatiniform rash may occur in streptococcal infection. The rash and shock of toxic shock can accompany staphylococcal cases. Intravascular haemolysis, jaundice and even crepitation of vulval tissues may indicate clostridial infection.

Management
Fever in the early puerperium should be taken seriously. If it is not obviously due to a local infection of the skin, breast or urinary tract, investigations should be carried out and empirical treatment begun without delay.

Cultures of blood, urine, lochia and genital swabs should be obtained. The possibility of retained products of conception should be considered; ultrasound examination may help to exclude this. Any retained products should be evacuated (by an experienced operator, as the uterus may be oedematous or friable). Evacuated products should be submitted for microbiological examination.

Initial antibiotic treatment should be active against Gram-negative rods, *Streptococcus pyogenes* and anaerobic organisms. A reasonable regimen would be a mixture of an aminoglycoside, high doses of benzylpenicillin (or ampicillin or amoxicillin) and metronidazole. A broad-spectrum cephalosporin could be substituted for the aminoglycoside and penicillin, but large doses are needed for an adequate antistreptococcal effect. Treatment may be modified when microbiological information becomes available. Some patients need intensive support, including treatment for shock, renal failure, adult respiratory distress and disseminated intravascular coagulation (see Chapter 19).

Prevention and control
Delivery should be as hygienic and atraumatic as possible. The placenta should be delivered without undue delay, and retained products should be promptly evacuated. Similar attention should be afforded to premature deliveries, stillbirths and terminations of pregnancy.

When a mother has a severe infection, her infant should be observed closely in the neonatal period as it is at increased risk of bacteraemia or meningitis caused by the same organisms.

Breast abscess during lactation

This is a distressing and painful condition, usually caused by *Staphylococcus aureus* infection. Bacteria probably ascend via the milk ducts, and replicate in an area of stagnation. The abscess often develops soon after breast-feeding commences when the milk flow is relatively intermittent, and difficulties with cracked or infected nipples are commonest. Avoiding breast engorgement and nipple trauma reduces the risk of breast abscess.

A typical hot, tender lesion is palpable in the affected breast, and may point towards the surface as a red fluctuant area. 'Blind' treatment with moderate oral doses of flucloxacillin is justified in early cases, and may permit early resolution without significant interruption of breast-feeding. The antibiotics are secreted in milk, but are safe for the infant, though rare cases of penicillin rash can occur. In more severe cases pain may prevent continued feeding; aspiration or drainage of fluctuant lesions may afford relief, and antibiotic treatment will clear residual infection. Lactation can be maintained by gently expressing milk from the affected side and allowing the infant to feed from the other.

Antibiotics during lactation
Few antibiotics are contraindicated during lactation. Sulphonamides are excreted in small amounts in breast milk, and could exacerbate kernicterus, or cause skin rashes or rare haemolysis in glucose-6-phosphate dehydrogenase deficiency. Chloramphenicol may cause neutropenia in the infant. Isoniazid is also excreted in breast milk; if it is given, both mother and baby should receive pyridoxine supplements and be monitored for possible adverse effects.

Antimicrobials safe to use in lactation

These include penicillins, cephalosporins, aminoglycosides, erythromycin, standard doses of metronidazole (but large single doses are not recommended), aciclovir, rifampicin, ethambutol, pyrazinamide and chloroquine (which is not excreted sufficiently for a prophylactic effect in the baby).

Antimicrobials best avoided in lactation

These include tetracyclines, ciprofloxacin, vancomycin, high-dose sulphonamides, imipenem and ganciclovir.

Rare zoonoses in pregnancy

These are brucellosis, Q fever and enzootic ovine abortion (a *Chlamydia psittaci* disease of sheep). All cause abortion in their primary bovine or ovine hosts, and severe infection with placental damage and abortion in pregnant women. However, if both mother and fetus survive, the baby is not permanently harmed.

Farmers, farmers' wives or women working with animals are the population at risk. Q fever has also spread from cats during the birth of kittens. The diagnosis may be suspected on epidemiological grounds. Urgent serodiagnosis should be sought (see Chapter 7).

Early treatment is essential, and should not harm the fetus. Q fever and enzootic ovine abortion may be treated with intravenous erythromycin, to which chloramphenicol can be added if a prompt response is not obtained. Q fever may also respond to monotherapy with chloramphenicol. Ciprofloxacin has been used in maternal Q fever, but has not been shown to abolish placental infection. It is not recommended in pregnancy, but its use may be justified in this serious situation.

Brucellosis in pregnancy is difficult to treat, as tetracycline, by far the best choice, may harm the fetus. The risk of co-trimoxazole plus rifampicin is small, and may be justified in this situation, but there are now a number of reports of successful treatment with oral rifampicin, 900 mg daily, which is considered to be the treatment of choice in endemic countries. Treatment should be continued for at least a month. Re-treatment with a tetracycline, given to the mother after delivery, is not thought to be necessary after successful treatment with rifampicin (see Chapter 25).

Part 4: Disorders Affecting More Than One System

Tuberculosis and Other Mycobacterial Diseases

18

Introduction

Tuberculosis is a granulomatous disease caused by infection with some species of mycobacteria. Depending on the portal of entry of the infection, and on the degree of haematogenous spread, many organs and systems of the body can be affected. The commonest sites of infection are the lungs or the lymph nodes. Other important sites are the bowel, the peritoneum, the meninges, the bones and joints, the renal system and the skin.

The disease is important because it is very common in many countries (Fig. 18.1), the pulmonary form is highly infectious, and tuberculosis can cause severe morbidity and mortality in people of all ages. Drug-resistant tuberculosis is a rapidly increasing problem worldwide.

Epidemiology

Tuberculosis is one of the commonest infections of humans and is responsible for more deaths than almost any other infectious disease. The World Health Organization declared TB a global emergency in 1993 and it is estimated that one-third of the world's population are infected, with nine million active cases and two million deaths worldwide in 2003. The numbers infected with tuberculosis are

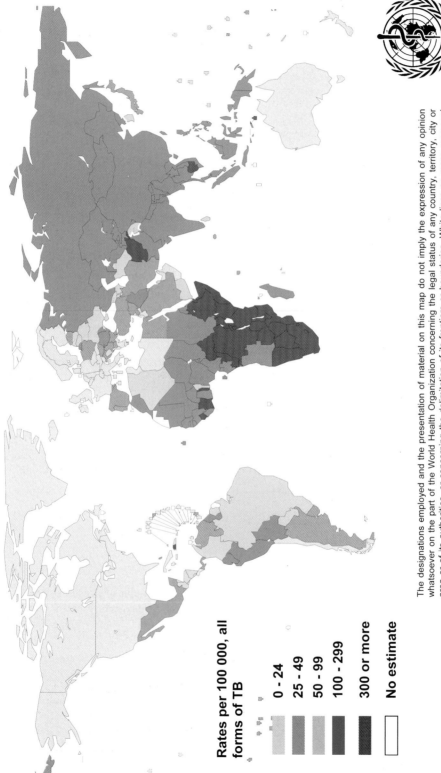

Rates per 100 000, all forms of TB

- 0 - 24
- 25 - 49
- 50 - 99
- 100 - 299
- 300 or more

No estimate

The designations employed and the presentation of material on this map do not imply the expression of any opinion whatsoever on the part of the World Health Organization concerning the legal status of any country, territory, city or area or of its authorities, or concerning the delimitation of its frontiers or boundaries. White lines on maps represent approximate border lines for which there may not yet be full agreement.

© WHO 2004

Figure 18.1 Incidence of tuberculosis (TB) reported in different regions of the world in 2003. Courtesy of the World Health Organization.

anticipated to increase further in the next few years, particularly in countries where most of the cases occur. This trend in low-income countries is explained by a combination of factors including increasing population, migration from rural to urban areas with associated poverty and poor social conditions, deterioration of the public health infrastructure needed to control tuberculosis, and the HIV epidemic.

In many high-income countries, after almost a century of steady decline, the incidence of TB has started to rise again (Fig. 18.2). In the UK the incidence has increased since the mid-1980s and in 2003 stood at 12.8 per 100 000 (slightly lower than the average for western Europe of 14.4 per 100 000). The disease in the UK is mainly concentrated in big cities (nearly 50% of all cases occur in London) and affects particular population groups. In most Western countries a high proportion of the burden of the disease falls on migrants. For example in 2003, 70% of all the TB cases in the UK occurred in foreign-born individuals. Other risk groups include the elderly (who experience reactivation of infection acquired many years ago), homeless people, those with drug and alcohol dependency, and prisoners. Overcrowded conditions facilitate transmission of infection and outbreaks can also occur in settings, such as hospitals and schools, where vulnerable people are collected together.

The prevalence of exposure to mycobacteria, as detected by tuberculin testing, rises with age. When the schools BCG programme operated in the UK, the tuberculin sensitivity among school children aged 10–13 years, tested prior to bacillus Calmette–Guérin (BCG) administration, was between 1 and 2%.

Most infections in developed countries are caused by *Mycobacterium tuberculosis* and infection due to *M. bovis* is rare. Risk factors for *M. bovis* infection include occupational contact with infected cattle and drinking un-pasteurized milk, both of which are more common in low-income countries.

Clinical and epidemiological classification of mycobacteria

There are more than 85 species of mycobacteria, and, with the availability of rapid sequencing facilities, many new species continue to be proposed. Most mycobacteria are environmental organisms that rarely, if ever, cause human disease. Their natural habitat is soil, and fresh and estuarine water. Only a minority of species are obligate human pathogens. The remaining human diseases arise when host defences are reduced by physical or immunological defect. The classification set out below is a simplified approach that is of clinical use.

Obligate pathogens

This important group includes those organisms whose isolation or detection implies that disease is present and requires specific chemotherapy. It includes the organisms of the *M. tuberculosis* complex: *M. tuberculosis*, *M. africanum*, *M. bovis*, and bacille Calmette–Guérin (BCG). *M. leprae* is in a group of its own.

Pulmonary opportunists

M. kansasii, *M. malmoense* and *M. xenopi* infection are found most commonly in patients who have lower respiratory tract abnormalities, such as chronic obstructive pulmonary disease or bronchiectasis. They are, rarely, isolated from patients with acquired immunodeficiency syndrome (AIDS). Infection with these organisms usually resembles indolent progressive pulmonary tuberculosis.

AIDS-related opportunists

Organisms of the *M. avium-intracellulare* (MAI) com-

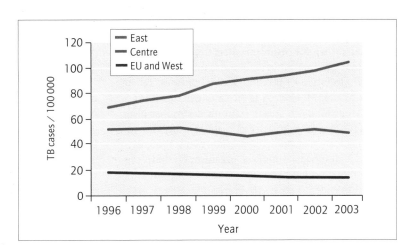

Figure 18.2 Tuberculosis notification rates by geographic area, Europe, 1996–2003.

plex are common causes of systemic infections late in the progress of AIDS when the CD4 count falls below 50×10^6/l. MAI are ubiquitous in the environment. Infection is probably acquired mainly by ingestion or inhalation. Cervical adenopathy is a common presentation, but systemic infections become increasingly common with waning immunity. They include infiltration of parenchymal organs, enteritis, meningoencephalitis, serositis with pericardial effusions or ascites, and systemic disease with mycobacteraemia. MAI infections are rare in immune-competent patients but have been associated with outbreaks of pulmonary infection in a hospital environment. Several 'new' mycobacteria have been isolated from patients with HIV infection including *M. genevense*, *M. haemophilum*, *M. cookei* and *M. hiberniae*. These remain uncommon.

Other opportunistic infections

In the past, opportunist infections with mycobacteria were rare. Nowadays, they occur in transplant patients, those treated for lymphomas and those with inherited cellular immunodeficiencies, as well as in HIV-infected patients. In the absence of normal inflammatory responses they often cause low-grade, persisting feverish illness, or disseminated infections (see above). Skin and tissue abscesses, or bacteraemias may be caused by 'rapid growers' (see below).

The most commonly isolated mycobacteria in the UK

1 *Mycobacterium tuberculosis*.
2 *M. avium-intercellulare*.
3 *M. kansasii*.
4 *M. bovis*.
5 *M. xenopi*.
6 *M. chelonei*.
7 *M. fortuitum*.

Skin pathogens

Mycobacterium marinum is found in old, concrete-lined swimming pools, rivers and fish tanks or aquaria. Infection occurs when the organism enters through broken or macerated skin, and causes chronic granulomatous infection and ulceration of the skin and, sometimes, subcutaneous tissues. *M. ulcerans* is the causative organism of buruli (tropical) ulcer. This is a chronic destructive ulcer, often of the foot or leg, which erodes the skin and the subcutaneous tissues, including bone. Infection is thought to arise through inoculation of the organism by sharp vegetation, though it has never been isolated from the environment in endemic areas. *M. tuberculosis* can also act as a skin pathogen, causing a chronic scarring infection of

the skin, principally on the face. The disease is called lupus vulgaris because of the wolf-like facies of sufferers in the pre-antibiotic period.

Rapid growers

These organisms include *M. fortuitum* and the *M. chelonei* complex, and they differ from other mycobacteria in the speed of their growth. Isolates grow on Löwenstein–Jensen and other selective media in 2–48 h. The organisms have low virulence but can be a problem in some circumstances. *M. chelonei* has caused systemic infections in neutropenic patients. Water (including natural ponds, hydrotherapy pools, or water used for injections), has been a source of infection in outbreaks. Both *M. chelonei* and *M. fortuitum* can cause abscesses in patients who have been injected with 'sterile' fluids contaminated by this organism (for example, at injection sites in diabetics whose needles or multidose insulin containers have become contaminated). *M. fortuitum* is a rare cause of opportunistic bone and joint infections.

Microbiology and pathogenesis of mycobacterial diseases

Mycobacteria are rod-shaped organisms possessing a high proportion of lipid and glycolipid in their cell walls, which makes them resistant to ordinary methods of staining (though they are related to Gram-positive organisms such as *Actinomyces* and *Listeria*). They can survive exposure to acid and alcohol, and are therefore not killed by the acidic gastric environment. They can be demonstrated by staining with auramine or hot carbolfuchsin stains, which resist discoloration by acid–alcohol solvents.

Most mycobacteria replicate more slowly than other pathogenic bacteria, requiring incubation for between 1 and 6 weeks before detectable growth occurs in artificial media. This means that definitive diagnosis is sometimes only confirmed many days after the commencement of empirical treatment. Also, sensitivity testing to antituberculosis drugs, which often relies on subculture of organisms in the presence of antibiotics, may take some weeks. A few species (called 'rapid growers') can produce detectable growth in a few days.

The pathogenesis of mycobacterial infection has proved difficult to research; however, a number of important pathogenic attributes of mycobacteria can be identified. These mechanisms are gradually being elucidated by the use of molecular cytology and immunology techniques.

Mainly due to the very lipid-rich cell wall, up to 40% of the dry weight of mycobacterial organisms is made up of lipid, and several lipid antigens have been identified as potential pathogenicity determinants.

Handling of mycobacterial antigens by the immune system

Once ingested by antigen-presenting cells, mycobacterial antigens are processed and presented in the classic way, in association with human leucocyte antigen (HLA) class II molecules. T lymphocytes bearing receptors that can recognize this complex bind to the macrophage, causing the release of interleukin 2 (IL-2), and the activation of T helper cells. IL-2 stimulates both antigen-specific T cells and antigen non-specific N cells. All of these cell types are capable of producing gamma-interferon, which is important in activating bactericidal mechanisms in macrophages.

Basis of the wasting effect of mycobacterial infections

The macrophage releases several cytokines while reacting to mycobacteria. These include tumour necrosis factor-alpha (TNF-alpha), IL-3 and granulocyte–macrophage colony-stimulating factor (GM-CSF), which probably have an important role in inducing symptoms such as fever and wasting. Mycobacterial lipoarabinomannan is a strong inducer of TNF.

Inhibition of cell-mediated immune responses

Macrophage activation is inhibited by mycobacterial lipid antigens such as lipoarabinomannan, and capsule-like materials such as the phenolic glycolipid of *M. leprae*. These are thought to act by scavenging reactive oxygen intermediates and interfering with the efficiency of the microbicidal oxidative system. Protein antigens of mycobacteria, such as catalase and superoxide dismutase, may also act in this way.

Mycobacterial antigens may also interfere with the activation of T-cell responses. Lipoarabinomannan has been implicated in blocking the stimulatory effect of gamma-interferon by inhibiting its interaction with macrophage surface molecules such as protein kinase C. Lymphocytes and macrophages that are recruited to the site of infection but cannot destroy the mycobacteria contribute to the progressive formation of granulomas.

Mechanisms of survival within macrophages

A key characteristic of the pathogenic mycobacteria is the ability to survive inside macrophages. Mycobacterial ingestion is mediated via the chemokine CR1 and CR3 receptors, which do not stimulate microbicidal oxidative responses. Mycobacteria are able to survive inside macrophages by arresting the maturation of the phagocytic vacuole and preventing its acidification.

Heat-shock proteins are produced by phagocytosed mycobacteria, and this may also be an adaptive mechanism for survival inside the macrophage.

Possible pathogenicity factors of *Mycobacterium tuberculosis*

1 Lipoarabinomannan (induces tumour necrosis factor and scavenges oxidizing molecules).
2 Catalase (scavenges oxidizing molecules).
3 Superoxide dismutase (scavenges oxidizing molecules).
4 Inhibitors of phagosomal–lysosomal fusion.
5 Heat-shock proteins.

Diagnosis of tuberculosis

Advice on the detection and management of tuberculosis, based on summation of the best available evidence, is available on the website of the National Institute for Health and Clinical Excellence (www.nice.org.uk).

The diagnosis of tuberculosis is suggested by the typical general features, and localizing symptoms and signs. Confirmation of the diagnosis is not always easy, as there are relatively few mycobacteria in the primary lesion, and they are encased in a dense granuloma (Fig. 18.3). Few cases of primary pulmonary tuberculosis have a positive sputum examination. Pleural, pericardial or ascitic fluids rarely yield positive smears or cultures. Only about 1 in 5 cases of tuberculous meningitis have positive cerebrospinal fluid (CSF) bacteriology. Bacteriological examination should always be performed, as it is diagnostic when positive. It is not reliable in excluding primary tuberculosis.

Tuberculin tests

The tuberculin test is an intradermal test for cell-medi-

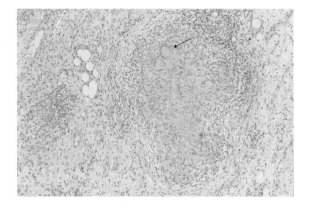

Figure 18.3 Haematoxylin and eosin-stained section of a granuloma, showing densely packed mononuclear infiltrate and multinucleate giant cells (arrow). Acid-fast bacilli are not seen, as they require special stains.

ated hypersensitivity to tuberculoprotein. It is not directly related to active tuberculosis, and has poor diagnostic sensitivity and specificity (see below). It is most often used for screening individual patients or contacts, to detect the development of a positive test. However, when microbiological diagnosis may be unavailable or delayed, it can provide evidence of an active immunological reaction to mycobacterial exposure.

Tuberculin sensitivity is usually demonstrated, and roughly quantified, by the Mantoux test. In some settings a 'tine test' is performed, using a disposable inoculation device, ready-impregnated with tuberculin (this test is subject to false-negative results, and is not recommended for use in Britain).

The Mantoux test

The test uses purified protein derivative (PPD), which is provided in solutions of 2 tuberculin units (2 TU) in a 0.1 ml dose (or a stronger preparation of 10 TU per 0.1 ml).

The standard Mantoux test is performed by inoculating 2 TU of PPD in 0.1 ml, intradermally, into the anterior surface of the forearm, to produce a small 'bleb' in the layers of the epidermis.

The result of the test is 'read' after an interval of 48–72 hours: the diameter of any resulting area of induration (*not* of any erythema) is marked and measured.

• *A negative test result* is induration reaching a diameter of 0–5 mm.

• *A positive result* is an area of induration of 6–14 mm in diameter (most positive reactions cause itching or irritation).

• *A strongly positive result* is induration of greater than 15 mm.

⚠ Very strong tuberculin reactions may produce extensive oedema, induration, ulceration at the injection site or lesions in the ascending draining lymphatics.

⚠ A further test, using 10 TU of PPD may be carried out if a 2 TU test produces negative results, and further testing is clinically indicated (this is usually in patients with impaired immune responses).

The tuberculin test is negative in those who have never been infected with tuberculosis. It becomes positive 3–5 weeks after infection. A strongly positive test, or conversion from negative to positive, indicates active tuberculosis unless the diagnosis is disproved.

Previously fit individuals who are reacting strongly to current infection with tuberculosis often develop strongly positive tuberculin reactions (Fig. 18.4). Some well individuals who are repeatedly exposed to mycobacterial proteins, including residents in endemic areas, and nurses

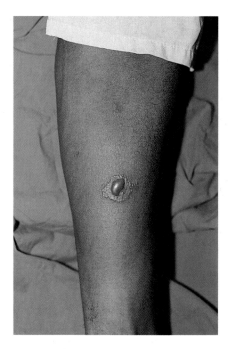

Figure 18.4 Strongly positive Mantoux test: this patient had a short history and high fever; the main reaction is 18 mm in diameter.

and doctors in chest or infectious diseases departments, may also display strongly positive tuberculin reactions. If such an individual is well and has a normal chest X-ray, treatment is not indicated, unless the positivity is related to a recent family contact (see section on prevention and control, below).

The tuberculin test may be misleadingly negative if performed too early after infection, before it has converted to a positive response. This can occur when primary tuberculosis presents as erythema nodosum or meningitis, soon after exposure to infection. It is worth repeating the test after 1 or 2 more weeks when tuberculosis is strongly suspected.

A debilitated patient or one with overwhelming tuberculous infection may have a falsely negative tuberculin test because of suppression of cell-mediated responses. This is most common in post-primary disease (see below) or in immunosuppressed patients. Improvement in the patient's general condition often restores the ability to mount a response.

Histology and cytology as diagnostic tools

Infected tissue such as pleura or lymph node is often accessible to needle biopsy or fine-needle aspiration, which

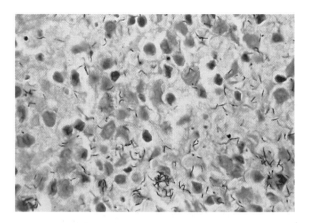

Figure 18.5 Ziehl–Nielsen-stained material obtained from a caseating mediastinal lymph node. Many acid–alcohol-fast bacteria are seen, with the typical appearance of *Mycobacterium tuberculosis*.

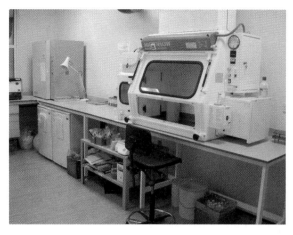

Figure 18.6 View of a CL3 laboratory, showing the exhaust protective cabinet that provides safety to the operator while manipulating cultures containing hazardous pathogens.

may permit demonstration of acid-fast bacilli (Fig. 18.5). Excision biopsy of cervical lymph nodes is a straightforward procedure, which contributes to cure. Histological examination of biopsy material shows granulomata. In the absence of caseation, a firm diagnosis of tuberculosis cannot be made, and sarcoidosis or other granulomatous conditions must be excluded by other means. Tissue sections can be stained to demonstrate acid–alcohol-fast bacilli, which are diagnostic if present. Culture of unfixed tissue should always be carried out, as this increases diagnostic yield by about 50%. It is important not to place all of the tissue obtained into formalin or other fixative, as this prevents its use for culture.

Microbiological diagnosis of mycobacterial disease

Specimens
Specimens that may contain *Mycobacterium tuberculosis* are often extensively manipulated during diagnostic procedures in the microbiology laboratory, and the pathogen is capable of causing severe, long-term disease. It must be handled in laboratory containment level 3 (CL3, Fig. 18.6: see also Chapter 23).

Sputum is the most important specimen for the diagnosis of pulmonary tuberculosis. Early-morning specimens are preferable. In some patients excretion of bacilli may be intermittent or scanty; other patients find it impossible to produce a satisfactory specimen. In these circumstances bronchoalveolar lavage is indicated, but morning aspiration of the gastric contents (to recover swallowed

organisms) is a low-cost alternative that may yield a positive diagnosis. Unfortunately, microscopic examination of gastric aspirate can give false positive results, so these specimens are only suitable for culture.

Cerebrospinal fluid is required for the diagnosis of meningitis. Other naturally sterile fluids may be examined, but the diagnostic yield is often disappointing: culture and histological examination of a pleural biopsy increases this yield in pleural tuberculosis. Early-morning urine specimens can be used for the diagnosis of renal tuberculosis. Twenty-four-hour collections are less useful because of frequent contamination with other bacteria. Pus may be submitted for acid-fast staining and mycobacterial culture if tuberculosis is suspected.

Blood culture has become more important in the diagnosis of disseminated mycobacterial infections following the HIV epidemic. *Mycobacterium avium-intracellulare* infection can readily be demonstrated in blood culture. In these patients faecal smear and culture can also yield a positive diagnosis, as intestinal infection is present before dissemination takes place.

In the laboratory, specimens that are normally sterile can be processed without decontamination, on non-selective media, while those that possess a normal bacterial flora require decontamination before inoculation. Cerebrospinal fluid, pus and blood do not require decontamination but sputum and faeces do. Urine, which may sometimes be contaminated, should be examined by Gram stain and, if non-mycobacterial organisms are seen, it is subjected to a decontamination procedure (see below).

Specimens that may be examined for mycobacterial infection

Key: D, decontamination; C, culture; H, histology; S, smear and acid-fast staining; PCR, polymerase chain reaction techniques.

Body fluids

1 Sputum (D, S, C, PCR).
2 Gastric aspirate (C).
3 Effusion fluids (S, C).
4 Early-morning urine specimens (S, C).
5 Cerebrospinal fluid (PCR, S, C).
6 Pus (D, S, C).
7 Blood (C).
8 Faeces (D, S, C).
9 Fine-needle lymph-node aspirate (S, C).
10 Bronchoalveolar lavage (BAL) specimen (PCR, S, C).

Tissues

1 Lymph-node biopsy (H, S, C).
2 Liver biopsy (H, S, C).
3 Pleural biopsy (H, S, C).
4 Uterine curettings (H, S, C).
5 Biopsy of affected skin (H, S, C).
6 Bone marrow (H, S, PCR, C).

Microscopy

The examination of sputum smears is central to the diagnosis of pulmonary tuberculosis. Early-morning samples are collected and, if not processed immediately, these should be refrigerated to prevent overgrowth of other bacteria. Since sputum is contaminated with mouth flora, it must be decontaminated before it can be used for culture. Several specimens from the same patient can be pooled and decontaminated using 4% sodium hydroxide. This acts by killing the other bacteria present in the specimen. This effect is not absolutely specific, so care is necessary to avoid killing too many mycobacteria, and causing false negative results. The treated specimen is centrifuged, and the deposit is examined microscopically and inoculated onto culture medium (see below).

Slides can be stained either by a modification of the Ziehl–Nielsen method, in which hot carbolfuchsin is used to stain the mycobacteria, or by the phenol auramine technique, which uses a fluorescent dye easily visible under ultraviolet illumination. Auramine staining enables large numbers of specimens to be screened quickly using a UV microscope. All positive slides are then overstained by the Ziehl–Nielsen method to confirm the identity of the fluorescent objects. This technique is particularly suited to laboratories with a large throughput. In smaller laboratories, or where an ultraviolet microscope is not available, Ziehl–Nielsen is the method of choice.

The lower limit of detection by stained smear is approximately 10^4 CFU/ml. Excretion of bacilli can be intermittent, so that a negative result does not absolutely exclude the diagnosis. However, the importance of microscopical diagnosis of tuberculosis cannot be over-emphasized. Not only is it a rapid and relatively sensitive diagnostic technique, but it also identifies the most infectious patients, who excrete large numbers of organisms.

Culture of mycobacteria

All but a few species of mycobacteria are slow-growing. This means that other bacteria present in the specimen would rapidly overgrow if not adequately suppressed. Specimens from sterile sites can be inoculated directly on to isolation media but those, such as sputum, from sites with a normal bacterial flora require decontamination. Growth of contaminating species is further inhibited by the incorporation of dyes, for instance malachite green in Löwenstein–Jensen medium, or antibiotics, as in Kirchner's selective medium. Culture must be performed in firmly closed containers to prevent the release of infectious organisms, and to prevent desiccation of the culture medium.

Many different media are used in the isolation of mycobacteria. All contain a source of fatty acids; fresh eggs in Löwenstein–Jensen, or purified oleic acid in Middlebrook's medium. Liquid media increase the diagnostic yield because they permit inoculation of a larger amount of specimen. The optimal approach is to use a combination of solid and liquid media. A positive diagnosis is made when colonies grow on solid medium, or liquid medium becomes cloudy. Colonies of *Mycobacterium tuberculosis* on solid media are usually rough and a buff colour, and the smear often shows 'cording' (see Fig. 18.5), but these characteristics are not sufficiently typical to allow a presumptive diagnosis.

More recently, the diagnosis of mycobacterial infection has been improved by the introduction of radiometric detection of mycobacterial growth. This technique utilizes a Middlebrook broth and a machine that continuously monitors the culture bottles to detect the production of CO_2 by one of various means. This has reduced the time taken to detect a positive isolate from about 21 days to less than 10 days.

Using polymerase chain reaction (PCR) nucleic-acid amplification techniques, mycobacterial DNA can be detected in clinical specimens. This technique has a similar sensitivity to that of culture, but is much quicker, giving a result in a matter of hours. By amplifying the *rpoB* gene and sequencing the product it is possible to detect the presence of rifampicin resistance.

Identification of mycobacterial species

Conventional methods of identifying mycobacteria include studying bacterial morphology using Ziehl–Nielsen stain. Organisms can then be subcultured on to identification and sensitivity media. Definitive diagnosis is based on biochemical tests, ability to grow at various temperatures, and ability to produce coloured pigments in the presence or absence of light. Most modern laboratories identify mycobacteria by molecular techniques including PCR, hybridization or direct sequencing of the 16S rRNA gene.

Typing mycobacteria

Several techniques are now available to type *M. tuberculosis* using analysis of restriction fragment length polymorphisms with several different probes. An alternative is a PCR-based method that amplifies repeat sequences. This has the advantage that the identity of the organism is in the form of a reproducible numerical code. These methods are very valuable, not only in worldwide epidemiological studies, but in episodes of transmission within a hospital, laboratory or community.

Sensitivity testing

Resistance determinants in mycobacteria are not located on transmissible plasmids, phages or transposons, as in other bacteria, but arise by spontaneous mutation. Mutation to rifampicin resistance occurs once in 10^8 cell divisions. In a patient with pulmonary tuberculosis there are approximately 10^{13} mycobacteria, so some bacteria will inevitably possess resistance factors. If more than one drug is given to the patient, the likelihood of generating resistance to all of the agents in the combination is multiplied (i.e. 10^8 rifampicin $\times$ 10^6 ethambutol and 10^6 pyrazinamide = 10^{20}, so the risk is once in 10^{20} divisions). Generally, if 1–10% of a patient's mycobacterial population is resistant to an antimicrobial drug, laboratory tests will indicate resistance to that drug. By using combinations of antituberculosis agents, the risk that an organism resistant to all of the prescribed agents will develop is very low. If patients do not receive appropriate medication, it is possible that mycobacteria will be exposed only to single effective agents, increasing the risk of induction of resistance.

Resistance is described as primary if it arises in patients never previously treated for tuberculosis. It implies that there has been transmission of a resistant strain in the community. Secondary resistance arises during therapy and indicates poor compliance or an inadequate prescription. Multiple drug resistance is defined as the presence of at least rifampicin and isoniazid resistance.

Resistance to rifampicin is caused by mutations in a particular area (resistance hot-spot) of the beta-subunit of the RNA polymerase gene *rpoB*. Resistance to isoniazid can arise via deletion of or mutation in the catalase *katG* gene, or mutation in the *inhA* or *aphC* genes. Studies of

isoniazid-resistant populations of *M. tuberculosis* suggest that there are other mechanisms still to be discovered. Although the mechanism of action of pyrazinamide is uncertain, resistance is probably associated with mutation in the pyrazinamidase gene *pncA*. Resistance mechanisms have also been determined for streptomycin, fluoroquinolones and ethambutol.

In countries where antituberculosis therapy is closely controlled, wild mycobacterial strains are usually sensitive to all agents. Where control of treatment is poor, patients may take only one drug (or one effective drug) for long periods of time, and resistance is more common. A high cost of medical care, poor advice and the choice of inadequate regimens contribute to this. Extreme poverty may deny treatment to many. Also, patients must understand the need to continue therapy after they return to apparently normal health. Where drugs are provided free by government schemes, they may be sold by patients to supplement inadequate incomes.

Methods of testing for antimicrobial resistance

The resistance ratio method is the method most commonly employed in UK reference centres. It uses Löwenstein–Jensen medium containing differing dilutions of antituberculosis drugs. Essentially, a minimum inhibitory concentration (MIC) is obtained for the test organism and for a range of simultaneously tested control strains. An organism with an MIC below that of the controls, by a factor of at least four, is considered sensitive.

The proportion method is a method in which cultures are inoculated on to plates containing test antibiotic or no antibiotic. A strain is considered resistant to a test antibiotic if the proportion of bacteria growing in the presence of drug, compared to the non-drug control, is greater than 1%.

Growth detection methods: all of the current automated mycobacterial culture methods can be used to perform rapid susceptibility testing, giving results in about 10 days. The rate of growth of the test organism is compared in the presence and absence of the antituberculosis agent.

Molecular detection methods use nucleic acid amplification techniques (NAAT) and sequence analysis of appropriate genes, to permit rapid determination of resistance for antibiotics, such as rifampicin or fluoroquinolones, to which resistance emerges through a small number of mutations. This approach cannot be used where there are several potential mechanisms of resistance, as in the case of isoniazid. There are now several commercial NAAT methods that allow detection of mutations in the *rpoB* gene. This approach is especially valuable as rifampicin resistance can be used as a marker for multiple-drug resistance.

The varying pathology of tuberculosis

Primary tuberculosis

Primary tuberculosis is the result of primary infection in a non-immune host. There is usually a small focus of inflammation, with a few mycobacteria surrounded by a dense granuloma (see Fig. 18.3). When this occurs in the lung it is called a Ghon focus. Regional lymph nodes are often enlarged, and the combination of primary granuloma and enlarged nodes is called a primary complex.

Primary tuberculosis is often clinically silent; indeed, naturally acquired immunity often follows healing of an inapparent primary complex. Healed lesions slowly calcify, leaving an irregular, easily identified X-ray-dense shadow.

Post-primary tuberculosis

Post-primary tuberculosis is a re-infection or reactivation illness. It can follow primary infection or immunization after a few weeks to many years. The resulting immune re-action causes the formation of large, exuberant granulomata, often with central, cheesy necrosis, called caseation. In pulmonary infection, necrotic tissue is coughed away, leaving cavities. In solid organs or soft tissues, the caseous material resembles pus, and may discharge. The abscess-like lesion is indurated rather than hot and inflamed, and is therefore called a cold abscess.

In post-primary tuberculosis the immune response is progressively suppressed by the mycobacteria, which survive within macrophages. Numerous organisms are released from caseating tissues. Post-primary, cavitating ('open') pulmonary tuberculosis can be highly infectious to susceptible close contacts because of the large numbers of mycobacteria released during coughing.

Miliary or disseminated tuberculosis

Miliary tuberculosis occurs when mycobacteria are disseminated via the bloodstream. It may follow rupture of an active granuloma into a blood vessel. Many organs are affected. The severity of disease varies from mild fever and malaise to severe, debilitating illness. Macroscopically detectable granulomata are visible in the organs as small white nodules rather like millet seeds (the origin of the expression miliary tuberculosis). They may be visible as widespread, small spots on the chest X-ray.

Mycobacterial infections in the immunosuppressed

When cell-mediated immunity is lost, granulomas tend not to form in response to mycobacterial infection. There is also a lack of cellular infiltrate in infected tissues. Histological diagnosis may then be difficult. A strong index of suspicion must be maintained to avoid overlooking the diagnosis. Techniques such as culture and PCR, used in examining specimens such as stools, tissue and blood, are then very important in confirmatory investigation.

Organism list

Human organisms

- *Mycobacterium tuberculosis*
- *M. africanum*
- *M. bovis*
- BCG

Other mycobacteria capable of causing typical tuberculosis

- *M. kansasii*
- *M. avium-intracellulare* complex
- *M. xenopi*

Mycobacteria that can cause cold abscesses

- *M. fortuitum*
- *M. chelonei*
- *M. scrofulaceum*
- BCG (if mistakenly injected subcutaneously)

Mycobacteria associated with skin lesions

- *M. marinum*
- *M. ulcerans*

Organism of leprosy

- *M. leprae*

Primary tuberculosis

Introduction

Primary tuberculosis characteristically affects certain organs and systems. The commonest affected organ is the lung, but the pleura, lymph nodes, peritoneum, pericardium and meninges are other sites where primary disease occurs.

Clinical features

The clinical presentation is a combination of the general features of tuberculosis and of signs and symptoms related to the affected site. The general features are persisting fever, night sweats, anorexia and weight loss. There is no predictable change in the blood count, the liver function tests or other blood biochemistry. The erythrocyte sedimentation rate and C-reactive protein are usually raised.

The clinical features relating to the affected site are often helpful in suggesting the diagnosis, because of both the typical symptoms and the typical type of lesion produced.

Erythema nodosum

Erythema nodosum sometimes accompanies primary tuberculosis, and is occasionally the first sign of the disease. Other associations with erythema nodosum, including sarcoidosis or streptococcal infections, are important differential diagnoses (see Chapter 5), but tuberculosis should always be actively excluded.

In rare cases, erythema nodosum develops early in the treatment of tuberculosis. This is similar to the situation in leprosy, where this is a well recognized complication of treatment, occurring with a surge of cell-mediated immune reactivity.

Primary pulmonary tuberculosis

Primary pulmonary tuberculosis, when it is clinically evident, often causes a persistent, dry cough. The site of the infection is usually the periphery of the midzone of one lung. The lesion is too small to produce abnormal signs on physical examination, but is visible on chest X-ray as a fluffy opacity with a diameter of 1–2 cm. The mediastinal lymph nodes on the affected side may be enlarged and, together with the primary focus, form the primary complex (Fig. 18.7).

Endobronchial tuberculosis

Endobronchial tuberculosis sometimes occurs as a primary infection in children. Granulomata develop in the bronchial mucosa of one of the larger airways, causing partial obstruction. There is a persistent, wheezy cough and often a fixed wheeze on auscultation over the affected airway. The narrowed section of bronchus may be visible on chest X-ray, and can be demonstrated on computed tomographic (CT) or magnetic resonance (MR) scanning.

Pleural tuberculosis

Pleural tuberculosis is commonest in young adults. Pleuritic pain is the characteristic clinical feature, and a pleural rub is occasionally heard. Pleural effusion may cause shortness of breath and typical physical signs on chest examination; the pleural rub disappears as the pleural surfaces are separated by effusion. The chest X-ray shows a typical opacity with an up-curved surface (Fig. 18.8).

Tuberculous pericarditis

In tuberculous pericarditis the features of tuberculosis are combined with those of pericarditis (see Chapter 12 and Fig. 18.9). Occasionally, the patient may present with heart failure due to tamponade. Late in the course of disease, fibrosis of the pericardium may lead to constrictive pericarditis.

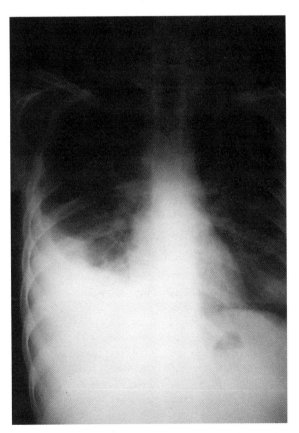

Figure 18.8 Primary tuberculosis: pleural effusion in an Indian teenager with malaise, night sweats, weight loss and low-grade fever.

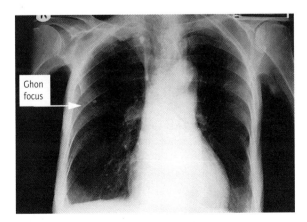

Ghon focus

Figure 18.7 Primary tuberculosis: Ghon focus on chest X-ray. The density of the lesion suggests healing and calcification.

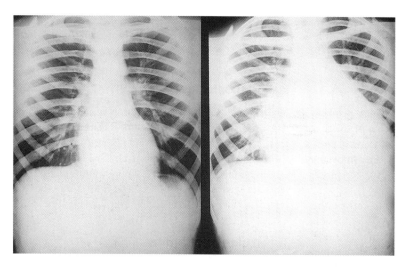

Figure 18.9 Tuberculous pericarditis in a 57-year-old Pakistani businessman: he presented with several days' increasing chest pain, anorexia and fever. Echocardiography revealed the small pericardial effusion, which rapidly enlarged over the next 10 days.

Lymph-node tuberculosis

Lymph-node tuberculosis occurs particularly in children and young adults. It usually affects the cervical or mediastinal nodes. In children, caseating or suppurating lymphadenitis may be due to *M. avium-intracellulare* or, less commonly, *M. scrofulaceum*. Swelling may develop slowly or alarmingly suddenly, when lymphoma must be urgently excluded. Cold abscess formation is nowadays rare, but can occur, even some weeks after starting treatment (Fig. 18.10).

Mediastinal lymph-node swelling may cause cough, due to extrinsic tracheal irritation. Dull, central chest pain is common. Physical signs are few, unless compression of an airway produces a fixed wheeze. The middle lobe bronchus is susceptible to compression by enlarged lymph nodes: this may lead to obstruction of the lobe, with secondary pyogenic infection (middle lobe syndrome). The chest X-ray often shows a nodular shadow widening the mediastinum, with or without narrowing of a central airway (Fig. 18.11).

On rare occasions a swollen lymph node ruptures into an airway, protruding into it and obstructing it. Complete tracheal obstruction may be fatal, but urgent bronchoscopy with bypass or resection is sometimes possible.

Tuberculous peritonitis

Tuberculous peritonitis is most common in Asians. It presents with abdominal discomfort and distension. Minor ascites is common, showing as separation of bowel loops on X-ray or as a fluid collection in the pelvis or peritoneal reflections on imaging. Massive ascites is extremely rare.

Post-primary tuberculosis

Introduction

Post-primary tuberculosis produces expanding, caseating granulomata that release numerous mycobacteria. The tissues affected are different from those involved in primary disease. The commonest site affected is the apex of the lung. Less common are bones and joints, the bowel, renal tract, and the male and female genital tracts. In contrast to primary tuberculosis, spontaneous healing is uncommon. Progressive tissue destruction occurs, causing increasing debility and, eventually, death.

Pulmonary tuberculosis

Introduction and epidemiology

Pulmonary tuberculosis is one of the most important epidemic respiratory infections worldwide. In 2003 44% of

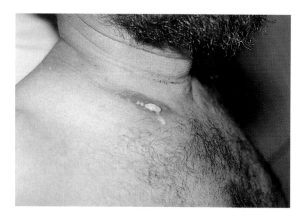

Figure 18.10 Tuberculous lymphadenitis: this 42-year-old man had typical symptoms and a swollen cervical lymph node for 4 months. Chemotherapy did not prevent the formation of a draining sinus.

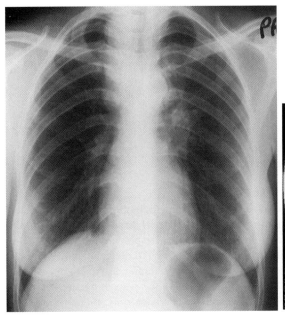

(a)

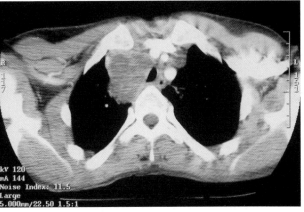

(b)

Figure 18.11 (a) Tuberculous lymphadenitis affecting a mediastinal node: this teenager presented with typical symptoms of tuberculosis, plus a cough and substernal chest pain. (b) CT scanning showed that the enlarged lymph node was compressing the trachea.

all TB cases globally were infectious (i.e. smear-positive) pulmonary TB.

Pathology

The usual cause is *M. tuberculosis* but, as discussed above, other mycobacteria can cause pulmonary disease, often in people with pre-existing pulmonary disease, other debility or immunosuppression. Pulmonary tuberculosis may exist alone or coexist with post-primary tuberculosis in other sites.

Clinical features

Presenting features include the typical quartet of fever, night sweats, anorexia and weight loss. Cough is usual and there is often sputum, which may appear purulent but is rarely copious. A few patients have episodes of blood-streaking in the sputum and the occasional case expectorates fresh blood or blood clot. A few patients have no sputum, and children rarely expectorate their sputum, as it is immediately swallowed.

Physical examination may be surprisingly uninformative. Most patients with pulmonary tuberculosis have few abnormal signs. Crepitations over the affected apex or dullness to percussion and perhaps bronchial breathing are the commonest findings.

The chest X-ray often shows typical changes that are virtually diagnostic. A ragged-edged opacity is seen in the apex of the lung, occasionally of both lungs. The opacity contains small and large lucencies, which represent cavities. Often one or more larger, thick-walled cavities are present (Fig. 18.12).

The erythrocyte sedimentation rate and C-reactive protein may be elevated, but are not always abnormal. As debility and weight loss progress, the blood albumin falls and anaemia of chronic inflammation often develops. Rarely, bizarre haematological abnormalities occur, including thrombocythaemia, leukaemoid reactions or pancytopenia. Their pathogenesis is poorly understood.

Diagnosis

The diagnosis is often evident from the history and the chest X-ray appearances. Confirmation is obtainable from examination of sputum, bronchoalveolar lavage or early-morning gastric aspirate. In early disease and in non-debilitated patients the tuberculin test is often strongly positive. However, in patients originating from highly endemic areas, tuberculin test results are difficult to interpret. The specificity of the tuberculin reaction is poor, with many false-negative results.

It is important to be alert to the possibility of multiple sites being affected since pulmonary disease can coexist with extra-pulmonary disease. In the UK in 2002, 59% of all reported cases had pulmonary TB with or without extra-pulmonary disease (earlier estimates suggest that about 6% of cases have coexisting pulmonary and extra-pulmonary infection), while 41% had only extra-pulmo-

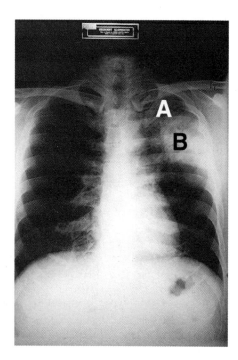

Figure 18.12 Post-primary pulmonary tuberculosis. Sputum microscopy and culture were both positive. A, apical opacity; B, large thick-walled cavity.

nary disease. This proportion has changed little in recent years.

Pulmonary tuberculosis responds readily to short-course, multi-drug chemotherapy, but disease in other sites may be slower to improve. For instance, a patient with pulmonary and articular disease may have an improving chest X-ray and a deteriorating joint (see Fig. 14.10). Similarly, a patient with pulmonary and renal involvement may have completely healed lung lesions, but persisting low-grade infection in a kidney. It is therefore important to examine cases of pulmonary tuberculosis for evidence of extra-pulmonary and disseminated disease. At the least, a urine examination and a full physical examination should be carried out.

Complications
Before treatment
Infected caseous material or sputum may overflow into other parts of the bronchial tree, causing tuberculous bronchopneumonia. This causes multinodular segmental or lobar opacities on chest X-ray and significant impairment of respiratory function.

A cavity may erode a pulmonary blood vessel, causing significant haemoptysis. The appearance of blood-streaked sputum or fresh blood clots is alarming. Bed rest, lying on the side of the lesion to minimize drainage into the opposite lung, and a check of the platelet count are

advisable. Many cases resolve, especially if antituberculosis treatment is in progress. A few cases have severe bleeding, requiring transfusion, emergency embolization of the bleeding vessel or even lobectomy. Catastrophic and fatal haemorrhage is rare.

Patients with severe, extensive pulmonary disease may develop tuberculous laryngitis, with harsh cough and hoarseness. Such patients are extremely infectious. However, they usually respond rapidly to antimicrobial chemotherapy, though some scarring and hoarseness may remain after healing.

Paradoxical reactions
After the start of treatment, immune function is rapidly restored as the mycobacterial load is reduced. The resulting sudden inflammatory response can cause high fever, anaemia, transaminitis, a rapid rise in inflammatory indices and a rapidly falling albumin, leading to hypotension and peripheral oedema. This is most intense in patients who are debilitated, or have coexisting depression of cell-mediated immunity. This reaction can be reduced by treatment with oral prednisolone or intravenous hydrocortisone. In situations where increased inflammation could be damaging (e.g. in meningitis or pericarditis, and in severe pulmonary disease with respiratory impairment), corticosteroids should be given from the onset of antituberculosis treatment. High doses (at least 30–40 mg daily of prednisolone) are required, as co-administration of rifampin effectively halves the blood levels of corticosteroids, due to microsomal enzyme induction. Doses may be stepped down as the patient gains condition over 3–6 weeks. Attention to nutrition, with adequate protein, vitamin and mineral intake, assists in recovery.

Inadequate or failing therapy
In patients with both pulmonary and extra-pulmonary disease, care should be taken to ensure that both have been adequately treated before chemotherapy is discontinued (see above). Resolution of symptoms, weight gain, signs of healing and fibrosis in X-ray or imaging appearances and return of inflammatory markers to normal levels are all helpful in confirming a treatment response.

Failure to respond should prompt a review of the antimicrobial sensitivities of the infecting organism, repeat of culture and sensitivity tests and a review of compliance with therapy.

Tuberculous meningitis

Introduction
Between 100 and 200 cases per year of tuberculous meningitis are reported in the UK. Meningitis is a relatively common manifestation of primary tuberculosis in children and young adults, in whom it can develop within 3–4

weeks of exposure. In adults and the elderly it is more often associated with extensive tuberculosis elsewhere in the body. It is a diagnostic opportunistic condition in AIDS, and is particularly seen in cases originating in Africa.

The disease is important because it can present insidiously, the diagnosis is often difficult, and early CSF changes are similar to those of viral meningitis. Delayed diagnosis or treatment can lead to severe, sometimes irreversible complications.

Clinical features

The disease can present as typical infectious meningitis, but more commonly causes meningitis of slow onset, progressive drowsiness or personality disorder (which can be mistaken for depression, or age-related dementia), or presents with a neurological complication. Headache may not be prominent, but mild or moderate meningism is often demonstrable by careful examination. Fever is almost always significant, but may be absent in immunosuppressed or debilitated patients. Older patients may have signs of pulmonary tuberculosis, or of disease elsewhere. Not all patients complain of night sweats, anorexia or weight loss.

Haematological and biochemical examination of the blood is often normal, though tuberculosis can produce unpredictable changes in the white cell count. If the meningitis is part of miliary or systemic disease, the liver function tests may be mildly abnormal. The erythrocyte sedimentation rate is not predictably altered, but the CRP is usually elevated.

The CSF appears clear in tuberculous meningitis, even though it contains excess white cells. From 10 to 2000 or more lymphocytes per cubic millimetre may be found. The CSF protein is elevated, sometimes so much so that it forms a 'spidery' white clot if the fluid is allowed to stand. The glucose level falls, and may be undetectable in advanced cases.

Diagnosis

The most important factor in diagnosis is suspicion, especially in immunosuppressed patients. Examination of the CSF is mandatory. In most cases imaging of the brain is indicated before lumbar puncture. This is often normal in early cases, but may show oedema or hyperaemia of the meninges, particularly at the base of the brain, or one or more enhancing lesions indicating tuberculomas.

Mild CSF changes, with slightly raised protein, a few excess lymphocytes and a slightly low glucose level, are difficult to distinguish from the changes of viral meningitis. A careful search should therefore be made for evidence of mycobacteria, by microscopy and culture. However, only 20–25% of cases have organisms identified in the CSF by these methods. Polymerase chain amplification of myco-

bacterial DNA in CSF samples is very important in making an early diagnosis and has been shown to be more sensitive than culture in some studies.

A strongly positive tuberculin test may support the diagnosis. A negative test is unhelpful (see above). Childhood tuberculous meningitis sometimes presents before the tuberculin test has converted to positivity. If other diagnostic methods remain negative, the tuberculin test can be repeated after 3–4 weeks (even after presumptive therapy has been commenced).

Management

Treatment should be commenced urgently if tuberculous meningitis is strongly suspected, without awaiting bacteriological confirmation. Initial quadruple therapy is appropriate, using isoniazid, rifampicin, pyrazinamide and ethambutol. Streptomycin does not cross the blood–brain barrier, and is not indicated. Continuation therapy is started after 2 months. Trials of short-course chemotherapy have not been completed for tuberculous meningitis. Treatment is usually continued to complete a total of one year's therapy.

Tuberculous meningitis causes granulomatous vasculitis of meningeal blood vessels, and meningeal oedema, which may increase significantly when treatment removes the inhibitory effect of mycobacteria on immune and inflammatory function (paradoxical reaction). Corticosteroids should therefore be given at the commencement of treatment. A typical dose would be prednisolone 40 mg daily for the first 7–14 days, and then tailing off as fever resolves. Corticosteroids may also aid improvement if neurological abnormalities are present when treatment is started. As long as effective antituberculosis treatment is given concurrently they have no adverse effect on the tuberculosis.

Complications
Tuberculomas

Space-occupying granulomas may already exist when the patient presents. As treatment improves cell-mediated immunity, they may enlarge, causing headache and/or neurological signs. Imaging of the brain demonstrates the lesions, which usually have surrounding oedematous changes. Tuberculomas can appear, or recur, many weeks after beginning antituberculosis treatment, and a minority of patients present with features of space-occupying lesions without previous symptoms of meningitis. Treatment with step-down dosage of dexamethasone or prednisolone, titrated against symptoms and signs, should be given until the tuberculomas have responded to chemotherapy. This may take many weeks or months. On rare occasions they fail to respond, and require neurosurgical intervention.

Focal lesions

Focal spinal cord or brain lesions can be caused directly by granulomas or by granulomatous vasculitis. Cranial nerve lesions, paraparesis and cauda equina syndromes are the commonest problems. Cauda equina lesions are often due to multiple granulomas, and tend to present with a mixture of leg weakness and bladder dysfunction. Spinal cord lesions can also be caused by extradural tuberculous abscesses, related to vertebral tuberculosis.

Hydrocephalus

Hydrocephalus may occur as a result of extensive fibrosis or as part of a poorly controlled granulomatous process. In either case, corticosteroids offer hope of prevention and improvement. Deteriorating consciousness after initial improvement should always prompt further brain imaging. Temporary or permanent CSF drainage may be needed, depending on the response to therapy.

Renal tuberculosis

Introduction

Tuberculosis of the renal tract is assumed to be bloodborne. The kidney is almost always involved; lesions in the collecting system, the ureters, the vas deferens and epididymis result from descending infection.

Initially, infection in the renal medulla affects one or more pyramids. Swelling deforms the adjacent renal calyx, and destruction of the pyramid may cause a bulbous enlargement of the calyx. In untreated disease, the kidney is slowly replaced by a tuberculous abscess. Ureteric obstruction by seedling granulomas, back pressure and hydronephrosis may complicate the infectious process.

Clinical features

Initially clinical features are minimal. Dull flank pain on the affected side occurs in established disease. Low-grade or swinging fever is also common. Many cases have typical symptoms of urinary tract infection and some have superimposed bacterial infections, as deformity and partial ureteric obstruction predispose to bacterial colonization. Some cases come to light because they present with a complicating epididymitis.

A small proportion of patients with renal tuberculosis have coexisting pulmonary disease. Chest X-ray should therefore be performed, and sputum, bronchoalveolar lavage or gastric aspirates examined.

Diagnosis

Diagnosis is by detecting typical renal lesions by imaging, and demonstration of mycobacteria in urine. The specimen of choice is the whole of an early-morning voiding of urine, in which acid-fast bacilli will have been concentrated overnight.

Imaging of the kidneys may show typical distortion of the calyces, expansion of a kidney by the inflammatory process, or areas of calcification related to partial healing. It is advisable to perform imaging that can show ureteric obstruction, as early relief of obstruction can avoid effective loss of the entire kidney.

Tuberculous epididymitis

This presents as a progressive, indurated swelling of the affected epididymis. The inflammation is subacute, with dusky red swelling and relatively little pain. Both sides may be affected, often asymmetrically. The diagnosis is usually confirmed by demonstrating the associated renal infection. If this proves difficult, biopsy with histology and culture will confirm the aetiology and exclude malignancy.

Tuberculosis of the female pelvis

Tuberculosis of the female genital tract usually causes subacute or chronic salpingitis. Granulomas develop in the mucosa of the Fallopian tube and 'seed' the endometrium, many of them being shed during the menses. Progressive distortion and occlusion of the tubes lead to infertility, which is often the only manifestation of the disease.

Clinical features are few in most cases. They include minimal vaginal discharge, mild to moderate suprapubic discomfort, or dyspareunia, in the presence of a large cold abscess. Occasionally, the condition presents subacutely with suprapubic pain and a moderate fever.

The diagnosis may be suggested by an ultrasound appearance of tubal swelling. Laparoscopy permits inspection and biopsy of the tubes. Premenstrual endometrium is obtained by dilatation and curettage for histological and microbiological examination. Typical granulomata or acid-fast bacilli may be seen in the tissue. Mycobacteria can be demonstrated by microscopy and culture.

Tuberculosis of bones and joints

Introduction

Tuberculosis of bones and joints is blood-borne. The joints most often affected are the spine and the hip. The knee and the wrist are less common sites, and involvement of other joints is rare. Synovitis without adjacent bony involvement is occasionally seen, mainly affecting the extensor tendons of the wrist and hand.

Infection begins in the cartilage of the joint, and spreads by a process of caseation and destruction into adjacent bone. The capsule of a joint may rupture, allowing a cold abscess to extend, sometimes erupting at the skin as a sinus. An example of this is the appearance of an abscess in the groin resulting from infection in the lumbar spine, with pus tracking along the psoas muscle from its spinal origin (see Fig. 20.4). Destruction of bone surfaces and cortex can lead to severe deformity. The classic example of this is Pott's disease of the spine, in which infection originates in a disc and spreads to the two adjacent vertebrae (Fig. 18.13). Collapse of the vertebral bodies produces an angular kyphosis at the level of the infection.

Clinical presentation

The clinical presentation is often with fever and pain. Soft-tissue swelling of the synovium and capsule follows, and effusion develops. Spinal tuberculosis may present with neurological features caused by extradural abscess formation and spinal cord compression. Magnetic resonance imaging is the investigation of choice, as it can demonstrate bone marrow oedema, destruction of cartilage and/or bone, soft-tissue changes and other features of inflammation. X-ray changes occur late, and considerable disease must exist before they are detectable. The X-ray changes are indistinguishable from those of chronic pyogenic infection.

Diagnosis

Diagnosis depends on suspicion and specific tests. Joint effusions usually have a high protein content, low glucose and pH, and a predominantly lymphocytic pleiocytosis. Synovial swellings may produce an exudate containing soft masses of inflammatory material, which look similar to melon seeds.

Demonstration of acid-fast bacilli or mycobacterial DNA, or culture of mycobacteria, may be possible from aspirated effusion or synovial biopsy. In advanced disease, bone biopsy may yield positive results.

Problems in treatment of bone and joint tuberculosis

There is often a considerable delay between the start of treatment and the resolution of inflammation and pain. Sinus formation and bone destruction may progress for some weeks after treatment is commenced. Nevertheless, evidence suggests that short-course therapy of 6 or 9 months is effective for bone and joint TB.

Surgical drainage and debridement may contribute to cure in extensive disease (see Fig. 14.10). There is little evidence that bed rest, spinal immobilization or splinting of joints affects the rate of healing or degree of final deformity. However, both may be useful in limiting initial pain and instability during treatment.

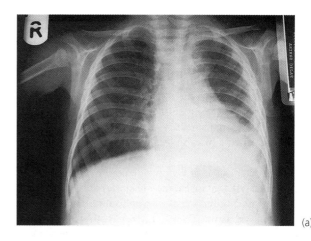

(a)

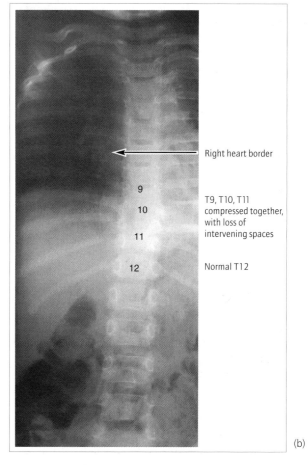

(b)

Figure 18.13 This Bosnian refugee was born in internment and did not receive bacillus Calmette–Guérin (BCG) vaccination. At the age of 18 months she had persistent cough, fever and back pain, with evidence of: (a) left-sided pleural disease on chest X-ray; and (b) spinal osteomyelitis, with loss of two disc spaces and vertebral volume in the lower thoracic spine.

Miliary or disseminated tuberculosis

In miliary tuberculosis, blood-borne dissemination of mycobacteria gives rise to small granulomas in many organs. They are often visible in the chest X-ray and pathologically on the surfaces and cut sections of the solid organs. In rare cases they can be seen in the retina on fundoscopy.

The diagnosis of miliary tuberculosis may be suggested by the chest X-ray appearances (Fig. 18.14). There is rarely significant sputum production, and sputum examination is usually negative. There is often a modest elevation of liver alkaline phosphatase levels in the blood, because of the many space-occupying lesions in the liver. The white cell count and erythrocyte sedimentation rate are not predictably abnormal, but the CRP is raised.

Liver or bone marrow biopsy material may show small granulomas. These are not always caseating, and acid-fast bacilli are not always demonstrable on Ziehl–Nielsen or auramine staining. Some of the material should be submitted unfixed for culture, as this may yield a diagnostic growth of mycobacteria, allowing speciation and sensitivity testing. A positive PCR test may be diagnostic.

Cryptogenic miliary tuberculosis

Disseminated tuberculosis may rarely present simply as a fever and raised inflammatory markers, often with weight loss, with or without an abnormal white cell count or liver function tests. Bone-marrow biopsy may be positive on PCR or culture. The tuberculin test may be strongly positive, but is negative in about 40% of cases. The condition must be suspected on epidemiological grounds or by exclusion.

Treatment of tuberculosis

Introduction

The mainstay of treatment in tuberculosis is effective, multidrug, antimicrobial chemotherapy. In the great majority of cases this will produce cure of the disease with a negligible chance of relapse. There are two phases of treatment.

Initial (intensive) phase (usually 2 months)

For the first 2 months of treatment, ideally, four first-line drugs are given, of which two should be rifampicin and isoniazid, and the third should be pyrazinamide. Ethambutol is the usual fourth drug, though in some countries, intramuscular streptomycin is a widely available, inexpensive alternative. This regimen aims to deliver at least two or three effective drugs, pending the results of sensitivity testing. It greatly reduces the load of mycobacteria by using drugs with a range of actions to attack rapidly replicating intracellular and extracellular mycobacteria. It also initiates therapy against more slowly replicating and dormant organisms.

Continuation phase (usually 4–7 months)

In this phase, rifampicin and isoniazid are continued and the additional drugs are stopped. In most Western countries the sensitivities of the patient's mycobacterium will be known before the beginning of the continuation phase. Provided that pyrazinamide has been used throughout the intensive phase, and the organism is sensitive to it, the continuation phase should be continued for 4 months. If pyrazinamide was not given, or the organism was pyrazinamide resistant, continuation therapy should be for 7 months. Most patients therefore receive a 6- or 9-month course of treatment, and for a sensitive organism this is curative in close to 99% of cases. Continuing follow-up after appropriate treatment of fully sensitive tuberculosis is therefore rarely indicated.

Approximately 5% of patients receiving rifampicin plus isoniazid have to stop one of these drugs because of unwanted effects. If unwanted effects or resistance of the organism prevent the use of rifampicin or isoniazid, then other drugs must be substituted. Rifampicin plus ethambutol can be given as continuation treatment. It must be given for 7 months, with regular monitoring for ethambu-

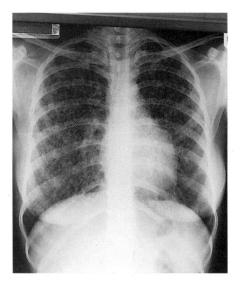

Figure 18.14 Miliary tuberculosis. Both lung fields contain numerous small, ill-defined, round opacities (the solid organs are similarly affected).

tol-related ocular toxicity (see below). If rifampicin cannot be given, treatment with three alternative drugs must usually be continued to complete an 18-month course.

Antituberculosis drugs must be taken as a continuous course, to avoid the emergence of resistance, with its risk of failed therapy or relapse. In many countries this is encouraged by directly observed therapy (DOT), using rifampicin-containing short-course regimens (DOTS: directly observed, short-course).

'Intermittent' regimens

Several different intermittent dosing regimens have been tested. These have used drugs daily for 2 months and then two or three times weekly for 4 months, or drugs three times weekly for the whole treatment period. All of these regimens appear equally effective.

Courses depending on weekly dosing have been used for immunocompetent patients, but are associated with an increased relapse rate in HIV-infected cases.

Drug-resistant tuberculosis

This is tuberculosis resistant to either rifampicin or isoniazid. The rates of these resistances vary in different countries; in England and Wales in 2002, isoniazid resistance occurred in about 7% of cases, and rifampicin resistance in about 1%.

Multidrug-resistant tuberculosis (MDRTB)

This is tuberculosis caused by *M. tuberculosis* resistant to rifampicin and isoniazid, with or without resistance to other antimycobacterial drugs. In England and Wales in 2002 just under 1% of cases had MDRTB but higher rates occur in other countries.

Risk factors for MDRTB

1 A past history of TB, especially if incompletely or intermittently treated.
2 Birth or exposure in higher-risk areas (e.g. Asia, Africa, southern and eastern Europe, Turkey, former USSR, Latin America).
3 Contact of a person with MDRTB.
4 Disease unresponsive to first-line therapy: fever continuing after 2 weeks, sputum smear positive after 2 months or sputum culture positive after 3 months.

Management of MDRTB

MDRTB is not more infectious than sensitive disease, but is more difficult to treat and more likely to relapse. Patients with positive sputum smears should avoid close association with new contacts, especially immunosuppressed individuals. When in hospital they should be nursed in a negative-pressure isolation room. Visitors to the room should wear a dust-mist (FFP3 or NIOSH95) mask.

Drug therapy should be supervised by an expert in MDRTB, as it is complex, carries an increased risk of toxicity, and may require modification during the course of treatment. For initial therapy, at least five drugs should be given, of which three have not been taken before. An optimum regimen can be chosen for continuation when sensitivities have been confirmed, and this treatment should be continued until cultures become negative. Following this, continuation therapy with at least three effective drugs should be followed for a time decided according to the individual's initial response. From a minimum of 9 months to 24 months or more is the common range. Some patients, particularly the immunosuppressed, may need lifelong treatment, to control low-grade disease or avoid recrudescence of dormant infection.

Surgery has a place in the management of MDRTB. Large cavities are difficult to 'sterilize', and may compromise respiratory function. Necrotic areas of lung may be inaccessible to antimicrobials. In these circumstances, CT scanning and lung function tests should be performed, to assess the patient's suitability for surgical excision of affected tissue. This not only reduces the burden of infection, but also ensures that remaining tissue is well-perfused and able to respond to chemotherapy.

Support and encouragement is needed if demanding drug regimens and isolation procedures are to be followed continuously for the required duration. DOT is particularly helpful in this situation. Further resistances may emerge during therapy; regular sputum cultures should therefore be monitored throughout therapy and, when possible, for at least 5 years after stopping drugs. Some patients never achieve permanently negative sputum cultures, but may stabilize their clinical condition. They must be educated about the risks their organism presents to their contacts and be warned of the danger to immunosuppressed contacts.

Mycobactericidal drugs used in tuberculosis (first-line drugs)

Rifampicin

Daily dose: 600 mg (450 mg if patient <50 kg), child 10 mg/kg; intermittent dose: 600–900 mg three times weekly, child 15 mg/kg three times weekly.

This is the most important drug in the treatment of tuberculosis, as it kills slowly replicating mycobacteria throughout the course of treatment. Its inclusion therefore permits short-course chemotherapy of 6–9 months' duration. It is well absorbed by mouth and widely distributed in the body. It penetrates moderately well into the CSF.

The unwanted effects of rifampicin include gastrointestinal distress and a red-orange discoloration of urine, tears and other body fluids. Soft contact lenses will also become discoloured. Urticarial rashes, mild eosinophilia or itching are rare effects.

The most important adverse effect is hepatocellular liver damage. There is always a temporary elevation of transaminases in the blood when antituberculosis treatment is commenced, but this should peak by 3–4 weeks, and slowly subside thereafter. Levels should not climb to more than three or four times the upper limit of normal. In patients with pre-existing hepatic impairment the rifampicin dose should not exceed 8 mg/kg daily.

In some cases of miliary tuberculosis, an inflammatory reaction in early treatment produces liver enzyme abnormalities unrelated to drug toxicity. If transaminases rise to over four times normal, or jaundice appears, all antituberculosis drugs should be withdrawn. When liver function tests have returned to normal specific treatment may be cautiously reintroduced (see below).

Rifampicin occasionally causes a syndrome of fever, myalgia and anorexia, particularly in patients on intermittent or interrupted therapy. This does not improve as therapy continues, and the drug must usually be withdrawn.

⚠ Rifampicin is a powerful inducer of microsomal enzymes. It increases the metabolism of the contraceptive pill, sulphonylureas, corticosteroids, azole antifungal drugs, warfarin and other coumarins, protease-inhibitor and some non-nucleoside reverse transcriptase inhibitor antiretroviral drugs. Rifampicin has been successfully used in combination with nucleoside reverse transcriptase inhibitors, and with efavirenz (with efavirenz dosage increased to 800 mg daily). It should not be used with protease inhibitors. Patients using oral contraception should be advised to use alternative methods of contraception during rifampicin therapy.

Rifabutin

Daily dose: for treatment of tuberculosis: 150–450 mg/day: for treatment of 'atypical' mycobacterial infections: 450–600 mg/day (not recommended for children).

This is a rifamycin, closely related to rifampicin. Cross-resistance between the two drugs is common. It is often used when treatment for tuberculosis or atypical mycobacterial infection must be combined with antiretroviral therapy, as its interactions are more easily managed than those of rifampicin. Its dose must be halved when it is co-administered with nelfinavir or amprenavir, and reduced with indinavir and atazanavir. Its dose must be increased if it is given with efavirenz. It should not be co-administered with ritonavir or saquinavir.

⚠ Rarely, patients develop uveitis during long-term therapy (particularly when drugs that elevate rifabutin levels are co-administered). The condition can often be controlled with non-steroidal anti-inflammatory treatment or step-down corticosteroid dosage; other adverse effects are similar to those of rifampicin.

Isoniazid

Daily dose: 300 mg, child 5 mg/kg; intermittent dose: for adult or child, 15 mg/kg three times weekly.

This drug is highly effective in killing rapidly replicating mycobacteria, allowing early dramatic reduction in mycobacterial load. It is given with rifampicin for the whole duration of chemotherapy, and contributes to the effectiveness of short-course treatments. It penetrates well into the CSF.

Unwanted effects include nausea and vomiting, hypersensitivity rashes and occasional cerebral disturbance or convulsions.

It can produce hepatocellular damage (at any stage of treatment), especially in middle-aged and elderly patients. Except when transient hepatocellular disorder occurs in the first 3–4 weeks of treatment, isoniazid must be discontinued if it causes hepatocellular toxicity.

Peripheral neuritis is an important side-effect, commoner in slow acetylators of the drug. The neuritis is painful and disabling, improving only slowly and often incompletely on withdrawal of isoniazid. This side-effect can be completely avoided by giving pyridoxine supplements to patients at risk of peripheral neuropathy (those with diabetes, alcohol dependence, chronic renal failure, malnutrition and HIV infection). The usual pyridoxine dosage is 10 or 20 mg/day.

Ethambutol

Daily dose: for adult and child, 15 mg/kg; intermittent dose: for adult and child, 30 mg/kg three times weekly or 45 mg/kg twice weekly.

This is a slightly less powerful drug than rifampicin and isoniazid, but is useful in combination with them for the intensive phase of therapy. It can also be used as one of two drugs for continuation therapy, but the total duration of treatment must then be 9 months, rather than 6.

Ethambutol can cause optic neuritis, loss of red-green colour discrimination and visual impairment. These effects are particularly likely in the elderly and in patients with renal impairment. Prolonged or high dosage also carries the risk of ocular effects. The patient notices loss of acuity and colour appreciation, both of which are reversible if the drug is promptly discontinued. Regular enquiry for visual symptoms and testing of acuity before and during treatment are advisable. The drug should not be given to pre-school children or others who cannot understand and report visual defects.

Pyrazinamide

> Daily dose: 2.0 g (1.5 g if <50 kg); intermittent dose three times weekly: 2.5 g (2.0 g if <50 kg), child 50 mg/kg; twice weekly: 3.5 g (3.0 g if <50 kg), child 75 mg/kg.

This is a bactericidal drug that is well absorbed and enters the CSF particularly well. It is highly effective against replicating intracellular organisms, but not against slowly metabolizing organisms later in the course of treatment. It is a useful addition to initial therapy, especially in the treatment of tuberculous meningitis, but is less useful after the first 2–3 months.

Pyrazinamide can be hepatotoxic, and also produces rashes, including urticaria. On occasions it can precipitate acute gout.

M. bovis is innately resistant to pyrazinamide.

Streptomycin

This is an effective drug whose usefulness is limited by ototoxicity, vertigo and nephrotoxicity, as well as the necessity for intramuscular administration. It is now rarely used in UK practice. It is most useful as part of the intensive phase of treatment, or as a reserve drug for resistant disease. The usual dose of 1 g daily should be reduced in those over age 40, in small patients and in those with renal impairment. Streptomycin can be given three times weekly once the sputum is smear-negative. It is highly advisable to check pre- and post-dose streptomycin levels at 2- to 4-weekly intervals.

The total dose given should not exceed 100 g, above which toxicity becomes much more likely.

> ## Protocol for cautious reintroduction of anti-TB treatment after early hepatocellular reaction
> British Thoracic Society recommendation
> - Step 1: isoniazid 50 mg daily for 3 days.
> - Step 2: isoniazid 100 mg daily for 3 days.
> - Step 3: isoniazid 200 mg daily for 3 days.
> - Step 4: isoniazid 300 mg daily.
> - Step 5: continue isoniazid 300 mg daily and add rifampicin 150 mg daily for 3 days.
> - Step 6: increase rifampicin to 300 mg daily for 3 days.
> - Step 7: increase rifampicin to 600 mg daily and continue.
> - Step 8: continue rifampicin and isoniazid and add ethambutol 15 mg/kg daily for 3 days.
> - Step 9: add pyrazinamide to the above regimen and continue.

> ⚠ Transaminase levels should be measured daily and the regimen should be interrupted if a significant elevation of transaminases, or jaundice occur: it will be evident which drug has caused the reaction, allowing it to be discontinued, and substituted by an alternative.

Additional and reserve (second-line) drugs

Ciprofloxacin, ofloxacin or moxifloxacin

These broad-spectrum antibiotics have proved effective against several types of mycobacteria, and are well tolerated by most patients. They are very useful in treating MDRTB. Moxifloxacin has the advantage of once-daily dosing.

Amikacin

This aminoglycoside may be effective when streptomycin is not. It must be given intramuscularly, and has the same side-effects as other aminoglycosides, so should not be given in combination with them. Like streptomycin, it can be given in a daily or a three times weekly regimen. Blood levels should be checked, as for streptomycin.

Prothionamide

This is a bacteriostatic drug that is useful in resistant *M. tuberculosis* infections. It penetrates well into the CSF. Side-effects are mainly those of gastrointestinal distress, but there is a long list of rarer side-effects, including rashes, blood dyscrasias and liver dysfunction. It is not marketed in the UK, so must be specially purchased.

Clarithromycin

This macrolide drug is particularly useful in treating *M. avium-intracellulare* infections, but may also be effective against *M. tuberculosis*. It is a broad-spectrum antibiotic, which can predispose to *Candida* infections. Its use is limited by nausea in some patients.

> ⚠ Clarithromycin increases plasma levels of rifabutin (with increased risk of uveitis): rifabutin dosage should be reduced.

Capreomycin

This aminoglycoside is only used in tuberculosis. The principles of its use are the same as for amikacin. As well as ototoxicity and vertigo, it can cause renal impairment, hepatotoxicity and skin reactions.

Para-aminosalicylic acid (PAS)

This is a mycobacteriostatic drug, well absorbed by mouth, but which must be given in voluminous doses, usually as numerous gelatin capsules three or four times daily. Its main problem is its frequent association with gastritis and dyspepsia, but it can also cause hepatotoxicity and occasional rashes. It is not marketed in the UK, so must be specially purchased.

Cycloserine

This is a rather toxic mycobacteriostatic drug that can be used in combination with other antimycobacterial agents. It has unpleasant side-effects, including headache, dizziness, depression, agitation, convulsions and allergic rashes. It tends to be a drug of last choice.

Non-tuberculous mycobacterioses

Introduction

These conditions are caused by accidental acquisition of environmental *Mycobacterium* species, commonly resulting in skin infections or inoculation abscesses. More extensive disease, or mycobacteraemic disease, can occur in immunosuppressed individuals (see Chapter 22).

MAI infections and HIV

Clarithromycin and ethambutol are accepted to be two first-line drugs in managing MAI infections. Studies suggest that the addition of rifabutin improves survival in patients not taking highly active antiretroviral therapy. Most specialists use it routinely in treating HIV-infected cases.

> ⚠ Co-administration of clarithromycin increases blood levels of rifabutin, increasing the risk of uveitis; the rifabutin dose should be reduced.

Mycobacterium chelonei

This organism is most commonly seen in cold abscesses at injection sites, particularly in insulin-dependent diabetics, who have many injections. Outbreaks of bacteraemia have occurred in immunosuppressed patients.

The infection can be treated with amikacin. Attempts at excision or drainage often result in extension or recurrence. Additional or alternative drugs include clarithromycin, cefoxitin, doxycycline and clofazimine.

Mycobacterium fortuitum

This organism also causes inoculation abscesses. It is a rare cause of bone and joint disease, especially in the diabetic foot. It is often sensitive to a wide range of antimicrobial agents, including amikacin, co-trimoxazole, clarithromycin, cefoxitin and ciprofloxacin.

Mycobacterium marinum

This organism typically causes 'swimming pool granuloma' in individuals who have contact with rivers, ponds and aquaria.

It is sensitive to a number of drugs, including rifampicin, ethambutol, ciprofloxacin, co-trimoxazole, streptomycin and doxycycline.

Mycobacterium ulcerans

This is the causative organism of tropical ulcer. Most cases respond slowly to multidrug treatment, but debridement and wound hygiene are important in managing extensive disease.

Prevention and control of tuberculosis

Introduction

Tuberculosis is rare in communities that enjoy good living conditions. The decline in tuberculosis that occurred during the latter half of the 19th century and the early part of the 20th century preceded other control measures and is attributed to improvements in housing conditions, nutrition and social deprivation (see Chapter 26). The most important means of prevention and control of TB is to find and appropriately treat infectious cases, thus reducing the reservoir of infection and the burden of disease in the community. To achieve this, patients and their health care providers must have good awareness of TB, and services appropriate for the population's needs must be readily available, particularly for those at high risk.

Reductions in infection due to *M. bovis* have been achieved in many countries by pasteurization of milk. In many countries, cattle are regularly tuberculin-tested and positive cases are removed from the herd, but there is not strong evidence that this has a major additive effect over pasteurization.

Management of the case and close contacts

Most patients with pulmonary tuberculosis can be treated at home and need not be separated from other household members, provided that chemoprophylaxis is given to young children in the household (see below). Where hospital admission is required because of severe disease or for social reasons, the patient should be nursed in a single room until no longer infectious. With modern antimicrobial therapy and fully sensitive organisms, this is usually achieved within 2 weeks, even though some bacilli may still be seen in sputum smears. Patients who are sputum-negative or who have non-pulmonary disease can be nursed in a general ward.

Close contacts of sputum-positive cases should be tuberculin tested and have a chest X-ray. Close contacts are defined as household members (and classroom contacts if the index case is a teacher or a school child). Where the tuberculin test is negative, it should be repeated 2–3 months after the last exposure to determine whether tuberculin conversion has occurred.

Chemoprophylaxis of mycobacterial infections is indicated for the following individuals

1 Children under 16 with a strongly positive tuberculin test, irrespective of BCG vaccination status.
2 Adults with a strongly positive tuberculin test and no previous BCG vaccination.
3 Adults who have tuberculin converted.
4 Young (<35 years) Asian adults with a strongly positive tuberculin test.
5 Immunosuppressed HIV-positive individuals, for primary or secondary prophylaxis of MAI disease.

 BCG vaccine should be given to unvaccinated contacts under 35 years of age who remain tuberculin-negative.

Isoniazid is the drug of choice for chemoprophylaxis of *M. tuberculosis*. It is usually given for 6 months. Isoniazid plus rifampicin can be given for 3 months. Longer-term or continuing chemoprophylaxis may be indicated for HIV-positive contacts.

Clarithromycin is often used for prophylaxis of MAI disease.

Drug-resistant tuberculosis

Outbreaks of drug-resistant disease can occur, often in populations that are socially or medically vulnerable or where ensuring treatment completion may be difficult. For example, a large outbreak of isoniazid-resistant tuberculosis was identified in London in 2001 associated with drug use and prison.

Control of such a situation requires energetic case finding, adequate isolation of cases in hospital (see Chapter 23), close supervision of TB treatment (directly observed therapy in the patient's own environment), and diligent contact tracing and follow-up.

Screening of migrants

Many developed countries operate a screening programme for new migrants from countries where tuberculosis is common. The aims of these programmes vary. Some include detection of active disease only. Others include identification of infected individuals who might benefit from chemoprophylaxis, or of uninfected and unvaccinated people who might benefit from immunization. New entrants from highly endemic countries have the highest rates of TB in the first few years after migration. Most cases, however, present several years after arrival in the new country and the rates of TB remain elevated in the foreign-born, compared with the indigenous population, for many years after resettlement. Ongoing awareness of TB in migrants is therefore important.

Bacille Calmette–Guérin (BCG) vaccine

BCG vaccine contains a live attenuated mycobacterium derived from *M. bovis*. It stimulates a protective immune response to *Mycobacterium tuberculosis*, *M. bovis* and probably other members of the human tuberculosis group. It also protects against the development of lepromatous leprosy. It is given as a single intradermal dose. A local reaction develops at the immunization site within 2–6 weeks, beginning as a small papule, which increases in size, often ulcerates and gradually heals, leaving a small, 'punched out' scar. The vaccine is contraindicated in patients with immunosuppression, including asymptomatic HIV-positive individuals.

Estimates of protection by BCG have varied in different field trials, from zero in one study in India, to 90%. Many studies, including earlier studies in the UK, have shown protection of approximately 70%, lasting for at least 20 years. Some of the variable results may be caused by a gradual change in the genome of BCG over the years, a phenomenon that has only recently been recognized. BCG is generally recognized to be better at preventing severe disease in young children than disease in adult populations.

Policies for the use of BCG vary considerably between countries. In countries where infection in young children is common, the vaccine is routinely administered at birth. Many developed countries only offer the vaccine to groups at particular risk of tuberculosis. These may include contacts of cases with respiratory tuberculosis, healthcare workers, teachers and immigrants from countries where tuberculosis is common. In the USA, the number of groups recommended for BCG is very limited; control of the disease rests on case detection and contact tracing (see below). In the UK, BCG used to be given routinely to all tuberculin-negative school children at 10–13 years of age. Since 2005, with the changing epidemiology of tuberculosis, BCG is now primarily targeted to infants (aged 0 to 12 months) living in areas where the incidence of TB is 40/100 000 or greater or with a parent or grandparent who was born in a country where the incidence of TB is 40/100 000 or greater.

Tuberculin testing before BCG immunization

BCG vaccine can safely be given without prior tuberculin testing to infants up to one year old. In older infants, children and adults, a tuberculin test should be performed first.

A negative tuberculin test indicates that the individual has not previously been infected or received BCG vaccination; such individuals can be given BCG. A positive test indicates past infection or previous vaccination and immunization with BCG is not required. A strongly positive reaction may indicate active disease; such individuals should be referred for further investigation.

UK Department of Health recommendations for BCG vaccination

1 All infants living in areas where the incidence of TB is 40 per 100 000 or greater.
2 Infants whose parents or grandparents were born in a country with a TB incidence of 40 per 100 000 or greater.
3 Previously unvaccinated new immigrants from high-prevalence countries.
4 Those at risk due to their occupation, e.g. healthcare workers, veterinary staff, staff of prisons, those intending to live or work in high-prevalence countries for 1 month or longer.
5 Contacts of known cases.

⚠ School children aged 10–13 will be reviewed for risk factors for tuberculosis, and will be Mantoux-tested and vaccinated if indicated.

Leprosy

Introduction

Leprosy is an indolent disease, mainly affecting the skin, nerves and mucosae, but also capable of infecting the eye, muscles and testicles. It is caused by *M. leprae*, which has low infectivity and is extremely slow-growing, with approximately one replication per fortnight. It has never been cultured in artificial media, but will grow slowly in some animals.

Spread is by close contact, usually among families and particularly from individuals with extensive multibacillary mucosal lesions.

The importance of leprosy is that it produces peripheral nerve lesions, leading to paralysis and anaesthesia of limbs, trophic ulcers and neuropathic joint disease. In untreated cases there is no means of preventing these effects, which can produce devastating deformities, including autoamputation.

Clinical features and grading of disease

The clinical features depend on the nature of the patient's immune response to the infection. For the purpose of determining treatment and prognosis, leprosy is graded into paucibacillary disease (with five or fewer lesions at presentation) and multibacillary disease (more than five lesions). Split skin smears, from skin lesions, are extremely helpful, as patients with detectable bacilli can be assigned to the multibacillary group.

The first presentation of leprosy is often rather mild, and is called indeterminate leprosy. An ill-defined area of skin gradually loses some of its pigmentation, and becomes hypoaesthetic. The face or hand is often affected, and the patient often presents because of a cut or burn of the site, resulting from diminished sensation.

Subtle nerve damage must be sought, as early treatment is essential to avoid the development of anaesthesia, and the resulting deformities. Nerves can be palpated to detect tenderness or thickening (the ulnar nerve, peroneal nerve, median and accessory nerves are all superficial and accessible to examination). Standard nylon monofilaments are nowadays used to detect loss of light touch sensation. Muscle weakness, particularly in the wrist, hand and ankle (e.g. early wrist- or foot-drop), must be carefully sought and documented.

Examinations should be repeated at intervals of a month, or more frequently when treatment has been commenced, as nerve damage can occur during and after treatment, due to prolonged immunological activity.

'Tuberculoid' disease

Tuberculoid (TT) disease is a clinical term used to describe the features seen in those who mount a strong cell-mediated response to the infection. It is characterized by one or two hypopigmented skin lesions, and localized, asymmetrical inflammation and thickening of peripheral nerves. Biopsy of affected skin shows typical granulomas, with palisades of epithelioid cells, surrounding collections of Langhans-type giant cells and active lymphocytes. Bacteria are rarely, if ever, seen in these lesions, which are termed paucibacillary.

Lepromatous disease

Lepromatous (LL) disease is associated with an absent cell-mediated response. There is intense oedema of affected tissues, which contain no granulomas and no lymphocytes. Only undifferentiated macrophages are seen, and these are packed with acid-fast bacilli. There is mycobacteraemia, with spread to distant areas of the skin, nerves, nasal and pharyngeal mucosa, eye, muscles, testicles and reticuloendothelial tissues, such as the spleen, liver and bone marrow (especially that of the phalanges). This is multibacillary disease.

Lepromatous disease is expressed as multiple, often symmetrical lesions. Skin lesions are nodular, and not hypoaesthetic because the small nerves are oedematous rather than inflamed. The nasal mucosa is affected early, and the resulting nasal discharge is highly infectious. Facial and lip swelling, often with a collapsed bridge of the nose, causes a typical, 'leonine' facies. Nerve trunks are swollen; the Schwann cells become packed with bacilli and proliferate, adding to physiological and anatomical disruption. There is gradual loss of sensation, starting with small-fibre functions and eventually affecting all modalities. Involvement of the eye can cause conjunctivitis or keratitis; muscle involvement causes weakness of small muscles, including the smooth muscles of the skin and superficial blood vessels; bone involvement leads to loss of alveolar bone from the jaw, of the nasal septum and of phalangeal joint surfaces.

Borderline leprosy

Borderline (BB) leprosy is an intermediate form in which lesions contain a lymphocytic infiltrate and epithelioid cells, but no giant cells are seen and bacilli survive within the epithelioid cells. There is a continuous spectrum of intermediate forms between tuberculoid, the intermediate borderline and the extreme lepromatous disease (TT–BT–BB–BL–LL). These are distinguished by histological features, by the degree of localization of skin, nerve and other lesions, and by the number of bacilli (on a scale of 1–6) detectable in lesions. Indeterminate, TT and BT cases are usually paucibacillary, while BB, BL and LL cases are multibacillary.

Borderline disease is unstable, and a small fluctuation in immune response or bacterial activity can cause a shift either way along the spectrum.

Diagnosis

The three aspects of diagnosis comprise:
1 physical examination for typical skin lesions, anaesthesia in lesions or nerve distributions, and nerve thickening;
2 biopsy of atypical or indeterminate skin lesions (this is also helpful for staging the disease);
3 examination of split-skin smears for acid-fast bacilli (a small cut is made through the epidermis, without drawing blood, and the blade is then used to scrape a little tissue fluid from the exposed dermal tissue and transfer it to a microscope slide, where it is allowed to dry before staining).

 PCR-based diagnostic methods have been described.

Treatment

The World Health Organization recommends standard multidrug therapy with rifampicin, clofazimine and dapsone. Rifampicin reduces the bacillary load in a few days, while dapsone kills residual organisms. Clofazimine is as effective as dapsone, and also has an intrinsic anti-inflammatory effect, which lessens the likelihood of reactive conditions complicating treatment. Occasionally prothionamide, minocycline, ofloxacin or clarithromycin are used when other drugs are contraindicated.

> **Treatment of leprosy**
> The optimal regimen depends on the bacillary load.
> **1** Paucibacillary cases are treated with rifampicin 600 mg monthly, plus dapsone 100 mg daily for 6 months.
> **2** Multibacillary cases receive rifampicin 600 mg and clofazimine 300 mg monthly, plus dapsone 100 mg and clofazimine 50 mg daily for 2 years (though it is hoped that trials in progress will confirm that around 1 years' treatment is sufficient to avoid relapse).
> The three-drug regimen is adequate, even for dapsone-resistant organisms, and prevents the emergence of further resistance.

Clofazimine has the disadvantage that it causes first reddening of the skin, viscera and body fluids, and eventually a grey skin colour. Prothionamide is a useful second-line drug, used if clofazimine is refused or if adverse reactions occur to other drugs.

Patients are followed up at least monthly during treatment and, ideally, 6-monthly after stopping. Paucibacillary patients are examined for new lesions until they have been disease free for 2 years. Over 40% of patients may still have lesions at the end of treatment for paucibacillary disease, due to the persisting immune response, but this does not indicate treatment failure. Multibacillary patients, with their high bacillary load, have a higher relapse rate than others, with a continuing risk of relapse for around 4 years. Multibacillary patients should therefore have split-skin smears during treatment, and treatment is continued until these are consistently negative. Follow-up of these patients continues until 5 years without disease.

Complications

Several types of immunological change can occur in leprosy; all are more common after the start of treatment.
1 Changes in the staging of the disease, either downgrading towards the lepromatous end of the spectrum or upgrading towards the tuberculoid end.
2 Type 1 (reversal) reactions, which represent increasing sensitivity to bacterial antigens, are associated with new or worsening nerve lesions, muscle weakness and newly

active skin lesions. Patients with multibacillary disease are at higher risk of new nerve lesions, and those with multibacillary disease and pre-existing nerve lesions have above 60% risk of developing further lesions.

3 Type 2 reactions, affecting about 20% of LL and some BL cases, with the development of nodular, inflamed lesions of erythema nodosum leprosum (ENL) on the face and limbs, and sometimes inflammation of the uveal tract, fingers, peripheral nerves and testicles.

In all cases multidrug treatment should be continued. Type 1 reactions are treated with oral prednisolone, beginning with a dose of 30–40 mg daily and stepping down over 4–6 months. Type 2 reactions usually respond to a short, step-down course of prednisolone. Some experts favour treatment of repeated or severe type 2 reactions with thalidomide (unlicensed for this use) in diminishing doses from 400 mg daily to 50 mg daily (but great care must be taken to avoid using this in a pregnant woman).

The lowest dose can be given for some months if necessary.

Many patients need physiotherapy, orthopaedic shoes or supportive limb braces to overcome established nerve lesions, and to protect anaesthetic extremities from injury. In some, tendon transplants may improve hand or wrist function. Emergency surgery to open nerve sheaths or surrounding fascia can limit damage caused by severe inflammatory swelling.

Prevention and control

This is based on: case finding, often by local health workers; working to abolish the social stigma of the diagnosis; treating infectious cases, who are the reservoir of infection; and educating the public about the curability of the disease and the needlessness of suffering nerve lesions and disfigurement.

Bacteraemia and Sepsis 19

Introduction

Bacteraemia is the condition in which bacteria circulate in the blood. Analogous pathology can occur with blood-borne fungi (e.g. *Candida* spp.), viruses (e.g. dengue) or parasites (e.g. *Plasmodium falciparum*). Such infections are not always associated with fever or illness. Repeated cultures of blood show that bacteria are detectable for 12–48 h after mild tissue trauma, such as dental extractions, and endoscopic examination of the bladder or biliary tree. Transient, self-limiting bacteraemia is also recognized in *Salmonella* and listerial infections, and may lead to the presentation of metastatic infections without preceding illness.

Conditions where transient bacteraemia is common, and usually clinically inapparent
1 Dental treatment.
2 Endoscopic procedures.
3 Urinary tract infections.
4 Bowel infections.

These bacteraemias and their damaging effects are controlled and terminated by the natural bactericidal and anti-inflammatory properties of blood components and vascular endothelium (see below). When the equilibrium between the pathogenicity factors of the blood-borne organism and the body's defensive systems is overcome, systemic effects of infection develop, and cause the features of sepsis.

Sepsis and sepsis syndromes

The definitions of sepsis and sepsis syndromes have been unified considerably by the recommendations of both European and American consensus groups. They define steps in a continuum of increasingly severe illness (Table 19.1). Sepsis is usually associated with infection, but the same sequence of events can follow non-infectious insults, such as trauma, burns, haemorrhage, pancreatitis and injury by chemical toxins. In the absence of infection, some experts call the response a 'systemic inflammatory response syndrome (SIRS)', and the severe form is often called 'multi-organ dysfunction syndrome (MODS)'.

The acute (or adult) respiratory distress syndrome (ARDS)

The lung is particularly susceptible to damage in sepsis, and can suffer a syndrome of non-hydrostatic pulmonary oedema, which may cause respiratory failure and threaten life. Its usual definition is:
1 a condition with acute onset (usually 2–4 days after the initiating insult);
2 bilateral interstitial infiltrates seen on chest X-ray;
3 an arterial PO_2 of <8 kPa, regardless of PEEP (positive end-expiratory pressure);
4 a pulmonary artery occlusion pressure of <18 mmHg (i.e. excluding congestive cardiac failure).

At the onset of ARDS, the excess interstitial fluid is a pure exudate, but this may progress to inflammatory alveolar exudate (containing neutrophils) and then to a fibroproliferative picture and eventual fibrosis. Mild cases can resolve at an early stage.

The pathogenesis of sepsis syndromes

This is a complex interaction of the immune response, priming and activation of neutrophils, and endothelial activation. Many contributory factors are now known, but the mechanisms of their actions and interactions are poorly understood.

Initiating factors

The most familiar of these is endotoxin (lipopolysaccharide, LPS). It is a component of Gram-negative bacterial cell walls that is taken in and immunologically presented by macrophages, which become activated. Activated macrophages interact with ligands on T-lymphocytes, initiating helper-cell and other immunological functions. In the blood endotoxin is bound by lipopolysaccharide-binding protein (LBP), and may then be cleared, or react with the CD14 ligand on macrophages, again activating them. Another protein, called bacterial permeability-increasing protein (BPI), is highly homologous with LBP, and also binds and inactivates endotoxin. Antigens other than LPS also interact with macrophage CD14. Superantigens can bind T lymphocytes with macrophages that are not presenting processed antigen, through the macrophage V-beta receptor (see Chapter 1).

The role of macrophages

Activated macrophages secrete tumour necrosis factor-alpha (TNF-alpha) and interleukin 1 (IL-1), leading to many responses such as activation of T cells, endothelial activation and 'priming' of neutrophils, which makes them rigid and adhesive, immobilizing them in the microcirculation. Activated macrophages also secrete IL-8, a chemokine powerfully attractive to neutrophils; it causes

Table 19.1 Definitions of sepsis-related conditions

Clinical disorder	Definition
Sepsis	Clinical evidence of infection plus evidence of a systemic response manifested by two or more of: • temperature >38 °C or <36 °C • heart rate >90 beats/min • respiratory rate >20 breaths/min or PCO_2 <4.3 kPa (<32 mmHg) • white cell count >12 × 10⁹/l or <4 × 10⁹/l (or >10% immature neutrophils)
Severe sepsis	Sepsis associated with organ dysfunction: Hypotension: a systolic blood pressure of <90 mmHg or a fall of >40 mmHg from baseline in the absence of other causes of hypotension Oliguria: <30 ml/h Lactic acidosis Confusion Hepatic dysfunction
Septic shock	Severe sepsis with hypotension despite adequate fluid resuscitation (refractory septic shock is present if hypotension persists for >1 h despite fluid resuscitation and pharmacological support)

primed neutrophils to release superoxide radicals and lysosymes, which are intensely inflammatory and further damage the endothelium to which they are bound.

The role of TNF-alpha

TNF-alpha alone can induce a sepsis response, and is probably a final signal in many of the processes of sepsis. As well as its immunological effects, it causes up-regulation of cellular adhesion molecules (ICAMs), which react with leucocyte beta integrins to allow endothelial attachment of neutrophils and macrophages.

The role of neutrophils

Endothelium-bound neutrophils, which have been primed and activated, produce many substances that damage the underlying endothelium. Agranulocytic individuals are not susceptible to severe sepsis syndromes or to ARDS. Early evidence suggests that antibody preparations binding neutrophil elastase can protect against ARDS or reduce its severity.

The range of abnormalities in sepsis

Endothelial damage leads to extravasation of intravascular fluid, with tissue oedema and an early fall in blood volume and albumin levels. Activated endothelium generates nitric oxide, a powerful vasodilator that causes a loss of peripheral resistance not fully compensated by increased cardiac output. These factors predispose to hypotension and shock. Poor tissue perfusion and poor lung function lead to widespread hypoxia (and hypoxic macrophages secrete more IL-8, recruiting more neutrophils).

Disseminated intravascular coagulation

Endothelial damage activates the coagulation cascade, causing fibrinogen consumption, fibrin deposition, activation and consumption of platelets and eventually disseminated intravascular coagulation (DIC). This obstructs the microcirculation, exacerbates hypoxia, further damages endothelium and brings the risk of haemorrhage. Endothelial damage may lead to vasoconstriction, further reducing tissue perfusion in areas already affected by the hypoperfusion of early sepsis.

Epidemiology of bacteraemia and sepsis

Bacteraemia is common. Approximately 30 000 blood isolates are reported from laboratories in the UK each year, and approximately 1% of hospital patients have bacteraemia on admission. Over 10% of hospital patients have features of sepsis, and sepsis affects over 50% of intensive care patients. This probably represents only a small frac-

tion of the true incidence, as many cases are not diagnosed and not all diagnostic laboratories make regular reports. The infections reported are biased towards more seriously ill patients and may not be representative of all bacteraemic infections. Nevertheless, it is clear that approximately 60% of reported bacteraemias are acquired by people in hospital, who are already ill. Neonates and the elderly are at greatest risk.

A wide range of pathogens cause bacteraemia. The recent reports of the most important organisms are shown in Fig. 19.1. Bacteraemias may be considered under four general categories:

1 staphylococci and streptococci;
2 Enterobacteriaceae;
3 anaerobic and aerobic opportunists; and
4 other, community acquired bacteraemias.

Staphylococci and streptococci

Staphylococci and streptococci are both community acquired and hospital acquired (Table 19.2). The epidemiology of staphylococcal bacteraemia is described later in this chapter (p. 400). Over 7000 streptococcal bacteraemias are reported each year, of which half are due to *Streptococcus pneumoniae*. Pneumococcal bacteraemia occurs mainly in the elderly and is often accompanied by pneumonia. It also affects patients of all ages with risk factors such as sickle-cell disease, asplenia, chronic renal, cardiac, liver or lung disease and diabetes mellitus. It is associated with an average mortality of greater than 20%.

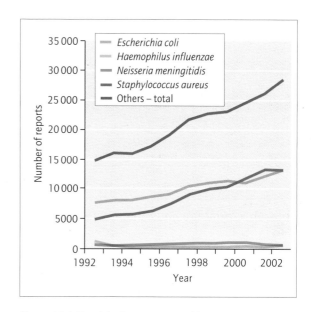

Figure 19.1 Trends in the occurrence of bacteraemias in the UK. Source: Health Protection Agency.

Table 19.2 Sources of coccal bacteraemias

Organism	May originate from
Staphylococcus aureus	Skin, nasopharynx (50% of all cases) Intravenous devices or pacemaker systems (recently implanted) (25% of all cases) Bone and joint disease or surgery or prosthesis (10% of all cases) Genitourinary disease or surgery (2% of all cases)
Streptococcus pneumoniae	Lung infection (50% of all cases) Intra-cranial or meningeal infection (10% of all cases) Patients with altered immunity, splenectomy or alcoholism (up to 10% of all cases) Rarely, intra-partum infection or peritonitis
Coagulase-negative staphylococci	Intravascular devices including pacing systems, haemodialysis and parenteral nutrition systems (over 50% of all cases) Peritoneal dialysis systems Intra-cerebral shunts and valves connected with the right atrium Intra-cardiac prostheses
Group A streptococci	Skin and soft tissues (most cases) Rarely, post-partum infections, genitourinary disease or surgery, bone and joint infection, pharyngitis
Group B streptococci	Neonatal infection (over 50% of all cases) Genitourinary infection or surgery, including termination of pregnancy
Group C and G streptococci	Skin and soft tissue infection
Streptococcus 'milleri'	Meningitis, ventriculitis, brain or liver abscess

Table 19.3 Sources of Gram-negative bacteraemias

Organism	May originate from
Escherichia coli (the cause of over 50% of all Gram-negative bacteraemias and about 25% of all bacteraemias)	Urinary tract colonization or infection, especially after instrumentation Intestinal disease or surgery Pancreaticobiliary disease, instrumentation or surgery
Klebsiella pneumoniae, subsp. *aerogenes*, subsp. *ozaenae*, *K. oxytoca* *K. pneumoniae* subsp. *pneumoniae* causes about 50% of all *Klebsiella* bacteraemias	As for *E. coli*, plus rare cases of suppurative lobar pneumonia
Enterobacter sp. and *Serratia* sp.	As for *E. coli*. Increasingly common in intensive care units and immunosuppressed patients
Proteus sp. (a member of the family Proteaceae, which also includes Morganella and Providencia)	Genitourinary and pelvic disease, surgery and instrumentation Renal or urinary tract stones (to which *Proteus* predisposes by metabolizing urea and forming alkaline ammonia)
Salmonella sp.: including *S. typhi* and *S. paratyphi*	Food-borne and water-borne intestinal infection (see also Chapter 8)

Enterobacteriaceae and other Gram-negative rods

Enterobacteriaceae include *Escherichia coli*, *Klebsiella*, *Citrobacter*, *Enterobacter*, *Proteus* and *Salmonella* spp (Table 19.3). These gastrointestinal organisms are often hospital acquired. The great majority (between 7000 and 8000 per year) are due to *E. coli* and occur in elderly, often surgical, patients. *Klebsiella* infections commonly complicate antibiotic treatment of chest or urinary tract infections in Western countries, but occur with a similar epidemiology to *E. coli* sepsis in some Far Eastern regions.

Anaerobic and aerobic opportunists

Anaerobic opportunists include *Bacteroides*, *Clostridium* and anaerobic streptococci, and aerobic opportunists include *Acinetobacter*, *Aeromonas*, *Pseudomonas* and *Serra-*

Table 19.4 Sources of 'hospital' bacteraemias

Organism	May originate from
Pseudomonas sp. and *Burkholderia* sp.	Immunocompromised patients (25% of all cases) Biliary tract disease, surgery or instrumentation
Acinetobacter sp.	Immunocompromised patients (33% of all cases) Intravenous devices Cardiac prostheses Rarely, neurosurgical procedures
Bacteroides sp. (especially *B. fragilis*)	Abdominopelvic disease or surgery Liver abscess Rarely, post-partum
Clostridium sp. (including *C. perfringens*)	As for *Bacteroides*, but also biliary tract sepsis, perineal or lower limb trauma, soil-contaminated or necrotic wounds
Anaerobic cocci (including *Peptococcus* and *Peptostreptococcus*)	Post-partum infections Abdominopelvic disease or sepsis Rarely, neonatal bacteraemia
Enterococci (including *Enterococcus faecalis*, *E. faecium* *Streptococcus durans*)	Intravascular cannulae, especially in intensive care unit patients (25% of all cases) Pancreaticobiliary disease, surgery or instrumentation Abdominopelvic disease or surgery

tia (Table 19.4). These are also mainly hospital acquired, affecting elderly, postoperative, debilitated or immunosuppressed patients.

Other community acquired bacteraemias

These include *Neisseria meningitidis* and *Listeria monocytogenes*. Meningitis is often an accompanying feature. Meningococcal bacteraemia occurs mainly in children under the age of 4. A second, smaller peak of meningococcal disease occurs in young teenagers (see Chapter 13). Asplenic patients and those with complement deficiencies are at increased risk of meningococcal disease. Listerial bacteraemia particularly affects adult immunocompromised patients and pregnant women. Infection in pregnancy often results in stillbirth or neonatal bacteraemia (see Chapter 17).

Defences of the blood

Phagocytes

Phagocytes are mobile cells that ingest particulate matter, including bacteria. The main phagocytes of the blood are the neutrophils, but cells of the monocyte–macrophage line are also phagocytic and play a vital part in antigen presentation in the immune system. Pathogens are ingested into vesicles called phagosomes. Within the neutrophil these fuse with lysosomes, forming phagolysosomes into which are liberated proteolytic enzymes and highly oxidative agents, including free radicals, which poison and destroy the bacteria. Neutrophils are particularly protective against staphylococci, yeasts and pseudomonads.

Neutrophils readily leave the bloodstream by rolling along the endothelial capillary wall, then adhering to endothelial cells, and finally passing between the cells into the tissues (Fig. 19.2). They are attracted to sites of complement activation by the chemotaxin C5a, and act as tissue and mucosal surface phagocytes. Many neutrophils are destroyed in the inflammatory process; others emerge on the surfaces of epithelia and are shed from the body.

Alternative complement pathway

The alternative complement pathway is a soluble defence mechanism. The third component of complement, C3, adheres to bacterial surfaces, where it is slowly broken down to C3a and C3b. In the presence of factor B, active C3bBb is formed, and is stable in the presence of properdin. It promotes the production of C5b from C5, and C5b induces the 'lytic pathway', through which the membrane attack complex is formed at the cell surface. This complex crosses and disrupts the cell membrane, destroying the bacteria by lysis (Fig. 19.3).

The alternative complement pathway does not depend on the presence of antibody, and therefore provides a non-specific means of destroying bacteria. It is particularly important in defending against Gram-negative cocci.

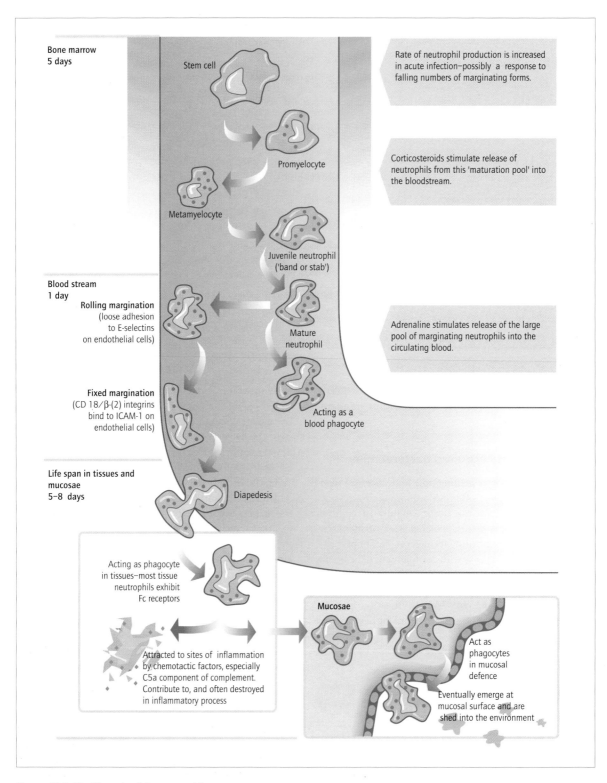

Figure 19.2 The life cycle of the neutrophil.

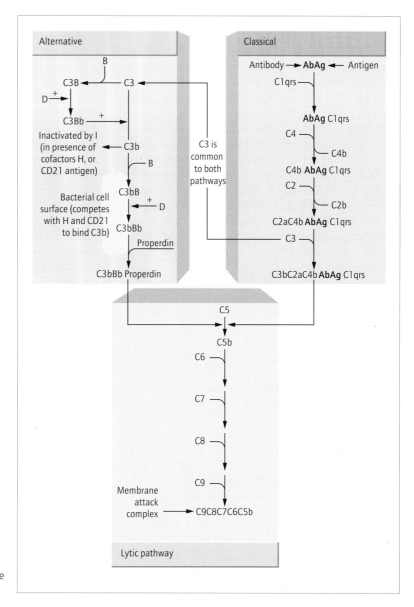

Figure 19.3 The actions of the alternative and classical complement pathways.

People with defects in this pathway are at increased risk of gonococcal and meningococcal bacteraemia. These bacteraemias are not more severe in complement-deficient patients, but in rare individuals who lack properdin they are aggressive and fulminating.

Iron binding

Iron binding depends on both specific and non-specific iron-binding proteins. Bacteria replicate inefficiently, and have reduced capacity to produce toxins when they lack iron. Iron-binding proteins such as ferritin (an acute-phase protein, see Chapter 1) increase greatly in infection and compete with bacteria for the iron they require.

Spleen

The spleen aids removal of bacteria from the blood. It not only provides conditions for phagocytosis in its sinusoids, but it removes engorged phagocytes from the bloodstream and promotes their destruction by other cells. People who lack a functioning spleen are at increased risk of pneumococcal bacteraemic diseases, and also to neisserial and *H. influenzae* type B diseases.

Antibodies

Antibodies circulate in the blood and can adhere to or agglutinate bacteria. Bound antibody or antigen–antibody immune complexes can activate the classical complement pathway (see Fig. 19.3). This is much faster than the alternative pathway, rapidly destroying bacteria and generating large amounts of chemotactic factors to summon phagocytes and cytotoxic cells. Phagocytes bear Fc receptors, which will bind to antibody on bacteria. This process, called opsonization, promotes rapid phagocytosis of the organisms. Thus, as well as directly disabling organisms, antibodies also enhance or recruit other defensive mechanisms of the blood. Defects of the alternative complement pathway are negated if the individual has antibodies to an organism, as the classical complement cascade can be rapidly activated.

The disadvantage of antibodies is that they take time to develop. Unless the patient is already immune, the nonspecific bactericidal mechanisms must 'buy time' while an antibody response is mounted.

Bactericidal properties of the blood

1 Phagocytes.
2 Alternative complement pathway activity.
3 Iron binding (deprives bacteria of iron).
4 Specific antibody.
5 Classical complement pathway activity.
6 Removal of capsulated bacteria by the spleen.

The natural defences of the blood are overcome when bacteria enter the circulation faster than they are removed, or when the bacteria replicate in the blood. There are three ways in which bacteria invade the bloodstream.

Escape from a site of natural occurrence

Typical sites of origin include the bowel, urogenital system or skin. When these tissues are damaged by trauma, inflammation or malignancy, natural skin or mucosal barriers to the passage of bacteria are disrupted. Manipulation, examination under anaesthesia or surgery all increase the likelihood that bacteria will be released into the bloodstream.

Some pathogens cause bacteraemia by colonizing surface mucosae, invading the epithelium and then escaping into the blood. This is the probable pathogenesis of pneumococcal, meningococcal and *H. influenzae* B bacteraemic diseases.

Release from a focus of infection

Abscess cavities are lined with granulation tissue, which is full of tiny, fragile blood vessels. These are easily disrupted and invaded by pathogens. The same is true of tissue surrounding a devitalized or necrotic area. The trauma of catheterization or endoscopy may provoke release of bacteria from an infected hollow organ, such as the urinary bladder or the gallbladder.

Inoculation of bacteria

Inoculation can occur by bite, scratch, trauma or 'needlestick'. Some bacteria, e.g. *Streptococcus pyogenes*, are so pathogenic and successful at evading blood and tissue defences, that a trivial injury can lead to overwhelming infection. Animal bites may inoculate organisms such as *Pasteurella multocida*, flea bites transmit plague, and inoculation accidents in hunters are an important means of transmission of tularaemia. Bacteria may replicate at the site of the inoculum or in a draining lymph node, from where bloodstream invasion proceeds. Bacteria can also gain direct access to the circulation when intravenous prosthetic devices become infected by skin organisms including *S. aureus*, *S. epidermidis* and *Corynebacterium jeikeium*.

Pathological accompaniments of bacteraemia

Pathology at the site of origin

Pathology at the site of origin is sometimes obscured by the effects of the bacteraemia itself. An asymptomatic gallstone or biliary stricture can cause overwhelming Gram-negative sepsis, which will defy treatment until the originating infection is removed. Renal or hepatic abscesses, infection behind a ureteric stricture, a small undrained empyema or an infected intracranial sinus are all capable of maintaining a bacteraemia, unless the loculated infection is drained. These predisposing conditions should always be actively sought.

The heart must always be examined, preferably by trans-oesophageal echocardiography (TOE), to exclude endocarditis. Apparent cure of bacteraemic disease after antibiotic therapy can be followed by recrudescence and cardiac damage because of inadequately treated endocarditis (Fig. 19.4).

Effects of exotoxins

Effects of exotoxins are widespread. Staphylococcal and streptococcal exotoxins can both cause rashes and toxic shock. Staphylococcal enterotoxins cause diarrhoea. Streptococcal toxins include leucocidins, haemolysins and hyaluronidase. Both Gram-positive and Gram-negative bacteria can produce toxins that cause tissue necrosis. *Clostridium perfringens* bacteraemia is sometimes associated with severe, toxin-mediated haemolysis. In sick neonates, clostridial toxins also probably contribute to necrotizing enterocolitis.

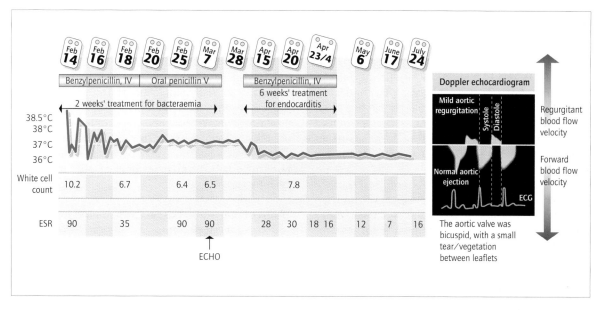

Figure 19.4 Endocarditis complicating bacteraemic disease: this 27-year-old presented with septic arthritis of the ankle and was found to have *Streptococcus pyogenes* bacteraemia; after completion of treatment at 2 weeks, the erythrocyte sedimentation rate (ESR) rose and a systolic murmur led to the diagnosis and treatment of endocarditis. ECG, electrocardiogram; ECHO, echocardiogram.

Effects of inflammatory responses

Effects of inflammatory responses contribute to fever and rigors. Complement activation, kinin activation and release of toxic agents from phagocytes are highly damaging, and add 'secondary mediators' to the classical pathways of fever production (see Chapter 1). They promote endothelial damage, platelet activation, vascular shunting and poor ventilation–perfusion matching.

Metastatic infections

Metastatic infections occur when tissues are seeded with bacteria from the bloodstream. Common sites affected include the lungs, the bones, the kidneys and the susceptible endocardium. In some bacteraemias, e.g. *Streptococcus constellatus* and *S. intermedius* of the *S. 'milleri'* group, the brain and liver are also at risk. Meningitis, arthritis and empyema are relatively frequent complications of bacteraemias caused by Gram-positive cocci. There is a very high incidence of endocarditis if *Staphylococcus aureus* bacteraemia is sustained for more than 48 hours.

Some bacteraemias are trivial or silent, but present with complications. *Salmonella* bacteraemias may not be accompanied by bowel symptoms, but result in osteomyelitis in sickle-positive individuals, or soft-tissue abscess in others. Staphylococcal joint infection is probably the result of silent bacteraemia, as is streptococcal pericarditis or pneumococcal peritonitis.

Diagnosis of bacteraemia

Clinical diagnosis

Suspicion of a bacteraemia can sometimes be aroused by the presence of characteristic physical signs. Typical rashes may accompany streptococcal, staphylococcal, meningococcal and gonococcal bacteraemias. Large-joint arthritis commonly accompanies gonococcal bacteraemia. Lobar pneumonia is often present in pneumococcal bacteraemia. Nodular lung opacities or lung abscesses may be seen in staphylococcal bacteraemia. Disease or obstruction of the urinary tract, biliary tree or bowel often leads to Gram-negative and/or enterococcal bacteraemia, often with a contribution from the anaerobic *Bacteroides fragilis*. Recent instrumentation of hollow organs, or surgery, makes bacteraemia more likely.

Early hypotension is much more common in Gramnegative bacteraemias, particularly those caused by Enterobacteriaceae. Shock is usually absent until the late stages of Gram-positive bacteraemias. An important exception to this is the early presence of shock in *Clostridium perfringens* bacteraemia.

Laboratory diagnosis

Blood cultures should immediately be obtained, and incubated without delay. A minimum of two sets of cultures is essential; more than three do not significantly increase the likelihood of a positive diagnosis. Other specimens that should be cultured include urine, and sputum if obtainable. Specimens such as pus, urinary catheter urine, wound swabs, drain aspirate, throat swabs, joint aspirate, cerebrospinal fluid and diarrhoea stool may be cultured in appropriate circumstances.

Blood cultures are most likely to be positive if collected when the fever is rising. This is less relevant when bacteraemia is continuous, as in endocarditis or cannula-related sepsis. The yield from blood culture increases with the volume cultured. However, with high volumes the dilutional effect of the culture medium is reduced, and the bactericidal properties of the blood may prevent a positive culture being obtained. A 1 in 10 dilution is usually optimal. In neonates and small children a positive culture can be obtained with a smaller volume of blood, as the degree of bacteraemia is usually higher.

The blood culture media should be sufficiently nutritious to support the growth of aerobic and anaerobic pathogens. More than one medium is required and microbiology laboratories provide blood culture 'sets'. These vary from laboratory to laboratory but usually consist of one bottle suitable for the growth of aerobic, capnophilic and facultatively anaerobic organisms and a second culture containing added reducing agents to aid the isolation of anaerobic species such as *Bacteroides* or *Fusobacterium*. Blood culture bottles can be subcultured within 6 h for a rapid result and the supernatant examined by Gram stain or tests for bacterial antigens such as pneumococcal capsular polysaccharides and streptococcal group antigen to yield a rapid presumptive diagnosis. Most laboratories now use automated blood culture systems that detect the presence of bacterial growth by radiometric, infrared or colorimetric detection of carbon dioxide production or changes in electrical impedance of the medium. These systems are continuously monitored, allowing rapid detection of positive cultures without the need for routine subculture (see Chapter 3).

Management of bacteraemia and sepsis

Antibiotic treatment

The mainstay of treatment is control of the underlying infection. All possible cultures should be obtained before initial antibiotic treatment is commenced. A rational choice of empirical treatment may be possible if the history and examination point to a particular pathogen or group of pathogens.

When treating critically ill patients, it is important to consider the possible adverse effects of the antibiotic treatment. Some antibiotics can damage renal or hepatic function, or interact with drugs that may be used to support the patient. Wherever possible, antibiotics should be used that will not cause further damage to the patient's organs or metabolism. Antibiotic dosage should be tailored to the patient's excretory or metabolic capacity (see Chapter 4).

Aims of managing the systems and life support of bacteraemic patients

The infected patient with sepsis needs support until antimicrobial therapy has time to act. The ambitions of supportive treatment are ideally to support life and prevent further deterioration by:

1 maintaining the blood pressure;
2 maintaining the cardiac output;
3 counteracting the effects of low peripheral resistance;
4 maintaining blood and tissue oxygenation;
5 minimizing the effects of endotoxin by modulating cytokine and inflammatory responses.

Supporting the circulation

The initial step is to optimize the intravascular fluid volume. Where peripheral resistance and tissue perfusion are reduced, a moderate over-replacement of lost intravascular fluid is often required, and is monitored by central venous pressure measurement. There is controversy as to whether crystalloid fluids (such as saline solutions) or colloids (such as polygels or albumin) are most effective. Crystalloids are widely distributed in the interstitial fluid as well as intravascularly; they provide filterable solute for the renal tubules and can replenish lost intracellular fluid, but they are also quickly excreted. Colloids tend to remain in the vascular space; they improve blood pressure and organ perfusion but are not filtered through glomeruli. Different colloid substances are broken down in the circulation at different rates, and the longer-lasting ones cannot be removed if the patient becomes fluid overloaded and develops congestive cardiac failure.

Current practice is to give crystalloids first for an immediate effect, and then to follow up with colloid. The most useful monitors of circulatory function are the blood pressure, the central venous pressures and the urine output. The blood pressure need not be completely normalized; systolic pressures of 50–60 mmHg are adequate if cerebral function, urine output and tissue perfusion are satisfactory. A right atrial pressure of 5–10 mmHg is usually adequate; higher pressures are not often helpful and may contribute to congestive cardiac failure and poor tissue perfusion. The pulmonary capillary wedge pressure may be measured, using a flotation balloon catheter. If it is higher than the right atrial pressure, the left ventricle is not adequately moving blood from the lungs, and systemic perfusion is probably poor.

The difference between the body's core temperature and skin temperature gives an idea of skin, and therefore tissue, perfusion. If the skin temperature (usually measured at the foot) is more than 2 °C lower than the core (measured rectally or at the eardrum), this suggests inadequate tissue perfusion. Equal or nearly equal temperatures indicate vasodilatation and increased cardiac output.

Drugs used to support the circulation

These are only effective if there is sufficient intravascular fluid volume, indicated by adequate central venous pressure. Fluid replacement should therefore be optimized when drugs are used to support haemodynamic function.

Adrenoceptor agonists: both alpha (norepinephrine, phenylephrine and methoxamine) and beta or mixed (epinephrine metaraminol and isoprenaline) adrenoceptor agonists are used for a broad-spectrum inotropic and chronotropic effect. The alpha adrenoceptor agonists cause a degree of vasoconstriction and are useful for maintaining blood pressure in severe vasodilatation.

DOPA agonists: these increase myocardial contractility via beta$_1$-adrenergic receptors. Dopamine is routinely used in infusions of up to 4 µg/kg/min for its inotropic effect on the heart. Higher doses cause vasoconstriction and tachycardia, and may contribute to heart failure. Dobutamine may be added in infusions of 2.5–10 µg/kg/min for extra inotropic effect. Dopexamine is believed to increase renal perfusion by acting on peripheral dopamine receptors. It is commenced at 500 ng/min and increased in 0.5–1.0-µg steps at 15-min intervals up to 6 µg/min for an optimal effect.

Phosphodiesterase inhibitors: these (milrinone and enoximone) increase cardiac output in the presence of raised filling pressures.

Vasoconstrictors: these (angiotensin, vasopressin and their analogues) raise blood pressure, but at the expense of intense vasoconstriction, which may compromise tissue perfusion.

Although blood pressure, cardiac output and peripheral vascular resistance can be manipulated by using combinations of these drugs, the evidence that tissue perfusion or oxygenation is improved by these effects is lacking. As each patient has a different severity of disease and different combination of problems, controlled trials of treatment outcomes are difficult to perform.

Managing hypoxia and ARDS

Increasing the percentage of inspired oxygen will increase the arterial oxygen saturation, and this can be monitored by pulse oximetry or arterial blood-gas estimations. Removal of carbon dioxide is not increased, however, and care must be taken not to allow the patient to become severely hypercapnic as this impairs cardiac function, increases acidosis and raises intracranial pressure.

Positive-pressure ventilation overcomes decreased lung compliance and may expel fluid from the oedematous lungs. Recent experience suggests that low pressure/low tidal volume ventilation aids alveolar patency and minimizes barotrauma to the lungs. Positive end-expiratory pressure or continuous positive airways pressure often improve gas exchange. They have the disadvantage of reducing venous return and cardiac output, so often cause a requirement for increased cardiovascular support.

Moderate doses of corticosteroids may be beneficial in improving lung function and preventing progression from the inflammatory to the organizing and fibrotic stage of ARDS. Prolonged courses of treatment may be required.

There is little evidence of benefit from extra-corporeal membrane oxygenation (ECMO), but it may support a patient with a short-term reversible oxygenation deficit. Neither inhaled vasodilators such as nitric oxide or prostacyclin, nor surfactant have proved beneficial.

Antagonists of the mediators of sepsis

Plasma exchange to remove toxins and inflammatory mediators has been trialled on several occasions in both Gram-positive and Gram-negative infections. Benefit has not been demonstrated. It is now generally accepted that high-dose corticosteroids have no demonstrable benefit in managing sepsis.

Direct inhibition of endotoxin, TNF-alpha or IL-1 is theoretically possible, using monoclonal antibodies. Animal studies show that this is effective if the antibodies are given before signs of sepsis are established, but this is impractical in human medicine, and controlled trials have shown either no benefit or slight disadvantage in antibody treatments. It is now recognized that beta adrenoceptor agonists have anti-inflammatory properties; they can reduce the production of TNF-alpha, IL-1 and IL-6, while those with alpha agonist action tend to have the opposite effect. A discriminating choice of adrenoceptor agonist may be more important in this respect than in direct haemodynamic support.

Activated protein C (drotrecogin) is available for the adjunctive management of severe sepsis. By enhancing the availability of the activated agent, it has a number of theoretical beneficial effects, including: fibrinolysis; dilatation of small blood vessels; endotoxin binding and clearance; and promotion of the complement cascade. It appeared to be significantly effective in one large trial, but only in a subgroup of patients with severe sepsis. Its major complication is severe haemorrhage. It is contraindicated in meningitis, thrombocytopenia, intracerebral tumour, severe chronic liver disease, and within one month of surgery. A later trial showed it to be ineffective in mild sepsis, and failed to confirm its effect in severe sepsis.

Nutrition

Patients with bacteraemic diseases have high calorie requirements, and rapid turnover of macronutrients and micronutrients. Controlled trials show that nutritional support improves the outcome of treatment for bacteraemia and sepsis. Enteral feeding is physiological and prevents atrophy of the intestinal mucosa. Nasogastric intubation is unpleasant and increases management requirements in intubated patients, so endoscopically placed gastrostomy catheters are increasingly used when long-term enteral nutrition is needed.

Patients with reduced bowel function cannot absorb enteral nutrients and must be fed parenterally. This is usually done via a dedicated lumen of a multilumen right atrial catheter. A gradual change is made to enteral feeding as soon as improving bowel function allows.

Immediate management of patients with sepsis and septic shock

1 Resuscitate: optimize intravascular and tissue fluid volumes using crystalloid and colloid infusions, and measure arterial and central venous pressures (target values: mean arterial pressure at least 65 mmHg; central venous pressure 8–12 mmHg; urine output at least 0.5 ml/kg hourly).

2 Obtain specimens for microbiological investigations: blood, urine, fluid, pus cultures, also, swabs, bronchial lavage or biopsy specimens where indicated.

3 Give empirical antibiotic therapy: to 'cover' for likely pathogens, modify when laboratory results are available.

4 Search for the origin of infection: by clinical examination, X-ray, echocardiography, imaging, cultures.

5 Consider recombinant activated protein C (rAPC): in patients with indications.

6 Maintain vigorous support: optimize tissue oxygen delivery by giving oxygen with or without non-invasive assisted ventilation; manage blood glucose; consider corticosteroid replacement therapy; give prophylaxis for deep vein thrombosis and gastroduodenal stress ulcers.

7 Refer to ITU if target values cannot be achieved or maintained: for intermittent positive-pressure ventilation; support of circulatory function, using inotropic and/or vasopressor agents; renal replacement therapy when indicated.

8 Maintain adequate nutrition: using enteral or parenteral feeding techniques.

Adverse effects of prolonged intensive care

Intensive, or critical, care support allows patients to survive during treatment of bacteraemia and its pathological effects. However, the modes of care involved are physically and physiologically stressful, and may result in medium or long-term adverse effects.

The lungs are subjected to pressure stress, and high oxygen levels. They may be subjected to one or more episodes of healthcare-acquired pneumonia. The respiratory muscles are disused during positive-pressure respiratory support, and consequently lose strength and bulk. This results in a need for medium-term respiratory rehabilitation, and may cause longer-term reduction in respiratory reserve.

The cardiovascular system is subjected to major pharmacological stimuli, in order to maintain perfusion of essential organs and tissues. To achieve this, intensive alpha-adrenergic stimulation may be necessary. The resulting peripheral vasoconstriction and pre-existing tissue hypoxia can result in the unavoidable loss of digits or, occasionally, the periphery of a limb, on order to preserve the patient's life.

Critical illness polyneuropathy

This condition is a mainly distal neuropathy that affects up to 70% of patients with sepsis, though it is not always clinically evident. It can cause difficulty in weaning from ventilation. It is recognized, and distinguished from the proximal radiculopathy of Guillain–Barré syndrome, by electromyography (EMG), which shows decreased compound muscle action potentials and denervation potentials in muscle. Sensory nerve action potentials are also reduced. It recovers in a few weeks with supportive management.

Critical illness myopathy

This is a rare atrophic or necrotic myopathy of unknown aetiology, which is more likely with long periods of intensive care. Reduced muscle action potentials are seen on EMG, and the muscle is unresponsive to direct stimulation. Muscle biopsy often shows changes, which vary from type II fibre atrophy to extensive muscle necrosis. Sensory action potentials remain normal. The condition tends to improve slowly, but muscle necrosis may be irreversible.

Staphylococcal bacteraemic disease

Introduction and epidemiology

Staphylococcus aureus is the commonest cause of Gram-positive bacteraemia. Nearly half of cases are community-derived and affect previously healthy people, in whom the infection often has an insidious onset, sometimes delaying the diagnosis because sufferers do not initially appear severely ill. A few strains of *S. aureus* carry the Panton–Valentine pathogenicity factor, a powerful leucocidin that predisposes to early, severe sepsis. In intravenous drug abusers, *S. aureus* originating from the skin is the commonest blood pathogen.

An increasing number of cases are hospital-acquired, related to intravenous devices, pacemakers and surgical wounds. In the UK, over half of such cases are due to methicillin-resistant strains of *S. aureus*. Healthcare-associated staphylococcal infection affects all age groups and carries a high mortality of around 25% (up to 70% in the elderly).

Staphylococcal bacteraemia is a grave threat to patients with biological implants and damaged heart valves. Prosthetic joints have around a 50% risk of becoming infected as a result of the bacteraemia. A similar proportion of susceptible heart valves are at risk if staphylococcal bacteraemia is sustained for more than 48 hours.

Pathology

Many patients with non-healthcare acquired staphylococcal bacteraemia have no obvious predisposing factor. Two serogroups of *S. aureus*, groups 5 and 8, predominate among bacteraemic isolates worldwide.

Phagocytosis is important in defence against staphylococci. Sufferers of cystic fibrosis have poor phagocyte function, and suffer severe staphylococcal chest infections but rarely septicaemia. Patients whose phagocytes cannot mount a bactericidal respiratory burst, as in chronic granulomatous disease, are at risk of repeated, severe staphylococcal infections.

Clinical features

The onset of illness is often slow, with increasing fever and malaise. Helpful physical signs include pain in a large joint (usually without effusion), severe periarticular tenderness of all joints, or numerous small pustules on the skin. Patients usually have high, swinging fever, but are haemodynamically normal. Rare cases have coexisting toxic shock syndrome (see Chapter 5).

The white cell count is often misleadingly normal, but the proportion of neutrophils is usually near 90%. After some days a neutrophil leucocytosis develops. The platelet count may be slightly reduced. Patients infected with Panton–Valentine leucocidin-possessing strains may maintain a neutropenia. Renal function is moderately impaired; proteinuria and microscopic haematuria are common. In half or more of cases the chest X-ray is abnormal, showing nodular opacities, abscesses, consolidation, effusion or pleural abscess. Intravenous drug abusers tend to suffer severe, embolic lung sepsis and, often, endocarditis, which may be right-sided.

In untreated cases, renal failure is progressive, because of toxaemia, and numerous microabscesses in the kidneys.

Disseminated intravascular coagulation is common and can cause peripheral gangrene. Metastatic infections are a real danger and include aggressive endocarditis, pneumonia, suppurative arthritis, soft-tissue abscesses and abscesses of the kidney, liver or brain.

Diagnosis

Even in patients who have received oral antibiotics, blood cultures are usually positive in all bottles within 18–24 h.

Management

For methicillin-sensitive strains, the treatment of choice is intravenous flucloxacillin, in divided doses totalling at least 6 g/day. Flucloxacillin does not reach bactericidal levels in all tissue sites.

Either fusidic acid or rifampicin may be added, to improve tissue antibiotic delivery in some circumstances. Both are well distributed in the body, but rifampicin is useful in controlling infection in the presence of implanted prosthetic devices. Fusidic acid penetrates well into bone. Fusidic acid is irritant to peripheral veins but both drugs are well absorbed orally or intragastrically. Neither fusidic acid nor rifampicin should be given alone, because staphylococci can make a one-step mutation to complete resistance within 4 or 5 days of exposure to them.

Clindamycin penetrates tissues well, and is effective in sites with poor perfusion and low redox potential. It is valuable in severe bone and soft tissue infections. It can be given orally with good effect, and is useful where intravenous access is critically difficult (for instance in longer-term treatment and in intravenous drug abusers).

In patients who have skin allergies to penicillins, cephradine or cefuroxime may be substituted, but should not be given to patients who have anaphylactic reactions to penicillin. In such difficult situations, ciprofloxacin, clindamycin, teicoplanin or vancomycin may be useful.

Many *S. aureus* are sensitive to aminoglycosides on laboratory testing. However, these have relatively poor penetration into abscesses, and often give disappointing results when used as antistaphylococcal monotherapy.

For methicillin-resistant *S. aureus* (MRSA), vancomycin is the treatment of choice. It must be given as an infusion over at least 90 min to avoid the severe hypersensitivity reaction of vasodilatation (red man syndrome), bronchospasm and acute hypotension. Infusion also avoids high peak levels, which predispose to nephrotoxicity and ototoxicity. Peak and trough levels at the end of infusion and immediately before the next dose should not exceed 30 and 10 mg/l, respectively. The dose should be reduced in renal impairment. Aminoglycosides and loop diuretics enhance vancomycin toxicity. Teicoplanin may be an effective and less toxic alternative, but some staphylococci have increased minimum inhibitory concentration (MIC) and minimum bactericidal concentration (MBC) values for this drug. Other alternatives may include trimethoprim, ciprofloxacin or, when sensitivity tests indicate likely effectiveness, fusidic acid, rifampin or tetracyclines. MRSA are often resistant to macrolides and clindamycin.

Important antistaphylococcal drugs

• Flucloxacillin i.v. 1–2 g 6-hourly (child under 2 years, 250–500 mg 6-hourly; 2–10 years, 500 mg–1 g 6-hourly); *or*
• Cloxacillin i.v. 500 mg–1 g 4–6-hourly (child under 2 years, 125–250 mg 6-hourly; 2–10 years, 250–500 mg 6-hourly); *plus*
• Fusidic acid orally 750 mg 8-hourly (child under 1 year, 50 mg/kg daily; 1–5 years, 750 mg daily; 5–12 years, 1.5 g daily, all in three divided doses) or i.v. 580 mg 8-hourly; *or*
• Rifampicin orally or i.v. 300–600 mg 12-hourly (child under 3 months, 5 mg/kg 12-hourly; 3 months–12 years, 10 mg/kg 12-hourly to maximum 600 mg/day); *or*
• Clindamycin orally 300–600 mg 6-hourly – may be increased to 900 mg 6-hourly in life-threatening disease (child over 1 month, 30–40 mg/kg daily – *minimum* 300 mg daily at any age); discontinue immediately if diarrhoea or colitis develops (may also be used as monotherapy).

Alternative drugs

1 Cephradine i.v. 1–2 g 6-hourly (child 50–100 mg/kg daily in four divided doses).
2 Cefuroxime i.v. 750 mg–1.5 g 6–8-hourly (child 60–100 mg/kg daily in three divided doses).
3 Trimethoprim by slow i.v. injection or infusion 150–250 mg 12-hourly (child 6–9 mg/kg daily in two or three divided doses).

Important drugs for treating MRSA

• Vancomycin i.v. infusion over 90 min, 500 mg 6-hourly (neonate up to 1 week, 15 mg/kg followed by 10 mg/kg 12-hourly; 1–4 weeks, 15 mg/kg followed by 10 mg/kg 8-hourly; child over 1 month, 10 mg/kg 6-hourly); plasma levels should peak (1 h after dose) not above 30 mg/l, trough not above 10 mg/l.
• Teicoplanin i.v. 400 mg 12-hourly for three doses, then 400 mg daily – may be reduced to 200 mg daily after good response (child over 2 months, 10 mg/kg daily – may be reduced to 5 mg/kg daily after good response).

⚠ Special drugs effective against Gram-positive organisms, such as linezolid, daptomycin and quinupristin/dalfopristin (Synercid), may be available, under the direction of a specialist in infection or clinical microbiology.

Coagulase-negative staphylococci

Coagulase-negative staphylococci are common causes of bacteraemia in neonatal intensive care, particularly associated with cardiothoracic surgery and the use of indwelling intravenous devices. Many are resistant to flucloxacillin. They must therefore be treated as methicillin-resistant staphylococci, with clindamycin, trimethoprim, doxycycline, gentamicin, fusidic acid, teicoplanin or vancomycin, depending ultimately on the results of sensitivity testing.

The duration of treatment will depend on response. Uncomplicated cases can be adequately treated with 7–10 days' antibiotics. Cases with multiple abscesses or loculated infections need longer courses. Drainage may be required for empyema, large lung abscesses or secondary bone infection. Endocarditis needs long-term treatment and follow-up (see Chapter 12).

Streptococcus pyogenes bacteraemia

Introduction and epidemiology

Streptococcus pyogenes (group A streptococcus, or GAS) is a commensal of the skin and upper respiratory tract. It can also colonize the female genital tract, particularly following examination or manipulation during pregnancy and delivery.

Many streptococcal bacteraemias originate from skin or throat infections, which may not themselves be severe. Examples include small cuts or abrasions, which may have healed by the time bacteraemia is recognized. Streptococcal puerperal fever may follow normal or assisted delivery, Caesarean section or other gynaecological procedures (see Chapter 17).

Although a rarer cause of bacteraemia than either *Staphylococcus aureus*, meningococci or pneumococci, *Streptococcus pyogenes* is a dangerous organism; the mortality of treated streptococcal septicaemia is 25–30% in most Western countries.

Pathology

No specific predisposition to streptococcal bacteraemia is recognized. There are many strains of *S. pyogenes*, some more virulent than others. Different strains predominate for several years at a time in slow epidemic fluctuations. M1 types have produced serious disease in the UK in the 1990s. These strains have shown a significant incidence of resistance to erythromycin.

Clinical features

The presenting complaint is usually of severe feverish illness with a short history of 1 or 2 days. Many patients lack any helpful physical sign, but enquiry for helpful preceding events, or warning signs may be rewarding. Mild recent skin lesions may be important, even if they have healed. Some patients complain of a severe sore throat with major systemic upset. A sinister warning feature is swelling and severe pain in lymph nodes draining the original site. This may be seen in the axilla, or the inguinal or tonsillar nodes. Spreading erythema, abscess formation, necrosis, ulceration or sloughing of the nodes is occasionally seen.

Occasionally, areas of erysipelas, cellulitis or a scarlet fever-like rash develop. The appearance of erysipelas, other than in the typical sites on the face or leg, is strongly suggestive of bacteraemia. Septic arthritis also indicates a risk of bacteraemia. Necrotizing fasciitis is a rare, life-threatening feature of *S. pyogenes* bacteraemia. Any fever in the puerperium should be actively investigated (see Chapter 17).

Non-specific features of severe illness may be prominent. These include watery diarrhoea, persistent vomiting, meningism or pain in the back or thighs. Confusion and reduced consciousness are rare.

The white cell count often shows a modest leucocytosis of 12 or 13×10^9/l. Renal and liver function is usually maintained, as is the blood pressure, until the preterminal stages. In later or untreated cases, an ill-defined pneumonia or a pleural effusion may occur. Shock and profound hypoxia can develop very quickly and are extremely difficult to reverse.

Diagnosis

Initial suspicion must often be based on clinical clues. Appropriate specimens from throat, skin or other lesions may produce a growth of beta-haemolytic streptococci within 24 h. Blood cultures are not reliably positive for up to 3 days in some cases, as *S. pyogenes* has a slow initial growth curve.

Management

Benzylpenicillin is the treatment of choice, except in the presence of streptococcal necrotizing fasciitis, when clindamycin is the drug of choice, as it is effective in low redox potentials, and has excellent tissue penetration.

A suitable dose of penicillin is 1.2 g 2-hourly. It is notoriously difficult to eradicate streptococcal bacteraemia, so the dose should not be reduced too soon after an apparent response. High dosage for 4–5 days may be needed before intermittent recurrences of fever and toxaemia are controlled. A further 7 days or more of conventional dosage is then advisable; longer in complicated cases.

Tissue damage caused by streptococcal toxins can progress for many hours after the start of treatment. Erythemas may appear; inspired oxygen requirements may increase; patients may need a period of positive-pressure ventilation or inotropic support. Close monitoring

should therefore be continued until a definite response is established.

Few drugs are as effective as penicillin. Broad-spectrum penicillins or cephalosporins may be effective, but not more so than penicillin. Erythromycin is ineffective against a proportion of virulent strains. Poor progress after an initial response may be better managed by increased penicillin dosage (limited only by nausea, drowsiness or other penicillin toxicity) than a change of drug. A change of therapy should be made only for compelling reasons, when clindamycin is a good alternative.

For a patient with anaphylactic reactions to penicillins, clindamycin or a glycopeptide may be given.

Complications

A common complication of treatment is a mild febrile reaction to the high penicillin dosage. So long as the patient's physical condition is improving, this is not an indication to change the treatment. Rarely, continued high-dose penicillin treatment results in profound, but reversible, agranulocytosis. No further beta-lactam agents must then be given (including oral agents), as there is complete cross-reactivity between them. Treatment should be continued with alternative drugs such as clindamycin (plus an aminoglycoside or a broad-spectrum quinolone during the neutropenic phase).

The classic poststreptococcal conditions of rheumatic fever, erythema nodosum and erythema multiforme are rarely seen, but scarlet fever may occur. When it does, it is often severe, with pleural effusion, ascites, renal impairment and hypotension (streptococcal toxic shock syndrome).

Tissue necrosis occasionally occurs in affected skin or lymph nodes even when the bacteraemia is relatively easily controlled.

Gram-negative septicaemia

Introduction and epidemiology

Gram-negative septicaemia usually originates from abdominal or pelvic pathology. In young age groups appendicitis, trauma or gastroduodenal surgery are the commonest predispositions. In middle age, gallbladder or biliary disease and bowel surgery are more common. In the elderly, or patients with indwelling urinary catheters, the urine is often colonized with organisms and is an important source of bacteraemia. Liver abscesses can accompany sepsis in all age groups, and must be actively excluded if other sources are not found. *Escherichia coli* is by far the commonest cause of Gram-negative bacteraemias.

Clinical features

While many Gram-negative bacteraemias are continuous or occur many times hourly or daily, some (often originating from loculated infection) are intermittent, causing brief episodes of fever once or twice daily or weekly (Fig. 19.5). In either case, rigors, tachycardia and a lowered blood pressure accompany the fever. In continuous bacteraemias the blood pressure can remain sufficiently low to compromise cerebral and renal perfusion, while in the intermittent type, it usually normalizes when the bacteraemia ceases after 30–90 min.

Significant prolongation of the prothrombin time tends to occur early in Gram-negative bacteraemias, and elevation of fibrin degradation products indicates DIC. There may be a raised white cell count, with neutrophilia, but in overwhelming sepsis the white cell count is often low.

Diagnosis

Blood cultures are essential, and should be combined with attempts to obtain specimens from infected sites of origin of the bacteraemias. Imaging of the liver, kidneys, pancreaticobiliary system or renal drainage system is often helpful. Disease of the bowel may reveal itself by causing pain or bowel dysfunction when ulcer disease, malignancy or diverticulitis are the origin of the infection. Patients with colonized urine, however, often deny urinary symptoms.

Management

Intensive supportive measures are often required. The choice of antimicrobial chemotherapy needs to include initial cover for Enterobacteriaceae, enterococci and anaerobes. Combinations of drugs are recommended. A broad-spectrum cephalosporin is effective against many coliforms, and can be combined with metronidazole to combat anaerobic infection. Meropenem is effective against both Gram-negative rods and anaerobes, and has some action against enterococci. If enterococcal sepsis is likely, co-amoxiclav or amoxicillin is also useful, and gentamicin may be added for synergistic effect. This combination is effective against many Gram-negative rods, and the co-amoxiclav also has an anti-anaerobic action.

> ### Typical regimens for treating Gram-negative bacteraemia
>
> • Cefotaxime i.v. 1–2 g 8-hourly (neonate, 100–200 mg/kg daily; child, 150–200 mg/kg daily, both in two to four divided doses); *or* ceftriaxone by i.v. infusion 2–4 g daily; neonate 20 to 50 mg/kg daily; child <50 kg 50–80 mg/kg daily *plus*
> • Metronidazole rectally 1 g 8-hourly or i.v. 500 mg 8-hourly; child, any route, 7.5 mg/kg 8-hourly.
>
> ### Alternative
>
> • Co-amoxiclav by i.v. infusion (expressed as amoxicillin) 1 g 6–8 hourly; infant up to 3 months 25 mg/kg 8-hourly (12-hourly in neonate or premature infant); 3 months to 12 years, 25 mg/kg 6–8 hourly *plus or minus*
> • Gentamicin by slow i.v. injection 2–5 mg/kg daily in three divided doses (child under 2 weeks, 3 mg/kg 12-hourly; 2 weeks–12 years, 2 mg/kg 8-hourly). Plasma concentrations should be: peak (1 h after dose) not above 10 mg/l, trough not above 2 mg/l.

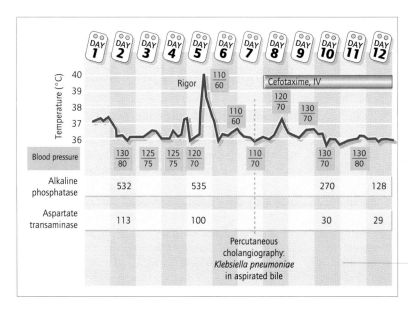

Figure 19.5 Intermittent Gram-negative bacteraemia: this patient's gallbladder was colonized with *Klebsiella pneumoniae*; his fevers were accompanied by rigors and hypotension; the fever and liver function improved with antibiotic treatment.

For Gram-negative infections when *Pseudomonas* spp. may be present

- Gentamicin or another aminoglycoside may be given *with* (but not mixed in the same syringe or infusion) azlocillin i.v. 2–5 g 8-hourly; neonate, 100 mg/kg 12-hourly; 7 days–1 year, 100 mg/kg 8-hourly; 1–14 years, 75 mg/kg 8-hourly; *or* piperacillin i.v. all ages, 200–300 mg/kg daily in four to six divided doses.
- For neutropenic patients (see also Chapter 22): piperacillin plus tazobactam (Tazocin®) by i.v. infusion adult and child over 12 years, 4.5 g 8-hourly *or* meropenem i.v. 1 g 8-hourly (2 g 8-hourly for meningitis); child 3 months–12 years, 10–20 mg/kg 8-hourly.

Alternative

- Ceftazidime i.v. 1–2 g 8-hourly (maximum 3 g/day in the elderly); child up to 2 months 30 mg/kg 12-hourly, over 2 months 60–100 mg/kg daily in 2 or 3 divided doses.

The antibiotic spectrum may be narrowed, or the least toxic drugs selected, when the results of culture are available.

Complications

The most important complication is failure to respond to appropriate antibiotic treatment. The usual reason for this is loculated infection, in which the organisms are inaccessible to the antibiotic. Image-guided, or surgical drainage is then necessary to obtain resolution of the infection.

Follow-up of patients with bacteraemia

Patients responding to treatment of their bacteraemia should be reviewed during convalescence as recurrence sometimes occurs. A particular danger is a persisting undetected originating condition (such as a carcinoma of the colon, biliary stricture or endocarditis). Examination should always be performed to detect abdominal tenderness, an enlarged liver or gallbladder or developing heart murmurs. It is also useful to review the C-reactive protein level, as a rise can give early warning of continuing infection.

Introduction: causes of pyrexia of unknown origin

A pyrexia of unknown origin (PUO) is a raised body temperature for which there is no obvious cause. Fever is defined as a body temperature reaching at least 37.8–38°C. Many trivial and self-limiting conditions cause short-lasting fevers. These include many infections, allergic and inflammatory reactions, whose formal diagnosis would not lead to management decisions and would therefore be uneconomical. They are excluded from consideration in PUO by making a detailed definition.

Definitions of pyrexia of unknown origin
1 The original definition was developed in the 1950s and 1960s, when modern imaging and molecular diagnostic methods were not available: PUO is a persisting or frequently occurring fever of 38.5°C or higher, which has continued for 3 weeks or more, and for which no diagnosis is evident after 1 week of in-patient investigation.

2 A more modern definition, developed at Duke University, takes account of modern diagnostic procedures: PUO is a fever of 38.5°C or higher, which has persisted or frequently occurred for at least 3 weeks, and whose aetiology remains unknown after 3 days of in-patient investigation, or three visits for out-patient investigation.

Because fever is a response to cytokine signals, generated via endothelial activation, infection is not its only cause. Non-infectious causes of PUO are common and important. Most causes of PUO are not exotic diseases; they are common diseases presenting without their usual symptoms and physical signs. With the development of new diagnostic techniques, the list of causes of PUO is gradually changing: for instance concealed soft-tissue infections, or connective tissue disorders such as systemic lupus erythematosus, now rarely defy early diagnosis.

The causes of fever can be described under six main headings.
- Infections (30–40% of cases; over 50% in children).
- Malignancies (20–30% of cases).

- Connective tissue disorders (10–20% of cases; over 30% in patients aged >50 years).
- Hypersensitivity reactions.
- Rare endocrine and metabolic conditions.
- Factitious fever (fever induced deliberately by the patient) comprising up to 10% of some series.

Low-grade fever is common in embolic or thrombotic conditions, such as recurrent pulmonary embolism.

Important infectious causes of pyrexia of unknown origin

1 Tuberculosis (usually non-pulmonary, miliary or cryptic).
2 Infections with a long time course (such as infectious mononucleosis, toxoplasmosis, bartonellosis or Q fever).
3 Infective endocarditis.
4 Sepsis (especially in a hollow organ) or abscess, e.g. liver, or dental abscess.
5 Imported diseases (such as typhoid fever or brucellosis).

Important malignant causes of pyrexia of unknown origin

1 Lymphomas (Hodgkin's and non-Hodgkin's).
2 Leukaemias (especially monocytic or myelomonocytic types).
3 Histiocytoses (usually in small children).
4 Renal adenocarcinoma.
5 Primary hepatic carcinoma.
6 Rare, atrial myxoma.

Connective tissue disorders that may present as pyrexia of unknown origin

1 Giant cell arteritis.
2 Polymyalgia rheumatica.
3 Polyarteritis nodosa.
4 Wegener's granulomatosis.
5 Sarcoidosis (and rare conditions such as idiopathic granulomatous hepatitis).
6 Crohn's disease.
7 Juvenile rheumatoid arthritis.
8 Adult Still's disease.

Common causes of hypersensitivity with fever
Drugs
1 Sulphonamides (including sulfasalazine).
2 Beta-lactam antibiotics.
3 Rifampicin, isoniazid and other anti-TB agents.
4 Phenytoin.
5 Quinidine.

Environmental factors
1 Fungus-infected hay (farmer's lung).
2 Bird proteins (pigeon-fancier's lung).

Rare metabolic causes of pyrexia of unknown origin

1 Uncontrolled hyperthyroidism or subacute thyroiditis.
2 Autoimmune adrenal failure (Addison's disease).
3 Porphyrias (acute intermittent or mixed types).
4 Familial relapsing polyserositis (familial Mediterranean fever).
5 Other familial fevers (disorders of acute inflammatory reactions; TNF receptor disorders).
6 Rare cases of vasoactive intestinal polypeptide-producing tumour (VIPoma) and glucagonoma.

Initial assessment

Assessment of a patient with PUO must include a search for both infectious and non-infectious conditions. Screening procedures work more efficiently if the patient's history, symptoms and physical condition are first assessed in detail. The physician can then choose initial and follow-up investigations in a structured way by using an epidemiological and clinical database to indicate which tests are most likely to be useful. Otherwise an almost infinite range of investigations could be performed, each with a different sensitivity and specificity, presenting the investigator with a hugely complex task when interpreting results.

The initial assessment of a patient with fever of unknown origin is the same as for any feverish patient (see Chapter 1). It is important to obtain as much information as possible about the onset and evolution of the prolonged fever. It is advisable to visit the patient on more than one occasion to do this, as few patients remember to mention every relevant point in a single interview.

It is also important to re-examine the patient a few days after the initial complete physical examination, as new physical signs, such as enlarged lymph nodes, rashes or cardiac murmurs, may appear as the condition progresses. Changes in already existing signs may also help to reveal recognizable characteristics of a disease.

Important points in the epidemiological history
Exposure
1 Exposure to infection; by contact with other cases, travel, food, water, occupation or recreation, or by association with animals, including farm animals or pets.
2 Exposure to allergens; antibiotics or other drugs, environmental agents such as bird proteins, organic dusts such as animal or bird dander, cotton or contaminated hay, industrial dusts and vapours, all of which can cause inflammatory or granulomatous lung disease.

Predisposition

1 Indicated by a family history, e.g. of connective tissue diseases, familial Mediterranean fever, acute intermittent porphyria or Reiter's syndrome.
2 Exposure to carcinogenic agents, such as radiation, including intensive radiotherapy, or sustained immunosuppressive therapy, for instance with cyclosporin or anti-TNF antibodies, which may indicate an increased likelihood of malignant disease because of impaired 'immune surveillance' or response.

Protection or resistance

1 The result of natural immunity following previous infection.
2 Induced by immunization or chemoprophylaxis.

⚠ These do not confer absolute protection from a condition, but reduce the likelihood of a particular disease and allow the investigator to place it lower in the differential diagnosis.

Important points in the clinical history

An earlier precipitating condition can suggest appropriate diagnostic tests (for instance: a sore throat in rheumatic fever, typically occurring 10–14 days earlier; an 'illness of infection' in typhoid fever; or erythema chronicum migrans in borreliosis).

The severity or pattern of the fever itself is not often helpful (see Chapter 1). The tertian fever of malaria is an exception, occurring when the disease is well established.

Developing viral infections are often marked by prostration, myalgia, arthralgia and shivering attacks.

Abscesses and loculated sepsis are often accompanied by intermittent bacteraemias, indicated by rigors – severe shaking chills that make speech and other movement difficult (Fig. 20.1).

Important aspects of physical examination

Localized bone or joint pain

Such symptoms precede X-ray changes by many days or weeks, and should therefore be further investigated by other imaging techniques (see Chapter 14).

Soft systolic murmurs

These may be signs of endocarditis, pericarditis or myocarditis. They should be confirmed and investigated by echocardiography, and reassessed regularly to detect changes.

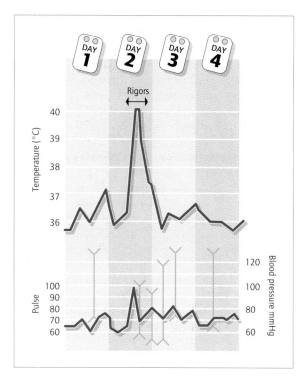

Figure 20.1 Isolated spike of fever accompanied by rigors. This patient had a *Klebsiella pneumoniae* infection of the obstructed biliary tree, with intermittent bacteraemias.

Lesions in skin, mucosae and nails

These may be signs of embolic or vasculitic phenomena in endocarditis or vasculitis. They include petechiae, Osler's nodes, splinter haemorrhages in the finger or toe nails, small, vasculitic lesions of the digits or retinal haemorrhages. Small retinal haemorrhages and cytoid bodies have a similar significance.

Macular or maculopapular rashes may accompany juvenile rheumatoid arthritis; livedo reticularis may occur in polyarteritis nodosa.

Mild meningism

Mild meningism may be a warning sign of subacute disease such as tuberculous meningitis or neuroborreliosis. Early investigation by imaging and lumbar puncture is essential.

Chest X-ray

A chest X-ray must always be performed without delay. Extensive pulmonary consolidation or cavitation can exist with few or no physical signs (Fig. 20.2). Granulomas, infiltrations, small pleural effusions and mediastinal

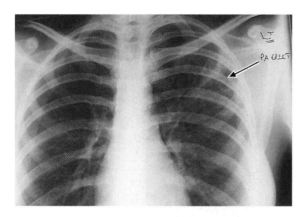

Figure 20.2 Chest X-ray of a midwife returning from work in Africa with persisting low-grade fever and cough. The chest was clinically clear, but this cavitating opacity (arrow) prompted investigation and treatment for tuberculosis.

swellings are other important abnormalities immediately detectable by X-ray.

Initial laboratory investigations

Haematological examination

1 Red cells (number, morphology,).
2 Stained blood film (intracellular inclusion bodies or parasites).
3 Granulocyte counts (immature granulocytes, inclusions or ingested pigment).
4 Lymphocyte count (subsets and morphology, e.g. activated mononuclear cells).
5 Platelet count.
6 Inflammatory indices, C-reactive protein (CRP) and erythrocyte sedimentation rate (ESR; see Chapter 1).

⚠️ In some connective tissue and granulomatous conditions, such as systemic lupus erythematosus and Crohn's disease, the CRP may be near normal while the ESR is markedly elevated.

Blood biochemical tests

1 Aspartate transaminase elevation indicates tissue damage.
2 Alanine transaminase is elevated if the liver is the damaged organ.
3 Alkaline phosphatase is elevated in cholestasis and space-occupying liver lesions (and rarely originates from sites of increased bone metabolism).

4 Other enzymes worth investigating include: blood amylase levels, which may be high in pancreatitis or inflammation of the salivary glands; creatine kinase elevations are seen in myositis, troponins in myocarditis; and lactate dehydrogenase levels may be high in lymphoma.
5 Serum ferritin is an acute-phase protein, often moderately elevated in severe inflammation; very marked elevation is characteristic of 'adult Still's disease' (adult presentations of juvenile rheumatoid arthritis).

Urine examination

1 Urine cytology and biochemical tests, including dip-tests, may give early indication of infection (see Chapter 2), and also detect products of haemolysis and bilirubinuria too mild for clinical detection, and proteinuria, which may indicate nephrosis. Urine microscopy, which demonstrates red-cell or white-cell casts, aids diagnosis of nephritis. Biochemical screening tests for porphyria are carried out on urine samples (preferably taken during episodes of fever).

Initial microbiological investigations

Microscopy of easily obtained specimens

This is a useful and rapid diagnostic procedure that may identify pathogens in urine (Fig. 20.3) and parasites in stools (see Chapter 2), while investigation of sputum can confirm a diagnosis of tuberculosis. Bronchial aspirate, drained effusions or cerebrospinal fluid (CSF) can be examined using various stains or polymerase chain reaction (PCR) techniques, to reveal fungal, herpesvirus or respiratory virus infections. Stained smears of splenic aspirate can reveal Leishman–Donovan bodies in visceral leishmaniasis, and fine-needle aspirate of lymph nodes may be stained to demonstrate granulomas, malignant cells, parasites, spirochetes or acid–alcohol-fast bacilli. Actinomycosis can be diagnosed by the Gram-stained appearance of 'sulphur granules' from pus.

Initial cultures (see also Chapter 2)

Blood cultures

Blood cultures are essential, and should be obtained before antimicrobial treatment whenever possible. At least two sets should be taken, at different times and preferably when the temperature is rising or has just risen, as fever is induced when pathogens are released into the blood. In pyrexia of unknown origin, prolonged incubation is advisable, to detect slow-growing or highly fastidious organisms.

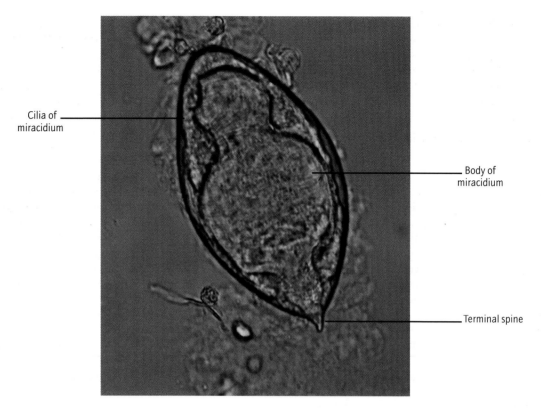

Cilia of miracidium

Body of miracidium

Terminal spine

Figure 20.3 Ovum of *Schistosoma haematobium* in the urine of an engineer with episodic haematuria: he had been working on a river estuary in Africa.

Other specimens for bacterial culture

Urine culture is also important, as both acute and chronic urinary infections can exist without localizing symptoms or signs. The early-morning urine (EMU) can be cultured for mycobacteria. Stool cultures may be positive in enteric fevers when blood cultures have failed for any reason. It may be important to obtain CSF (usually after brain imaging) if meningeal or neurological signs are present.

Pus, discharge or vesicle fluid must always be collected for culture. It is more helpful to the microbiologist if a significant volume can be obtained, rather than just a smear on a swab. However, smears dried onto glass slides, or swabs, can be used to prepare material for PCR techniques. To preserve fastidious anaerobes, pus may be inoculated into blood culture media. A pathogen can still be identified in some cases where previous antibiotic treatment or fragility of the pathogen result in negative cultures. Bacterial 16S (or fungal 18S) ribosomal RNA can be amplified by PCR-based techniques, sequenced, and identified by comparison with a sequence database (see Chapter 3).

If appropriate specimens are not accessible, or the patient cannot, for instance, expectorate, specimens should be obtained by specific techniques, such as fine-needle aspiration, image-directed aspiration of loculated material, or bronchoalveolar lavage.

Cultures for virological investigation

Cultures for virological investigation are equally important, especially in the immunosuppressed. For respiratory viruses, nose and throat swabs, nasopharyngeal aspirate, bronchial aspirate and lavage specimens may be processed. For enteroviruses, throat swabs, stool and (if appropriate) CSF may be examined. Urine can be cultured for cytomegalovirus and paramyxoviruses. Respiratory specimens, vesicle scrapings and vesicle fluid are appropriate for herpesvirus cultures.

Nucleic acid amplification techniques as screening and diagnostic tests

It is now possible to amplify microbial ribosomal RNA directly from body fluid or tissue samples. If 16S-rRNA (of bacterial origin) or 18S-rRNA (of fungal origin) is detected in a sample, its nucleotide structure can be defined by sequencing. The organism can then be identified by comparing this sequence with a reference database of sequences (see Chapter 3). This technique may be successful in identifying organisms that do not grow in standard cultures, or whose growth has been inhibited by prior antibiotic therapy. It can be successful even using tissue samples that have previously been 'fixed' in preservatives, depending on whether sufficient RNA has survived to be adequately amplified.

Initial serological investigations

These need not be elaborate, unless the clinical situation demands it, but serum must always be obtained as early as possible, for comparison with later specimens. The need for paired sera, taken 10–14 days apart, means that some serological diagnostic tests must be deferred until the second serum is obtained. Many tests, however, are useful screening tests and can provide early results. A positive monospot test is virtually diagnostic of Epstein–Barr virus infection, and remains positive for some weeks after the onset of fever. As well as immunoglobulin M (IgM) and other antibody tests for infectious agents, autoantibody tests for connective tissue diseases are useful early tests.

Antigen tests for hepatitis B, or *Cryptococcus neoformans*, and urinary antigen tests for *Legionella pneumophila* are also examples of early diagnostic serology.

Tuberculin test

Tuberculosis is common among PUO patients, so a tuberculin test should be considered, despite its known poor specificity (see Chapter 18). Although not diagnostic, a strongly positive Mantoux test supports a diagnosis of active tuberculosis.

Initial work-up plan for pyrexia of unknown origin

1 Detailed clinical and epidemiological history.
2 Physical examination.
3 Chest X-ray.
4 Blood count, differential count and morphology.
5 C-reactive protein (and/or erythrocyte sedimentation rate).
6 Liver and renal function tests.
7 Dipstick urine tests.
8 Microscopy of blood film and urine (CSF, stool and lesion fluid if indicated).
9 Culture of blood, urine and respiratory specimens (CSF, stool, pus and lesion fluid if indicated).
10 Acid-fast stain and tuberculosis cultures (sputum, gastric aspirates, early morning urine and CSF) if indicated.
11 Heterophile antibody detection test.
12 Tuberculin test.
13 Immediate serological tests for specific pathogens.
14 Save baseline serum.

Interpretation of initial findings

The results of initial tests will become available during the first week of investigation. Many diagnoses will be made from these results, including bacteraemias, culture-positive infective endocarditis, pneumonias, urinary tract infections, pulmonary tuberculosis, Epstein–Barr virus infection and common parasitic diseases. Some malignancies such as leukaemias will also be revealed.

While results are accumulating, the physical examination should be reviewed a number of times, both to check on doubtful signs and to detect new or changing ones. A cardiac murmur can suddenly become obvious, finger clubbing can develop or lymph nodes enlarge.

Further investigation is indicated if no diagnosis is apparent or if initial assessment suggests a line of investigation. In some cases the patient's condition is poor, and a wider range of tests must be completed without delay. Three main types of investigation are possible:
1 further serological tests for infections;
2 extensive tests for connective tissue or immunological diseases;
3 tissue diagnosis (imaging and/or biopsy and culture).

Serological tests (and probes) for infections

These tests can be used to follow on from initially negative microscopy and culture tests, and to investigate a wider range of possible infectious conditions. The choice of investigations will depend on the results of epidemiological and clinical investigations already carried out. Here are some examples of useful clinical indicators.

- Anaemia;
  - haemolytic: consider malaria, *Mycoplasma pneumoniae* infection, *Escherichia coli* O157 infection with haemolytic uraemic syndrome, clostridial sepsis and rarities such as thrombotic thrombocytopenic purpura, bartonellosis or babesiosis;
  - of chronic disease: consider tuberculosis, human immunodeficiency virus (HIV) infection, or persistent sepsis;
  - with haemophagocytosis: consider tuberculosis, brucellosis, leishmaniasis, also cytomegalovirus infection and, rarely, Epstein–Barr virus infection (which can occur associated with immunosuppression).
- Eosinophilia: consider any tropical helminth infection, schistosomiasis or liver fluke, rare trichinellosis (also occurs in Europe), drug allergy.

• Hepatocellular disorder: consider viral hepatitis, leptospirosis, rickettsioses, Epstein–Barr virus infection, primary cytomegalovirus infection and rarities such as bartonellosis and viral haemorrhagic fevers.

• Cholestatic changes: consider granulomatous conditions such as tuberculosis, brucellosis, Q fever and rare histoplasmosis or cryptococcosis; also consider single or multiple liver abscesses, or cystic conditions and rare schistosomiasis and liver flukes.

• Sterile pyuria: consider tuberculosis, brucellosis, or rare *Chlamydia trachomatis* or *Mycoplasma hominis* infection.

• Persistent cough or respiratory symptoms: consider tuberculosis, enteric fever, tularaemia, respiratory viral infections, particularly paramyxoviruses, adenoviruses and rare hantavirus pulmonary syndrome, as well as pulmonary eosinophilia due to migrating parasites including *Strongyloides*.

• Persisting meningism, encephalopathy or fits: consider tuberculosis, borreliosis, any arboviral encephalitis, cysticercosis and rare listeriosis, syphilis or neurobrucellosis.

• After travel abroad: consider zoonoses such as brucellosis, Q fever, tularaemia, hantavirus infections, leptospirosis, schistosomiasis, also rare melioidosis and malarial tropical splenomegaly.

Serological and nucleic acid amplification tests for infection in PUO

1 Antigen detection (blood antigen for hepatitis B or D, CSF antigen for cryptococcosis, tissue antigen for cytomegalovirus).

2 Single-step antibody tests (immunoglobulin M) for acute viral infections, borreliosis, toxoplasmosis or malarial tropical splenomegaly, gel precipitin test for amoebiasis, diagnostic titres of *Legionella* antibodies.

3 Blood PCR-based specific tests for hepatitis C RNA, Epstein–Barr virus, cytomegalovirus HHV6 or HHV8 DNA, or viral haemorrhagic fever RNAs.

4 CSF PCR tests for herpes simplex virus, varicella zoster virus or cytomegalovirus DNA, enterovirus or rabies RNA, and mycobacterial DNA.

5 Paired-serum antibody tests (for atypical pneumonias, leptospirosis, yersiniosis).

6 Search for bacterial or fungal ribosomal RNA, using PCR techniques on blood or CSF specimens; the RNA can be sequenced, and compared with a library of rRNA examples, to identify its pathogen of origin.

Some patients take longer than 10–14 days to produce diagnostic titres or elevations of antibody levels. A serological diagnosis cannot be made before 6–9 weeks in some cases of legionellosis, borreliosis and leptospirosis. Even the anti-streptolysin O titre (ASOT) may take 3 or 4 weeks to reach diagnostic levels. It is always worth taking a late serum specimen if a serological diagnosis has not been evident in the first month of fever.

Tests for connective tissue and granulomatous diseases

Connective tissue and granulomatous diseases are good mimics of infection. Some, such as systemic lupus erythematosus, are particularly like viral infections, presenting with fever, neutropenia and sometimes rash. Others may cause neutrophilia, for example polyarteritis nodosa and Wegener's granulomatosis. These further mimic bacterial disease by producing focal inflammatory lesions in the respiratory system. The granulomatous disease sarcoidosis must be distinguished from tuberculosis, and non-respiratory infections of several types. Crohn's disease can mimic intestinal infections.

Elevated inflammatory indices (CRP and ESR)

An elevated ESR is a prominent finding in vasculitic and granulomatous conditions. Whilst a very high ESR is not exclusive to connective tissue diseases, it is unwise to dismiss such a diagnosis unless the finding can be otherwise explained. The CRP may also be elevated but in some conditions it is not raised in proportion to the increase in the ESR.

Autoantibodies

Autoantibodies are detectable in many connective tissue diseases, and may be diagnostic. Sarcoidosis is not associated with autoantibodies but if there is lung involvement the serum angiotensin-converting enzyme (SACE) level is elevated, occasionally even when pulmonary function and chest X-ray appearances are normal.

Serological tests for common connective tissue disorders

1 Rheumatoid factor.

2 Antinuclear factor, double-stranded-DNA antibodies.

3 Other specific antibodies (smooth-muscle, mitochondrial, thyroid, etc.).

4 ANCAs (anti-neutrophil cytoplasmic antibodies: p-ANCA and c-ANCA).

5 SACE (serum angiotensin-converting enzyme; elevated in pulmonary sarcoidosis).

Tissue diagnosis (imaging and biopsy)

The availability of accurate imaging techniques has greatly simplified the investigation of PUO. Imaging can be used to demonstrate the anatomy of tissues and organs, to test their function and to detect inflammation within them. Imaging can identify lesions suitable for biopsy, and assist in guiding the biopsy needle.

Ultrasound scans

Ultrasound scans are non-invasive and relatively inexpensive. The technique depends on showing differences in the sonic density of tissues and will delineate the anatomy of organs, and lesions within organs, especially if these are outlined by thin fatty planes. It is particularly useful for demonstrating enlargement of the abdominal organs, and for detecting abscesses, cysts or space-occupying lesions. It can demonstrate soft-tissue swelling and periosteal elevation in early osteomyelitis. It can also define pelvic lesions, particularly in and around the female genital tract, and trans-vaginal techniques are particularly helpful in this area. It is regularly used in investigation of PUO cases, and is often diagnostic.

Its limitation is its requirement for expert operation and interpretation, as anatomical definition is not perfect and the ultrasonic beam makes shadows that can be confused with lesions.

Echocardiography

Echocardiography is ultrasonic imaging of the heart that can show a two-dimensional, real-time picture of the beating heart and its valves. Doppler echocardiography demonstrates the direction, velocity and turbulence of blood flow (see Figs 12.4 and 12.5) These techniques can show vegetations on heart valves, abscesses of the valve rings and septum, pericardial effusions and even dilatations of the coronary arteries. The only limitation of echocardiography is its occasional inability to demonstrate small lesions, giving rise to false negative results. Trans-oesophageal echocardiography is significantly superior to trans-thoracic studies in demonstrating small valvular vegetations.

Computed tomographic scans

Computed tomographic (CT) scans are high-definition computed tomograms derived from axial X-rays of the body. They accurately reveal the anatomy of organs and can demonstrate lesions of 0.5 cm or less. Contrast media can demonstrate increased blood supply to inflamed lesions by enhancing the radiodensity of affected tissue. CT-guided biopsy and aspiration can be performed (Fig. 20.4).

Magnetic resonance scans

Magnetic resonance (MR) scans detect the density of mobile magnetic atoms in the tissues. The most abundant such atom in the body is hydrogen, present in body water. MR scans are therefore good for showing vascularity of tissues, but can also demonstrate subtle variations of the water content. Oedema is easily seen, allowing detection of inflammation by MR imaging, even before there is a change in radiodensity (i.e. in X-ray or computed tomographic appearance). This makes MR scanning the investigation of choice for early detection of bone marrow oedema in osteomyelitis. Its sensitivity makes it an investigation of choice for detection of posterior fossa lesions in the brain and meninges. It also permits imaging of vascular anatomy and blood flow, which can detect the presence of vasculitic change in small and medium-sized arteries. So-called 'dynamic' MR scanning techniques can image moving organs such as the heart, though this has not proved superior to high-quality echocardiography. It is possible to enhance MR images with media containing magnetic atoms, some of which are radioisotopes, but radioisotopes are potentially damaging to some tissues, while magnetic fields are not. A disadvantage of magnetic resonance scans is that the scanners' enormously powerful magnets can disturb moveable magnetic implants, such as pacemakers, haemostatic clips or metal stents. Patients whose magnetically susceptible implants are unstable in the tissues cannot undergo MR scans.

Isotope localization scans

Isotope scans involve the intravenous injection of radioisotopes, which become concentrated in abnormal tissues. Subsequent scanning with an appropriate detection system, usually a gamma camera, will produce a picture demonstrating the affected area. The most commonly used isotopes are technetium, which demonstrates increased blood flow to inflamed tissues, and gallium, which accumulates in areas rich in inflammatory mediators. Technetium is used for bone scans, which can demonstrate inflammatory lesions in bone long before X-rays can show altered bone anatomy. Gallium can reveal abscesses or foci of inflammation in many organs and tissues (Fig. 20.5). Neither of these types of scans can distinguish between changes due to infection or other causes of inflammation or cytokine activation.

Isotope scans can also be used to demonstrate the relationship between structure and excretory function in the kidneys and liver. Appropriate isotope scans can show whether the isotope is cleared from the organ and concentrated in the excreted fluid. The concentrated isotope can then be used to generate an image of the renal collecting system, bladder or bile ducts.

Labelled neutrophil scans

In vitro radio-labelling of the patient's own neutrophils, followed by re-injection and appropriate imaging after a

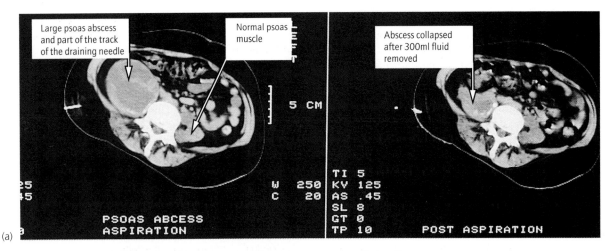

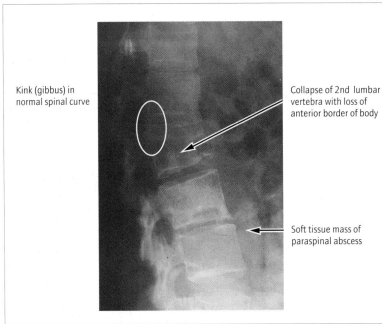

Figure 20.4 (a) Computed tomographic-guided aspiration of pus from a large psoas abscess: as well as confirming tuberculosis this was a therapeutic procedure – 300 ml of infected material was removed; (b) X-ray appearance of the same abscess.

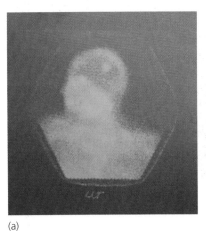

Figure 20.5 Gallium scan: (a) left lateral and (b) coronal view showing concentration of activity in the left parietal bone: this teenager with fever, anaemia and weight loss had a localized lymphoma.

period of distribution, allows the demonstration of sites where large numbers of neutrophils are congregated. These sites will include pyogenic foci such as abscesses, effusions or loculated infection. Foci of non-infectious inflammatory reaction may be demonstrated, such as the focal lesions of polyarteritis, or the pericarditis of rheumatic fever or Dressler's syndrome. A disadvantage of this kind of technique is its poor ability to demonstrate widespread small lesions (because no concentration of isotope is sufficient to be individually detectable), or very widespread inflammation, such as generalized peritonitis, in which the isotope is too widely distributed to be visible as a focus.

The original labelled neutrophil scans used a technique of exposing the patient's blood to tertiary amine-complexed indium oximes. These complexes are phagocytosed by the neutrophils, which are re-injected into the patient after time has been allowed for adequate labelling. The labelling and scanning process may take one or more days to complete.

Newer techniques use radionuclide-labelled monoclonal antibodies, which label neutrophils more rapidly than indium complexes, and whose concentration is detectable much sooner after re-injection. One of these labels is a Fab label, and a promising development is an anti-CD15 antibody, which preferentially labels neutrophils that have migrated out of the blood circulation (because tissue-located neutrophils express a higher level of CD15 than circulating neutrophils). Using this agent, the labelling and scanning technique can be completed in a few hours, rather than one day or more.

PET and SPECT scanning

Positron emission tomography and single photon emission spectrometry are emerging methods of demonstrating metabolic activity (or its absence) in living tissues, by targeting chemicals such as neurotransmitters or glycolytic pathway components, using radionuclide substrates. CT-PET scanning allows the functional scan to be presented in the format of a CT image, detecting the functional and structural context together. The images produced can indicate sites of tissue hyperactivity when there is insufficient inflammatory reaction for detection by other imaging methods, and where tissue anatomy is macroscopically normal. They have been used to demonstrate areas of encephalitis, or small foci of tissue activity in lymph nodes or parenchymal organs in such conditions as chronic brucellosis or borreliosis.

X-ray techniques

X-ray techniques are still important in the investigation of PUO. This is particularly true in bowel disease, where scanning images may not be able to distinguish between

the fluid-filled lumen of the bowel and an abscess in the folds of the peritoneum. Contrast studies can demonstrate mucosal disease (Fig. 20.6), perforations and fistulae of the bowel and of other hollow organs.

Laparoscopy

Laparoscopy allows direct examination of organs in the abdomen or pelvis. It is useful for detecting abnormalities of the abdominal lymph nodes, peritoneum, liver and female genital tract. Biopsies can be obtained under laparoscopic control. Other procedures for examining tissue planes or the lumen of the organs include mediastinoscopy, bronchoscopy, and endoscopy of the bowel, biliary tract and urinary system. Biopsies, brushings or washings can be obtained during these procedures.

Laparotomy and thoracotomy

Laparotomy and thoracotomy are rarely necessary in the investigation of PUO. They can be helpful when signs or symptoms point to pathology, or when lesions are demonstrable by imaging but are too small or inaccessible to approach by guided biopsy. A mini-laparotomy or mini-

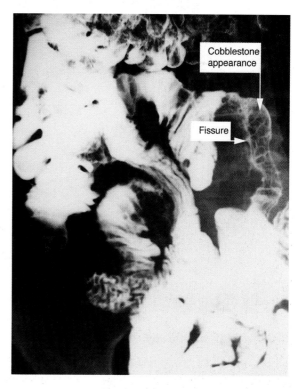

Figure 20.6 Barium meal examination showing cobblestone change and fissuring in the terminal ileum: this Indian patient had ileal tuberculosis.

thoracotomy can then be performed, to examine the area directly, and obtain tissue specimens. This may permit distinction between, for instance, tuberculosis, sarcoidosis and malignancy.

Making the most of biopsy material

It is usual to fix biopsy material in formalin for histological examination, but portions should also be retained in sterile water or saline for bacteriological and virological examination, or for special staining, for instance to demonstrate tumour antigens, immune complexes or enzyme activities. Bone marrow, or aspirates from abscesses and cysts, can be cultured like blood cultures or directly inoculated on to culture media. Solid tissues must be treated in the laboratory before inoculation. Smears may be prepared for later virological or bacteriological study. Specimens for viral culture may be refrigerated, but should not be frozen. Although cultures cannot be performed on fixed specimens and serology is rarely possible, DNA survives fixation poorly but may still be detectable by molecular techniques.

> **Imaging and biopsy procedures in the diagnosis of PUO**
>
> Key: B, biopsy may be performed; C, standard and tuberculosis culture should be performed, in addition to bacterial/fungal culture or PCR-based studies.
> 1 Ultrasound scans (especially abdominal and pelvic; B, C).
> 2 Computed tomographic scan of the head or body (B, C).
> 3 Magnetic resonance scan of the head or body (B, C).
> 4 Gallium scan (whole body).
> 5 Technetium bone scan.
> 6 Laparoscopy (B, C).
> 7 Bronchoscopy (B, C).
> 8 Mini-laparotomy and mini-thoracotomy (B, C).
> 9 Liver biopsy (C).
> 10 Lymph-node biopsy (C).
> 11 Bone marrow biopsy (C).
> 12 Temporal artery biopsy.
> 13 Skin lesion biopsy (C).
> 14 Functional tissue scanning (PET and SPECT).

Trials of therapy

Risks of trials of therapy

Ideally, trials of therapy should be avoided in most cases of PUO. Anti-inflammatory agents, particularly corticosteroids, may mask untreated infection and allow it to advance, leading to complications such as perforation of hollow organs or progression to bacteraemia. Antimicrobial agents can cause hypersensitivity reactions and other adverse effects, which might cause increased fever, rash, blood disorder or organ failure, which could be mistaken for features of the original condition. They may also prevent recovery of organisms by culture while failing to treat the disease adequately. This is common in enteric fevers, listeriosis, brucellosis and *Streptococcus pyogenes* infections. Antimicrobial therapy alone may fail to resolve abscesses or loculated infections, which must still be identified and drained to cure the condition. Therapy for tuberculosis includes broad-spectrum agents such as rifampicin or quinolones, which can inhibit culture of pyogenic organisms. Tetracyclines and co-trimoxazole may partly inhibit malarial parasites or other protozoa, masking parasitic infections.

Finally, corticosteroids can inhibit the immune responses that make serodiagnosis possible, and can abolish a positive tuberculin test response.

> **Risks of trials of therapy**
> 1 Reduced usefulness of diagnostic cultures.
> 2 Modification of infection without cure.
> 3 Adverse reaction to therapy complicating the illness.
> 4 Corticosteroids may reduce the usefulness of immunological tests.
> 5 Corticosteroids may permit progressive infection with reduced signs of inflammation.

Uses of trials of therapy

In spite of the risks, trials of therapy are sometimes useful. This might be the case when a diagnosis is strongly suspected after investigation, but cannot be proved. If the suspected diagnosis is a specific infection, treatment may be attempted, using the narrowest possible antimicrobial spectrum. Obtaining the expected response is then supportive of the presumed diagnosis.

Care must be taken not to treat another condition inadvertently. For instance, trial of antituberculosis treatment should be performed using isoniazid (INAH), pyrazinamide, ethambutol, etc., rather than drugs also active against pyogenic organisms. Patients undergoing a trial of corticosteroids should be examined frequently, and X-rays or imaging results should be reviewed, in case infection is potentiated by immunosuppression. Failure to improve on reducing doses of steroids should cause review of the presumed diagnosis.

When the patient's condition is critical a 'blind' trial of therapy may be unavoidable. This may be very important if a life-threatening infection, such as listerial rhombencephalitis is suspected, or a severe vasculitis could cause pulmonary haemorrhage or cerebral ischaemia. All possible specimens should first be obtained for investigation. The physician must then make a best-guess decision (assisted by expert colleagues in relevant specialities) on the treatment or treatments to try. If possible, treatments should be introduced sequentially but this is not always feasible. Once started, treatment should not be stopped before it has had a chance to produce results; while many pyogenic infections improve promptly, enteric fevers may take up to a week and tuberculosis as much as a month to respond by improvement of fever.

Reasons to attempt trials of therapy

1 To gain further evidence for a strongly suspected diagnosis by obtaining the expected response to therapy.
2 In an emergency when the patient is severely ill.

When treatment is apparently effective it is difficult to decide how long it should be continued. This will depend on the type of condition suspected and on any laboratory data that may become available. The pulse and temperature chart gives the earliest indication of success. Improvement or normalization of the C-reactive protein level is a good indicator of reduced inflammation. Other helpful data include improvement in X-ray abnormalities, reduction in lymphadenopathy or organomegaly, and restoration of albumin or haemoglobin levels.

Successful treatment of tuberculosis may improve cell-mediated immunity, causing the tuberculin test to become strongly positive and confirming the diagnosis. Treatment can then be continued for the appropriate 6 or 9 months. In connective tissue disorders the ESR is a useful guide to response. Non-specific indicators of response include improving serum albumin levels and regained body weight.

When no improvement can be obtained in PUO

This occurs in 5–15% of published series. It is most common in cases with mild or subacute fevers. About half of these cases eventually recover and most of the others remain feverish but do not deteriorate. A mortality rate of 7–12% is often found in undiagnosed PUO, with the higher rates in elderly patients. The commonest cause of prolonged high fever, defying diagnosis, is occult lymphoma.

Investigation of PUO is therefore well worthwhile, usually leading to diagnosis and cure, and rarely ending in failure.

Post-infectious Disorders

21

Introduction

Most people expect to make a steady improvement after suffering an acute infection. This process of convalescence is complex, and depends on a number of overlapping physiological events.

Normal features of convalescence

Suppression of active infection
In many cases, as in the cure of tonsillitis, meningitis or endocarditis, this is probably achieved by eradication of the causative pathogen by the action of the innate and the adaptive immune system, plus or minus the effect of antimicrobial drugs. In other cases the infected cells are shed, as in influenza, or infected material is discharged when loculated infection is drained.

In some diseases the pathogen is suppressed rather than destroyed. In toxoplasmosis, tissue cysts survive and reactivation is prevented by a continuing immune reaction. Viruses of the herpes group remain latent unless the immune system is impaired sufficiently for recrudescent infections to occur. It is now suspected that some viral pathogens are suppressed by selecting inactive mutants from the range of mutations that occur during rapid replication. Some cases of hepatitis B e antigenaemia may cease when an e-antigen-negative mutant is selected, while e-antigen-positive viral products are destroyed by anti-e antibodies.

Methods of terminating active infection
1 Destruction of pathogens and their antigens.
2 Shedding or destruction of virus-infected cells.
3 Making pathogens inaccessible to the immune system (e.g. in pseudocysts).
4 Selecting antigenically inert mutants of the pathogen.
5 Establishment of latency.

Return of immune responses to resting states
As the amount of microbial antigen falls, antigen presentation ceases and the stimulus to further immunological

activation subsides. However, many activated immune cells have been generated, including macrophages, helper T cells, cytotoxic T cells and clonally proliferating B and T cells, as well as suppressor T cells. Many activated T cells are programmed to self-destruct by an enzymic process (called apoptosis) that limits their lifetime. The presence of increased levels of interleukin-2 while infection persists tends to delay this destruction and maintain immune activation.

As antigen levels decline, the production of immune complexes slows down, slowing the activation of the classical complement pathway and the coagulation cascade, leading to a reduction of chemotactic and pro-inflammatory factors in the tissues.

Events favouring reduced immune activity and inflammation

1 Falling levels of antigen and immune complexes.
2 Reduced antigen presentation.
3 Reduced cytokine production.
4 Reduced immune complex-mediated complement activation.
5 Apoptosis of activated T cells.

Repair of tissue damage

The catabolic state of acute infection must be reversed to allow protein construction and repair of damaged tissue. This in turn permits the restoration of normal bodily functions; for instance, absorption of nutrients after bowel infection or glucose metabolism after severe liver infection. This process can be surprisingly prolonged; for instance, there is mild but measurable hypoxaemia for several weeks after acute bronchiolitis in infants. During this period the metabolic rate remains raised, with concurrent tachycardia and easy fatigability.

Features of post-infectious disorders

In post-infectious disorders the process of convalescence is interrupted, or complicated by unwanted effects of the immune response. Inflammation may arise in different tissues from those affected by the original infection. In some cases the post-infectious condition overlaps the acute infection in time, as when erythema nodosum complicates primary tuberculosis. The post-infectious condition may follow the acute infection with little or no interval, as when erythema multiforme complicates episodes of herpes simplex. This is typical of a vasculitic condition stimulated by the infection. Post-infectious conditions caused by the actions of antibodies or immune complexes tend to follow the acute infection with an interval of 10–14 days

or more, as in reactive arthritis or post-streptococcal nephritis (Table 21.1).

Possible pathogenesis of post-infectious conditions

There are several mechanisms by which post-infectious conditions may arise. Molecular techniques for detecting pathogens and developments in understanding of immunology continue to elucidate these.

Persisting low-grade infection

Persisting low-grade infection has been demonstrated in post-infectious encephalitis, which is often histologically indistinguishable from acute infectious encephalitis. Early evidence for persistence of infection came when measles virus was recovered from the brain tissue of cases of subacute sclerosing panencephalitis (SSPE). Immunoglobulin M (IgM) antibodies can persist for weeks or months in some infections such as hepatitis A, infectious mononucleosis and toxoplasmosis, which can all be followed by a debilitating and prolonged convalescence. More recent work on erythema multiforme associated with herpes simplex has used polymerase chain amplification to demonstrate that herpes simplex virus DNA is present in epidermal lesions but not in adjacent, normal skin.

Molecular mimicry

Molecular similarity between antigens of the pathogen and the host is the cause of some conditions. In the Guillain–Barré syndrome there is evidence of immunological attack against components of myelin. Experimental demyelination of nerve roots very similar to that seen in Guillain–Barré syndrome can be produced in animals by anti-myelin antibodies, and about 25% of patients with the condition have anti-ganglioside antibodies. Nephritogenic strains of *Streptococcus pyogenes* produce a unique antigen that cross-reacts with glomerular structures.

Immune complex disease

Immune complex disease is probably often the cause of post-infectious arthritis. In the synovitis that often follows meningococcal disease, meningococcal antigen has been demonstrated in biopsies of affected synovium. Similarly, in Reiter's syndrome, chlamydial antigen has been found in synovium.

Persisting inappropriate immune reaction

Persisting inappropriate immune reaction is probably a mechanism in some of the cytopenias. It is known that haemolysis in *Mycoplasma* pneumonia and Epstein–Barr infections is related to the inappropriate production of anti-I and anti-i antibodies to blood-group antigens. Thrombocytopenia is likely to be caused by a similar

Table 21.1 Some post-infectious disorders and associated pathogens

Disorder	Common associations	Rare associations
Aplastic anaemia		Non-A/non-B hepatitis
Arthritis	Rubella *Meningococcus* *Yersinia*	*Salmonella* *Shigella* *Campylobacter* Mumps
Encephalitis	Varicella Measles Influenza Mumps	Rubella Yellow fever vaccine
Acute disseminated encephalomyelitis	*Mycoplasma* Many viral infections	Legionellosis
Transverse myelitis	Viral and 'atypical' infections	
Erythema multiforme	Herpes simplex	*Mycoplasma pneumoniae*
Erythema nodosum	Tuberculosis Leprosy	*Yersinia*
Glomerulonephritis	*Streptococcus pyogenes*	Hepatitis B Mumps
Guillain–Barré	*Campylobacter* Cytomegalovirus	Hepatitis A or B Respiratory viruses
Haemolysis	*Mycoplasma pneumoniae*	Epstein–Barr virus Syphilis
Haemophagocytic syndrome		Epstein–Barr virus Cytomegalovirus
Reiter's syndrome	Chlamydial genital infections	*Shigella* infections
Reye's syndrome	Varicella	Aspirin as a cofactor Influenza
Rheumatic fever	*Streptococcus pyogenes*	
Serositis	*Meningococcus*	
Thrombocytopenia	Rubella Mumps Varicella	Epstein–Barr virus Tuberculosis
Thrombotic thrombocytopenic purpura (TTP)		HIV infection; other infections

mechanism; indeed, anti-platelet antibodies can be demonstrated in some cases.

A condition like systemic lupus erythematosus occasionally occurs after infectious mononucleosis. This is associated with positive anti-DNA antibodies, joint pains and a raised erythrocyte sedimentation rate (ESR), but has a limited duration of weeks or months. Acquired thrombotic thrombocytopenic purpura (TTP) is caused by antibody-mediated inhibition of the metalloproteinase that cleaves 'giant' von Willebrand factor to its active form.

Failure of an effective immune response

In rare cases of post-infectious haemophagocytosis, a subtle immune deficit delays or prevents the clearance of viruses such as Epstein–Barr virus from host cells, resulting in the persistence of immortalized, infected lymphocytes, which are not destroyed by apoptosis. A mounting cell-mediated immune response results, with infiltration of parenchymal organs and uncontrolled phagocytosis of host cells by activated macrophages. The immune deficit can be inborn, or acquired, often as a result of transplant immunosuppression or cytotoxic therapy.

Susceptibility of the patient

Special attributes of the patient may influence the occurrence of post-infectious conditions, by influencing the antigens displayed by host cells, or the efficiency of various components of the immune response. It is known that non-secretors of blood group substances are more likely than others to develop rheumatic fever. Reiter's syndrome is almost exclusive to patients with the human leucocyte antigen (HLA) B27 tissue type, but non-Reiter's arthropathies have a weaker association with this tissue type. A total of 88% of patients who have recurrent erythema multiforme after herpes simplex episodes have tissue type DQw3, and 71% have DRw53.

Possible mechanisms of post-infectious disease

1 Persistent low-grade infection.
2 Molecular mimicry.
3 Immune complex disease.
4 Persisting inappropriate immune activity.
5 Failure to terminate the infection and the resulting immune response.
6 Immunological susceptibility of the patient.

Erythema multiforme

Epidemiology

This skin disorder affects mainly teenagers and young adults. It has been described after streptococcal infections, primary tuberculosis, *Mycoplasma pneumoniae* infections and after exposure to drugs such as antibiotics, anti-epilepsy drugs and some antiretrovirals (notably, nelfinavir or abacavir). It sometimes accompanies Epstein–Barr virus infection in toddlers. Rarer causes include nelfinavin, diphenoxylate and traumas such as radiotherapy.

Repeated episodes of erythema multiforme occur in some patients who have cold sores. The skin rash appears 2 or 3 days after the onset of the mucocutaneous lesions, and lasts up to a week. The severity of the rash is related to the extent of the herpetic lesions on each occasion.

Pathology

The skin lesions are associated with oedema and necrosis of epidermis with dilatation and inflammatory infiltration of the subepidermal blood vessels. The oedematous epidermis may become raised, forming bullae. In herpes simplex-associated disease, viral antigen and viral genome can both be demonstrated in the keratinocytes of the affected skin, but are not present in normal skin.

Clinical features

Illness begins abruptly, with fever and developing rash. Typically, many of the lesions are target- or iris-shaped with an erythematous halo surrounding an oedematous or dusky centre (Fig. 21.1). The rash is densest on the extremities, including the palms and soles (Fig. 21.2), but is also exaggerated in areas exposed to sunlight or trauma. Untreated, the lesions last for 2–6 weeks, some of them coming and going all the time. There may be arthralgia, especially of the knees and/or ankles. There is no characteristic change in the white cell count. The ESR is usually high: 70 mm/h or more.

Stevens–Johnson syndrome

Stevens–Johnson syndrome is the term used when erythema multiforme affects mucosae as well as skin. The disease is usually severe, with many skin bullae and severe,

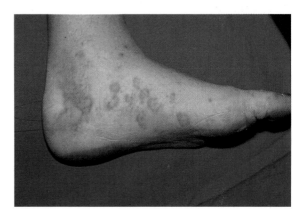

Figure 21.1 Target lesions of erythema multiforme.

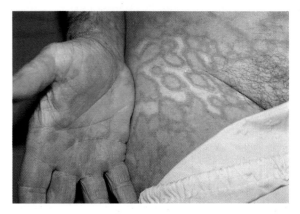

Figure 21.2 Erythema multiforme: showing the rash, which also involves the palms.

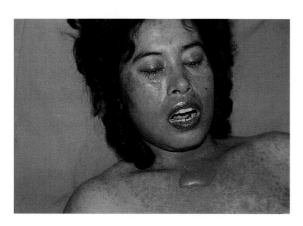

Figure 21.3 Stevens–Johnson syndrome: severe involvement of the oral and conjunctival mucosae and epidermal necrolysis.

sometimes haemorrhagic inflammation of the oral, conjunctival and genital mucosae. There is danger of widespread epidermal necrolysis, in which the keratinized layer of the epidermis becomes loose, and can be moved over the lower epidermis, or rubbed off it (Nikolsky's sign; Fig. 21.3).

Management

The rash almost always resolves spontaneously. Precipitating causes should be removed. Severe cases are often treated with corticosteroids, though there is no good evidence of benefit, and the risk of infection may be increased. The ESR can be used as an objective measure of improvement.

Patients with epidermal necrolysis should be nursed in a warm room and kept well hydrated, as heat and moisture are rapidly lost from the exposed lower epidermis. There is risk of secondary infection, particularly with *Staphylococcus aureus*, which may cause bacteraemia. If a patient develops fever, tachycardia or shock, blood cultures should be obtained and early antimicrobial treatment commenced.

Complications

These are few, once the patient has passed the severe part of the illness. The inflammation is superficial, and the epidermal layers are replaced, with only an occasional atrophic scar. After severe mucosal ulceration, adhesions may develop, for instance between the lip and gum, or in the conjunctival sac. These may require division and split-skin grafting.

Guillain–Barré syndrome (ascending polyneuritis) and other nervous system syndromes

Epidemiology

This is a condition in which there is demyelination of the nerve roots. It has an incidence of around 2/100 000 in the UK and often follows a viral-type respiratory infection; 25% follow *Campylobacter* enteritis or other gastrointestinal illness, with an interval of 1–3 weeks. About 10% follow cytomegalovirus infections, and occasional cases are associated with infectious mononucleosis, *Mycoplasma pneumoniae*, legionnaires' disease or immunizations.

The disease affects all age groups, but tends to be more prolonged in middle-aged individuals and the elderly.

Pathology

Inflammation and demyelination in the spinal and cranial nerve roots are presumed to be of immunological origin, but responsible antigens have not been identified. If inflammation is sufficiently severe, axonal degeneration in nerve trunks may follow.

Dysfunction of other nerves, such as the facial nerve, brachial plexus or individual nerve trunks, also occasionally occurs.

Clinical features

Guillain–Barré syndrome usually begins with paraesthesia, followed by weakness in the feet and lower legs. Muscle pains may be significant. Initially, objective signs are lacking and the patient may be thought hysterical. The disease progresses up the spinal levels, with weakness of the trunk and arms, and facial paralysis. As the paralysis intensifies, tendon reflexes are lost and the patient becomes immobile. Respiratory failure may result from intercostal and diaphragmatic paralysis. Early facial paralysis warns of sudden respiratory failure. Sensory features include loss of light touch and joint position sense. There are no upper motor neurone signs; the plantar responses, when they can be elicited, are down-going. There is little involvement of sphincters, and this is transient if it occurs. A rare form of demyelination syndrome causes descending paralysis, and is difficult to distinguish from botulism.

Progression may cease at any stage, but usually halts within 2 or 3 weeks. Some cases progress to complete paralysis within 2–3 days. After stabilization there is progressive recovery. This may be rapid, taking 2 or 3 weeks until the patient can walk independently. In elderly patients, or when significant axonal degeneration has occurred, it can take 3–12 months or more, and the recovery of muscle strength may be compromised by disuse atrophy and a degree of denervation. There is a 10–20% mortality, largely from complications of intensive care. Of those who recover, 90% return to full motor activity; the remainder have variable residual weakness.

Abnormal laboratory findings are confined to the cerebrospinal fluid (CSF). Early in the course of the illness there is an excess of lymphocytes. The CSF protein level gradually rises, out of proportion to the lymphocytosis, and may reach levels of several grams per litre.

Electrophysiological investigation in established disease shows marked slowing of conduction velocities and diminished evoked muscle and sensory nerve action potentials. These are scattered and mild in the first few days, and may be difficult to demonstrate.

Management

Monitoring of the vital capacity and the arterial oxygen saturation are important, to warn of the need for ventilation. The paralysed patient also needs skin care and attention to pressure areas.

Intravenous immunoglobulin treatment may halt the advance of the disease, or produce remission. Plasmapheresis has resulted in remission of prolonged paralysis, but may require supplementation with immunosuppressive therapy for a variable time to prevent relapse.

Henoch–Schönlein disease

Epidemiology

This is a vasculitic condition that presents with fever, purpuric rash, arthralgia, synovitis, abdominal pain and glomerulonephritis. There is no unique precipitating condition, but there has often been a preceding respiratory infection and some sufferers have evidence of recent streptococcal disease. Children and young adults are most often affected, though cases occasionally occur in the middle-aged.

Clinical features

Rash

The rash affects mainly the extensor surfaces of the feet, ankles and legs, and commonly affects the buttocks. The lesions are raised and of variable size. A variable proportion are purpuric, and those on the feet may be haemorrhagic and bullous (Fig. 21.4). Capillaries in affected skin contain IgA immune complexes and inflammatory infiltrate.

Synovitis

Synovitis most often affects the ankles and knees, but any joint can be involved. Tendon sheaths around the ankle, and sometimes the wrist, are often swollen and painful.

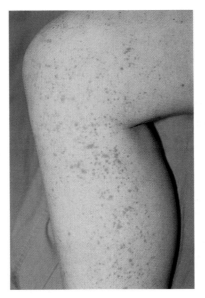 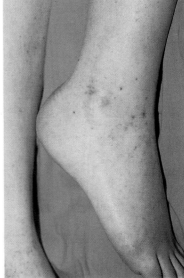

Figure 21.4 Henoch–Schönlein disease: characteristic rash with raised and haemorrhagic elements.

Abdominal pain

The abdominal pain is caused by swelling and haemorrhagic lesions of the bowel. Severe cases can develop vomiting and abdominal rigidity. Varying amounts of blood may be seen in the stools. Nodular mucosal swelling occasionally predisposes to intussusception.

Nephritis

The nephritis is a focal glomerulitis pathologically identical to IgA nephropathy. Haematuria and proteinuria are common. Renal failure can occur, with hypertension and oedema.

Main features of Henoch–Schönlein disease

1 Rash on the legs and/or buttocks.
2 Joint pain.
3 Abdominal pain.
4 Proteinuria (and often microscopic haematuria).
5 Reduced renal function.

Management

Management is symptomatic, as the disorder is usually self-limiting in children. Corticosteroids do not alter the course of the disease. About a quarter of children who develop nephritis are at risk of deteriorating renal function later in life. A minority of patients require dialysis for renal failure and a few do not recover. Early consultation with a nephrologist is advisable, as cases with significant renal failure, particularly adults, may benefit from early treatment with corticosteroids and/or other immunosuppressive drugs.

Rare complications are bowel haemorrhage, rarely bowel necrosis, or intracerebral haemorrhage.

Post-streptococcal glomerulonephritis

Epidemiology

This disorder affects children of age 1–5 years, and is rare in other age groups. It can follow *Streptococcus pyogenes* infection of the skin or throat. It usually occurs 10–14 days after the initiating infection, but can appear after as long as 4 weeks.

Clinical features

The onset is abrupt, with fever, malaise and often loin pain. There is puffiness of the feet and face. Haematuria and oliguria are often noticed by the child's mother.

Physical findings are of mild loin tenderness, oedema and hypertension. There is mild to moderate haematuria and proteinuria; significant protein loss is rare. Urine microscopy shows epithelial casts that contain red blood cells, confirming the glomerular origin of the problem.

Main features of post-streptococcal nephritis

1 Fever and loin pain.
2 Facial and dependent oedema.
3 Haematuria and oliguria.
4 Impaired renal function.
5 It follows 10–21 days after streptococcal infection.

Diagnosis

Transient microscopic haematuria is common in acute streptococcal infections and does not indicate glomerulonephritis. Glomerulonephritis can complicate other infections, including viral hepatitis, infectious mononucleosis, atypical pneumonias and childhood viral infections. Drugs, including beta-lactams, occasionally cause interstitial nephritis. Care should be taken to show that the glomerulonephritis is truly associated with streptococcal infection, especially if it is severe or prolonged.

Evidence of recent streptococcal infection is provided either by persisting positive culture from throat or skin, or by significantly elevated anti-streptolysin O titre (ASOT) or other antibody titres (see Chapter 6).

Renal biopsy is rarely performed, as the natural course of the disease is short, with diuresis and improvement in creatinine levels within 7–10 days. Histology shows swollen glomeruli with endothelial proliferation and neutrophil infiltration. Irregular deposits of immune complexes can be demonstrated on the glomerular basement membrane.

Management

The throat or skin infection should be treated with penicillin or a suitable alternative (see Chapter 6). The glomerulonephritis almost always resolves without specific treatment. Diuretics may be indicated to reduce oedema; a few patients require protein and potassium restriction, or control of severe hypertension. Dialysis is rarely indicated. Once the kidneys have recovered, renal function usually remains normal.

Reye's syndrome

Introduction and epidemiology

This is a rare childhood disease consisting of encephalopathy, cerebral oedema and fatty infiltration of the liver, with a characteristic picture of liver dysfunction. It is associated with transient but severe damage to mitochondria. Disrupted and necrotic mitochondria are demonstrable on liver biopsy in affected cases.

The syndrome most often follows infection with varicella or influenza B, less commonly echovirus infections, atypical pneumonias and adenovirus infections. Aspirin treatment for the preceding feverish illness predisposes to Reye's syndrome. A steady reduction in the incidence of Reye's syndrome followed the discontinuation of aspirin use in children. Aspirin is therefore not recommended for children below the age of 16 years.

Clinical features

Illness begins a few days after the onset of the precipitating infection. There is increasing drowsiness, vomiting and often convulsions. Hepatomegaly is common, but jaundice is rare. There is biochemical evidence of liver dysfunction, with a raised blood ammonia and low blood urea. The blood glucose may fall, leading to hypoglycaemic convulsions. There is sometimes significant impairment of blood clotting.

The most important aspect of the disease is raised intracranial pressure and reduction of cerebral perfusion, which can cause permanent cerebral damage.

Main features of Reye's syndrome
1 Drowsiness and vomiting.
2 Raised intracranial pressure.
3 Hepatomegaly.
4 Biochemical signs of liver failure.
5 Onset within 1 week of respiratory or gastrointestinal illness.

Diagnosis

There is a clinical overlap with several inborn errors of fatty acid oxidation and urea synthesis. These conditions should be sought by expert investigation, particularly in children below the age of 15 months.

Management

Management is directed at maintaining adequate cerebral perfusion pressure, with the aid of intracerebral pressure monitoring. Admission to a paediatric intensive care unit is a matter of urgency. The disease is self-limiting; recovery occurs slowly as the mitochondria recover from the temporary insult. If cerebral damage is prevented, full recovery can be achieved.

Rheumatic fever

Introduction and epidemiology

Rheumatic fever is a multisystem inflammatory disease that follows throat infection with a wide range of *S. pyogenes* serotypes. It does not occur after streptococcal skin infections. It is commonest between the ages of 5 and 16 years, but is reported up to age 25 or 30. After this age, susceptibility seems to be lost. Rheumatic fever has been rare in Western countries for some decades, but in the 1980s large epidemics occurred in North America. This may be because of the reappearance of epidemic types of *S. pyogenes* with an antigen resembling those of cardiac and other tissues. The streptococci recovered from North American cases produced mucoid colonies, and most were of M type 18.

The throat infection preceding rheumatic fever is variable in severity. Only half of the North American cases sought a medical consultation, and only a quarter of all cases received antibiotic treatment.

Clinical features

There is an interval of 2–3 weeks between the acute infection and the onset of rheumatic fever. The various manifestations of the disease have been classified into major and minor features, or criteria, to aid clinical diagnosis (Table 21.2).

The commonest presentation is with fever and flitting or migratory polyarthritis.

Arthritis
The arthritis is an acute, painful synovitis, often with effusion. It affects large joints, particularly the knees and ankles, but the joints of the arm can also be involved. Typically, one joint is affected for a few days, then improves rapidly while another becomes inflamed (flitting arthritis).

Table 21.2 Major and minor manifestations of rheumatic fever

Major manifestations	Minor manifestations
Arthritis (70%)	Fever
Carditis (50%)	Previous episode(s)
Chorea (20%)	Raised ESR or CRP
Rash (12%)	ECG abnormalities
Nodules (5%)	Arthralgia

The percentages shown are for recent North American cases. CRP, C-reactive protein; ECG, electrocardiogram; ESR, erythrocyte sedimentation rate.

Carditis

About 10% of cases have clinical evidence of carditis at presentation. All tissues of the heart are affected. There is acute pericarditis, with a degree of purulent exudate. The myocardium is inflamed, and may contain typical Aschoff bodies: inflammatory bodies consisting of degenerative epithelioid central areas surrounded by haloes of inflammatory cells. As a post-mortem finding these are pathognomonic of rheumatic fever. Myocardial inflammation is often reflected by cardiographic changes, such as prolongation of the P-R interval, axis changes or other conduction changes. The commonest findings on examination of the heart are a soft first heart sound, a third heart sound or a short systolic or diastolic (Carey–Coombs) murmur.

Endocarditis

The endocarditis is usually macroscopic, with verrucous vegetations on the affected valves. The mitral valve is most often affected, often leading to mitral regurgitation, but vegetations can cause stenotic features. The aortic valve is affected in boys more often than in girls. The right heart valves, especially the pulmonary, are rarely affected. Heart failure is usually related to severe valvular disease, but is exacerbated by myocardial inflammation. Echocardiography is useful to demonstrate the valvular lesions and to evaluate left ventricular function.

Chorea

Chorea often affects older children and girls, and can be the sole major manifestation of the disease. The abnormal movements may be mild and partly disguised, making the patient appear restless, or they may be more gross and obtrusive. They can develop late in the illness, when other manifestations are subsiding, and persist for weeks or months.

Rash

The rash is classically erythema marginatum (see Fig. 1.11), a rash of raised, serpiginous lesions that waxes and wanes with the fever from hour to hour. More often there is an urticarial, multiform or even erythema nodosum-like rash, which is similarly changeable.

Nodules

Nodules are usually less than 1 cm in diameter. They are subcutaneous and often found over tendons on the extensor surfaces of the arms. They occur most often in patients with severe carditis.

Diagnosis

This is based on clinical assessment, using the major and minor criteria, and on demonstrating evidence of recent streptococcal infection (see Chapter 6). Two major criteria are sufficient to make the diagnosis; if only one is found, one minor criterion plus evidence of streptococcal infection should be demonstrated.

The presence of Aschoff bodies in biopsy material such as skin nodules may be helpful, but the inflammatory changes in tissue other than myocardium are rarely specific.

Management

Termination of the inflammation is important, to minimize the risk of scarring in affected heart valves. The treatment of choice is aspirin, but other non-steroidal anti-inflammatory agents may be preferred in younger children, if sufficiently effective. Corticosteroids are not effective.

Bed rest and/or diuretic treatment may be indicated if there is heart failure. Fatalities from acute rheumatic fever are rare, but severe congestive heart failure or pericardial effusion can sometimes be life-threatening.

Prevention

This is important, as a first attack is often followed by recurrences with each successive streptococcal throat infection. With each attack there is further fibrosis of the valve rings, shortening of the chordae tendinae and increasing deformity of heart valves, leading to valve failure. Deformed valves are at increased risk of acquiring infective endocarditis.

Phenoxymethylpenicillin 250 mg twice daily is safe and effective prophylaxis. Erythromycin or a first-generation cephalosporin may be given to penicillin-allergic patients. Prophylaxis is usually continued until the individual leaves school. For those working in schools or hospitals it may be continued for longer, even lifelong, but the risk of recurrence is small after the age of 30.

Post-streptococcal movement disorders

It is now recognized that abnormal movements can follow streptococcal infections. Children of school age are most often affected. The movements may be choreiform, or be represented by grimacing or posturing of the head. Functional brain scanning has been used to demonstrate the presence of antibodies in the basal ganglia of affected individuals. The condition usually lasts for many months, and may persist for years.

Reiter's syndrome

Introduction and epidemiology

Reiter's syndrome is a multisystem disorder that affects almost exclusively individuals with tissue type HLA B27. In its commonest form it follows non-specific genital infection (usually chlamydial) in men. It can also affect patients of either sex after an attack of bacillary dysentery. It has similarities with features of ankylosing spondylitis, with which it shares features of pathogenesis.

Clinical features

The syndrome consists of synovitis, sacroiliitis, conjunctivitis, mucocutaneous rashes and aortitis, beginning 10–14 days after the precipitating condition. Fever is common, as is polymorph leucocytosis in the peripheral blood. The ESR is usually raised to 70 mm/h or more and the C-reactive protein is also very high. Chlamydial antigens have been demonstrated in the affected synovium but this, and tissue typing, are expensive and not usually necessary for diagnosis.

Synovitis with associated conjunctivitis

This is the commonest presentation. The synovitis typically affects tendon sheaths, especially of the hand and foot, but also of the elbow and shoulder. There is also inflammation of large and small joints (Fig. 21.5), commonly affecting the knee, ankle and shoulder. Sacroiliitis is prominent, causing low back pain, local tenderness and discomfort on 'springing' the pelvis. Sacroiliitis is often radiographically demonstrable, showing as soft-tissue swelling and perisacral sclerosis. This does not occur in rheumatic fever, so helps to differentiate between that condition and Reiter's syndrome with aortic valve involvement.

Circinate balanitis

Circinate balanitis is the typical mucocutaneous rash. This is often a red, roughly circular, slightly weeping lesion with scaly margins, affecting the periurethral part of the glans penis. Non-specific papular or scaly rashes may also occur on the glans and at the angles of the mouth.

Aortitis

Aortitis is not always clinically evident. Sometimes a soft systolic murmur is audible, but in rare, severe cases aortic incompetence may occur and lead to sudden heart failure.

Keratoderma blennorrhagica

Keratoderma blennorrhagica is a thickened, scaling rash of the palms and soles, in which flat vesicles are often seen. It is characteristic of late Reiter's syndrome, often occur-

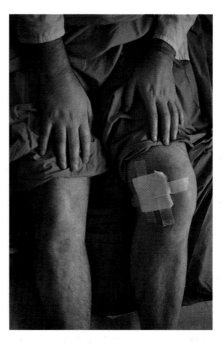

Figure 21.5 Reiter's syndrome: synovitis of the left knee and tendinitis of the dorsum of the right hand.

ring after the presenting features have responded somewhat to treatment.

Uveitis and cardiac dysrhythmias

Uveitis and cardiac dysrhythmias are rarer manifestations.

> **Main features of Reiter's syndrome**
> 1 Synovitis.
> 2 Sacroiliitis.
> 3 Conjunctivitis.
> 4 Aortitis.
> 5 Circinate balanitis.
> 6 Late keratoderma blennorrhagica.

Management

This is symptomatic. The initiating illness should be treated if still active, but the Reiter's syndrome is probably related more to antigen already fixed in the affected tissues than to continuing active disease.

Non-steroidal anti-inflammatory drugs are the mainstay of treatment. Systemic corticosteroid treatment is avoided, if possible, but corticosteroid eye drops may be necessary to control uveitis. Rare cases of aortic valve disruption may need surgical treatment, but control of inflammation is important to provide a good basis for any implanted valve structure.

The disease has a prolonged and relapsing course, often requiring several weeks or months of treatment with anti-inflammatory agents.

Thrombotic thrombocytopenic purpura (TTP)

Introduction and pathology

Thrombotic thrombocytopenic purpura is a multisystem syndrome characterized by a rapid onset of purpura or bleeding with a low platelet count, microangiopathic anaemia, usually renal failure, and variable neurological disturbances often leading to severe encephalopathy and coma. Pathological investigations show a mixture of endothelial damage, platelet activation and microthrombi in the vasculature of many organs. Although similar in appearance to haemolytic–uraemic syndrome (HUS; see Chapter 8), acquired TTP is not toxin-mediated. It is associated with dysfunctional von Willebrand factor. Native, ultra-large von Willebrand factor is not cleaved to its func-tional form, because the metalloprotease enzyme required to perform this cleavage is damaged by an acquired autoantibody. TTP can complicate many infections, and has now also been described as a complication of HIV infection, probably related to the polyclonal B-cell activation that is characteristic of HIV infection, with overproduction of unusual autoantibodies.

TTP can also occur associated with systemic lupus erythematosus, other collagen disorders, or the use of cytotoxic agents such as cyclosporin or mitomycin.

Management

Successful management depends on early recognition of the disease, permitting treatment before irreversible organ damage is established. Fresh, or fresh frozen plasma can reconstitute the deficient enzyme, and is more effective if combined with plasmapheresis to remove the abnormal antibody. Serum should be collected from the patient before plasmapheresis and fresh plasma infusion, so that tests for HIV infection and other predisposing conditions can be carried out.

Part 5: Special Hosts, Environments and the Community

Infections in Immunocompromised Patients 22

Introduction

Earlier chapters have described how organisms can overcome host defences and cause infection. This chapter discusses the infections that occur when host defences are reduced by disease, medical treatment or inherited disorders. Each or all of the components of the immune system can be compromised. Differing patterns and degrees of immune compromise result in different patterns of infection as different opportunities are opened to invading organisms. Immunocompromised patients may be infected by organisms with little pathogenicity that are usually incapable of causing primary infection, for example, organisms that form part of their bacterial flora or are derived from the environment. The interpretation of microbiological culture results is therefore more difficult in these cases. Also, pathogens that affect non-compromised patients infect the immunocompromised, often causing severe or persisting disease. The absence of an immune response may modify the clinical picture; typical clinical features may not occur (for instance, the typical rash in varicella or measles).

The recognition and treatment of infection in immunocompromised patients is very important, as any delay can allow the establishment of potentially fatal disease. The classic features of high fever, sepsis and shock are often absent, as the development of these depends on immune functions. Hospital admission, extensive investigation and vigorous therapy are usually indicated.

Classification of infections in immunocompromised patients

Immunodeficiencies are classified into primary/congenital immunodeficiency (including defects in B cells, T cells, complement and phagocytes) and secondary or acquired immunodeficiency resulting from disease, malignancy or immunosuppressive therapy.

Immune deficiencies can also be classified into seven groups:
1 disorders of the innate immune system, e.g. phagocytosis;
2 neutropenia and neutrophil dysfunction, e.g. congenital granulomatous disease (CGD);
3 T-cell deficit;
4 hypogammaglobulinaemia;
5 complement deficiencies;
6 splenectomy;
7 broad-spectrum immunodeficiency related to haematological or other malignancy, chemotherapy or immunosuppression for transplants.

This is a simplification, as deficiency in one component of the system leads to imbalance and failure of other components (for example, T cells are the main component of cell-mediated immunity, but T-cell help is also essential for the optimal activity of the humoral response).

Disorders of innate immunity are often found in hospital practice, when, for example, treatment breaches the physical barriers to microbial invasion (see Chapter 23).

Table 22.1 Common deficits in immune function and the infections with which they are associated

Immune deficit	Caused by	Bacterial infections	Other infections
Complement	Congenital deficiency	*Neisseria* spp. *Streptococcus pneumoniae*	
Spleen	Surgery, trauma, sickle-cell anaemia (functional)	*S. pneumoniae* *Haemophilus influenzae* (type b)	*Plasmodium* spp. *Babesia* spp.
Gamma-globulin	Congenital deficiency Multiple myeloma CLL AIDS	*S. pneumoniae* *H. influenzae* (non-capsulate)	*Pneumocystis jiroveci* *Giardia intestinalis* *Cryptosporidium parvum*
Neutrophils	Chemotherapy of leukaemia Bone marrow transplantation dysfunction, e.g. CGD	Enterobacteriaceae Oral streptococci *Pseudomonas aeruginosa* *Enterococcus* spp.	*Candida* spp. *Aspergillus* spp.
T cells	Marrow and other transplantation AIDS Cancer chemotherapy Lymphoma Steroids	*Listeria monocytogenes* *Mycobacterium tuberculosis* *M. avium-intracellulare* *Salmonella* spp. *Rhodococcus equi*	*P. jiroveci* *Toxoplasma gondii* *Cryptosporidium parvum* *Leishmania* spp. Herpesvirus CMV Varicella-zoster virus *Cryptococcus neoformans* *Histoplasma* spp. and other systemic yeast infections

AIDS, acquired immunodeficiency syndrome; CGD, congenital granulomatous disease; CLL, chronic lymphocytic leukaemia; CMV, cytomegalovirus.

Infections associated with acquired immunodeficiency syndrome (AIDS) are discussed in Chapter 16. The other immunodeficiencies and the common infections associated with them are summarized in Table 22.1.

With the exception of some congenital abnormalities, immune deficits are rarely single. Patients undergoing chemotherapy for leukaemia or bone marrow transplantation are primarily neutropenic, but also have impaired cell-mediated immunity, predisposing, for example, to cytomegalovirus (CMV) infection. Treatment involves the use of many skin-piercing cannulae, predisposing to *Staphylococcus epidermidis* and other skin-derived infections. Fungal infections are also common in these patients, facilitated by the combination of neutropenia and decreased cell-mediated immunity.

Neutropenia

Neutropenia can be an adverse reaction to treatment with a number of drugs, but also accompanies acute leukaemia or its treatment. The risk of infection increases significantly once the neutrophil count falls below $0.5 \times 10^9/l$, and is proportional to the period of neutropenia. More than half of all patients suffering an episode of neutropenia will develop an infection. The mortality from these infections is high if not promptly treated.

Epidemiology

The epidemiology of infection in neutropenic patients is complex. Not only are patients and their underlying diseases diverse, but treatment protocols that lead to neutropenia are constantly evolving and this results in a changing pattern of infection, both between patients and at different stages of treatment in the same patient.

Bacterial infections

Bacteraemia occurs in 20–30% of neutropenic patients. The principal bacteria implicated in these patients are Gram-negative rods, but Gram-positive cocci are also important. The frequency with which these organisms cause infection changes with developments in treatment and antimicrobial prophylaxis. Hospital patients are susceptible to colonization with resistant organisms because of both the administration of antibiotics and the effects of serious underlying disease. Hospital organisms may be transmitted on the hands of medical and nursing attendants or ingested in food, notably washed vegetables.

Enterobacteriaceae and *Pseudomonas* spp. are the most commonly isolated Gram-negative pathogens. They are usually derived from the patient's own intestinal flora, gaining access to the circulation when the rapidly multiplying intestinal epithelium is damaged by antineoplastic agents or X-irradiation. Improvements in drug treatment of these pathogens have reduced both the incidence and mortality associated with them. Although they remain the commonest Gram-negative bacilli isolated, the prophylactic use of broad-spectrum 4-fluoroquinolones has diminished the incidence of these infections. A decreasing incidence of *Klebsiella* spp., *Serratia* spp. and *Enterobacter* spp. has also been noted in many centres.

Infections with Gram-positive organisms have become more common in neutropenic patients, rising from around 30% of bacterial infections in the mid-1960s to 60% in the mid-1990s. In most centres, the main organisms reported are *Staphylococcus epidermidis*, the oral streptococci *Streptococcus mitis* and *S. oralis*, *Enterococcus* spp., *Staphylococcus aureus* and *Corynebacterium jeikeium*. Methicillin-resistant *S. aureus* (MRSA) strains are an increasing problem in hospitals where these organisms have established themselves. The almost universal use of long-term intravascular access devices such as Hickman and Portacath catheters favours infection with organisms derived from the skin. Infection with oral streptococci may follow mucositis secondary to chemotherapy.

Although uncommon, enterococcal infections are becoming more important because of the use of 4-fluoroquinolones for prophylaxis and cephalosporins for treatment of fevers; enterococci are resistant to both of these agents. Aminoglycoside- and vancomycin-resistant enterococci (VRE) are an increasing problem in intensive care medicine.

> ### Bacteria associated with infections in neutropenia
>
> 1 Enterobacteriaceae (*Klebsiella*, *Serratia* and *Enterobacter* spp.).
> 2 Skin-derived organisms: *Staphylococcus epidermidis*, *S. aureus*, including methicillin-resistant *S. aureus*, and corynebacteria, including *Corynebacterium jeikeium*.
> 3 Other streptococci including oral streptococci and enterococci.
> 4 *Pseudomonas* spp.
> 5 *Bacillus cereus*.
> 6 Other coagulase-negative staphylococci.

Fungal infections

Patients with prolonged neutropenia are at risk of invasive fungal infection. The frequent use of potent antibacterials encourages colonization by *Candida albicans*, the most commonly isolated organism. Infection with yeasts such as *C. dublini* and *C. krusei*, which are naturally resistant to antifungal prophylaxis, can cause a difficult problem. Infections with *Candida glabrata* and *C. parapsilosis* have been associated with intravenous cannulae.

Aspergillus spp. are the most common filamentous fungi causing invasive disease, although infections with *Fusarium* spp. and *Pseudoallerchia boydii* have rarely been reported. *Aspergillus fumigatus* is the most commonly isolated species, but *A. flavus* is increasing and *A. niger* is also reported. Infection is acquired by inhalation. *Aspergillus* spores are normally present in the air, and more are released during building work. They pose a grave threat to severely neutropenic patients, causing a progressive pneumonia associated with a high mortality.

> ### Fungi associated with infections in neutropenia
>
> 1 *Candida albicans*.
> 2 *Candida parapsilosis*.
> 3 *Candida dublini*.
> 4 *Candida krusei*.
> 5 *Candida glabrata*.
> 6 *Aspergillus fumigatus*.
> 7 Other *Aspergillus* spp.

Prevention of infection during neutropenia

Protective isolation (see Chapter 23) is intended to reduce the risk of infection in neutropenic patients. The patient is nursed in a side room, and is given sterile water and food with a low microorganism content. Attendant staff should observe hand-washing routines and may use sterile gloves when working with patients. *Aspergillus* infection can be reduced if the air entering the patient's room has been high-efficiency particulate air (HEPA)-filtered. These precautions are designed to prevent the replacement of the normal flora with potential pathogens or resistant organisms.

As the major infecting agents in neutropenic patients come from their own indigenous flora, various suppressive antimicrobial regimens have been used to prevent infection. Non-absorbable antibiotics have been given orally to reduce the numbers of bacteria in the bowel. A combination of framycetin (later replaced with neomycin), colistin and nystatin ('FRACON'), or of gentamicin, vancomycin and nystatin have both been used. The efficacy of these regimens has never been clearly established, but they appear to select for aminoglycoside-resistant organisms, notably strains of *Klebsiella* spp.

Current antimicrobial prophylaxis is based on the concept of colonization resistance, which suggests that the obligate anaerobic flora prevent colonization by facultative Gram-negative organisms, by competing for intestinal attachment sites and nutrients, and by production of bacteriocins and toxic free fatty acids. Antibiotics should therefore be targeted to facultative organisms, sparing the protective obligate anaerobic flora. Meta-analysis of trials of prophylaxis indicate that this approach is beneficial.

Early protocols using co-trimoxazole prophylaxis had some success, but patients still became colonized with resistant organisms. This was prevented by the addition of colistin to the regimen. Side-effects such as skin rashes and neutropenia were common. More recently the 4-fluoroquinolones norfloxacin, ofloxacin or ciprofloxacin have become the main antibiotics used for prophylaxis in many centres. Quinolone prophylaxis is usually combined with agents active against fungi (see below). Such regimens are superior to neomycin–colistin and co-trimoxazole–colistin, both in preventing infection with Gram-negative bacilli and in avoiding selection of resistant organisms. Ofloxacin and ciprofloxacin may be more effective than norfloxacin, which is less well absorbed. This suggests that these agents have a role both in suppressing Gram-negative bacilli in the gut and in inhibiting the early stages of systemic infection. The quinolones are less active against Gram-positive infections, which are now becoming the main cause of bloodstream infections.

Antifungal prophylaxis

Oral nystatin alone or in combination with oral amphotericin B is a useful non-absorbed prophylaxis, as it reduces fungal colonization of the mouth and gut. Fluconazole, which is well absorbed orally, has been successful in preventing yeast infections, but it has no useful activity against *Aspergillus* spp. or some yeasts such as *C. krusei* and *C. glabrata*. Itraconazole has useful activity against *Aspergillus* spp. Patients on prophylaxis with fluconazole may become infected with *Candida* spp. that are naturally resistant to it. Alternative agents include itraconazole, which has a wide spectrum of activity including filamentous fungi such as *Aspergillus*. The newer agent caspofungin may also be useful in preventing infection.

Antiviral prophylaxis

Latent viruses may reactivate during neutropenia. This most commonly occurs when cytomegalovirus (CMV) infection reactivates soon after bone marrow transplantation. Patients at risk of reactivation include those with documented seropositivity for the virus before transplant, and those whose donor was seropositive. Most centres adopt a prevention strategy in which quantitative CMV PCR is used to estimate viral copy number after transplantation. When an increase in copy numbers is detected, intervention is then begun with ganciclovir treatment. Antiviral treatment is continued until the immune system is reconstituted and the viral load has reduced to undetectable. Foscarnet is used for patients who cannot tolerate ganciclovir.

> **Prevention of infection in neutropenic patients**
> 1 Protective isolation.
> 2 Ward hygiene, including steam-pressed linen (to kill spores).
> 3 Filtered air supplies.
> 4 Suppressive antimicrobial regimens.
> 5 Antifungal prophylaxis.
> 6 Surveillance of viral load and treatment of reactivated viral infections.

Treatment of fever in neutropenic patients

Fever in neutropenic patients is assumed to indicate infection unless proved otherwise. Untreated bacterial infections progress rapidly; more than half of neutropenic patients with *Pseudomonas* bacteraemia will die within 24 hours unless appropriate treatment is prescribed.

Many trials of empirical therapy have been performed in search of a best treatment. Consideration of locally prevalent organisms and their resistance pattern is important. Aminoglycosides alone have not proved effective in neutropenic patients and are therefore usually combined with a beta-lactam antibiotic. The importance of *Pseudomonas* as a pathogen means that the regimen should have optimal activity against this agent. A ureidopenicillin, such as azlocillin or piperacillin, plus amikacin, has obtained response rates above 60% in neutropenic patients, or greater if Gram-negative bacteria were isolated. A combination of ceftazidime and amikacin has been a popular regimen but, following a large multicentre trial, the European Organization for Research and Treatment of Cancer (EORTC) proposed that a carbapenem, such as meropenem or imipenem, is likely to be successful in initial therapy. It may be necessary to include caspofungin or amphotericin, and/or a glycopeptide to deal with fungal infection, *Corynebacterium jeikeium*, MRSA or *S. epidermidis*. Empirical treatment is later modified, according to the results of culture or response to initial therapy. Empirical regimens need to be adjusted in the light of hospital epidemiology where this indicates the presence of multiple drug-resistant organisms.

In this environment of empirical, broad-spectrum treatment, several organisms that possess innate resistance to many antimicrobials are recognized as major opportunistic pathogens. They include MRSA, *S. aureus*, *Stenotrophomonas maltophilia* and *Enterococcus faecium*. Glycopeptide therapy is often necessary to treat the Gram-positive organisms, Failure to respond to antibacterial agents may indicate the need for antifungal therapy with caspofungin (now increasingly used to treat suspected fungal infections in neutropenic patients) or amphotericin B. Liposomal preparations of amphotericin enable patients to tolerate effective doses of this agent.

(see Chapter 16)

Treatment of fever in neutropenic patients (IDSA recommendations)

Initial therapy
- *Either* meropenem *or* ceftazidime; *plus or minus* vancomycin.
- *Alternative*: aminoglycoside *plus* piperacillin *or* azlocillin.

After reassessment at 3 days
- *If patient is now afebrile and low risk* (cultures negative; neutrophils >100 × 10⁹/l): change to oral therapy with a quinolone or beta-lactam for a further 4 days, or until neutrophils are approaching 500 × 10⁹/l.
- *If patient remains febrile*: take further cultures, investigate further, and change to alternative empirical regimen.

Reassessment at 5–7 days
- *If patient is afebrile and low risk*: continue treatment for 5–7 days more, then stop and review.
- *If fever continues and neutrophils are >500 × 10⁹/l*: stop treatment and review.
- *If fever continues and neutrophils are <500 × 10⁹/l*: add caspofungin *or* amphotericin B and continue existing regimen for 2 weeks after fever subsides and there are no signs of infection.

The outcome of therapy is related to the degree and duration of neutropenia. The cytokines GM-CSF (granulocyte–macrophage colony-stimulating factor) and G-CSF can be used to stimulate neutrophil production. This significantly shortens the period of neutropenia, reducing the number of infections and the days of fever.

Patients with chronic granulomatous disease are nowadays treated with interferon, which restores the defect in granulocyte function.

Fungal infection, especially with *Aspergillus* spp., can be a difficult problem for immunosuppressed patients. Liposomal amphotericin B is tolerated in much larger doses than other forms of the drug, and amphotericin B colloidal dispersion may have similar tolerability. Surgery can be useful for pulmonary *Aspergillus* infection; infected segments of lung can be excised after recovery of neutrophil numbers, to remove a locus of infection. Newer agents such as caspofungin, itraconazole or voriconazole are useful and relatively non-toxic alternatives, which are particularly useful when amphotericin use is undesirable because of toxicity or resistance.

T-cell deficiency

T-cell deficiency is increasingly common, with increasing use of corticosteroids, cyclosporin, serolimus, tacrolimus and other immunosuppressive agents in the chemotherapy of malignancies, and in transplantation medicine. The human immunodeficiency virus (HIV) epidemic has also contributed to the number of patients with T-cell deficiency, but the recent introduction of HAART (highly active antiretroviral therapy) has enabled physicians to reverse the deficiency and immunocompromised patients to recover immune function (see Chapter 16). Congenital T-cell deficiencies are rare, and can present as isolated T-cell dysfunction or combined with a hypogammaglobulinaemia.

Opportunistic pathogens in T-cell deficiencies

The main pathogens affecting T-cell-deficient patients are those that, in the human host, have an intracellular location. Among viruses, the naturally latent herpesviruses are common opportunistic pathogens. Mycobacteria and *Listeria* are well-recognized intracellular bacterial pathogens. Fungal and parasitic infections are more common and severe in affected patients.

Viral infections
Many viral infections occur in the course of leukaemia treatment, or transplantation immunosuppression (see below), often caused by the herpesviruses: herpes simplex, cytomegalovirus (CMV) and varicella zoster virus. Many of these infections can be prevented by prophylactic aciclovir. Unfortunately, a few herpes simplex and CMV isolates are now resistant to aciclovir and ganciclovir, respectively. Hepatitis B, hepatitis C, adenovirus, papillomavirus, polyomavirus and Epstein–Barr virus (EBV) may also cause clinical problems through increased activity and progression of chronic infections. Chickenpox infection, or herpes zoster, caused by varicella zoster virus, can be life-threatening. Patients with no immunity to chickenpox can be offered post-exposure prophylaxis with varicella zoster immunoglobulin (VZIG).

In children, measles can be life-threatening, complicated by giant-cell pneumonia and encephalitis. Predisposed children who have been exposed to measles virus should be protected passively with human normal immunoglobulin (HNIG). Live attenuated vaccines are generally contraindicated in children with immunodeficiencies, and should not normally be given to their household contacts, because of the risk of vaccine virus transmission. However MMR vaccine can be given to the sibs of affected children, as transmission of these vaccine viruses is exceptionally rare. Varicella vaccine may also be given (including to the patient if a sufficient non-immunocompromised interval occurs during a remission, or between episodes of chemotherapy).

Bacterial infections
The principal bacteria associated with T-cell deficiency are

mycobacteria, including *Mycobacterium tuberculosis*, *M. kansasii*, *M. avium-intracellulare* and *M. chelonei*. *M. tuberculosis* and *M. kansasii* are primarily respiratory pathogens, and often cause typical granulomatous lung disease, but in severe T-cell deficiency they can cause disseminated or miliary disease. *M. avium-intracellulare* is acquired by the gastrointestinal or respiratory route, and may cause bowel infection, lung infection or disseminated disease. *M. chelonei* is often acquired by inoculation and then causes local abscesses, but it has also been found to cause bacteraemia and fever of unknown origin. *Listeria monocytogenes* is an important cause of meningitis, and occasionally of peritonitis or bacteraemia. Often the only clue to the listerial aetiology is a history of immunosuppression.

Treatment of mycobacterial infections in T-cell deficiency

1 *Mycobacterium tuberculosis* and *M. kansasii*: treat as routine with triple or quadruple therapy, guided by antibiotic sensitivity testing (see Chapter 18).

2 *M. avium-intracellulare*: useful drugs include rifabutin orally 450–600 mg/kg daily; ethambutol orally 15 mg/kg daily; clarithromycin orally 250–500 mg 12-hourly for 2–4 weeks, then 250 mg 12-hourly; and amikacin i.m. or i.v. 7.5 mg/kg 12-hourly.

3 *M. chelonei*: co-trimoxazole 960 mg 12-hourly, reducing to 480 mg 12-hourly after 2–4 weeks.

Fungal infections

In contrast to the effect of neutropenia, superficial fungal infections are uncommon in patients with deficient T-cell function, but deep-seated or disseminated infection can be caused by organisms such as *Histoplasma capsulatum*. Cryptococcal infections, although increasing, are usually found only among patients on high doses of immunosuppressive drugs or with HIV infection. Patients with HIV disease often suffer from severe oropharyngeal and oesophageal candidiasis. *Pneumocystis jiroveci* pneumonia (PCP) is the commonest infection in AIDS patients (see Chapter 16). Before the HIV epidemic, it was usually reported in patients with leukaemia (particularly lymphoblastic leukaemia), congenital T-cell defects and corticosteroid therapy.

Treatment of systemic yeast infections in the immunosuppressed

1 Amphotericin by i.v. infusion starting with 250 μg/kg daily and increasing over 3–4 days to 1 mg/kg daily, adjust dose to minimize fever, nausea, hypokalaemia, renal impairment (which may be ameliorated by giving prednisolone before each dose) and rare neurological or haematological effects; some experts stop at a total dose of 1.0 g, others are guided by clinical cure.

Liposomal amphotericin (must be made up in warmed solution) by i.v. infusion; tolerated in doses of 3–4 mg/kg daily. It is the preferred agent when it is available, otherwise it is indicated for severe infections or when standard preparation is not tolerated.

Alternative: Fluconazole orally or i.v. 400 mg initially, then 200–400 mg daily *or* for resistant *Candida* infections: voriconazole orally 400 mg 12-hourly for 2 doses, then 200 mg 12-hourly; or i.v. 6 mg/kg 12-hourly for 2 doses, then 4 mg/kg 12-hourly (or 3 mg/kg 12 hourly if the higher dose is not tolerated).

2 For histoplasmosis or sensitive cryptococcal meningitis: fluconazole orally or i.v. 400 mg daily reducing after response to 200 mg daily; treat until clinical features are abolished (at least 6–8 weeks for cryptococcal meningitis).

Prophylaxis of PCP

Prophylaxis initiated before immunosuppression can prevent the development of disease in patients at risk of *P. jiroveci* (*carinii*) infection. It is particularly valuable when immunosuppression is related to bone marrow or organ transplantation. It can also be used in patients with AIDS, in whom PCP is uncommon until the T-cell count falls below 200×10^9/l. Monitoring of CD4 counts can indicate the appropriate time for institution of prophylaxis. Oral co-trimoxazole 480–960 mg daily is the prophylaxis of choice, but many patients are unable to tolerate this regimen. Alternatives include substitution of the sulfamethoxazole with dapsone, which provides equivalent protection with a lower incidence of adverse events and has the useful side-effect of decreasing the likelihood of reactivation of toxoplasmosis. Another effective prophylactic is aerosolized pentamidine 150 mg every 2 weeks or 300 mg 4-weekly, although this is less effective in children or in those who have already had an episode of pneumocystis infection.

Treatment of PCP

Co-trimoxazole orally or intravenously, 120 mg/kg per day in divided doses, is the treatment of choice. Pentamidine is an alternative, indicated for patients with a history of adverse reaction to co-trimoxazole. It is potentially very toxic and can cause hypotension during or immediately after intravenous administration. The dose is 4 mg/kg daily for at least 14 days; inhaled pentamidine 600 mg daily may be effective, and avoids severe systemic side-effects. Other regimens, used when patients fail to respond to the primary therapies, include atovaquone, a ubiquinone that interferes with parasite cytochrome metabolism, and trimetrexate, an antifolate drug related to methotrexate, which is given with folinic acid.

Parasitic infections

Toxoplasma gondii infection

Most *Toxoplasma gondii* infections in immunocompromised hosts result from reactivation of quiescent bradyzoite cysts. They usually present as multiple space-occupying lesions of the central nervous system. Rare, primary infections can cause fulminating meningitis or encephalitis. As with PCP, HIV infection is now the commonest predisposition. Hodgkin's disease, cardiac transplantation and acute leukaemia are non-HIV conditions associated with *Toxoplasma* infection.

Treatment and prophylaxis of toxoplasmosis

The treatment of choice is a combination of pyrimethamine 50 mg daily and sulfadiazine 4 g daily in divided doses. This is well absorbed orally and crosses the blood–brain barrier, though sulfadiazine can be given intravenously if necessary. Patients with cerebral toxoplasmosis usually respond by lysis of fever within 48 hours. Failure to respond should prompt a search for an alternative diagnosis. In patients with persisting immunocompromise, suppressive treatment is often indicated after successful therapy. Dapsone is useful for this purpose. A number of salvage treatments have been devised for cerebral toxoplasmosis, including clindamycin plus pyrimethamine, or trimetrexate plus folinic acid.

Cryptosporidiosis

Cryptosporidium parvum infection is very difficult to manage in immunosuppressed patients, who suffer continuing profuse, watery diarrhoea with abdominal pain. There is no tendency to natural resolution, as in immunocompetent children and adults.

C. parvum is naturally resistant to a wide range of disinfectants and antibiotics. A small proportion of patients appear to gain some benefit from spiramycin and others from paromomycin. Recent reports suggest that azithromycin in daily dosage produces a significant benefit. At present, treatment is mainly symptomatic, with antidiarrhoeal agents and antispasmodic drugs. Severely immunocompromised patients are advised to boil all drinking water, to reduce their risk of infection.

Isospora belli infection

Isospora belli is a coccidian parasite that, unlike *Cryptosporidium*, is susceptible to antimicrobial therapy. Patients should be treated with co-trimoxazole. Alternatives include metronidazole, furazolidine, quinacrine, nitrofurantoin and newer macrolide antibiotics. Patients may be maintained on suppressive doses of co-trimoxazole or a weekly dose of azithromycin.

Strongyloides stercoralis hyperinfection

Strongyloides stercoralis, a nematode infection, is acquired by direct penetration of infective larvae through intact skin followed by invasion of the small bowel by adults, which produce further larvae, perpetuating the infection (see Chapter 8). Infection can remain largely asymptomatic for more than 40 years. When cellular immunity is reduced, uncontrolled multiplication of the parasite can develop. This is known as the hyperinfection syndrome, which may be complicated by Gram-negative septicaemia, pneumonia or meningitis, as larvae deposit bacteria in the tissues.

Management of strongyloidiasis

Treatment with albendazole 400 mg daily for 3 days results in eradication in up to 80% of immunocompetent patients, but courses of 400 mg 12-hourly for up to 4 weeks may be needed in the immunosuppressed. Ivermectin 200 µg/kg (available on a named patient basis) is a highly effective alternative. The complication of Gram-negative septicaemia or meningitis, which may arise during uncontrolled hyperinfection syndrome, should be treated vigorously with third-generation cephalosporins such as cefotaxime.

Microsporidia have recently been recognized as important pathogens of patients with T-cell deficiency, particularly in AIDS.

Microsporidian species causing infection in immunosuppressed patients

1 *Encephalitozoon cuniculi.*
2 *Enterocytozoon beneusii.*
3 *Encephalitozoon intestinalis.*
4 *Nosema connori.*
5 *Vittaforma corneum.*
6 *Nosema* spp.

Diagnosis of opportunistic infection in patients with immunodeficiencies

Because of the diversity of potential infectious agents and the need for urgent therapy, rapid, accurate diagnosis is important in all immunosuppressed patients.

Blood cultures

All patients should have at least two blood cultures taken from different sites. If these are taken via an indwelling intravenous cannula, parallel cultures should be taken from a peripheral vein, to check whether any growth originated only from the catheter or also from the bloodstream. Blood should also be drawn for mycobacterial culture by conventional liquid medium, or in an automated detection system (which gives the advantage of earlier positive results; see Chapter 18). In patients potentially exposed to the systemic mycoses, cryptococcal antigen tests may be performed and cultures should be made to media designed to support fungal growth (in a containment level 3 laboratory: see Chapter 23).

Bronchoalveolar lavage

Pulmonary lesions and respiratory symptoms and signs should be investigated by bronchoalveolar lavage (see Chapter 7). The washings are examined by Gram, Ziehl–Nielsen and silver methenamine methods for the diagnosis of bacteria, mycobacterial pathogens and PCP. Herpesviruses may be detected by electron microscopy, and by rapid CMV culture methods. PCR and other DNA amplification techniques are now available for several diagnoses (e.g. CMV, HSV, VZV and *M. tuberculosis*).

Urinary antigen tests for *Legionella* infection can be performed.

Cultures from susceptible sites

Whenever fever occurs, sites such as long lines, peripheral cannulae, ventricular shunts and drains, tracheostomy sites and urinary catheters must be reviewed as potential origins of infection. Skin wounds, ulcers or multiple papules should also be investigated. Regular sets of cultures for surveillance or screening are taken in some settings, particularly haematology services. The intention is to give early warning of colonization with potentially dangerous or resistant opportunistic pathogens. The results of recent surveillance cultures may indicate which empirical therapy may succeed when fever occurs. However, the agent causing a fever is not always derived from superficial sites that can be routinely screened.

Early imaging in suspected localized infections

Cerebral toxoplasmosis can be suggested by the clinical picture of focal neurological deficit, and appearance of a ring-enhancing lesion on computed tomographic scanning. Serology is not diagnostic, as only immunoglobulin G antibodies are usually detectable, indicating either past infection or reactivation. Definitive diagnosis depends on brain biopsy, but this is usually only undertaken if the patient fails to respond to anti-toxoplasma therapy. Nocardiosis can produce cerebral and lung abscess. The rarer, subacute cryptococcosis can cause lung, skin and bone abscesses.

Strongyloides should not be forgotten

The diagnosis of strongyloidiasis should be attempted for all patients with a history of exposure before immunocompromising therapy is commenced (this includes locally immunosuppressive therapy for inflammatory bowel disease, which may also allow exacerbation of undiagnosed infection with *E. histolytica*). It can be made by serology and/or jejunal sampling by string test. Diagnosis of amoebiasis is made by examining stools and serology (see Chapter 8).

> ### Diagnosis of opportunistic infection
> 1 Blood culture.
> 2 Bone marrow culture.
> 3 Bronchoalveolar lavage.
> 4 Cerebrospinal fluid culture.
> 5 Brain scan or other imaging.
> 6 Surveillance cultures: mouth swabs, sputum, urine, sites of indwelling cannulae, areas of skin inflammation.
> 7 (Tests for strongyloidiasis or amoebiasis if indicated.)

Hypogammaglobulinaemia

Congenital hypogammaglobulinaemia

Congenital hypogammaglobulinaemia has two forms:
1 X-linked agammaglobulinaemia, in which patients become susceptible to infection after the first 6 months of life when maternal antibody is lost; and
2 common variable immunodeficiency, which can occur at any age, most commonly the third decade.

Infants aged 3–7 months always show a dip in total immunoglobulin levels, as maternal antibodies are lost, and generation of the child's own antibodies takes over. Infants may temporarily have gammaglobulin levels well below the accepted lower limit, but are not susceptible to opportunistic infections.

Functional hypogammaglobulinaemia

Functional hypogammaglobulinaemia develops in patients with multiple myeloma, due to arrested B-cell maturation. It also occurs in patients with chronic parasitic infections such as leishmaniasis and trypanosomiasis, due to polyclonal B-cell activation and overproduction of low-affinity, non-specific antibody at the expense of specific high-affinity antibody.

Patients with deficiency in T-cell function are also susceptible to pyogenic, particularly pneumococcal and other respiratory, infections due to the loss of T-cell help in antibody production.

Opportunistic infections

The main impact of hypo- or agammaglobulinaemia is on the respiratory and gastrointestinal tracts, with resulting failure to thrive. Patients suffer recurrent and chronic respiratory infections, particularly sinusitis and pneumonias. *Streptococcus pneumoniae* and non-capsulate *Haemophilus influenzae* are the commonest pathogens. *Mycoplasma pneumoniae* and chlamydial pneumonias are also more common and persistent in affected adults. In the intestinal tract, infections with *Campylobacter*, *Giardia* and *Cryptosporidium* may be more persistent than in normal subjects. Rare cases of progressive enteroviral infection with meningoencephalitis occur in agammaglobulinaemic patients. Enteroviruses are demonstrable by culture and PCR techniques in brain, cerebrospinal fluid, muscle and other tissues.

Recurrent suppurative lung infections lead inevitably to bronchiectasis if immunoglobulin is not replaced regularly. Intravenous immunoglobulin is readily available and regular infusions are effective in preventing recurrent and progressive infection. It is given to all infection-prone patients with congenital agammaglobulinaemia. The role of immunoglobulin in myeloma and other malignant diseases is less clear.

Complement deficiency

Hereditary complement deficiencies are rare but give rise to recurrent pyogenic infections, depending on which component of the complement pathway is deficient. Deficiency in the later components of the complement cascade, C7–C9, results in reduced ability to generate the membrane attack complex and achieve lysis of Gram-negative bacteria. The clinical consequence is recurrent infection with Gram-negative cocci, usually *Neisseria meningitidis*.

Kindreds with deficiency of components of the alternative complement pathway suffer more frequent and more serious *S. pneumoniae* infections, including meningitis. This is due to the importance of the alternative complement pathway in opsonizing pneumococcal cell wall components for clearance (see pp. 393–5, Chapter 19). Defects in this pathway increase the likelihood, but not the severity, of meningococcal and gonococcal bacteraemias. Rare individuals with properdin deficiency can develop devastating meningococcal infection. Immunization against the organisms carrying high risk will reduce the likelihood of severe infections in patients with alternative pathway defects, by allowing the recruitment of the classical complement cascade (see Chapter 1).

Deficiencies of each of the individual complement components have been described. Acquired complement deficiency occurs in immune disorders such as systemic lupus erythematosus.

Splenectomy

Following splenectomy there is a continuing risk of serious sepsis, with an incidence of approximately 0.5–1.0% per year. The risk varies with age and is particularly high in infants and children. Risk also varies with the indication for splenectomy; high mortality is associated with splenectomy for lymphoma and thalassaemia. The increased risk diminishes with time after splenectomy, but is never eliminated. Patients with sickle-cell disease have functional asplenia and suffer similar susceptibility to sepsis.

The most important infecting organism for splenectomized patients is *S. pneumoniae*, which causes approximately two-thirds of infections in most series. Other important bacteria are *H. influenzae* and *Escherichia coli*. Malaria may follow a fulminant course in patients with splenectomy. Splenectomy is an important predisposition to rare *Capnocytophagia canimorsis* infections, which usually follow dog bites.

Preventing infections in asplenic patients

Prophylactic vaccination with 23-valent capsular polysaccharide pneumococcal vaccine should be offered 2 weeks before an elective splenectomy. Antibody responses are reduced in magnitude and duration after splenectomy, so vaccination should be repeated 3–6 years later. Local reactions are common on revaccination. Improved responses may be obtained with conjugate pneumococcal vaccine, though the range of pneumococcal serotypes included in the vaccine is smaller (seven capsular types). *Haemophilus influenzae* type b (Hib) conjugate vaccine and meningococcal vaccine should also be given.

Antimicrobial prophylaxis with penicillin V should also be prescribed but its value may decrease as the prevalence of penicillin-resistant pneumococci increases. Patient education is important in encouraging patients to consult their physician quickly at the onset of a fever, or possibly to use an antibiotic regimen prescribed for early treatment without consultation. In reality many patients who undergo splenectomy are lost to follow-up, occasionally with fatal consequences.

> **Prevention of infection in splenectomized patients**
> 1 Immunization against *Streptococcus pneumoniae* and *Haemophilus influenzae*, preferably before splenectomy.
> 2 Antimicrobial prophylaxis.
> 3 Patient education and information.
> 4 Use of alerting card or bracelet.

Infections in transplant patients

Organ transplant patients

Organ transplant patients have usually received their

transplants because of underlying disease of the organ concerned. This may be non-infectious, as in terminal renal failure or ischaemic heart disease. Some transplants, however, are performed for diseases caused by severe or persisting infection; examples include liver transplants for acute or chronic liver failure caused by viral hepatitis, and heart transplant for myocarditis of infective origin. Fortunately, the original infection has often been terminated by the same immune and inflammatory response that damaged the affected organ. In a few cases, the infecting agent is not entirely removed with the infected organ, and recrudescence of local or generalized infection can occur. Thus hepatitis B virus can exist in the pancreas, and possibly other tissues, and will eventually re-infect a transplanted liver.

Anti-rejection therapy

Anti-rejection therapy produces major suppression of cell-mediated immune responses, designed to disable the effects of cytotoxic and natural killer cells on the transplanted organ. In the early post-transplant period, aggressive immunosuppression is obtained with corticosteroids, azathioprine, cyclosporin or other immunosuppressive drugs. Corticosteroids and azathioprine are broad-spectrum immunosuppressants that affect both B- and T-cell function, and also depress phagocytes and eosinophils. Cyclosporin, mycophenolate, sirolimus and tacrolimus have a narrower spectrum but strongly affect cell-mediated immune responses. As the transplant stabilizes and the early risk of acute rejection passes, the degree and spectrum of immunosuppression can be reduced. Many patients take maintenance doses of cyclosporin, tacrolimus or azathioprine, sometimes with a small supplement of corticosteroids.

There is therefore an evolution of susceptibility to infection after a transplant.

Evolution of infection risks in transplant patients
1 Risk derived from original disease (e.g. renal failure or haematological malignancy).
2 Risk of hospital admission and surgery.
3 Risk of disease contracted from transplanted tissue (e.g. toxoplasmosis).
4 Early risk of opportunistic infections during strong immunosuppression.
5 Later risk of opportunistic infections due to chronic suppression of cell-mediated immunity.

The patient will begin with an increased susceptibility to infection simply because of chronic underlying disease. Hepatic cirrhosis causes splenic dysfunction and increased risk of pneumococcal disease. Hypersplenism may cause pancytopenia. Renal failure causes a broad-spectrum susceptibility to infections.

Since the transplant itself is a surgical procedure, the patient will bear the infection risks of hospital admission, anaesthesia and surgery, and sometimes also those of intensive care.

On rare occasions, the transplanted organ contains infective agents related to the past infection history of the donor. These may be persisting or latent viruses, such as CMV, human herpesvirus type 6 (HHV6) or HHV8, or dormant organisms such as *Toxoplasma*. Donors from tropical countries may have migrating parasites such as *Strongyloides* in their organs. Infections originating in this way usually become evident soon after the transplant. Efforts are made to eliminate the possibility of transmissible infection in the donor, or to ensure that the recipient has antibodies to agents such as *Toxoplasma* and CMV if the donor is also positive.

The effects of strong immunosuppression are usually important for about 3 months. During this time a wide range of viral, fungal, parasitic and atypical infections may occur. Bacterial infections are also likely, and are vigorously investigated and treated. Skin and bowel suppressive treatment may be given with topical disinfectants and non-absorbable antibiotics and antifungal agents. Aciclovir prophylaxis is often given, and protects against herpes simplex infections and also against CMV disease (though the mechanism for the latter is poorly understood). Anti-CMV immunoglobulin is also used in some cases, and has an additive effect with antivirals.

Importance of cytomegalovirus in post-transplant patients

One of the most difficult infections to treat in the first 3 months after a transplant is CMV infection. Unlike AIDS patients, whose pulmonary CMV infections are rarely clinically important, more than half of transplant patients who develop CMV pneumonitis in the first 3 months post-transplant will die. The treatment of CMV pneumonitis depends on the use of ganciclovir. This drug can cause neutropenia, thrombocytopenia and oncogenesis. In combination with other bone-marrow suppressive drugs, it can cause severe bone marrow depression. Foscarnet, used for CMV retinitis in AIDS, is not recommended for other CMV infections. It is extremely toxic, causing renal impairment in half of those treated and making cyclosporin therapy difficult. The decision to use ganciclovir is usually dependent on monitoring the CMV viral load in blood by quantitative PCR methods, to indicate when therapy is required.

Ganciclovir treatment of cytomegalovirus pneumonitis

- Ganciclovir by i.v. infusion over 1 h, 5 mg/kg 12-hourly for 14–21 days; may be continued at 5 mg/kg daily (or 6 mg/kg daily on 5 days per week), if risk of recurrence exists.
- Oral valganciclovir 900 mg daily is an alternative to i.v. continuation therapy.

The later risks of reduced cell-mediated immunity are largely related to infections caused by reactivation of latent infections, and include herpes zoster, toxoplasmosis or progressive multifocal leucoencephalopathy (caused by JC virus). Infection with environmental agents of low pathogenicity, such as *Listeria monocytogenes* or *Nocardia asteroides*, is also a continued risk (Table 22.2).

Bone marrow transplant patients

Bone marrow transplant patients have slightly differ-ent problems because, as well as the immunosuppressive effects of their treatment, lymphoma, leukaemia or myeloma will have caused immunosuppression before any cytotoxic therapy or transplant is undertaken. The transplant itself is preceded by ablative chemotherapy and radiotherapy, causing virtually complete suppression of immune responses. The transplanted bone marrow is also often treated to suppress its cytotoxic cell population, in an effort to avoid severe graft-versus-host disease. As the transplanted bone marrow gradually becomes established, many of its functions will recover (Table 22.3).

The need for immunosuppressive maintenance therapy will depend on the source of the transplanted marrow, and its degree of tissue match with the recipient. Autologous transplants (the patient's own bone marrow, harvested during remission of malignancy) rarely need immunosuppressive support. Transplants from close relatives may require little or no anti-rejection treatment once they are established. Transplants from unrelated donors can require permanent immunosuppression, and may

Table 22.2 Range of infection risks following organ transplantation

Time after transplant	Disease risk
First 2 weeks	Infectious complications of anaesthetic and surgery: *Staphylococcus aureus* or enterobacterial infection of wound, enterobacterial, staphylococcal or candidal infection of organ or deep tissues Pneumococcal or other chest infection
First 3 months	Chest infections with CMV, *Legionella, Aspergillus, Mycobacterium, Candida* or other fungus Skin and mucosal infections with herpes simplex, *Candida* or other yeast Infection of long-term indwelling cannula Cerebral infections with *Toxoplasma, Aspergillus* or *Candida*
After 3 months	Skin infection (herpes zoster) Pneumonia with *Nocardia, Aspergillus* or CMV Cerebral infection with *Listeria, Cryptococcus, Nocardia, Toxoplasma* or JC virus

CMV, cytomegalovirus.

Table 22.3 Spectrum of immune dysfunction following bone marrow transplant

Duration of effect	Type of defect
First month	Lack of natural killer cells (and other non-CD8 cytotoxic cells)
3–4 months	Total T-cell count deficiency Decreased interleukin-2 production Decreased neutrophil chemotaxis
4–6 months	Decreased CD4 (helper/inducer) cells Decreased CD8 (cytotoxic) cells Decreased response to polysaccharide antigens and reduced immunoglobulin A responses to antigen
1 year or more	Decreased secondary antibody responses Decreased alveolar macrophage functions Reduced proliferative responses by CD4 cells

also cause graft-versus-host disease, which itself carries a risk of infection and/or rejection.

As bone marrow transplants are becoming more successful, and are increasingly used to treat metabolic and other non-malignant disease, immunosuppressive therapy is likely to evolve, and possibly become narrower in spectrum. Expertise in long-term follow-up and the treatment of late manifestations of opportunistic infection (which may include slowly progressive or chronic infections) will become increasingly important.

Hospital Infections

23

Introduction

The hospital is an ideal environment for the transmission of pathogens, because patients with similar diseases and susceptibilities are housed in an enclosed community. Patients share contact with many healthcare workers each day. In this setting, the ward, patients and workers become colonized by organisms adapted to the special environment. New susceptible individuals are frequently added to the population, and are at risk of colonization and infection. In Britain, approaching 15% of all patients admitted to hospital develop a hospital-acquired infection, with the risk increasing for a longer hospital stay.

Respiratory tract, urinary and wound infections are common in all hospital patients. In addition, immunocompromised patients readily develop infection with organisms of low virulence. Patients in the intensive therapy unit, where many antibiotics are used, can be colonized with naturally resistant organisms, which may cause pneumonia or bacteraemia.

Organisms with multiple antibiotic resistances can cause problems on general wards as well as in special units. So-called methicillin-resistant *Staphylococcus au-reus* (MRSA) strains are an example. Many are resistant not only to antistaphylococcal penicillins, but also to a range of other antistaphylococcal agents. They readily colonize skin wounds, ulcers and indwelling devices such as intravenous cannulae and urinary catheters. If they cause infections, prolonged treatment may be necessary, using expensive or toxic drugs, such as teicoplanin or vancomycin.

Hospital infection not only imposes a burden of illness and prolonged admission on the patient, it also imposes the cost of investigation and treatment on the hospital, as well as preventing the use of the bed for other patients. It carries the risk of spread, particularly to other patients, and demands time-consuming and expensive control measures.

The concepts of host, organism and environment have already been discussed (see Chapter 1). The same approach can be applied to the natural history of hospital infection. In this chapter the main hospital pathogens will be described. We will also review the special susceptibilities of patient populations to organisms that often have low pathogenicity in the community, and discuss features of the hospital environment that influence the transmission of these pathogens.

Patient susceptibilities to hospital infection

Although it seems glib to say that patients are in hospital because they are sick, the need for admission implies an alteration in host defences. This is obvious in patients immunocompromised by an illness such as leukaemia, or by treatment such as cytotoxic chemotherapy or high-dose corticosteroids. It is less obvious in fit patients admitted for routine surgery. However, the effects of anaesthesia and post-operative pain may inhibit coughing, leading to post-operative hypostatic pneumonia, or they may make micturition difficult, leading to urinary infection. The surgical wound itself presents another potential site of entry for infection. It is essential, therefore, to assess each patient carefully for such factors.

Patient predispositions to hospital infection

1 Pre-existing condition (chronic chest disease, obstructed urinary outflow or previous immunosuppression).
2 Need for invasive devices (intravenous cannulae, urinary catheters, etc.).
3 Effect of surgery (skin wound, tissue trauma, opening colonized viscus, anaesthesia, immobilization, introduction of foreign material such as joint prosthesis or arterial graft).
4 Effect of antibiotic treatment (antibiotic-associated diarrhoea, colonization by resistant organisms, predisposition to superficial fungal infections).
5 Effect of immunosuppressive treatment (corticosteroids, cancer chemotherapy or transplant immunosuppression).
6 Exposure to healthcare workers and other patients who may carry or transmit pathogens.
7 Exposure to pathogens in the environment, especially bedding and food.

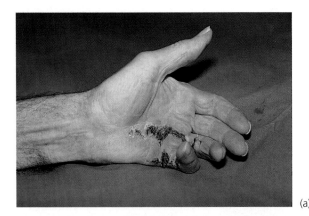

(a)

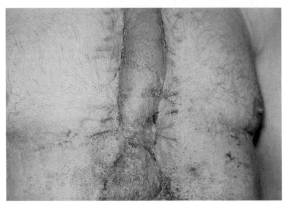

(b)

Figure 23.1 (a) An infected minor operation wound on the hand. Methicillin-resistant *Staphylococcus aureus* (MRSA) was recovered from swabs. (b) A patient in the same ward required skin grafting after MRSA infection of a sternotomy wound.

Infection due to intravenous cannulae

Intravenous devices of many kinds can be placed in the vascular system for varying periods. The time for which they can be maintained in place depends significantly on the likelihood of infection in each site. Peripheral venous catheters are readily colonized by organisms of the skin flora. Trivial infection, with mild inflammation, is common but more invasive disease with organisms such as *S. aureus* can cause significant morbidity and mortality (Fig. 23.1).

Many patients now have venous catheters that enter the right atrium. These may be used for intravascular monitoring, as with Swan–Ganz catheters, or to give intravenous feeding or drugs that are irritant to peripheral veins. When long-term intravenous therapy is required, long lines, such as Hickman or Portacath catheters, can

be inserted via a subcutaneous track or tunnel. Infection in these devices is serious as not only must the infection and its complications be managed, but the line must also be replaced using a new tunnel. This requires a second operative procedure, and is also costly in terms of time and resources.

The common pathogens of intravenous catheters are flora from the skin of the host, particularly *S. epidermidis* and *S. aureus*. More rarely, corynebacteria may be implicated, especially the naturally multidrug-resistant species *Corynebacterium jeikeium*, which may cause line-related sepsis in leukaemic patients. *Acinetobacter* spp. and enterococci can cause line infections in intensive care.

Clinical features

These are usually mild unless septicaemia supervenes. Vigilance is important in detecting this early stage, when treatment is likely to be successful, and complications few. There may be signs of inflammation at the site of the skin entry, with tenderness, cellulitis or slight purulent exudate. As in infective endocarditis, bacteraemia is usually continuous. Fever is often present but is usually mild; around 37.5–38.5 °C. When line-related sepsis is likely, the patient should be examined for signs of metastatic infection or endocarditis.

Investigation and diagnosis

Whenever a patient with an intravascular device develops fever, the device should be suspected as being the origin of the symptoms. The insertion site should be inspected and swabs taken if there are signs of inflammation or infection. Blood cultures should be collected, ideally through the suspect line and also from peripheral venous sites. Venous cultures may be negative if the bacteraemia is scanty, but positive results from blood taken from the device will point strongly to this origin of infection. Multiple cultures are required as the organisms are often derived from normal flora, and it is only when repeated isolates of an organism with similar biochemical activity and antibiotic susceptibility profile are made that infection is confirmed.

The tip of the cannula can be cultured and a standardized method can be used to facilitate the interpretation of the results (the Maki roll method). The main problem with this approach is that the results are semiquantitative and the cannula must be removed to examine it. This difficulty can be overcome by the introduction of brush sampling methods that allow the tip of the cannula to be sampled while it is still *in situ*. This is especially valuable for patients with a long-standing intravenous access device.

Prevention and control

The control of line-related sepsis must start with education and training in medical and nursing schools. A strict protocol of skin disinfection and sterile technique must always be used when inserting intravenous access devices. The choice of device is also important. Those with side ports are prone to colonization at this point, where no flow occurs. Giving sets may also provide a nidus for colonization. This is especially true if multiple access points are available, each with a dead space where fluids can become static. Contamination can be introduced into intravenous fluids and giving sets by repeated addition of drugs to the intravenous system. Ideally additive drugs should be incorporated into intravenous fluids in the manufacturing pharmacy, under sterile and controlled conditions.

The most important factor in preventing line-related sepsis is the regular review of inserted lines. To facilitate this the date when lines were inserted must be documented in the patient's case record. Peripheral intravenous lines should ideally be re-sited every 48 h. Central lines should be changed if there is evidence of infection. The life of tunnelled lines is much longer than that of peripheral ones, but clinicians must be aware of the risks of infection, and intervene to remove the line whenever it occurs. Special situations, such as intravenous feeding, encourage infection by providing a rich supply of nutrients within the catheter lumen. They are best managed by a specialist team that includes clinicians, pharmacists and microbiologists.

Measures to prevent line-related sepsis

1 Choice of device (avoiding side ports and dead spaces).
2 Aseptic and atraumatic insertion.
3 Preparation of additive drugs and parenteral feeds in the pharmacy.
4 Maintenance of adequate hygiene and dressing of insertion site.
5 Regular review of insertion site.
6 Replacement of giving set (and cannula when indicated) at appropriate intervals.
7 Removal of cannula from inflamed site.
8 Removal or changing of cannula in a bacteraemic patient.

Treatment

When bacteraemia is associated with an intravenous access device the device should be removed. The skin insertion site and the cannula tip should both be cultured, and compared with the results of blood cultures. When the infecting organism is of low virulence, such as *S. epidermidis*, these measures should be sufficient but if the infection is severe a glycopeptide antibiotic can be given for 48 h. When *S. aureus* is the infecting organism, 2 weeks' intravenous antistaphylococcal therapy is needed to minimize mortality and complications. After completion of treatment, physical review should be performed to exclude persisting focal infection, such as endocarditis or osteomyelitis. When *S. epidermidis* colonizes a 'precious' cannula, an attempt is sometimes made to eradicate the organisms by instilling a glycopeptide into the cannula. Even if successful, this should be followed by vigilant review to detect recrudescence.

Management of line-related sepsis

1 Removal of the affected device.
2 Culture of site, cannula and blood.
3 Short course of glycopeptide antibiotic (active against staphylococci and corynebacteria).
4 Full treatment if *Staphylococcus aureus* is isolated.
5 Physical examination and/or investigation to exclude endocarditis and other complications.
6 Review of the need for insertion of new device.

Infection associated with urinary catheters

Indwelling urinary catheters provide easy access for ascending infection of the urinary tract. After a number of days, organisms will reach the bladder, often by ascending between the catheter and the urethral wall. Permanent bladder catheterization is always associated with bacterial colonization of the urine.

Gram-negative organisms are the commonest colonizers of the catheterized bladder. *Escherichia coli*, *Klebsiella pneumoniae* and *Pseudomonas* spp. are often seen. *Proteus* spp. are also found in chronically stagnant urine. Their ability to metabolize urea and produce alkaline ammonia predisposes to the deposition of calcium as stones or 'sand'. In turn, these deposits can act as a reservoir of infection.

After trans-urethral prostatectomy, *S. aureus* or coagulase-negative staphylococci can cause urinary colonization, occasionally complicated by epididymitis.

Catheter-related urinary colonization is often asymptomatic, but there is a risk of ascending infection or bacteraemia if the catheter becomes blocked or is vigorously manipulated. This colonization is inevitable and does not require treatment unless there is evidence of infection. If fever, urinary tract symptoms or rigors occur, appropriate antibiotic therapy should be given. The catheter should usually be replaced before chemotherapy is discontinued, to ensure adequate drainage of the bladder and to remove a possible reservoir of infection.

Closed drainage systems, in which the catheter is never opened directly to the environment, delay the entry and ascent of organisms, and afford a barrier to the introduction of hospital pathogens from attendants' hands. The risks can also be minimized by careful attention to aseptic technique when the catheter is inserted, to strict hand hygiene by healthcare workers who may manipulate the catheter and urinary drainage system, and to the personal hygiene of the catheterized patient.

Avoiding urinary catheter-related sepsis

1 Sterile, atraumatic insertion.
2 Appropriate choice of catheter type and size.
3 Use of closed drainage systems.
4 Hand hygiene of healthcare workers who handle the catheter.
5 Maintenance of good patient hygiene.
6 Replacement of catheter at appropriate intervals.
7 Removal of calcific deposits from the bladder (if they form).
8 Avoidance of excessive catheter manipulation or unnecessary bladder washouts.
9 Treatment of bacteriuria only when symptomatic.

Susceptibilities of intensive therapy patients

Patients in the intensive therapy unit (ITU) are susceptible to infection for several reasons. Immune responses are often diminished by the stress and metabolic effects of existing disease. Many patients in intensive care have recently undergone anaesthetic and surgical procedures. Additionally, many of the barriers to infection provided by innate immunity (see Chapter 1) are breached because of the need for complex intravenous therapy, invasive monitoring, urinary catheters, artificial ventilation and extracorporeal procedures such as dialysis or haemofiltration.

Tracheal intubation and artificial ventilation

Bacterial lung infections are the commonest complication of admission to intensive care. The endotracheal tube provides a means for organisms in the pharynx to bypass the defensive mucociliary blanket and gain direct access to the lower respiratory tract. Where a patient has been in hospital for some time and may have received several courses of antibiotics the normal upper respiratory flora has often been replaced with multidrug-resistant Gram-negative organisms such as *Acinetobacter*, *Stenotrophomonas* or *Enterococcus faecium*. These organisms are naturally resistant to many first-line antibiotics and can easily invade the respiratory tract. Artificial ventilation usually imposes the need for muscular paralysis; this inhibits the normal sighing and coughing reflexes, further reducing the ability of patients to resist bacterial invasion of the lungs.

Common causes of lung infections in hospital settings

1 *Streptococcus pneumoniae* (often local strains, may be penicillin-tolerant).
2 Methicillin-resistant *Staphylococcus aureus* (usually needs glycopeptide treatment).
3 *Moraxella catarrhalis* (usually produces beta-lactamase).
4 *Klebsiella pneumoniae* (always resistant to ampicillin).
5 *Escherichia coli*.
6 Enterococci (need beta-lactam/aminoglycoside, alternatively aminoglycoside–glycopeptide combination, meropenem or imipenem; a few are aminoglycoside- or glycopeptide-resistant).
7 *Pseudomonas* or related organisms (need aminoglycoside, extended-spectrum penicillin or antipseudomonal cephalosporin).
8 *Candida* (needs fluconazole as itraconazole is unpredictably distributed in critically ill patients).
9 Other yeasts (may need amphotericin treatment).

Other susceptibilities

The patient in the ITU has a multiplicity of intravenous and intra-arterial cannulae. The principles of cannula management, described above, must be vigorously applied. All ITU patients have indwelling urinary catheters, which must also be managed with meticulous attention to hygiene.

In addition to the special risks of intensive care procedures, many patients also have reduced innate and specific immunity because of organ failure, underlying malignancy, previous or pre-existing infection or chronic airways disease.

It is now known that early and adequate nutrition greatly improves the prognosis for intensive care patients. Enteral feeding often requires the insertion of percutaneous gastric cannulae, producing a further susceptible skin puncture. The feed itself must be hygienically prepared, as spoilage can introduce enteric infection. Intravenous feeding carries a high risk of cannula-related infection (see above).

The types of organisms causing local and bacteraemic infections in the ITU depend on local environmental factors and antibiotic usage. Common pathogens causing problems of treatment include *Klebsiella*, and other antibiotic-resistant Gram-negative rods, an increasing number of which are resistant to many antibiotics because they produce extended-spectrum beta-lactamase enzymes (ESBLs). Enterococci are also relatively common causes of infection, and a minority of these are tolerant, or resistant to amoxicillin, gentamicin or glycopeptides. MRSA can cause intermittent outbreaks of colonization and infection.

Surgery and its contribution to infection

Modern anaesthetic and surgical techniques make extensive and complex surgery possible, but the impact of infection greatly influences the outcome. Lister, the pioneer of antiseptic surgery, said that each operation was an experiment in bacteriology. This remains true today, though more control can be exerted over the experiment and its adverse effects. Patients must often be admitted to the ITU after a complex operation. This adds to the range of infections to which they are susceptible.

Several factors influence the occurrence of infection in surgical patients: the patient, the operation, the antimicrobial prophylaxis, the surgical team, the hospital environment and the post-operative care.

The patient

Patients often come to surgery with pre-existing health problems. Minimizing the time between admission and the surgical procedure will limit the opportunity to acquire resistant hospital pathogens. Wherever possible, existing infection should be treated before surgery is undertaken, and preferably before hospital admission. Patients with respiratory infections should receive appropriate antibiotics and physiotherapy. Antimicrobials should not be prescribed for trivial reasons in advance of surgery, as these may allow the replacement of sensitive normal flora with multidrug-resistant hospital strains.

Prophylaxis of surgical infections

The introduction of antimicrobial prophylaxis has done much to reduce the incidence of surgical infection. There are a number of basic principles that guide its use. The agents used should be bactericidal and active against the organisms likely to cause infection. To ensure that they are available at the susceptible site at the time of operation, the first dose is often given intramuscularly as part of premedication or at induction if an intravenous preparation is used. There is no evidence that additional benefit is gained by continuing prophylaxis for more than 1–3 days.

Choosing appropriate surgical prophylaxis

For this purpose, operations can be classified into three categories: clean, contaminated and infected.

Clean operations

In clean operations only the skin, or a site such as a joint, which is normally bacteriologically sterile, is breached. In this case, the commonest organisms implicated in postoperative infection are staphylococci from the skin. Postoperative infection after clean operations is almost always mild wound infection, and affects less than 2% of patients. Antimicrobial prophylaxis is not usually indicated. The exception is when a prosthetic device, such as a vascular graft or hip prosthesis, is to be inserted. In these circumstances the consequences of infection are catastrophic and prophylaxis is indicated.

Systems designed to minimize the transmission of skin bacteria into the operation site include filtered air supplies to the operating theatre, impermeable, ventilated suits for surgeons, adequate skin preparation at the operation site, and sterile surgical drapes or dressings.

Attempts to eradicate those bacteria that enter the wound include the use of antibiotic-impregnated orthopaedic cement, and even antibiotic-impregnated intravascular prostheses. Nevertheless, the most common organisms infecting implanted devices remain staphylococci from the patient's own skin.

Neurosurgical operations in which the meninges are opened carry a risk of post-surgical meningitis. This is rare, but when it occurs it is often caused by *S. aureus* or

Gram-negative organisms, including *Pseudomonas* spp. and *Acinetobacter* spp.

Contaminated operations

In these operations the surgeon opens an organ, such as the large bowel, that possesses a normal flora. Without prophylaxis the risk of infection varies between 10 and 40%. When the bowel is opened, a mixture of facultative and obligate anaerobes is released, and prophylaxis active against these organisms should include metronidazole and a broad-spectrum antibiotic such as a second-generation cephalosporin. In the upper gastrointestinal tract obligate anaerobes are uncommon and prophylaxis with a second-generation cephalosporin alone is adequate. After gastric and duodenal surgery, candidal infections occasionally occur. When obstruction is present, obligate anaerobes may accumulate, for example in the stagnant biliary tree or stomach, and the prophylaxis must be adjusted accordingly.

Instrumentation through a colonized or infected hollow organ is similar to a contaminated operation. Examples are cystoscopy or ureteroscopy of the infected urinary tract and endoscopic retrograde cholangiopancreatography, when the endoscope must enter the sterile biliary tree after passing through the colonized duodenum. Prophylaxis must be given in these circumstances, as the risk of infection approaches 100%, and bacteraemia is common.

Infected operations

Infected operations are those in which infection already exists, and contamination with pathogens is inevitable. Drainage of an intra-peritoneal abscess and excision of perforated bowel are good examples. Appropriate antimicrobial therapy, rather than prophylaxis, should be prescribed in this situation.

Examples of prophylactic regimens for surgical procedures

1 For upper gastrointestinal tract, endoscopic retrograde cholangiopancreatography of infected biliary tree: single dose of gentamicin or broad-spectrum cephalosporin 2 h before surgery.
2 For colonic and rectal surgery: single dose of gentamicin or cefuroxime *plus* metronidazole 2 h before surgery.
3 For hysterectomy: single dose of metronidazole rectally or i.v. 1–2 h before surgery.
4 For high amputations of the leg: benzylpenicillin i.v. or i.m. 300–600 mg 6-hourly for 5 days or metronidazole rectally 1 g or i.v. 500 mg 8-hourly for 5 days.
5 For joint replacement: cefuroxime or co-amoxiclav in standard parenteral doses for not more than 3 days (bone cement also contains aminoglycoside to inhibit skin-derived staphylococci).

The surgical team

The surgical team, or scrubbed team, is intimately involved with the operation. If one of the team has an infected or colonized skin or respiratory site, the pathogen involved, often a *Streptococcus* or *Staphylococcus*, may be shed into the patient's wound. The number of bacteria released into the operating theatre air depends on the number of persons in the theatre and their movement. The surgical team should therefore be as small as possible, and movements should be limited as much as possible.

The number of organisms shed can also be related to the design of surgical gowns and the type of material used. A direct relationship between this and post-operative infection is less clear. Impervious materials reduce shedding to a minimum but are very uncomfortable to wear. In the Charnley system, where the surgical team wear impervious 'space suits', these have individual air supplies and are cooled.

Surgical infections in immunocompromised patients

Modern medical practice means that there are now many patients with severe immunocompromise (see Chapter 22). These patients, however, do not form a homogeneous group. In addition to their underlying conditions, which impose differing infection risks, the nature of their immunocompromise will alter the range of infections to which they are subject. Renal and liver transplant patients require less immunosuppression to prevent rejection and are therefore subject to fewer opportunist infections than cardiac or allogeneic bone marrow transplant patients, in whom infection is a more important determinant of outcome. In understanding the infection risks to which patients are exposed it is convenient to discuss these under treatment considerations and immunological factors (below).

Treatment considerations

In many patients, transplantation or other therapeutic intervention imposes particular surgical risks. In cardiac transplantation the chest wall wound is a common site of infection, as is the urinary tract in renal transplant patients. For patients receiving radiotherapy to the abdomen or pelvis, radiation damage to the intestinal epithelium makes bacterial escape across the gut wall more likely. After liver transplantation, post-operative infections such as cholangitis and biliary peritonitis are related to the surgery on the biliary tract.

Immune considerations

Different diseases have a differing impact on the various components of the immune system. In some, such as human immunodeficiency virus (HIV) infection, the major effect falls on T-cell function. Cyclosporin treatment also has a major immunosuppressive effect on T cells. In multiple myelomatosis the main effect is on humoral immunity. Neutrophils are principally affected in patients undergoing induction chemotherapy for leukaemia. Other conditions such as splenectomy can cause poor clearance of blood-borne parasites, and a defect in control of capsulate organisms. Immunodeficiency is seldom purely cellular or humoral: for instance, T-cell disorder also leads to humoral immunodeficiency, because of the associated lack of T-helper function.

Environmental factors in hospital infection

The hospital environment is very different from the general environment, and poses particular dangers to patients ill-equipped to resist infection. Hospitals are often crowded, with much movement of patients, healthcare workers, volunteers and visitors, providing many opportunities for person-to-person transmission by contact. Hospital food may transmit food-borne disease if kitchen hygiene or food handling is unsatisfactory. Food is also often a reservoir of local *Pseudomonas* strains. The air supply in the hospital environment is controlled and may allow for the transmission of aerosol-borne organisms such as legionellae if systems are not properly maintained. Similarly, airborne organisms such as varicella zoster virus can spread rapidly in the hospital environment if the appropriate control measures are not taken. *Aspergillus* spores always exist in unfiltered air, and can infect neutropenic patients.

Hospital water supplies

The water supply of a hospital is complex. Unlike a domestic building or office system, there are great demands for water which must be delivered to a very large number of sites and for widely differing uses. In addition to wash-hand basins and showers, there is a need for central heating and air-conditioning. Hydrotherapy pools and birthing pools exist in many general hospitals. Several departments also require the delivery of steam for heating or disinfection. Most governments have issued minimum standards for maintenance of air-conditioning systems, cooling towers and hospital water supplies.

In the life of a hospital, water use will evolve with changing demands, and changing use of ward and laboratory areas. This affords many opportunities for lengths of pipework to be extended, or to go out of use. Cold water may become stagnant in these areas, or hot water may lose its heat. A great danger is that *Legionella* spp. will colonize warm water, especially when stagnant water has given up its protective chlorine. Sporadic cases or outbreaks of legionellosis may then occur. Inadequately cleaned and disinfected air-conditioning systems or cooling towers are also common sources of legionellosis (see p. 155). *Legionella pneumophila* serogroup 1 tends to cause classic legionnaires' disease, while other strains have reduced virulence, usually only causing disease in immunocompromised patients. Any area where fast-running water may cause aerosols is a potential source of infection. Shower heads and spray-type taps may harbour legionellae in rubber washers, and disperse large numbers of organisms.

Frequent use, adequate maintenance, ensuring adequate chlorination, and cleaning are all important in minimizing risk. Legionellae cannot survive at temperatures above 55 °C, or replicate below 20 °C. The risk of scalding by adequately hot water can be reduced by installing mixer taps. Redundant or overlong stretches of pipework, where legionellae could multiply, must be avoided. Pasteurizing the hot water system, by applying extra heat, may help to clear legionellae from the system. In addition it may be necessary to add additional chlorine to the cold water system. These extra measures may be necessary after pipework has been decommissioned for repair or alteration. The addition of water filters and treatment of the water with chlorine dioxide may also be employed. Despite this it may prove impossible to completely eradicate environmental *Legionella* from the water supply. It may then become necessary to provide sterile water to prevent infection in immunocompromised patients.

Control of legionellae in hospitals
1 Minimize long or redundant runs of pipework.
2 Ensure adequate chlorination of cold water.
3 Ensure adequate temperature of hot water.
4 Avoid spray taps and rubber washers.
5 Maintain taps, shower heads and cooling systems meticulously.
6 Consider pasteurization of colonized hot water systems.
7 Use filters or additional disinfection in critical areas.

Hospital air supplies

The piping and trunking of hospital air-conditioning may gradually accumulate much dust and building rubble. Sporing organisms and fungi such as *Aspergillus* spp. can thrive in these conditions. They can be discharged into the air and cause respiratory or systemic disease in susceptible individuals. Operating theatres, laboratories and individual patient isolation rooms are sites of particular risk.

Operating theatres

The construction and maintenance of operating theatres

is a specialist subject. The emphasis is on easy-to-clean, impermeable surfaces, including those of movable equipment. Design is intended to minimize the need for movement of staff, and to direct the movement of patient, staff and theatre waste away from the operating or clean areas, rather than towards or through them.

There are guidelines that set out the maximum number of organisms tolerable in the air of an operating theatre. Air is supplied to the theatre through filters. Before the theatres may be used, and following repairs to the filters or other decommissioning, air quality should be tested. A special air sampler (e.g. Casella slit sampler) is used, which draws a known volume of air through a slit and deposits particles on solid bacteriological medium. Colonies of organisms can be counted after the medium is incubated, and speciation performed if indicated.

Hospital equipment and the spread of infection

Disposable hospital equipment such as syringes, needles, blood lancets, scalpels, intravenous cannulae and urinary catheters makes the introduction of infection by often-repeated invasive procedures very rare. Blood-borne infections can still be transmitted, however, by inoculation accidents with used sharp instruments. Most hospitals have strict protocols for the handling and disposal of sharps. Since the occurrence of variant Creutzfeldt–Jakob disease (vCJD), and the discovery of prion protein in human tonsillar and bowel tissue, there is now a policy to use disposable surgical instruments and drapes wherever possible, and to apply strict protocols to the decontamination of all operating theatre equipment.

Some equipment is too complex and expensive for single use. Examples include endoscopes and associated biopsy equipment, complex surgical instruments, positive-pressure ventilators and high air-loss beds. These all have internal channels that can come into contact with body fluids, wound exudates or infected tissues. All have been involved in hospital outbreaks of infection. Automated cleaning systems, in which a sequence of cleaning solutions and disinfectants is pumped through the equipment for a fixed time, have reduced mishaps in decontamination after use. Such systems are expensive, and can only be used properly if the hospital possesses enough equipment to allow a satisfactory cleaning cycle before the next use.

Even such apparently simple equipment as mattresses, linen and beds can contribute to infection. The impermeable covers of mattresses can develop holes, allowing moisture and bacteria to enter. Linen that is washed at too low a temperature, or not pressed, can be contaminated by spore-bearing organisms. Low skin-pressure beds may be made of foam components, plastic beads or air-filled rubber bubbles, all of which can accumulate fluid and pathogens if not properly maintained and cleaned. All of these have caused outbreaks of skin infection, sometimes with systemic extension, often in patients with burns, skin grafts or immunosuppressive conditions.

Isolation facilities in hospitals

There are two main types of patient isolation, designed for different purposes.
• In source isolation the aim is to ensure that the organisms infecting or colonizing the patient are not transmitted to other patients or staff.
• In protective isolation the aim is to prevent organisms being transmitted to patients with special susceptibility to infection.

Source isolation

There are several types of source isolation depending on the infection that is being controlled. Until recently, isolation protocols were divided into groups depending on the routes of transmission (including blood, wound and enteric, and respiratory isolation). It is simpler, and mistakes are less likely, if a universal type of isolation is practised for all infected patients, and this is now becoming the rule in many Western hospital settings.

More stringent, high-security isolation is occasionally indicated for dangerous infections, whose treatment is difficult and which may be transmitted to medical carers. Only a few viral haemorrhagic fevers, rare bacterial pneumonias such as plague, and some fungal pneumonias fall into this category. Patients with infectious multidrug-resistant tuberculosis or suspect SARS or pandemic influenza are also cared for in special facilities.

Blood, wound and enteric isolation

The purpose of blood, wound and enteric isolation is to prevent the transmission of organisms normally spread by contact or ingestion. Patients are nursed in a side room that contains a wash-hand basin and preferably a separate toilet facility. Nursing and medical staff remove white coats before entering the room and put on an apron or gown. The apron is discarded and hands are washed before leaving the room. Gloves are worn when handling the patient's body fluids or excreta.

Infections that can be contained by blood, wound and enteric precautions

1 Most bacteraemias.
2 Localized infections and wounds of the skin.
3 Infections with methicillin-resistant *Staphylococcus aureus*.
4 Viral hepatitis.
5 Infectious diarrhoeas.

Respiratory isolation

The precautions taken here are similar to blood, wound and enteric precautions, but the patient's room should be ventilated by a system that extracts the air to the exterior, and does not permit air flow into other ward areas. If the patient is transferred to another department of the hospital for further treatment or investigations, such as the radiology department, the patient should wear a face mask. The mask does not prevent contact with organisms suspended in droplet nuclei, but it does provide a barrier against gross contamination by fragments or large droplets of sputum that may be produced during coughing. Nurses or physiotherapists having close facial contact with a patient may wear face masks for similar reasons. The recent SARS epidemic illustrated the need for effective masks and gowns when managing these patients in addition to care in removing potentially contaminated personal protective equipment. As a result of this, many hospitals now have a policy to provide dust-mist respirators (face masks made of material that filters inspired and expired air) for use in respiratory isolation.

Contamination of the air outside the patient's room can be minimized by keeping the door closed. This will also ensure that the ventilation system is not overloaded by extraneous air flows.

Infections that can be contained by respiratory isolation

1 Pulmonary tuberculosis (sensitive organisms, for first 2 weeks of treatment; resistant organisms, until sputum smear is negative).
2 Varicella and herpes zoster.
3 Measles.
4 Pertussis.
5 Diphtheria.
6 Skin infections in patients with exfoliative conditions.
7 Respiratory infections such as pandemic influenza or those caused by unusually resistant organisms, e.g. penicillin-resistant *Streptococcus pneumoniae*.

Control of methicillin-resistant *S. aureus*

Infection with MRSA seriously limits options for antimicrobial therapy, as methicillin resistance is usually associated with resistance to other agents. Many studies show that the mortality rate associated with MRSA is higher, and the length of hospital stay is longer, than for patients with susceptible *S. aureus* strains (resulting in significantly increased healthcare costs). Acquisition of MRSA colonization carries the highest relative risk for developing a hospital infection. Staff hands are the main vehicle of transmission but some individuals ('staphylococcal dispersers') release air-borne infectious particles, heavily contaminating the environment. Infected patients and, where possible, carriers should be managed in single-room accommodation with blood, wound and enteric precautions (see above). An isolation ward is often required in hospitals where MRSA numbers are substantial. MRSA transmission is likely to occur if wards are overcrowded and antibiotic prescribing is uncontrolled. Infection control policies such as hand-washing may be neglected if staffing levels are too low or many temporary staff are employed. The stringency of control activities will vary with the nature of the hospital (e.g. nursing home or high-dependency environment) and the degree of patient susceptibility.

Hospitals with rare MRSA cases

In hospitals in which MRSA emerges rarely, efforts should be made to prevent it being established. Central to this is the universal acceptance and implementation of the infection control policies (staff must actually wash their hands between each patient contact). Alcohol gels may be useful in encouraging staff to decontaminate their hands between patients as the gel causes less trauma than soap to the carer's skin, and its use takes less time than hand washing. The implementation of this approach has caused a reduction in MRSA transmission in some hospitals. There is good evidence that uncontrolled antibiotic prescribing provides ideal conditions for MRSA to become established. An emphasis on appropriate prescribing is therefore an essential part of controlling the transmission of resistant organisms. Patients admitted from other hospitals or from overseas should be isolated and screened, and those found to be colonized or infected must remain isolated.

Hospitals with endemic MRSA

When MRSA is endemic, a graded approach may be employed depending on the degree of risk to the patient.
• In minimal risk areas, such as psychiatric units or long-stay care of the elderly units, isolation is not necessary.
• In low-risk areas, including most medical wards, acute care of the elderly wards and non-neonatal paediatric units, colonized patients should be isolated and eradication of MRSA from the ward should be attempted.
• In moderate risk areas, including general surgery, urology, dermatology, neonatal, obstetrics and gynaecology wards, screening of patients and staff should be performed if there are more than two cases, carriers should be isolat-

ed and antiseptic detergent used for patient washing.

• In high-risk areas such as intensive care units, special care baby units, burns units, transplantation services, and orthopaedics, trauma and vascular surgery wards, patients should be screened on admission, transfer to another ward area, and on discharge. All colonized patients should be isolated, and staff should be screened. If there is evidence of transmission, surgical antibiotic prophylaxis regimens may be changed to include a glycopeptide. When there is evidence of continuing transmission, ward closure may be necessary to interrupt the chain of contacts, but this should be managed in consultation with clinicians, managers and the infection control team. Physical control measures must be supplemented with effective antimicrobial prescribing policies.

Screening procedures for MRSA

Patients may be screened for MRSA to investigate possible transmission in the hospital environment. The nose, perineum and any susceptible skin site, such as a venepuncture site or tracheostomy, as well as any infected area or burn, should be sampled. Staff should be screened if there is evidence of continuing transmission in the face of effective physical control measures. Exfoliated skin areas, due to psoriasis, dermatitis or eczema, are at high risk of persistent colonization by MRSA.

There are a number of different laboratory approaches to the detection of MRSA (see also Chapter 3). Selective media often use a high salt content, which *S. aureus* can tolerate, to reduce the growth of commensal organisms. An indicator system such as mannitol and phenol red may be used to indicate colonies for further identification. Definitive identification of MRSA is made by confirming the presence of coagulase and DNase together with a test of methicillin resistance.

Some laboratories have introduced DNA amplification techniques based on the *mecA* gene to speed up detection and identification. A simple rapid alternative is a latex agglutination technique that detects the *mecA* gene product PBP2′. A new real-time PCR method enables MRSA to be identified in carriers within hours of presentation and it is being evaluated in routine use.

It is important that MRSA are typed so that the epidemic can be monitored and new strains identified. First isolates from a new outbreak can be referred to the Health Protection Agency reference laboratory for detailed typing.

Control of drug-resistant pulmonary tuberculosis in hospitals

Patients who are sputum-positive for resistant *M. tuberculosis* should be managed in single-room isolation until two or three successive sputum smears are negative for acid–alcohol bacilli. The patient's room should be at neg-

ative pressure relative to the adjacent hospital areas. To reduce the bacterial load in the room, air should be extracted to the exterior to be replaced by clean air that will dilute organisms in the room (the US Centers for Disease Control recommend at least six complete air changes per hour).

Susceptible visitors to the room should wear close-fitting, dust/mist filter masks that cover nose and mouth. A 'negative-pressure' room should be used for procedures such as obtaining induced sputum specimens or performing bronchial lavage, bronchoscopy or upper GI endoscopy.

Strict isolation methods have recently been applied to the control of multidrug-resistant tuberculosis, after the occurrence of outbreaks in patients and care workers (Table 23.1). Patients have been nursed in negative-pressure single rooms and this has assisted in the control of epidemics. In the UK, BCG vaccine is offered to all tuberculin-negative healthcare workers, and this may have helped to protect them from occupational infection (see Chapter 18).

High-security isolation

Highly secure isolation is provided by rooms in which the air flow is strictly controlled, and all air leaving the room is filtered to remove droplet or viral particles. Healthcare staff wear protective clothing, including face protection, gloves, trousers and boots, which provide a barrier to personal contamination. In exceptional cases, mechanically filtered respiratory protection is indicated, or the patient may be cared for in a filtered environment such as a bed isolator. All infectious waste from such units is decontaminated by heat or chemical methods before leaving the area.

These facilities are only available in specialist units. The only conditions that might normally require such strict isolation are viral haemorrhagic fevers transmissible from person to person (Crimean-Congo, Ebola, Lassa and Marburg virus infections, and some rarer, related arenavirus infections; see Chapter 24). Small-scale, negative-pressure filtered isolators are available for the laboratory handling of specimens containing dangerous pathogens (Fig. 23.2).

Protective isolation

Protective measures are used for patients who are highly susceptible to infection (see Chapter 22), though many of the infections that they acquire originate from their own colonizing flora or latent organisms. Patients with severe neutropenia are nursed in protective isolation when their neutrophil count falls below 0.5×10^9/l. Protection is not only physical, in the form of single-room isolation, with filtered air to reduce the risk of *Aspergillus* infection. It must also include arrangements to control the risk of in-

Table 23.1 Minimum requirements for the isolation of patients with suspected or proven tuberculosis

Type of disease	Type of patient/contacts	Infectious	Potentially infectious*	Non-infectious
Drug-sensitive disease	Other patients immunocompetent	Single room	Open ward	Open ward
	Other patients immunocompromised	Negative-pressure room	Single room	Open ward
Drug-resistant disease	Other patients immunocompetent	Single room	Open ward	Open ward
	Other patients immunocompromised	Negative-pressure room	Single room	Open ward
Multidrug-resistant disease	Other patients immunocompetent	Negative-pressure room	Single room	Open ward
	Other patients immunocompromised	Negative-pressure room	Negative-pressure room	Single room

*Three negative consecutive smears but one or more cultures positive or cultures awaited. From the Interdepartmental Working Group On Tuberculosis 1998; available from the Department of Health.

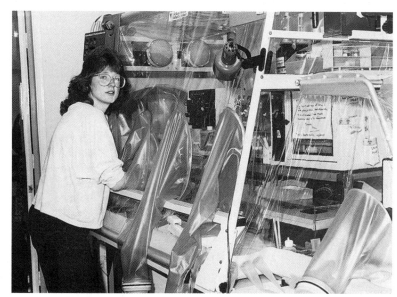

Figure 23.2 Negative-pressure filter isolator for high-security laboratory work. This laboratory contains an incubator, an automated blood count machine, a cassette-based biochemistry analyser, a coagulometer, a blood mixer, a microscope and a centrifuge in a bench area of 2 m².

fection from such sources as Gram-negative organisms in food or gifts of flowers, or *Listeria* in soft cheeses.

Maintaining effective isolation

Often the most difficult aspect of all these forms of isolation is ensuring that healthcare staff adhere to the effective policies. Large ward rounds, many ancillary staff, numbers of untrained voluntary workers and crowds of visitors all contribute to transmission of infection, and busy staff may often neglect the simple precautions of hand washing when work pressures become intense.

Prevention of infection in laboratories

In European legislation, pathogens come under the heading of biological agents and are classified into categories depending on the hazard that they present to the people who work with them. Where they are handled in the workplace, an assessment of the resulting risk must be made, and protocols to minimize this are designed in accordance with the provisions of the Control of Substances Hazardous to Health (COSHH) regulations.

Definition of a biological agent

A biological agent is: any microorganism, cell culture or human endoparasite, including any that have been genetically modified, that may cause any infection, allergy, toxicity or otherwise create a hazard to human health.

In the UK the Advisory Committee on Dangerous Pathogens (ACDP) is a committee of the Health and Safety Executive, which advises on all aspects of occupational and public health in relation to dangerous biological agents. This includes advising on the design and operation of laboratories and other workplaces, depending on the type of organisms handled. Pathogens may be classified according to the recommendations of the ACDP, and other countries have similar systems. There are four hazard groups.

Summary of the hazard groups for microbial pathogens (Table 23.2)

- Category 1: an organism that is unlikely to cause disease in immune-competent humans.
- Category 2: an organism that may cause human disease, and that might be a hazard to laboratory workers, but is unlikely to spread in the community. Laboratory exposure rarely produces infection and effective prophylaxis or treatment is usually available.
- Category 3: an organism that may cause severe human disease and presents a serious hazard to laboratory workers. It may present a risk of spread to the community, but there is usually effective prophylaxis or treatment available.
- Category 4: an organism that causes severe human disease and is a serious hazard to laboratory workers. It may present a high risk of spread to the community and there is usually no effective prophylaxis or treatment.

Laboratory safety requires a well-designed laboratory suite. Attention must be paid to the materials employed in floors, walls and benching, and to the provision of services, including water supply and ventilation. All units where work on hazard Category 2, 3 or 4 organisms is intended must inform the Health and Safety Executive before work commences. For hazard Category 4 agents, the agents must be named.

Containment level 3 facilities must be provided for any laboratory likely to isolate and handle organisms in hazard group 3 or that examines specimens that might contain such organisms (e.g. sputum). A hospital laboratory suite may contain many laboratories and facilities, such as media preparation rooms, autoclave facilities, an incubator room and a cold room. Each of the individual laboratories within the clinical microbiology suite should conform to containment level 2 or containment level 3 (COSHH regulations).

A containment level 3 laboratory should be sited away from the main work of the department and access must be limited to authorized personnel who are trained in the use of the rooms and in the manipulation of hazard group 3 organisms. The doors should be locked when the laboratory is not in use. A continuous air flow through the laboratory, and away from other laboratory areas, must be maintained when work is in progress. A system must operate to prevent positive pressurization of the room if the extraction fans fail. Reverse air flows through the ventilation system must also be prevented. A microbiological safety cabinet of class 1 or class 2 must be available and all procedures where cultures or specimens are manipulated must be performed in this cabinet, which should be exhausted through a high-efficiency particulate air (HEPA) filter to the outside air. The laboratory should be sealable so that it can be fumigated. Ventilation, filters and cabinets

Table 23.2 Examples of the categorization of pathogens and their laboratory handling.

Category	Pathogen	Laboratory procedures
Category 2	*Staphylococcus aureus* *Streptococcus pyogenes* *Streptococcus pneumoniae* *Escherichia coli* Cytomegalovirus	Open, easily cleanable bench with adequate workspace; dedicated working overalls; separate rest area; no eating, drinking, smoking etc. in the laboratory; hand-washing facilities available in the laboratory
Category 3	*Salmonella typhi* *Shigella dysenteriae* *Brucella* spp. *Mycobacterium tuberculosis* Hepatitis B virus	As above, plus separate room dedicated to this category; manipulations must be performed in class 1 safety cabinets; dedicated overalls; hand-washing facilities in room; ventilation by air extraction to exterior
Category 4	Rabies virus Lassa virus Marburg virus Ebola virus	As above, but separate unit away from general circulation; HEPA-filtered, negative-pressure ventilation; all work in class 3 cabinets or laboratory isolator; all laboratory waste and effluent disinfected before leaving unit

HEPA, high-efficiency particulate air.

must be regularly maintained and tested, to ensure continuing adequate function.

Safety cabinets

Class 1 cabinets
Class 1 cabinets, or exhaust-protective cabinets, are simple in design, with air being drawn through the face of the cabinet and out through the HEPA filter to the outside air.

Class 2 cabinets
Air is drawn into the cabinet through a HEPA filter and directed downwards on to the work surface. A portion of the filtered air is extracted. Class 2 microbiological safety cabinets provide protection to the operator and also to the work. They are therefore suitable for tissue culture, where contamination of cell lines must be minimized.

Class 3 cabinets
These are similar in clinical design to class 1 cabinets, except that they are fully enclosed and the operator works through glove ports (Fig. 23.3). Air is drawn into the cabinet and exhausted through a HEPA filter. This type of cabinet provides maximum protection to the worker from aerosol hazard. Some argue that the need to manipulate all materials and equipment with gloved hands increases the risk of accidental self-inoculation when needles and other sharp implements must be used.

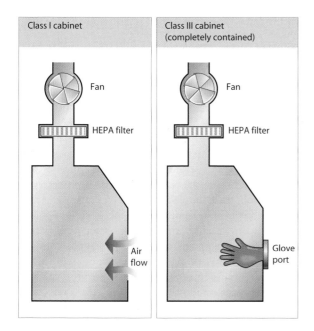

Figure 23.3 Class I and class III exhaust-ventilated cabinets. HEPA, high-efficiency particulate air.

Containment level 4

Containment level 4 laboratories are rare and are usually sited at national reference or research laboratories. They are operated on the basis of complete security of the material used in the laboratory. Laboratory workers completely change their clothes before entering the laboratory via an air-lock while work is contained in class 3 cabinets and there is a negative pressure between the laboratory, the air-lock and the outside. Air enters through HEPA filters and is extracted through a pair of HEPA filters. A double-sided interlocked autoclave ensures that all material leaving the laboratory is rendered safe. The worker should be visible within the laboratory through glass panels and an intercom or telephone system provided with an additional competent person available to assist in emergencies. Respirators must be available for emergency use.

Control of infection in hospitals

Appropriate organization and effective management protocols are essential to the control of hospital infection. These arrangements promote good clinical practice to minimize the occurrence of infection, promote the teaching and use of universal precautions and other infection control methods, and allow a coordinated response to outbreaks when they arise. The control of infection organization has two main strands: the Control of Infection Committee and the Control of Infection Team.

Control of Infection Committee

The Control of Infection Committee is generally a subcommittee of the senior medical committee or chief executive of the hospital, and is empowered to develop and implement infection control policies and procedures. The chairperson of the committee is usually a consultant microbiologist or a consultant of equivalent expertise, such as a consultant in infectious diseases or communicable disease control. Other key staff include a hospital manager, and senior representatives from hospital services involved in infection control procedures (operating theatres, sterile supplies, nursing staff, cleaning and catering services, maintenance services and medical, intensive care and surgical departments). The committee should be able to co-opt those professionals whose expertise or executive action are required.

Outbreak control committee
When an outbreak occurs, the Control of Infection Committee may meet to coordinate a response, although in general a smaller group is often more practical, and can

include clinicians and nursing staff from the affected area, non-clinical hospital staff, and a community liaison or press officer.

Control of Infection Team

The Control of Infection Team is the core group whose role is to implement the control of infection policy, and to monitor its effectiveness. This team usually contains the Control of Infection Officer (CIO), who is also the chairperson of the Control of Infection Committee. National authorities are increasingly insisting that the CIO has completed a diploma course in hospital infection control. In addition there are control of infection nursing staff, and these often include ward-based control of infection 'link' nurses. They work together to manage the control of infection policy on a day-to-day basis, to collect statistics on infection rates and to identify any problems. The team monitors compliance with the policy at ward level and implements emergency control measures when indicated. It undertakes specific surveillance of particular organisms, such as MRSA in surgical units, *Klebsiella* endemicity in ITUs and antibiotic-resistant urinary pathogens, and organizes screening of suspected carriers. It also provides assistance and advice to hospital staff on current policy, on appropriate infection control policies for any new planned activities or facilities and on developments made necessary by changes of activity or procedures in various hospital departments.

The remit of the Control of Infection Team and Committee is necessarily wide and impacts closely on the activity of many clinical and other departments. This is because apparently trivial factors can have a profound effect on transmission of microorganisms in the hospital environment.

Control of infection standards

Examples of these have been drawn up by the Association of Medical Microbiologists, the Hospital Infection Society and the Infection Control Association in the Health Protection Agency. Governments issue criteria for management structure and responsibilities related to control of infection and often use this as a mechanism to guide funding or establish standards. It recommends policies and procedures for appropriate microbiological services, surveillance for the control of infection, and relevant education policies.

This requires that an infection control structure with sufficient resources and clear lines of responsibility should be set out. This responsibility lies directly with the senior management of the hospital. A Control of Infection Committee and effective Control of Infection Team should be constituted. The plans for controlling outbreaks of infection should be laid down by the Control of Infection Committee. The remit is very wide, involving all aspects of the hospital's work, including mortuary services, sterile supply services, hotel services, engineering, disposal of waste products and purchase of equipment. None of these can be supported without an adequate microbiological laboratory. Regular surveillance for infections should be in place. Education is the main method of preventing transmission of infection, and the Infection Control Team should be central to the education of medical, nursing and paramedical staff. Adequate resources and staffing must be provided by the hospital management to ensure that this important task can be carried out.

Control of an outbreak

With effective control of infection procedures in place, outbreaks ideally should be the exception. They are inevitable, however, because of the nature of the hospital environment, admitting patients from the community who may be incubating diseases such as *Salmonella* infections or chickenpox, or receiving specialist referrals from other hospitals where particular 'problem' organisms may be common. Each department must have a plan for responding to outbreaks that can be foreseen, such as *Salmonella* infections, MRSA, legionnaires' disease and multidrug-resistant Gram-negative organisms such as multidrug-resistant *Acinetobacter*.

There are three main strands to controlling outbreaks within the hospital. Once a true outbreak has been confirmed, they are:

- to identify the source or reservoir;
- to halt the transmission; and
- to modify the host risk.

Identifying sources and reservoirs

Reservoirs of infection in hospitals are identified by screening patients within the hospital environment or before admission to hospital. Patients coming from hospitals with known epidemics, e.g. MRSA, may be screened for the presence of this organism and isolated until shown to be negative. Where the environment is a likely reservoir of infection, for example in legionnaires' disease, it may be necessary to sample the water supply at various points. Gram-negative bacteria may have their reservoir in particular hospital equipment, such as water baths, mattresses or endoscopes. When blood-borne viruses are involved, intravenous infusion, sampling or injection protocols may have been violated. Outbreaks such as MRSA infections may continue by person-to-person spread. Testing other patients and staff then becomes important in identifying the source of the problem.

Halting transmission

This is the most difficult aspect of infection control as it involves improving routine procedures such as hand washing and asepsis, isolating colonized and infected patients, and modifying the way in which patients are nursed. Cohort nursing or targeted nursing may allow individual groups of infected patients to be nursed by a team not involved with caring for other, non-infected patients. It may be necessary to stop admitting new patients until the situation is controlled.

Modifying host factors

It is also difficult to modify host risks. Some of these change naturally as patients recover from operations or can cease immunosuppressive treatment. Where epidemics of multidrug-resistant organisms are a problem, it may be possible to modify the host risk by controlling antibiotic usage. This will require liaison between the prescribing clinicians and the microbiology and infection control teams. Several studies show that strict control of antibiotic prescribing can tip the balance against MRSA and halt the spread of this organism.

Hospital cleaning and disinfection

These are two important aspects of the Control of Infection Committee's work. If the hospital is clean and its equipment and special areas adequately disinfected, staff can work safely and the risk of spreading infection is minimized.

Disinfection

Disinfection is the removal of sufficient microbial contamination from equipment to allow its safe use. This may range from the cleaning and disinfection of a vacated bed to the removal of all microbial contamination from a reusable surgical instrument.

Disinfection by cleaning

This is an extremely effective way of decontaminating floors, furniture and ordinary work surfaces. It entails the removal of dust and organic matter by wiping or washing with detergent solution, and then wiping dry with a clean cloth. Machine-washing cutlery and crockery at an adequate heat (above 80 °C) is a good method of decontamination, as is washing linen at an adequate temperature (above 75 °C). Steam-pressing bed sheets and towels is sufficient to remove many bacterial spores.

In some circumstances additional treatment with disinfectants (various types of antibacterial agents) is necessary to reduce bacterial contamination further. There is a vogue for using commercial cleaning systems using hydrogen peroxide vapour to try to interrupt transmission of multidrug-resistant organisms.

Types of disinfectants used for different purposes

1 Chloros (sodium hypochlorite or bleach) is an oxidizing agent that kills most vegetative bacteria and many viruses. A 1% solution is used for cleaning surfaces. For disinfecting spillages, a 10% solution is poured over, before they are wiped up, or absorbed into granules and swept up. Chloros is toxic to humans and corrodes many metals.

2 Halogen disinfectants (chlorides and iodides) have low toxicity to human skin. They kill many bacteria, and iodine can kill spores. They are used as skin washes, scrubs and disinfectants. Chlorhexidine (containing chloride) and Betadine (containing iodine) are extensively used in intensive care units and operating theatres. Chlorine dioxide is useful for decontaminating surfaces in the laboratory, including category 3 containment facilities, in view of its excellent activity against *M. tuberculosis*. It can be used to replace glutaraldehyde, to which staff can develop hypersensitivity. It works by the creation of metastable chlorous acid in solution with release of chlorine dioxide upon contact with the microorganism. Due to structural differences between bacteria and mammalian cells it is not toxic to human cells.

3 Alcohols act rapidly to kill vegetative bacteria, many viruses and fungi. They may be used as sprays to disinfect surfaces, such as trolley tops, and also as rubs for rapid hand disinfection. Halogen disinfectants can be dispensed as alcoholic solutions, for additional bactericidal effect. Alcohols are easily diluted to below effective concentrations by evaporation, and penetrate organic matter poorly. Their flammability makes them too dangerous to use near diathermy units and other operating theatre equipment.

4 Aldehydes are non-corrosive and kill a wide range of organisms. Glutaraldehyde (Cidex) solution is widely used in automated decontamination systems. Sealed rooms and large equipment are occasionally decontaminated by fumigation with formaldehyde dissolved in water vapour.

5 Phenol-based disinfectants such as Hycolin are non-corrosive, and highly toxic to microorganisms but are not very active against viruses. They are used to disinfect contaminated surfaces such as bed frames, bathroom equipment and the floors of ambulances. Suspensions of tarry phenolics, such as Sudol, are used for cleaning highly contaminated stone and ceramic floors, such as mortuary areas. They are highly microbicidal, but leave a residue that is hard to remove from instruments and equipment.

6 Ethylene oxide is a gas that kills bacteria, viruses, fungi and bacterial spores. It can penetrate complex instruments, and is used for cleaning laparoscopes, arthroscopes, reusable cardiac catheters and delicate surgical instruments. It is applied at low pressure, usually mixed with carbon dioxide to avoid risk of explosion, and at a temperature of 55 °C. Instruments must be aired after treatment to remove the irritant gas.

7 Peroxygen compounds are disinfectants with a broad spectrum of activity, including bacteria, fungi and viruses. They exert their effect through strong oxidative activity. They exert an adequate effect on a broad range of surfaces.

⚠ Disinfectants do not work adequately in the presence of organic matter, which may degrade them. Debris can also prevent the disinfectant from reaching its target surface. All equipment should be cleaned or washed before being disinfected. Recent research suggests that treatment with protease-containing solutions can destroy residual contamination with prions.

Physical methods of disinfection

Heat is the most commonly used physical disinfection method in hospitals. Superheated and pressurized steam is used in autoclaving. Porous material such as operating gowns and drapes are prepared for use by this method. Stainless steel and some plastic and rubber equipment can also be autoclaved.

Adequate sterilization is a function of temperature and time of exposure. In an autoclave, air is drawn out of the load by creating a vacuum, and is replaced by pressurized steam. This process is repeated in five to eight cycles. The temperature in the vessel is then held, usually at 134 °C, for 3–20 min. An extended cycle can be used for potentially contaminated material from patients with CJD.

The effectiveness of the cycle can be tested by including heat-stable bacterial spores in the load, and showing that they do not germinate after treatment. The attainment of adequate temperatures can be shown by using autoclavable tape to wrap the load; the tape shows a colour change at the correct temperature–time combination. Equipment can be prepacked in semipermeable paper or plastic wrappers, and the whole package autoclaved. The equipment is then sterile until the pack is broken.

Closed jars and bottles would rupture in the waves of heat and pressure produced by standard autoclaves. Special autoclaves with balanced pressure and heat cycles must be used for these.

Gamma radiation is widely used in industry to remove microbial contamination after manufacture from plastic, silicon and rubber equipment, usually after packaging. As the gamma rays penetrate deeply, the method can be used to sterilize fragile and complex equipment such as cardiac catheters, pacemaker controllers and complex interventional radiology equipment.

Hospital waste disposal

A hospital produces a huge amount of waste. This includes:
• domestic-type waste from kitchens, washrooms, dining facilities and public areas;
• clinical waste such as used dressings, wound drainage, used disposable equipment and even organs and limbs from the surgical department;
• discarded used 'sharps' such as injection needles, lancets, stitch-removers and scalpels.

Domestic waste can be removed by local authority services and disposed of in various ways without hazard to disposal workers or the public. Clinical waste must be safely packed and clearly identified. In the UK it is put into strong, yellow plastic sacks at the site where it is generated. These sacks are stored in strong bins or skips until they are removed intact for incineration locally, or by a licensed disposal firm.

Used sharps are disposed of directly into strong, leak-proof bins at the site where they are generated. These bins are sealed when full, and are removed intact for storage and incineration.

Laboratory waste contains high concentrations of pathogens. All used containers, media and disposable equipment are autoclaved before they are either disposed of in the hospital waste system or recycled for laboratory use. Laboratories use the same sharps disposal system as the other hospital areas.

Case study 23.1: Hernia causes a pain in the neck

History

A 69-year-old man was admitted to hospital for a routine herniorrhaphy. His general health was good, except for moderate spinal osteoarthritis. He had not smoked tobacco since giving up cigarettes 10 years earlier. The anaesthesia and operation proceeded uneventfully, but 3 days later he noticed soreness and a slight discharge of pus from his wound. Swabs were taken, and a growth of methicillin-resistant *Staphylococcus aureus* (MRSA) was identified on culture. The wound was cleaned twice daily with chlorhexidine lotion, the discharge ceased and healing continued.

Further progress and investigation

One week later, he complained of intense aching in his lower neck, and pain in the medial side of his right upper arm. Examination showed temperature 38 °C, limitation of neck rotation, and pain on neck movement. Neurological examination was normal. X-rays of the cervical spine showed only osteoarthritic changes. On the following day the patient remained febrile. He complained of weakness of the legs, and developed acute retention of urine, with bladder volume 750 ml. An urgent MR scan of the cervical and thoracic spine was carried out (Fig. CS.6). The white blood cell count was $15.6 \times 10^9/l$, with 85% neutrophils.

Questions

- What is seen on the MR scan?
- What bacterial aetiology must be included in the differential diagnosis?
- What immediate treatment should be commenced?
- What further management is indicated?

Further management and progress

The MR scan showed oedema and a soft-tissue mass surrounding the second thoracic vertebra, with partial destruction of the vertebral body, and loss of the T1/T2 disc space (arrowhead). A pale mass, representing an extradural abscess, is surrounding and compressing the spinal cord (arrow). Blood cultures were taken, and antimicrobial therapy commenced with teicoplanin, cefotaxime and metronidazole. (The glycopeptide antibiotic was considered essential, because of the recent isolation of MRSA from this patient.) The neurosurgical team arranged immediate surgical decompression and drainage of the abscess.

Culture of pus from the abscess cavity produced a pure, heavy growth of MRSA. The blood culture remained negative, and the patient made a complete recovery after a total of 6 weeks parenteral therapy.

Comment

Although colonization and infection of skin and small wounds is often a trivial problem, hospital-acquired resistant organisms can cause severe and life-threatening disease if they invade systemically. This patient must have acquired his vertebral abscess by blood-borne spread from a brief bacteraemia, probably with lodgement of MRSA in the mildly traumatized osteoarthritic tissues of the neck. He was fortunate that earlier investigation had revealed that he was colonized by a resistant organism, so that effective empirical treatment could be chosen.

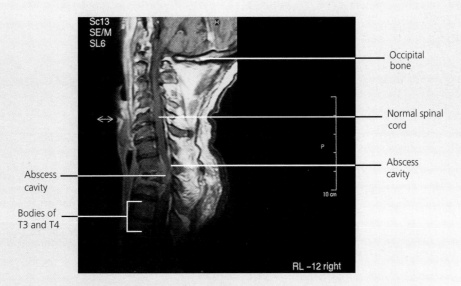

Figure CS.6 MR scan showing oedema and partial destruction of second cervical vertebra with loss of the T1/T2 disc space, and a pale abscess mass surrounding and compressing the spinal cord.

Travel-associated and Exotic Diseases

24

Introduction

Worldwide travel has increased enormously in the last 25 years, as travel has become easier, quicker and more affordable (Fig. 24.1). Air travel is rapid, allowing movement around the world in a shorter time than the incubation period of almost any disease. Only the most remote or hostile areas are inaccessible to tourism, but even these may be entered by journalists, aid workers and researchers such as geologists, anthropologists and ecologists. Overland travel is popular, as is the experience of sharing unfamiliar living conditions with local people. Infectious diseases are found in all areas of the world, depending on local conditions of ecology, epidemiology and public health. Outbreaks of respiratory or intestinal infection can also occur in the partly closed environment of large hotels or cruise ships.

Travellers may be vulnerable to infections unfamiliar to their home-based medical services, presenting diagnostic problems, difficulties in management and unexpected complications. Contagious diseases are also hazardous to contacts and can be serious public health problems.

Unfamiliar features of imported diseases
1 Epidemiology.
2 Presenting features.
3 Diagnostic methods.
4 Management requirements.
5 Unexpected complications.
6 Unexpected infectiousness.

Factors that increase the vulnerability of travellers
1 The temptation to take risks with food, water, animals and sexual contacts when relaxing away from the conventions of home.

2 The different epidemiology of some diseases in different environments (e.g. heterosexual versus homosexual transmission of HIV, prevalence of open pulmonary tuberculosis or existence of epidemic diseases such as diphtheria).
3 The incomplete understanding of health hazards and protective measures with which travellers often arrive at a destination.
4 The stress that accompanies long journeys across time zones, which may make travellers unusually susceptible to some diseases.
5 In the case of refugees: privation, malnutrition and preexisting disease or injury, which may widen the range of infections to which they are vulnerable.

Disease list

Rather than being listed by organism, diseases of travel can conveniently be considered in aetiological or epidemiological groups.

Diseases common worldwide
- Influenza
- Community-acquired pneumonias
- Urinary tract infections
- Meningococcal disease
- Sexually transmitted diseases

Infectious diseases related to climate and environment
- Dermatophyte infections
- Folliculitis
- Skin infections related to arthropod and other types of bites
- Skin infections caused by marine organisms

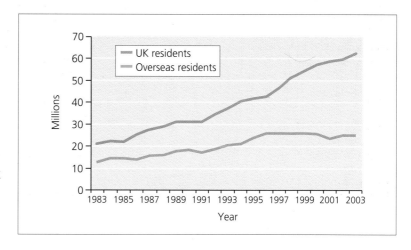

Figure 24.1 Journeys to and from the UK 1983 to 2003. Source: National Statistics website: www.statistics.gov.uk. Crown copyright material is reproduced with the permission of the Controller, HMSO.

Diseases controllable by public health measures

Sanitation, food hygiene and safe drinking water

- Hepatitis A
- Hepatitis E
- Viral gastroenteritis
- Traveller's diarrhoea
- Bacterial food poisoning
- Bacillary dysentery
- Enteric fevers
- Cholera
- Giardiasis
- Amoebiasis
- Cryptosporidiosis
- Helminth infections

Immunization

- Poliomyelitis
- Diphtheria

Education

- Sexually transmitted diseases
- HIV infection

Risks of contact with mud and water

- Leptospirosis
- Hookworms
- Strongyloidiasis
- Schistosomiasis
- Liver flukes
- Guinea worms (increasingly rare)

Diseases with arthropod vectors

- Dengue fevers
- Arboviral encephalitides
- Other arboviral infections, e.g. phlebotomus fever
- Rickettsial infections
- Plague
- Lyme disease
- Malaria
- Leishmaniasis
- Trypanosomiasis
- Filariasis
- Onchocerciasis

Some important zoonoses

- Borreliosis
- Toxoplasmosis
- Brucellosis
- Hantavirus infections
- Plague
- Tularaemia
- Melioidosis
- Rabies
- Anthrax

Viral haemorrhagic fevers

- Yellow fever
- Dengue haemorrhagic fever
- Lassa fever and other arenavirus infections
- Marburg fever
- Ebola fever
- Crimean-Congo haemorrhagic fever

Many of these diseases are discussed in some detail in other chapters. Important or common diseases not mentioned elsewhere will be described in detail in this chapter. Brief information on presentation, diagnosis and management will be given for the remainder.

Diseases that are common worldwide

Introduction

However exotic a traveller's destination, common infections still pose a risk. Indeed, crowding, stress, insect bites, altered personal hygiene facilities, fatigue and altered patterns of hydration may predispose to clinical expression of common infections.

Some epidemic diseases may be active at the destination while in abeyance at home, as epidemic cycles wax and wane around the world. Influenza, enteroviral meningitis, and meningococcal disease are examples.

Influenza prophylaxis

Pandemics of influenza A often begin in eastern Asia and spread westwards across the world, following the winter season, first southwards to Australasia and then northwards to Europe and North America. World Health Organization laboratories maintain surveillance during the influenza season every year, to warn of the emergence of new epidemic strains. Appropriate vaccines can quickly be constructed.

Immunization is useful for travellers who will enter areas of influenza activity, and should always be offered to elderly individuals, or those with special susceptibilities who would be given the vaccine routinely in the UK. It affords around 75% protection to healthy recipients and, although less protective against infection, reduces the risk of severe disease and death in debilitated and elderly recipients.

Prophylaxis of meningococcal disease

Group B meningococci, for which there is still no effective vaccine, remain the prevalent epidemic organisms of

Europe and North America, but in the Middle East, Africa and South America most epidemics are of group A. Since the late 1980s, Muslim pilgrims visiting crowded sites in Saudi Arabia for the Hajj and Umrah pilgrimages have been at risk of group A and, more recently, W135 disease. Local outbreaks of group A and C meningococci also occur in many countries, particularly west and central Africa.

Meningococcus vaccine, containing polysaccharide antigens of group A and C, affords good protection to travellers to endemic areas for at least 3 years after a single dose. Conjugate group C vaccine is available for children below the age of 2 years. For pilgrims to the Hajj, polyvalent A/C/Y/W135 vaccine is recommended, regardless of any previous immunization against meningococcal types. This recommendation includes children below the age of 18–24 months, though the vaccine may have reduced immunogenicity in this group.

Other respiratory infections

Elderly travellers, those with impaired immunity or preexisting disease are at risk of *Streptococcus pneumoniae* infection. In some parts of the world, penicillin and multidrug-resistant strains of *S. pneumoniae* are common. All travellers over the age of 5 years who are in groups at increased risk of pneumococcal disease (see Chapter 7) should be offered 23-valent polysaccharide pneumococcal vaccine. Susceptible children below age 5 may be offered 7-valent conjugated pneumococcal polysaccharide vaccine.

Other travellers may consider immunization, especially if they will experience predisposing situations such as dusty environments, frequent underwater diving, high altitude exposure, or work in healthcare environments in affected countries.

Legionnaire's disease is commonly contracted during travel to warm countries, where cooling towers and domestic water systems are susceptible to colonization by legionellae (see Chapter 7).

Diseases related to climate and environment

Skin problems in hot climates

Skin disorders are common infection-related problems reported by travellers. In hot climates the skin is constantly moist and easily becomes macerated or traumatized by the friction of moist clothing. Hair follicles or sweat glands may be blocked by soft keratin plugs, causing hyperaemia and papular swelling, often called a sweat rash.

Cuts from stones, sharp vegetation or wood splinters should be avoided by wearing suitable clothing when walking or working in areas of risk.

A healthy skin can also be maintained by avoiding insect attack and wearing loose, light clothes, preferably made of absorbent natural fibres. Both the clothes and the skin should be regularly washed. Antiperspirants are not as effective in hot climates as in temperate ones, and may contribute to blocking of hair follicles and sweat ducts.

In cold climates, or at high altitude, the skin can become dry and fissured. The use of gloves, attention to skin hygiene, and the application of emollient or 'barrier' creams, can help to avoid infection-susceptible lesions.

Infected arthropod bites

Insect bites or, occasionally, the bites of bugs or fleas can become the focus of spreading superficial infections. Impetigo lesions are common (Fig. 24.2), and can easily spread to other sites of mild trauma, or other family members. Staphylococcal folliculitis can also occur. Staphylococcal or streptococcal cellulites easily affect traumatized skin areas. Both conditions can cause systemic illness if they become extensive. Both streptococcal and staphylococcal lesions can lead to toxin-mediated diseases, such as scarlet fever or toxic shock syndrome, which are more common in children.

Fungal infections

Yeast and dermatophyte infections easily occur in moist or traumatized skin. Many of these are caused by pathogens

Figure 24.2 Three sisters stayed for the summer in the Caribbean with their grandmother, who was horrified by these skin lesions. They are, however, simple secondary *Staphylococcus aureus* infection (impetigo) in insect bites.

such as *Candida* spp. or common dermatophytes, which are easily treated with topical imidazole antifungal drugs, such as clotrimazole cream.

Rarely, subcutaneous infections can follow inoculation of fungal elements from vegetation, wood or soil. These tend to produce granulomatous, crusting lesions that grow slowly after a prolonged incubation period. Diagnosis depends on demonstration of the fungi, which rarely grow in artificial culture, in biopsy specimens. Treatment should be supervised by an expert. Oral itraconazole or terbinafine, sometimes with surgical excision of large lesions, is the usual mode of treatment. A few infections respond only to oral saturated potassium iodide. Treatment courses must usually be continued for weeks or months.

Other skin infections

Cuts from rock, fish spines or corals can become infected by marine organisms, such as vibrios. *Acinetobacter* spp. are also relatively common causes of skin infection in the tropics. These Gram-negative infections do not respond to oral penicillins and cephalosporins, but can often be treated orally with tetracyclines or quinolones.

In areas where diphtheria exists, skin ulcers and abrasions may be colonized or infected with *Corynebacterium diphtheriae*. Infected lesions often have a greyish membrane at the base, and a slight serosanguineous discharge. In local child populations the small dose of toxin produced in the lesion often induces natural immunity. Troublesome lesions respond rapidly to treatment with penicillins or erythromycin, but it should be remembered that such lesions are infectious. Immunity declines slowly after childhood immunization, so adult travellers may be susceptible. Visitors to Western countries from overseas may never have been immunized.

Skin infections in hot climates

1 Staphylococcal folliculitis.
2 Staphylococcal and streptococcal infections of insect bites.
3 Infections with *Acinetobacter*, pseudomonads or marine vibrios.
4 Colonization or infection with *Corynebacterium diphtheriae*.
5 Yeast and dermatophyte infections.
6 Rare, subcutaneous fungal infections.
7 Skin invasion by arthropods or their larvae.

Insect and other arthropod bites: importance and prevention

Mosquitoes, fleas and ticks not only inflict bites that may cause painful lesions in their own right; they can also become infected with aggressive pathogens, or introduce arthropod-borne pathogens, including malaria, rickettsioses and severe viral infections.

Loose clothing is a useful barrier, which can protect against the bites of mosquitoes, flies and ticks; the arms and legs should be covered at dusk when mosquitoes are highly active. The lower legs can be protected by boots and socks. Window and door screens keep insects out of buildings; the use of 'knock-down' sprays or 'mosquito coils' kills any that have entered.

In malarious areas mosquito-proof bed-nets offer important protection, which is much increased by impregnation with permethrin.

Insect repellents can be used on skin not covered by clothing (and can be used to impregnate clothing, to protect against lice and fleas). The most effective of all insect repellents is DEET (*N*,*N*-diethylmetatoluamide). This is available as creams, sprays and liquid formulations in concentrations of up to 50%. It can be used safely on the skin of individuals of all ages, and does not need to be applied thickly. It must be re-applied if sweating, swimming or rubbing remove the protective surface application. DEET is toxic if ingested. Small children should not, therefore apply the preparations to their own skin; and the containers should be kept out of their reach. Permethrin can be used to impregnate clothes, which should be dried without creasing or folding after application.

Preventing other insect attacks

Female chigger or jigger fleas lodge in the soft skin of the feet, usually between the toes, and enlarge to up to 1 cm in diameter as they fill with eggs, causing painful, nodular lesions that are susceptible to secondary infection. When small, they can be teased out with a needle. The use of footwear in sandy areas affords good protection.

Tumbu flies deposit eggs on drying laundry, and emerging larvae penetrate the skin of the wearer. As the larva grows and develops, a boil-like lesion results, in which wriggling movement can often be felt. Covering the lesion with Vaseline causes the maggot to emerge for air, when it can be grasped and removed. Ironing all laundry prevents this problem by killing the eggs before the garment is worn. Insect repellents are available that will persist in clothing after several washes.

Diseases controllable by public health measures

Traveller's diarrhoea

Introduction

Traveller's diarrhoea is the commonest infection problem reported by travellers; it is defined for research studies as the occurrence of at least three abnormally loose stools in any day, or one or more loose stools with features such as vomiting, abdominal cramps or fever. It is common, affecting from 8 to 50% of travellers, depending on local sanitary standards. Around 85% of investigated cases are proven to be caused by bacteria or respond to treatment with antibiotics; 10% are caused by parasites, and 5% of cases are probably viral. The majority of bacterial cases are caused by enterotoxigenic *Escherichia coli* (ETEC); other bacterial pathogens include entero-aggregative *E. coli*, campylobacters, salmonellae, shigellae, non-cholera vibrios and *Yersinia enterocolitica* (see Chapter 8). There have been persisting outbreaks of viral gastroenteritis, transmitted from person-to-person, affecting cruise ships. Up to half of traveller's diarrhoea cases are fatigued or bed-bound for at least 1 day.

Clinical features
Symptoms begin most often in the first week after reaching the destination. The illness lasts an average of 4 days and few patients have more than five or six diarrhoea stools per day. An attack caused by ETEC is followed by immunity to the local ETEC strain.

Management
The most important treatment is adequate rehydration, using locally available safe drinks, rehydration salts, or home-made rehydration solution (see Chapter 8). Short-term use of anti-diarrhoea agents may help to reduce inconvenience, colic and distress.

Most cases are brief and not significantly shortened by treatment with antibiotics. In highly endemic areas 40–60% of ETEC are susceptible to agents such as amoxicillin, trimethoprim, co-trimoxazole or ciprofloxacin. Elderly and debilitated patients, who may suffer prolonged or complicated illness, may be advised to take a 1- to 3-day course of oral ciprofloxacin at the first sign of diarrhoea, stopping the antibiotic when diarrhoea ceases.

Prevention
Although several antibiotics and also bismuth preparations can reduce the incidence of diarrhoea in travellers, resistant ETEC soon emerge where antibiotics are frequently used. Chemoprophylaxis is therefore not recommended unless the necessity is exceptional. Vaccines against enterotoxins are being developed, mainly for the prevention of gastroenteritis in children, but they may become a useful safety measure for travellers. Dietary precautions can contribute to prevention, especially in the short term. Travellers who suffer from diarrhoea have more often than not taken chopped fresh fruit, sandwiches with mixed fillings, raw or lightly cooked seafood and untreated water (including ice cubes).

Water safety
It is safest to avoid untreated water, including fruits and salads washed in it, and ice cubes made from it. Commercial brands of mineral water and carbonated drinks are usually safe, as are tea and coffee made with boiling water, or wines and beers sold in cans and sealed bottles. Untreated water may be purified by boiling but this is not always convenient. Chemical disinfectants such as iodine and chlorine preparations will usually kill bacteria and viruses but filtration is additionally required for some parasites including *Giardia* and *Cryptosporidium*. Some reverse-osmosis filtering systems can remove bacteria, viruses and parasites, but they are quite bulky compared with ordinary filters.

Typhoid and paratyphoid fevers

Introduction and epidemiology
These diseases, caused by *Salmonella enterica* var typhi and paratyphi A or B, are uncommon but often severe infections of travellers. They occur where sanitation is poor or drinking water is insufficiently safe. Even with modern treatment, morbidity can be considerable, and an increasing tendency to antibiotic resistance means that some cases are difficult to treat. In 2000 there were an estimated 22 million cases of typhoid worldwide with over 200 000 deaths. The UK sees between 200 and 300 cases of typhoid and paratyphoid every year, the majority of which are imported, particularly from the Indian subcontinent. Studies suggest that those people travelling to visit friends and relatives may be at particular risk.

Pathology of enteric fever
For a few days after infection, salmonellae replicate in the gut, and are excreted in the faeces. This is followed by a primary bacteraemia, which is usually asymptomatic. The organism then replicates in reticuloendothelial cells, including within the Peyer's patches. A secondary bacteraemia heralds re-invasion of the gut and the onset of symptoms. During this period blood cultures are positive, and salmonellae can be isolated from the urine in some patients. Treatment of the infection may render the blood sterile, but bone marrow cultures may remain positive until late in the course of antimicrobial treatment.

Clinical features
Typhoid fever is the typical enteric fever. The incubation period varies from 6 days to 4 weeks, but averages 2 weeks. Symptoms begin insidiously, with fever (which increases daily), headache, abdominal discomfort, constipation and often a dry cough; early typhoid fever is often mistaken for 'flu'. The pulse rate often fails to rise in proportion with

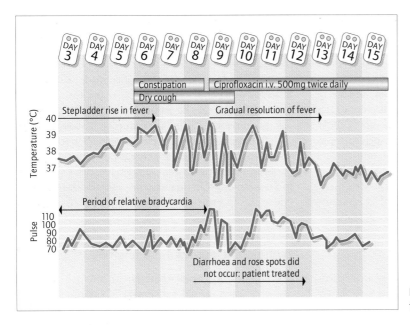

Figure 24.3 Clinical chart of the course of typhoid fever.

the temperature (relative bradycardia: Fig. 24.3). There may be a neutrophilia at this stage. Confusion is common, varying from taciturnity or bad dreams to frank delirium or apparent psychosis. Confused or psychotic patients are often very restless and may hurt themselves. Rarely, they may be mistakenly referred to a trauma clinic or psychiatric service.

After 7–10 days the fever reaches its peak; a handful of rose spots often appears on the flanks, buttocks or costal margins (Fig. 24.4), and diarrhoea begins. At this stage tachycardia develops and the white blood cell count usually shows a neutropenia.

In untreated cases complications can be expected from the second week of illness. The commonest are intestinal bleeding or perforation, usually from deeply ulcerated Peyer's patches. Bleeding may be slight, and mixed with greenish diarrhoea stools, but can be catastrophic. Similarly, small perforations can become walled off by omentum, causing temporary local signs of peritonism, which resolve with continued treatment. Large or multiple perforations require emergency surgery. Bleeding and perforation are the main causes of fatalities from enteric fevers.

Late features and complications of typhoid fever

1 Bowel haemorrhage.
2 Bowel perforation.
3 Acute cholecystitis.
4 Osteomyelitis (especially spinal).
5 Other, rare metastatic infections.
6 Relapse of the acute illness.
7 Prolonged *Salmonella typhi* excretion.

Relapse is an important feature of typhoid, with an incidence of 10–15% occurring after either treatment or spontaneous recovery. Fever recurs and blood cultures are again positive. While often less severe than the original illness, relapse is occasionally severe or fatal. It is more likely after inadequate treatment, and rare after successful treatment with ciprofloxacin or ceftriaxone.

Less common presentations

Less common presentations of enteric fevers are important. Paratyphoid A closely resembles typhoid, except that rose spots are rarely seen. Paratyphoid B may have an incubation period of 4–5 days. It is usually a bacteraemic and diarrhoeal disease from its onset, the greenish watery stools becoming bloody as the feverish illness progresses.

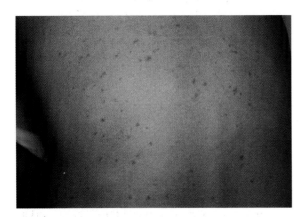

Figure 24.4 Rose spots on the ninth day of typhoid fever.

In this disease a widespread rash of rose spots often develops. Some patients with enteric fevers have only persisting fever, fatigue and malaise but remain ambulant; they are at risk of late complications.

Typhoid fever is often atypical in children. The fever does not follow the classic evolution but is often high, swinging and persistent, while bowel signs and symptoms may be slight or absent. Splenomegaly is common in late-presenting cases. Signs of pneumonia or sepsis may predominate in babies.

Possible presentations of typhoid fever in infants and children

1 Complications of high fever.

2 Persisting fever with splenomegaly.

3 Acute respiratory disease and/or sepsis in babies.

Respiratory features are common in all age groups. Cough is a frequent symptom and the chest X-ray sometimes shows segmental or nodular pneumonitis; *S. typhi* may be isolated from sputum. Rare features of typhoid include acute cholecystitis, or osteomyelitis, which particularly affects the lumbar spine.

Clinical diagnosis

The evolution of clinical features often suggests the diagnosis, particularly if typical rose spots appear.

Laboratory diagnosis

In common with other Enterobacteriaceae, salmonellae can be readily cultured on simple nutrient-selective media.

In cases of suspected enteric fever blood, urine and faeces should be submitted for bacteriological culture. In difficult cases, or those recently treated with antibiotics, bone marrow culture can yield positive results. Blood culture has a sensitivity of approximately 60%, and bone marrow 80%. Detection of positive isolates may be delayed using conventional culture systems, but in automated systems, positives can be detected within 24 h and subculture commenced. Isolation of *S. typhi* in faeces alone must be interpreted with caution, as it may indicate asymptomatic carriage rather than true infection.

Laboratory diagnosis of typhoid fever

1 Blood culture.

2 Bone marrow culture.

3 (Urine or faeces culture in appropriate clinical illness.)

Antigen detection

S. typhi lipopolysaccharide antigen can be detected in the serum and the urine of patients with typhoid. Techniques reported include counterimmunoelectrophoresis and enzyme-linked immunosorbent assay (ELISA). Results can be obtained more rapidly than from bacterial culture, and remain positive after chemotherapy has been initiated.

Management

In some areas of the world, widespread use of antibiotics for the treatment of diarrhoeas has led to antibiotic resistance of enteric fever pathogens. Typhoid and paratyphoid A, acquired in India and Pakistan, are rarely susceptible to quinolone antibiotics, and are likely to be resistant to amoxicillin. The treatment of choice in these circumstances is ceftriaxone, which must be given intravenously. For enteric fevers acquired in Africa, and other regions, quinolones are usually effective and have the advantage of oral administration, with a very low risk of relapse after recovery. In Sri Lanka, typhoid fever is usually successfully treated with oral co-trimoxazole or high oral doses of amoxicillin.

The temperature does not always fall immediately when enteric fevers are treated with effective drugs. It may take two or three days to fall. However, the absence of any fall, or the persistence of positive blood cultures after 48 hours should lead to review of the choice of antibiotic. Severely ill patients may suffer a 'Herxheimer-like' exacerbation of fever, prostration and hypotension at the start of specific treatment. This can be treated with a step-down course of intravenous hydrocortisone, starting with 100 mg three times daily, and rapidly reducing according to the patient's response.

Antibiotic treatment of typhoid fever

1 Oral ciprofloxacin 500 mg twice daily for 10 days (child 7.5 mg/kg twice daily).

2 Intravenous ciprofloxacin may be given in a dose of 400 mg twice daily (child 5 mg/kg twice daily) until oral therapy is possible.

3 Alternatives:
- Co-trimoxazole orally 960–1440 mg twice daily (child 6 weeks to 5 months, 120 mg; 6 months–5 years, 240 mg; 6–12 years, 480 mg; all 12-hourly for 2 weeks).
- Amoxicillin orally 500 mg–1 g 8-hourly for 2 weeks (child up to 10 years, 250 mg 8-hourly).

4 For antibiotic-resistant infection: ceftriaxone i.v. 1–2 g daily (can also be given i.m., if i.v. route not available).

⚠ Chloramphenicol used to be widely used but carries a small risk of agranulocytosis, and up to 15% of cases suffer relapse after initial response; it may still be effective in difficult cases: orally or i.v., 2–3 g daily in divided doses (child 50–100 mg/kg daily) for 2 weeks.

Paratyphoid A and B are less predictable than typhoid in their response to antibiotics. Ciprofloxacin or ceftriaxone are the two drugs of choice. Ciprofloxacin-resistant paratyphoid B has been reported.

Prevention and control

Avoiding high-risk foods and drinking water can much reduce the chance of exposure to enteric fevers, though it is difficult for most travellers to adhere completely to safe eating principles. Travellers visiting areas where enteric fevers occur can be protected by vaccination. Available vaccines offer around 70–75% protection from typhoid fever.

Immunization against typhoid fever

1 Vi-polysaccharide vaccine (Typhim Vi or Typherix), given in a single 0.5 ml i.m. dose. May produce irritation and redness at the injection site. A booster is recommended after 3 years.
2 Combined inactivated hepatitis A with Vi-polysaccharide vaccine is available as a single 1 ml i.m. dose for adolescents and adults (Hepatyrix for individuals over age 15 years; ViA-TIM for individuals over age 16 years). A booster against typhoid fever is recommended after three years. These vaccines may be given subcutaneously to individuals using anticoagulants, or suffering from thrombocytopenia or haemophilia.

Cholera

Introduction

In spite of its rarity in developed countries, cholera is still an important infection worldwide. It may occur as part of an epidemic or arise sporadically in the developing world. It is occasionally found in travellers returning from areas where cholera transmission occurs.

Epidemiology

Cholera is spread mainly through drinking faecally contaminated water. Food, especially shellfish, may also be a vehicle of infection. Large epidemics occur in countries lacking safe drinking water and adequate facilities for the disposal of sewage. Cholera is a pandemic infection, capable of causing epidemics that spread simultaneously in many countries around the globe. Smaller epidemics and outbreaks have been reported in Mediterranean and east European countries.

During the 19th century, several cholera pandemics spread from India throughout Asia, Europe and the Americas. During the first half of the 20th century, the disease was largely confined to Asia. The seventh cholera pandemic spread from Indonesia in 1961, and has now reached all continents, including South America. In 2002, 142 000 cases of cholera including 4500 deaths were officially reported to the World Health Organization, although this is almost certainly an underestimate. Ninety-seven per cent of reported cases were reported from Africa. In most countries, infection is due to the E1 Tor biotype, although the classic biotype has re-emerged in Bangladesh. A new serotype, O139, appeared as a cause of epidemic illness in the Bay of Bengal in 1992, but it has remained confined to Asia – where it is responsible for approximately 15% of laboratory-confirmed cases of cholera.

Cholera is rare in developed countries. There is a small focus of infection in Texas and Louisiana due to a unique strain of *Vibrio cholerae* O1. There has been no indigenous cholera infection in the UK this century. Travellers to endemic areas are only occasionally affected. Between 1990 and 2001, 126 cases of cholera have been imported into England and Wales. It has been estimated that the risk of infection in a traveller is about 1 in 500 000.

Pathogenesis

Cholera is a toxin-mediated disease caused by the O1 (and, potentially, the related O139 serotype) of *V. cholerae*, which exists in the El Tor and classical biotypes. Cholera toxin is very closely related to the heat-labile toxin of *Escherichia coli* (see Chapter 8). In the future, the same vaccine may be effective against both heat-labile *E. coli* and cholera toxin.

Clinical features

The usual incubation period is 3 or 4 days. The severity of illness is extremely variable, many patients simply having a gastroenteritis-type disease. In classic cholera there is an abrupt onset of severe diarrhoea, at first watery and brown, but quickly changing to pale fluid stools containing only a little mucus and cell debris – the so-called rice-water stools. Fever is not prominent. Continuous fluid loss quickly leads to shock. The diarrhoea contains many organisms and is highly infectious. This, coupled with the difficulty of maintaining personal and domestic hygiene, contributes to the rapid spread of the disease.

Diagnosis

Clinical suspicion will be alert in an epidemic or outbreak situation. There are few diseases that cause such sudden dehydration in adults, though child cases may be hard to distinguish from other types of severe gastroenteritis. Examination of the diarrhoea stools will give an early result but, even so, treatment should not be delayed pending diagnosis.

Laboratory diagnosis
Microscopy
V. cholerae organisms have characteristic darting motility, and can be seen in freshly passed stool specimens from patients with acute disease. This rapid diagnosis can be confirmed by demonstrating inhibition of motility with specific antiserum.

Isolation
Specimens should be transported to the laboratory immediately but, if this cannot be done, specialized transport media such as that of Cary–Blair can be employed.

Media for the isolation of pathogenic vibrios have a high pH (8.6) and contain bile salts that together inhibit the growth of other enteric bacteria. The most commonly used medium is thiosulphate–citrate–bile-salt–sucrose (TCBS) agar. *V. cholerae* ferments sucrose after 24 h incubation producing yellow colonies due to a colour change of bromothymol blue indicator with acid production. Other vibrios, including food-poisoning species such as *V. parahaemolyticus*, grow well on this medium. Most, including *V. parahaemolyticus*, are non-sucrose fermenters whose colonies appear blue/green.

Suspect colonies are subcultured and identified using the conventional biochemical tests employed for the identification of Enterobacteriaceae (see Chapter 3). There are more than 70 serotypes of *V. cholerae* based on the lipopolysaccharide 'O' antigen. Only O1 and O139 have been associated with human disease. Confirmed colonies of *V. cholerae* are therefore serotyped, by a slide-agglutination test using anti-O1 and anti-O139 antisera. Toxin production by the organism must then be confirmed, as only toxigenic O1 and O139 strains are pathogenic, and only these strains require public health action. Rapid diagnostic techniques including latex agglutination and DNA amplification techniques have been described but have only limited availability in developing countries where the disease is common.

Laboratory diagnosis of cholera

1 Comma-shaped bacteria with darting motility in the faeces.
2 Sucrose-fermenting organisms identified on thiosulphate–citrate–bile-salt–sucrose agar.
3 *Vibrio cholerae* confirmed using biochemical tests.
4 Serotype O1 or O139 identified by slide agglutination.
5 Toxigenicity confirmed serologically.

For epidemiological purposes, the O1 and O139 strains can be typed, using agglutination, biochemical or phage reactions, as El Tor or classic types, which each also exist in three biotypes (Inaba, Ogawa and Hikojima).

In areas where culture is not possible, serological surveys using techniques to detect antibodies to the O1 lipopolysaccharide antigen can be useful in epidemiological surveys.

Management

Because the mucosal cells remain intact in cholera their absorptive function is undamaged. Oral rehydration is therefore successful in over 90% of cases. Intravenous rehydration should be given to exhausted or shocked patients. Very large initial volumes of 4–6 l may be needed, followed by several litres per day while diarrhoea lasts. Spontaneous recovery is usual once hydration is controlled, but the diarrhoeal illness may last for 4–5 days. Chemotherapy may be indicated in severe or prolonged disease, or in the elderly and debilitated. Tetracycline is often effec-

tive; ciprofloxacin may also be given, and is easier to use if parenteral treatment is necessary.

Treatment of cholera

1 Oxytetracycline orally or via nasogastric tube 500 mg 6-hourly for 3–5 days.
2 Alternative: ciprofloxacin orally 500 mg 12-hourly *or* i.v. 200 mg 12-hourly, both for 3–5 days.

While asymptomatic carriage of classic cholera strains is unusual, the El Tor strains may be excreted by asymptomatic carriers and by convalescent patients. Patients and their close contacts should therefore have follow-up stool examination before being released from medical supervision. A 4- or 5-day course of chemotherapy will eradicate excretion in most cases.

Prevention and control

The most useful measure in preventing the spread of cholera is provision of safe drinking water and sanitary disposal of human faeces. Food likely to be contaminated, especially fish and shellfish, should be thoroughly cooked before eating. Travel and trade restrictions between countries are not effective.

Cholera vaccine is of limited use. The protection afforded is limited and lasts for 6–24 months, so that recipients are advised to take precautions with food and water despite immunization. The vaccine has no effect on carriage and is not useful in preventing spread. The World Health Organization has no requirements in the International Health Regulations for a certificate of vaccination against cholera. Nevertheless, evidence of vaccination in travellers from infected areas may occasionally be required. The vaccine may be appropriate for those who are unable to take adequate precautions in highly endemic or epidemic settings. This would include aid workers assisting in disaster relief or refugee camps, and more adventurous backpackers who do not have access to medical care.

Oral cholera vaccine

Inactivated Inaba and Ogawa (including El Tor biotype) *V. cholerae* O1 plus recombinant B-subunit of cholera toxin

• Adult and child over 6 years: 2 doses with an interval of 1 week.
• Child aged 2–6 years: 3 doses each separated by an interval of 1 week.
• Food should not be taken for 1 hour before and after dosing.
• Boosters recommended after 2 years for individuals aged over 6 years, after 6 months for younger individuals.

⚠ Injectable cholera vaccines are available in some countries, but they offer only limited and short-lasting protection; they are not available in the UK.

Diseases with arthropod vectors

Introduction

Arthropod bites may transmit serious vector-borne diseases. Travellers to affected areas often do not anticipate the intensity of arthropod activity, or understand the day- or night-biting habits of locally prevalent arthropods.

Lyme disease (borreliosis)

Introduction and epidemiology

Lyme disease is caused by *Borrelia* spp., which are transmitted from animals to humans by the bite of hard ticks of the genus *Ixodes* (Fig. 24.5). The natural hosts of the *Borrelia* species are small rodents, but human disease is often acquired via deer. Dogs can also be infected. The main organism causing borreliosis in the USA is *B. burgdorferi* and in various parts of Europe is *B. afzelii*, *B. garinii* or *B. vientiana*. They all cause disease with similar clinical characteristics; US strains may be more likely to produce large-joint arthritis and carditis, while European strains may more often produce early and late central nervous system and dermatological manifestations.

Borreliosis is most common following occupational or leisure exposure to open forest or parkland, in the season when ticks are active. The presence of large animals such as deer increases the likelihood of infection. Transmission has been recorded in the New Forest, the deer parks of London and south-east England, and in parts of Scotland. Imported cases may originate in European countries such as Germany or Austria, in Scandinavian countries and in eastern and central areas of the USA. A tick must remain attached for about 20 h to transfer an infective dose of borrelias in its saliva.

Early Lyme disease and erythema chronicum migrans

After an incubation of 1–3 weeks, about 80% of infected patients develop a characteristic rash surrounding the tick bite. A disc of erythema expands, often clearing in the centre (Fig. 24.6). This may encompass an entire limb before disappearing. Multiple erythema chronicum migrans (ECM) lesions are occasionally seen. Patients often have fever, muscle and joint aches and malaise during the eruption, which can last for 2–4 weeks. There may be a mild leucocytosis, and the inflammatory indices are mildly to moderately elevated.

Borrelial lymphocytoma

Borrelial lymphocytoma is a rarer skin manifestation that can occur at the same time as ECM, after it, or occasionally in late disease. It is usually a dusky nodule or plaque often affecting the ear (especially in children) or the breast. It may reach 1–5 cm in diameter and ulcerate, appearing similar to a cigarette burn, and can last for several months if untreated. Few patients have associated constitutional symptoms. Histology shows non-specific lymphocytic infiltration with germinal centres, and must be differentiated from other granulomatous disorders and lymphomas.

Early disseminated Lyme borreliosis

These disorders can follow the original infection after an interval varying from 2 or 3 weeks to 2 or 3 months.

Lyme arthritis

Lyme arthritis is common in the USA, affecting about half of patients whose early infection is untreated. Its onset varies from a few days to 2 years after exposure. It is usually an asymmetrical large joint arthritis, but is occasionally palindromic. There is synovitis and often moderate effusion of affected joints. The effusion has a high protein

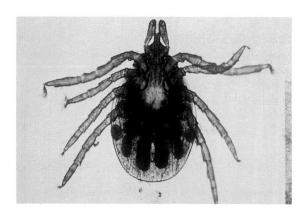

Figure 24.5 Hard ixodid tick, the vector of borreliosis. Hard ticks are also vectors for tick-borne encephalitis and some rickettsial infections. Courtesy of the Ministry of Defence.

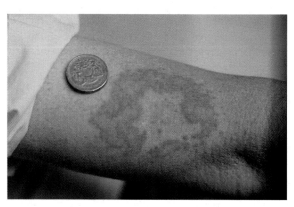

Figure 24.6 Erythema chronicum migrans. Courtesy of Dr M. G. Brook.

level and contains neutrophils. Seronegative rheumatoid arthritis and rheumatic fever are important differential diagnoses. Untreated Lyme arthritis tends to recur progressively less frequently over 2–4 years. Few patients have permanent or erosive joint disease.

About 1 in 10 untreated patients develop intermittent arthralgias and periarticular pain without synovitis or effusion. In some cases these symptoms persist for 5 years or more.

Peripheral neuropathies

Peripheral neuropathies are common in European infections, and may accompany other manifestations. Facial paralysis is one of the most common neuropathies, but others, including unilateral phrenic nerve palsy (Fig. 24.7), have been described. They tend to resolve spontaneously over a period of weeks.

Neuroborreliosis

Relapsing lymphocytic meningitis was recognized long before Lyme disease. It is now known that a significant proportion of these cases are caused by neuroborreliosis.

Polyradiculitis is a disabling and progressive feature of Lyme disease, more common in European types of infection. It presents as localized pain in the distribution of the affected nerve roots, with dysfunction of the associated nerves. A typical presentation would be low back or sacral pain with a weak knee or foot drop. Paraesthesiae, loss of sensation and absent reflexes are common findings.

The radiculitis is often accompanied by meningism and lymphocytosis in the cerebrospinal fluid (CSF). Occasional plasma cells are also seen, and the CSF glucose level may be slightly lowered. This syndrome of relapsing meningoradiculitis (Banwarth's syndrome) was described before

the aetiology was understood. Occasionally, encephalitis or encephalopathy also occurs.

Cardiological effects

The cardiological effects are a result of myocarditis, often with conduction defects. Complete heart block is common (Fig. 24.8). Prolonged and progressive cardiomyopathy has been described in occasional cases.

Late (chronic) Lyme disease

This occurs as a peripheral, glove-and-stocking neuropathy, associated with a typical, asymmetrical violaceous inflammation of the skin called atrophic acrodermatitis. It most commonly affects a foot or heel or sometimes the elbow or hand. Histology of affected skin shows oedema and a lymphocytic infiltrate, often with many plasma cells. After months or years without treatment the lesions become thin and atrophic. Even after months or years, it is amenable to treatment.

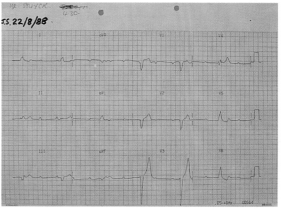

(a)

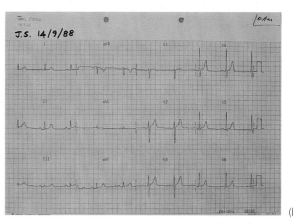

(b)

Figure 24.8 (a) Electrocardiogram (ECG) showing a complete heart block in a young patient with secondary Lyme disease. (b) The second ECG after treatment shows that conduction is normal (same patient as in Figure 24.7).

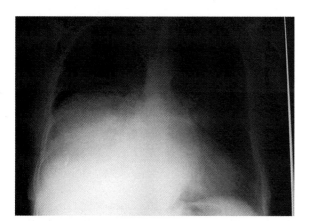

Figure 24.7 Lyme disease: this patient had a transient phrenic nerve palsy, with raised right diaphragm.

Clinical manifestations of borreliosis

Early localized
- Flu-like illness.
- Erythema chronicum migrans.
- Borrelial lymphocytoma.

Early disseminated
- Arthritis.
- Myocarditis.
- Neuropathies.
- Relapsing lymphocytic meningitis.
- Polyradiculitis.
- Relapsing meningoradiculitis

Late
- Peripheral neuropathy.
- Atrophic acrodermatitis (acrodermatitis chronica et atrophica).

Diagnosis

Clinical and epidemiological suspicion is important. Careful examination may show a very wide ring of inflammation due to ECM.

Initial laboratory tests include IgM and IgG ELISA tests (using *B. burgdorferi* flagellar antigens in the USA, and *B. afzelii* in Europe). Positive results are verified by Western blotting, to confirm that the antibodies are specific for the antigens of these borrelias, rather than cross-reacting species. The confirmation of borreliosis is not straightforward, as patients may not display IgM antibodies until up to 3 weeks after the onset of illness, and antibody results may remain permanently negative after early antibiotic treatment.

In specialist laboratories, *B. burgdorferi* can be cultured from active skin lesions or blood. Polymerase chain reaction tests can demonstrate *Borrelia* spp. DNA in the synovium and CSF, and in blood in the stage of ECM, but this test is not generally available.

When there are neurological features, locally produced antibodies or borrelial DNA can be demonstrated in the CSF, and this is useful evidence of infection.

Treatment

The treatment of choice is doxycycline 200 mg daily for 14 days (or 200 mg initial dose, then 100 mg daily for 21 days). This is probably equally effective for all stages of infection. Amoxicillin is also effective. Erythromycin is an alternative that may carry a slightly higher risk of relapse.

Cefotaxime and ceftriaxone have both been used successfully in neuroborreliosis, but in half or more of patients symptoms persist after a 2- or 3-week course. Even long courses are not always followed by cure, and it is then worth trying oral doxycycline, or amoxicillin. Longer continuation courses may have some benefit in late disease.

Treatment of borreliosis

1 (a) Early: doxycycline orally 200 mg daily for 14 days *or* 200 mg initial dose, then 100 mg daily for 21 days. (b) *Alternative*, or for child: ampicillin orally 500 mg 6-hourly (child under 10 years, 250 mg 6-hourly); *or* amoxicillin orally 500 mg 8-hourly (child under 10 years, 250 mg 8-hourly) for 3 weeks. (c) *Second choice*: erythromycin orally, 500 mg 6-hourly (child up to 2 years, 125 mg 6-hourly; 2–8 years, 250 mg 6-hourly) for 3 weeks.
2 Alternative for disseminated or late disease: cefotaxime i.v. 1–2 g 8-hourly *or* ceftriaxone i.v. 1–2 g daily, both for 2–4 weeks; *followed by* doxycycline orally 100 mg daily *or* ampicillin 250 mg 6-hourly *plus* probenecid 500 mg twice daily (child under 10 years, half adult dose) for 4–8 weeks.

Prevention of borreliosis

Avoidance of tick contact, by appropriate use of clothing and insect repellents, is important. Inspection of the skin and early removal of ticks is also helpful, as around 20 hours' attachment is necessary before an infective dose of borrelias is transferred from tick to host.

Many primary care physicians offer post-exposure prophylaxis for prolonged tick attachment, or multiple tick bites in endemic areas. The usual regimen is a 1-week course of oral amoxicillin.

Dengue fevers

Introduction and epidemiology

Dengue virus is a flavivirus, whose four serotypes are widely distributed in tropical areas. The highest burden of disease occurs in SE Asia and the Western Pacific, but over the last few years there has also been a rising trend in South America and the Caribbean. Large epidemics can occur and travellers in affected areas are at considerable risk of exposure. The UK has seen around 50 confirmed cases imported each year recently, with another 200 probable or suspected. Infection is transmitted from person to person by *Aedes* sp. mosquitoes, which are common in rural and urban areas, and bite in the daytime. The feverish illness can be severe and debilitating. Second infection with a different serotype can lead to the enhanced illnesses, dengue haemorrhagic fever (DHF) or dengue shock syndrome (DSS: see below).

Virology

Flaviviruses are spherical enveloped RNA viruses 40–60 nm in diameter. The envelope is covered with viral glycoproteins: the M (membrane) and E (envelope) antigens. With the capsid C protein, seven other non-structural proteins required for replication are transcribed from the viral genome. Attachment, endosomal-membrane fusion and haemagglutination are mediated through the E antigen. Several of the non-structural antigens also mediate viral serotype and leucocyte responses. Flaviviruses are adapted to grow in insect and vertebrate cells at the wide range of temperatures that they will encounter through their life cycle.

Clinical features

The incubation period is usually 5–7 days, but can be up to 10 days. Illness begins abruptly with high fever, often severe arthralgia and frontal headache (this has been called breakbone fever). At this stage there is neutropenia and there may be mild hepatocellular disturbance in liver function tests. Symptoms often abate after 4–6 days but in many cases the fever continues or recommences, there is a generalized lymphadenopathy, and a macular rash may appear on the trunk and proximal limbs (Fig. 24.9). The platelet count often falls and the rash may become petechial, but significant bleeding does not occur and the condition is transient; fever abates in another 5 or 6 days. Convalescence is moderately rapid, taking 2 or 3 weeks.

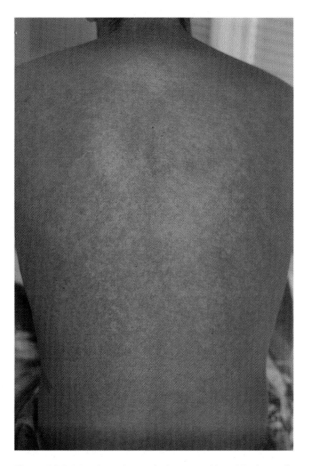

Figure 24.9 Macular rash seen in the second feverish phase of dengue fever. Courtesy of Dr M. G. Brook.

Diagnosis and management

Typical illness is readily recognized, but many cases are mild or atypical. When dengue fever is locally active, any fever continuing for more than 2 days (especially with a 'relative bradycardia') is likely to be caused by dengue infection. Virus can be demonstrated by polymerase chain reaction (PCR) techniques in blood during the acute phase. Serological testing in the first 3 weeks shows high antibody titres and the presence of immunoglobulin M (IgM) antibodies. IgG antibodies also appear early, and persist long-term. Management is symptomatic, as there is no specific treatment.

Complications

Immunity to dengue fevers is type-specific. Second attacks can occur with different serotypes, when antibody-bound virus is not neutralized, but can attach to macrophages via their Fc receptors, enhancing both viral entry into cells and cell activation. The immunopathological features are then enhanced, producing DHF. A typical feverish onset develops into severe disease with profound thrombocytopenia. Haemorrhage, pleural effusions and associated secondary pneumonitis may occur, with up to 15% mortality. Shock, presaged by haemoconcentration, is a serious development requiring vigorous supportive treatment; the mortality of dengue shock syndrome (DSS) is near 40%. Children are most often affected, but the disease has been recorded in adults.

> **Stages of dengue haemorrhagic fever**
> - Stage I (thrombocytopenia): positive tourniquet test (Fig. 24.10).
> - Stage II (capillary leak): haematocrit increased by 20% or more compared with initial level.
> - Stage III (circulatory failure): pulse pressure equal to or less than 20 mmHg.
> - Stage IV (profound shock): hypotension unresponsive to rehydration.

Other arboviral infections

Introduction

There are many arboviral infections, transmitted by mosquitoes, sandflies and ticks. They are all characterized by short incubation periods and can cause intense feverish syndromes, some with lymphadenopathy or rash. Important among these are the zoonotic encephalitides, which cause severe meningoencephalitis with a high risk of long-term sequelae (see Chapter 13). The reservoirs of infection are birds, rodents or other small mammals and transmission to larger animals and humans is by arthropod bites. In the USA and south America horses are often affected (by eastern and western equine encephalitis and Venezuelan equine encephalitis). Birds may be a reservoir (as in West Nile virus infection, long recognized in parts

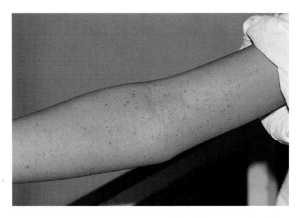

Figure 24.10 Dengue haemorrhagic fever: positive tourniquet test. Courtesy of Dr D. Lewis.

of Africa, the Middle East and eastern Mediterranean, now widespread across eastern and central USA). St Louis encephalitis (in Florida and central America) has a reservoir in birds, and California encephalitis is harboured by rodents. Transmission of these infections is by mosquito vectors. In Nepal, India and Far Eastern countries, culecine mosquitoes carry Japanese B encephalitis from pigs to humans. European tick-borne encephalitis, transmitted by hard ticks, is endemic in wooded areas of central Europe and eastern Scandinavia, and eastern strains exist in eastern Europe, Mongolia and China.

Prevention and precautions
For many of these diseases the main preventive measure is control of mosquitoes in endemic areas, and avoidance of mosquito or tick bites by humans. Effective vaccines are available against tick-borne encephalitis and Japanese B encephalitis. They should be offered to travellers who will have rural exposure in endemic areas. They are not necessary unless the traveller remains close to the reservoir of infection, as the vectors have a very limited range.

Tick-borne encephalitis vaccine
Contains inactivated tick-borne encephalitis virus of a European strain. Three doses by i.m. or deep subcutaneous injection; the first 2 doses separated by 3 weeks to 3 months, the third dose after 9–12 months.

Boosters recommended at 3-yearly intervals, if antibody levels indicate the need.

Adverse events include high fever, and rare arrhythmias.

Indicated for travellers who will be hiking, camping or working in wooded areas in central and eastern Europe and southern Scandinavia, especially in the warm season from April to September.

Japanese B encephalitis vaccine (for named patient use in the UK)
Contains inactivated JE virus. Three subcutaneous doses are given on days 0, 7 and 28–30. Adverse events include hypersensitivity reactions: recipients should be monitored for 30 minutes before leaving the clinic. Recommended for travellers who will spend 1 month or more in rural or farming areas of north India, the Far East, China, Korea and Japan.

Yellow fever

Introduction
Yellow fever is a flavivirus infection, transmitted by mosquitoes of the *Aedes* sp. It is endemic in parts of the tropics, including Africa and tropical South America. It is hardly ever imported into European countries because of the high acceptance of vaccination by tourists, assisted by the International Health Regulations, which require vaccination of travellers who enter endemic areas.

Epidemiology
Two patterns of disease transmission occur. In the *sylvatic cycle*, monkeys are the reservoir of infection, and the virus is transmitted from monkey to monkey by mosquito bites. The disease can be transmitted from monkeys to humans, and from human to human, by the bite of the *Aedes* sp. of mosquito (principally *A. aegypti*). These mosquitoes breed mainly in small water containers such as discarded coconut shells, and inhabit villages and towns. Periodic explosive outbreaks of *urban yellow fever* then occur, often in association with a breakdown in mosquito control measures.

In South America, the vector is a forest-dwelling mosquito of the genus *Haemagogus*. The disease occurs endemically among young adult male forest workers.

Yellow fever has never been reported outside Africa or mainland South America, with the exception of Trinidad, which is now free of the disease (Fig. 24.11).

The incidence of yellow fever has risen sharply in recent years. Official figures represent only a small fraction of the total cases and the WHO currently estimates the global figure to be 200 000 cases every year with 30 000 deaths. The vast majority of cases and deaths occur in sub-Saharan Africa but Latin America is now at greater risk of urban epidemics than at any time in the last 50 years.

Clinical features
After an incubation period of 6 or 7 days, symptoms appear abruptly, with fever, myalgia and prostration. As with other flavivirus infections, a remission of fever and symptoms often occurs after 4 or 5 days. Sometimes the illness ends here, but severe cases progress, with further fever,

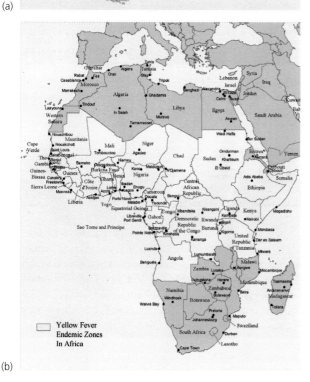

clinical jaundice, bleeding characterized by haematemesis, and a progressive sepsis syndrome with multiorgan failure.

Diagnosis

The diagnosis can be suspected in unvaccinated patients who have recently visited endemic areas. Virus can be demonstrated in the blood by PCR and culture in the first few days of illness. Serological diagnosis is complicated by difficulty in distinguishing antibody generated by immunization from that resulting from infection. It may be offered by arbovirus reference laboratories.

Prevention and control

Yellow fever vaccine is prepared from the attenuated 17D strain of yellow fever virus. One dose protects for at least 10 years. As with other live vaccines, it may not be given to pregnant women or the immunocompromised. Several countries in endemic zones of Africa and South America require certification of vaccination. A rare syndrome of vaccinaemic disease has been described, causing neurological signs and symptoms after vaccination of immunocompetent recipients.

In endemic areas, the disease is preventable by mass immunization and vector control programmes.

Rickettsioses

Introduction

Rickettsioses are systemic diseases that are common in many countries. Rocky Mountain spotted fever (caused by *Rickettsia rickettsii*) is tick-borne, and is endemic in the Rocky Mountains and in several rural areas on the eastern seaboard of the USA. Epidemic typhus (*R. prowazekii*) affects many populations infested by lice, which are the reservoir of infection. Both of these diseases are life-threatening if untreated. Less grave but still severe illnesses are tick typhus (*R. conorii*), endemic typhus (*R. typhi*, transmitted by fleas from mouse to humans) and scrub typhus (*Orientia tsutsugamushi*), transmitted by mites. Similar but rare diseases are trench fever, caused by *Bartonella quintana*, and the ehrlichioses, caused by *Ehrlichia* and *Anaplasma* spp. The agents of these two diseases are members of the family Rickettsiaceae. Q fever is caused by the sheep pathogen, *Coxiella burnetii*, another member of the Rickettsiaceae.

Diseases caused by Rickettsiaceae
Typhus group
• Epidemic typhus (*Rickettsia prowazekii*)
• Murine or endemic typhus (*R. typhi*)

Figure 24.11 Endemic areas for yellow fever in (a) South America; (b) Africa. Courtesy of the US Centers for Disease Control and Prevention.

Spotted fever group
- Rocky Mountain spotted fever (*R. rickettsii*)
- Tick typhus (*R. conorii* and *R. africae*)
- Scrub typhus (*R. tsutsugamushi*)
- Rickettsial pox (*R. akari*)
- Many other species of *Rickettsia* cause mild to moderate illness in various ecologies: examples include *R. africae, R. australis, R. felis, R. sibirica, R. honei, R. microtimonae*, etc.

Q fever
- *Coxiella burnetii*

Bartonelloses
- Trench fever (*Bartonella quintana*)
- Cat-scratch fever (also pelious hepatitis and pelious dermatitis) (*B. henselae*)

Ehrlichioses
- Monocytic ehrlichiosis (*Ehrlichia chaffeensis*)
- Granulocytic ehrlichiosis (*Anaplasma phagocytophilum*)
- (*Neorickettsia sennetsu*, possibly a zoonosis with dog-tick-man transmission).

Clinical features

All of the rickettsial diseases have an incubation period of around 13 days. They begin abruptly with swinging fever and frontal headache. Confusion is common during the peaks of fever. The main pathology is an endovasculitis, which causes a rash and bleeding diathesis. The rash of Rocky Mountain spotted fever begins as macules on the hands and feet, then spreads over the body, becoming petechial or haemorrhagic. That of epidemic typhus begins in the axillae, and also becomes purpuric as it spreads. Rashes, often with a petechial element, may be seen in the other rickettsioses (Fig. 24.12). That of rickettsial pox is variegate, having macular, purpuric and pustular elements. It often closely resembles chickenpox.

About half of all imported rickettsial diseases in the UK are tick typhus, originating from the Mediterranean, the Arabian Gulf or Africa. The eschar of the originating tick bite may still be visible on admission as a black scab surrounded by inflammation (Fig. 24.13). The rash is generalized and the conjunctivae are suffused. The blood count shows a slight neutrophilia and transaminases are usually elevated; sometimes clinical jaundice is present. Mild abnormalities of clotting and of renal function are often detectable (a milder version of the intravascular coagulation, bleeding and systems failure of the more severe rickettsioses).

Untreated tick typhus lasts for about 3 weeks before the fever and rash resolve. Although rarely fatal, it is a severe disease followed by debility and prolonged convalescence.

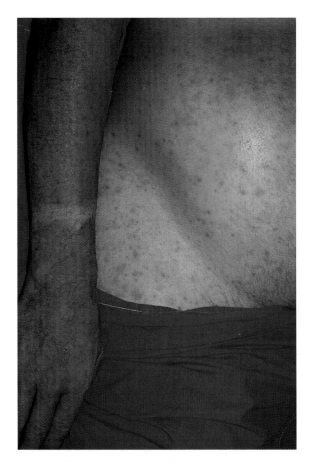

Figure 24.12 Tick typhus: the patient has a maculopapular rash with a petechial element, conjunctival injection, severe headache and myalgia.

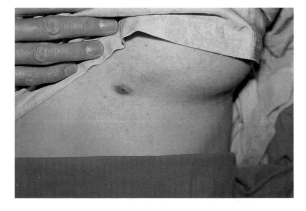

Figure 24.13 Tick typhus: a black eschar at the site of the infecting tick bite.

Endemic typhus may be more severe, causing renal failure or intravascular coagulation.

The ehrlichioses cause persisting febrile disease without rash, but usually with moderately elevated liver transaminases. The history of tick-exposure or tick-bite, and the presence of inclusion bodies in the monocytes or granulocytes, are helpful indicators of this diagnosis.

Diagnosis

The travel history, arthropod exposure and clinical features often suggest the diagnosis of rickettsiosis. Severe cases with purpura must be distinguished from meningococcal disease, or viral haemorrhagic fevers.

The diagnosis is usually made serologically using immunofluorescence, complement fixation or specific IgM enzyme immunoassay (EIA). PCR-based diagnosis is also available.

Q fever is usually diagnosed by a complement fixation method using acute and convalescent serum. *Coxiella burnetii* expresses different antigens at different phases of infection. Concentrations of antibodies to antigens known as phase 2 antigens rise in acute infection but antibodies to phase 1 antigens are only elevated in chronic granulomatous Q fever or endocarditis. EIA and PCR-based methods are available in reference laboratories.

Bartonella henselae is demonstrated in lymph nodes by Warthin–Starry silver staining (Fig. CS.1) and less effectively by Gram staining. It has been isolated in culture: freshly prepared brain–heart infusion agar containing 5 or 10% rabbit or horse blood should be used and the plates should be incubated in a humid atmosphere for up to 3–4 weeks. *Bartonella* spp. grow best on solid or semisolid media and do not produce turbidity or convert enough carbon dioxide for ready detection in automated systems. Colonies on blood agar are pleomorphic. Immunofluorescence and EIA techniques have been described for the detection of IgM and IgG antibodies to *Bartonella henselae*. PCR-based techniques have proved to be the most sensitive diagnostic tests.

Management

The treatment of choice is doxycycline. Chloramphenicol, orally or intravenously, is an alternative. Erythromycin may be effective, but resistance sometimes emerges during treatment. Quinolones might be an alternative option. A 10–14-day course is usually required. As in typhoid fever, the temperature may not fall to normal for 3–4 days.

Treatment of rickettsial infections
1 Doxycycline 200 mg daily for 10–14 days.
2 Alternative: chloramphenicol orally or i.v. 500 mg 6-hourly for 10–14 days.

Malaria

Introduction and pathology

Malaria is one of the most important imported diseases, which can be caused by four *Plasmodium* species, one of which, *P. falciparum*, causes severe, 'malignant' disease. The benign malarias are debilitating diseases with a relapsing course. Falciparum malaria may be life-threatening and should be treated as a medical emergency. Malaria should be actively excluded in every feverish traveller from the tropics. Around 2000 cases are reported each year in the UK, with up to a dozen fatalities. Most cases affect travellers who have not taken prophylaxis, particularly people travelling to visit friends and relatives.

Malarial sporozoites attach to and enter red blood cells, using surface antigens such as blood group determinants. There they multiply by binary fission, digesting haemoglobin to derive energy and producing a waste haem pigment, haemozoin, which is deposited in endothelium and surrounding tissues, causing an intense inflammatory response and activation of platelets and complement. The daughter parasites rupture the red blood cell to enter the plasma and continue the blood infection cycle by infecting new cells. Haemolysis is increased by an immune response to malarial antigens on the surface of infected cells.

Plasmodium falciparum parasites can enter red cells at all stages of maturity, whereas *P. vivax* infects only reticulocytes and *P. malariae* favours senescent cells. As *P. falciparum* parasites multiply to the schizont stage they cause rigidity of the infected cell, with the appearance of 'sticky' surface projections. This leads to sludging of parasitized red cells in the microcirculation, particularly of parenchymal organs. *P. falciparum* causes a more intense infection, with more vascular damage and inflammatory response than the other malaria types. Vascular shunting, tissue hypoxia, endothelial damage and disseminated intravascular coagulation occur in severe infections.

Causes of the four types of malaria
1 *Plasmodium falciparum* (malignant tertian malaria).
2 *P. vivax* (benign tertian malaria).
3 *P. ovale* (benign tertian malaria).
4 *P. malariae* (benign quartan malaria).

⚠ Recently, some cases of apparent *P. vivax* malaria have been shown by genotyping to be caused by *P. knowlesii*, formerly thought to be exclusively a pathogen of monkeys; the infections responded to conventional antimalarial treatment.

Clinical features

The only consistent clinical features of malaria are fever and rigors. Patients present initially with a chaotic, swinging fever; rigors occur when the temperature rises. The fever becomes periodic when synchronous release of parasites is established after 7–14 days. Fevers tend to occur every third day (tertian fever) in vivax and ovale malaria, or every fourth day (quartan fever) in malariae malaria. In falciparum malaria the fevers are less regular, but may approximate to a tertian pattern. Many non-specific symptoms may be present, including abdominal pain, headache, diarrhoea, dysuria and frequency, sore throat and cough. Physical examination may be normal or splenomegaly may be detectable. In chronic or relapsing malaria the spleen can be very large. Hepatomegaly and mild jaundice may also be present.

Cerebral malaria presents with encephalopathy. It is largely a disease of the non-immune and in endemic areas it mainly affects children below the age of 4 years, in whom it must be distinguished from childhood bacterial meningitis. Hypoglycaemia, convulsions and hypoxia readily occur and worsen the prognosis considerably. Blackwater fever results from severe intravascular haemolysis. Profound anaemia, jaundice, haemoglobinuria and acute renal failure quickly develop in untreated cases. Acute respiratory distress syndrome (ARDS) is common, and frequently coexists with cerebral disease. The absence of bile in the urine distinguishes haemolytic jaundice from that of viral hepatitis.

Important features of malignant malaria

1 Encephalopathy (cerebral malaria).
2 Pulmonary oedema (acute respiratory distress syndrome).
3 Acute renal failure.
4 Severe intravascular haemolysis, with profound anaemia.
5 Haemoglobinuria (blackwater fever).

Diagnosis

The only reliable means of diagnosis is the demonstration of parasites in the red blood cells or parasite antigens in the blood.

Parasite detection is best done by preparing thick and thin blood films, which are stained with Giemsa or Field's stain, respectively. Rapid Romanowsky-type stains are also suitable for thin films. Scanty parasites are easier to detect in thick films, while thin films are often additionally helpful in speciation (Fig. 24.14). Some automated blood-counting machines can also detect malarial parasites, but have not proved as reliable as tests done by an experienced observer.

Stick or card tests can be used to detect the histidine-rich surface antigen or lactate dehydrogenase enzymes of *P. falciparum* or *Plasmodium* sp. in blood. These are very sensitive tests, but false positive results can occur in patients with rheumatoid factor.

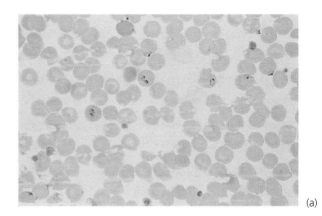

(a)

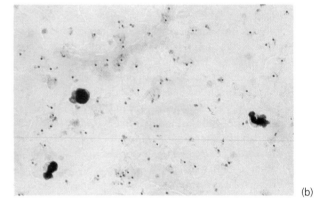

(b)

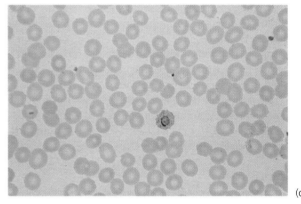

(c)

Figure 24.14 Malaria: (a) thin blood film, stained with a Romanowsky stain, shows numerous *Plasmodium falciparum* trophozoites, some red cells with double parasitization and accroche forms (parasite applied to the rim of the red cell); (b) thick blood film, Field's stain, shows numerous *P. falciparum* trophozoites and one white blood cell; (c) thin blood film shows *P. vivax* trophozoite in a cell with typical Schuffner's dots. (*Continued.*)

Management

Benign malarias

Benign malarias should be treated promptly with chloroquine, to which they are rarely resistant. Chloroquine

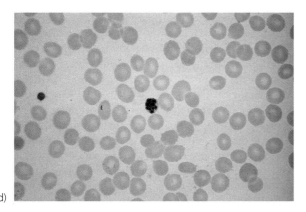

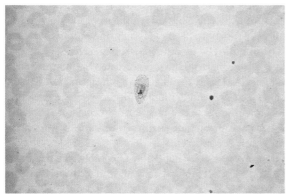

Figure 24.14 (*Continued.*) (d) thin blood film shows *P. vivax* schizont; (e) thin blood film shows *P. ovale* trophozoite.

clears the parasites from the blood, and terminates the acute feverish illness.

Chloroquine acts quickly and the fever should be abolished within 12–24 h. Nausea or intestinal irritation is rarely severe enough to prevent completion of the treatment. Pruritus particularly affects dark-skinned people and may make continued treatment intolerable, in which case quinine may be substituted.

Chloroquine exacerbates skin conditions such as psoriasis. Alternative agents such as quinine, or possibly Malarone®, should be used in this situation.

Eradication of liver parasites

Eradication of liver parasites (hypnozoites) is necessary after treatment of acute vivax or ovale malaria, as these 'dormant' parasites are not killed by chloroquine or quinine, which mainly affect developing schizonts. The only treatment currently available for this is oral primaquine (see box).

Treatment of acute non-falciparum malaria

1 Chloroquine (as base) 600 mg immediately (four tablets).
2 Next dose, after an interval of 6 h, 300 mg (two tablets).
3 Two more doses of 300 mg, 24 and 48 h after second dose.

⚠ Equivalent doses of chloroquine preparations are: chloroquine base 150 mg = chloroquine sulfate 200 mg = chloroquine phosphate 250 mg.

Clearing hypnozoites from the liver

1 *For P. vivax eradication:* primaquine, orally 30 mg daily for 14 days (500 μg/kg for children).
2 For *P. ovale eradication:* primaquine, orally 15 mg daily (250 μg/kg daily for children). This may be taken in two divided doses, or as 15 mg daily for 4 weeks if nausea is a problem with a 30 mg dose.

⚠ Primaquine is contraindicated in individuals who have glucose-6-phosphate dehydrogenase deficiency, in whom it may cause severe haemolysis. Although experts in malaria management may cautiously treat patients with mild deficiencies, in most cases, primaquine cannot be given. Primaquine should not be given to pregnant women, in case it leads to severe haemolysis in the fetus. The strategy for managing patients who cannot take primaquine must be to treat relapses of malaria promptly, until they cease to occur. Most patients cease to suffer relapses after 18–30 months.

Difficult cases

Chloroquine resistance has been well-documented in some cases of vivax malaria contracted in Indonesia and Papua New Guinea. Quinine is then the treatment of choice. Malarone® (atovaquone plus proguanil) or coartem (artemether plus lumefantrine) are alternatives.

For malaria cases in which the parasites cannot be speciated (often because their morphology is altered by the effects of partial or suboptimal treatment), treatment should be with quinine plus either doxycycline or fansidar, as for falciparum malaria (see below).

Uncomplicated falciparum malaria

Patients with falciparum malaria should be admitted to hospital to initiate treatment.

An increasing proportion of *P. falciparum* parasites are chloroquine resistant in all endemic areas, so that chloroquine is now unreliable therapy. Uncomplicated cases respond to oral quinine 600 mg 8-hourly. The dose interval can be increased to 12-hourly, or the dose reduced to 400 mg 8-hourly, if features of cinchonism (nausea, tinnitus or deafness) occur. On treatment the temperature drops slowly or erratically, often remaining normal only after 2–3 days. Parasitaemia should become undetectable after 48 h of treatment.

With quinine monotherapy 7–10 days' potentially toxic treatment is needed. This can be shortened by giving quinine for 4 or 5 days, then discontinuing the quinine and giving doxycycline 200 mg daily for 10 days or, in patients who cannot take doxycycline, Fansidar®, three tablets, as a single dose. Fansidar® resistance is common in the Far East and some east African countries, but failures with quinine/Fansidar® regimes rarely occur in UK practice. Clindamycin is also an effective antimalarial drug, especially useful for children, and in pregnancy, when doxycycline is contraindicated.

Other oral treatments for uncomplicated falciparum malaria include Malarone®, a fixed-dose combination of proguanil hydrochloride 100 mg and atovaquone 250 mg, and co-artem (Riamet®), a fixed-dose combination of artemether 20 mg and lumefantrine 120 mg. Malarone® has few adverse effects, other than nausea, but occasional cases of recurrence after treatment are related to a mutation that confers resistance to atovaquone. Co-artem must be taken with a lipid-rich snack or drink (such as milk) to maximize absorption.

Mefloquine can be used to treat uncomplicated falciparum malaria, but is usually only used when other drugs are unavailable. Psychotic side-effects are relatively common at therapeutic doses; mild ataxia and nausea are common; the drug is contraindicated in epileptics, in severe liver disorder and in pregnancy. Rare side-effects include skin rashes and cardiac conduction defects. This drug can also be used to treat chloroquine-resistant vivax malaria.

Treatment of uncomplicated falciparum malaria

1 Quinine, orally, 600 mg 8-hourly for 5 days *followed by* doxycycline 200 mg daily for 7 days *or* Fansidar® single dose of three tablets (child: quinine, orally 10 mg/kg 8-hourly for 7 days, *followed if quinine resistance possible by* Fansidar®: up to 4 years, half tablet; 5–6 years, 1 tablet; 7–9 years 1.5 tablets; 10–14 years, 2 tablets).
Alternative for treatment after initial quinine: clindamycin orally 300 mg 6-hourly for 5 days (child 20 mg/kg, maximum 900 mg daily).
2 Atovaquone/proguanil (Malarone®) orally, four tablets daily for 3 days (child 11–20 kg, one tablet daily; 21–30 kg, two tablets daily; 31–40 kg, three tablets daily).
3 Co-artem (Riamet®) for adult, and child over 35 kg, or 12 years, orally, four tablets initial dose, followed by four tablets at 8, 24, 36, 48 and 60 hours (total 24 tablets).
Alternative: Mefloquine: two doses of 10 mg/kg, 6 or 8 h apart; maximum dose 1500 mg (child 20–25 mg/kg in two divided doses 6 hours apart).

Severe and complicated falciparum malaria

Well-recognized severe forms of malignant malaria include cerebral malaria and severe haemolysis with dark urine (blackwater fever). Other important features include: acute respiratory distress syndrome (ARDS); renal failure; acidosis; and disseminated intravascular coagulation. These clinical features may occur separately or together. Although patients vary in their susceptibility to severe disease, and residents of endemic areas may be less likely to suffer severe disease, any patient with malaria can deteriorate rapidly and without warning. For this reason, cases with a parasitaemia above 2% of all red cells, must be treated immediately with intravenous therapy.

During the early phase of treatment, blood glucose, respiratory function, haemoglobin, urine output and renal function must be monitored. Hypoglycaemia occurs easily in severe malaria, and is exacerbated by quinine, which inhibits the release of glucose from the liver. The Glasgow coma scale is useful in monitoring recovery from cerebral malaria in adults. A combination of hypoxia and hypoglycaemia may precipitate seizures, which greatly worsen the prognosis in cerebral malaria.

The average case-fatality rate of complicated malaria is around 15%, but long-term sequelae in recovered patients are rare.

Treatment of severe and complicated falciparum malaria

Intravenous quinine is the mainstay of treatment in many countries. This cannot be given by bolus injection because of the high risk of cardiac depression, cerebral toxicity, nausea and vomiting. The safest procedure is to give an infusion of 10 mg/kg in 5% dextrose over 4 h. This dose can be repeated 12-hourly until oral therapy is possible (see above). The first dose of quinine should be 20 mg/kg in very ill patients, and in children. The pharmacokinetics of quinine show that the drug is concentrated in parasitized red cells, so in the initial phase of high parasitaemia, little free drug exists in the blood, and toxicity is rare; however, as the parasite count falls, most of the drug remains free in the plasma and symptoms of cinchonism may then develop.

Quinidine intravenously, 10–15 mg/kg 12-hourly, is effective if quinine is unavailable.

Other important aspects of treatment include maintenance of adequate blood glucose levels, correction of anaemia by transfusion if necessary, and avoidance of convulsions. Many experts advocate the use of prophylactic phenytoin in patients with coma.

Pulmonary oedema may require vigorous treatment; diuretics and fluid restriction are not highly effective (and in excess could contribute to reduced cerebral blood flow). High inspired oxygen tension and intermittent or continuous positive-pressure ventilation are required in many cases.

Acute renal failure may require support therapy with haemoperfusion or dialysis.

In spite of severe cerebral disturbance and difficulties with fluid handling, there is little evidence that treatment for cerebral oedema is helpful in cerebral malaria, except possibly in very young children. Indeed, it has been shown that dexamethasone therapy may *prolong* coma without improving the outcome.

There is increasing evidence that artesunate is superior to quinine in the treatment of severe falciparum malaria. It clears the parasitaemia more quickly, and reduces the duration of fever. It has fewer toxic effects and, although it may cause mild cerebral irritation, this does not affect the progress of the patient's recovery.

Drug treatment of severe and complicated falciparum malaria

1 Quinine i.v. 10 mg/kg 12-hourly as an infusion over 4 hours in 5% dextrose (first dose should be 20 mg/kg (maximum 1.4 g) in children or if there is risk of quinine resistance). Change to treatment for uncomplicated malaria when the patient's condition permits.

2 Artesunate (named patient treatment in the UK). *Recommended dose:* 2.4 mg/kg i.v. (or 120 mg) loading dose, followed by eight doses of 1.2 mg/kg (or 60 mg), given 12-hourly.

Exchange transfusion is strongly advocated by some specialists, but hardly ever employed by others. It rapidly reduces parasitaemia in peripheral blood, but its effect on parasites sequestered in the microcirculation, and on their pathological effects, is less certain. The trials that have been reported did not demonstrate an improved outcome with exchange transfusion.

Prophylaxis of malaria

Key steps in avoiding morbidity from malaria

A Awareness of the presence of malaria risk.

B Bite avoidance, using appropriate behaviour, clothing, knock-down insecticide and insect repellent.

C Chemoprophylaxis.

D Diagnosis and treatment of fever without delay.

UK Advisory Committee on Malaria Prevention

The risk of malaria is greatly reduced by avoidance of mosquito bites. The anti-mosquito measures already described should always be used by travellers to endemic areas to enhance the effect of chemoprophylaxis.

Recommendations for chemoprophylaxis of malaria

1 For benign malarias: chloroquine 300 mg weekly; second choice: proguanil 100 mg daily, both starting 1 week before exposure and taken continuously until 4 weeks after the last exposure.

2 For falciparum malaria in all endemic regions of the world: mefloquine 250 mg (one tablet) weekly, starting 2–3 weeks before exposure and continuing for 4 weeks after last exposure. Mefloquine resistance occurs in a number of areas, but particularly in the borders of Thailand with Cambodia and Myanmar (Burma). Mefloquine can cause ataxia and nausea, which are reduced if the dose is taken after an early evening meal. Agitation and intrusive dreams, and rare cases of severe depression or psychosis have been reported. At prophylactic doses severe adverse effects are uncommon (about 1 in 10 000 recipients). This drug is contraindicated in epileptics, and those with a history of psychiatric disease. Accumulating data suggest that it is safe in the first trimester of pregnancy.

3 Alternative: Malarone®: one tablet daily starting 3 days before exposure and taken until 1 week after last exposure (child 11–20 kg, one-quarter of adult dose; 21–30 kg, half adult dose; 31–40 kg, three-quarters of adult dose).

4 For falciparum malaria in the Cambodian and Myanmar borders of Thailand (where mefloquine resistance is very common): doxycycline 100 mg daily, starting 3 days before exposure and continuing until 4 weeks after last exposure. This drug must be taken with at least half a glass of water, as it may cause oesophageal irritation. A minority of users suffer from an increased tendency to sunburn: the use of hats, other clothing and sunscreen should be recommended. *Alternative*: Malarone® (see above).

Malarial prophylaxis for pregnant women and for children

Falciparum malaria threatens life, and in pregnancy endangers both the mother and the pregnancy itself. Neonates and infants are susceptible to severe disease. Malarial prophylaxis is much less risky than the disease itself and should never be omitted.

Malarone® is not recommended for use in pregnancy or breastfeeding, as there are insufficient data on the safety of atovaquone; however, it is available for prophylaxis in children from 11 kg (see above). Tetracycline must not be used in infancy, pregnancy or during breastfeeding.

The best option for prophylaxis of falciparum malaria in infancy, pregnancy and breastfeeding is mefloquine. If mefloquine cannot be tolerated and other drugs are contraindicated, weekly chloroquine 300 mg *plus* daily proguanil 100 mg may be used, but is much less effective, and vigilance must be maintained in case malaria occurs. Quinine is the drug of choice for malaria treatment in pregnancy.

UK recommendations for child doses are shown in Table 24.1.

Even if prophylaxis has been taken continuously, malaria should be actively excluded if a traveller becomes feverish after returning home; falciparum malaria occasionally occurs up to a year later, and vivax up to 2 or 3 years.

Standby treatment for malaria

Even when prophylaxis is properly used, there is a small risk of malaria. Travellers in remote areas who cannot obtain timely investigation and treatment, may be given a supply of emergency treatment. Suitable drugs for this are quinine (600 mg 8-hourly for 5 days), *plus* Fansidar (single dose of three tablets), or Malarone® (four tablets daily for 3 days). Medical attention should be sought as soon as possible after their use, so that the blood can be checked for parasites and anaemia.

Leishmaniasis

Introduction

Leishmaniasis is disease caused by protozoa of the genus *Leishmania*. There are a number of different species, all transmitted by sandflies, causing several types of either cutaneous or visceral (systemic) disease.

The natural history of the disease involves the transmission of flagellate parasites via the bite of a sandfly. The parasites then lose their flagella (becoming 'amastigote' forms) and invade macrophages. Many parasites survive within macrophages by inhibiting the elaboration of nitric oxide by the activated phagocyte. Although subclinical infections probably often occur, established disease has a prolonged time course, and visceral infections cause progressive reticuloendothelial disease with increasing immunosuppression and, if untreated, will end fatally.

Cutaneous leishmaniasis is common in tropical countries, the Middle East and many Mediterranean areas, and also occurs in tropical areas of central and south America. It is caused by local invasion of dermal macrophages, and usually affects exposed sites such as the face, forearm or legs. Small rodents, and sometimes dogs, are often the reservoir of infection. Visceral leishmaniasis has a more patchy distribution within the same tropical areas, affecting tropical areas of Africa (but is rare in West Africa) and Bangladesh. It is a rare disease in the northern and central Mediterranean area, but is absent from the central areas of the Arabian peninsula.

Cutaneous leishmaniasis (CL)

Pathogens of 'old world' cutaneous leishmaniasis

- *Leishmania tropica*: transmitted from human to human; incubation period of up to 4 months. Typically produces an ulcerated lesion, 1.5–5 cm diameter, with an indurated (rolled) margin, slowly healing to leave a 'tissue paper' scar.
- *L. major*: transmitted from rodents (gerbils) to humans.
- *L. infantum*: rare cause of CL in Europe.

Table 24.1 Malarial prophylaxis: UK recommendations for child doses

| Weight in kg | Drug and tablet size | | | | Age |
	Chloroquine 150 mg base	Proguanil 100 mg	Mefloquine 250 mg	Doxycycline 100 mg	
Under 6.0	0.125 dose	0.125 dose			
	1/4 tablet	1/4 tablet	NR	NR	Birth to 12 weeks
6.0 to 9.9	0.25 dose	0.25 dose	0.25 dose		3 months to
	1/2 tablet	1/2 tablet	1/4 tablet	NR	11 months
10.0 to 15.9	0.375 dose	0.375 dose	0.25 dose		1 year to
	3/4 tablet	3/4 tablet	1/4 tablet	NR	3 years 11 months
16.0 to 24.9	0.5 dose	0.5 dose	0.5 dose		4 years to
	1 tablet	1 tablet	1/2 tablet	NR	7 years 11 months
25.0 to 44.9	0.75 dose	0.75 dose	0.75 dose	Adult dose	8 years to
	1 ½ tablets	1 ½ tablets	¾ tablet	from age 12 years 1 tablet	12 years 11 months
45 kg & over	Adult dose	Adult dose	Adult dose	Adult dose	13 years and over
	2 tablets	2 tablets	1 tablet	1 tablet	

When both are available, weight is a better guide than age for children over 6 months.

Caution: In some countries tablet size may vary. NR: Not recommended.

Note: These doses do not always correspond with the dosages stated in the Statement of Product Characteristics (Product 'data sheet'), but are recommended by the HPA Advisory Committee on Malaria Prevention in UK Travellers. For Malarone paediatric® the manufacturer's recommendations for dosages should be followed.

• *L. aethiopica*: occasional cause of CL in Ethiopia and Kenya highlands.

Pathogens of 'new world' cutaneous leishmaniasis

• *L. mexicana* complex: var *mexicana*; var *amazonensis*: transmitted from forest rodents to humans; var *venezuelensis*: has an unknown reservoir. Typically causes scabbing, ulcerating lesions, e.g. *L. m. mexicana* causes 'chiclero's ulcer' on the ear lobe of gum-tree workers.
• *L. braziliensis* complex: var. *braziliensis*: transmitted by forest rodents and domestic animals; var. *panamensis* and var. *guyanensis*: mainly carried by sloths; var. *peruviana*: transmitted from dogs to humans. Most cause both CL and mucocutaneous leishmaniasis. *L. b. braziliensis* can also cause a visceral disease.

In the old world CL is mostly caused by *L. tropica* or *L. major*. A typical oriental sore is caused by var. *major*, and has an incubation period of around 2–6 weeks. The ragged, punched-out ulcer usually appears on the face or extremities, and is accompanied by regional lymphadenopathy. A scab may repeatedly form and separate, and the edges of the lesions may be thickened (see Fig. CS.8 in Case Study 24.2). The lesion reaches up to 2.5 cm in diameter and heals slowly over 3–6 months, leaving a depressed, tissue-paper scar.

A more indolent, granulomatous type of lesion can rarely occur with an incubation period as long as a year. It appears as a purplish nodule, which gradually breaks down and slowly heals over several months. Lymphadenopathy is rare.

The diagnosis can be made clinically in typical cases, or confirmed by demonstrating the protozoa in scrapings or split skin smears from the edge of a lesion or in biopsy material.

In the new world, ulcerating lesions can be caused by *L. mexicana* (chiclero's ulcer, usually affecting the pinna of the ear), or *L. peruvia*, which causes ulcers on the face of children, and leaves atrophic scars.

Specific treatment of localized cutaneous lesions is not often required, as spontaneous healing occurs in a few weeks or months. Disfiguring facial lesions may be treated as for visceral leishmaniasis (see below). Intralesional meglumine may be effective in controlling extending or frequently recurring lesions.

Oral fluconazole and itraconazole may be effective against old world cutaneous leishmaniasis.

Mucocutaneous leishmaniasis (ML)

Leishmania (Viannia) braziliensis causes extensive ulceration, usually of the face. If not vigorously treated, this metastasizes along lymphatic pathways to cause disfiguring mucocutaneous disease, with granulomas and tissue destruction, which includes the nasal septum in about 5% of sufferers. Secondary infection commonly exacerbates the condition. Vigorous and prolonged drug treatment is necessary for ML, using liposomal amphotericin, sodium stubogluconate and/or paromomycin. Secondary infection may complicate the lesion and should be treated promptly to avoid further scarring. Reconstructive surgery may be necessary to restore the cosmetic appearance and function of the face and mouth.

Visceral (systemic) leishmaniasis (VL)

VL is a systemic disease associated with dissemination of parasites throughout the reticuloendothelial system.

Causes of visceral leishmaniasis

The *L. donovani* complex:
• *L. donovani*: human to human transmission; in India, Bangladesh, Burma, Sudan, Kenya, Somalia.
• *L. infantum*: a reservoir in dogs; in dry rocky terrain across the Mediterranean, Middle East, Asia and western China.
• *L. chagasi*: a reservoir in dogs, foxes and opossums; in northern South America and southern Mexico.

L. infantum causes opportunistic visceral leishmaniasis in HIV-positive patients, often following exposure in southern Europe.

The incubation period averages about 3 months, but can range from 10 days to two years. Fever is constant for the first few weeks, but then becomes intermittent. The typical features of massive splenomegaly, pancytopenia and increased skin pigmentation (which gives the disease the name 'kala-azar' – the black sickness) develop slowly over many months. One-third of patients have moderate liver enlargement. The erythrocyte sedimentation rate is often very high, and is related to a polyclonal elevation of immunoglobulin G (IgG). Most untreated cases die of uncontrolled bleeding, due to thrombocytopenia and hypersplenism, or secondary bacterial infections, including dysentery and tuberculosis.

The diagnosis may be made by demonstrating protozoa (Leishman–Donovan bodies) packed into the mononuclear phagocytes of the spleen (Fig. 24.15), liver or bone marrow. Splenic aspiration is most likely to give positive results. Serodiagnosis by ELISA test and genome detection by PCR are also possible. Culture of blood, bone marrow or splenic aspirate, or biopsy, is performed in reference laboratories.

Atypical forms of infection have been reported in US soldiers exposed to cutaneous leishmaniasis infection in Afghanistan, Iraq and the Middle East. These included low-grade fevers (but no abnormal blood or skin signs), arthralgia, diarrhoea, or mildly elevated transaminases. Cultural and nucleic acid detection techniques demon-

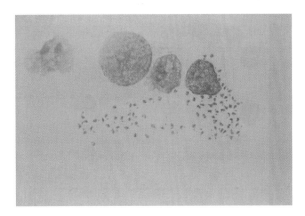

Figure 24.15 Systemic leishmaniasis: this patient had typical hepatosplenomegaly, pancytopenia and persisting fever. Splenic aspirate revealed Leishman–Donovan bodies – macrophages packed with the protozoan parasites.

strated the presence of *Leishmania* and specific DNA in affected tissues or bone marrow.

The treatment of systemic leishmaniasis should be supervised by a specialist. Where it is available, liposomal amphotericin is the treatment of choice, as it has low toxicity.

Management of visceral leishmaniasis

1 Liposomal amphotericin B 1 mg/kg daily (up to 3 mg/kg as initiation therapy), used in a 3-week course.

2 Miltephosin, orally, 50 mg twice daily for 28 days, has a high response rate, but may be followed by relapse in a significant number of patients (contraindicated during pregnancy; avoid pregnancy for 6 months after treatment).

3 Organic pentavalent antimony (sodium stibogluconate, Pentostam), i.m. 20 mg/kg daily for 28 days (may be given by i.v. infusion over 20 minutes in thrombocytopenic patients). Side-effects such as vomiting, coughing and substernal pain are common. Pancreatitis is common in patients with associated HIV infection. Injections of this drug are themselves painful. Failure of pentavalent antimony treatment is increasingly common in India.

4 Paromomycin (an aminoglycoside) i.m. 15 mg/kg daily for 21–28 days, can be used as an alternative treatment, or combined with antimonial treatment.

⚠ PKDL (post-kala-azar dermal leishmaniasis) can follow resolution of *L. donovani* (but not other VL) infections. It usually consists of the appearance of hypopigmented nodules and plaques, but can cause severe desquamating lesions also affecting mucosae. It can occur months or years after the initial illness, and is more common, with shorter latency, in central Africa. Scanty amastigotes are present in the lesions. PKDL should be treated with a 3-month course of VL therapy.

Lymphatic filariasis

Lymphatic filariasis is a disease caused by the nematode worms *Wuchereria bancrofti* and *Brugia malayi*. Adult worms live coiled together in the lymphatics of humans. Pregnant female worms release large numbers of microfilariae, which reach the peripheral blood and must then be ingested by a biting mosquito to complete their life cycle. Several species of mosquito can transmit the disease.

Clinical manifestations begin 9–12 months after infection. They are probably due to a hypersensitivity reaction to the release of larvae into the blood. There is usually a series of episodic fevers, each of which lasts a few hours. The fever is rarely high, but shivering and sweating are common. Lymphangitis may be visible if it affects lymphatics near to the skin surface; common sites are the lower leg or the thigh. Pain and redness of the skin mimic cellulitis or erysipelas. Abdominal pain or scrotal inflammation are rarer manifestations.

The filariae die within about 5 years of a single infecting episode, and no further symptoms occur. In rare cases with many re-infections, accumulating fibrosis of affected lymphatics can eventually produce lymphoedema (elephantiasis).

Diagnosis is best made serologically by an IgG antifilaria ELISA test. Microfilariae can be demonstrated in the peripheral blood of patients in endemic areas, but the sensitivity of this test is only about 70%. Midday and midnight blood films should be examined, to allow for differing periodicities of microfilaraemia.

The treatment of choice is ivermectin, which clears many parasites without significant toxicity, and without causing febrile reactions. Diethylcarbamazine kills both adult worms and microfilariae. The dose is 6 mg/kg daily in divided doses. Febrile allergic reactions are common at the onset of treatment, which is commenced under close medical supervision, starting with 1 mg/kg daily and working up to a therapeutic dose. Full dosage is then continued for 21 days. Antihistamines and/or corticosteroids may be needed until febrile reactions cease.

Treatment of filariasis (excluding onchocerciasis)

1 Ivermectin, 150 mg, single dose.

2 Diethylcarbamazine orally:
- Day 1: 1 mg/kg as a single dose.
- Day 2: 3 mg/kg as a single dose.
- Day 3: 6 mg/kg as a single dose.
- Next 20 days: 6 mg/kg daily.

Loiasis

Loiasis, caused by *Loa loa*, is often asymptomatic but may produce localized allergic skin swellings (Calabar swellings), or is visible if an adult worm migrates across the eye. If it requires treatment, diethylcarbamazine is effective.

Onchocerciasis (river blindness)

Onchocerciasis is transmitted by *Simulium* flies, which deposit their eggs near fast-flowing water. Adult *Onchocerca volvulus* live in the skin, often forming macroscopic nodules that contain convoluted worms. Illness is caused by inflammatory reactions to the millions of microfilariae that invade the skin and eye. Treatment with diethylcarbamazine is dangerous because of severe inflammatory reactions with possible destructive involvement of the eye. Ivermectin 150 mg produces gradual and sustained reduction in microfilariae with little allergic reaction, and the dose can be repeated annually until infection is eradicated. Surgical excision of skin nodules has also been used as a means of reducing microfilarial numbers.

Some important zoonoses

A zoonosis is an animal disease that can be accidentally transmitted to humans. Several such diseases, including Q fever, ornithosis, hydatid disease, hantavirus infections and Lassa fever, have already been mentioned in other chapters of this book. A few diseases, such as *Campylobacter* infections and salmonelloses, are common to animals and humans and are relatively readily transmitted between humans as well as between animals. True zoonoses are not easily transmitted from person to person, save for cases of exceptional pulmonary disease, as in pneumonic plague.

Most zoonoses are acquired when humans intrude into the animals' environment, or handle animals or their carcasses. This occurs during farming, handling raw animal products such as hides, meat or bones, and during hunting, camping or trekking. Consumption of untreated or uncooked products such as milk, cheese or preserved meat can also be a means of acquiring a zoonosis.

Ongoing improvements in animal husbandry, occupational health measures and food hygiene have made many zoonoses uncommon in Britain and other western countries. Reported cases of zoonotic disease therefore often affect travellers or migrants.

Common activities that predispose to zoonoses
Consuming untreated milk, cream, yoghurt and curd cheese
1 *Salmonella*, *Escherichia coli* O157 or *Campylobacter* infections.
2 Brucellosis.
3 Q fever.
4 Tick-borne encephalitis (when the animal is infected).

Hunting, trapping, skinning and butchering wild animals
1 Plague.
2 Tularaemia.
3 Rabies.

Butchering farm animals
1 Q fever.
2 Streptococcal skin infections.
3 Brucellosis.
4 *Erysipelothrix* skin infections.
5 *Streptococcus suis* systemic infections.
6 Anthrax.
7 Crimean-Congo haemorrhagic fever (including from ostriches).

Eating undercooked meat
1 Toxoplasmosis.
2 Trichinellosis (from pork, boar, horse).
3 Pork or beef tapeworms.
4 *Salmonella*, *Escherichia coli* O157 or *Campylobacter* infections.

Handling dead animals, untanned hides or unpasteurized bonemeal
1 Q fever.
2 Tularaemia.
3 Plague.
4 Anthrax.

Ingestion or inoculation of animal urine
1 Leptospirosis.

Camping, hiking or forestry working in warm climates
1 Tick-borne encephalitis.
2 Borreliosis.
3 Arboviral encephalitides.
4 Hantavirus infections (haemorrhagic fever with renal syndrome (HFRS) or hantavirus pulmonary syndrome).

Toxoplasmosis

Introduction and epidemiology

Toxoplasma gondii is a protozoan parasite of the phylum Sporozoa. It is found in the tissues of almost all warm-blooded creatures, but the only hosts for its definitive life cycle are the cat family. Cats are infected by predation on other infected creatures, or by consumption of oocysts derived from the faeces of recently infected cats.

In the cat, oocysts develop in the intestinal mucosa, and release parasites into the bloodstream and tissues. These replicate rapidly by fission, and are called tachyzoites. Tissue tachyzoites form pseudocysts: masses of organisms within expanded cells. The organisms in tissue cysts are quiescent, as they are controlled by cell-mediated immunity, and are called bradyzoites. Meanwhile some parasites re-enter the enterocytes and develop into oocysts, which are shed in the faeces. Shedding of oocysts begins about 10 days after infection and persists for 2–3 weeks.

Tachyzoites, bradyzoites and oocysts are all infectious by the alimentary route. Humans are usually infected by contact with uncooked meat. Bradyzoites can be demonstrated in many types of butcher's meat, and in the types of ham that are preserved and eaten without cooking in many countries. Cats, particularly kittens, who become infected as they begin to predate, excrete infectious oocysts for a limited period and can then infect their human owners.

Toxoplasmosis has occurred when a transplanted heart contained bradyzoites. In the immunosuppressed recipient the presentation of the disease may be modified.

In the UK the peak age for seroconversion to toxoplasmosis is 15–35 years, and about half of all adults have evidence of past infection. Immunocompetent individuals rarely have a recognizable illness. The importance of the disease lies in its ability to cause trans-placental infection (see Chapter 17) and opportunistic infection in acquired immunodeficiency syndrome (AIDS) sufferers (see Chapter 22).

Clinical features

Clinically expressed toxoplasmosis tends to present in one of three ways.

1 A mononucleosis syndrome is common in young adults. The features are fever, malaise and one or more enlarged lymph nodes. There is atypical mononucleosis in the peripheral blood but the heterophile antibody test is negative. The illness is self-limiting, with a variable duration, up to many weeks.

2 A single, persistently enlarged lymph node, or occasionally a skin or soft-tissue nodule. The differential diagnosis of lymphadenitis is large, and excision biopsy is often needed to exclude tuberculosis, sarcoidosis or lymphoma.

3 Acute choroidoretinitis is a feature of late-stage toxoplasmosis. Although there may have been a feverish illness some weeks or months before the onset of eye symptoms, such a history may be remote, and is rarely elicited. Ocular features of toxoplasmosis can develop months or years after congenital infection.

Rare manifestations include myocarditis, encephalitis, encephalomyelitis and pneumonitis (all more common in the immunosuppressed).

⚠ In immunosuppression due to advanced HIV infection, bradyzoites may become activated. This is particularly common in the brain. The patient may present with seizures, reduced consciousness level or focal neurological signs, or a combination of these. CT scans typically show one or more ring-enhancing lesions.

Diagnosis

Several serological tests are useful in the diagnosis of toxoplasmosis. A latex agglutination test is widely used for screening sera. Titres rise rapidly during acute disease; titres of 1 : 4000–1 : 64 000 are common, and fall slowly over 18–24 months to reach background seropositivity.

Confirmatory tests carried out in reference laboratories include IgM ELISA tests, the *Toxoplasma* dye test (which uses live trophozoites to measure antibody-mediated inhibition of dye uptake by live parasites), a complement fixation test and a haemagglutination test. The dye test is the 'gold standard' against which other tests are compared.

The *Toxoplasma* dye test rises and falls in parallel with the latex agglutination test, but is less prone to false-positive results. The complement fixation test has a similar time course. Dye-test antibody titres may reach 1 : 32 000–1 : 64 000 in acute disease.

IgM antibodies indicate recent infection. They persist for 6 months or more, suggesting that the initial infection subsides slowly as bradyzoite pseudocysts form and are controlled by the immune response. IgM antibodies sometimes persist for over a year. This is a problem for women who wish to conceive although, if the dye test and latex agglutination titres are falling, the risks of parasitaemia and trans-placental infection are almost certainly negligible. Antibody affinity studies can indicate whether the infection is at a late stage; these are reference laboratory tests.

Culture of CSF may be helpful in acute brain infection. This is a specialist procedure, and must usually be prearranged with a reference laboratory. Positive cultures can also be obtained from infected placenta, products of conception, CSF and brain in congenital infection.

PCR-based tests are particularly useful in investigating CSF in neurological disease, or for biopsy material.

Late toxoplasmosis: eye disease

Choroidoretinitis is a late manifestation of toxoplasmosis, and occurs when acute antibody titres have declined. Dye test and latex agglutination titres of 1 : 256–1 : 512 (dye test) or 1 : 128 (latex agglutination) are typical. The haemagglutination test has a slower response to acute infection, and may show significantly elevated titres at the time that eye disease occurs.

Histological diagnosis

A typical histological picture in excised lymph nodes is strongly suggestive of toxoplasmosis. Giemsa-stained tissue smears, myocardial biopsy or brain biopsy preparations can occasionally be shown to contain cysts or typical crescent-shaped trophozoites.

Diagnosis of toxoplasmosis

1 Serology: IgM ELISA, latex agglutination, complement fixation test (haemagglutination test in choroidoretinitis).
2 Histological appearance of biopsied nodes.
3 Giemsa-stained tissue specimens.
4 Culture of cerebrospinal fluid or affected tissue (rarely performed).
5 Polymerase chain reaction-based tests.

Treatment

Toxoplasmosis is self-limiting in most cases, and primary infection is often diagnosed relatively late. The risks of treatment must therefore be balanced against the likely benefit. However, treatment is always justified in myocardial or central nervous system disease, and in symptomatic immunosuppressed patients.

The treatment of choice is a combination of sulphonamide and pyrimethamine. Sulfadimidine is the best choice of sulphonamide available in the UK. The dose of sulphonamide should not be less than 1 g 6-hourly. Pyrimethamine is given in a dose of 50 mg daily.

Sulphonamides and pyrimethamine are both inhibitors of folate synthesis, and in these doses can cause a significant fall in the white cell count. The blood count should be closely monitored during treatment, which must be continued for at least 4 weeks. Folic acid or folinic acid may be given during treatment.

Sulphonamides are contraindicated in pregnancy. The treatment of choice is then spiramycin. This is not available in the UK, and must therefore be purchased for the individual patient (see Chapter 17).

For patients who cannot tolerate sulphonamides, pyrimethamine plus either clindamycin or azithromycin is a suitable alternative regimen.

Brucellosis

Introduction and epidemiology

Brucella spp. infect many warm-blooded animals, particularly hoofed animals, such as goats, cattle and camels, in which they often cause infectious abortion. Inapparent infection and excretion of the organisms also occur. The animal's milk may be contaminated, either by excretion in the milk itself or by contamination during unhygienic milking processes. Control programmes have made the infection extremely rare in the UK, and most herds are nowadays *Brucella*-free. Almost all cases are now imported. Brucellosis is still common in farming areas of Turkey and in rural parts of Africa and Asia. In some tropical countries milk is still 'preserved' by adding the animal's urine.

Humans usually acquire infection by consuming unpasteurized dairy products or by contact with infected meat or products of conception, and many cases follow occupational exposure. The infective dose is low, and aerosol-transmitted laboratory infections can occur. Chronic brucellosis or undulant fever was described some years after recognition of acute disease, again after consumption of cows' milk.

Clinical features

Brucellosis has a variable clinical picture. The incubation period ranges from 5 days to several weeks, and the onset may be acute or insidious.

Acute brucellosis

Acute brucellosis has an abrupt onset, with high swinging fever, often rigors and sometimes myalgia and arthralgia. The patient feels severely unwell, but may display few physical signs. About half have enlarged lymph nodes in the cervical chain, and about a quarter have splenomegaly. The white cell count is often, but not always, raised and the liver function tests may be slightly abnormal because of granulomatous hepatitis. Almost any organ or tissue can be involved, although granulomatous inflammation of reticuloendothelial organs is the most constant feature. This causes lymphadenopathy, splenomegaly and abnormalities of the liver function tests.

Infective arthritis affecting a large joint, or chronic osteomyelitis of the spine are relatively common manifestations of brucellosis. Differentiation of vertebral arthritis from tuberculosis is important. Brucellosis tends to affect the cervical spine, but accurate diagnosis depends on clinical suspicion and appropriate diagnostic tests (see also Chapter 18).

Orchitis is relatively common. It develops after several days of fever, and may be associated with chills and increased malaise. Physical signs vary from aching pain in the testicles to acute tenderness and swelling. Renal involvement is usually evidenced by proteinuria and 'sterile' pyuria.

Neurobrucellosis is uncommon, but can be severe. Meningoencephalitis is the commonest feature, and can sometimes be life-threatening. Encephalopathy is a rare feature. Depression or acute psychosis can occur; only the accompanying fever, or local disease elsewhere, will suggest the diagnosis.

Endocarditis is a rare manifestation of brucellosis, often causing rapid valve destruction.

Untreated acute brucellosis is rarely fatal, but causes severe morbidity for 5 or 6 weeks. Brucellae are intracellular pathogens that are able to survive inside cells of the reticuloendothelial system using a superoxide dismutase and nucleotide-like substances to inhibit the intracellular killing mechanisms of the host.

Chronic brucellosis

Chronic brucellosis is uncommon. It may follow on from an acute attack or commence insidiously. Typically, it causes malaise, depression and a fever which waxes and wanes over periods of 2 or 3 weeks (undulant fever). The white cell count is often low, because of neutropenia.

Clinical features of brucellosis

  1 Fever (undulant in chronic infection).
  2 Pyogenic arthritis.
  3 Spinal osteomyelitis.
  4 Lymphadenopathy.
  5 Splenomegaly (especially in acute disease).
  6 Abnormal liver function.
  7 Orchitis or testicular pain.
  8 Endocarditis.
  9 Meningoencephalitis.
  10 Depression or psychosis.
  11 Sterile pyuria.

Diagnosis

Diagnosis depends on epidemiological or clinical suspicion, and on appropriate laboratory tests.

Blood cultures are often positive, but *Brucella* spp. are relatively slow-growing, and cultures must be maintained for 4 weeks until they are declared negative if conventional culture methods are used. Modern automated blood culture systems allow positive detection within a few days. Culture of aspirated joint effusions, bony abscesses, or liver or bone marrow biopsies is much more sensitive than blood culture, so as many different specimens as possible should be obtained whenever the diagnosis is considered.

Serological tests are useful, and often more rapid than blood cultures. Traditional agglutination reactions (standard agglutination test, SAT) are often used but can show a strong anamnestic response after re-exposure to *Brucella* antigens, giving rise to a false-positive diagnosis. Enzyme-linked immunosorbent assay (ELISA) tests are widely performed, and can detect immunoglobulin M (IgM) antibodies in acute disease, or high levels of IgA and IgG in chronic infection.

Liver biopsy will often show multiple granulomata, especially in established disease. These are rarely caseating, unless caused by *B. suis*, and must be distinguished from the granulomata of sarcoidosis or miliary tuberculosis.

Treatment

The treatment of choice is a combination of doxycycline and rifampicin, which should be continued for 28 days. Relapse is rare after this regimen. Doxycycline plus an aminoglycoside is almost equally effective, but is less convenient than doxycycline plus rifampicin because of the need for intramuscular injection and blood-level monitoring.

For endocarditis or neurobrucellosis, doxycycline plus at least one additional drug should be continued for a total of 3 months.

Co-trimoxazole produces early improvement, but is followed by relapse with positive blood cultures in 35–50% of cases. It is not recommended as monotherapy. Quinolones have not been found as effective as tetracycline plus rifampicin. Both drugs are useful as components of multidrug therapy.

Depression or psychotic symptoms do not always respond readily to antimicrobial therapy. Psychiatric support may be required, and additional treatment with psychotropic drugs may be needed.

Brucellosis in pregnancy is likely to cause severe placental infection and abortion. Tetracyclines are contraindicated, but treatment with rifampicin 900 mg daily has proved successful, and successful continuation of pregnancy with this regimen has been reported. Children, in whom doxycycline is contraindicated, may be treated with a combination of co-trimoxazole and rifampicin.

Treatment of brucellosis

  1 Doxycycline orally 200 mg daily *plus* rifampicin 600 mg 12-hourly for 4 weeks.
  2 Alternative: doxycycline 200 mg daily 4–6 weeks *plus* gentamicin 2–5 mg/kg daily in three divided doses (with blood-level monitoring) for 4 weeks.
  3 In pregnancy: rifampicin orally 900 mg daily for 4–6 weeks.
  4 For a child under 12 years; co-trimoxazole orally; 6 months–5 years, 240 mg twice daily; 6–12 years, 480 mg twice daily for 8 weeks; plus rifampicin 10 mg/kg daily for the first 4 weeks.

Recurrences can occur, even after full tetracycline plus aminoglycoside treatment. They should be treated with a second course of the same therapy.

Rabies (see also Chapter 13)

Rabies is a lyssavirus infection of many warm-blooded animals including birds, squirrels, skunks, cats, horses and cattle, but is most often transmitted to humans by dogs, wolves, insectivorous bats and mongooses. The virus causes myeloencephalitis and is excreted in tears and saliva. It is inoculated by bite, scratch or mucosal contamination from animals excreting the virus. Virus enters the peripheral nerves via which it travels to the central nervous system.

The incubation period is only 1–2 weeks after bites on the head or face, but averages 60–90 days, and can be over a year, for bites of the lower leg, remote from the central nervous system. Illness often begins with paraesthesiae at the inoculation site. Extreme anxiety is common at this stage. The main clinical features are then either ascending polyneuritis or rapidly developing encephalitis. The patient may be obtunded or speechless (dumb rabies), or excitable and agitated (furious rabies; Fig. 24.16) with intermittent spasms of the pharynx precipitated by stimulation of the face or mouth (hydrophobia). There is no specific treatment.

Early diagnosis depends on clinical suspicion, and exclusion of more treatable causes of myelitis or encephalitis. Rabies virus can be demonstrated by PCR or immunofluorescence, and cultured from saliva, tears, nuchal biopsies and corneal scrapings. Rising titres of antibodies can be demonstrated. Negri bodies are seen in some neural cells in both humans and animals.

Prevention of rabies (see also Chapter 13)

1 Most travellers do not need pre-exposure prophylaxis: education about risk avoidance and post-exposure measures is sufficient.

2 Post-exposure prophylaxis with cell-culture derived inactivated rabies vaccine plus human rabies immunoglobulin (HRG) is highly effective.

3 Pre-exposure prophylaxis with a three-dose course of vaccine should be offered to those who will work with animals and for travellers visiting remote endemic areas where immediate post-exposure treatment may not be readily available.

Plague

Introduction

Plague, caused by *Yersinia pestis*, is naturally a flea-borne disease of rodents that exists in many rural and wooded areas throughout the world. Plague is transmitted from rat to rat by the rat flea, whose pharynx becomes blocked by oedema and replicating bacteria. When an infected flea bites, bacteria are regurgitated through its mouthparts, inoculating a large infective dose into its host. As infected animal hosts die and flea populations lose access to them, other animals and humans are increasingly bitten, and human cases of plague occur. Urban foci of transmission also exist, where feral animals, humans and rats share the environment (Fig. 24.17). Most human cases are acquired from inoculation via flea-bite transmission or by handling infected rodents. A few cases are acquired by inhalation of organisms, often from an infected human or companion animal.

Classical plague causes infection arising from the site of inoculation and spreading to the draining lymph nodes, frequently leading to severe bacteraemic disease. Infection

Figure 24.16 A cat with furious rabies attacks an object which has entered its cage. Courtesy of Dr D. Lewis.

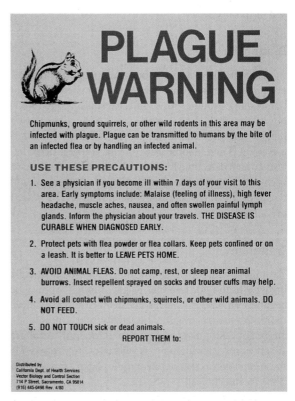

Figure 24.17 Plague hazard warning in an urban setting.

by the inhalational route is very efficient, and causes a severe and highly transmissible pneumonitis.

Plague is an infection that could be employed, via airborne infection, by bioterrorists.

Clinical considerations

The incubation period is usually 2–4 days, but can be up to 12 days. The onset of disease is abrupt, with fever, prostration and rigors. The characteristic clinical features of the local manifestations are inflammation or eschar at the site of inoculation and enlarged, suppurating regional lymph nodes (buboes). Inguinal nodes are more often affected than others. The mass of enlarged nodes is surrounded by boggy, often haemorrhagic oedema, and in untreated cases will often point and discharge pus after a week or two. Toxaemic and bacteraemic features soon appear. Haemorrhage, due to disseminated intravascular coagulation, is a major feature, causing a widespread bruising, and haemorrhage from the skin and mucosae.

Pneumonitis

Pneumonitis is a less common feature, but is often rapidly fatal. Extensive lung involvement and respiratory failure, with expectoration of bloody sputum, can develop in 24–36 h. Patients with pneumonitis are highly infectious, and readily transmit pneumonic plague to family members and health workers.

Diagnosis

A high index of suspicion is necessary to obtain the important epidemiological history. Gram-stained smears of lymph-node aspirate, pus or infected sputum often show numerous small Gram-negative rods with characteristic bipolar staining. The organism grows readily in standard cultures and blood culture media. Antigen-detection tests (which detect the F1 antigen), for use on blood, have been developed for use in endemic areas. *Y. pestis* should be handled using laboratory containment level 3 (CL 3) precautions.

Treatment

Treatment with broad-spectrum antibiotics is usually highly effective for bubonic plague, though it must be begun promptly, as the disease evolves quickly. Pneumonic plague has a high case fatality rate. Quinolones, aminoglycosides, tetracyclines and chloramphenicol are all effective. Ciprofloxacin is effective prophylaxis for those exposed to cases of pneumonic plague (doxycycline is the second choice).

Prevention of plague

There is no safe and highly effective vaccine for plague, though inactivated bacterial vaccines are available in some countries. Prevention depends on knowledge of risks and risk avoidance.

> Plague is an internationally notifiable disease. Countries in which human plague cases exist are listed weekly in the World Health Organization *Weekly Epidemiological Report*.

Tularaemia

Introduction

This is a disease of rodents and birds caused by *Francisella tularensis*. It is usually transmitted to humans by inoculation, either by bite or scratch from live animals, or by injuries acquired when handling or skinning carcasses. Ticks may also transmit tularaemia. It is usually a sporadic disease affecting hunters, trappers and tourists to rural or forested areas. European strains of *F. tularensis* can cause a pharyngeal or alimentary infection through ingestion of infected meat or animal products. Tularaemia has never been endemic in Britain.

The disease may be localized to the skin and local lymph nodes (ulceroglandular form) or cause systemic infection with fever, variable rashes and sometimes severe pneumonia (typhoidal form). Rare cases of ocular, pharyngeal or abdominal infection also occur. Patients with pneumonitis have a significant mortality. The infective dose is extremely low, and laboratory-acquired infection easily occurs if there is a breach of laboratory protocol.

This agent is recognized as an organism with potential for use in terrorist attacks.

Cutaneous–lymphatic (ulceroglandular) tularaemia

In cutaneous–lymphatic tularaemia, a nodular or suppurative lesion develops at the inoculation site, with extension along lymphatic channels. There is marked enlargement of draining lymph nodes, which are often tender and painful. Occasionally, the primary lesion appears as a painful conjunctival ulcer. Lymph-node pathology can exist without a detectable skin lesion.

Bacteraemic ('typhoidal') tularaemia

The patient presents with persisting high fever, with few local features. The severity of the illness varies from persisting fever to prostrating and debilitating disease. Splenomegaly and a transient rash are sometimes seen.

Some typhoidal cases develop a widespread pneumonitis, which can lead to respiratory failure. Although the infective dose by inhalation is very low, pneumonitis cases do not appear to excrete infectious bacteria, and human to human transmission by the inhalational route has not been reported.

Diagnosis

Diagnosis depends on recognizing the history of exposure, and excluding similar diseases, such as brucellosis and typhoid fever, as a cause of the symptoms. Blood cultures are rarely positive, but organisms can be cultured from skin lesions and lymph-node aspirate or biopsy. Sputum may be positive in cases of pneumonitis.

Serological and nucleic-acid-based tests are available at reference laboratory level. False positives and cross-reactions occur, so that interpretation of results is based on expert knowledge.

Treatment and prevention

Treatment is always warranted, as there is a mortality of 5–8% in systemic cases. Quinolones or aminoglycosides are the treatments of choice, producing a rapid response in lung and systemic disease. The skin and lymph-node lesions often heal more slowly, even when treated vigorously. Tetracyclines or chloramphenicol are also effective, but relapse is more common after treatment with these drugs.

There is currently no effective vaccine. Those with potential exposure should be educated about risks. The organism should be handled in laboratory Containment Level 3. Post-exposure chemoprophylaxis with effective antibiotics may be offered after inhalational exposure (e.g. in laboratory accidents).

Melioidosis

This is caused by *Burkholderia pseudomallei*, whose reservoir is probably in the water buffalo and its wetland habitat. Infection is commonest in Asian and Far Eastern farming areas, notably in Vietnam and Thailand. Intense rural contact is necessary to acquire infection.

Subclinical infection is probably common. Human infection presents as severe, rapidly progressive bacteraemic disease, sometimes years after exposure to infected areas. Pyogenic features, such as pneumonia, empyema or skin and soft tissue infection are seen in around half of cases. Over half of cases have pre-existing disorders such as diabetes, alcohol or drug abuse, immunosuppression or liver disease.

A high index of suspicion is required to recognize the possible diagnosis. The organism can be recovered from pus or blood cultures.

The organism is resistant to aminoglycosides and many beta-lactam antibiotics, so that initial empirical treatment for sepsis is often ineffective. Also, resistance tends to emerge during treatment. Carbapenems, ceftazidime and intravenous co-trimoxazole are often successful. A combination of doxycycline plus co-trimoxazole has been used successfully in endemic areas. Co-amoxiclav is also often effective, and is used for treating children. Prolonged medication, combined with effective drainage of loculated infection, is often necessary.

Prolonged follow-up is advisable, to detect relapses.

Relapsing fever

This borrelial infection exists in a range of ecologies throughout the world, depending on the interaction between several species of *Borrelia* and their life cycles. Tick-borne relapsing fever is transmitted between hosts by the bite of Argacid soft ticks of *Ornithodorus* species. Louse-borne relapsing fever is transmitted between humans by the human lice (*Pediculus humanus* and *P. capitis*).

Relapses occur because the surface antigen expression of the parasites varies frequently, depending on periodic variation in the expression of genes residing in extrachromosomal linear DNA. The 'spontaneous' crises and Jarisch–Herxheimer reactions seen in these diseases are caused by sudden release of TNF, mediated by a pyrogenic variable major lipoprotein released from the borrelial outer membrane.

• *B. duttonii* relapsing fever: human to human spread via ticks. Found in east, central and southern Africa.

• Other tick-borne relapsing fevers: *B. crocidurae*; *B. hermsi*; *B. parkeri*; *B. turicatae*; *B. persica*; *B. tholozani*; *B. hispanica*, etc.: spread to humans from rats, mice, chipmunks, gerbils, squirrels etc. in dry, bushy terrain or within dwellings. Found in central and western USA, central and south America, north and east Africa, the Middle East, and a belt from Uzbekistan to western China.

• Louse-borne relapsing fever (*B. recurrentis*): human to human spread by the human body louse or head louse. Remains endemic in highland areas of Ethiopia, Yemen, Peru and Bolivia: can establish a transmission cycle in overcrowded situations where poor hygiene prevails (e.g. war zones, refugee camps and populations affected by natural disasters).

Clinical features

All types of disease have an incubation period of 3–18 days (average around 7–10 days).

Tick-borne relapsing fever begins abruptly with fever, prostration, headache, muscle and joint pains, and usually severe sweats. Skin and mucosal haemorrhages appear as petechiae or subconjunctival lesions. Abdominal pain and diarrhoea are common in established fever. Up to 10% of patients develop transient cranial nerve dysfunctions, particularly facial paralysis, and also peripheral pareses and paraesthesiae. Up to seven or eight increasingly mild relapses then occur, every week or two.

Louse-borne relapsing fever can cause simply a mild fever, but around half of cases are severe, with delirium and nightmares plus all of the abrupt manifestations of the tick-borne types, often progressing to severe illness with multi-organ failure. There may then be hepatomegaly, obvious jaundice, splenic enlargement and severe bleeding into body surfaces. Pneumonia, transient myocarditis or, rarely, ARDS occur in up to 15% of cases. Infection in pregnant women often results in fetal loss.

The features of relapsing fever episodes often increase markedly on day four or five of the episode, and then gradually subside in a phase of severe sweating and hypotension (spontaneous crisis).

Diagnosis

The pink-staining coiled borreliae can be seen on thick and thin blood films; they are more plentiful and therefore easier to demonstrate in severe illness. Serology will demonstrate antibodies to *Borrelia* spp., but these can be confused by cross-reaction with the borreliae of 'Lyme' disease, if the patient has been exposed to both infections. *B. duttonii* and *B. recurrentis* can be cultivated in a reference laboratory setting.

The differential diagnoses of recurring fever with haemorrhagic manifestations and jaundice include malaria, rickettsial diseases, trench fever and rat bite fever.

Treatment

Doxycycline and erythromycin are the antibiotics of choice in relapsing fever. Penicillins and chloramphenicol are also effective.

Treatment of relapsing fever

1 *Tick-borne relapsing fever:* doxycycline orally, 100 mg daily for 10 days (child or pregnant woman, erythromycin orally, in a standard dose for 10 days).

2 *Louse-borne relapsing fever:* doxycycline 100 mg single dose (*for intravenous therapy* in severe disease: erythromycin i.v. 300 mg single dose, or *for a child* 10 mg/kg to a maximum of 300 mg).

⚠ An acute, potentially severe exacerbation of disease can follow 1–3 days after commencing therapy (Jarisch–Herxheimer reaction); this does not respond to corticosteroid treatment, but may be prevented or lessened by pre-treatment with an anti-TNF antibody.

Rat bite fever

Introduction

Two organisms may be transmitted by the bites of rats – *Streptobacillus moniliformis*, with an incubation period of 7–10 days, and *Spirillum minus*, with an incubation period of 1–4 weeks. *Streptobacillus moniliformis*, the cause of Haverhill fever, can also be transmitted by contaminated food or milk; milk-borne outbreaks have been described.

Both pathogens cause fever and peripheral rash. The fever tends to be relapsing when caused by *Spirillum minus* and swingeing when caused by *Streptobacillus moniliformis*. The rash may be papular or petechial and occasionally contains small pustules. Arthritis is common in *S. moniliformis* infections. Both types of infection are rare causes of pyrexia of unknown origin with a rash. They are both usually accompanied by neutrophilia. The liver function tests may be slightly abnormal, and prolonged prothrombin times can be demonstrable.

S. moniliformis is a rare cause of endocarditis.

Diagnosis

Diagnosis depends largely on cultures of pus, joint fluid and blood. *S. moniliformis* can be fastidious, but in appropriate culture media produces tangled chains of Gram-negative bacteria. *Spirillum minus* does not grow well in artificial media.

Treatment

Treatment with parenteral benzylpenicillin is effective against both organisms. A course of 1.2 g 6-hourly for 7–10 days is usually sufficient. Streptobacillary endocarditis requires 4–6 weeks' treatment.

Anthrax

Introduction and epidemiology

Anthrax is a disease particularly of hoofed animals, caused by *Bacillus anthracis*, an aerobic, spore-bearing, Gram-positive rod related to *B. cereus* and *B. subtilis*, but with a wider range of antibiotic sensitivities than either. In its natural hosts it causes a fatal septicaemic disease in which the animal's blood becomes packed with bacilli. After the host's death, the bacilli form spores that remain viable in the soil for decades. Farm animals are susceptible; an outbreak in the UK in the 1980s affected pigs. Humans become infected by close contact with infected or dead animals, with bones, bonemeal, hides, hooves or meat. Badger-hair shaving brushes occasionally caused infection in the past. The usual route of infection is inoculation into the skin, less often by inhalation or ingestion of spores.

A highly dispersible preparation of anthrax spores, mailed in sealed envelopes, has been successfully used in a terrorist attack. An industrial accident in Russia in the 1980s caused many human and animal deaths after release of highly dispersible spores in a crowded city.

Clinical features

An oedematous skin lesion (malignant pustule) is the commonest presentation of anthrax. An inflamed nodule ulcerates and forms a black scab, surrounded by a halo of vesicles or pustules. A common site is the neck at the collar, the dorsum of the wrist or the arm. Local draining lymph nodes are enlarged and tender. A helpful diagnostic feature is the enormous extent of oedema that surrounds the lesion (Fig. 24.18).

Inhalational anthrax can follow inhalation of massive spore loads, often from hides or dusty bonemeal. It is a haemorrhagic mediastinitis and mediastinal lymphadenitis which is often rapidly fatal leading to death in 2 or 3 days, but is not transmitted from person to person.

Gastrointestinal anthrax is probably acquired by ingestion of large spore loads. It causes abdominal pain and severe watery diarrhoea, which contains many sporing organisms.

Septicaemia is usually fatal. It can occur with untreated skin infection, but is common with pneumonitis and gastrointestinal disease. Death is often due to pulmonary embolism or cerebral vein thrombosis, even after antibiotic treatment has reduced fever.

Diagnosis

Culture of lesion swabs or vesicle fluid will produce filamentous colonies in blood agar, which are composed of tangled chains of square-ended Gram-positive rods

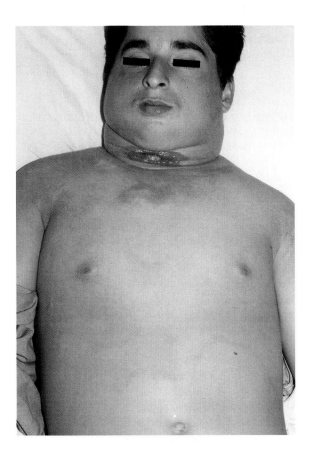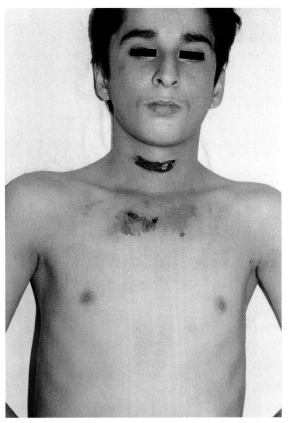

Figure 24.18 Anthrax, showing (a) the lesion with its halo of vesicles at the centre of the neck, and extensive odema, affecting the face and the whole torso. (b) After recovery, the extent of the earlier oedema is dramatically evident. (Reproduced with permission from Felek *et al.* (1999) *Journal of Infection* **38**: 201.)

(medusa head colonies). Spore stains demonstrate central spores in bacteria from mature colonies. Blood cultures readily produce a heavy growth; indeed, methylene blue-stained blood smears from septicaemic animals and humans will often demonstrate bacilli (without spores, which do not form in the living host).

In inhalational anthrax, the chest X-ray rarely shows lung opacity, but CT scanning may demonstrate mediastinal inflammation and distortion; small pleural effusions are common, and are often cellular and bloodstained. Blood cultures are usually positive.

Treatment

Cutaneous anthrax

Treatment should be commenced while the diagnosis is being confirmed, to minimize the risk of bacteraemia. Ciprofloxacin intravenously 200 mg 12-hourly or orally 500 mg 12-hourly is the treatment of choice. Intravenous benzylpenicillin 2.4–3.0 g 6-hourly is an alternative. Oral ampicillin 500 mg–1 g 6-hourly should be used for continuation. Treatment should be continued until the lesion is healed, and the oedema completely resolved.

Inhalational anthrax

Although regarded as almost uniformly fatal, experience in recent incidents of deliberate release showed that early, vigorous treatment with effective antibiotics can be effective in over 60% of cases.

> ⚠ Anthrax is not susceptible to treatment with broad-spectrum cephalosporins; an empirical regimen for the treatment of severe chest infections, which includes co-amoxiclav, ciprofloxacin or moxifloxacin, is preferable if admitted patients with inhalational anthrax are to benefit from early, effective treatment.

Antibiotic treatment should be continued for 6 weeks for inhalational anthrax, as germination of remaining spores can cause illness for some time after the initial exposure. The treatment of choice is ciprofloxacin orally, 500 mg twice daily.

Prevention and control

Inactivated anthrax vaccines, which stimulate the production of antibody to the protective antigen (PA), are available for animal and human use. Inactivated vaccine is offered to workers at occupational risk; few adverse effects are recorded other than local inflammation. Hides, and hoof and bone products must be heat treated or chemically treated to destroy spores before they are moved between countries or used in manufacturing processes. The tanning process renders hides and leather safe.

Zoonotic streptococcal infections

Introduction

Some animals are colonized and may become infected by pyogenic streptococci that rarely affect humans. Two well-recognized examples of this are *Streptococcus suis*, of pigs, and *S. zooepidemicus*, which can affect horses.

Streptococcus suis

S. suis has an epidemiology in pigs similar to that of the meningococcus in humans. The organism is carried in the nasopharynx, particularly of piglets. When subjected to crowding, or the stress of transport, the animals may develop clinical meningitis. Humans are infected through close contact, usually with pig carcasses, and tend to develop meningitis with bacteraemia.

Gram-positive cocci may be seen in the CSF of human cases. A beta-haemolytic *Streptococcus* is demonstrated on culture, but is Lancefield group R or S rather than A (C or G), as expected. Treatment is with benzylpenicillin, with or without an aminoglycoside. A course of 10–14 days is usually required.

Streptococcus zooepidemicus

S. zooepidemicus causes bacteraemic or soft-tissue infections, usually in people who have close contact with horses. It appears in culture as a beta-haemolytic *Streptococcus* of Lancefield group C. The treatment of choice is benzylpenicillin plus an aminoglycoside.

Herpesvirus simiae infection

Herpesvirus simiae inhabits the mouth and mucocutaneous borders in monkey species, in a similar way to herpes simplex virus in humans. It causes cold-sore lesions in a proportion of affected animals. Monkey-handlers can become infected if the animal's saliva is inoculated via a bite or scratch.

Human infections may be local, producing persistent herpetic vesicles, but there is a severe risk of potentially fatal viral encephalitis. Treatment with aciclovir is effective in suppressing the skin lesions, but relapse often follows cessation of therapy. Affected individuals therefore often require long-term aciclovir treatment.

The nature of a herpetic lesion in a monkey-handler can be confirmed by isolation and characterization of the virus from vesicle or skin scrapings. *H. simiae* is a hazard category 4 virus for which there is no reliable treatment or prophylaxis. Diagnostic culture is therefore carried out in reference laboratories such as the Virus Reference Division of the HPA Centre for Infection in London, or the Centers for Disease Control in Atlanta, Georgia, USA, where containment level 4 facilities exist.

Zoonotic paramyxoviruses

Hendra virus

Hendra virus caused severe systemic and respiratory disease in horses in two sites in Australia in the late 1990s. Human stable workers were infected, probably by contact with respiratory secretions of sick horses, with at least one death. The reservoir of virus was found to be in fruit bat colonies.

Nipah virus

Epidemics of encephalitic disease in western Malaysia in 1999 were at first thought to be Japanese encephalitis but, after 80% of sufferers were found negative for this virus, cultures revealed a paramyxovirus related to Hendra virus. This virus was found to be widespread in herds of pigs, on which the area's economy was based. Many patients were treated with ribavirin, though the effectiveness of this is not proven. Many herds of infected pigs were slaughtered, with a consequent reduction in human cases. Smaller outbreaks of Nipah virus infection have been described or suspected in Singapore and Indonesia. The reservoir of infection is in fruit bats living in trees within the pig farms.

Viral haemorrhagic fevers

Introduction

The term viral haemorrhagic fevers (VHFs) describes several groups of severe viral infections in which haemorrhage is part of the clinical picture. These include some diseases already discussed in this book.

Disease list

Arboviruses
- Dengue haemorrhagic fever (flavivirus)
- Yellow fever (flavivirus)
- Crimean-Congo haemorrhagic fever (CCHF; nairovirus)
- Others (e.g. Rift Valley fever)

Hantavirus infections
- Haemorrhagic fevers with renal syndrome
- Hantavirus pulmonary syndrome

Arenaviruses
- Lassa fever
- Argentine haemorrhagic fever (Junin virus)
- Bolivian haemorrhagic fever (Machupo virus)
- Other pathogenic arenaviruses, endemic in South America (Guanarito, Sabia)

Filoviruses
- Marburg disease
- Ebola virus haemorrhagic fever

> ⚠ Yellow fever is an internationally notifiable disease. Countries where human cases exist are listed in the World Health Organization *Weekly Epidemiological Report*.

These viral infections tend to cause severe systemic diseases with fever, malaise, variable sore throat and headache, arthralgia and increasing prostration. Multi-organ damage is common, with evidence of liver dysfunction, bone marrow depression, renal impairment and evidence of widespread tissue damage (falling sodium levels, elevation of non-liver transaminases, falling blood pressure, encephalopathy and extreme lassitude). These problems may be accompanied by specific features such as rash, diarrhoea or renal failure. Haemorrhage is usually due to platelet deficiency or dysfunction, and significant disseminated intravascular coagulation is a late feature, except in CCHF.

Haemorrhagic fevers with renal syndromes

These diseases are caused by viruses of the hantavirus family, whose reservoir is in asymptomatic mice and voles. Transmission to humans is either by inoculation or by inhalation of body fluids or excreta from host animals. Severe disease, caused by Hantaan and Seoul virus (in the Far East and Korea) and Dobrava–Belgrade virus (in eastern Europe), cause eye pain, moderate haemorrhage and severe uraemia, with a case fatality rate of up to 15%. Milder diseases of the same type occur in western Europe and forested parts of Scandinavia. These are often caused by Puumala virus.

The disease occurs in three main phases:
- an acute influenza-like syndrome;
- an intermediate stage of hypotension or shock, accompanied by haemorrhagic features and thrombocytopenia; and
- a late stage of oliguria and renal failure.

Diagnosis is based on demonstrating IgM antibodies, usually by ELISA. The viruses can be recovered in cell cultures of blood or serum; viral genome can be demonstrated in blood by PCR techniques.

Management is mainly supportive. Early treatment with ribavirin can abort the hypotensive and nephropathic phases of the disease.

Hantavirus pulmonary syndrome

This is a severe, haemorrhagic or infiltrative pneumonitis with a short incubation of around 3–4 days, and fulminant course. It is caused by a variety of hantaviruses, transmitted by inhalation of dense aerosols of infected mouse excreta (typically from mouse-infested dwellings), and is endemic throughout the Americas. It can also occur in other hantavirus-endemic areas, if aerosolized infected material is inhaled. Human to human transmission has rarely been described.

Many infections (caused by Sin Nombre and New York viruses) do not cause renal disease, but others (Bayou, Andes and Black Creek Canyon virus) can cause renal failure. The case fatality rate is 40% or more, due to respiratory and multi-organ failure. Diagnosis is by PCR of blood, by culture of blood or respiratory secretions, or by serology in late disease. High-dose ribavirin treatment has not proved to be effective.

Major viral haemorrhagic fevers (transmissible from person to person)

Although all haemorrhagic fevers are serious for affected patients, most are not transmitted from person to person. However, some have caused significant nosocomial transmission and do not reliably respond to antiviral treatment. They are therefore subject to special precautions when diagnosed or suspected in non-endemic countries. These are (in approximately increasing order of infectiousness): Lassa fever, Marburg disease, Ebola virus haemorrhagic fever and Crimean-Congo haemorrhagic fever. As the maximum incubation period of viral haemorrhagic fevers is 3 weeks, they can be excluded if more than 21 days have elapsed between leaving the endemic area and the onset of fever. In the UK *The Management and Control of Viral Haemorrhagic Fevers* (HMSO, 1996) sets out the recommendations of the Department of Health for obtaining advice and arranging management for suspected cases. High-security infectious diseases units (HSIDUs) have special clinical and laboratory facilities for handling patient management, while diagnostic tests are performed in special pathogens reference laboratories at laboratory containment level 4.

Lassa fever

Introduction

Lassa fever is an arenavirus infection whose natural reservoir is the multimammate rat *Mastomys natalensis*, which only carries Lassa virus in West Africa (other arenaviruses less pathogenic to humans exist in other parts of Africa). Transmission is by inoculation or mucosal contamination of infectious excreta from asymptomatic rats. Small numbers of medical and laboratory staff have been infected when handling cases or specimens.

Many cases may be mild or self-limiting, as seropositivity is common in endemic areas. However, Lassa fever is a common cause of hospital admission in endemic areas and many patients have severe illnesses of 2–3 weeks' duration. Those who have high transaminases (more than 10 times the upper reference level) usually have high viraemias (more than 4 $\log_{10}$ tissue culture infective dose 50 (TCID50)), and have a mortality approaching 80%, if not treated.

The disease has three phases:

1 an insidious onset with fever, malaise, muscle aches, dry cough, sore throat, moderate gastrointestinal symptoms and increasing prostration;

2 rising transaminases, minor haematuria, blood streaking of sputum, reduced blood pressure, extensive non-pitting oedema of the lower face and neck and often mild to moderate encephalopathy and/or nerve deafness;

3 shock, multi-organ failure and haemorrhage, loss of capillary integrity with widespread severe oedema – peripheral blood neutrophilia reflects extensive tissue damage.

The progress of the infection may cease at any stage.

Diagnosis

Diagnosis depends on clinical and epidemiological suspicion. Patients with Lassa fever usually have: persisting fever above 38.8 °C; a history of rural exposure in highly endemic areas; a history of activities that bring them into contact with infected rats or rat-contaminated environments.

Laboratory diagnosis is based on demonstrating viral RNA in blood by PCR techniques. IgM antibodies are usually positive by ELISA after 5–7 days of illness. IgG antibodies persist long-term.

Viruses are readily cultured in vero, or other common cell-lines, and produce a cytotoxic effect or are detectable by immunofluorescence. Blood cultures are known to remain positive in afebrile patients in early convalescence (probably because neutralizing antibodies are not established until 9–14 weeks of convalescence). Urine cultures may remain positive for several weeks.

> ⚠ Serological diagnosis may be impossible in severely ill patients, who may not display an antibody response; a negative antibody test in PCR-proven infection is a very poor prognostic sign.

It is important to exclude immediately life-threatening diseases such as malaria, and no patient should be referred as a case of viral haemorrhagic fever before malaria has been adequately considered and investigated.

> ### Haemorrhagic conditions that should be considered before a diagnosis of viral haemorrhagic fever is assumed
> 1 Malignant malaria.
> 2 Meningococcal disease.
> 3 Severe rickettsial infections.
> 4 Gram-negative septicaemia with disseminated intravascular coagulation.

Treatment

Treatment is based on intensive support and early ribavirin therapy, which ameliorates disease but does not abolish viraemia or viruria. Ribavirin is not effective if commenced later than 7 days after onset; it is more effective if given intravenously.

Ribavirin for intravenous use is available in referral centres, and can reduce mortality in severe disease. Follow-up is necessary to confirm eventual clearance of virus from blood and urine.

> ### Ribavirin therapy for Lassa fever
> Ribavirin, by slow intravenous injection, 2 g loading dose, followed by 1 g 6-hourly for 4 days, followed by 0.5 g 6-hourly for 6 days.
>
> Dose-related haemolysis is a common adverse effect, which recovers spontaneously after treatment.

Prevention and control

Prevention and control measures are concentrated on close family members and medical staff who have contact with the patient and his or her blood. Casual and social contacts are not at risk. In most cases surveillance of health and temperature during the possible incubation period is sufficient. Oral ribavirin chemoprophylaxis has been given to high-risk contacts such as sexual partners or persons contaminated with the patient's blood, but there is no firm evidence of its effectiveness.

Argentine haemorrhagic fever

Argentine haemorrhagic fever is caused by Junin virus, an arenavirus whose natural hosts are harvest mice. Epidemics occur in northern Argentina, particularly during the corn harvest. The disease is like Lassa fever; petechial rashes and platelet dysfunction are common. Patients with severe disease may have encephalopathies. The mortality is 10–15%.

Convalescent plasma, containing neutralizing antibodies to Junin virus, can reduce mortality to 3%, but a late mild encephalopathy is common in treated patients. An effective vaccine is now available.

Other arenavirus haemorrhagic fevers

Machupo virus causes occasional cases and outbreaks in rural Bolivia. Guanarito virus causes outbreaks in Venezuela, and Sabia virus has been identified in Brazil, as well as in two laboratory-associated infections.

Marburg disease and Ebola haemorrhagic fever

Epidemiology

The epidemiology and reservoir of these filovirus infections is not understood.

The original outbreak of Marburg infections followed the importation of infected vervet monkeys into a scientific laboratory. Transmission occurred from monkeys and their tissues to laboratory workers, from patients to medical attendants, and from one convalescent patient to his wife (the patient's semen was found to contain virus). The main route of transmission is from blood and body fluids and the case fatality rate is about 50%. One patient in a South African outbreak had Marburg virus isolated from the anterior chamber of the eye many weeks after her illness.

Ebola infection has caused large community outbreaks with fatality rates of 75–80%. The first outbreak was complicated by extensive hospital transmission, perhaps associated with re-use of needles. Countries affected by cases include Congo, Sudan, Gabon and Côte d'Ivoire (one case only) (Fig. 24.19). A nurse in Johannesburg was fatally infected when caring for a patient from Gabon in 1996. Most outbreaks appear to originate from human exposure to an infected monkey or chimpanzee, though there is no evidence that these are the natural reservoir of disease. A laboratory-acquired case occurred in England in 1976. Three strains of virus are pathogenic for humans: the Zaire strain (EBO-Z), the Sudan strain (EBO-S), and the Côte d'Ivoire strain. The Reston strain is highly pathogenic and has caused large outbreaks in captive monkeys, particularly in the Philippines, but is not pathogenic for humans.

Clinical features

The clinical picture is similar in Marburg and Ebola infections. After 3 or 4 days' insidious onset with high fever and prostration, diarrhoea occurs and a measles-like rash is common. Mild or gross haemorrhagic features may quickly develop, including bloody diarrhoea with ab-

Distribution of Lassa, Ebola and Marburg viruses in Africa

- ● Lassa virus transmission
- · Ecology favouring Lassa virus
- ● Ebola virus transmission
- · Ecology favouring Ebola virus
- ● Marburg virus transmission
- · Ecology favouring Marburg virus

Figure 24.19 Distribution of Lassa, Ebola and Marburg viruses in Africa. Reproduced with permission from B. Bannister. Viral haemorrhagic fevers. *Medicine* 2005; **33**: 7–16.

dominal pain (which may lead to inappropriate abdominal surgery). Many cases progress to multi-organ failure, severe haemorrhage and death. Leucocytosis occurs and transaminases rise as tissue damage proceeds. Platelet dysfunction is the main cause of haemorrhage.

Diagnosis

The diagnosis may be suspected on clinical and epidemiological grounds, especially during outbreaks. Early diagnosis is by demonstration of viral RNA in blood. The viruses grow readily in cell cultures and can be demonstrated by immunofluorescence. IgM and IgG antibodies in the patient's blood can be demonstrated at appropriate stages of the disease.

No antiviral agent, including interferon, has a significant beneficial effect; vigorous supportive treatment must be given. Strict patient isolation is more urgent than with Lassa fever, as both the patient and his or her body fluids are more infectious, and there is a small possibility of droplet infection from either source.

Crimean-Congo haemorrhagic fever

Epidemiology

The natural epidemiology of this nairovirus disease depends on tick-borne transmission between warm-blooded animals. Cattle and farmed ostriches are important reservoirs in many African countries but camels and rodents may be important elsewhere. Crimean-Congo haemorrhagic fever (CCHF) is not restricted to tropical Africa, but is endemic in the Middle East and parts of Bulgaria, former Yugoslavia, Pakistan, Afghanistan and the former southern USSR. A large outbreak affected Albania in the late 1990s. Secondary cases are relatively common, particularly among clinical care personnel, partly because of the massive haemorrhage that patients may suffer.

Clinical features

Many strains of CCHF exist in its wide epidemiological range. In southern Africa, the resulting disease may not be so severe as in Pakistan. The clinical picture follows an incubation of 4–10 days, with a maximum of 13 days. Abrupt onset of severe fever, muscle pains and prostration is followed in 2 or 3 days by collapse, extensive bruising and purpura, and massive epistaxis, haematemesis and melaena. A rapid rise in transaminases and profound drop in the platelet count occurs during the first week of illness. In these patients there is evidence of intravascular coagulation, and ARDS may develop. Cases of intermediate severity can occur, and must be differentiated from meningococcal or rickettsial diseases or from haemorrhagic chickenpox.

Ribavirin reduces the severity of CCHF, and causes a rise in the platelet count and reduction of haemorrhage. Vigorous supportive treatment is also needed in severe cases.

Public health measures for viral haemorrhagic fever cases

Viral haemorrhagic fevers are unfamiliar diseases in the UK, where they are extremely rare. There are several important aspects to their safe and effective management.

1 Recognition of exposure and risk of the diagnosis.

2 Ensuring that more common severe infections are not neglected. Over 90% of viral haemorrhagic fever suspects have a final diagnosis of falciparum malaria, which can cause disseminated intravascular coagulation (see text note, p. 478). Blood films must be examined to exclude malaria before the question of viral haemorrhagic fever is seriously considered in most cases. Seriously ill patients can be commenced on empirical quinine treatment. Suspicion of and empirical treatment for meningococcal or rickettsial disease may also be important.

3 Seeking specialist advice and support. Regional infectious and tropical disease specialists will have up-to-date information on endemic areas, and will often be able to exclude the diagnosis of viral haemorrhagic fever on the basis of epidemiological and clinical data.

Cases where the diagnosis cannot readily be excluded fall into two main categories:

• Medium-risk cases, exposed to endemic areas but not to known sources of infection. If malaria studies and initial investigations do not suggest an alternative diagnosis, these patients' specimens should be handled in laboratory containment level 3 until definitive diagnostic tests have been completed (the patient may be transferred to a specialist infectious diseases unit for investigation and management).

• High-risk cases, when the diagnosis of viral haemorrhagic fever is known, obvious or strongly suspected, or is confirmed after investigation as a medium-risk case. Such cases should be managed in a high-security infectious diseases unit (HSIDU), where special control of infection measures are employed during investigation and management. The staff at the HSIDU will accept such cases, and will liaise with ambulance services to arrange transfer.

Viral haemorrhagic fevers are notifiable diseases. They should be notified urgently by telephone to the appropriate Proper Officer, with follow-up in writing. The names of face-to-face healthcare contacts should be recorded for health surveillance. Contacts of high-risk cases are followed up for 3 weeks after the last contact, or until the diagnosis is disproved.

Standard hospital procedures are usually adequate for cleaning and disinfection of the patient's environment.

Summary of personal health measures recommended for travellers

1 Vaccinations for epidemic diseases. The desirability of these will depend on the current prevalence of epidemics in the destination country. Vaccines available include:
• influenza vaccine;
• meningococcal vaccine;
• measles mumps rubella vaccine for infants.
(Cholera vaccine is no longer required by any country.)

Other vaccines for personal protection will depend on whether a traveller will have other occupational or social exposures:
• tetanus, diphtheria and polio vaccines or boosters;
• hepatitis B vaccine.

2 Protection from food- and water-borne diseases is recommended for travellers to countries where efficient sanitation and safe water supplies cannot be guaranteed. The following measures may be considered:
• typhoid immunization;
• hepatitis A prophylaxis with inactivated vaccine;
• human normal immunoglobulin can confer immediate, short-term protection from hepatitis A if urgently indicated;
• chemoprophylaxis for diarrhoea (special circumstances only);
• water filter/purification kits (campers and independent travellers).

3 Malaria prophylaxis is essential for everybody who travels to an endemic area. Measures to avoid insect bites should also be emphasized.

4 Yellow fever vaccination is recommended for all travellers who will enter an endemic area. Certificates of vaccination are often required for travellers passing from endemic areas to other countries. Certification is subject to International Health Regulations, and is available only from accredited centres.

5 Immunization against local infections may be advisable for travellers having rural or community exposure, especially during certain seasons. Diseases to consider include:
• Japanese B encephalitis (stays of more than 1 month in rural farming areas of affected countries);
• tick-borne encephalitis (camping, walking or rural work in forested areas of affected countries during late spring and summer);
• rabies (independent travel in remote areas of affected countries; relief work, especially if involving animal contact in those countries; caving in areas where bat rabies occurs).

Assessment of fever in a returning traveller

Different diseases may become evident at different times after a traveller returns from an overseas visit, depending on the incubation periods of the various infections (Table 24.2).

Table 24.2 Intervals between a traveller's return home and the presentation of imported diseases

Time after return home	Disease
During first week	Viral gastroenteritis Traveller's diarrhoea Bacillary dysentery, sexually transmitted diseases, influenza, dengue (and other arboviral infections)
1–2 weeks	Malaria Hepatitis A Typhoid fever Paratyphoid fever Rickettsial infections
2–4 weeks	Typhoid fever Amoebiasis Hepatitis C Katayama fever
1–6 months	Hepatitis B HIV seroconversion illness Hepatitis E Amoebiasis Rabies Cutaneous leishmaniasis Systemic leishmaniasis
More than 6 months	Relapses of vivax or ovale malaria Reactivation of malariae malaria Strongyloidiasis (larva currens) Rabies Systemic leishmaniasis AIDS

AIDS, acquired immunodeficiency syndrome; HIV, human immunodeficiency virus.

Case study 24.1: Young men with itchy feet

History

Three young men shared a 2-week holiday at a Caribbean resort, where they spent most of their days barefoot, participating in beach activities. Near the end of their holiday, they each developed localized, red, intensely itchy lesions on their feet (Fig. CS.7). These lesions persisted for 3 weeks, despite treatment with clotrimazole cream, and a referral to the travellers' clinic was made.

Questions
- What is the diagnosis?
- What is the pathology of the lesion?
- What treatment would you recommend?

Management and progress

Based on the typical lesions, a clinical diagnosis of cutaneous larva migrans was made. This is due to skin invasion by the larvae of dog hookworms. Hookworm ova are deposited on sand or soil in the faeces of infected dogs. The ova release larvae, which undergo a soil cycle, becoming invasive, and are destined to adhere to and burrow through the skin of animals walking on the affected ground. Larvae adhering to human skin will invade locally, producing an irritating allergic reaction. They cannot, however, complete their life cycle, and die after several weeks. The larvae can be killed by a systemic course of albendazole, 400 mg twice daily for 3 days *or* ivermectin 200 µg/kg, single dose. An alternative is a paste of 10% thiabendazole (which can be made by a pharmacist), applied under occlusion to aid skin penetration. Topical corticosteroid cream may help to reduce itching during the next week or so, as larval antigens disperse.

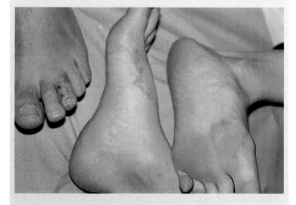

Figure CS.7 Intensely itchy lesions in the thinner skin of the feet, one with an obvious inflammatory 'track'.

Case study 24.2: Insect bites that will not heal

History

A 26-year-old woman went for a 2-week holiday to a coastal resort in a north African country. Three weeks after returning home, she became aware that two small 'insect bites' on her right leg were enlarging and becoming nodular. During the following 2 weeks, three further lesions appeared, and the original two lesions had become ulcerated, with scabs which repeatedly separated and re-formed.

Physical examination

This showed the scabbed and nodular lesions (Fig. CS.8). There was no constitutional complaint, the temperature was normal and there was no hepatosplenomegaly or regional lymphadenopathy.

Questions

- What is the diagnosis?
- How can the diagnosis be confirmed?
- How should the condition be treated?

Investigation and management

A clinical diagnosis of cutaneous leishmaniasis was made, based on the slow evolution, the size (1.0–2.0 cm diameter) and the chronic ulceration and scabbing with a typical 'rolled' epithelial border. This type of lesion is probably due to *Leishmania tropica* var. *major*. Scrapings from the granulomatous mound at the base of the lesion were examined histologically for the presence of parasites within mononuclear phagocytes. This did not prove positive. A more reliable test would be to biopsy the lesion, including part of the central area. In view of

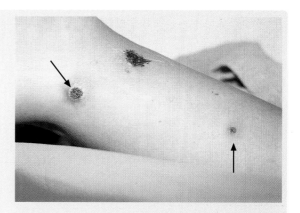

Figure CS.8 Lesions interpreted as insect bites on the right shin and calf (one has been biopsied, and is bleeding). The large, intact lesion (upper left arrow) shows a typical 'rolled' epithelial edge and central scab. A smaller, more recent lesion is seen on the lower calf (lower right arrow).

the typical clinical appearance, it was considered unjustified to produce a scar by undertaking biopsy. Serology is negative in cutaneous leishmaniasis.

There is no evidence that topical anti-leishmanial drugs influence the course of cutaneous leishmaniasis. Systemic therapy may be indicated for large or disfiguring lesions, but is toxic, and not recommended for uncomplicated lesions. Spontaneous healing should occur in 3–5 months, with a small, atrophic skin area remaining.

There is some evidence that oral fluconazole or itraconazole in a 1-week course can shorten the course of the lesions.

Role of national agencies

In the UK, a number of agencies currently have responsibility for health protection of the population, from infections, and from environmental, chemical and radiological hazards. In England, the 'Health Protection Agency' (HPA) has this responsibility, in Wales, the 'National Public Health Service for Wales' and in Scotland 'Health Protection Scotland'. In Northern Ireland the responsibility is shared by the HPA and the four Health and Social Service Boards that manage health provision. Most countries have similar organizations that are particularly focused on infection hazards, for example the Centers for Disease Control and Prevention in the USA. In some countries the functions are performed by the Ministry of Health. In Europe there is increasing emphasis on joint working between countries and a new European Centre for Communicable Disease Surveillance was established in 2005.

The Health Protection Agency is used here as an example of how various functions are conducted by national agencies.

The Health Protection Agency has several divisions whose professionals work together on infectious diseases at national, regional and local levels (Fig. 25.1). Nationally, the Centre for Infections combines microbiological and epidemiological expertise in health protection programmes for specific diseases (e.g. tuberculosis), as well as for populations (e.g. travellers and migrants), and settings (e.g. health care associated infections) of particular risk.

The microbiological functions include:
- reference laboratories for bacterial and viral pathogens;
- antibiotic sensitivity determination;
- molecular sub-typing;

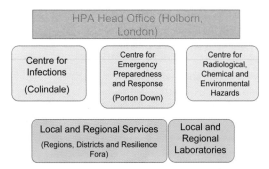

Figure 25.1 The divisions of the HPA.

- microbiological research and development of new technologies;
- vaccine development;
- training in the microbiology of infectious disease.
  The epidemiological functions include:
- surveillance of infectious diseases and of immunization programmes in England;
- provision of advice on prevention and control, for development of national policy and to assist local health professionals;
- investigation of national outbreaks;
- monitoring emerging and re-emerging infections globally;
- liaison with international colleagues;
- epidemiological research;
- training in the epidemiology and control of infectious disease.

Another centre of the HPA, the Centre for Emergency Preparedness and Response (CEPR), provides diagnosis and reference services for 'special pathogens', most of which must be handled at higher laboratory containment levels. It also provides research into vaccines for dangerous pathogens, and expertise in managing large biological emergencies. Its role in infectious disease control is to ensure that through planning, other parts of the agency and the health services more widely are capable of a speedy and effective response to public health emergencies.

The HPA also provides chemical and radiological protection services through the Centre for Radiological, Chemical and Environmental Hazards (CRCE).

The Local and Regional Services division of the HPA collaborates with local authorities, primary care providers, hospitals and emergency services, to provide operational functions through several regional offices that relate to the Government Offices of the Regions. Regional functions include the general oversight of local infection-related events, routine surveillance of infections and outbreak control activities. HPA Regions also participate in planning for national coordinated responses to large-scale biological events, such as epidemics.

Each regional team coordinates the activities of several local Health Protection Units (HPUs). There are currently 39 HPUs in England, each covering an area broadly corresponding to a county or police boundary. The task of each HPU is to engage with the National Health Service primary care trusts (PCTs), acute hospital trusts and local authorities in their area and agree with them how health protection should be delivered locally. The HPA maintains a regional laboratory within each of the regions that provides specialist services to neighbouring NHS hospitals. In addition, there are a number of collaborating laboratories that are managed by the health services but that have been given additional resources to provide public health outputs. These include enhanced surveillance and reporting, collaboration in special studies and food, water and environmental microbiology testing. The microbiological component of Local and Regional Services is coordinated by a Regional Microbiologist who aims to provide a link with the Centre for Infection and the local HPUs.

Health Protection Unit functions for infectious diseases control

1 Local disease surveillance.
2 Access to local/reference laboratories.
3 Alert systems.
4 Investigation and management of incidents and outbreaks.
5 Ensuring local delivery and monitoring of national action plans for infectious diseases, including immunization programmes.

Other agencies are also important in the surveillance and control of infectious diseases in the UK.

• The National Health Service: health care professionals working in the NHS are vital to surveillance and control of infectious diseases; they report cases to surveillance systems and enact the health interventions that limit spread. More specifically within hospitals there is a Control of Infection Officer, usually the local microbiologist.
• The Office for National Statistics (ONS) collects vital statistics (births, marriages and deaths), and provides population estimates and projections for England and Wales. The equivalent organization in Scotland is the General Register Office (Scotland) and in Northern Ireland it is the Northern Ireland Statistics and Research Agency.
• The Department for the Environment, Food and Rural Affairs (DEFRA) is a ministry responsible for the control of infection in animals. Close collaboration between DEFRA and the various health protection agencies is important in relation to the control of zoonotic infections.
• The Food Standards Agency (FSA) is responsible for overseeing the safety of food production, manufacture and handling up to the point of sale.

Communicable disease and the law

In the UK the legal framework for control of infection in the community has developed over the past 150 years. Much of the old legislation was brought together under the Public Health (Control of Disease) Act (1984) and the Public Health (Infectious Diseases) Regulations (1988). These acts do not apply in Scotland, although similar legislation does exist.

Responsibility for the control of communicable disease was vested in local authorities during the 19th century, when the post of Medical Officer of Health (MOH) was established. When the National Health Service was created, the MOH, who was employed by the local authority, retained responsibilities for control of infection and related activities such as immunization. The post of MOH was abolished in 1974; however, some residual functions, including communicable disease control, were retained by local authorities.

Each local authority has a designated proper officer (in Scotland, medical officer) with statutory powers for prevention and control of infection. These statutory powers relate to the notifiable diseases (see below) and to certain other diseases, usually defined. The proper officer is usually (but not always) a consultant in communicable disease control (CCDC) who in England is a HPA employee mainly based in a HPU, but acting on behalf of the local authority with respect to these statutory powers. In Scotland, the equivalent post is the consultant in public health medicine (communicable disease and environmental health).

International Health Regulations

As well as national legislation related to infectious diseases there is also international legislation. The most important are the International Health Regulations (IHR). These are a legally binding code of practices and procedures designed to prevent the international spread of infectious diseases, while not interfering unnecessarily with the international movements of people and goods. The current regulations include: procedures for notification of certain diseases (plague, cholera and yellow fever); health-related rules for international trade and travel; health organization (including procedures and practices at airports, seaports and borders), and documentation requirements. The IHR were first developed by the World Health Organization in 1969 and member states agree on the regulations by consensus. They were revised in 1981, and in 2005 a further revision was ratified by the World Health Assembly. The revised IHR have a broadened remit with emphasis on building surveillance and response capacity in member states, as well as international communication.

Communicable disease surveillance

Principles and practice of surveillance

Surveillance of disease is traditionally considered to be the continuous systematic collection, collation and analysis of data, and the timely dissemination of public health information for assessment and action as necessary.

The primary aim of such formal surveillance is to identify trends or clusters of disease that require preventive action; thus surveillance is *information for action*. Surveillance data are also used to evaluate control measures such as vaccination programmes, to plan resource allocation and to provide baseline epidemiological information for research workers.

While such formal surveillance is the cornerstone of infectious disease epidemiology and control, other non-routine functions are also increasingly seen as surveillance today. Surveillance for public health includes all activity that provides information requiring responses or actions to protect or promote health. It therefore also includes healthcare staff and other people noticing and reporting the unusual, and conversely public health professionals alerting healthcare staff when they become aware of new threats. Since surveillance should never be divorced from the responses or actions it demands, it is more correct to speak of 'surveillance and response'.

The process of formal surveillance involves:
1 collection of data;
2 analysis of data to provide statistics;
3 interpretation of statistics to provide meaningful information;
4 dissemination of narrative reports to those who need to know.

An ideal surveillance system should meet the following criteria.
1 The disease under surveillance must be of sufficient importance to justify the resources required to undertake effective monitoring.
2 The surveillance should be timely – the data should be collected and interpreted sufficiently quickly to enable effective control measures to be taken. This is particularly important for communicable diseases, where large outbreaks can occur very suddenly and cause considerable morbidity.
3 The data collected should be representative of the total population. In practice this is difficult to achieve as surveillance systems usually build on routine data sources from atypical populations such as hospital inpatients. The role of the epidemiologist is to understand and take account of the biases inherent in surveillance systems when interpreting the data.
4 It should be consistent over time and between geographical areas. If the proportion of cases detected by a surveillance system is not constant, it becomes difficult to interpret any changes in reported disease incidence. During periods of increased disease incidence the efficiency of reporting tends also to increase. A change in the criteria for reporting a disease can have a large effect on reported incidence. For example, the inclusion of patients with a CD4 count of less than $200/\mu l$ in the case definition for acquired immunodeficiency syndrome (AIDS) resulted in a 200% increase in the cumulative total of cases reported in the USA.
5 Reporting should be complete wherever possible.
6 It should be simple, in both structure and ease of operation.
7 It should be flexible enough to adapt to changing information needs (for example, emergence of a new disease).

Criteria for an effective surveillance system
1 Disease under surveillance is of sufficient public health importance.
2 Timeliness.
3 Representative of the total population.
4 Consistency over time and between geographical areas.
5 Completeness of reporting.
6 Simplicity.
7 Flexibility.

A surveillance system may be passive, stimulated-passive or active.

• A *passive* system relies on routinely collected data, e.g. notifications of infectious disease, where cases are reported without any specific encouragement or inducement. When special efforts are made to improve reporting to a passive surveillance scheme, e.g. by written reminders or following up non-reporting centres, the scheme becomes *stimulated-passive*.

• An *active* surveillance system is where all potential reporting units are contacted at regular intervals and specifically asked to report the condition under surveillance; in addition, if no cases have been seen, a 'negative' report is requested. The British Paediatric Surveillance Unit (BPSU; see below) is a good example of an active surveillance system.

In general, passive surveillance systems are used for common or less severe diseases such as food poisoning, where it is not essential, or possible, to ascertain every case. Stimulated-passive or active reporting is usually only necessary for rare or serious conditions, or those where a public health programme of elimination is planned, for example poliomyelitis.

Sources of data

Reports of outbreaks and other infectious disease incidents are often provided on an *ad hoc* basis. There are however, many sources of data, routine and non-routine, that constitute the formal communicable disease surveillance network.

Routine sources of data on infectious disease in the UK

1 Statutory notifications to the consultant in communicable disease control or equivalent.
2 Laboratory reports to Centre for Infection (CfI).
3 General practitioner reporting schemes.
4 Hospital in-patient and out-patient statistics.
5 Returns from sexually transmitted disease clinics.
6 Death certificates.
7 School medical officers.
8 Occupational health departments.

Routine data sources

There are several routine sources of data on infectious disease in the UK. The most important are notifications, laboratory reports, general practitioner surveillance schemes, hospital data, clinic returns and death certificates.

Notification

Notification is a statutory requirement. All doctors are required to notify any cases of specified infections seen by them to the designated proper officer for the relevant local authority. Notification is based on clinical suspicion and does not require laboratory confirmation, although if the diagnosis is altered as a result of laboratory investigations, the notification can be corrected by the notifying doctor. Urgent cases, such as suspected diphtheria, should be notified immediately by telephone (but must still be formally notified in writing). A small fee is payable for each notification.

Notifiable diseases in England

1 Acute encephalitis.
2 Acute poliomyelitis.
3 Anthrax.
4 Cholera.
5 Diphtheria.
6 Dysentery (amoebic or bacillary).
7 Leprosy.
8 Leptospirosis.
9 Malaria.
10 Measles.
11 Meningitis.
12 Meningococcal septicaemia (without meningitis).
13 Mumps.
14 Ophthalmia neonatorum (neonatal ophthalmia).
15 Paratyphoid fever.
16 Plague.
17 Rabies.
18 Relapsing fever.
19 Rubella.
20 Scarlet fever.
21 Smallpox.
22 Tetanus.
23 Tuberculosis.
24 Typhoid fever.
25 Typhus.
26 Viral haemorrhagic fever.
27 Viral hepatitis.
28 Whooping cough.
29 Yellow fever.

⚠ The notification system has developed over more than 100 years. Many of the conditions are of historical interest only, whereas others, e.g. legionellosis, are noticeable by their absence. A local authority has the discretion to add conditions to the list that occur in that authority; addition of a disease to the national list requires approval by the Secretary of State.

Weekly summaries of notifications are forwarded from local authorities to Health Protection Scotland, the Communicable Disease Surveillance Centre, Northern Ireland, and the Health Protection Agency's Centre for Infections, where they are collated into weekly, quarterly and annual reports.

Laboratory reports

Laboratory reports provide reliable information on many infections. These are provided on a voluntary basis by National Health Service and other laboratories to national authorities including Health Protection Scotland, The National Public Health Service for Wales, and the Health Protection Agency's Centre for Infections. The main limitation of laboratory data is that they are biased towards patients with severe infections or where laboratory confirmation is likely to influence management. For some conditions therefore, especially many viral infections, laboratory reports do not provide a representative picture of infection in the community. Additional, mandatory schemes have been developed to follow trends in hospital acquired infections notably methicillin-resistant *Staphylococcus aureus* and *Clostridium difficile*. In these mandatory schemes health service hospitals are required by the government to report monthly on, for example, the number of MRSA bacteraemias.

General practitioner surveillance schemes

The Royal College of General Practitioners (RCGP) operates a national network of 'spotter' practices, serving a population of approximately 750 000, who provide weekly reports of first-time consultations on a variety of infectious and non-infectious conditions. It provides a particularly sensitive index of influenza activity.

In addition to the RCGP scheme, local general practitioner surveillance networks operate in several parts of the country.

Hospital data

Hospital data provide useful information on infections that usually require admission, such as meningococcal meningitis. They tend, however, to be incomplete and out-of-date.

Clinic data

Clinic data are useful for certain conditions, particularly sexually transmitted diseases. Annual returns of numbers of diagnoses from sexually transmitted disease clinics are collected by Health Protection Scotland, The National Public Health Service for Wales, Department of Health, Social Services and Public Safety (DHSS&PS, Northern Ireland) and the Health Protection Agency's Centre for Infections.

Death certificates and other sources of data

Death certificates make a limited contribution to surveillance, as deaths from infection are rare. However, the ratio of deaths to cases (the case fatality ratio) can provide useful information on the changing severity of some diseases and the effectiveness of new treatment measures.

Other useful routine sources of data include the Medical Officers of Schools Association (MOSA), which reports incidents in boarding schools, NHS Direct (which collects information based on calls to the telephone helpline), and some occupational health departments.

Special surveillance schemes

For some infections, routine data are insufficient to inform action, so enhanced surveillance systems have been set up. For example, in the UK, enhanced surveillance systems have been developed for tuberculosis and for antibiotic-resistant gonorrhoea.

For other infections, routine data are not available and special surveillance systems have been established. Examples in the UK include: the confidential reporting of AIDS and human immunodeficiency virus (HIV)-related disease (to Health Protection Scotland and the Health Protection Agency's Centre for Infections); the National Congenital Rubella Registry at the Institute of Child Health; and the British Paediatric Surveillance Unit (BPSU). Under the BPSU scheme, all consultant paediatricians are sent a monthly card with a menu of around 12 reportable conditions. Any cases seen by the physicians are then followed up by the lead investigator for that condition. Conditions may be added to or deleted from the menu on application to the BPSU. The BPSU scheme has been particularly useful for ascertaining cases of rare infectious disorders such as subacute sclerosing panencephalitis.

Dissemination of information

Laboratory and notification data are published electronically on a weekly basis by the Health Protection Agency's Centre for Infections (*Communicable Disease Report*) and by Health Protection Scotland. Similar bulletins are published in other countries, for example the *Morbidity and Mortality Weekly Report* (MMWR) in the USA. In addition, special reports are produced by these agencies on specific diseases or at-risk populations at regular intervals. These are generally available via their websites. Websites are particularly useful in presenting up-to-date information about emerging situations, or new advice on the diagnosis, management or containment of infectious diseases. An example is the regular presentation of data and advice about the SARS epidemic in 2002.

Surveillance in other countries

Most countries operate a notification system for communicable diseases, although the list of reportable conditions and the method used for reporting varies widely from one country to another. Surveillance is, however, not robust in many parts of the world. The new International Health Regulations will aim to encourage the establishment of stronger surveillance systems in World Health Organization member states. Increasingly, surveillance is being conducted at supra-state level. Pan-European surveillance has been developed for a number of infections, including tuberculosis, AIDS, travel-associated legionellosis, meningococcal infection, enteric pathogens and influenza.

Prevention and control of infectious disease

A wide range of measures may be used to prevent or control infection in the community and in hospitals (Table 25.1). Prevention may be classified as primary, secondary or tertiary.

• Primary prevention aims to prevent or reduce exposure to the infectious agent. This is the most effective method, but also the most difficult to achieve.

• The purpose of secondary prevention is to detect infection at an early stage, so that control measures to prevent serious disease or further spread can be taken.

• In tertiary prevention, the aim is to minimize the disability arising from infection.

Procedures designed to control infection in the community are discussed in this chapter; for the control of infection in hospitals see Chapter 23.

For communicable diseases the potential for spread from person to person is determined by the frequency of contacts in a population, the proportion of that population already immune, how long a person remains infectious and the probability of transmission in a contact between an infected and a susceptible person. These determinants predict the range of measures that can be used to control the spread of a communicable disease through a population.

Measures to prevent the spread of communicable diseases

1 Preventing contact between susceptibles and the source of infection, e.g. isolation.

2 Altering the number of immune individuals in a community, e.g. immunization.

3 Reducing the period of infectivity of a case, e.g. by treatment with antibiotics.

4 Reducing the risk of transmission in a contact between an infected individual and a susceptible, e.g. condom use for sexually transmitted infections.

Social and environmental factors

Although infection is still an important clinical problem in high-income societies, many of the more serious diseases that were common in the past have largely been brought under control. In contrast, infectious disease remains a major cause of morbidity and mortality in low-income countries.

The main factors in reducing the burden of infectious disease in Western societies were the improvements in social and environmental conditions that took place in the late 19th and early 20th century. Public health legislation forced industry and local government to spend money on sanitation and better housing. The average family size shrank rapidly during the early part of the 20th century, resulting in less crowding. Better nutrition meant that the population was less susceptible to disease.

The most important environmental measures are provision of adequately treated drinking water and safe disposal of faeces. Water-borne infection accounts for more deaths in low-income countries than any other disease (Fig. 25.2). The recent spread of cholera throughout South America illustrates the vital importance of basic sanitation. Even in the UK, water-borne outbreaks, for example of cryptosporidiosis, are relatively common, due to

Table 25.1 Prevention and control of infection

Primary prevention	Secondary prevention	Tertiary prevention
Immunization (pre-exposure)	Immunization (post-exposure)	Effective treatment of acute infection*
Improved housing and sanitation	Contact tracing	Management of post-infectious disorders†
Provision of safe food and pasteurization of milk	Screening of food handlers, health workers, etc.	Physiotherapy, speech therapy
Vector control	Chemoprophylaxis	
Behaviour modification (sexual, hygiene, etc.)	Effective surveillance	
Isolation, barrier nursing‡	Outbreak investigation and management	
Disinfection, sterilization‡		
Laboratory safety‡		

* See Chapter 4. † See Chapter 21. ‡See Chapter 23.

Figure 25.2 Consumption of untreated water in a developing country: a major source of infection. Courtesy of SmithKline Beecham.

treatment failures and post-treatment contamination of water supplies.

Better housing conditions have made a major contribution to controlling respiratory infections such as tuberculosis that are spread by close person-to-person contact. The reduction in tuberculosis illustrates the relative importance of environmental conditions compared to more high-tech prevention methods. Cases of respiratory tuberculosis have markedly declined since reliable records began last century. The introduction of mass radiography, chemotherapy and bacillus Calmette–Guérin (BCG) vaccination in the second half of the 20th century has made virtually no difference to the rate of decline (Fig. 25.3).

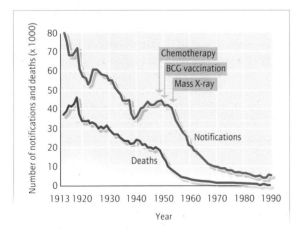

Figure 25.3 Respiratory tuberculosis in England and Wales from 1913 to 2002. The disease started to decline long before chemotherapy and other medical interventions became available, due to improvements in living conditions. Source: Health Protection Agency.

Paradoxically, the burden of some infectious diseases actually rises as living conditions improve. This applies to conditions where the rate of complications is greater in adults than in children. Epidemics of paralytic poliomyelitis occurring in the 1940s and 1950s are attributed to improved sanitation with a consequent reduction of wild virus circulation in young children (Fig. 25.4). This led to an increase in the average age at which infection occurred. Because paralytic poliomyelitis is the more likely clinical presentation in older age groups, the number of paralytic cases actually increased. A similar phenomenon has recently been observed for hepatitis A in some countries, where the number of cases with jaundice is rising as the average age at infection increases.

Another infection to emerge as a result of modern living is legionnaires' disease. The causal agent, *Legionella pneumophila*, thrives in microenvironments such as showerheads, cooling towers and air-conditioning units. Outbreaks of legionnaires' disease occur when bacterially contaminated aerosols are generated from these water systems, particularly when the systems are not properly maintained (Fig. 25.5, and Chapter 26).

Health education

Many health education programmes have been conducted at local and national levels with the aim of reducing exposure to infectious diseases. These include safe sex campaigns, needle exchange schemes, advice to pregnant women, guidance on food hygiene and advice to travellers. These campaigns have been conducted by government-funded bodies, voluntary agencies and industry (especially in relation to food hygiene). At the local level many individuals and agencies may be involved, including consultants in communicable disease control (CCDCs), general practitioners, health promotion departments, health visitors and voluntary groups.

With a few notable exceptions, health education has achieved only limited success in preventing exposure to communicable disease. Research has shown that campaigns often succeed in raising public awareness, but seldom result in behaviour modification, and where this does occur it is often short-lived. The smaller the perceived risk of infection, the less likely it is that an education programme will succeed. Newly emergent diseases such as AIDS and new variant Creutzfeldt–Jakob disease, together with the high media profile for diseases such as food poisoning, legionellosis, meningococcal meningitis, and methicillin-resistant *Staphylococcus aureus* has shifted the public perception of infectious disease in recent years. However, in general, the education programmes most likely to succeed are those that involve the local community in their design and implementation, and are on-going rather than one-off campaigns.

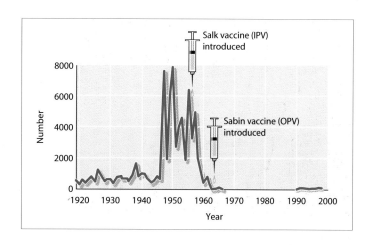

Figure 25.4 Paralytic poliomyelitis in England and Wales from 1914 to 1998. The outbreaks in the 1940s and 1950s were associated with an increasing age at infection, with a corresponding increase in the ratio of paralytic to non-paralytic cases. IPV, inactivated poliovaccine; OPV, oral poliovaccine.

Figure 25.5 Inside view of a cooling tower – a potential source for *Legionella pneumophila*. This badly maintained tower was the source of a large outbreak of legionnaires' disease.

Food safety

Reports of salmonellosis, *Campylobacter* and other bacterial causes of food poisoning have increased considerably

in recent decades (see Chapter 8). This has attracted considerable public attention and led to a major review of the legislation governing food safety, culminating in the Food Safety Act (1990). Food law has now been harmonized across the European Community since the introduction of the single European market.

Much of the responsibility for enforcement of food safety legislation lies with environmental health officers (EHOs), who are employed by Local Authorities to inspect food premises. The Department for the Environment, Food and Rural Affairs (DEFRA) is responsible in England for the enforcement of regulations that apply to farms. This includes the extensive legislation that was introduced to control salmonella infection. The Food Standards Agency was set up in 2000 to avoid the previous situation where the same body was responsible for both food production and safety. It is an independent food safety watchdog to protect the public's health and consumer interests in relation to food.

Pasteurization of milk greatly reduces the risk of exposure to many pathogens such as *Mycobacterium bovis* and *Campylobacter* sp. In Scotland it is illegal to produce or sell unpasteurized milk, whereas in England and Wales it is not. The ban in Scotland has led to a substantial reduction in the number of milk-associated outbreaks compared to England and Wales (Table 25.2).

Vector control

Vector control is particularly important in tropical countries where arthropods play an important role in many infections, both as vectors and as primary causal agents (Table 25.3). Attempts to control insect populations by the regular use of pesticides have usually been unsuccessful due to the emergence of resistance. Control of arthropod vectors is better achieved by altering the human environment than by widespread, untargeted use of insecticides.

Time period	England and Wales		Scotland	
	No. of outbreaks	No. of cases	No. of outbreaks	No. of cases
1980–1982	40	540	21	1090
1983–1984	22	518	8	46

Table 25.2 Milk-borne outbreaks of salmonellosis in England, Wales and Scotland, from 1980 to 1984: impact of compulsory pasteurization in Scotland in 1983. Source: Health Protection Agency and Scottish Centre for Infection and Environmental Health

Table 25.3 Arthropods of medical importance

Arthropod	Diseases
Mosquito	Malaria, dengue fever, filariasis, yellow fever
Sandfly	Leishmaniasis, sandfly fever
Fly	Trypanosomiasis
Flea	Plague, rickettsial fevers, tungiasis
Tick	Relapsing fever
Mite	Scabies, typhus
Maggot	Myiasis
Louse	Pediculosis, relapsing fever, typhus

• Reducing the availability of breeding grounds: successful measures have included the drainage of marshes where mosquitoes can breed, alteration of watercourses to reduce areas of stagnation, clearance of exposed refuse dumps to reduce the number of old motor tyres, cans and containers in which water can collect, and to reduce the population of rats and rat fleas. Covering bowls and tanks used to store water in households is important in denying access to breeding mosquitoes. The availability of hygienic laundry and personal bathing facilities, good quality clothing, and cosmetic insect-repellents supports personal protection from fleas, lice and mosquitoes.

• Keeping vectors out of homes: improved house construction removes areas for nesting and resting of vectors in human communities. Metal, wood or tile roofing is less insect-friendly than thatch or palm. Window and door screens keep insects out, and air-conditioning systems also keep buildings insect-free.

• Sentinel systems: in areas where seasonal or occasional outbreaks of vector-borne disease occur, surveillance can give warning of increasing case-numbers. Observation and serological testing of chicken flocks is used to detect increasing cases of St Louis encephalitis virus infection in the USA. Virological investigation of birds found dead or sick is used to identify West Nile virus activity in many countries. Reporting of encephalitis deaths in horses is also useful. When sentinel systems indicate a rising incidence of infections, targeted anti-mosquito campaigns, and public advice, can be commenced. Following surveillance for Japanese encephalitis virus activity, targeted use of environmental insecticide spraying was used to protect visitors to South Korea during the Seoul Olympic Games.

• Travellers to the tropics should be educated and advised to reduce the risk of infection by taking personal measures to avoid insect bites using, for example, repellents such as diethyltoluamide, mosquito nets and protective clothing.

Immunization

The introduction of effective immunization programmes was one of the most significant public health achievements of the 20th century. The control of smallpox, polio and diphtheria would not have been possible without the development of immunizing agents against these diseases.

Vaccines and immunoglobulins

Immunization may be achieved passively by administration of an immunoglobulin preparation, or actively by use of a live or non-replicating vaccine.

Vaccines

Vaccines are derived from whole viruses and bacteria, or their antigenic components. Live vaccines are prepared from attenuated strains that have minimal pathogenicity but are capable of inducing a protective immune response. They multiply in the human host and provide antigenic stimulation over a period of time. This results in durable immunity, usually after a single dose. Vaccine failures are uncommon, and are usually the result of inadequate handling or administration. The main disadvantage of live vaccines is that they occasionally cause a full-blown infection in the recipient. This is most likely to occur in the immunosuppressed patient; live vaccines are usually contraindicated for this group. Their use in pregnancy should also be avoided because of the potential risk of fetal infection.

Non-replicating vaccines contain either inactivated whole organisms or antigenic components. Agents that enhance the immunogenic effect of the vaccines (adjuvants) may be added. These agents (aluminium salts are commonly used) often help to keep the antigen in a localized or particulate form. Increasingly sophisticated methods are being used to produce these vaccines, such as protein–polysaccharide conjugation (*Haemophilus influenzae*) and genetic expression of protective antigens (hepatitis B). Because replication cannot take place in the human host, such vaccines are safe for use in pregnancy and in the immunosuppressed. Their disadvantage is that more than one dose is usually required for protection.

Local reactions are relatively common with non-replicating vaccines; this is related to the quantity of antigen they contain.

Types of vaccines, their modes of action and contraindications

Live
Examples: bacillus Calmette–Guérin (BCG), measles/mumps/rubella, oral polio, yellow fever.
1 Multiply inside the human host and provide continuous antigenic stimulation over a period of time.
2 Provide durable immunity, usually after a single dose.
3 Contraindicated in pregnancy and the immunosuppressed.

Non-replicating
Examples: pertussis, influenza, *Haemophilus influenzae* type b, rabies, typhoid.
1 Do not multiply inside the human host.
2 Antibody response is related to the antigen content and potency.
3 Multiple doses are often required, with subsequent booster doses.
4 No general contraindications.

Immunoglobulins
Immunoglobulins are extracted from plasma by ethanol fractionation. They provide short-term protection against certain infections. They are also used in the management of immune disorders and to supplement antiviral therapy, especially in the immunocompromised (see Chapter 22). The preparations available in the UK are either from pooled plasma or from hyperimmune donors.

Pooled plasma
Pooled plasma contains antibodies to viruses that are prevalent in the general population. Human normal immunoglobulin (HNIG) was widely used in the past for pre-exposure and post-exposure prophylaxis of hepatitis A. However, an active hepatitis A vaccine, effective both pre-exposure and post-exposure, is now available. The uses of HNIG are now limited. Here are some examples of where it is sometimes used.
• Protecting household contacts in outbreaks of hepatitis A, but only where there has been a delay in identifying cases and contacts may be at particular risk of severe disease.
• Post-exposure prophylaxis of measles in immunosuppressed contacts and some pregnant women and infants.
• Post-exposure prophylaxis of rubella in early pregnancy (where the risk of intrauterine transmission and serious sequelae is high) if termination is not acceptable to a non-immune woman.
• Preventing or attenuating an attack of polio in immunosuppressed persons who have either inadvertently received live polio vaccine themselves or whose contacts have been given it (this is now only relevant in situations such as polio outbreaks, where live polio vaccine may be used).

Hyperimmune donors
• Hepatitis B immunoglobulin, in combination with active immunization, may be used for post-exposure prophylaxis following accidental exposure to infected blood, and for babies born to acutely or chronically infected mothers.
• Varicella zoster immunoglobulin may be indicated where a susceptible person at high risk of severe varicella (e.g. immunosuppression, pregnancy, neonates) has had a significant exposure to chickenpox or herpes zoster.
• Tetanus immunoglobulin may be used in the management of tetanus-prone wounds in patients who have not been immunized or in whom the last dose of vaccine was given more than 10 years previously.
• Tick-borne encephalitis immunoglobulin is available in countries where the disease is endemic, notably Austria, for prophylaxis following tick bites.
• Rabies immunoglobulin may be indicated for prophylaxis following warm-blooded animal bites in countries where the disease is endemic in the animal population.

The most up-to-date information on indications and dosage schedules for the use of immunoglobulins in the UK is available on the Health Protection Agency website.

Strategic aspects of immunization programmes
The aim of an immunization programme may be eradication, elimination or containment of a disease.
• *Eradication* is total absence of the organism in humans, animals and the environment. Once a disease has been eradicated, the immunization programme can be discontinued. The only disease that has been eradicated by immunization is smallpox. Smallpox had many features that favoured eradication: it was an easily recognizable illness with no subclinical or latent infection; long-term carriage of the virus did not occur; visible evidence of immunity (a characteristic scar) was available; there were no non-human hosts; infectivity was low and the incubation period was relatively long. Poliomyelitis shares many of these characteristics, and the World Health Organization set a target to eradicate polio globally by the year 2000. Although this target was not reached, significant progress has been made in most parts of the world. By the end of 2003 polio had been eliminated from all but six countries worldwide. The main challenge to eradicating polio has been the reintroduction of polio to several African countries and Yemen, which followed reduced vaccination levels in West Africa.
• *Elimination* is where the disease has disappeared, but the organism remains in animal hosts, the environment or

causing subclinical infection in humans. The persistence of *Clostridium tetani* in the environment, and of *Bordetella pertussis* in the older age-groups of human populations are examples. Unlike eradication, it is not possible to discontinue immunization.

• *Containment* is the point at which a disease, although not eliminated, is no longer considered to be a significant public health problem. The current situation with *Haemophilus influenzae* type b infections is an example of this.

Possible aims for an immunization programme

1 Eradication: removal of the causal agent (e.g. smallpox).

2 Elimination: absence of disease, although the causal agent remains.

3 Containment: reduction of disease to the point at which it is no longer a public health problem.

There are two basic approaches to immunization programmes: universal or selective.

• Universal immunization has been adopted for most of the childhood vaccines.

• A selective programme aims to protect only those at risk from disease. This is less expensive than universal immunization, and tends to be used for the more costly vaccines such as hepatitis B. In practice it is often difficult to identify and immunize those who are genuinely at risk.

Immunization schedules

The ages at which vaccines are given, and the preparations used, vary considerably from one country to another. The approach in the UK has been to minimize the number of clinic visits and to secure protection as early in life as possible, without compromising efficacy. The currently recommended schedule is shown in Table 25.4. The British schedule is similar to that used in many European countries and in the USA.

Surveillance of immunization programmes

The ingredients for a successful immunization programme are a safe, effective vaccine and high coverage (uptake) in the target population. The safety and efficacy of vaccines are established in clinical trials before they are licensed. After licensing, batches of all vaccines are regularly tested for potency and toxicity at the National Institute for Biological Standards and Control (NIBSC) before release. Any severe or unusual reactions to vaccines should be reported on a yellow card to the Commission on Human Medicines. Additional surveillance schemes have been established to monitor the efficacy and safety of some vaccines, e.g. BCG. Annual serological surveys of age-specific antibody prevalence to measles, mumps and rubella (MMR) are undertaken by the Health Protection Agency's Centre for Infections to monitor the impact of the MMR vaccine. Coverage of vaccines is assessed from annual returns to the health departments and the COVER (cover of vaccination evaluated rapidly) scheme, which is run by the Centre for Infections.

The target coverage for childhood vaccines is 95% at 2 years of age. Most immunizations are given by general practitioners, who are paid according to whether they achieve targets. Each health district has a designated immunization coordinator, who is usually either a community paediatrician or the CCDC, with local responsibility for management of the programme. Vaccine coverage has improved considerably in recent years (Fig. 25.6) although there was a decline in MMR coverage in the late 1990s,

---

Table 25.4 British childhood immunization schedule 2006

Vaccine(s)	Age	Comment
Diphtheria/tetanus/acellularpertussis/inactivated polio/haemophilus influenzae b (DTaP?IPV/Hib) **plus** conjugated pneumococcal vaccine	2 months	Primary course
DTaP/IPV/Hib **plus** meningococcus C vaccine	3 months	
DTaP/IPV/Hib **plus** pneumococcal **and** meningitis C vaccines	4 months	
Haemophilus influenzae b/meningitis C	12 months	Booster
Measles/mumps/rubella **plus** conjugated pneumococcal vaccine	13 months	First dose Booster
Adult diphtheria/tetanus/inactivated polio	13–18 years	Booster
Measles/mumps/rubella	3–5 years	Second dose
Tuberculosis (BCG)	Neonates at high risk and individuals aged 10–14 years at increased risk (see Chapter 18)	

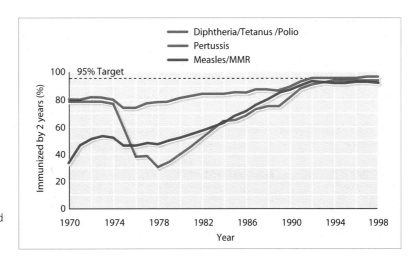

Figure 25.6 Vaccine coverage in England and Wales from 1970 to 1998. Source: Health Protection Agency.

following claims of a link between the MMR and autism. Large scientific studies have, however, demonstrated no such link.

Surveillance of immunization programmes

1 Clinical trials before vaccine introduction.
2 Regular 'batch' testing for vaccine antigen content and purity.
3 Surveillance of disease before and after vaccine introduction.
4 Reporting of vaccine adverse effects for investigation ('yellow cards').
5 Surveillance of vaccine coverage.

An example of the use of vaccine surveillance was the finding that the incidence of *Haemophilus influenzae* type b infections rose in the late 1990s, despite excellent coverage in the childhood vaccination programme. Investigation revealed that one of the available vaccines had low immunogenicity in some settings, and also that a pre-school booster dose of vaccine was necessary to maintain protection against infection at entry to pre-school in the at-risk age group.

Contact tracing

Contact tracing is used to identify those who have been in contact with an infectious disease, in order that preventive measures can be taken. These measures may include screening for evidence of infection and subsequent treatment, active or passive immunization and/or chemoprophylaxis (see below).

For contact tracing to be successful, prompt notification of the index case and follow-up of contacts are of the essence. This depends on the index case being able to identify their contacts. It is most likely to be effective for infec-

tious diseases with longer incubation periods (allowing time for intervention) and where adequate chemotherapy is available. Contact tracing is an important control method for tuberculosis. In this context recent developments in the ability of laboratories to distinguish between strains has enabled contact tracing and outbreak investigation to be more focused by indicating those individuals who are genuinely related to the index case by proving a close similarity by molecular means (see Chapters 2 and 18).

Special considerations apply to contact tracing for sexually transmitted diseases (in this context, called 'partner notification'). In the UK, the National Health Service [Venereal Diseases] Regulations in 1968 first outlined the partner notification process and gave a standard for good practice, including maintenance of confidentiality. Existing guidance on partner notification dates from 1980. Anonymised contact slips are used that can be cross referenced to cases. Partners may either be contacted by the patient directly or by Health Advisors attached to genitourinary medicine clinics (without naming the patient concerned).

Chemoprophylaxis

Chemoprophylaxis is the name applied to a course of anti-infective treatment, given to someone who is infected by a pathogen but who does not have active disease. It is generally used for control of more serious infections such as diphtheria and meningococcal disease. It may also be used during influenza outbreaks among high-risk patients, such as elderly nursing-home residents. Chemoprophylaxis aims either to eliminate carriage of pathogenic organisms, reducing the risk of progress to clinical infection and preventing transmission to those not yet exposed, or to treat contacts who are non-immune and may be incubating the disease. Elimination of carriage requires prophylaxis of a large contact network, whereas individual protection can be achieved by treating only those who have been in close contact with the index case. Both strategies can be difficult to implement as they require healthy individuals to take antibiotics that may produce side-effects. Single-dose regimens are the most effective. It is important to determine the antibiotic sensitivity of the strain from the index case, as this may influence the agent selected for chemoprophylaxis.

Example of a chemoprophylaxis regimen in the UK

Meningococcal disease

1 Rifampicin 600 mg orally twice daily for two days; (child 1–12 years, 10 mg/kg twice daily; child less than 1 year, 5 mg/kg twice daily).
2 Alternative for adults and children aged 5 years and above (also useful when large numbers of contacts need prophylaxis; ciprofloxacin: single oral dose of 500 mg (child 5–12 years 250 mg).
3 Alternative, also useful in pregnancy and breastfeeding: ceftriaxone i.m. 250 mg single dose (child under 12 years, 125 mg).

⚠ Rifampicin is contraindicated in jaundice or known hypersensitivity to rifampicin. Interactions with other drugs such as anticoagulants, anticonvulsants and hormonal contraceptives should be considered. Side-effects should be explained, such as staining of urine and contact lenses.

⚠ Ciprofloxacin is contraindicated in pregnancy, and the manufacturers do not recommend using it in children or growing adolescents unless the risks outweigh the possible benefits. Ciprofloxacin has been associated with anaphylactic reactions.

Screening for infectious diseases

Screening in its widest sense is the detection of disease in its pre-symptomatic phase. Identifying those who are infected before they develop disease, or become infectious themselves, can allow action to prevent on-going transmission. Screening in communicable disease control most usually goes hand in hand with contact tracing. For example: once contacts of a case of tuberculosis have been identified, they may be investigated to see whether they have become infected; if they have, appropriate measures may be taken to treat them and to manage their contacts. In this scenario the history of contact makes the risk of infection high compared to the general population, and intervention benefits both the individual and public health.

A different, more formal, 'screening programme' is one in which populations or risk groups with no identified con-

tact with a case of communicable disease are investigated. An example is antenatal screening for HIV infection.

Criteria for an effective and ethical communicable diseases screening programme

1 The disease should be an important public health problem.
2 There should be a recognizable latent or early symptomatic stage.
3 The screening test should be harmless, sensitive and specific.
4 There should be clear benefits to the individual from identifying the infection.
5 Effective measures should be available to prevent the on-going transmission of infection.

In the UK the most effective screening programmes have been those aimed at preventing vertical transmission of rubella, syphilis and HIV (see Chapter 17).

It is rare that all these criteria can be met and screening for infectious disease is sometimes adopted in response to public and political pressure rather than on the basis of sound public health.

Outbreak investigation

Prompt identification and control of outbreaks can prevent substantial morbidity. For example, it was estimated that 30 000 cases of salmonellosis were averted following the investigation of a single outbreak that was caused by contaminated imported chocolate (Fig. 25.7).

Methods

When an outbreak occurs, the fundamental tasks that must be performed are: describing the incident; analysing data from investigations to determine the cause; and giving appropriate public health advice to control the outbreak. There are a series of basic steps in the investigation of an outbreak. These are not necessarily sequential and control measures may be instituted at any stage.

Steps in the investigation of an outbreak of infectious disease

1 Establish that an outbreak exists.
2 Confirm the diagnosis.
3 Define the population at risk.
4 Define, collect and count cases.
5 Describe the cases according to time (onset), person (age/sex) and place.
6 Formulate a hypothesis to explain the source, mode of transmission and duration of outbreak.
7 Test the hypothesis by an analytical study.
8 Evaluate effectiveness of control measures.
9 Communicate findings.

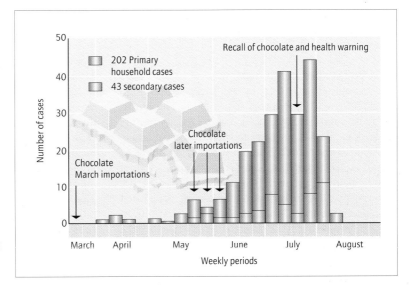

Figure 25.7 The epidemic curve for an outbreak of *Salmonella napoli* caused by contaminated chocolate, showing the impact of rapid identification and removal of the source. Source: Health Protection Agency.

Recognizing an outbreak

An outbreak may be recognized in different ways. Laboratories may detect greater than expected numbers of isolates of a common pathogen, or begin to identify isolates of an unusual pathogen. There may be increased reports of clinical cases of disease, or new reports of a disease not recognized as endemic in the area.

In a widespread outbreak, individual cases may be dealt with by different clinicians and laboratories in different regions. Effective collaboration of local, regional and national surveillance is essential in detecting such outbreaks, as they may only be recognized when all cases are counted together. For example, an outbreak of serious illness and death amongst injecting drug users was recognized in Glasgow, Scotland, in 2000, because a series of similar and unusual deaths were examined at post-mortem by one pathologist. When similar cases were looked for in England a large number were found but because each had been dealt with by a different team in widely separate geographical areas the connection had not been made. The cause of the outbreak was found to be an unusual pathogen, *Clostridium novyii*, associated with subcutaneous heroin injection. This example highlights the fact that effective surveillance is more than simply the collection of routine data; it depends equally on healthcare workers noticing and reporting the unusual.

Confirming the existence of an outbreak

When apparent case-clusters are identified, it is important to distinguish between true outbreaks and pseudo-outbreaks. Sudden increases in reports of illness may be caused by the availability of a new or more sensitive laboratory test, the appointment of a physician with a particular disease interest, or increased public awareness, perhaps because of media reports about a disease. An important first step is therefore to confirm the diagnosis. This may seem obvious, but may be overlooked. In 1985, a large outbreak of legionellosis associated with a hospital outpatient department was initially labelled as epidemic influenza until appropriate laboratory investigations were undertaken.

Characterizing the outbreak

Once the diagnosis has been confirmed, a case definition should be developed. Initially this should be non-specific so that all suspected cases can be identified and investigated; later on in the investigation a more rigorous definition, often with laboratory confirmation, will be required. This case definition should then be the basis for collecting and counting cases. Collection of cases will often involve active case finding; for example, by directly contacting general practitioners, laboratories and hospitals.

At this stage of the investigation, the aim is to gather basic epidemiological information to describe the 'time, person, place' characteristics of the outbreak and to formulate a hypothesis to explain the source of infection, the mode of transmission and the duration. The basic epidemiological information, collected for all cases, must include date of onset of symptoms, age/sex and place of residence. This information should be tabulated and an epidemic curve drawn. The nature of the curve will indicate whether a point source, continuing source, or person-to-person transmitted outbreak is involved (see Chapter 1). Additional information required will depend on the nature of the outbreak. For example, a salmonella incident will involve obtaining food consumption histories, whereas in an outbreak of legionnaires' disease details of likely exposures should be sought, such as foreign travel. These preliminary enquiries should be carefully conducted with cases affected early in the outbreak, ideally as face-to-face interviews.

Analytical studies of outbreaks

The hypothesis generated by the early investigation should be tested by a formal analytical study. This is usually a case-control study. In a case-control study, exposure histories are sought from cases and healthy controls, using a structured questionnaire administered by telephone or face-to-face interview; in larger investigations a postal questionnaire may be used. The relative risk of exposure to the postulated source of the outbreak is then calculated for cases and controls. The controls must be drawn from the same population as the cases, so that they have had the same opportunity for exposure to the source. The main disadvantage of such studies is the potential for biases. These may arise from selection of controls but the major difficulty is 'recall bias'. Interviews are often conducted several days or even weeks after the event. Patients may not remember their exposures or may bias their responses towards their perception of the source of the outbreak. For example, patients affected by food poisoning frequently ascribe their symptoms to the meal consumed immediately before the onset of illness, rather than in the period 24–48 h previously; their recall for the earlier meals may therefore be less accurate. This can be overcome by giving prompts, such as asking patients to consult their diary during the interview. It is also important to conceal the exposure variable under suspicion in the questionnaire by including questions about other exposures that are not under investigation.

Organization and management

At an early stage an outbreak control team should be convened, which should meet frequently to review progress and plan control measures. This group may include: the local CCDC, microbiologist, environmental health officer, director of public health, and hospital control of infection officer, a representative of the suspected source (e.g. a water company) and, where appropriate, regional or national representatives of e.g. the Health Protection Agency or equivalent. It is important that all communication with the press is through a single source, and that the outbreak control team should agree the contents of all press statements.

All stages of the outbreak investigation should be carefully documented. This is particularly important where legal action is likely to ensue. Regular reports of the progress of the investigation should be produced that are approved by the outbreak control team. The final report should be a comprehensive account of the investigation and include an evaluation of the control measures that were implemented as well as recommendations for the prevention and management of future incidents

Useful websites

- Health Protection Agency; http://www.hpa.org.uk/
- Health Protection Scotland; http://www.hps.scot.nhs.uk/
- National Public Health Service for Wales; http://www2.nphs.wales.nhs.uk/icds/
- Communicable Disease Surveillance Centre, Northern Ireland; http://www.cdscni.org.uk/
- Centers for Disease Control; http://www.cdc.gov/
- Office for National Statistics; http://www.statistics.gov.uk/
- Department for the Environment, Food and Rural Affairs; http://www.defra.gov.uk/
- Food Standards Agency; http://www.food.gov.uk/
- World Health Organization; http://www.who.int/en/

Emerging and Re-emerging Infections

26

Introduction

Recognition of emerging viral infections

• 1937 West Nile fever: 1950s: culture from humans and mosquitoes, Uganda, emerged in USA 1998.
• 1957 Marburg virus: culture from sick African laboratory primates and humans in German laboratory: re-emerged 1999, Congo.
• 1970 Monkeypox virus recognized in West Africa as smallpox became rare.
• 1976 Ebola virus: culture from cases in Congo and Sudan: re-emerged 1995, Congo.
• 1980 HTLV-I: by lymphocyte culture.
• 1983 HIV: by lymphocyte co-culture.
• 1988 Human herpes virus type 8 identified in Kaposi's sarcoma tissues.
• 1989 Hepatitis C virus: by detecting virus mRNA in serum.
• 1993 Hantaviruses causing pulmonary syndromes: cell culture and serodiagnosis, USA.
• 1994 Hendra virus: culture of material from humans, horses and bats in Australia.
• 1995 H5N1 avian influenza epidemic: culture from poultry and humans: Hong Kong.
• 1997 Australian bat lyssavirus: first indigenous rabies-like virus in Australia, by serology and culture from a bat.
• 1998 Nipah virus: serology, culture, PCR of material from humans, pigs, bats: Malaysia.
• 1998 Enterovirus 71: culture from affected children in Taiwan.
• 2003 SARS coronavirus: culture from humans and civet cats: China.
• 2004 Monkeypox cases occur in the USA.

Emerging bacterial infections

• 1951 Penicillin-resistant *S. aureus* identified, London, UK.
• 1976 *Legionella*: histology and culture from lung tissue, in large epidemic, USA.
• 1982 *Borrelia*: culture from ticks, and humans with Lyme arthritis, USA.
• 1983 *Helicobacter pylori*: spiral organisms visualized in gastric biopsies
• 1988 Anisakis: *Ehrlichia*-like organisms recognized in granulocytes in sick humans.
• 1989 *Bartonella henselae* propagated from blood.
• 1992 *Vibrio cholerae* O139 cultured from faeces of patients in India.
• 1982 Epidemic methicillin-resistant *S. aureus* strain

identified in Australia.
- 1998 Vancomycin-resistant enterococci identified in USA.
- 2003–4 Multidrug-resistant *Acinetobacter baumanii* increasingly identified in ITUs, UK.

Emerging parasitic infections

- 1976 Cryptosporidia: microscopy of faeces from sick humans.
- 1977 Cyclospora: microscopy of faeces, 1993 life cycle demonstrated in human enterocytes.
- 1990s and 2000s Large expansion in fox population (following rabies control) supports spread of *Echinococcus multilocularis* in foxes and voles across Western Europe, the Balkans, Eastern Europe: now found along western end of the Danube valley: human cases occurring outside recognized endemic areas.

Emergence of novel agents

- 1980s Endemicity of *Pneumocystis jiroveci* recognized when it causes fatal opportunistic infection in AIDS cases: now identified by rRNA analysis as a fungus with some similarities to protozoa.
- 1986 Bovine spongiform encephalopathy recognized by brain histopathology of affected animals: human form (vCJD) first suspected 1995.

Some examples of deliberately released pathogens

- 1987 *Salmonella enterica*: epidemic of salmonellosis from a salad bar later found to have been deliberately caused when a political group was investigated by police for other issues, USA.
- 2001 *Bacillus anthracis*: caused outbreak of pulmonary anthrax, USA.
- In invasions, wars and sieges through the ages: reports of attempting to infect enemies by throwing the bodies of plague victims, or donating blankets used by smallpox cases.
- In reminiscences of former cold-war scientist: reports of development of weapons containing smallpox, Ebola viruses, plague bacilli and *Francisella tularensis*.

The reasons why infections emerge or re-emerge

Overview

Both pathogens and human populations continuously change and evolve in parallel with each other. Pathogens change by mutation or genetic re-assortment, and successful organisms expand their population through natural selection. Human populations change by expansion, movement and migration, urbanization, and changing practice in, among other areas, commerce, agriculture and food production (currently exemplified by increasing dependence on intensive agriculture and mass production).

Globalization is the name given to the process of increasing the connectivity and interdependence of the world's markets and businesses. This process now means that, for instance, fresh food from the other side of the world is available in supermarkets, and large groups of people may visit several countries per year for business. In addition to the globalization of commerce, ever increasing numbers of people are now travelling around the world for holidays and education. These changes offer a continuing opportunity for humans to encounter new pathogens, which may become established in human populations for the first time. The pathogens may then spread contiguously within a population, as well as by travel and movement of the affected population. With the speed of modern transport epidemic or pandemic diseases can spread around the globe more quickly than before. For example, the pandemic of Hong Kong influenza in 1957 took 4 months to spread from the Far East but in 2003 SARS was distributed from Hong Kong to many different countries in a few weeks.

From time to time over the ages, humans have artificially introduced disease into another community as an aid to invasion, or an act of war or terrorism. This action, officially termed 'deliberate release', is currently a concern in some parts of the world, particularly as modern biotechnology theoretically offers the opportunity to maximize the infectiousness, pathogenicity, antimicrobial resistance or transmissibility of an introduced pathogen.

> ### Reasons for the emergence and re-emergence of pathogens
> **1** Alteration of pathogens by mutation, evolution or genetic change.
> **2** Human or animal population movements.
> **3** Alteration of the human contacts and environment by urbanization, globalization, mass manufacturing, climate change or social breakdown.
> **4** Deliberate release of specific pathogens as a hostile act.

The development of novel or altered pathogens

Novel pathogens arise by genetic mutation
Existing, known pathogens all gradually change in genetic and antigenic structure, each at a different rate, and those with rapid multiplication can evolve particularly rapidly. Rapidly dividing, single-stranded RNA viruses lack built-

in processes for repairing mistakes or damage to their RNA sequences, and can produce many mutations in a few dozen generations. Most of the mutations will result in unfit viruses that die out but, occasionally, an altered strain will succeed in establishing a replicating population. The new strain may become a more infectious, virulent, transmissible pathogen or have an additional host tropism. The antigenic structure of influenza A virus changes progressively in this way. Every few years a virus evolves that is both an efficient pathogen and sufficiently different from previous virus strains that it is not suppressed by existing immunity in human populations. This process of continuing evolution is called 'antigenic drift'. It is the cause of annual seasonal epidemics of influenza A, which vary in size, severity and duration from year to year.

Even slowly replicating pathogens, such as *Mycobacterium tuberculosis*, can achieve significant change in genotype and phenotype by a process of mutation and selection. Resistance to anti-tuberculosis drugs occurs by mutation. The genes that determine sensitivity to the various drugs each mutate at a different rate. In a patient with a high bacterial load a significant number of mutated organisms will already be present. If a patient is treated with combination chemotherapy the mutated strains will be eradicated because they remain susceptible to most components of the regimen. It is highly unlikely that a patient will posses a strain resistant to two or more components of the regimen. However, if a patient is treated sub-optimally, and is only receiving one effective drug, the wild-type organisms may be suppressed by it while the newly mutated, resistant strain has a selective advantage and will grow rapidly. Exposure to another single drug by adding a new agent to a failing regimen allows this process to be repeated, leading to the emergence of a strain resistant to both drugs. This is how multidrug resistant tuberculosis is thought to emerge.

New phage types of salmonellas frequently arise from existing types by slight genetic change. They easily spread in crowded flocks and herds of farm animals, and enter the human food chain. *Salmonella enteritidis* phage type 4 arose in this way in the 1980s and caused widespread, persistent infection and vertical transmission in egg-producing chicken flocks. Apparently well, colonized chickens produced eggs that contained salmonellae, leading to outbreaks of severe disease in humans who ate the contaminated eggs.

Vibrio cholerae serotype O139 arose in 1992 in India and Bangladesh by mutation from the pathogenic serotype O1 and, like O1, was able to carry the cholera toxin gene. *V. cholerae* O139 was officially designated as the organism responsible for the eighth pandemic of cholera. However, the O139 strain has not progressed to countries outside Asia (where it causes about 15% of all cases of cholera). The original O1 serotype has remained prevalent, and is still the major epidemic-causing organism (Fig. 26.1).

Novel pathogens can suddenly emerge by genetic hybridization or re-assortment

Organisms can evolve very rapidly by genetic hybridization. It is thought that humans were readily infected by a novel SARS virus whose genome was largely that of a mammalian strain of coronavirus, but which contained a significant length of RNA derived from a strain that naturally infected birds.

The RNA genome of influenza A virus is divided into eight separate segments. If two types of influenza A virus simultaneously infect a single host, the segmented genes of the two viruses easily become mixed during replication, when the viral genomes are assembled within infected host cells (Fig 26.2). This process of genetic re-assortment may generate a novel virus strain, which possesses some genes from each of the two original viruses. This is the basis of sudden 'antigenic shift' in influenza A viruses. Avian influenza virus A H5N1 occasionally causes fatal infections in humans who handle affected birds, though human-to-human transmission is still extremely rare. If a human was simultaneously infected by avian H5N1 and a 'human' influenza virus during an influenza outbreak, a new hybrid virus could emerge, with avian H5 haemagglutinin. This could be both pathogenic and transmissible in humans, causing a pandemic of severe influenza. The devastating pandemic of influenza in 1917–18 was caused by the combination of a swine (pig) H1 component with a human N1.

Severe acute respiratory syndrome (SARS): the life and death of a pandemic

Epidemiology

SARS was recognized in March 2003 following a global alert from the World Health Organization (WHO) of outbreaks of unexplained severe respiratory illness among healthcare workers in Hanoi, Vietnam, and subsequently Hong Kong. The virus probably originated from Guangdong Province, China, where the first recognized cases occurred in November 2002. SARS rapidly spread worldwide (Fig. 26.3). Between March and July 2003 over 8400 cases, with over 900 deaths, were reported to the WHO from 32 countries. The virus was probably mainly transmitted through droplet spread, requiring reasonably close contact. It appeared to be less infectious than influenza. Most cases were close family members and hospital workers who cared for SARS patients. The case fatality rate was age dependent, ranging from less than 1% in persons under 24 years to over 50% in those above 65 years.

Figure 26.1 Epidemic status of *Vibrio cholerae* seventh (serotype O1) and eighth (serotype O139) pandemics in 1993.

Virology and pathogenesis

The SARS coronavirus is a single-stranded positive-sense RNA virus. It encodes a single polyprotein that is cleaved by a virus-encoded protease into several non-structural proteins including RNA-dependent RNA polymerase, ATP helicase and several structural proteins including a haemagglutinin esterase. The spike glycoprotein forms the petal-like surface projections. There are three groups of coronavirus, each with a different animal host range. SARS coronavirus is distantly related to all three of these groups. Its origin is not clear but serological evidence suggests that it has not previously circulated in human populations.

Infection was probably acquired through the respiratory tract. The incubation period of 2–7 days was followed by high fever and increasing features of pneumonia, often with a lung opacity on X-ray. The lung was the main focus of pathology but other systems, notably the gastrointestinal tract, were involved. In the lung there was interstitial lymphocyte and mononuclear cell infiltration, with giant cells of macrophage origin. Desquamation of alveolar pneumocytes and hyaline membrane formation are followed by the development of an organizing pneumonia.

Laboratory diagnosis

Laboratory research into the SARS coronavirus continues. No reliable early test exists. RT-PCR is the most sensitive rapid test available, but its utility is limited by the low titre of virus found in many clinical specimens. The SARS coronavirus grows in diploid cell lines but blind passage is

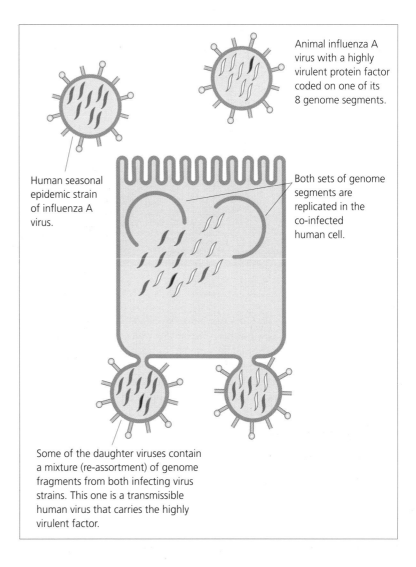

Animal influenza A virus with a highly virulent protein factor coded on one of its 8 genome segments.

Human seasonal epidemic strain of influenza A virus.

Both sets of genome segments are replicated in the co-infected human cell.

Some of the daughter viruses contain a mixture (re-assortment) of genome fragments from both infecting virus strains. This one is a transmissible human virus that carries the highly virulent factor.

Figure 26.2 Diagram of the mechanism of reassortment during simultaneous infection by two related viruses.

sometimes required for isolation. An IgM EIA is available for detecting seroconversion.

Prevention and control

The WHO announced that the chain of human-to-human transmission had been broken in May 2003. This was achieved using conventional public health measures: early detection of cases, tracing the contacts of cases, isolation of cases and quarantine of contacts, infection control in hospitals, informing and educating health professionals and the public. No vaccine or specific treatment has yet been developed. Since the outbreak in 2003, occasional sporadic cases have been reported from China. Vigilance for suspected outbreaks continues in public health systems around the world.

The recently revised International Health Regulations are intended to encourage effective recognition of, and response to, events such as the SARS pandemic.

Human travel, migration and population changes

Human travel for business and leisure

Today, many millions of people and their families travel for leisure, business and adventure. They may be exposed to diseases that they would not encounter at home. Almost everyone knows that intestinal infections are common when travellers encounter a new range of bowel pathogens. However, families travelling to tropical countries, from which they emigrated years before, may not

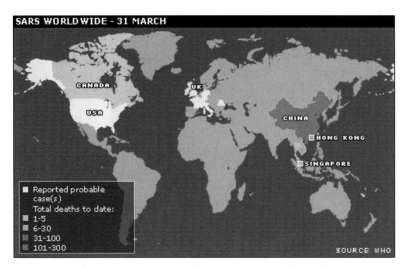

Figure 26.3 The worldwide reports of suspected SARS cases in early 2003: these cases were identified by an early clinical definition and included many individuals whose illness was found to be influenza, or other infections, once laboratory tests could be performed – nevertheless, they provided a basis for the effective isolation and quarantine of potentially infected cases, which interrupted transmission and halted the pandemic. Map reproduced with permission from BBC Asia-Pacific news website: http://news.bbc.co.uk/2/hi/asia-pacific, 29 June 2003.

realize that any immunity that they may once have had to malaria will have waned. Few northern Europeans realize that cutaneous or visceral leishmaniasis can be contracted in some areas of southern Europe and the Mediterranean. Few British travellers are aware that tick-borne encephalitis is widespread in some parts of central Europe and Scandinavia. Adventure travellers might be exposed to mammals or bats that carry rabies viruses. Business and adventure travellers may have sexual exposure in a part of the world where the epidemiology of HIV and other sexually transmitted diseases is quite different from that which they recognize at home, and can become the focus of outbreaks of exotic sexually transmitted diseases in a new ecology. In 2004–5, increasing numbers of men who have sex with men in the UK were diagnosed with lymphogranuloma venereum, a STD that is usually found in tropical countries (Fig. 26.4). In 2003, the newly emerged SARS coronavirus was carried rapidly from mainland China to Hong Kong, and from there to Canada, by business and family travel between the large Asian communities in these countries.

Human population movement and social change

Human populations tend to expand, develop and move. The reasons for these changes include enlarging population, changes in farming practices, commercial and housing pressures and the imperatives of war and famine.

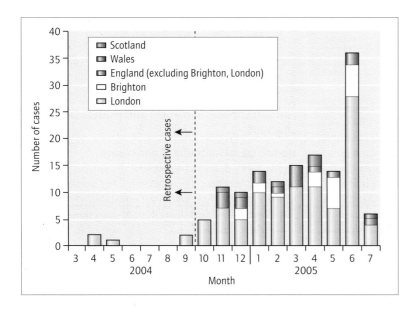

Figure 26.4 The outbreak of lymphogranuloma venereum in men who have sex with men in the UK. Figures from HPA CDR weekly 2005, no. 30.

When humans move into new environments, they often encounter new infection hazards. An example of this occurred when the Panama Canal was built. To accommodate the numerous construction workers, sprawling shanty towns grew up, with many water tanks and latrines, in a previously sparsely inhabited region of tropical rain forest. In this jungle, a 'sylvatic cycle' of yellow fever existed, in which mosquitoes carried the virus from monkey to monkey. The human settlements attracted the mosquitoes, and a human epidemic of yellow fever occurred, which eventually halted work on the canal, until adequate sanitary and anti-mosquito measures were instituted.

Displaced and deprived populations

Refugees and displaced populations often suffer from crowding, poor sanitary facilities, lack of facilities for bathing and laundry, and lack of basic healthcare. In these circumstances, skin, respiratory and intestinal infections may become epidemic or endemic. This can occur within just a few days in a population affected by natural disaster and breakdown of services, as occurs with floods and earthquakes. In the longer term with breakdown of health services, failure of immunization programmes and the advent of malnutrition, epidemic diseases, such as childhood measles, may become established, and tuberculosis often emerges and spreads. Even small-scale deprivation and homelessness can produce a large public health problem. In New York in the early 1990s, a large community of homeless drug-abusers, many of whom were HIV-infected, became a reservoir of untreated and infectious pulmonary tuberculosis. Many hundreds of cases occurred, and the rising incidence of new cases was only halted by a large and aggressive campaign of case-detection and case-isolation and free, directly observed treatment, centred on the patients' localities and needs.

Where social and family breakdown occur, the incidence of sexually transmitted diseases may also rise dramatically. Diseases of deprivation and social breakdown may later be carried to regions where the affected individuals eventually settle.

Natural disasters take away the opportunity for safe disposal of human excreta and often destroy irrigation and drainage systems. Populations of potential disease vectors, such as rats, feral dogs and mosquitoes, often increase dramatically, followed by an upsurge in diseases such as leptospirosis, plague, dengue, malaria or rabies.

Aid workers, the military and peacekeepers may be exposed to local diseases when they enter areas of social and environmental disturbance, where domestic and community hygiene and health facilities have been disrupted. Many of them are obliged to live in rented accommodation or simple dwellings. In West Africa, a number of such workers contracted Lassa fever in the first years of this century, while earlier, following civil wars in the Balkans, cases of hantavirus infection occurred in peacekeeping troops. More recently, a number of cases of cutaneous leishmaniasis have been diagnosed in British armed forces (Fig. 26.5).

Changes in human environments and commerce

Zoonotic infections arising due to occupation, farming and food production

Humans may come into new contact with animals when they keep or farm animals. It is thought that the international epidemics of SARS possibly resulted from infection

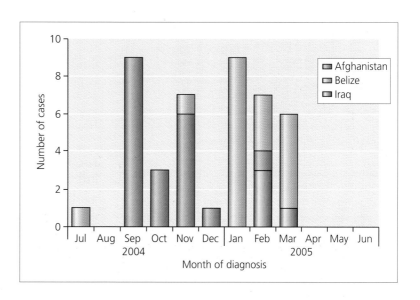

Figure 26.5 Cases of cutaneous leishmaniasis diagnosed in UK military personnel after deployment in endemic areas: an increasingly frequent situation.

of humans by a coronavirus strain acquired from close contact with civet cats. Civet cats are a food source in a number of Far East countries, and are kept and sold for the food market. They are often kept together with poultry, providing an opportunity for the coexistence of avian and mammalian coronaviruses. The first cases of SARS arose in rural China where this practice is common.

An avian influenza A (H5N1) virus has caused large epidemics among domestic poultry in China, Cambodia, Vietnam and some other Asian and east European countries in the early 21st century. Some humans in close contact with infected birds, especially young children, have become infected, with a case-fatality of around 55%. The disease has been carried westwards across the Eurasian continent by migrating wild birds. To the time of writing, no human-to-human transmission or co-infection with a human influenza virus has been reported, but plans to avert an epidemic or pandemic of an altered strain have been made worldwide.

In South Africa and other countries, ostriches are now farmed for their meat. Ostriches are subject to severe tick infection, and ticks in South Africa may carry the virus of Crimean Congo haemorrhagic fever (CCHF). Ostriches infected by tick bites are asymptomatically infected with the virus, and can support a viraemia. In the 1990s, an outbreak of CCHF occurred in workers at an ostrich-packing plant near to Tygeburg, as a result of exposure to infected ticks and ostriches.

Changes and movement in animal populations

Emerging zoonoses may be disseminated by animal movements. Nipah virus infection emerged in the rural pig-farming industry of Western Malaysia in the 1990s, causing many deaths in pig farmers. Pig exports carried the disease to Singapore, where more human deaths resulted.

Diseases spread by travel are not new

'The plague of Athens' 430 BC originated in Ethiopia, and spread via Egypt and Libya to enter the port of Piraeus. '*Headache, cough, retching......diarrhoea and a rash of small blisters, starting at the top. Some who recovered were blind*' (Thucydides). A quarter of the Greek army, recently returned from overseas, was destroyed. Most deaths occurred on the 7th to 9th day. Survivors developed immunity.

Animal populations themselves wax and wane, often in step with fluctuations in weather and food supply. In the 1990s, an expansion in the muskrat population in Siberia resulted in increased numbers of cases of Omsk haemorrhagic fever in humans who had rural contact with the virus-carrying animals.

Animal movement may introduce zoonotic infection into a new region. A recent example is the emergence of West Nile virus in the USA. West Nile virus has been rec-ognized as a zoonotic pathogen for many years, in Africa, mainland Europe and the eastern Mediterranean, where it is carried by migrating birds and transmitted between birds, animals and humans by biting mosquitoes. In 1999, in New York, deaths in wild and captive birds, particularly crows, were followed by the occurrence of encephalitis cases in humans (see Chapter 13). West Nile virus was identified in both birds and humans.

The virus was almost certainly introduced to the USA by the accidental or illegal importation of an infected bird. Since introduction, the virus has spread, both along bird migration routes and contiguously by local transmission, affecting birds, mosquitoes and humans in almost every US state (Fig. 26.6).

The virus has caused sporadic cases and outbreaks of disease in horses and humans in Europe since the 1960s. There was a cluster of cases in the south of France in 2003, and in 2004 two human cases were seen in Ireland, affecting travellers who had recently returned from Portugal.

The importation of exotic animals for sale as pets can result in the importation of disease. It has long been recognized that imported parrots and parakeets can carry *Chlamydia psittaci*. In 2005, a parrot in quarantine after arrival in the United Kingdom became sick, due to infection with influenza A H5N1. The parrot and a number of other birds were euthanased, and human quarantine technicians were given prophylactic antiviral drugs. No further cases arose.

A monkeypox outbreak in the USA

In 2004, giant Gambian rats were popular pets in the USA, and prairie dogs (rodents indigenous to the USA) were often kept with them in pet stores and households. Imported Gambian rats carried the monkeypox virus and passed it to prairie dogs, which became sick. A number of human cases of monkeypox occurred in children and adults who had close contact with sick prairie dogs. Human infection resulted from inoculation of infected material through bites or scratches from infected animals. In humans, a local lesion of inflammatory swelling and pocks existed for several days before the occurrence of systemic disease with marked fever and a generalized rash. No deaths or human-to-human transmission occurred.

Food-borne zoonoses and other food-borne infections

An emerging infection in animals may be transmitted to human populations if infected animal products enter the human food chain. In 1988 a novel neurological disease of cattle, diagnosed as a bovine spongiform encephalopathy (BSE), was recognized in cattle in the United Kingdom. A large BSE epidemic in cattle was amplified greatly by the practice of including recycled bovine material in meat-

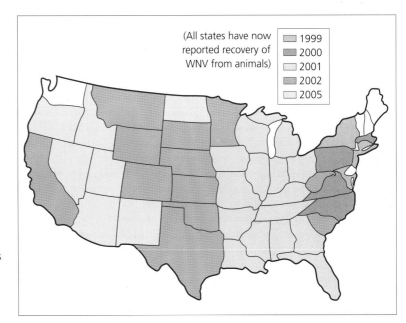

Figure 26.6 The spread of West Nile virus across USA states, after its introduction into New York State in 1999. Based on figures from the Centres for Disease Control and Prevention, Atlanta, USA.

and bone-meal products used as high-protein cattle feed. BSE-contaminated nervous system and reticuloendothelial tissue was used to manufacture many processed meat products for human consumption, including hamburgers, sausages and pates. Eventually, a novel neurological disease, variant Creutzfeldt-Jakob disease (vCJD) was recognized in humans and its cause was shown to be the infectious BSE agent. The incidence of vCJD declined following the eradication of BSE in cattle (Fig. 26.7 and Case study 26.1).

Fresh and chilled fruits, vegetables and manufactured foods are nowadays transported long distances, and can carry human pathogens with them. The resulting out-

breaks of human infection have included shigellosis (lettuces from European countries), enteric fever (sugar cane from Asia), cyclosporiasis (soft fruits from Central America), listeriosis (unpasteurized soft cheeses from France and the USA), and salmonellosis (eggs from northern and southern Europe, chocolates containing unpasteurized dairy products from Italy, salami snacks from Europe, and rice snacks from Israel).

Environmental change affects the epidemiology of a disease

The human environment may be changed by deforesta-

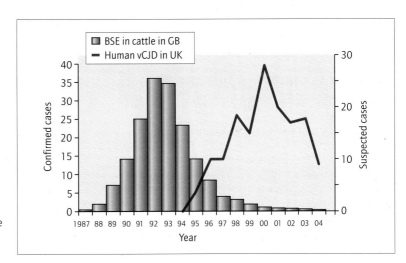

Figure 26.7 The epidemic curves for bovine spongiform encephalopathy in cattle and vCJD in humans, 1987–2004. Based on data from the Department of the Environment, Food and Rural Affairs and the Department of Health.

tion, urbanization, drainage or irrigation of land, or by climate change. These changes influence the ecology of the affected region, thus altering the prevalence of pathogens, reservoirs of infection and vectors. Deforestation may remove the habitat of monkeys that are a reservoir for yellow fever virus, bringing them closer to human dwellings. Heavy urbanization destroys the pools and lakes where mosquito vectors of malaria can breed, but towns are often the site of numerous cooling towers and air-conditioning systems, which may become reservoirs and sources of infection with *Legionella* spp.

Cattle farming and fish farming favour the activities of terrestrial and aquatic rats and voles, which are the reservoirs of *Leptospira* spp. responsible for Weil's disease and other leptospiroses. The use of wet 'paddy' fields and ploughs drawn by water buffaloes for rice farming is associated with the incidence of melioidosis. This is a disease caused by *Burkholderia pseudomallei*, an environmental organism associated with wetlands and water buffalo.

The distribution of some diseases and vectors has already changed as a result of global warming. Sandfly species capable of transmitting cutaneous leishmaniasis have become established further north in Europe than previously. Occasional cases of malaria currently occur in hot weather near airports in northern Europe, probably as a result of the importation and survival of malaria-carrying mosquitoes. However, it would only take a small increase in average temperatures to enable *Plasmodium vivax* to complete its life cycle in the anophelene mosquitoes that already exist in this region.

A large epidemic of vivax malaria occurred in East Anglia (which was then a wet fen) during a hot summer in the First World War, after repatriated soldiers convalescing from malaria carried the infection to the UK. It is estimated that around 70% of the population of the Fens suffered from malaria before the winter weather eradicated the parasites. In the El Nino year of 1998, exceptionally high rainfall in east Africa was accompanied by a large increase in the incidence of malaria.

Deliberate release of biological agents

Introduction

The strategy of assaulting opponents with infectious diseases is not new. Infection has been introduced by throwing the bodies of plague victims into besieged cities (siege of Kaffa), by making presents of blankets contaminated by smallpox cases (American-Indian wars), by attempt-

ing to develop bombs containing anthrax spores (British Second World War experiments), by contaminating salad-bar foods with salmonella (USA, 1980s), and by sending anthrax spores through the postal system (USA, 2003).

In the 21st century there is widespread concern about the risk of terrorist attacks involving chemical, biological or radiological materials, and most countries have systems for the detection of such attacks and for responses designed to avoid or minimize their damaging impact.

There is general agreement about the infectious agents most likely to be effectively used.

> **Biological agents considered most likely to be used as weapons**
> 1 *Bacillus anthracis*
> 2 *Yersinia pestis**
> 3 *Francisella tularensis*
> 4 Variola virus (smallpox agent)*
> 5 Viral haemorrhagic fever viruses (e.g. Ebola)*
>
> The agents marked * are capable of person-to-person transmission.

These agents are all moderately or highly infectious if dispersed by the airborne route, and are likely to be readily adaptable to spread by small- or large-scale aerosol generation. Medical practitioners and emergency-room staff are encouraged to be vigilant for the clinical features of these infections, and to seek help from their local Public Health teams if they suspect them.

Many other pathogens, such as food-poisoning pathogens, *Brucella* spp. or rickettsiae, could potentially be adapted for deliberate release, or 'weaponized'. First-line medical teams are therefore advised to maintain awareness of the likely presentation of any disease of deliberate release.

> **Unusual presentations of infections**
> These could indicate that a deliberate release has occurred
> 1 Infection caused by a rare strain of pathogen, or one with unusual antibiotic resistances.
> 2 An outbreak of an infection not endemic to the affected region.
> 3 A pattern of epidemic not consistent with natural behaviour of the pathogen (e.g. many cases of a zoonosis in people with no animal exposure, or many cases of a vector-borne infection when the vector is not prevalent in the region).
> 4 More than one outbreak occurring simultaneously in unconnected areas or communities.

How emerging and re-emerging infections spread

Introduction

Emerging pathogens extend their range by the same methods of transmission and spread as established and recognized pathogens (see Chapter 1). In addition, some of the means by which humans encounter new pathogens, travel, population movements, mass production of food, are also means by which rapid dissemination of a pathogen can occur.

Local, person-to-person transmission

Infections transmitted only by close personal contact may often be slow to spread, and can be limited by good conditions of hygiene and housing. An example is diphtheria, which re-emerged in homeless, unemployed communities following the rapid break-up of the former Soviet Union in the 1980s. Homeless individuals crowded together in large shopping complexes, where they could avoid the cold weather, and attempt to make a living by begging or small-scale selling of cigarettes and other goods. Once diphtheria emerged in these groups, many homeless people became carriers and cases. Childhood vaccination programmes had been interrupted, and this increased the susceptibility of the local communities to spread of the disease. A slowly extending, smouldering epidemic became established in the region. Detection of cases, and antibiotic treatment, limited the impact of the epidemic, but did not prevent the occurrence of new cases.

Tourists from Scandinavian and European countries were unaware of the risk, and some became infected during begging or shopping transactions. However, in their own countries, good living conditions, medical care and public health measures prevented further spread of the disease.

Airborne transmission

Airborne transmission of disease, by the movement of infectious aerosols (particles of dust or fluid with diameters below 1 μm, in contrast to person-to-person spread by larger droplets of diameter usually greater than 5 μm), is a common cause of local outbreaks of disease. Examples of airborne transmission include the transmission of varicella between patients in a large hospital ward, or transmission of hanta virus pulmonary syndrome in a large, rat-infected warehouse. However, large community outbreaks of legionellosis have been caused by infected aerosols emitted from municipal or industrial cooling towers. When the air is still, often on warm evenings or sultry days, aerosols can remain suspended for many hours, expanding into a large, invisible plume or cloud that deposits infectious particles beneath it. One summer in London, UK, an infected aerosol spread from the contaminated cooling tower of a large broadcasting company building. An area of more than one square kilometre of busy city streets was affected for several days, resulting in over 90 reported cases of legionnaires' disease. Many affected individuals had only briefly visited the area. Thus, from an urbanized environment, many cases of illness were generated, presenting in many parts of London and the UK, even though the disease was not transmissible from person to person.

International spread of infections

Although the SARS coronavirus was not highly infectious, and was also mainly transmitted by close personal contact, it did not cause a localized problem of the kind resulting from the diphtheria epidemic. It was able to spread rapidly across the world because of several key features of the infection and the population affected.

Features of SARS permitting rapid worldwide spread
1 Virus excretion continued for up to 3 weeks.
2 Infection could not be treated by antibiotic therapy.
3 There was no effective vaccine to protect travellers or contacts.
4 Some infected individuals were affluent, and travelled widely by international air routes while incubating or suffering from the disease.
5 The affected population existed in several large communities in different continents, with much travel between them.
6 SARS cases often infected carers and close contacts in their households.
7 Hospital admission and respiratory care procedures for severe cases greatly increased the transmission of infections, particularly to healthcare workers.

Sexual transmission ensures that a pathogen can spread in all human populations.

The human immunodeficiency virus (HIV) was recognized in the late 1970s when outbreaks of *Pneumocystis* pneumonia and severe candidiasis affected American men who were not known to be immunocompromised. The infecting HIV virus almost certainly emerged in Africa by evolution from a similar simian virus. Initially a rare disease, the epidemic was amplified by sexual transmission in the rapidly urbanizing population centres of Africa, affecting both men and women. From there it was carried by travellers to the USA. As soon as it entered a population that was mobile across the world, the virus spread more rapidly. In the USA a genetically adapted strain spread rapidly in sexually active communities of homosexual men. Later, it was carried by travellers to other Western countries. Asian and Far Eastern countries were among the last to recognize large outbreaks. Blood-borne spread and vertical transmission were recognized early, and vertical transmission from pregnant women to their newborn remains a major means of transmission, especially in developing countries. The number of people who need diagnosis and treatment continues to rise worldwide.

Dissemination of infection in the food chain

An example of infection carried by mass-produced foods is the national outbreak of illness due to the newly recognized toxigenic *E. coli* O157 in the USA, in the early 1980s. Hundreds of cases of haemorrhagic colitis, and associated cases of haemolytic uraemic syndrome in children, were traced to the consumption of hamburgers from a popular nationwide fast-food chain.

Investigation revealed that the hamburger patties were all manufactured at a single, central point, where ground beef products were mixed on an enormous scale, shaped into patties, frozen, and distributed daily to outlets all over the USA. A small amount of contaminated beef product could therefore produce a nationwide epidemic.

A similar, large epidemic of food poisoning occurred in Japan in the late 1990s when, as part of a national school meals manufacturing and delivery system, *E. coli* O157-contaminated radish salad was distributed to schoolchildren. Over 9000 cases of infection were reported.

Infection transmitted through non-food biological products

Nowadays human and animal cells, tissues and manufactured biochemicals are placed in human bodies on a large scale. Infections resulting from these practices do not cause outbreaks, because they affect only the recipients of the transferred tissues. However, they are now common problems associated with the expansion of complex medical practice, and are one of the significant costs of medical advance.

Blood products

Blood transfusion has been practised for over a century. More recently, blood products such as platelets, neutrophils, lymphocytes, immunoglobulins and coagulation factors have become widely used. Some of these are manufactured from the pooled blood derivatives of dozens or hundreds of donors. It is now known that some such products can potentially carry and transmit blood-borne infections. Blood cell products, on the whole, cannot be treated to destroy living pathogens, as this would also destroy the viable cells needed by the recipient. It is possible, however, to test donated blood for the presence of known blood-borne pathogens, and to discard contaminated donations. It is also usual practice to defer taking blood from donors who may have been in contact with pathogens until the risk of infection has passed, for instance after potential tropical exposure to malaria. Some blood products, such as factor IX concentrate, are pasteurized before use. Pure, infection-free factor VIII is manufactured by a genetic modification process. Other blood products, such as intramuscular immunoglobulin preparations, appear to be naturally decontaminated by the manufacturing process and have never been shown to transmit infection.

Pooled blood products are no longer manufactured from blood donated by individuals who have lived in the United Kingdom because of the potential risk that they could contain an infectious amount of the vCJD agent. For the same reason, in 2004 the Department of Health for England announced that people who had received a blood transfusion in the UK since 1980 would no longer be able to give blood. Similarly, bovine serum albumin from countries where BSE has been endemic is not used in the manufacture of pharmaceuticals and vaccines.

Hormones and other bioactive agents

Until the early 1980s, hormone preparations for medical use were often manufactured from human tissues. This practice came under close scrutiny when several cases of classical CJD occurred in patients who had been treated with human growth hormone (HGH). At the time, HGH was extracted from human pituitary glands harvested in post-mortem rooms. Despite the policy of harvesting only from cadavers with no history of neurological disease, it became clear that CJD had been transmitted by HGH injections in both Europe and the USA. The situation was saved by the development of genetically engineered, pure HGH. Human gonadotrophins, for fertility treatment, were also originally pituitary-derived, but are now manufactured by alternative means.

Tissue transplants

Blood products have only a limited life in the body of the recipient, but other cell- and tissue-transplants are intended to last for the life-span of the patient. They can potentially transmit all of the blood-borne infections that are known to be transmitted by blood (for instance, islet cell transplants could theoretically transmit hepatitis viruses). Some unexpected pathogens, which can reside long term in tissues, have also been transmitted by organ transplants. Examples include strongyloidiasis, contracted from transplanted kidneys, and toxoplasmosis, from transplanted hearts. Such infections can be difficult to treat in patients who are significantly immunosuppressed, so donors and recipients are serologically tested, to ensure that the donor does not have an infection to which the recipient is susceptible. In some settings, the recipient can be partly protected by infusions of specific immunoglobulin, or by chemoprophylaxis (for instance against severe cytomegalovirus infection).

Cadaveric dura mater grafts are no longer used after the potential for CJD transmission was recognized. Corneal grafts have been the means of rabies transmission in a handful of incidents, when corneas were harvested from unrecognized rabies cases.

In the late 1980s and early 1990s, the use of animal organs for human transplantation was seriously contemplated. The organs of pathogen-free pigs were thought to be of the correct size and function. However, the discovery that replication-competent retroviruses (porcine endogenous retroviruses: PERVS) are encoded in the genome of all pigs was a major factor in the abandonment of xenotransplantation projects.

How emerging infections are controlled

Strategies to prevent the spread of emerging infections

1 Detection and surveillance in countries, in Europe and by the World Health Organization. The World Health Organization global observation, alert and response network (GOARN), which assists with outbreak investigation and control in low-income countries, and gives international situation updates and advice during pandemics.

2 Communication and assessment systems nationally and internationally to alert to and assess the significance of any emerging infections, and to facilitate collaboration in their management.

3 Planning; national and international planning for the eventuality of emerging infections.

4 Control measures, e.g.:
- public and health worker education;
- national and international health regulations;
- control of animal, plant or food transport;
- health and hygiene for transported animals, animal products and cargoes;
- control of human travel and transport systems;
- vector control;
- isolation of cases and suspected cases;
- hospital and primary care infection control procedures;
- early, effective treatment.

5 Quarantine of exposed people.

6 Prophylaxis programmes.

Detection and surveillance

As discussed in Chapter 25, most countries have routine systems of reporting and counting infectious diseases, mainly based on statutory notifications or laboratory reports. These permit an ongoing knowledge of the usual occurrence of common and uncommon conditions, and recognition of any changes. This is valuable when a changing pattern of disease is suspected, particularly in detecting re-emerging diseases or new variants of old diseases. These systems are inefficient at detecting and following rapidly developing situations, though they are relatively good for documenting the appearance and behaviour of more slowly advancing outbreaks.

For some situations routine surveillance systems are inadequate and other methods have to be used, e.g. 'syndromic' surveillance systems. These may either be used for the prompt detection of recognizable transmissible infections with short incubation periods and distinctive clinical features (e.g. influenza) or may aim to detect unusual clinical syndromes that may presage a new disease, for example the USA-based global sentinel surveillance system 'Geosentinel'. They rely on prompt reporting by astute 'first responders': family practitioners, walk-in centres, accident and emergency departments and 'phone-in' medical advice services where ill people first seek advice.

An increase in the occurrence of acute respiratory, gastrointestinal, headache or rash-associated illness can quickly be detected by these systems. Such systems do not rely on definitive diagnosis, but are very responsive to an emerging epidemic of, for instance, influenza in the winter season or viral gastroenteritis in the summer. Some systems, such as the NHS Direct telephone advice service in England, have records that can be reviewed in real time, as often as hourly in an urgent situation, and analysed on a regional basis. Syndrome reporting systems allow a public health response to be immediately set up when case numbers rise above a specified alert level. Where a new emerging infection has been detected (e.g. SARS), surveillance systems then become very important in monitoring the new disease. Systems may need to be developed for novel diseases at very short notice, depending upon the urgency and seriousness of the situation.

Laboratory identification of a new pathogen

When a new infectious syndrome is detected, microbiologists must quickly determine whether this represents a new presentation of an old pathogen or is caused by a completely novel agent. Effective identification requires an international consortium of clinical microbiology laboratories that are able to apply a range of conventional diagnostic techniques to the new agent. These may include conventional culture methods, for example the SARS coronavirus was first detected in tissue culture, and *Legionella* was first isolated by a research group who had previously worked with intracellular bacteria. Molecular diagnostic methods are especially useful, using primers that will amplify more than one organism. Thus, 16S rRNA genes or primer sets are targeted against various genes in a wide range of pathogens. However, they contain altered bases that reduce the specificity of the assay and allow amplification of closely related, but not identical, pathogens. There are still many human pathogens for which we do not have a delineated clinical syndrome or diagnostic method, and new pathogens frequently emerge. As methods of culture and molecular diagnostics develop we will identify many new agents.

Communication and assessment; nationally and internationally

Many countries have systems that alert the national responsible agencies to the possibility of an emerging infection. Communication of information to those who must implement appropriate actions is an essential part of surveillance activities. The public health significance of any event also needs to be assessed. The Health Protection Agency, for example, holds weekly 'Infection Update' meetings, which provide a forum for professionals at regional and national levels to share information about outbreaks and incidents of infectious diseases occurring nationally. A summary of the information produced from this meeting is shared with professionals who require it.

Many countries report important cases and outbreaks of infection to the World Health Organization. This activity will be strengthened in the revised International Health Regulations, which should come into force in 2007. The WHO can warn of emerging problems, and may declare endemic areas or countries. It makes regular reports on certain diseases such as yellow fever, cholera, plague and winter influenza cases. It would issue an immediate warning if, for instance, a novel influenza serotype or an avian influenza strain was shown to be transmitted from person-to-person. In the SARS pandemic of 2003, the WHO declared some affected cities or countries to be zones of transmission, to and from which travel should be minimized.

In 2000, four fatal cases of Lassa fever were reported from countries in Western Europe, when the most recent previously diagnosed case had been in 1986. This alerted the world to the fact that travellers, peacekeepers and aid workers were increasingly entering endemic areas in West Africa.

The WHO also manages the Global Observation, Alert and Response Network (GOARN), which assesses and reports on emerging events. It will publish data on the progress of the event, and collate expert assessments of the means of spread of the outbreak and possible means of control. It will also advise on and help to organize responses to infection outbreaks in countries that cannot provide the full range of expert resources locally: examples include outbreaks of Ebola fever in Congo in the mid-1990s, Marburg fever in Uganda in the late 1990s and SARS in China in 2003.

National and international specific response plans

Some diseases are of such significance that individual countries have defined stepwise alert levels with specific planned responses. In the UK, plans exist for pandemic influenza, avian influenza, SARS and deliberate release events. These plans permit the issue of information, the use of vaccination programmes, the provision of prophylaxis or treatment, and the adjustment of routine work in the hospitals and clinics of a country, to free-up beds and healthcare staff for the care of infected individuals. Response algorithms and written information are kept available for future use.

In an extreme situation, schools in an affected region might be closed for a period of time or large gatherings such as sports events postponed, to reduce the risk of person-to-person transmission of a disease, or its transport away from the venue by visiting spectators.

In the international arena, the progress of the situation across countries can be documented and reported, accumulating scientific knowledge can be made available, the experience of countries most affected can be provided to others, and advice based on International Health Regulations can be given (for instance, advice on travel restrictions, vaccination requirements or quarantine).

Education of healthcare workers, emergency services providers and the public

Healthcare workers are the first, and front-line, responders in any infection-related emergency. Their contributions to controlling the situation include:
- recognition of suspicious cases or outbreaks;
- care of cases and contacts;
- collection of data about the cases;
- participation in prophylaxis and vaccination programmes;
- support for those suffering side-effects of prophylaxis or vaccination;
- support and advice to the 'worried well'.

To perform these tasks, healthcare workers require appropriate training and exercises, to practice and improve basic skills. These are provided by including appropriate educational material in routine training, within personal development programmes, and by conducting specific training and exercises for selected front-line staff. Newsletters and information handbooks, CD-ROMs and websites provide updates. A large resource of web-based information is provided and promoted.

In the event of a pandemic, such as SARS or influenza, the emergency services need to support the response, interact with the public and ensure safety by the use of effective infection control and personal protective measures. In most countries, they are trained to do this, and carry out regular small- and large-scale exercises. The largest exercises, such as major incident responses, often take place in the public arena.

If a deliberate release of infectious material is suspected, the police, or state investigation services have a major role. Protocols are developed to ensure effective collaboration between them and other responders. Regional, government and international offices also develop protocols and exercise them, to ensure effective national and international communication and coordination of responses.

After real emergencies, such as the SARS epidemic, as much audit and review as possible is carried out, to identify areas where further learning is required, and to inform and improve any future responses.

Modelling the behaviour of outbreaks and epidemics

High-quality outbreak modelling programs are now available, which can help to predict the size and spread of emerging outbreaks. They can be used to evaluate the likely effectiveness of different control measures in hypothetical situations. Models can be based on accumulating knowledge of a new disease, as with SARS. If a similar pathogen has caused previous epidemics or pandemics,

archived data from these can be combined with current knowledge, to make predictive models for new outbreaks. This has been done for such diseases as smallpox, and pandemic influenza. Figure 26.8 shows the epidemic curve for the influenza pandemic of 1918–19, with three waves of influenza infection affecting Great Britain. Superimposed on this curve are the predicted curves, assuming that either 10% or 20% of the population could have been treated with an antiviral drug (evidence suggests that, in most pandemics, 15–25% of the population develop clinical disease).

Isolation, quarantine and treatment of cases and contacts

Predictive models show that isolation of infected and infectious cases at the onset of an epidemic has a large effect in limiting the size and spread of an epidemic where the agent is spread person-to-person. In large-scale outbreaks, such as smallpox and SARS, it is clear from review of previous events that uncontrolled hospital admission of sick cases greatly increases the rate of transmission. Infection control measures should be employed from the outset, and should be applied to suspected cases, rather than only those with a proven diagnosis. Minimizing the number of infected cases means that drug stockpiles are more likely to be adequate for treating those who need it, and that vaccine supplies can be used economically, to target individuals and communities at risk.

> **Isolation** is the care of patients in individual accommodation with appropriate control of infection measures, in order to avoid contact between infected and susceptible individuals.

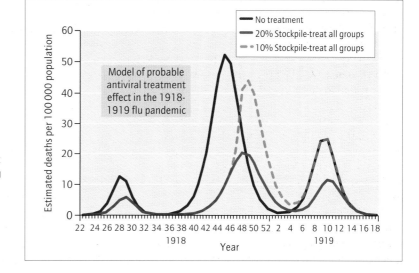

Figure 26.8 Modelling the probable effect of antiviral treatment on the influenza pandemic of 1918–19 in Great Britain. From: Gani R, Hughes H, Fleming D, Hedlock J, Leach S. Potential impact of antiviral drug use during influenza pandemic. Emerging Infections Diseases [serial on the internet], 20 September 2005. Available from www.cdc.ncidod/EID/vol11no0904-1344.htm.

Isolation, and effective infection control measures, do not usually require complex or high-technology arrangements but must be done correctly and without fail. They are done best by trained teams. Urgent training in patient care and infection control techniques may be needed to ensure that enough personnel are available. In the SARS pandemic of 2003, healthcare facilities had to be designated and cleared for use at short notice. Where they were not available, they were installed in temporary buildings, improvised or modified. As a result of this experience, some national healthcare providers now pre-designate emergency healthcare facilities or arrangements.

> **Quarantine** is the separation of individuals or groups, who are at risk of developing an infectious condition, until there is no longer a risk that they will become infectious and transmit infection to others.

Quarantine is intended to prevent contact between individuals susceptible to the infection and those who are infectious, or may become so. Potentially infectious individuals include those who may be incubating the infection because they are contacts of known cases or have been in an infected area. Cases who have recovered from the acute infection may continue to excrete the infectious agent for some time, and may also require quarantine. The period of quarantine is determined by knowledge of the infectious period of the infection, or by testing for infectiousness. If this is not possible, a reasonable arbitrary period may be chosen until more information emerges.

Home quarantine, when a person remains in his or her own household, is economical and preferred by quarantined individuals. It is usually necessary to perform checks to ensure that the quarantine is being observed, and that the quarantined person has not become unwell. When an individual is at high risk of illness, or when significant infectiousness may precede symptoms, quarantine may have to be supervised in an observation unit or healthcare facility.

The press and the broadcasting media

Both the public and healthcare workers often regard the press and the media with scepticism or distrust. However, radio, television and the newspapers are major sources of information and opinion in times of emergency. Many people also seek information from the world wide web, which provides a range of useful advice and some that is harmful and eccentric.

In local, national and international emergencies, the media are often asked to provide rapid and effective routes of communication to the public. They can disseminate authoritative and continuously updated information and advice on the risks to members of the public and their families. They can inform on sensible means of personal protection from risk, on travel, and on how to care for cases of some infections (for instance, for influenza: 'stay at home, keep away from other people, keep warm, drink plenty of fluids, use proprietary antipyretics and analgesics, only seek medical advice for severe respiratory symptoms or declining health').

The media can also broadcast information and instructions on more specific subjects, such as the availability of healthcare, vaccines or medicines and where to get them. Most importantly, they can remind people that a risk still exists or inform that the danger is over. They can also provide expert opinion and trusted 'talking heads', such as established news and investigative journalists, or popular scientists and doctors, to whom people are likely to listen.

Nevertheless, they will freely criticize perceived inefficiency, failure and low-quality information and they will always seek out public interest stories, whether these are helpful or not.

The transmission of information, using modern information technology and electronic media, is virtually instantaneous. Healthcare agencies aim to use this resource, to provide good quality information quickly, to pre-empt the emergence of inaccurate or malicious 'expertise', and to engage the support of the public in combating disease threats.

Case study 26.1: Variant Creutzfeldt–Jakob disease: establishing causality of an emerging infection

History

In 1996, the UK Creutzfeldt–Jakob disease (CJD) surveillance unit in Edinburgh recognized 10 cases of a new variant of human spongiform encephalopathy disease, with a different clinical course and pathological features from classical CJD. The average age at diagnosis was around 27 years, in contrast to classical CJD, which is rarely diagnosed in individuals aged under 50. It was thought possibly to be a human manifestation of infection with the agent of bovine spongiform encephalopathy (BSE), a recently recognized disease of cattle. BSE is a transmissible spongiform encephalopathy (TSE), naturally acquired by cattle by the oral route, and transmissible to other species. Classical CJD is also transmissible from person- to-person; thus transmission of BSE from cattle to man seemed plausible.

Question

■ What epidemiological procedures could be used to investigate this new condition?

Epidemiological investigation

Cases were actively sought, using a case definition based on:

■ the distinct clinical course (onset with psychiatric symptoms, followed by abnormal sensation at 2 months, ataxia at 5 months, myclonus at 8 months, akinetic mutism at 11 months and death between 12 and 24 months);

■ the unique pathological findings, including formation of plaques of abnormal prion protein (PrP^{sc}) surrounded by a halo of spongiform change, distributed throughout the cerebrum and cerebellum.

By the end of 1998, 35 cases had been diagnosed in the UK and one in France.

Relationships of time, place and person (including personal exposure) were sought by case-note studies and interviews. A number of epidemiological criteria indicated that the association with BSE was causal:

■ all but one of the cases were identified in Britain, the only endemic country for BSE;

■ there was a temporal relationship between the onset of BSE cases in cattle and the recognition of human vCJD 10 years later;

■ there was a pattern of likely exposure – although there was a case outside Britain, this was a body builder who had probably injected himself with a growth hormone preparation derived from bovine brain material;

■ macaque monkeys infected with BSE developed brain pathology identical to that of human vCJD.

Question

■ What next steps in controlling the outbreak could confirm the origin of vCJD?

Outbreak control measures

1 Transmission in cattle was interrupted by banning the use of animal feed that contained bovine and other animal proteins.

2 The preparation of carcasses was strictly controlled, to ensure the removal of all potentially infectious tissues before entry into the food chain.

3 The cohort of cattle that could have been incubating the disease was slaughtered.

Following these measures, the numbers of vCJD cases have declined. Surveillance continues, however, to ensure that longer incubation cases are not overlooked (the longest recorded incubation for a human spongiform encephalopathy is 40 years).

Index